# Exponential and Natural Logarithm

| x | $e^x$ | $e^{-x}$ | ln x | x | $e^x$ | $e^{-x}$ | ln x |
|---|---|---|---|---|---|---|---|
| 0.0 | 1.0000 | 1.0000 | — | 5.0 | 148.41 | 0.0067 | 1.6094 |
| 0.1 | 1.1052 | 0.9048 | −2.3026 | 5.1 | 164.02 | 0.0061 | 1.6292 |
| 0.2 | 1.2214 | 0.8187 | −1.6904 | 5.2 | 181.27 | 0.0055 | 1.6487 |
| 0.3 | 1.3499 | 0.7408 | −1.1204 | 5.3 | 200.34 | 0.0050 | 1.6677 |
| 0.4 | 1.4918 | 0.6703 | −0.9163 | 5.4 | 221.41 | 0.0045 | 1.6864 |
| 0.5 | 1.6487 | 0.6065 | −0.6931 | 5.5 | 244.69 | 0.0041 | 1.7047 |
| 0.6 | 1.8221 | 0.5488 | −0.5108 | 5.6 | 270.43 | 0.0037 | 1.7228 |
| 0.7 | 2.0138 | 0.4966 | −0.3567 | 5.7 | 298.87 | 0.0033 | 1.7405 |
| 0.8 | 2.2255 | 0.4493 | −0.2231 | 5.8 | 330.30 | 0.0030 | 1.7579 |
| 0.9 | 2.4596 | 0.4066 | −0.1054 | 5.9 | 365.04 | 0.0027 | 1.7750 |
| 1.0 | 2.7183 | 0.3679 | 0.0000 | 6.0 | 403.43 | 0.0025 | 1.7918 |
| 1.1 | 3.0042 | 0.3329 | 0.0953 | 6.1 | 445.86 | 0.0022 | 1.8083 |
| 1.2 | 3.3201 | 0.3012 | 0.1823 | 6.2 | 492.75 | 0.0020 | 1.8245 |
| 1.3 | 3.6693 | 0.2725 | 0.2624 | 6.3 | 544.57 | 0.0018 | 1.8405 |
| 1.4 | 4.0552 | 0.2466 | 0.3365 | 6.4 | 601.85 | 0.0017 | 1.8563 |
| 1.5 | 4.4817 | 0.2231 | 0.4055 | 6.5 | 665.14 | 0.0015 | 1.8718 |
| 1.6 | 4.9530 | 0.2019 | 0.4700 | 6.6 | 735.10 | 0.0014 | 1.8871 |
| 1.7 | 5.4739 | 0.1827 | 0.5306 | 6.7 | 812.41 | 0.0012 | 1.9021 |
| 1.8 | 6.0496 | 0.1653 | 0.5878 | 6.8 | 897.85 | 0.0011 | 1.9169 |
| 1.9 | 6.6859 | 0.1496 | 0.6419 | 6.9 | 992.27 | 0.0010 | 1.9315 |
| 2.0 | 7.3891 | 0.1353 | 0.6931 | 7.0 | 1096.63 | 0.0009 | 1.9459 |
| 2.1 | 8.1662 | 0.1225 | 0.7419 | 7.1 | 1211.97 | 0.00083 | 1.9601 |
| 2.2 | 9.0250 | 0.1108 | 0.7885 | 7.2 | 1339.43 | 0.00075 | 1.9741 |
| 2.3 | 9.9742 | 0.1003 | 0.8329 | 7.3 | 1480.30 | 0.00068 | 1.9879 |
| 2.4 | 11.023 | 0.0907 | 0.8755 | 7.4 | 1635.98 | 0.00061 | 2.0015 |
| 2.5 | 12.182 | 0.0821 | 0.9163 | 7.5 | 1808.04 | 0.00055 | 2.0149 |
| 2.6 | 13.464 | 0.0743 | 0.9555 | 7.6 | 1998.20 | 0.00050 | 2.0281 |
| 2.7 | 14.880 | 0.0672 | 0.9933 | 7.7 | 2208.35 | 0.00045 | 2.0412 |
| 2.8 | 16.445 | 0.0608 | 1.0296 | 7.8 | 2440.60 | 0.00041 | 2.0541 |
| 2.9 | 18.174 | 0.0550 | 1.0647 | 7.9 | 2697.28 | 0.00037 | 2.0669 |
| 3.0 | 20.086 | 0.0498 | 1.0986 | 8.0 | 2980.96 | 0.00034 | 2.0794 |
| 3.1 | 22.198 | 0.0450 | 1.1314 | 8.1 | 3294.47 | 0.00030 | 2.0919 |
| 3.2 | 24.533 | 0.0408 | 1.1632 | 8.2 | 3640.95 | 0.00027 | 2.1041 |
| 3.3 | 27.113 | 0.0369 | 1.1939 | 8.3 | 4023.87 | 0.00025 | 2.1163 |
| 3.4 | 29.964 | 0.0334 | 1.2238 | 8.4 | 4447.07 | 0.00022 | 2.1282 |
| 3.5 | 33.115 | 0.0302 | 1.2528 | 8.5 | 4914.77 | 0.00020 | 2.1401 |
| 3.6 | 36.598 | 0.0273 | 1.2809 | 8.6 | 5431.66 | 0.00018 | 2.1518 |
| 3.7 | 40.447 | 0.0247 | 1.3083 | 8.7 | 6002.91 | 0.00017 | 2.1633 |
| 3.8 | 44.701 | 0.0224 | 1.3350 | 8.8 | 6634.24 | 0.00015 | 2.1748 |
| 3.9 | 49.402 | 0.0202 | 1.3610 | 8.9 | 7331.97 | 0.00014 | 2.1861 |
| 4.0 | 54.598 | 0.0183 | 1.3863 | 9.0 | 8103.08 | 0.00012 | 2.1972 |
| 4.1 | 60.340 | 0.0166 | 1.4110 | 9.1 | 8955.29 | 0.00011 | 2.2083 |
| 4.2 | 66.686 | 0.0150 | 1.4351 | 9.2 | 9897.13 | 0.00010 | 2.2192 |
| 4.3 | 73.700 | 0.0136 | 1.4586 | 9.3 | 10938.02 | 0.00009 | 2.2300 |
| 4.4 | 81.451 | 0.0123 | 1.4816 | 9.4 | 12088.38 | 0.00008 | 2.2407 |
| 4.5 | 90.017 | 0.0111 | 1.5041 | 9.5 | 13359.73 | 0.00007 | 2.2513 |
| 4.6 | 99.484 | 0.0101 | 1.5261 | 9.6 | 14764.78 | 0.00007 | 2.2618 |
| 4.7 | 109.95 | 0.0091 | 1.5476 | 9.7 | 16317.61 | 0.00006 | 2.2721 |
| 4.8 | 121.51 | 0.0082 | 1.5686 | 9.8 | 18033.74 | 0.00006 | 2.2824 |
| 4.9 | 134.29 | 0.0074 | 1.5892 | 9.9 | 19930.37 | 0.00005 | 2.2925 |
| 5.0 | 148.41 | 0.0067 | 1.6094 | 10.0 | 22026.47 | 0.00005 | 2.3026 |

Note: $e^{17.4} = e^{7.4}(e^{10.0})$

$e^{-12.9} = e^{-2.9}(e^{-10.0})$

Note: $\ln 2700 = \ln(2.7 \times 10^3) = \ln 2.7 + 3 \ln 10$

$\ln(0.00051) = \ln(5.1 \times 10^{-4}) = \ln 5.1 - 4 \ln 10$

A BRIEF CALCULUS FOR BUSINESS, ECONOMICS, SOCIAL AND LIFE SCIENCES

# A BRIEF CALCULUS
FOR BUSINESS, ECONOMICS, SOCIAL AND LIFE SCIENCES

**JOSEPH N. FADYN**
Southern College of Technology

**WEST PUBLISHING COMPANY**
St. Paul   New York   Los Angeles   San Francisco

DEDICATED TO MY WIFE KAREN

| | |
|---|---|
| Text design: | Janet Bollow |
| Technical Illustrations: | Illustrious, Inc., Ross Rueger |
| Copyediting: | Susan Gerstein |
| Composition: | Jonathan Peck Typographers, Ltd. |
| Cover design: | Diane Beasley Design |
| Cover image: | Joseph Stella, *The Voice of the City of New York Interpreted: The Bridge*, 1920–22<br>Collection of The Newark Museum<br>Purchase 1937 Felix Fuld Bequest Fund<br>(Detail of above painting used on spine) |

COPYRIGHT ©1991    By WEST PUBLISHING COMPANY
50 W. Kellogg Boulevard
P.O. Box 64526
St. Paul, MN 55164–0526

All rights reserved

Printed in the United States of America

98 97 96 95 94 93 92 91    8 7 6 5 4 3 2 1 0

LIBRARY OF CONGRESS CATALOGING-IN-PUBLICATION DATA

Fadyn, Joseph N.
  A brief calculus for business, economics, social and life sciences
  Joseph N. Fadyn.
     p.  cm.
  Includes index.
  ISBN 0-314-77292-8
  1. Calculus.   I. Title.
QA303.F276   1991
515—dc20

90–47914
CIP

# CONTENTS

Preface ix

## CHAPTER 1 — FUNCTIONS AND GRAPHS 1

- **1.1** Introduction to Functions 2
- **1.2** Graphing Functions 9
- **1.3** The Algebra of Functions 22
- **1.4** Linear Functions and Applications 30
- **1.5** Quadratic Functions and Applications 47
- **1.6** Chapter 1 Review 60

## CHAPTER 2 — LIMITS AND AN INTRODUCTION TO THE DERIVATIVE 65

- **2.1** An Introduction to Limits 66
- **2.2** Limits at Infinity and Infinite Limits—Asymptotes 79
- **2.3** Continuity 95
- **2.4** The Tangent Line to a Curve 106
- **2.5** The Derivative of a Function 118
- **2.6** The Derivative as a Rate of Change 132
- **2.7** Chapter 2 Review 148

## CHAPTER 3 — THE DERIVATIVE 155

- **3.1** Basic Differentiation Formulas 156
- **3.2** The Chain Rule 166
- **3.3** Higher Derivatives 177
- **3.4** Implicit Differentiation 186
- **3.5** Related Rates 196
- **3.6** Differentials 206
- **3.7** Chapter 3 Review 216

## CHAPTER 4 — APPLICATIONS OF THE DERIVATIVE 221

- **4.1** Increasing and Decreasing Functions 222
- **4.2** Finding Maxima and Minima: The First Derivative Test 235
- **4.3** Concavity, Points of Inflection, and the Second Derivative Test 246
- **4.4** Absolute Maxima and Minima 260

| | | |
|---|---|---|
| | 4.5 | Curve Sketching  271 |
| | 4.6 | Optimization Problems  286 |
| | 4.7 | Business and Economics Applications  300 |
| | 4.8 | Chapter 4 Review  318 |

## CHAPTER 5 — EXPONENTIAL AND LOGARITHMIC FUNCTIONS  327

- 5.1 Exponential Functions  328
- 5.2 Logarithmic Functions  342
- 5.3 Derivatives of Exponential and Logarithmic Functions  353
- 5.4 Exponential Growth and Decay  364
- 5.5 Chapter 5 Review  378

## CHAPTER 6 — INTEGRATION  385

- 6.1 Antiderivatives and Indefinite Integrals  386
- 6.2 Integration by Substitution  397
- 6.3 The Definite Integral as a Net Change  406
- 6.4 Integration by Parts  415
- 6.5 Integration Using Tables  424
- 6.6 Chapter 6 Review  431

## CHAPTER 7 — ADDITIONAL INTEGRATION TOPICS  437

- 7.1 The Definite Integral as an Area  438
- 7.2 The Definite Integral as the Limit of a Sum  449
- 7.3 Applications of the Definite Integral  456
- 7.4 Numerical Integration  470
- 7.5 Improper Integrals  483
- 7.6 Chapter 7 Review  490

## CHAPTER 8 — FUNCTIONS OF MORE THAN ONE VARIABLE  497

- 8.1 Introduction and Examples  498
- 8.2 Partial Derivatives  506
- 8.3 Maxima and Minima  517
- 8.4 Constrained Optimization: Lagrange Multipliers  527
- 8.5 Regression: The Method of Least Squares  539
- 8.6 Double Integrals  551
- 8.7 Chapter 8 Review  568

## CHAPTER 9 — DIFFERENTIAL EQUATIONS  577

- 9.1 Introduction to Differential Equations  578
- 9.2 Separation of Variables  586

- 9.3 First-Order Linear Differential Equations  593
- 9.4 Applications of Differential Equations  599
- 9.5 Chapter 9 Review  611

## CHAPTER 10  APPLICATIONS OF CALCULUS TO PROBABILITY AND STATISTICS  617

- 10.1 Discrete Random Variables  618
- 10.2 Continuous Random Variables and the Uniform Distribution  631
- 10.3 Expected Value, Variance, and Median of Continuous Random Variables  641
- 10.4 The Normal Distribution  650
- 10.5 Chapter 10 Review  657

### ANSWERS TO ODD-NUMBERED EXERCISES  A1

Index  I1

# PREFACE

This book introduces students of business, economics, and the social and life sciences to important concepts of calculus directly applicable to their fields of interest. The prerequisites assumed are two years of high-school algebra or a single course in college algebra. No trigonometry is included in this book. It contains more than an ample amount of material for a one-semester or two-quarter course.

I have written this book using a motivational approach. Many sections of the text begin with an applied example that is later solved using techniques developed in the section. Virtually every section of the book contains one or more examples of applications from the fields of business, economics, or the sciences. I feel that the book is especially strong in applications to business and economics. Perhaps this is due in part to the fact that my educational background includes training both in formal mathematics and in quantitative methods.

## FEATURES

**EXAMPLES** The text contains approximately 370 examples. Of these approximately 295 are standard worked out examples, and approximately 75 are *participative examples*. The participative examples require the student to participate actively in the solution of the example by working out certain steps of the solution on his or her own. Virtually every section contains at least one participative example; answers follow immediately in the text.

**FORMAT** Important definitions, theorems, and formulas are titled and boxed for emphasis and ease of reference. In many cases, step-by-step summaries of important techniques are provided to assist the student.

**GRAPHS** The book contains about 350 figures, and a second color is used functionally. Wherever feasible, intuitive graphical interpretations of important ideas and results are provided.

**GRAPHING CALCULATOR/COMPUTER MATERIAL** The book contains a small amount of optional material for those instructors or students who wish to use a graphing calculator or graphing software for a microcomputer. This material is not calculator- or software-specific.

**EXERCISES** A great strength of this text is the number and variety of exercises. The book contains approximately 3,750 exercises. These exercises fall into three categories:

a. *Short Answer Exercises* Virtually every section contains a set of short answer exercises. These exercises precede the regular exercise set and their purpose is to clarify and reinforce the definitions, theorems, and concepts covered in the section. There are approximately 700 short answer exercises in the book.

b. *Regular (Computational) Exercises* About 2,900 regular exercises are included in the book. Of these, approximately 725 are applications exercises, and many of the applications exercises have multiple parts.

c. *Optional Graphing Calculator/Computer Exercises* The book contains about 150 optional graphing calculator/computer exercises. These exercises reinforce the concepts covered in the text and extend the student's ability to use a graphing calculator or a computer as a tool for solving more difficult calculus problems that might not normally be attempted in a course at this level.

CHAPTER REVIEWS   Each chapter of the book ends with an extensive Chapter Review Section. The Chapter Review contains a list of Key Terms, a summary of Key Formulas and Results, a set of Short Answer Review Questions, and a set of Review Exercises. A subset of the Review Exercises serves as a Chapter Test.

## SUPPLEMENTARY MATERIALS

A *Student Solutions Manual* contains detailed solutions to all odd-numbered regular and graphing calculator exercises. It is available for purchase by students from college bookstores.

An *Instructor's Solutions Manual* written by Ross Rueger of the College of the Sequoias contains solutions to all even-numbered exercises in the text and is available to instructors from the publisher.

A printed and bound *Test Bank* contains three versions of open-ended Chapter Tests as well as two versions of multiple-choice Chapter Tests for each chapter in the book. A total of about 950 questions are included in the *Test Bank*, which is available to instructors from the publisher.

A *Computerized Test Bank* is available from the publisher.

A *Graphing Calculator and Computer Graphing Guide* by Mickey Settle of Pensacola Junior College is available to students and instructors for the Casio fx-7000G, the Texas Instruments TI-81, and Macintosh and IBM computers.

A set of more than 70 *overhead transparencies*, which include many key figures and graphs from the text, is available to instructors.

## ACKNOWLEDGMENTS

I wish to thank the many instructors who reviewed the manuscript and helped to improve it. Thanks to Richard Brualdi, University of Wisconsin—Madison; Tom Caplinger, Memphis State University; David Dudley, Phoenix College; John Garlow, Tarrant County Jr. College; Ronnie Khuri, University of Florida; Peter Livorsi, Oakton Community College; David McKay, California State University—Long Beach; Ann Megaw, University of Texas—Austin; Dan Scanlon, Orange Coast College; and William Smith, University of North Carolina—Chapel Hill.

I would also like to thank the following professionals of West Publishing Company for their encouragement, support, and effort during the production of the text: Peter Marshall, Executive Editor; Rebecca Tollerson, Developmental Editor; Mark Jacobsen, Senior Production Editor; and Susan Gerstein, Copyeditor.

I am grateful to Ross Rueger and Charles Heuer of Concordia College in Moorhead, Minnesota for checking solutions to the exercises and for their valuable advice and suggestions.

Thanks to Annette Rohrs, who typed the manuscript with incredible speed and accuracy.

On a personal level, I would like to offer a special thanks to my wife Karen for her love, patience, and encouragement during the writing process.

# 1 FUNCTIONS AND GRAPHS

CHAPTER CONTENTS

1.1 Introduction to Functions
1.2 Graphing Functions
1.3 The Algebra of Functions
1.4 Linear Functions and Applications
1.5 Quadratic Functions and Applications
1.6 Chapter 1 Review

## 1.1 INTRODUCTION TO FUNCTIONS

The idea of a function is a pivotal concept in the study of mathematics. The function concept is particularly important in calculus. In fact, on virtually every page of this book you will be encountering functions in one way or another. Therefore we will begin our study of calculus with this fundamental concept: that of a function.

Throughout Chapter 1, we will be employing a common business-type example to illustrate various ideas. This will be a "perfume company" example, and you should imagine that it is *your* small business for the purposes of Chapter 1.

Suppose that you are in the business of manufacturing an exclusive new type of perfume which you sell for $200 an ounce. If you sell ten ounces of perfume this week, what is your revenue for the week? The answer is ($200)(10) = $2000. What is the revenue corresponding to a weekly sales of 15 ounces? To find this, we compute ($200)(15) = $3000. Notice that in each computation we multiplied the revenue per ounce ($200) by the number of ounces sold (10 and 15, respectively) to obtain the total revenue for the week. A listing of various weekly sales and the corresponding revenues is given in Table 1.1. The revenue generated depends upon weekly sales, and for each entry of weekly sales in this table there corresponds a single value of revenue. For example, the revenue corresponding to a weekly sales of 18 ounces is equal to $3600 = ($200)(18).

The relationship between weekly sales and revenue can be generalized as follows. If $x$ represents the weekly sales (the number of ounces of perfume sold in a particular week), then the revenue generated from the sale of these $x$ ounces is equal to $200x$ dollars. If we designate revenue by the variable $y$, then:

$$y = 200x$$

Summarizing, the formula $y = 200x$ means that to obtain the revenue ($y$) for a particular value of weekly sales ($x$), we multiply weekly sales by 200.

| WEEKLY SALES (ounces) | REVENUE |
|---|---|
| 5 | $1000 |
| 10 | $2000 |
| 15 | $3000 |
| 18 | $3600 |
| 20 | $4000 |

TABLE 1.1

EXAMPLE 1

Given that $y = 200x$, where $y$ represents revenue and $x$ represents weekly sales, find $y$ when (a) $x = 7$; (b) $x = 9.6$.

Solution

(a) When $x = 7$ the value of $y$ is $y = 200(7) = 1400$.
(b) When $x = 9.6$ the value of $y$ is $y = 200(9.6) = 1920$. ∎

In the preceding example, we say that revenue ($y$) is a function of weekly sales ($x$), where the functional relationship is given by the formula $y = 200x$. In other words, revenue $y$ depends upon (is a function of) weekly sales ($x$), and the rule defining this relationship that allows us to find revenue ($y$) given sales ($x$) is $y = 200x$.

**DEFINITION OF A FUNCTION**

A *function* is a rule that defines an association between two sets $X$ (called the *domain*) and $Y$ (called the *range*). In this correspondence, each element $x$ in the set $X$ is associated with one and only one element of $y$ in the set $Y$. In many (although not all) cases, the rule or functional relationship can be summarized by a formula.

If a function is defined by a formula, then it is understood that, unless otherwise specified, the domain of the function is the set of all real numbers that, when substituted into the formula, produce a real number as a result. Thus, *unless otherwise specified*, the domain of the function $y = 200x$ is all real numbers since, when any real number is substituted into this equation for $x$, the resulting value of $y$ is a real number. In certain applied situations, however, it may be necessary to restrict the domain of a function to values that "make sense" for the particular application.

In our perfume company example, the domain $X$ represents the set of possible values of weekly sales in ounces of our perfume. Thus, we might take $X$ to be the set of all nonnegative numbers; that is, $X = \{x \mid x \geq 0\}$. The symbolism $\{x \mid x \geq 0\}$ is read, "the set of all $x$ such that $x$ is greater than or equal to zero." Without further information about our perfume business, this would indeed be the domain of the function. Note, however, that this set $X$ implies that we can make and sell arbitrarily large amounts of perfume each week, since there is no upper bound on the values of $X$. In practice, if we are given the additional information that our small factory can only produce, say, 20 ounces of perfume per week, and if we do not stock any inventory from week to week but manufacture only enough perfume to cover sales for any particular week, then $X = \{x \mid 0 \leq x \leq 20\}$. The set $Y$ represents all the corresponding values for weekly revenue. Thus $Y = \{y \mid 0 \leq y \leq 4000\}$. We conclude that the domain of our revenue function is $X = \{x \mid 0 \leq x \leq 20\}$ and that the range of the function is $Y = \{y \mid 0 \leq y \leq 4000\}$.

Note also that for each value of weekly sales $x$ in $X$, there corresponds one and only one value of revenue $y$ in $Y$. For example, if weekly sales are $x = 3$ (ounces), the corresponding revenue is $y = 600$ (dollars) and no other value of $y$. As another example, if weekly sales are $x = 10.2$ (ounces), the corresponding revenue is $y = 200(10.2) = 2040$ dollars.

Another way to think about a function is as a "machine" into which we input $x$-values. The output from the function machine is the corresponding $y$-value. See Figure 1.1 for the general idea, Figure 1.2 for our specific perfume company example, and Figure 1.3 for an input value of 5 and the corresponding output of $1000 in our perfume company example.

To indicate that $y$ is a function of $x$, we write $y = f(x)$, and read this as "$y$ is a function of $x$" or "$y$ equals 'eff' of 'ecks.'" In this case, where the functional relationship can be summarized by a formula, $f(x)$ may be thought of as the formula in which the variable $x$ appears. The symbol $f(x)$ *does not* mean "$f$ times $x$." For example, we might write $y = f(x) = 200x$. Thus, $y$ is a function of $x$ (that is, $y$ depends on $x$) and the relationship between $y$ and $x$ is expressed by the formula $f(x) = 200x$. The notation $y = f(x)$ is referred to as

**FIGURE 1.1**

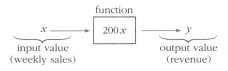

**FIGURE 1.2**

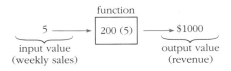

**FIGURE 1.3**

**FIGURES 1.1–1.3** A function viewed as an input-output machine.

"functional notation" and is quite useful. In this situation $y$ is called the *dependent variable* and $x$ is called the *independent variable*. This is so, because the value of $y$ depends upon the value of $x$, which we choose "independently" to input into the formula. Note that for any $x$ in the domain, $f(x)$ is actually the output (or $y$) value corresponding to an input value of $x$. As examples, consider $f(5) = 200(5) = 1000$, which is the output $y$-value corresponding to input value $x = 5$, and $f(18) = 200(18) = 3600$, which is the output $y$-value corresponding to input value $x = 18$. For any number $n$ in the domain, $f(n) = 200n$, which is the output $y$-value corresponding to input value $x = n$.

You should be aware that there are formulas that do not define functions. For example, in the formula $y^2 = x^2$, every input value of $x$ (except $x = 0$) produces *two* distinct output values of $y$. For example, if we input $x = 3$ into this formula, then:

$$y^2 = 3^2$$
$$y^2 = 9$$
$$y^2 - 9 = 0$$
$$(y - 3)(y + 3) = 0$$
$$y - 3 = 0 \quad \text{or} \quad y + 3 = 0$$
$$y = 3 \quad \text{or} \quad y = -3$$

Thus, an input value of $x = 3$ produces two different output values of $y$ (3 and $-3$), so we conclude that the formula $y^2 = x^2$ does *not* define $y$ as a function of $x$. Remember, the definition of a function requires that each element $x$ in the domain is associated with *one and only one* element $y$ in the range.

In calculus, we deal with many different types of functions. Usually the functional relationship will be defined by some formula. Although the variables $x$ and $y$ and the letter $f$ will be used most commonly to denote the functional relationship, other letters may also be used. For example, $A = g(t)$ indicates that "$A$ is a function of $t$," where the name of the function is $g$. We might use this notation, for example, if money were invested over a period of time and $A$ represents the amount of money accumulated after $t$ years.

Let us now consider some additional examples of functions.

### EXAMPLE 2

Find the domain and range of $f(x) = 2x^2 + 5$. Also, find the functional values $f(-2), f(0), f(3), f(n),$ and $f(x + h)$.

### Solution

Since any real value of $x$ may be substituted into this formula with a resulting unique (that is, one and only one) value of $y$, the domain of this function is $X = \{x \mid x \text{ is a real number}\}$. Note also that for any $x$, since $2x^2 \geq 0$, we have $2x^2 + 5 \geq 5$, so that the range of this function is $\{y \mid y \geq 5\}$. Also,

$$f(-2) = 2(-2)^2 + 5 = 13$$
$$f(0) = 2(0)^2 + 5 = 5$$
$$f(3) = 2(3)^2 + 5 = 23$$
$$f(n) = 2n^2 + 5$$
$$f(x + h) = 2(x + h)^2 + 5 = 2x^2 + 4xh + 2h^2 + 5$$

## FINDING THE DOMAIN OF A FUNCTION

> In finding the domain of many functions, our main concerns will be to *eliminate* from the domain of $y = f(x)$:
>
> i. any $x$'s that will produce the square root (or any other even root) of a negative number, and
> ii. any $x$'s that will produce a division by zero.

**EXAMPLE 3**  Find the domain of $f(x) = \sqrt{2x - 6}$. Also find the functional values $f(3)$, $f(5)$, $f(10)$, $f(n)$, and $f(x + h)$.

**Solution**  To find the domain of $f(x)$, note that to avoid trying to find the square root of a negative number (which is not a real number), we must have $2x - 6 \geq 0$, or $2x \geq 6$, or $x \geq 3$. Thus the domain is $X = \{x \mid x \geq 3\}$. The required functional values are:

$$f(3) = \sqrt{2(3) - 6} = \sqrt{0} = 0$$
$$f(5) = \sqrt{2(5) - 6} = \sqrt{4} = 2 \quad (not\ -2)$$
$$f(10) = \sqrt{2(10) - 6} = \sqrt{14} \approx 3.7417$$

(where the symbol $\approx$ means "is approximately equal to")

$$f(n) = \sqrt{2n - 6}$$

(if $n$ is in the domain of $f$, that is, $n \geq 3$)

$$f(x + h) = \sqrt{2(x + h) - 6}$$
$$= \sqrt{2x + 2h - 6}$$

(if $x + h$ is in the domain of $f$)

**EXAMPLE 4**  Find the domain of $z = h(t) = (2t + 1)/(t^2 - 16)$.

**Solution**  To find the domain of $h(t)$, note that to avoid division by zero, we must have $t^2 - 16 \neq 0$. Thus, consider solving the equation $t^2 - 16 = 0$:

$$t^2 - 16 = 0$$
$$(t + 4)(t - 4) = 0$$
$$t + 4 = 0 \quad \text{or} \quad t - 4 = 0$$
$$t = -4 \quad \text{or} \quad t = 4.$$

Since both $t = -4$ and $t = 4$ satisfy the equation $t^2 - 16 = 0$, and since we must have $t^2 - 16 \neq 0$, the domain must *exclude* the numbers $-4$ and $4$. So the domain of the function is $I = \{t \mid t \neq -4 \text{ and } t \neq 4\}$.

*Note: In the next example you should participate in finding the solution by filling in the blanks. Correct answers are given after the example.*

**6** CHAPTER 1 FUNCTIONS AND GRAPHS

---

**EXAMPLE 5 (Participative)**

**Solution**

For the function $f(x) = \sqrt{5x}/(x - 1)$, let us determine **(a)** the domain; **(b)** $f(0), f(5), f(7), f(n),$ and $f(x + h)$.

(a) To find the domain, note that we must have $5x \geq$ _____, so that $x \geq$ _____. In addition, we must have $x - 1 \neq$ _____, so that $x \neq$ _____.
We conclude that the domain is $\{x \mid x \geq 0 \text{ and } x \neq 1\}$

(b) We compute:

$$f(0) = \frac{\sqrt{5(0)}}{0 - 1} = \underline{\hspace{1cm}}$$

$$f(5) = \frac{\sqrt{5(5)}}{5 - 1} = \underline{\hspace{1cm}}$$

$$f(7) = \underline{\hspace{1cm}}$$

$$f(n) = \underline{\hspace{1cm}}$$

$$f(x + h) = \underline{\hspace{1cm}}$$

■

---

**EXAMPLE 6**

Suppose you own a small business and your supplier supplies you with component parts needed in the assembly of your product. The cost you must pay for the parts depends on the number you purchase. If you purchase fewer than 500, your cost is $2.00 per part. If you purchase at least 500 but fewer than 1000, your cost per part is $1.95, and if you purchase at least 1000 parts, your cost per part is $1.90. This information can be summarized by the following function:

$$C(x) = \begin{cases} 2.00x & 0 \leq x < 500 \\ 1.95x & 500 \leq x < 1000 \\ 1.90x & x \geq 1000 \end{cases}$$

This function gives the cost to you of purchasing $x$ component parts from this particular supplier. Find and interpret $C(400), C(650),$ and $C(1000)$.

**Solution**

This type of function is sometimes called a "multipart" function because there are really three different formulas given: (i) $2.00x$, (ii) $1.95x$, and (iii) $1.90x$. When we input a value of $x$, which of these three formulas do we use? The answer is that the appropriate formula depends upon the value of $x$. In fact, we

i. use the formula $C(x) = 2.00x$ if $0 \leq x < 500$,
ii. use the formula $C(x) = 1.95x$ if $500 \leq x < 1000$,
iii. use the formula $C(x) = 1.90x$ if $x \geq 1000$.

---

**SOLUTIONS TO EXAMPLE 5**

(a) 0, 0, 0, 1, $\{x \mid x \geq 0 \text{ and } x \neq 1\}$

(b) $0, \dfrac{5}{4}, \dfrac{\sqrt{35}}{6} \approx 0.9860, \dfrac{\sqrt{5n}}{n - 1}, \dfrac{\sqrt{5x + 5h}}{x + h - 1}$

Therefore,

$$C(400) = (2.00)(400) = \$800, \quad \text{the total cost for } x = 400 \text{ parts}$$
$$C(650) = (1.95)(650) = \$1267.50, \quad \text{the total cost for } x = 650 \text{ parts}$$
$$C(1000) = (1.90)(1000) = \$1900, \quad \text{the total cost for } x = 1000 \text{ parts}$$

Can you explain why it does not make too much sense to order exactly 499 parts from this company? If you cannot, see Exercise 29 for a hint. ■

## SECTION 1.1
### SHORT ANSWER EXERCISES

1. In the function $y = 250x$, the independent variable is ____ and the dependent variable is ____.
2. The set of all allowable input values to a function $y = f(x)$ is called the ____ of the function.
3. The set of all obtainable output values from a function $y = f(x)$ is called the ____ of the function.
4. (True or False) The domain of all functions is all real numbers. ____
5. For the function $y = f(x) = 50x + 600$, the symbol $f(7)$ represents the value of $y$ corresponding to $x =$ ____. What is $f(7)$ equal to numerically? ____
6. Given that $f(x) = 250x - 500$, then $f(2) =$ ____.
7. (True or False) The formula $y^2 = x^2 + 3$ defines $y$ as a function of $x$. ____
8. Since division by zero is not defined, the number 5 must be ____ from the domain of $f(x) = 1/(x - 5)$.
9. Since the square root of a negative number is not a real number, the domain of $g(x) = \sqrt{x - 1}$ is all real numbers $x$ such that $x - 1 \geq 0$ or $x \geq$ ____.
10. For the function $f(x) = x^2$, since the square of a real number is always nonnegative, the range is ____.

## SECTION 1.1
### EXERCISES

In Exercises 1–10, determine (a) $f(0)$, (b) $f(3)$, (c) $f(5)$, (d) $f(-2)$, (e) $f(n)$, and (f) $f(x + h)$. In parts (e) and (f), assume that $n$ and $x + h$ are numbers in the domain of $f$.

1. $f(x) = 100x$
2. $f(x) = 12x - 1$
3. $f(x) = x^2 - 3x + 1$
4. $f(x) = 3x^2 - 7x + 1$
5. $f(x) = x^3 - 7x + 1$
6. $f(x) = 1 - 2x - x^3$
7. $f(x) = 5$
8. $f(x) = 22$
9. $f(x) = \dfrac{-x^2 + 3}{x - 2}$
10. $f(x) = \dfrac{x - 3}{x^2 + 2x + 1}$

In Exercises 11–14, find the domain and the range of the function.

11. $f(x) = 10x$
12. $C(x) = 2x + 1, \quad x \geq 0$
13. $f(x) = 75$
14. $g(x) = 2x^2 + 3$

*In Exercises 15–19, find the domain of the function.*

**15.** $f(x) = \dfrac{x}{x-5}$

**16.** $f(t) = \dfrac{2t+3}{t^2 - 5t - 6}$

**17.** $f(x) = \sqrt{3x - 5}$

**18.** $f(x) = \dfrac{\sqrt{x+1}}{x-1}$

**19.** $R(x) = \dfrac{2x+3}{\sqrt{5x - 10}}$

*In Exercises 20–23, determine whether the formula defines y as a function of x.*

**20.** $y = 3x^2 - 10$

**21.** $|y| = 2x + 1$

**22.** $y^2 = 7x^2 + 1$

**23.** $y^3 = 3x^2$

**24.** Given a revenue function $R(x) = 400x$ where $x \geq 0$, determine the revenue for $x = 5$, $x = 10$, and $x = 20$.

**25.** (Cost function) The cost in dollars of producing $x$ ounces of a certain perfume is $C(x) = 50x + 600$ dollars per week. Find the weekly cost of producing 10 ounces of this perfume. Find the weekly cost of producing zero ounces of the perfume. Can you explain why this last cost is not equal to zero?

Business and Economics

**26.** (Demand function) Let $p = D(x) = 1000 - 2x$ represent the demand per month of a certain product. That is, in order for a total quantity $x$ of this product to be sold per month, the price (in dollars) must be $p$. For example, in order to sell 100 units of the product per month, the price must be $D(100) = 1000 - 2(100) = \$800$ per unit. Find and interpret $D(200)$, $D(300)$, and $D(400)$.

Business and Economics

**27.** (Volume discounts) A local company, SUPPLYIT, will supply your company with $x$ component parts at a cost in dollars determined by the following function:

$$C(x) = \begin{cases} 10x & \text{if } 0 \leq x < 12 \\ 9.5x & \text{if } 12 \leq x < 24 \\ 9x & \text{if } x \geq 24 \end{cases}$$

(a) Find $C(6)$, $C(13)$, and $C(30)$.
(b) State in words SUPPLYIT's policy on volume discounts on sales of parts to your company.

**28.** Some students argue that $f(x) = x^2$ is not a function in the following way: "$f(2) = 4$, and $f(-2) = 4$, so we get the same output value, 4, for two different input values, 2 and $-2$. Therefore $f$ is not a function." Of course, this argument is incorrect: the formula $f(x) = x^2$ defines a perfectly good function of $x$. Explain why the argument is fallacious.

**29.** (Cost function) For the cost function given in Example 6 of this section, compute $C(499)$ and $C(500)$. Can you now explain why it does not make too much sense to order exactly 499 parts from this supplier?

Social Sciences

**30.** (Psychological testing) After a fully rested individual has been awake for $x$ hours, the average amount of time required for her to complete a certain task is

$$T(x) = 2 + \sqrt{x}, \quad 0 \leq x \leq 72$$

where $T$ is measured in minutes. Find and interpret $T(0)$, $T(24)$, $T(48)$, and $T(72)$.

31. (Profit function) For the years 1990–1994, the annual profit of your small company has been given by

$$P(x) = x^2 - x + 2$$

where $P(x)$ is the profit in thousands of dollars and $x = 0$ represents 1990, $x = 1$ represents 1991, and so forth.
   (a) Compute the annual profit for each of the years 1990, 1991, 1992, 1993, and 1994.
   (b) Use the formula to predict the annual profit for 1995.

32. (Employee learning) After a new employee has been on the job for $t$ days on an assembly line, he can assemble approximately

$$N(t) = \frac{2t^2 + 3t}{t + 1}, \qquad 0 \le t \le 10$$

units of a certain product. Find and interpret $N(0)$, $N(1)$, $N(5)$, and $N(10)$.

33. (Advertising) The marketing department of a company has determined that for a certain product, each dollar of additional advertising per month produces two additional dollars in profit for advertising levels from $0 to $50,000. If monthly profit with no advertising is $20,000, find the function $P(x)$ that represents profit as a function of advertising expense $x$. Specify the domain of your function clearly.

34. (Packaging) A rectangular box (with no top) is to be made from a piece of cardboard 10 inches by 12 inches by cutting equal squares from the four corners and folding up the sides. Suppose $x$ is the length (in inches) of the squares that are cut out.
   (a) Find a formula that expresses the volume of the box in terms of $x$. Call your formula $V(x)$. *Hint:* The volume of a rectangular box is $V = lwh$ when $l$, $w$, and $h$ are the length, width, and height of the box, respectively.
   (b) Find and interpret $V(1)$, $V(2)$, $V(3)$, and $V(4)$.
   (c) Compute $V(10)$.
   (d) Does the answer for part (c) make physical sense for this problem?
   (e) Find the domain of the function for which you found the formula in part (a).

## 1.2  GRAPHING FUNCTIONS

You are probably already familiar with the fact that a pair of numbers $x$ and $y$ written in the form $(x, y)$ is called an *ordered pair* because the order of the numbers is important. As an example, $(2, 5)$ is an ordered pair, and $(2, 5)$ is *not* equal to $(5, 2)$. Ordered pairs of numbers can be graphed as points in a two-dimensional system called the *rectangular coordinate plane*. Figure 1.4 shows a rectangular coordinate plane on which we have plotted the ordered pairs $(1, 3)$, $(-1, 2)$, $(-2, -3)$, and $(2, -1)$ as points.

A graph of a function is a picture of the function. If $y = f(x)$, the graph describes the function $f$ geometrically and is drawn in the rectangular $xy$-coordinate plane. To obtain the graph of the function $y = f(x)$, we simply plot all ordered pairs $(x, y) = (x, f(x))$ for which $x$ is in the domain of $f$.

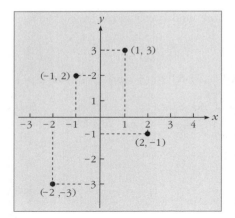

FIGURE 1.4  The graphs of several points in the rectangular $xy$-coordinate plane.

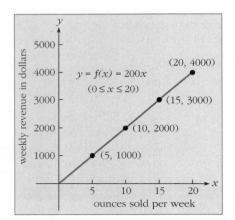

FIGURE 1.5  The graph of the revenue function $y = 200x$.

As an example, let us consider the revenue function from our perfume business of Section 1.1. Recall that this function is $y = 200x$, where $x$ represents the number of ounces of perfume sold in a particular week and $y$ represents the weekly revenue in dollars. Let us also recall that the restricted domain of this function is $X = \{x \mid 0 \leq x \leq 20\}$. To obtain a graph of this function, we first construct a table showing various values of $x$ and the corresponding values of $y$, where each $y$ is computed using the functional relationship $y = 200x$. The choice of values for $x$ is somewhat arbitrary (recall that $x$ is the "independent" variable). Let us select the values $x = 0$, $x = 5$, $x = 10$, $x = 15$, and $x = 20$ from the domain for inclusion in the table: see Table 1.2. Note the similarity of Table 1.2 to Table 1.1 of the previous section.

| $x$ | 0 | 5 | 10 | 15 | 20 |
|---|---|---|---|---|---|
| $y = 200x$ | 0 | 1000 | 2000 | 3000 | 4000 |

TABLE 1.2

Each value of $x$ represents weekly sales of perfume in ounces and the corresponding value of $y$ represents weekly revenue in dollars. For example, the ordered pair $(x, y) = (10, 2000)$ can be associated with the idea that if our weekly sales of perfume is 10 ounces, then our weekly revenue is $2000. Note that each ordered pair $(x, y)$ can also be interpreted as a point on the graph of $y = 200x$. Thus we have computed a total of five points on the graph of our function. These points are $(0, 0)$, $(5, 1000)$, $(10, 2000)$, $(15, 3000)$, and $(20, 4000)$. To draw the graph of this function, we simply plot our points in the $xy$-plane and then connect them by a smooth curve. In this case, as shown in Figure 1.5, the "smooth curve" is apparently a straight line. We will see in Section 1.4 that this is actually true. You should also observe that in order to sketch this graph, we used a different scale on the $y$-axis than on the $x$-axis. This was necessary because of the different sizes of the $x$'s and $y$'s in this example.

Graphical methods can be very powerful tools in the analysis of business and social science problems. To take a simple example, suppose you are able to obtain information about the profit your company receives from the sale of one of its products for various levels of sales. Using the figures from the last two months, you observe the data summarized in Table 1.3. These data are then graphed in Figure 1.6 and a smooth curve is drawn through the data points.

| $x$ (sales) | 10 | 20 | 30 | 40 | 50 | 60 | 70 | 80 |
|---|---|---|---|---|---|---|---|---|
| $y$ (profit) | 1000 | 2500 | 3500 | 4000 | 5000 | 5800 | 5500 | 5000 |

TABLE 1.3

The graph in Figure 1.6 shows clearly that our profit is largest when weekly sales are about 60 units. As we pass from sales of 60 units to sales of 70 units, our profit from this product decreases. This could be due, for example, to the fact that, to sell additional units beyond 60, the extra advertising expense is so large it decreases our profit.

It should be emphasized that the method of "connecting the dots" that we just used to derive the graphical relationship in this problem is very crude and

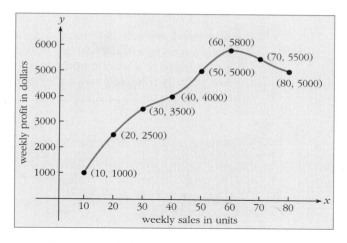

**FIGURE 1.6** A graph of the data points of Table 1.3, where a smooth curve has been drawn through the points.

imprecise. More formal and accurate methods of fitting curves to data are studied in statistics. We will consider one of these methods later in this book—the method of "least squares." It should be clear, however, that even a simple graphical data analysis such as ours can lead to useful insights.

### EXAMPLE 1

Graph $f(x) = x^2 - 2x$.

### Solution

First, we note that the domain of $f(x)$ is all real numbers. We construct a table of values by choosing as $x$-values (say) $x = -2$, $x = -1$, $x = 0$, $x = 1$, $x = 2$, $x = 3$, and $x = 4$. We find the corresponding $y$- (or $f(x)$-) values by substituting each of these $x$-values into the formula $f(x) = x^2 - 2x$. We obtain the table shown in Figure 1.7. By plotting these points in the $xy$-plane and joining them with a smooth curve, we obtain the graph of Figure 1.7. ∎

Once we have a graph of a function available, the range of the function is easy to "read off" from the graph. For example, in the graph of Example 1, the low point on the graph is $(1, -1)$ and apparently for every value of $y \geq -1$, we may find some $x$ such that $y = f(x)$. To verify this for a particular $y$, say, $y = 3$, draw a horizontal line through $(0, 3)$ on the $y$-axis as indicated in Figure 1.8. At the points where this horizontal line intersects the graph, draw vertical lines to intersect the $x$-axis. The values of $x$ where the vertical lines intersect the $x$-axis (which are $x = -1$ and $x = 3$) are the input values that will produce an output value of $y = 3$. That is, $f(-1) = 3$ and $f(3) = 3$. This shows that $y = 3$ is in the range of this function, since we can find at least one $x$-value that will produce a $y$-value of 3. Applying a similar argument for every $y \geq -1$, we find that the range of $f(x) = x^2 - 2x$ is $Y = \{y \mid y \geq -1\}$.

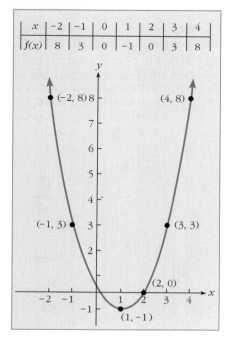

**FIGURE 1.7** The graph of $f(x) = x^2 - 2x$.

In general, we may determine whether a number $c$ is in the range of $y = f(x)$ by passing a horizontal line through the point $(0, c)$. If this line intersects the graph in at least one point, then $c$ is a number in the range. Further, if one point of intersection is $(a, c)$, then $f(a) = c$.

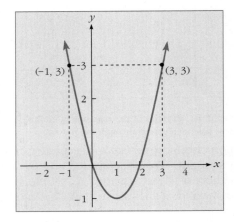

**FIGURE 1.8** The graph of $y = x^2 - 2x$. The horizontal line $y = 3$ intersects the curve in $(-1, 3)$ and $(3, 3)$, showing that $y = 3$ is a member of the range.

---

**EXAMPLE 2**    Graph $f(x) = \sqrt{x - 2}$.

**Solution**    Notice that because of the square root, the domain of $f(x)$ is all real numbers $x$ such that $x - 2 \geq 0$, or $x \geq 2$. Thus we will not have any $x$ values less than 2 in our table. The table and graph appear in Figure 1.9. Note that the range of $f(x)$ is $\{y \mid y \geq 0\}$. In sketching the graph we used the approximations $\sqrt{2} \approx 1.41$ and $\sqrt{3} \approx 1.73$.

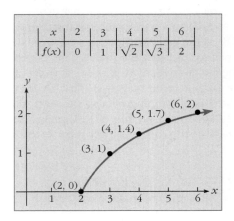

**FIGURE 1.9** The graph of $f(x) = \sqrt{x - 2}$.

## EXAMPLE 3

A local company, SUPPLYIT, supplies you with a part needed in the production of your product. The price SUPPLYIT charges depends on the number of parts $x$ that you order. SUPPLYIT charges $10 per part if you order fewer than 400 parts, $9.75 per part if you order between 400 and 700 parts (inclusive), and $9.50 per part if you order more than 700 parts.
(a) Determine the amount $C(x)$ that you must pay to SUPPLYIT to purchase $x$ parts from them.
(b) Graph the function $C(x)$.

### Solution

(a) Note that the cost $C(x)$ of $x$ parts depends upon the number of parts we order. As examples, consider:

If we order 200 parts, then we must pay $10 per part, so $C(200) = 10(200) = \$2000$.

If we order 550 parts, then we must pay $9.75 per part, so $C(550) = 9.75(550) = \$5362.50$.

If we order 750 parts, then we must pay $9.50 per part, so $C(750) = 9.50(750) = \$7125$.

We conclude that if $x < 400$ then $C(x) = 10x$; if $400 \leq x \leq 700$, then $C(x) = 9.75x$; and if $x > 700$, then $C(x) = 9.50x$. We can write $C(x)$ as follows:

$$C(x) = \begin{cases} 10x & \text{if } 0 \leq x < 400 \\ 9.75x & \text{if } 400 \leq x \leq 700 \\ 9.50x & \text{if } x > 700 \end{cases}$$

(b) We construct a table by choosing several $x$-values in each of the three ranges $0 \leq x < 400$, $400 \leq x \leq 700$, and $x > 700$; then we compute the $C(x)$-values using the formula appropriate to each range. Thus, if $0 \leq x < 400$, to find $C(x)$ we use $C(x) = 10x$; if $400 \leq x \leq 700$, to find $C(x)$ we use $C(x) = 9.75x$; and if $x > 700$, to find $C(x)$ we use $C(x) = 9.50x$. The table and the corresponding graph of $C(x)$ appear in Figure 1.10. Several things about this graph should be noted. First, the graph consists of three separate straight-line segments. The hollow circle (○) indicates that the corresponding point is not included on the graph, whereas the dark circle (●) indicates that the corresponding point is included. Even though we can purchase only whole units—for example, 399 units or 400 units and not 399.5 units—we have drawn the graph as a solid curve. In many cases we will assume that our graph is solid in order to do our analysis using calculus, since calculus is primarily used in the analysis of such curves. This will not cause us any difficulties so long as we realize (in drawing conclusions, for example) that in certain instances only *discrete* amounts (for example, 399 parts or 400 parts, and not 399.5 parts) may be involved. Getting back to our example, we can see from the table and the accompanying graph in Figure 1.10 that it would be rather foolish to purchase 399 parts from SUPPLYIT. The reason for this, of course, is that 399 parts will cost $3990, whereas 400 parts will cost only $3900.

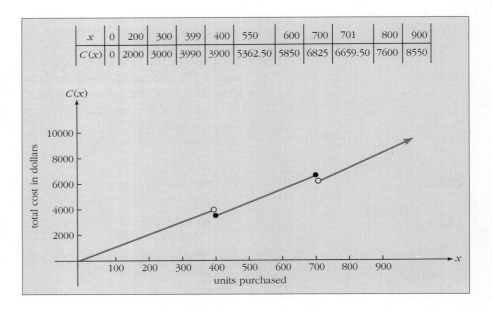

FIGURE 1.10  A sketch of the cost $C(x)$ of purchasing $x$ parts from SUPPLYIT.

*Note:  In the next example, you should participate in finding the solutions. Correct answers are given after the example.*

## EXAMPLE 4 (Participative)

Solution

Let us sketch the graph of

$$f(x) = \frac{1}{x-1}.$$

First, note that the domain of $f(x)$ is _____. In many cases, behavior of the graph "close to" an isolated number not in the domain of a function is interesting and important. Accordingly, in our table we should include some $x$-values that are "close to" but not equal to 1. Why can't we use an $x$-value of exactly 1? _____. Complete the following table by computing the $y$-values for the given values of $x$, and sketch the graph on the coordinate system provided in Figure 1.11.

| $x$ | −3 | −2 | −1 | 0 | $\frac{1}{2}$ | $\frac{3}{4}$ | $\frac{7}{8}$ | $\frac{9}{8}$ | $\frac{5}{4}$ | $\frac{3}{2}$ | 2 | 3 | 4 | 5 |
|---|---|---|---|---|---|---|---|---|---|---|---|---|---|---|
| $y = \frac{1}{x-1}$ | | | | | | | | | | | | | | |

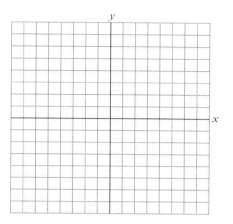

FIGURE 1.11

What is the range of $f(x)$? _____.

---

**SOLUTIONS TO EXAMPLE 4**

All real $x \neq 1$, since $x = 1$ produces division by 0. Because 1/0 is not allowed.

| $x$ | $-3$ | $-2$ | $-1$ | $0$ | $\frac{1}{2}$ | $\frac{3}{4}$ | $\frac{7}{8}$ | $\frac{9}{8}$ | $\frac{5}{4}$ | $\frac{3}{2}$ | $2$ | $3$ | $4$ | $5$ |
|---|---|---|---|---|---|---|---|---|---|---|---|---|---|---|
| $y = \dfrac{1}{x-1}$ | $-\frac{1}{4}$ | $-\frac{1}{3}$ | $-\frac{1}{2}$ | $-1$ | $-2$ | $-4$ | $-8$ | $8$ | $4$ | $2$ | $1$ | $\frac{1}{2}$ | $\frac{1}{3}$ | $\frac{1}{4}$ |

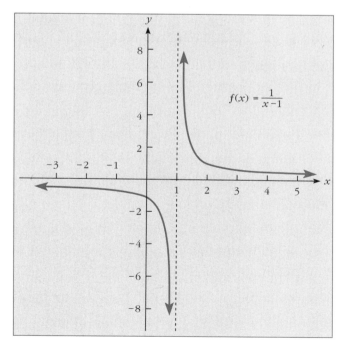

FIGURE 1.12

The range is all real numbers $y$ except $y = 0$.

## EXAMPLE 5

The total cost per week for a small company to produce $x$ custom children's rocking horses is $C(x) = 20x + 400$, where $C(x)$ is measured in dollars. The average cost per rocking horse is then defined by

$$\text{average cost per unit} = \frac{\text{total cost}}{\text{number of units}}$$

$$\overline{C}(x) = \frac{C(x)}{x}$$

Sketch the graph of $C(x)$ and $\overline{C}(x)$ on the same coordinate axes.

### Solution

First, we compute $\overline{C}(x) = (20x + 400)/x = 20 + (400/x)$. Since $x$ is the number of rocking horses produced per week, we must certainly have $x \geq 0$. So we take as the domain of $C(x)$ all real numbers $x \geq 0$ and as the domain of $\overline{C}(x)$, all $x > 0$. Observe that $\overline{C}(0)$ is undefined. We construct tables for $C(x)$ and $\overline{C}(x)$. Note that in constructing the table for $\overline{C}(x)$, we have chosen some values of $x$ "near" $x = 0$, which is not in the domain of $\overline{C}(x)$.

| $x$ | 0 | 1 | 5 | 10 | 15 | 20 |
|---|---|---|---|---|---|---|
| $C(x) = 20x + 400$ | 400 | 420 | 500 | 600 | 700 | 800 |

| $x$ | $\frac{1}{3}$ | $\frac{1}{2}$ | 1 | 5 | 10 | 15 | 20 |
|---|---|---|---|---|---|---|---|
| $\overline{C}(x) = 20 + \frac{400}{x}$ | 1220 | 820 | 420 | 100 | 60 | 46.67 | 40 |

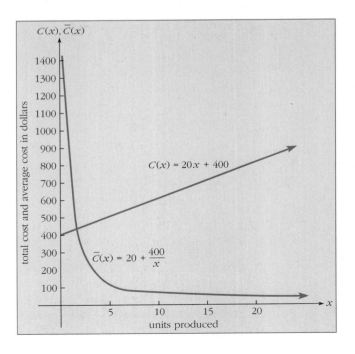

**FIGURE 1.13** The total cost function $C(x)$ and the average cost function $\overline{C}(x)$ of Example 5.

The graphs of $C(x)$ and $\overline{C}(x)$ are sketched on the same coordinate axes in Figure 1.13.

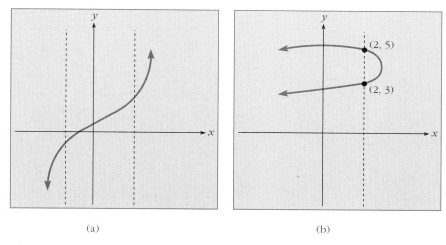

FIGURE 1.14  The graph in (a) is the graph of some function $y = f(x)$, whereas the graph in (b) is not the graph of any function of $x$.

We now address an interesting question that has an easy answer. Given a curve in a plane, does this curve represent the graph of some function? Let us consider the two curves in Figure 1.14. Curve (a) has the property that every vertical line intersects the graph in at most one point. Curve (b) lacks this property, since the dashed vertical line shown in the figure intersects the curve in *two* points, (2, 3) and (2, 5). It is therefore impossible for the graph in (b) to be the graph of some function $y = f(x)$. For if it were, then, since (2, 3) is on the graph, $f(2) = 3$. But also, since (2, 5) is on the graph, $f(2) = 5$. This clearly contradicts the definition of a function in that two different $y$-values ($y = 3$ and $y = 5$) are assigned to the same $x$-value ($x = 2$). This contradiction indicates that no such function $y = f(x)$ can exist. We conclude that the graph in (b) is not the graph of any function, while the graph in (a) *is* the graph of some function of $x$. This discussion motivates the *vertical line test*, which is stated here.

**VERTICAL LINE TEST**

> A curve represents the graph of some function of $x$ if and only if every vertical line intersects the graph in at most one point.

### Optional Graphing Calculator/Computer Material

Throughout this text, optional exercises for students with graphing calculators or access to microcomputer graphing software are included. A graphing calculator or graphing software can take much of the drudgery out of graphing routine functions and will allow you to graph more complicated functions relatively easily.

One important aspect of using a graphing calculator or graphing software is setting the range of $x$ and $y$ correctly so that you see the picture you want through your "graphing window." This usually involves setting a minimum value for $x$ ($x$ min), a maximum value for $x$ ($x$ max), and a scale for $x$ ($x$ scale); a minimum value for $y$ ($y$ min), a maximum value for $y$ ($y$ max), and a scale for $y$ ($y$ scale).

For example, if we attempt to graph $y = \sqrt{x - 4}$ using the following graphing window:

$x$ min $= -5$, $x$ max $= 5$, $x$ scale $= 1$,
$y$ min $= -5$, $y$ max $= 5$, $y$ scale $= 1$

which appears graphically as Figure 1.15(a), we then obtain the graph in Figure 1.15(b). It is clear that if we want to see more of the picture, we must move

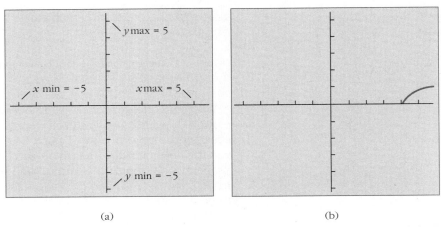

**FIGURE 1.16** A graphing "window" that might be used in an attempt to sketch $y = \sqrt{x - 4}$.

our graphing window over to the right. This is because the domain of this function is $x \geq 4$, so for $x < 4$ there is no graph! If the graphing window is changed to

$x$ min $= 2$, $x$ max $= 10$, $x$ scale $= 1$,
$y$ min $= -2$, $y$ max $= 4$, $y$ scale $= 1$

we obtain the graph in Figure 1.16, which gives us a much better picture of the function.

Unfortunately, adequate graphing windows vary from function to function and are not always easy to determine. Finding the domain of the function can help you choose an appropriate window, but expect a bit of trial and error in all but the simplest graphing problems that you do on a graphing calculator or computer. There are some graphing exercises at the end of this section on which you can practice.

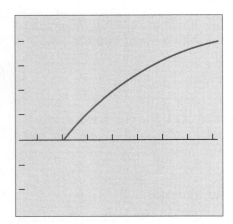

**FIGURE 1.15** The graph of $y = \sqrt{x - 4}$ using a more appropriate graphing "window" than that of Figure 1.14.

## SECTION 1.2
## SHORT ANSWER EXERCISES

1. Three points on the graph of $f(x) = x^2 + 1$ are $(-1, \_\_\_)$, $(0, \_\_\_)$, and $(1, \_\_\_)$.

2. Four points on the graph of $g(x) = x/(2 - x)$ are $(0, \_\_\_)$, $(1, \_\_\_)$, $(\frac{3}{2}, \_\_\_)$, and $(\frac{7}{4}, \_\_\_)$.

3. (True or False) Every graph represents the graph of some function $y = f(x)$. _____

4. A certain graph contains (among others) the points $(1, 2)$, $(-1, 3)$, $(0, 4)$, and $(1, -1)$. Which is true? \_\_\_\_
   (a) This graph represents the graph of some function of $x$.
   (b) This graph does not represent the graph of some function of $x$.
   (c) This graph may or may not represent the graph of some function of $x$; we can't tell from the information given.

5. When making up a table of values in preparing to graph a function $y = f(x)$, the values of $x$ we choose to include in the table must be in the _____ of the function.

6. Consider the following graph:

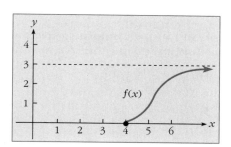

**FIGURE 1.17**

The domain of $f(x)$ is _____. The range of $f(x)$ is _____.

7. Which of the following four curves is the graph of some function $y = f(x)$?

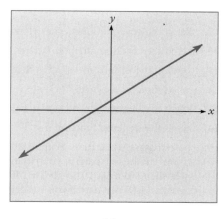

(a)

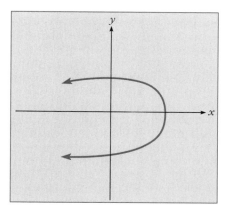

(b)

*(continued)*

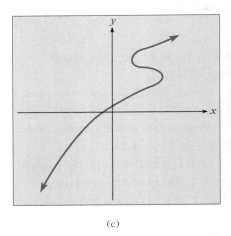
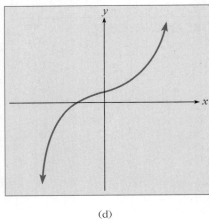

(c)  (d)

**FIGURE 1.18**

8. A horizontal line through (0, 6) intersects the graph of a certain function $y = f(x)$ at the point (4, 6). Is $y = 6$ included in the range of this function? _____ What is $f(4)$? _____

## SECTION 1.2
## EXERCISES

In Exercises 1–18, sketch the graph of each function. Use your graphs to determine the domain and range of each function.

1. $f(x) = 10x + 3$
2. $f(x) = -4x + 5$
3. $f(x) = x^2 + 2$
4. $f(x) = 2x^2 - 3x$
5. $f(x) = x^2 - 4x + 3$
6. $f(x) = x^2 + x - 6$
7. $f(x) = x^3$
8. $f(x) = x^3 + 4$
9. $f(x) = \sqrt{2x + 6}$
10. $f(x) = \sqrt{2 - x}$
11. $f(x) = \sqrt{3 - 5x}$
12. $f(x) = \frac{1}{2}\sqrt{4x - 8}$
13. $f(x) = 1/(x - 2)$
14. $f(x) = 4/(x + 2)$
15. $f(x) = 2/(3x + 6)$
16. $f(x) = 5/(20 - 4x)$
17. $f(x) = 2x + (1/x)$
18. $f(x) = 3x - (4/x)$

19. Sketch the revenue function $R(x) = 400x$ for $x \geq 0$.
20. Sketch the revenue function $R(x) = 5x$ for $x \geq 0$.
21. Sketch the cost function $C(x) = 50x + 600$ for $x \geq 0$.
22. Sketch the cost function $C(x) = 3x + 35$ for $x \geq 0$.
23. Sketch the demand function $p = D(x) = 1000 - 2x$ for $x \geq 0$.
24. Sketch the demand function $p = D(x) = 600 - 3x$ for $x \geq 0$.
25. You purchase some parts from SUPPLYIT, and the price SUPPLYIT charges depends on the number of parts $x$ you order. SUPPLYIT charges $10 per part if you order fewer than a dozen, $9.50 per part if you order at least a dozen but fewer than two dozen, and $9 per part if you order at least two dozen.
    (a) Determine the amount $C(x)$ that you must pay SUPPLYIT for your purchase of $x$ parts for $x > 0$.
    (b) Graph the function $C(x)$.

**Social Sciences**

**26.** (Psychological testing) Sketch the function
$$T = f(x) = 2 + \sqrt{x}, \quad 0 \le x \le 72$$
which measures the average amount of time in minutes for an individual to perform a certain task after being awake for $x$ hours.

**27.** (Average cost function) The total weekly cost to a company of manufacturing $x$ power screwdrivers is $C(x) = 8x + 1000$. If the average cost per screwdriver is $\overline{C}(x) = C(x)/x$, sketch the graphs of $C(x)$ and $\overline{C}(x)$ on the same coordinate axes.

**28.** (Postal rates) The cost of mailing a letter in 1990 was $0.25 for a letter weighing at most an ounce and $0.20 for each additional ounce or fraction thereof up to 11 ounces. For example, a letter weighing 2.4 ounces cost $0.65 to mail. Let $C(x)$ be the cost of mailing a letter that weighs $x$ ounces. Sketch the graph of $C(x)$ for $0 \le x \le 11$.

**Business and Economics**

**29.** (Supply and demand) If the price of a certain product is $p$ dollars, the number of units demanded per week by consumers (in thousands) is $d(p) = -p^2 - 2p + 10$ and the number of units supplied by manufacturers (in thousands) is $s(p) = p^2 + p + 1$.
(a) Graph $d(p)$ and $s(p)$ on the same coordinate axes.
(b) From your graph, estimate the coordinates of the point where the supply and demand curves intersect. Economists call this point the *market equilibrium point*.

**Life Sciences**

**30.** (Pollution) The concentration of a certain pollutant $x$ miles from a large factory in parts per million is given by the function
$$C(x) = \frac{500}{x^2 + 1}$$
(a) Sketch the graph of $C(x)$.
(b) Use your graph to estimate the minimum distance from the factory at which the concentration of the pollutant is less than or equal to 100 parts per million.

**31.** (Packaging) To construct a rectangular box without a top from a rectangular piece of cardboard 12 inches by 15 inches, equal squares are cut from the four corners and the sides are folded up. If $x$ is the length (in inches) of the squares that are cut out, the volume of the resulting box is
$$V(x) = (12 - 2x)(15 - 2x)(x) \quad \text{for } 0 \le x \le 6$$
(a) Sketch the graph of $V(x)$.
(b) Use your graph to estimate the length of the square to be cut out that will make the volume as large as possible. What is the approximate maximum volume?

**OPTIONAL**

**GRAPHING CALCULATOR/ COMPUTER EXERCISES**

Graph the following functions using a graphing calculator or a graphing utility on a microcomputer. Estimate the domain and range of each function by using your graph.

**32.** $f(x) = x^3 - 7x^2 + 3x + 5$

**33.** $f(x) = 3x^4 - 7x^3 - 10x^2 + 5x - 4$

**34.** $f(x) = \dfrac{x^2 - 3x + 5}{x^2 + 4}$

**35.** $f(x) = \dfrac{\sqrt{3x^2 + 5}}{x - 3}$

**36.** $f(x) = \dfrac{x^3 - 3x^2 + 2x + 5}{2x - 1}$

**37.** $f(x) = \dfrac{x^2 + x - 5}{2x^2 + 7}$

## 1.3 THE ALGEBRA OF FUNCTIONS

Returning to our perfume company example, suppose that we have the following revenue function:

$$R(x) = 200x, \quad 0 \le x \le 20$$

Note that we have replaced the "f" by an "R" in this revenue function. Also, suppose that the fixed costs of our small factory come to $600 per week. This is the amount of money (for rent, insurance, etc.) that it costs us to keep our operation in business even if we were to produce no perfume at all. In addition, suppose that the variable cost (for materials, labor, etc.) of producing one ounce of our perfume is $50. We can now write our cost function as follows:

$$C(x) = 50x + 600, \quad 0 \le x \le 20$$

Thus $R(x)$ represents the revenue generated by selling $x$ ounces of perfume and $C(x)$ represents the cost to us of producing $x$ ounces of perfume. It is a well known fact that a company's profit is equal to the revenue it generates (the money it "takes in") minus the cost incurred in generating this revenue (the money that "goes out" for the production of its product). In symbols,

$$P(x) = R(x) - C(x)$$

where $P(x)$ is the *profit function* and represents, in general, the profit to the company from manufacturing and selling $x$ units of its product.

For example, our small perfume company has the profit function

$$P(x) = 200x - (50x + 600)$$
$$= 150x - 600, \quad 0 \le x \le 20$$

Thus the profit in dollars that our company makes from the sale of 10 ounces of perfume is

$$P(10) = 150(10) - 600 = \$900$$

The purpose of the preceding example is to convince you that given two functions $y = f(x)$ and $y = g(x)$, it is sometimes important in business and social science applications to consider certain combinations of these functions. Indeed, such combinations of functions will also be important in our study of calculus. Basic combinations may be formed using addition, subtraction, multiplication, and division, and these combinations of functions are considered to be *new* functions. Let us consider some other examples.

EXAMPLE 1

Let $f(x) = 2x + 1$ and $g(x) = x^2 + 3x$. Then

(a) $f(x) + g(x) = (2x + 1) + (x^2 + 3x) = x^2 + 5x + 1;$
(b) $f(x) - g(x) = (2x + 1) - (x^2 + 3x) = 1 - x - x^2;$
(c) $f(x) \cdot g(x) = (2x + 1)(x^2 + 3x) = 2x^3 + 7x^2 + 3x;$
(d) $\dfrac{f(x)}{g(x)} = \dfrac{2x + 1}{x^2 + 3x}.$

∎

We use the notation $f^2(x)$ to indicate the product $f(x) \cdot f(x)$. Thus

$$f^2(x) = f(x) \cdot f(x) = [f(x)]^2$$

Similarly,

$$f^3(x) = f^2(x) \cdot f(x) = [f(x)]^3$$

and so forth.

## EXAMPLE 2

Let $f(x) = \sqrt{x-2}$ and $g(x) = 2/(x-3)$. Then

**(a)** $f(6) - g(6) = \sqrt{6-2} - \dfrac{2}{6-3} = 2 - \dfrac{2}{3} = \dfrac{4}{3}$;

**(b)** $\dfrac{f(11)}{g(11)} = \dfrac{\sqrt{11-2}}{\frac{2}{11-3}} = \dfrac{3}{\frac{1}{4}} = 3(4) = 12$;

**(c)** $f^2(5) = [f(5)]^2 = [\sqrt{5-2}]^2 = [\sqrt{3}]^2 = 3$.

## EXAMPLE 3 (Participative)

**Solution**

Given that $f(x) = 3x - 5$ and $g(x) = 2/x$, let's find **(a)** $f(x) \cdot g(x)$; **(b)** $f(5) - g(5)$.

For (a), consider:

$$f(x) \cdot g(x) = (3x - 5)(\underline{\quad}) = \underline{\quad\quad}$$

For (b), we have

$$f(5) - g(5) = [3(5) - 5] - (\underline{\quad}) = \underline{\quad}$$

Given two functions $f(x)$ and $g(x)$, we now consider another important type of function, called *the composition of f with g*, denoted by the symbol $f \circ g$. By definition,

## COMPOSITION OF FUNCTIONS

$$(f \circ g)(x) = f(g(x))$$

It is important to distinguish between the composition $f \circ g$ and the product $f \cdot g$. In general, they are *not* the same.

---

**SOLUTIONS TO EXAMPLE 3**      (a) $2/x$, $6 - (10/x)$; (b) $\frac{2}{5}$, $\frac{48}{5}$

**EXAMPLE 4**

Let $f(x) = x^2 + 1$ and $g(x) = 2x - 1$. Find (a) $(f \circ g)(x)$; (b) $(f \circ g)(3)$; (c) $(g \circ f)(x)$; (d) $(g \circ f)(3)$; (e) $f(3) \cdot g(3)$.

**Solution**

(a) $(f \circ g)(x) = f(g(x)) = f(2x - 1) = (2x - 1)^2 + 1 = 4x^2 - 4x + 2$
(b) $(f \circ g)(3) = f(g(3)) = f(2(3) - 1) = f(5) = 5^2 + 1 = 26$

Alternatively, using the result of part (a):

$(f \circ g)(3) = 4(3)^2 - 4(3) + 2 = 26$

(c) $(g \circ f)(x) = g(x^2 + 1) = 2(x^2 + 1) - 1 = 2x^2 + 1$
(d) $(g \circ f)(3) = g(f(3)) = g(3^2 + 1) = g(10) = 2(10) - 1 = 19$

Alternatively, using the result of part (c):

$(g \circ f)(3) = 2(3)^2 + 1 = 19$

(e) $f(3) \cdot g(3) = [3^2 + 1][2(3) - 1] = (10)(5) = 50$

Note that $(f \circ g)(x) = 4x^2 - 4x + 2$, whereas $(g \circ f)(x) = 2x^2 + 1$. We conclude that, in general, $f \circ g$ is *not* equal to $g \circ f$. Note also that the product $f(3) \cdot g(3)$ is *not* the same as either of the compositions $(f \circ g)(3)$ or $(g \circ f)(3)$. ∎

---

**EXAMPLE 5 (Participative)**

Given that $f(x) = x^2 + 2x - 2$ and $g(x) = \sqrt{x + 2}$, let's find (a) $(f \circ g)(x)$; (b) $(g \circ f)(x)$.

**Solution**

(a) We have

$(f \circ g)(x) = f(g(x)) = f(\sqrt{x + 2}) = (\underline{\phantom{xx}})^2 + 2(\underline{\phantom{xx}}) - 2$
$= \underline{\phantom{xxxx}}$

(b) For this part of the problem, note that

$(g \circ f)(x) = g(f(x)) = g(x^2 + 2x - 2)$
$= \sqrt{(\underline{\phantom{xxxxx}}) + 2} = \sqrt{\underline{\phantom{xxxx}}}$

To reiterate an important point, note that this problem again bears out the fact that, in general, $f \circ g \neq g \circ f$. ∎

We must take some care concerning the domain of the function $f \circ g$. Note that to compute $f(g(x))$, we must first take a number $x$ and substitute it into the formula for $g$ to obtain the output number $g(x)$. Then this output number $g(x)$ must be substituted into the formula for $f$ to obtain, finally, $f(g(x))$. This process can be illustrated as follows:

---

**SOLUTIONS TO EXAMPLE 5**

(a) $\sqrt{x + 2}, \sqrt{x + 2}, x + 2\sqrt{x + 2}$; (b) $x^2 + 2x - 2, x^2 + 2x$

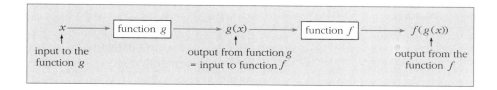

We conclude that to be in the domain of the composition $f \circ g$, a number $x$ must be in the domain of $g$ (since it must be used as an input value to the function $g$). But also, $g(x)$ must be in the domain of $f$, since $g(x)$ is used as an input value to the function $f$. Thus:

**THE DOMAIN OF THE COMPOSITION OF FUNCTIONS**

> The domain of the composition $f \circ g$ is the set of all numbers $x$ in the domain of $g$ such that $g(x)$ is in the domain of $f$.

**EXAMPLE 6**

Let $f(x) = \sqrt{3x - 12}$ and $g(x) = x^2$. Then

$$(f \circ g)(x) = f(g(x)) = f(x^2) = \sqrt{3x^2 - 12}$$

The domain of $f$ is the set of all real numbers $x$ such that $3x - 12 \geq 0$, or $x \geq 4$. The domain of $g$ is the set of all real numbers. Therefore, the domain of $f \circ g$ is

the set of all real numbers $x$    such that    $x^2$    $\geq$    4
↑                                ↑            ↑
in the domain of $g$                $g(x)$    in the domain of $f$

Since $x^2 \geq 4$ if and only if $x \geq 2$ or $x \leq -2$, the domain of $f \circ g$ is the set of all real numbers $x$ such that $x \geq 2$ or $x \leq -2$. ∎

**EXAMPLE 7**

An English teacher has found that the average grade $G$ on a certain placement test is related to the average I.Q. of $x$ for the class by the formula

$$G = f(x) = 6\sqrt{x} + 15$$

If, over the next five years, the average I.Q. of students attending this English class is projected to be

$$x = g(t) = 102 + 3t, \quad 0 \leq t \leq 5$$

where $t$ is time in years and $t = 0$ designates the present, find the average grade on this test three years from now.

Solution

We want to find $f(g(3))$. We have

$$f(g(3)) = f(102 + 3(3)) = f(111) = 6\sqrt{111} + 15 \approx 78.2$$

Therefore the average grade on this test three years from now will be 78.2. ∎

EXAMPLE 8  Suppose that the profit (in dollars) per week that your company makes is a function of the number $x$ of units sold and is given by the formula $P(x) = -x^2 + 850x - 40000$. On the other hand, the number of units that can be sold per week certainly depends upon (that is, it's a function of) the price $p$ that you charge per unit. Thus we conclude that profit ultimately depends upon the price you charge. This should make good sense if you just think about it! The profit is therefore expressible as a composition of functions. Note that the number $x$ of units of the product that you can sell at a given price is the *demand* for your product at that price. For the purposes of this example, suppose that the demand function is

$$x = q(p) = -2p + 1000, \quad \text{where } p \text{ is measured in dollars}$$

This function is graphed in Figure 1.19. Note that, for example, if you charge $200 per unit, then the quantity demanded will be

$$q(200) = -2(200) + 1000 = 600 \text{ units}$$

Now let us express your profit as a function of the price you charge per unit. This is done by computing the composition of $P$ with $q$, or $P \circ q$, where

$$\begin{aligned}(P \circ q)(p) &= P(q(p)) = P(-2p + 1000) \\ &= -(-2p + 1000)^2 + 850(-2p + 1000) - 40000 \\ &= -(4p^2 - 4000p + 1000000) - 1700p + 850000 - 40000 \\ &= -4p^2 + 2300p - 190000\end{aligned}$$

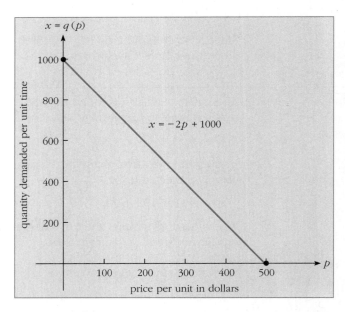

FIGURE 1.19  The graph of the demand function $x = q(p) = -2p + 1000$.

Thus, for example, if you charge a price of $200 per unit this week, your profit for the week will be

$$(P \circ q)(200) = -4(200)^2 + 2300(200) - 190000 = \$110,000$$ ∎

It would be very valuable for the company in Example 8 to know the price to charge per unit in order to maximize profit. This is exactly the kind of question that calculus can help us to answer, and we will take up this kind of question again after we have a bit of calculus under our belts.

## SECTION 1.3
### SHORT ANSWER EXERCISES

1. If $f(x) = x^2$ and $g(x) = 2x$, then $f(x) + g(x) = $ _____.
2. If $f(x) = x - 1$ and $g(x) = 3x$, then $f(x) - g(x) = $ _____.
3. If $f(x) = x^3$ and $g(x) = x - 1$, then $f(x) \cdot g(x) = $ _____.
4. If $f(x) = 2x^2$ and $g(x) = 3x^2 - x$, then $f(2)/g(2) = $ _____.
5. If $f(x) = 4x^2 - 2x + 1$, then $f^2(-1) = $ _____.
6. If $f(x) = 3x - 5$ and $g(x) = 4x + 1$, then $(f \circ g)(1) = $ _____.
7. (True or False) The domain of the composition $f \circ g$ is equal to the domain of the function $g$. _____
8. If $f(x) = 2x + 1$ and $g(x) = 1 - x$, then $(f \circ g)(3) = $ _____, whereas $f(3) \cdot g(3) = $ _____.
9. If $f(x) = 4x^2$ and $g(x) = 3x$, then $(f \circ g)(2) = $ _____, whereas $(g \circ f)(2) = $ _____.
10. (True or False) In general the composition $f \circ g$ is different from the product function $f \cdot g$. _____
11. From the result of Exercise 9, what can you conclude about the functions $f \circ g$ and $g \circ f$? _____
12. Given revenue function $R(x) = 300x$ and the cost function $C(x) = 200x + 5000$, find the profit function $P(x) = R(x) - C(x)$. _____

## SECTION 1.3
### EXERCISES

For each pair of functions in Exercises 1–6, find and simplify $f(x) + g(x)$, $f(x) - g(x)$, $f(x) \cdot g(x)$, and $f(x)/g(x)$.

1. $f(x) = 3x - 5$, $g(x) = 2x + 1$
2. $f(x) = 7x - 3$, $g(x) = x + 7$
3. $f(x) = \sqrt{x^2 + 1}$, $g(x) = -\sqrt{x^2 + 1}$
4. $f(x) = 4/x$, $g(x) = -2/(x - 1)$
5. $f(x) = 1/x$, $g(x) = 2x^3$
6. $f(x) = x^2 + 2x + 3$, $g(x) = -5x + 4$

**Exercises 7–12:** *for each of the Exercises 1–6, find $f(3) + g(3)$, $f(3) - g(3)$, $f(3) \cdot g(3)$, and $f(3)/g(3)$.*

For each pair of functions in Exercises 13–18, find and simplify $(f \circ g)(x)$ and $(g \circ f)(x)$. Also, find the domain of $f \circ g$.

13. $f(x) = 2x$, $g(x) = 3x + 5$
14. $f(x) = 7x - 5$, $g(x) = 3x - 2$

15. $f(x) = \sqrt{2x - 5}$, $g(x) = 1/x$ 
16. $f(x) = 2/(x - 4)$, $g(x) = \sqrt{1 - x}$
17. $f(x) = 4x/(x - 1)$, $g(x) = x^2$ 
18. $f(x) = 2x/(x + 1)$, $g(x) = x^3 + 1$

**Exercises 19–21:** for each of the Exercises 13–15, find $(f \circ g)(5)$, $(g \circ f)(5)$, and $f^2(5)$.

**Exercises 22–24:** for each of the Exercises 16–18, find $(f \circ g)(0)$, $(g \circ f)(0)$, and $f^2(0)$.

25. For Exercise 15, find $(f \circ g)(\frac{1}{5})$.
26. For Exercise 16, find $g^2(-8)$.
27. For Exercise 17, find $f(g(2))$.
28. For Exercise 18, find $g(f(1))$.
29. (Revenue function) Given a revenue function $R(x) = 75x$ and a cost function $C(x) = 25x + 300$, where $x$ is units sold and $R(x)$ and $C(x)$ are measured in dollars,
    (a) find the profit function $P(x)$.
    (b) If you sell 30 units of your product, what is your profit?
30. (Profit) You are in the business of manufacturing stereo speakers. You sell the speakers for $100 each, your fixed cost of production is $5000 per week, and your variable cost is $40 per speaker.
    (a) Find the function that expresses your profit per week in terms of the number of speakers sold per week.
    (b) If you sell 300 speakers next week, what profit will you make?

**Business and Economics**

31. (Profit and demand) Given a profit function $P(x) = -x^2 + 600x - 30000$ and a weekly demand function $x = q(p) = -3p + 900$, where $x$ is units sold and $p$ is the selling price in dollars per unit,
    (a) express profit as a function of the price per unit.
    (b) If we charge a price of $250, what is our profit?
    (c) If we charge a price of $250, what is our weekly demand?

**Social Sciences**

32. (Homeless people) The number of homeless people in a certain urban area is given approximately by

    $$n = f(x) = \sqrt{x} - 3\sqrt[3]{x}, \quad \text{for } x \geq 10000$$

    where $x$ is the population of the area. If

    $$x = 200000 + 20000t, \quad \text{for } 0 \leq t \leq 10$$

    where $t$ is time measured in years from the present,
    (a) find the present population of this area;
    (b) find a general formula for the number of homeless people in $t$ years;
    (c) find the approximate number of homeless people five years from now.

33. (Manufacturing output) The output $O$ of a certain maufacturing process depends upon the number of workers $w$ according to the formula

    $$O = 10\sqrt{w}$$

    If the number of workers who can be hired depends upon the hourly wage $x$ offered to them, where

    $$w = 5x, \quad 0 \leq x \leq 12$$

    (a) find the output as a function of the hourly wage $x$.
    (b) If the hourly wage is $5 per hour, find the output $O$.

**34.** (Sales and advertising)  Weekly sales $S$ for a certain product in one of the product lines of a company is a function of advertising $v$, where

$$S = 10\sqrt{v} + 2$$

Here advertising and sales are measured in thousands of dollars. The level of advertising in turn depends on the weekly budget $b$ for the product line according to the relationship

$$v = 0.1b - 10$$

where $b$ is measured in thousands of dollars.
(a) Find weekly sales as a function of the weekly budget $b$.
(b) If the weekly budget is $100,000, determine weekly sales.

Social Sciences

**35.** (School tax and teachers' salaries)  The amount of school tax $T$ collected per year in a certain school district depends upon the total assessed property value $V$ in the district:

$$T = 0.005V$$

The total amount of money $S$ available yearly for teachers' salaries is

$$S = 0.40T - 200000$$

(a) Find a formula that gives $S$ as a function of $V$.
(b) If the total assessed property value is 4 billion dollars, find the amount of money available for teachers' salaries.

Business and Economics

**36.** (Set-up and storage costs)  In manufacturing a certain product, a company incurs a set-up cost of

$$S = \frac{7500000}{x} \text{ dollars}$$

where $x$ is the number of units it produces in each production run and $1 \leq x \leq 50000$. Once the product is manufactured, the total storage cost (in dollars) is given by

$$R = 0.50x$$

(a) If total cost $T$ is the sum of set-up cost and storage cost, find a formula for $T$ as a function of $x$.
(b) Find the total cost if $x = 5000$.

OPTIONAL

**GRAPHING CALCULATOR/ COMPUTER EXERCISES**

**37.** Using a graphing calculator or computer, estimate the price we should charge, given the profit and demand functions of Exercise 31, in order to obtain the maximum weekly profit. What is the (approximate) maximum profit? What is the (approximate) weekly demand if we charge the price that will maximize weekly profit?

**38.** Graph the function $n = f(t)$ that you found in Exercise 32 and use your graph to estimate the time $t$ when the number of homeless people will exceed 400.

**39.** Graph the function $T = f(x)$ that you found in Exercise 36 and use your graph to estimate the value of $x$ that will yield the smallest total cost $T$.

# 1.4 LINEAR FUNCTIONS AND APPLICATIONS

**Linear Functions and Slope**

Let us begin the discussion of what are called linear functions by considering the profit function of our perfume company:

$$y = P(x) = 150x - 600, \quad 0 \leq x \leq 20$$

A table of values and the graph of $P(x)$ are shown in Figure 1.20.

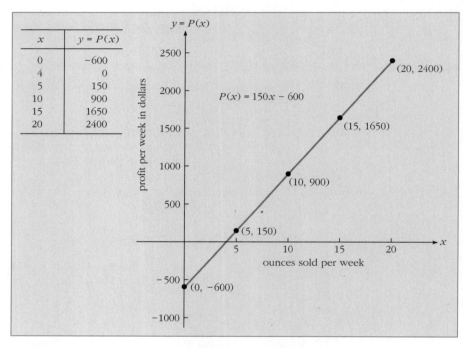

FIGURE 1.20  The graph of the profit function $P(x) = 150x - 600$.

Recall that $x$ is the number of ounces of perfume that we sell per week and $P(x)$ is the corresponding profit from the sale. Observe that a negative profit corresponds to a loss, and that your "break-even" point is $x = 4$ units, since if you sell less than 4 units per week you suffer a loss, whereas if you sell more than 4 units per week you enjoy a profit. This break-even point at which profit is equal to zero is the point at which our graph crosses the $x$-axis; it is called the *x-intercept* of the graph. The graph of our profit function $P(x)$ is apparently a straight line. In fact, $P(x)$ is a type of function that is called "linear."

**LINEAR FUNCTION** | If $m$ and $b$ are constants, then the function $f(x) = mx + b$ is a *linear function*.

It can be shown that the graph of every linear function is a straight line. In our profit function $P(x) = 150x - 600$, we have $m = 150$ and $b = -600$. Some other examples of linear functions are

$$f(x) = 12x + 3, \quad m = 12, b = 3$$
$$y = 100 - 17x, \quad m = -17, b = 100$$

Examples of functions that are *not* linear are

$$f(x) = 5x^2 + 3x + 7$$
$$g(x) = \sqrt{x - 3}$$
$$y = \frac{x - 5}{x + 1}$$

Since the graph of every linear function is a straight line, it is sufficient when graphing such a function to find two points on the graph and to connect them with a straight line that extends indefinitely in both directions beyond each point. In our profit function example, the line is not extended indefinitely in both directions since the domain is restricted to $0 \leq x \leq 20$. However, in graphing our profit function $P(x) = 150x - 600$, it would have been sufficient to determine two points on the graph—say, $(0, -600)$ and $(5, 150)$—and then to draw the straight line through these two points, taking care not to extend the graph beyond the domain interval $0 \leq x \leq 20$. This easy method works because we know that $P(x) = 150x - 600$ is a linear function and hence its graph is a straight line. Such a simple procedure would *not* be applicable if $P(x)$ were not linear—say, for example, if profit were given by $P(x) = -4x^2 + 2300x - 190000$.

## EXAMPLE 1

Sketch the graph of $f(x) = -3x + 5$.

### Solution

Since this function is linear with $m = -3$ and $b = 5$, we can easily sketch this graph by determining two points on the graph and drawing a line through these points. This is done in Figure 1.21.

| $x$ | $f(x)$ |
|---|---|
| 0 | 5 |
| 1 | 2 |

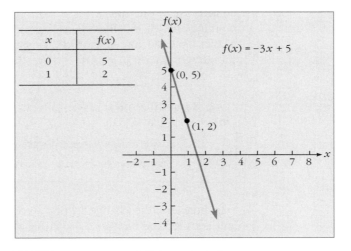

**FIGURE 1.21** The graph of $f(x) = -3x + 5$.

Note that although two points are sufficient to determine a straight line, it is prudent to find a third point on the graph before sketching as a check on your work. If your points don't "line up," check your computations again. If they still don't line up, perhaps your function is not linear. Check again to see that it satisfies the definition of a linear function.

In the definition of the linear function $f(x) = mx + b$, the numbers $m$ and $b$ have a special significance when you graph $f(x)$. If you inspect the graph in Figure 1.20 of $P(x) = 150x - 600$ (for which $b = -600$), you will see that the point $(0, -600)$ lies on the graph. Also, looking at Figure 1.21, which represents the graph of $y = -3x + 5$ (for which $b = 5$), we find that $(0, 5)$ is on this graph. Note that $(0, 5)$ is a point on the $y$-axis and represents the point at which the line corresponding to $y = -3x + 5$ intersects the $y$-axis. We say that the *y-intercept* of this line is equal to 5.

In general, if $y = mx + b$, then by setting $x$ equal to zero, we see that

$$y = m(0) + b = b$$

But setting $x$ equal to zero corresponds to finding the $y$-intercept of the graph, since any point at which the graph intersects the $y$-axis is of the form $(0, y)$ —that is, any such point has an $x$-coordinate equal to zero. We conclude that for the linear function $y = f(x) = mx + b$, *the number $b$ is the y-intercept* and represents the $y$-coordinate of the point at which the graph of $f(x)$ crosses the $y$-axis.

To determine the significance of the number $m$ in the linear function $f(x) = mx + b$, let us consider a table of values for the profit function $P(x) = 150x - 600$ that is somewhat different from the one we saw in Figure 1.20.

| $x$ | 0 | 1 | 2 | 3 | 4 | 5 | 6 | 7 | 8 | 9 | 10 |
|---|---|---|---|---|---|---|---|---|---|---|---|
| $y = P(x)$ | −600 | −450 | −300 | −150 | 0 | 150 | 300 | 450 | 600 | 750 | 900 |

Note that when $x$ increases from 0 to 1—an increase of one unit—our profit $P(x)$ increases from −600 to −450, an increase of 150. This is true since $-450 - (-600) = 150$. As another example, when $x$ increases from 4 to 5— an increase of one unit—we see that $P(x)$ increases from 0 to 150, which again is an increase of 150. Indeed, as you can check for any two consecutive $x$-values in this table, an increase of one unit in $x$ is always accompanied by a corresponding increase in $y$ of $m = 150$.

**INTERPRETATION OF $m$ IN $y = mx + b$**

One way of interpreting the number $m$ in the linear function $y = f(x) = mx + b$ is that

*$m$ represents the increase (or decrease—see Figure 1.21)*
*in $y$ corresponding to an increase of one unit in $x$.*

In business, the change in profit corresponding to increasing sales by one unit is called the *marginal profit*. So in the case of our profit function $P(x) = 150x - 600$, the marginal profit is $150 and corresponds to the amount of additional profit we will earn if we sell one more ounce of our perfume.

Inspecting the preceding table more carefully, we can see another pattern. For example, if $x$ increases by two units (say, from $x = 4$ to $x = 6$), then $y = P(x)$ increases by $2(150) = 300$ units (from $y = 0$ to $y = 300$). Indeed, as you can verify from the table, an increase in $x$ of $n$ units will produce an increase in $y$ of $150n$ units. Let us consider the points $(4, 0)$ and $(6, 300)$, and note that

$$\frac{300 - 0}{6 - 4} = \frac{300}{2} = 150 = m$$

For any two points on the graph of $P(x) = 150x - 600$, if we divide the difference in the $y$-coordinates of these points by the difference in the $x$-coordinates, we obtain the number $m = 150$. This is illustrated for the points $(0, -600)$ and $(10, 900)$ in Figure 1.22, where we have used the notation $\Delta x$ to refer to the *difference* between the $x$-coordinates of the points.

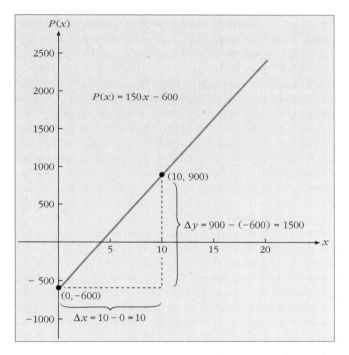

**FIGURE 1.22** The graph of $P(x) = 150x - 600$, showing $\Delta x$ and $\Delta y$ for the points $(0, -600)$ and $(10, 900)$.

The symbol "$\Delta$" is the Greek letter *delta*, and the symbol $\Delta x$ is read as "delta $x$," or sometimes as the "change in $x$" or the "difference in $x$." Do *not* think of $\Delta x$ as meaning delta multiplied by $x$. For example, as we pass from the point $(0, -600)$ to the point $(10, 900)$ along the graph, the "change in $x$" or the "difference in $x$" is $10 - 0 = 10$. Thus, we see that $\Delta x = 10$ for these two points. Similarly, $\Delta y$ is the "difference in $y$" or the "change in $y$." For example, as we pass from the point $(0, -600)$ to the point $(10, 900)$ on the graph, the "change in $y$" or the "difference in $y$" is $900 - (-600) = 1500$, so that $\Delta y = 1500$ for these two points. Also, observe that

$$\frac{\Delta y}{\Delta x} = \frac{900 - (-600)}{10 - 0} = 150 = m$$

**EXAMPLE 2**  Suppose that we can lower the variable cost (for labor, materials, etc.) of producing our perfume to $10 per ounce while maintaining a fixed cost (for rent, insurance, etc.) of $600 per week. Find the new cost function, the new profit function, and graph the old and new profit functions on the same coordinate axes.

**Solution**  Our new cost function is

$$C_N(x) = 10x + 600, \quad 0 \leq x \leq 20$$

Also, since $P(x) = R(x) - C(x)$, our new profit function is

$$P_N(x) = 200x - (10x + 600)$$
$$= 190x - 600, \quad 0 \leq x \leq 20$$

Note that our new marginal profit is $m_N = \$190$. In Figure 1.23 we have graphed both $P(x) = 150x - 600$ and $P_N(x) = 190x - 600$ on the same set of coordinate axes.

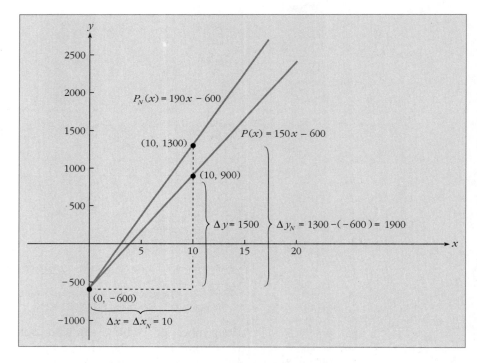

**FIGURE 1.23**  The graph of $P_N(x)$ yields a steeper line than the graph of $P(x)$.

In Example 2, the line for $P_N(x)$ is "steeper" than the line $P(x)$ as we travel from left to right. To see why this is true, consider the points $(0, -600)$ and $(10, 1300)$ on the graph of $P_N(x)$. We have

$$\frac{\text{change in } y}{\text{change in } x} = \frac{\Delta y_N}{\Delta x_N} = \frac{1300 - (-600)}{10 - 0} = \frac{1900}{10} = 190 = m_N$$

Recall that for $P(x)$, we had

$$\frac{\text{change in } y}{\text{change in } x} = \frac{\Delta y}{\Delta x} = \frac{900 - (-600)}{10 - 0} = \frac{1500}{10} = 150 = m$$

In other words, as $x$ changes from 0 to 10, the corresponding change in $y$ is larger ($= \Delta y_N = 1900$) along the line $P_N(x)$ than it is along the line $P(x)$, where $\Delta y = 1500$. The result is that $P_N(x)$ is a steeper line than $P(x)$. This measure of "steepness" or "inclination" is called the *slope* of the line and is equal to $m$ in the general linear equation $y = mx + b$.

**SLOPE OF A LINE**

> If $(x_1, y_1)$ and $(x_2, y_2)$ are any two points on a straight line, with $x_1 \neq x_2$, then the slope $m$ of the line is given by
>
> $$m = \frac{\text{change in } y}{\text{change in } x} = \frac{\Delta y}{\Delta x} = \frac{y_2 - y_1}{x_2 - x_1}$$

See Figure 1.24 for a graphical illustration of this definition.

Note that it is possible for the slope of a line to be negative. For example, if $f(x) = -3x + 5$ (see Figure 1.21), then the slope is $m = -3$. How do we interpret a negative slope? Observe that two points on this line are $(0, 5)$ and $(1, 2)$, so that

$$\frac{\text{change in } y}{\text{change in } x} = \frac{\Delta y}{\Delta x} = \frac{2 - 5}{1 - 0} = \frac{-3}{1} = -3 = m$$

The interpretation is that as we pass from the point $(0, 5)$ to the point $(1, 2)$ along this line, $y$ *decreases* 3 units for each 1 unit increase in $x$. This is so because $\Delta y / \Delta x = -3/1$. See Figure 1.25.

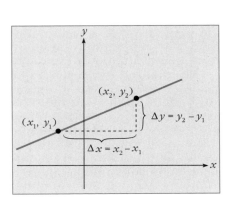

FIGURE 1.24 An illustration of the definition of the slope of a line.

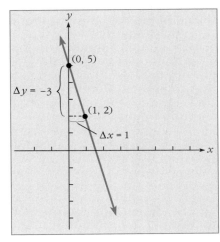

FIGURE 1.25 The slope of the line with equation $y = -3x + 5$ is $-3$.

From this example we can see that a line with negative slope will be falling as we move from left to right along the graph. On the other hand, if the slope is

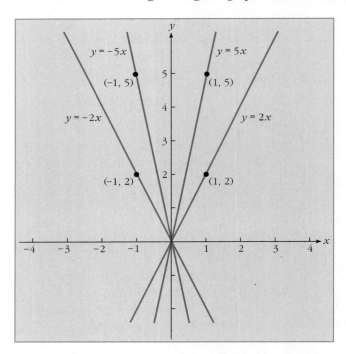

FIGURE 1.26 A comparison sketch of several straight lines that pass through the origin.

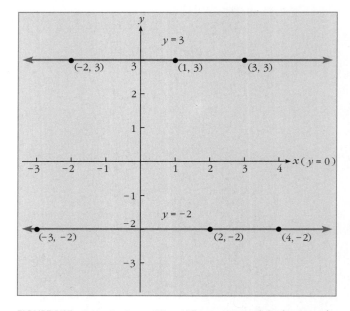

FIGURE 1.27 Some horizontal lines. The equation of the horizontal line passing through $(x_0, y_0)$ is $y = y_0$.

positive (as in our profit function, $P(x) = 150x - 600$, where $m = 600$; see Figure 1.20) then the line will rise as we move from left to right along the graph. In Figure 1.26 we illustrate several lines passing through the origin.

### Horizontal and Vertical Lines

A horizontal line has slope zero, since the change in $y$ is 0 for any change in $x$. Thus $m = \Delta y/\Delta x = 0/(x_2 - x_1) = 0$ if $x_1 \neq x_2$. A horizontal line, therefore, has the form $y = 0x + b$, or $y = b$. Thus the $y$-coordinate of any point on the graph of a horizontal line is equal to the $y$-intercept $b$. Figure 1.27 depicts several horizontal lines.

The slope of a vertical line is undefined, since for a vertical line the change in $x$ is 0 for any change in $y$. Therefore $m = \Delta y/\Delta x = (y_2 - y_1)/0$, and division by zero is undefined. A vertical line does *not* define a function since it fails the vertical line test, and therefore it cannot be written in the form $y = mx + b$, that of a linear function. It is not difficult to see, however, that the equation of a vertical line passing through the point $(x_0, y_0)$ is equal to $x = x_0$. See Figure 1.28 for the graphs of some vertical lines.

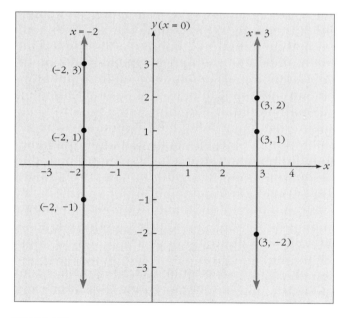

**FIGURE 1.28** Some vertical lines. The equation of the vertical line passing through $(x_0, y_0)$ is $x = x_0$.

### Forms of the Equation of a Straight Line

Let us now consider some forms for the equation of a straight line. The first form is already familiar as the form of the general linear function, and it is called the *slope–intercept form* of the equation of a straight line.

**SLOPE–INTERCEPT FORM**

$$y = mx + b$$

where $m$ is the slope, $b$ is the $y$-intercept.

Now, suppose that we know the slope $m$ of a straight line and that the line passes through the point $(x_1, y_1)$. Let $(x, y)$ be an arbitrary point on this line different from $(x_1, y_1)$. See Figure 1.29.

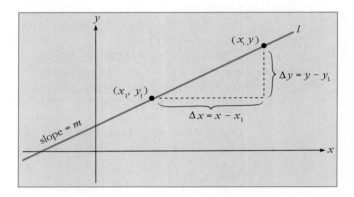

FIGURE 1.29 The derivation of the slope–intercept form of the equation of the straight line $l$.

Then

$$m = \frac{\Delta y}{\Delta x} = \frac{y - y_1}{x - x_1}$$

so that

$$y - y_1 = m(x - x_1).$$

We have just derived what is called the *point–slope form* of the equation of a straight line.

**POINT–SLOPE FORM**

$$y - y_1 = m(x - x_1)$$

where $m$ is the slope, $(x_1, y_1)$ is a point on the line.

Another form of the equation of a straight line is obtained by replacing $m$ in the point–slope form by its definition, $(y_2 - y_1)/(x_2 - x_1)$. We obtain the *two-point form* of the equation of a straight line.

**TWO-POINT FORM**

$$y - y_1 = \frac{y_2 - y_1}{x_2 - x_1}(x - x_1)$$

where $(x_1, y_1)$ and $(x_2, y_2)$ are two points on the line.

Finally, we have what is called the *general form* of the equation of a straight line, which is

**GENERAL FORM**

$$Ax + By = C$$

Any equation of a straight line can be put into general form. This form usually serves as a standard way to express the final equation when we are working problems.

**EXAMPLE 3** Find an equation of the line with slope $\frac{2}{3}$ and $y$-intercept $-2$.

Solution  We use the slope–intercept form $y = mx + b$ with $m = \frac{2}{3}$ and $b = -2$ to obtain

$$y = \frac{2}{3}x - 2$$

To put this equation into general form, we first clear it of fractions by multiplying both sides of the equation by 3:

$$3y = 2x - 6$$

Finally, we write the equation in general form:

$$2x - 3y = 6$$

**EXAMPLE 4** A line with slope $-4$ passes through the point $(-1, 2)$. Find an equation of the line and sketch the graph.

Solution  Since we are given the slope $m = -4$ and a point on the line, $(x_1, y_1) = (-1, 2)$, to find an equation of the line, it is convenient to use the point–slope form: $y - y_1 = m(x - x_1)$. Making the appropriate substitutions for $m$, $x_1$, and $y_1$, we get

$$y - 2 = -4[x - (-1)]$$
$$= -4(x + 1)$$

To put this equation into general form, we proceed as follows:

$$y - 2 = -4x - 4$$
$$4x + y = -2$$

To sketch the graph of the equation, note that we are given one point on the line, namely, $(-1, 2)$. To obtain another point, let $x = 0$ (say) in the equation $4x + y = -2$. We obtain

$$4(0) + y = -2$$
$$y = -2$$

So another point on the line is $(0, -2)$. See Figure 1.30 for the sketch.

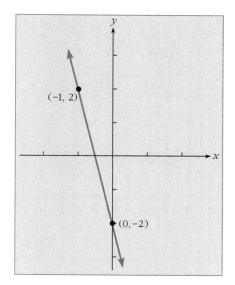

**FIGURE 1.30** The graph of the straight line $4x + y = -2$ of Example 4.

EXAMPLE 5 (Participative)

Solution

Let's find an equation of the line that passes through the points $(-2, 3)$ and $(2, -1)$, and let us also find the $y$-intercept of this line.

Since we are given two points on the line, $(x_1, y_1) = (-2, 3)$ and $(x_2, y_2) =$ _____, it is most convenient to use the two-point form of the equation of a straight line:

$$y - y_1 = \frac{y_2 - y_1}{x_2 - x_1}(\underline{\qquad})$$

Substituting the appropriate values, we get

$$y - 3 = \underline{\qquad}[x - (-2)]$$
$$= \underline{\qquad}$$

In general form, this equation is _____. To find the $y$-intercept of the line, we equate $x$ to zero in this equation to obtain $y =$ \_\_\_\_ as the $y$-intercept. ∎

### Parallel and Perpendicular Lines

Here are some other useful facts about the slopes of straight lines (proofs are omitted):

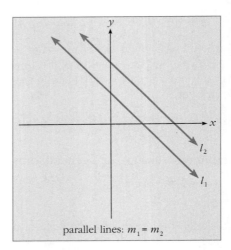

parallel lines: $m_1 = m_2$

FIGURE 1.31 Two parallel lines $l_1$ and $l_2$ have slopes $m_1 = m_2$.

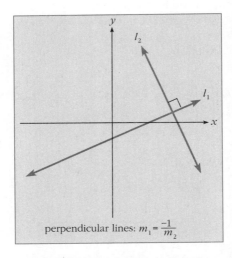

perpendicular lines: $m_1 = \frac{-1}{m_2}$

FIGURE 1.32 Two perpendicular lines $l_1$ and $l_2$ have slopes $m_1$ and $m_2$ such that $m_1 = -1/m_2$.

---

**SOLUTIONS TO EXAMPLE 5**

$(2, -1)$, $x - x_1$, $\dfrac{-1 - 3}{2 - (-2)}$, $-(x + 2)$, $x + y = 1$, $1$

i. Two nonvertical lines $l_1$ and $l_2$ with slopes $m_1$ and $m_2$ are *parallel* if and only if the slopes of these lines are equal; that is, $m_1 = m_2$ (see Figure 1.31).
ii. Two lines $l_1$ and $l_2$ (neither of which is vertical) with slopes $m_1$ and $m_2$, respectively, are *perpendicular* if and only if their slopes are negative reciprocals of one another; that is, $m_1 = -1/m_2$ (see Figure 1.32).

### EXAMPLE 6

In the production of a certain product, your monthly cost in dollars of manufacturing $x$ units of the product is $C(x) = 50x + 400$. If your fixed cost increases to $600 per month, find your new cost function, assuming that your variable cost remains the same. Graph both the old and the new cost functions on the same coordinate axes.

### Solution

First, observe that in the equation $C(x) = 50x + 400$, the fixed cost is $400 and the variable cost is $50. The fixed cost corresponds to the $y$-intercept, while the variable cost corresponds to the slope of the line $C(x) = 50x + 400$. If your fixed cost increases to $600 per month while your variable cost remains the same at $50, your new cost function is

$$C_N(x) = 50x + 600$$

Because these lines have the same slope (namely, 50), the lines $C(x)$ and $C_N(x)$ are parallel. These lines are graphed in Figure 1.33.

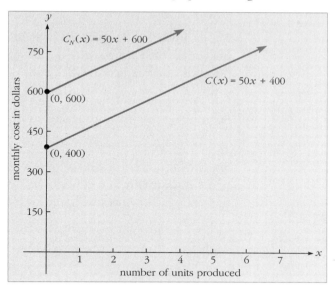

**FIGURE 1.33** The two cost lines of Example 6 are parallel because their slopes are both equal to 50.

### EXAMPLE 7 (Participative)

Let's find an equation of the line that passes through the point $(-1, 2)$ and is perpendicular to the line $2x + 3y = 7$. Also let us find the point of intersection of these two lines.

**42** CHAPTER 1 FUNCTIONS AND GRAPHS

Solution   In order to determine the slope, we first put the given line's equation into the slope–intercept form, $y = mx + b$. Solving the equation for $y$, we obtain

$3y = $ _____

$y = $ _____

The slope of the given line is the coefficient of $x$, which is _____. Since the line we want is perpendicular to this line, its slope is the negative reciprocal of $-2/3$, which is _____. Since we now know the slope of the perpendicular line and a point on the perpendicular line (namely, _____), we will use the point–slope form of the equation of a line to complete the problem. We have

$y - y_1 = m(x - x_1)$   (point–slope form)

Substituting $x_1 = -1$, $y_1 = $ _____, and $m = $ _____, we obtain

$y - 2 = $ _____

Writing this in general form, we get _____.

Next we are required to find the point of intersection of the two straight lines

$2x + 3y = 7$
$-3x + 2y = 7$

This is a problem in solving a "system" of two linear equations in two unknowns. One way to solve such a system is to solve the first equation for $y$. This was done earlier and the result is

$y = $ _____

Substituting this into the second equation gives us

$-3x + 2(\text{_____}) = 7$

Simplifying and solving for $x$ gives $x = $ _____. To obtain $y$ we may substitute this value of $x$ into either of the two original equations. Substituting into the first equation gives us

$2(\text{_____}) + 3y = 7$

so that

$y = $ _____

The point of intersection of the two lines is therefore _____. ■

---

**SOLUTIONS TO EXAMPLE 7**   $-2x + 7$, $-\frac{2}{3}x + \frac{7}{3}$, $-\frac{2}{3}$, $\frac{3}{2}$, $(-1, 2)$, $2$, $\frac{3}{2}$, $\frac{3}{2}(x + 1)$, $-3x + 2y = 7$, $-\frac{2}{3}x + \frac{7}{3}$, $-\frac{2}{3}x + \frac{7}{3}$, $-\frac{7}{13}$, $-\frac{7}{13}$, $\frac{35}{13}$, $\left(-\frac{7}{13}, \frac{35}{13}\right)$

EXAMPLE 8

In order to manufacture component parts, we purchase a machine that costs $24,000. If we use straight-line depreciation over a period of ten years, find an equation for the value $V$ of our machine after $t$ years, where $0 \leq t \leq 10$. Graph the function $V(t)$.

Solution

Using the straight-line method of depreciation implies that the value of our machine decreases by the same amount every year. Thus, over a ten-year period, the value must decrease from $24,000 to $0 (assuming that there is no "salvage value" at the end of ten years). The decrease per year must then be

$$\frac{\$24{,}000}{10 \text{ years}} = \$2400/\text{year}$$

Since every increase of one year for $t$ produces a corresponding decrease of $2400 for $V$, our function is linear: $V(t) = mt + b$, with $m = -2400$. To find $b$, note that $V(0) = b$, where $V(0)$ is the value of the machine at time $t = 0$ years. This of course is its value when new, which is $24,000. We conclude that $b = 24000$, so that

$$V(t) = -2400t + 24000, \quad 0 \leq t \leq 10$$

To graph $V(t)$, note that, say, $V(5) = -2400(5) + 24000 = \$12{,}000$, which represents the value of our machine after five years. So one point on the graph is $(5, 12000)$. Also, since $V(0) = 24000$ and $V(10) = 0$, other points on our straight-line graph are $(0, 24000)$ and $(10, 0)$. The graph of $V(t)$ is drawn in Figure 1.34. Can you see why this method of depreciating an asset is called "straight-line depreciation"?

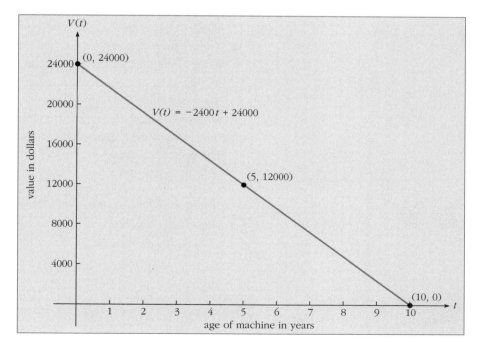

FIGURE 1.34  The value $V(t)$ of the machine after $t$ years graphs as a straight line.

## SECTION 1.4
### SHORT ANSWER EXERCISES

1. Given a profit function $P(x) = 20x - 600$, the slope of the straight line represented by this function is _____ and the $y$-intercept is _____.
2. For the profit function $P(x) = 20x - 600$, if sales ($x$) increase by 50 units, then profit (increases, decreases) by _____ dollars.
3. In a linear function $y = 2x + b$, if $(1, 0)$ is a point on the line represented by this function, then $b = $ _____.
4. (True or False) The points $(0, 0)$, $(1, 1)$, and $(2, 4)$ all lie on the same straight line. _____
5. An equation of the straight line with slope $-4$ and $y$-intercept 5 is $y = $ _____.
6. If $y = C(x) = 35x + 600$, then $C(5) = $ _____ and $C(10) = $ _____. Thus, $\Delta y/\Delta x = (C(10) - C(5))/(10 - 5) = $ _____, which is the slope of this line.
7. A company has a linear profit function with marginal profit equal to \$20 per unit and a break-even point equal to 50 units. The profit function is $P(x) = $ _____.
8. On a straight line, as we pass from the point $(5, 300)$ to the point $(15, 800)$ along the line, $\Delta x = $ _____ and $\Delta y = $ _____, hence $m = \Delta y/\Delta x = $ _____.
9. Four straight lines have slopes $-6$, $4$, $7$, and $10$ respectively. The "steepest" line among these is the line with slope equal to _____, while the "shallowest" line among these is the line with slope equal to _____.
10. The only straight line that does not define $y$ as a function of $x$ is a _____ line.
11. If the fixed cost of production decreases by \$500 per month while the variable cost remains constant, how much less does it cost per month to produce 100 units of our product? _____
12. Three lines $l_1$, $l_2$, and $l_3$ have slopes $-4$, $\frac{1}{4}$, and $-4$, respectively. We conclude that $l_1$ and $l_2$ are _____, $l_1$ and $l_3$ are _____, and $l_2$ and $l_3$ are _____.
13. The slope of the line with equation $2x - 3y = 5$ is _____.
14. Given the straight-line depreciation formula $V(t) = -500t + 2500$, where $t$ is time in years and $V(t)$ is value in dollars after $t$ years, what is the original price of this asset? _____

## SECTION 1.4
### EXERCISES

*In Exercises 1–8, find an equation of the line having the properties listed.*

1. slope $= 2$, $y$-intercept $= 7$
2. slope $= -3$, $y$-intercept $= 4$
3. slope $= \frac{2}{3}$, $y$-intercept $= -\frac{1}{3}$
4. slope $= -\frac{1}{5}$, $y$-intercept $= 1$
5. slope $= -1$, passes through $(3, 4)$
6. slope $= 4$, passes through $(-1, -2)$
7. slope $= \frac{1}{2}$, passes through $(0, -3)$
8. slope $= -\frac{2}{7}$, passes through $(-3, 5)$

*In Exercises 9–16, identify each function as linear or nonlinear.*

9. $f(x) = 2x + 7$
10. $f(x) = -2x - 3$
11. $f(x) = 12 - \frac{3}{5}x$
12. $f(x) = x^2 + 5$

13. $f(x) = \sqrt{3} + 5x$
14. $f(x) = \sqrt{x} + 2$
15. $f(x) = \dfrac{2x - 3}{x + 5}$
16. $f(x) = \dfrac{2x - 3}{8}$

*In Exercises 17–28, sketch the graph of the linear function and find the slope and the y-intercept. Interpret the slope by showing $\Delta x$ and $\Delta y$ in your diagram as x changes from 0 to 1.*

17. $f(x) = 3x + 2$
18. $f(x) = -3x - 3$
19. $f(x) = -2x + 1$
20. $f(x) = 5x - 4$
21. $f(x) = \frac{2}{3}x - 1$
22. $f(x) = -\frac{3}{4}x + 2$
23. $P(x) = 100x - 400, \quad x \geq 0$
24. $P(x) = 350x - 900, \quad x \geq 0$
25. $C(x) = 50x + 200, \quad x \geq 0$
26. $C(x) = 70x + 500, \quad x \geq 0$
27. $V(x) = -1000x + 5000, \quad 0 \leq x \leq 5$
28. $V(x) = -80x + 800, \quad 0 \leq x \leq 10$

*In Exercises 29–35, determine which of the given pairs of lines are parallel, which are perpendicular, and which are neither parallel nor perpendicular.*

29. $x + 2y - 7 = 0, y - 2x = 4$
30. $2x - 3y = 5, 6y - 4x = 1$
31. $4x + 5y = -3, 4y - 5x = 3$
32. $x - y = 3, y = 2x - 4$
33. $C(x) = 50x + 300, C(x) = 50x + 500, \quad x \geq 0$
34. $P(x) = 150x - 600, P(x) = 190x - 600, \quad x \geq 0$
35. $x = 6, y = 2$

36. Find an equation of the line that passes through the point $(1, -3)$ and is parallel to the line $2x - 3y = 4$.

37. Find an equation of the line that passes through the point $(-2, -5)$ and is parallel to the line $3x - 5y = 3$.

38. Find an equation of the line that passes through the point $(0, -4)$ and is perpendicular to the line $4x - 5y = 8$.

39. Find an equation of the line that passes through the point $(-4, 2)$ and is perpendicular to the line $7x - 3y = -2$.

40. Find an equation of the line that passes through the origin and is perpendicular to the line that passes through $(1, -2)$ and $(-3, -5)$.

41. (Linear profit function) Given a linear profit function $P(x) = 250x - 2000$, where $P(x)$ is dollars of profit derived from the sale of $x$ units of product:
    (a) find the profit for the sale of ten units;
    (b) find the number of units $x$ corresponding to a profit of $0. How do we usually refer to this number of units?
    (c) If sales increase by 15 units from their current level, by how much will profit increase?

**Business and Economics**

42. (Fixed and variable costs) If it costs $1500 to manufacture 75 wrist watches, and it costs $2000 to manufacture 100 wrist watches,
    (a) determine the cost function $C(x)$, assuming $C(x)$ is linear;
    (b) find the fixed cost of production per unit time period;
    (c) find the variable cost of production.
    (d) How much will it cost to manufacture 200 wrist watches?

**43.** (Demand function)  If you charge $5 per unit of a certain product, demand will be 100 units. If you charge $2 per unit, demand will increase to 550 units. Assuming that the demand function is linear,
  (a) determine the demand function $q(p)$, where $q$ is the quantity demanded at price $p$;
  (b) find the demand if you charge $3 per unit.

**44.** (Revenue and profit)  You are in the business of making and selling homemade ceramic items. For each ceramic piece, you charge $40. You estimate that your small shop costs you about $200 a month for fixed expenses. Material and labor for the production of each ceramic piece costs about $10.
  (a) Find your revenue function $R(x)$ and your cost function $C(x)$, assuming these functions are linear and all revenue is derived from the sale of your ceramic items.
  (b) Find your profit function $P(x)$ and graph this function.
  (c) Find your marginal profit and your break-even point.
  (d) If your cost for material and labor increases to $20 per ceramic piece, find and graph your new profit function $P_N(x)$.

**45.** (Depreciation)  We purchase a piece of heavy equipment for $75,000 for our business and we decide to depreciate it over a period of 15 years using the straight-line method.
  (a) Determine an equation for $V(t)$, the value of our asset at time $t$ (years).
  (b) What is the value of our asset after seven years?
  (c) Graph the straight-line value function $V(t)$.

## Social Sciences

**46.** (Population)  The current population of school-age children in a certain county is 78,000. If the number of school-age children in this county is increasing at the rate of 5,000 children per year,
  (a) how many school-age children will there be in this county next year? How many the year after next?
  (b) If $y$ represents the number of school-age children $t$ years from now, then $y = mt + b$. Find $b$ and $m$ in this linear relationship.
  (c) Find the number of school-age children in this county eight years from now.

## Business and Economics

**47.** (Market equilibrium)  If the price of a certain commodity is $p$ dollars per unit and if $x$ is the number of units of the commodity supplied to the market per week by producers, then

$$p = 2x + 300$$

If the number of units of this commodity demanded by consumers per week is $x$ when the price is $p$ dollars, then

$$p = 800 - 3x$$

Find the *equilibrium point* at which supply and demand are equal by finding the point of intersection of these two straight lines.

**48.** (Salary)  A salesperson's weekly salary $S$ consists of a base pay of $150 and 8% of the dollar amount of her weekly sales $x$.
  (a) Find the salesperson's weekly salary $S$ as a linear function of her weekly sales $x$.
  (b) If weekly sales amount to $1200, compute the salesperson's weekly salary.

**49.** (Temperature)  The relationship between degrees Fahrenheit, °F, and degrees Celsius, °C, is linear. If 0°C is equal to 32°F and 20°C is equal to 68°F,

(a) find the formula for converting degrees Celsius into degrees Fahrenheit.
(b) What is the Fahrenheit equivalent of 100°C (the boiling point of water)?

**50.** (Profit) The annual profit of a certain company has been decreasing at a constant rate since 1988, when profit was $20 million. In 1989, profit was $17.5 million. Assume that this trend continued through 1994.
(a) Express the profit $P$ of this company as a function of time $t$. Let $t = 0$ correspond to 1988, $t = 1$ correspond to 1989, and so forth. What is the domain of your function?
(b) What is the profit of this company in 1993?

Social Sciences

**51.** (Placement testing) Entering freshmen at a major university are given a mathematics placement test and these results are compared with their numerical grade in the freshman mathematics course Math 101. It has been found that students who score 50 on the placement test average 62 in Math 101, and students who score 65 on the placement test average 70 in Math 101. Assume that the relationship between placement test scores and grades in Math 101 is linear.
(a) If $p$ is the placement test score and $g$ is the final grade in Math 101, determine $g$ as a function of $p$.
(b) Use the relationship you found in part (a) to predict the grade in Math 101 of a student who scores 75 on the placement test.
(c) According to the relationship you found in part (a), what is the lowest grade on the placement test that will yield a predicted grade of C or better (70 or more) in Math 101?

OPTIONAL

**GRAPHING CALCULATOR/ COMPUTER EXERCISES**

*In Exercises 52–55, graph each of the straight lines and estimate the coordinates of the point of intersection using a graphing calculator or computer. Then find the exact coordinates of the point of intersection by solving the system of equations algebraically (see Example 7).*

**52.** $2x - 5y = 7$
$3x + 4y = 10$

**53.** $7x - 25y = 30$
$12x - 9y = 42$

**54.** $0.2x - 0.7y = 0.83$
$-0.7x + 0.12y = -0.37$

**55.** $0.92x + 0.73y = 1.24$
$1.45x - 6.70y = 2.31$

## 1.5 QUADRATIC FUNCTIONS AND APPLICATIONS

Besides the linear function we studied in Section 1.4, another important class of functions is the *quadratic* function. Let us see how quadratic functions might arise by returning to our perfume company example.

During the past few months we have noticed a decrease in the sales of our perfume. Looking over our records, we notice that sales have averaged about 12 ounces per week for the past three months, while previously we were always able to sell from 18 to 20 ounces per week. Since our profit function is $P(x) = 150x - 600$ (see Figure 1.20), selling 6 to 8 fewer ounces per week has decreased our weekly profit by between $900 and $1200. Since this is serious money for our small company, we begin to look for reasons for the decrease in sales, and ways to increase our profit.

Since we know that a new low-cost competitor began operations in the area some months ago, we feel that price may be a factor that we should consider. Specifically, if we lowered the price, would that increase our profit? To investigate this question, we decide to survey some of our customers, inquiring about the number of ounces of our perfume they would buy per week at various prices. Our survey specified prices of $150, $200, and $250 per ounce. A summary of the data we obtained from this survey appears in Table 1.4.

The information in Table 1.4 is called a "demand schedule" because it gives the number of ounces of perfume that our customers will demand (that is, buy) at various prices. As we might expect, as price increases, the demand decreases. We note further that *equal* price increases of $50 (for example, from $150 to $200, or from $200 to $250) produce *equal* decreases in demand of 8 ounces (for example, from 20 ounces to 12 ounces, or from 12 ounces to 4 ounces). Thus there is evidence that $x$ is a *linear* function of $p$, and that the slope of the line determined by this linear function is

$$m = \frac{\Delta x}{\Delta p} = \frac{12 - 20}{200 - 150} = \frac{4 - 12}{250 - 200} = -\frac{8}{50} = -\frac{4}{25}$$

Further, using the point–slope formula with $(p_1, x_1) = (150, 20)$,

$$x - x_1 = m(p - p_1)$$

$$x - 20 = -\frac{4}{25}(p - 150)$$

or, upon simplification,

$$x = -0.16p + 44, \quad 150 \leq p \leq 250$$

This demand function is sketched in Figure 1.35. Note that we have restricted the domain of our function to $150 \leq p \leq 250$, since this is the interval in which our survey data lie. Going beyond the range of our data by extending this line would amount to *extrapolation*. That is, since we have no data for prices less than $150 or greater than $250, it would be dangerous to assume that the linear relationship that holds on the range of our survey data, $150 \leq p \leq 250$, is valid outside that range.

We did a similar survey six months ago, and the results of the earlier survey are listed in Table 1.5. Once again, we note that these data are linear, since equal price increases of $50 are accompanied by equal decreases in demand of 8 ounces. In Figure 1.36, the data of Table 1.4 and Table 1.5 are graphed as two straight lines. Apparently, in the last six months, our firm's demand curve has shifted downward and to the left (as indicated by the direction of the arrows in Figure 1.36). The result is that less perfume is now demanded at each price than six months ago. We suspect that some of our customers are substituting the perfume of our low-cost competitor for our perfume.

Since revenue is equal to the price per unit multiplied by the quantity sold, we obtain our current revenue function:

revenue = (price)(quantity)

$$R = px$$
$$= p(-0.16p + 44)$$
$$= 44p - 0.16p^2, \quad 150 \leq p \leq 250$$

| PRICE (p) | WEEKLY DEMAND (x) |
|---|---|
| $150 | 20 ounces |
| $200 | 12 ounces |
| $250 | 4 ounces |

**TABLE 1.4** Demand schedule

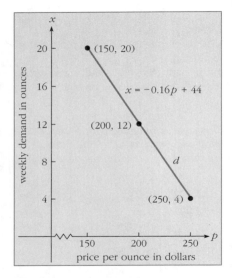

**FIGURE 1.35** The straight line $d$ is the demand function that has equation $x = -0.16p + 44$ for $150 \leq p \leq 250$.

| PRICE (p) | WEEKLY DEMAND (x) |
|---|---|
| $150 | 26 ounces |
| $200 | 18 ounces |
| $250 | 10 ounces |

**TABLE 1.5** Old demand schedule

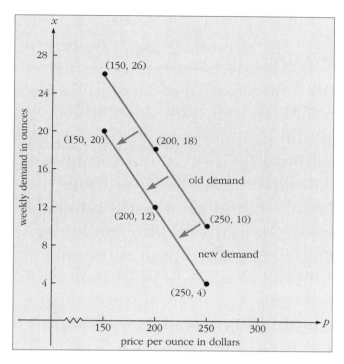

**FIGURE 1.36** A comparison of our old demand curve and our new demand curve shows that our firm's demand curve has shifted downward and to the left.

Since this revenue function contains a term involving $p^2$, our revenue function is no longer linear. This revenue function is the type of function we will study in this section: a *quadratic* function.

---

EXAMPLE 1

Determine the profit $P$ as a function of the price $p$ that we charge per ounce of perfume in the perfume company example.

Solution

Since our cost function is $C(x) = 50x + 600$ (see Section 1.3), where $x$ is the number of ounces sold per week, we find that

profit = revenue − cost
$$P = 44p - 0.16p^2 - (50x + 600)$$
$$= 44p - 0.16p^2 - [50(-0.16p + 44) + 600], \quad 150 \leq p \leq 250$$

Upon simplification, this becomes

$$P = -0.16p^2 + 52p - 2800, \quad 150 \leq p \leq 250$$

As shown in Example 1, profit $P$ is also a *quadratic* function of price $p$ in the perfume company example. We would now like to graph our profit function $P$ and determine the price $p$ we should charge to maximize our profit. Before completing this analysis we will first consider some general facts about quadratic functions that will assist us.

**QUADRATIC FUNCTION**

> If $a$, $b$, and $c$ are constants, then the function $f(x) = ax^2 + bx + c$ is a *quadratic function* if $a \neq 0$.

Note that if $a = 0$ in this definition, then $f(x) = bx + c$, which is a linear function—*not* a quadratic function. Here are some examples of quadratic functions.

$$f(x) = 3x^2 - 2x + 1, \quad a = 3, b = -2, c = 1$$
$$y = 5 + 7x - x^2, \quad a = -1, b = 7, c = 5$$
$$P = -0.16p^2 + 52p - 2800, \quad a = -0.16, b = 52, c = -2800$$

Here are some examples of functions that are *not* quadratic.

$$f(x) = 7x + 3 \quad \text{(This is a \emph{linear} function)}$$
$$g(x) = \sqrt{x^4 - 4} \quad \text{(This is \emph{not} equal to } x^2 - 2\text{)}$$
$$y = \frac{x^2 + 3x + 5}{3x - 4}$$

Just as the graph of every linear function is a straight line, the graph of every quadratic function is a U-shaped curve called a *parabola*. To graph a straight line, recall that it suffices to determine any two points on the line. Although a parabola is uniquely determined by any three points on its graph, we usually do *not* follow the procedure of determining any three points on the parabola and drawing the curve through these points. The reason for this is that, depending on the points we find, it may be difficult to obtain a good graph using only these three points. However, if we are selective in determining which points on the graph we find, and if we keep a few other simple facts in mind, we can sketch a parabola quickly and efficiently by determining only a few points on the graph.

In Figures 1.37, 1.38, 1.39, and 1.40 we have drawn some "typical" parabolas and we have labeled some key features of each curve.

The key features of a parabola are these:

(a) whether the parabola opens upward (∪) or downward (∩);
(b) the *x*-intercept(s) and the *y*-intercept;
(c) the vertex (which is the low point on the graph when the parabola opens upward, and the high point on the graph when the parabola opens downward);
(d) the line of symmetry (which is the line through the vertex and parallel to the *y*-axis). If we think of the line of symmetry as a "hinge," we can fold the graph over on itself using this hinge and the two halves will match perfectly.

A few more comments about the line of symmetry are in order. First, note that once the vertex is known, the line of symmetry is easy to find. Also, the line of symmetry is not part of the graph of the parabola. It is usually drawn as a dashed line to indicate this fact. The line of symmetry simply serves as a guide in drawing the actual parabola.

1.5 QUADRATIC FUNCTIONS AND APPLICATIONS   51

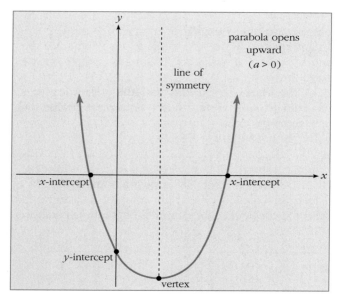

FIGURE 1.37

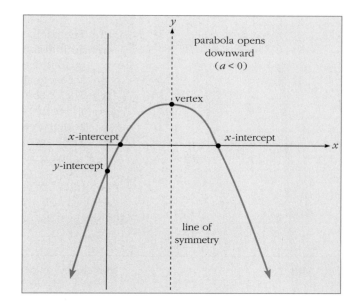

FIGURE 1.38

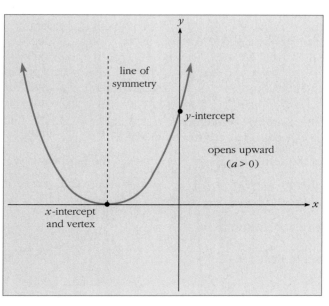

FIGURE 1.39

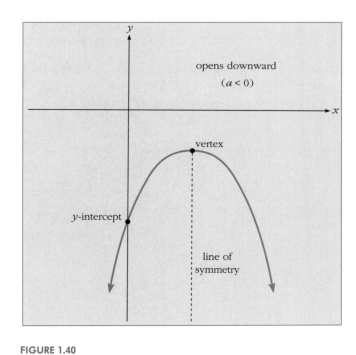

FIGURE 1.40

Figures 1.37–1.40 show some typical parabolas and their key features.

One procedure for graphing a parabola is to determine the key features of the graph and perhaps a few points on the graph other than the vertex or intercepts and then to draw the appropriate U-shaped curve. Note the following facts:

(a) The parabola opens upward if $a > 0$ and downward if $a < 0$.
(b) To find the $x$-intercept(s) (if any), we set $y$ equal to zero and solve the quadratic equation $ax^2 + bx + c = 0$ for $x$. To find the $y$-intercept we set $x$ equal to zero to find $y = c$.
(c) The $x$-coordinate of the vertex is $-b/2a$, and the $y$-coordinate is $f(-b/2a)$. The $x$-coordinate of the vertex is always midway between the $x$-intercepts if any $x$-intercepts exist.
(d) The line of symmetry has equation $x = -b/2a$.

Properties (a) and (c) can be proven easily by applying some concepts of calculus, which we will learn in Chapter 4. For right now, let us assume that properties (a)–(d) are valid and consider some examples of graphing parabolas.

## EXAMPLE 2

Sketch the graph of $f(x) = x^2 - x - 6$.

### Solution

First, observe that this function is quadratic with $a = 1$, $b = -1$, and $c = -6$.

(a) Since $a = 1 > 0$, the parabola opens upward ($\cup$).
(b) To find the $x$-intercept(s), we set $y$ (or $f(x)$) equal to zero to obtain

$$x^2 - x - 6 = 0$$

Solving for $x$ (by factoring), we get

$$(x - 3)(x + 2) = 0$$
$$x - 3 = 0 \quad \text{or} \quad x + 2 = 0$$
$$x = 3 \quad \text{or} \quad x = -2$$

We conclude that there are two $x$-intercepts: $x = 3$ and $x = -2$. To find the $y$-intercept, we set $x$ equal to zero to obtain $y = -6$.

(c) The $x$-coordinate of the vertex is $-\dfrac{b}{2a} = -\dfrac{(-1)}{2(1)} = \dfrac{1}{2}$. The $y$-coordinate of the vertex is $f(-b/2a) = f(\tfrac{1}{2}) = (\tfrac{1}{2})^2 - \tfrac{1}{2} - 6 = -\tfrac{25}{4}$. The vertex is located at the point $(\tfrac{1}{2}, -\tfrac{25}{4})$.

(d) The line of symmetry has equation $x = -\dfrac{b}{2a} = \dfrac{1}{2}$.

This parabola is sketched in Figure 1.41. ■

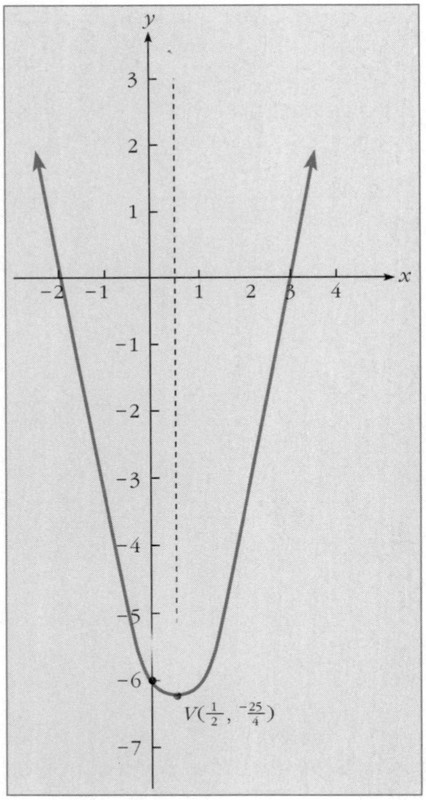

**FIGURE 1.41** The graph of $f(x) = x^2 - x - 6$ is a parabola.

Before we attempt the next example, recall that the solutions of the quadratic equation $ax^2 + bx + c = 0$ are given by the *quadratic formula*:

$$x = \frac{-b \pm \sqrt{b^2 - 4ac}}{2a}$$

EXAMPLE 3

Sketch the graph of $y = 2 + 4x - 3x^2$.

Solution

This function is quadratic with $a = -3$, $b = 4$, and $c = 2$.

(a) Since $a = -3 < 0$, the parabola opens downward.
(b) To find the $x$-intercept(s), we set $y$ equal to zero to obtain
$$2 + 4x - 3x^2 = 0.$$
Using the quadratic formula, we get:
$$x = \frac{-4 \pm \sqrt{4^2 - 4(-3)(2)}}{2(-3)}$$
$$= \frac{-4 \pm \sqrt{40}}{-6}$$
$$= \frac{-4 \pm 2\sqrt{10}}{-6} = \frac{2}{3} \pm \frac{1}{3}\sqrt{10}$$

The roots are $x_1 = \frac{2}{3} + \frac{1}{3}\sqrt{10} \approx 1.72$ and $x_2 = \frac{2}{3} - \frac{1}{3}\sqrt{10} \approx -0.39$, so there are two $x$-intercepts. To find the $y$-intercept, we set $x$ equal to zero to obtain $y = 2$.

(c) The $x$-coordinate of the vertex is $-\frac{b}{2a} = -\frac{4}{2(-3)} = \frac{2}{3}$. The $y$-coordinate of the vertex is $f(-b/2a) = f(\frac{2}{3}) = 2 + 4(\frac{2}{3}) - 3(\frac{2}{3})^2 = \frac{10}{3}$. The vertex is located at the point $(\frac{2}{3}, \frac{10}{3})$.

(d) The line of symmetry has equation $x = -\frac{b}{2a} = \frac{2}{3}$.

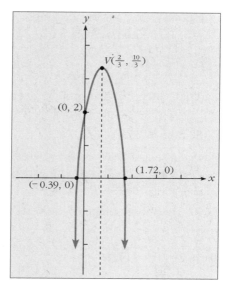

FIGURE 1.42 The graph of $y = 2 + 4x - 3x^2$.

The parabola is sketched in Figure 1.42.

EXAMPLE 4 (Participative)

Let's sketch the graph of $f(x) = 2x^2 + 2x + 1$.

Solution

We begin by noting that the given function is quadratic with $a =$ ____, $b =$ ____, and $c =$ ____.

(a) Since $a = 1 > 0$ the parabola opens _____.
(b) To find the $x$-intercept(s), we set $2x^2 + 2x + 1 = 0$. Using the quadratic formula with $a = 2$, $b = 2$, and $c = 1$, we find that
$$x = \frac{-2 \pm \sqrt{2^2 - 4(2)(1)}}{2(2)} = \frac{-2 \pm \sqrt{-4}}{4}$$

Since the quantity under the square root is negative, we conclude that there are no real solutions to the equation, so there are ____ $x$-intercepts. The $y$-intercept is $y =$ ____.

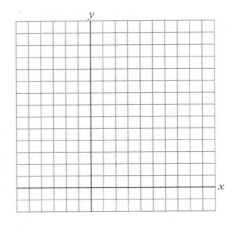

**FIGURE 1.43**

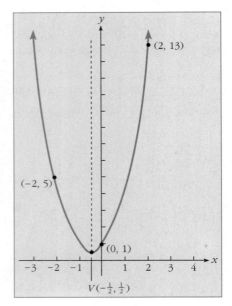

**FIGURE 1.44**

(c) The x-coordinate of the vertex is $x = -b/2a =$ ____. The y-coordinate of the vertex is $f\left(-\frac{1}{2}\right) =$ ____. The vertex is located at the point ____.

(d) The equation of the line of symmetry is ____.

(e) Since the graph has no x-intercepts, it is useful to determine at least one point on the graph to the left of the vertex and one point on the graph to the right of the vertex (besides the y-intercept) to assist us in drawing the sketch. Since the x-coordinate of the vertex is $x = -\frac{1}{2}$, we may choose to find the points on the graph having, say, x-coordinate $-2$ and x-coordinate 2. Since $f(-2) =$ ____, the point $(-2, 5)$ is on the graph. Also, since $f(2) =$ ____, the point $(2, 13)$ is on the parabola.

Sketch the parabola on the coordinate axes provided here. ■

Let us return to our perfume company example, where we had profit $P$ given as a quadratic function of price:

$$P = f(p) = -0.16p^2 + 52p - 2800, \quad 150 \leq p \leq 250$$

Since the graph of this function is a parabola that opens downward ($a = -0.16 < 0$), the high point on the graph (which corresponds to the maximum profit) will be the vertex. Then because $b = 52$, we conclude that the p-coordinate of the vertex is $p = \dfrac{-b}{2a} = -\dfrac{52}{2(-0.16)} = \$162.50$. The P-coordinate of the vertex is

$$f(162.50) = -0.16(162.50)^2 + 52(162.50) - 2800 = \$1425$$

So given our current demand schedule, we should charge a price of $162.50 per ounce for our perfume in order to maximize our profit per week at $1425. A graph of $f(p)$ is given in Figure 1.45. Note that we have drawn only that portion of the parabola that lies between $p = 150$ and $p = 250$, which was the range of the survey data from which we derived our demand curve of Figure 1.35.

Note that, although we can increase our profit from $1200 to $1425 by reducing our price from $200 per ounce to $162.50 per ounce, we are still suffering under the effects of our new competitor. That is, when our old demand schedule (Table 1.5) was in effect, we were able to sell more ounces of perfume per week and at a higher price. We had been charging $200 per ounce* and were able to sell 18 ounces at that price, obtaining a profit of $150(18) - 600 = \$2100$ per week. We are therefore still making $2100 - $1425 = $675 less per week than before. The reason for this is that our demand curve has shifted downward and to the left (see Figure 1.36), so that even if we price at the optimal level of

**SOLUTIONS TO EXAMPLE 4**

2, 2, 1; (a) upward; (b) no, 1; (c) $-\frac{1}{2}, \frac{1}{2}, \left(-\frac{1}{2}, \frac{1}{2}\right)$; (d) $x = -\frac{1}{2}$; (e) 5, 13; see figure 1.44.

*See Exercise 17 of this section for a discussion of our old pricing strategy.

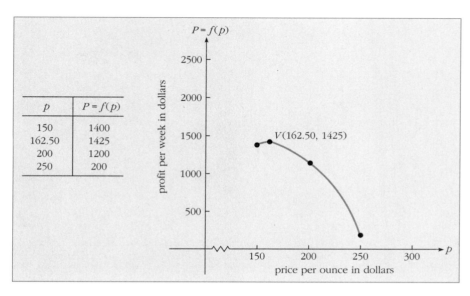

**FIGURE 1.45** The graph of the profit function $P = f(p) = -0.16p^2 + 52p - 2800$ for $150 \leq p \leq 250$.

$162.50 per ounce, we will still make less profit than before. Exercise 18 of this section asks you to consider alternative ways (other than pricing alone) to regain more of our profit.

---

EXAMPLE 5

Your friend Jim has recently purchased a small bakery that specializes in cakes. The previous owner told Jim that the cost function of the bakery is

$$C(x) = \frac{1}{500}x^2 - 2x + 1500$$

where $x$ is the number of cakes baked and $C(x)$ is the weekly cost in dollars of making $x$ cakes. Since Jim knows you are taking a calculus course, he asks you how many cakes he should bake per week in order to minimize his cost, assuming that the cost function he was given is still valid.

Solution

Note that the cost function is quadratic with $a = \frac{1}{500}$, $b = -2$, and $c = 1500$. The $x$-coordinate of the vertex gives the number of cakes to be baked per week in order to minimize cost. This number is

$$x = -\frac{b}{2a} = -\frac{-2}{2\left(\frac{1}{500}\right)} = 500$$

The minimum cost is therefore

$$C(500) = \frac{1}{500}(500)^2 - 2(500) + 1500 = \$1000$$

The cost function is graphed in Figure 1.46.

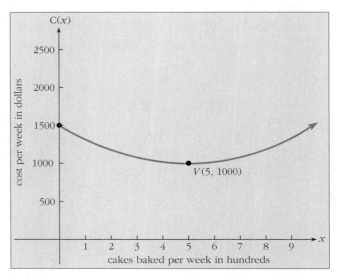

FIGURE 1.46 The graph of the bakery cost function, $C(x) = \frac{1}{500}x^2 - 2x + 1500$.

## EXAMPLE 6 (Participative)

$y = C(x)$

FIGURE 1.47

If your cost function is $C(x) = 0.04x^2 - 6x + 2000$, where $x$ is the number of units produced and $C(x)$ is the monthly cost in dollars of producing $x$ units, then

(a) your cost function is quadratic with $a =$ _____, $b =$ _____, and $c =$ _____.

(b) To minimize cost, your output should be $x = -b/2a =$ _____ units.

(c) When you produce 75 units, your minimum cost is $C(75) =$ _____.

(d) To graph the cost function, note that since $a = 0.04 > 0$, the parabola opens _____. Also, from parts (b) and (c), the vertex of the parabola is located at the point _____.

(e) To find the $y$-intercept, we set $x$ equal to zero to obtain $y =$ _____.

(f) To find the $x$-intercept(s), we set $y = C(x)$ equal to zero to obtain $0.04x^2 - 6x + 2000 = 0$. Using the quadratic formula, we find that, since $b^2 - 4ac =$ _____ $< 0$, there are no $x$-intercepts.

(g) Using the symmetry of the parabola, since $(0, 2000)$ is a point on the graph and since the line of symmetry is $x =$ _____, another point on the graph is $(150,$ _____$)$.

(h) Graph the cost function on the coordinate axes provided.

**SOLUTIONS TO EXAMPLE 6**

(a) 0.04, −6, 2000; (b) 75; (c) 1775; (d) upward, (75, 1775); (e) 2000; (f) −284; (g) 75, 2000; (h) see figure 1.48.

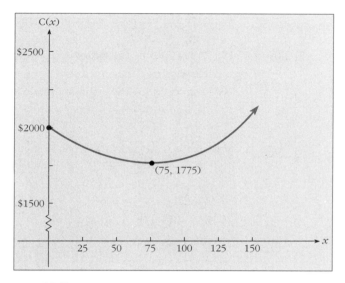

**FIGURE 1.48**

## SECTION 1.5

**SHORT ANSWER EXERCISES**

1. The function $f(x) = 7 - x + x^2$ is (a) linear; (b) quadratic; (c) neither linear nor quadratic.
2. The graph of every quadratic function is a curve called a _____.
3. The $y$-intercept of the parabola represented by $y = 7x^2 + 3x - 100$ is _____.
4. If our firm's demand curve shifts downward and to the left, then at each price, _____ units are demanded.
5. If your firm's demand is $q = 0.25p + 500$, then your revenue function is $R = pq = $ _____.
6. The $x$-intercepts of the parabola represented by $y = 2x^2 - 5x - 3$ are _____ and _____.
7. The $x$-coordinate of the vertex of the parabola represented by $y = 7 - 4x + 2x^2$ is _____.
8. If your revenue function is $y = R(p) = 60p - 0.30p^2$, where revenue is given in dollars, you can maximize revenue by charging a price of _____ per unit.
9. The vertex of a parabola is either the high point or the low point on the graph. If $y = ax^2 + bx + c$ and $a = 5$, the vertex will be the (high, low) point on the curve. _____
10. If $(-2, 4)$ is one point on a parabola having line of symmetry $x = 0$, then another point on the graph is _____.
11. For the parabola represented by $y = ax^2 + bx + c$, the line of symmetry is perpendicular to the _____-axis.
12. In the cost function $C(x) = \frac{1}{500}x^2 - 2x + 1500$, the fixed cost is _____ per time period.
13. (True or False) The graph of $y = x^3 - 7x^2 + 3x + 1$ is a parabola. _____

14. If the x-intercepts of a certain parabola are $-1$ and $4$, the x-coordinate of the vertex is ___.

15. (True or False) The parabola represented by $y = 4x^2 + 4x + 1$ has only one x-intercept. ___

16. (True or False) The graph of every quadratic function has a y-intercept. ___

17. (True or False) The graph of every quadratic function has an x-intercept. ___

## SECTION 1.5 EXERCISES

In Exercises 1–12, sketch the graph of each function. Be sure to locate the x-intercept(s), y-intercept, vertex and line of symmetry of each parabola.

1. $f(x) = x^2 - x - 2$
2. $f(x) = x^2 + 7x + 6$
3. $f(x) = 2x^2 - 9x - 5$
4. $f(x) = 9x^2 - 9x - 4$
5. $f(x) = x^2 - 7x + 3$
6. $f(x) = 2x^2 + 6x + 1$
7. $f(x) = 2x^2 + 3x + 5$
8. $f(x) = 3x^2 - 2x + 4$
9. $f(x) = -x^2 + 3x + 10$
10. $f(x) = -x^2 + 2x + 15$
11. $f(x) = 6 - x - 3x^2$
12. $f(x) = 12 - 3x - 2x^2$

13. (Demand and revenue) The demand for a certain product that our firm produces is given by the function $x = f(p) = 100 - 5p$, where $p$ is the price in dollars per unit and $x$ is the number of units demanded per week.
    (a) Graph the demand function.
    (b) Find revenue as a function of price.
    (c) Determine the price that will maximize revenue, and the maximum revenue.
    (d) Determine demand for our product at the price found in part (c).
    (e) Graph the revenue function.

14. (Demand and revenue) Given a linear demand function $x = f(p) = 500 - 4p$, where $p$ is the price per unit in dollars and $x$ is the quantity demanded per month in units,
    (a) graph the demand function;
    (b) find revenue as a function of price;
    (c) determine the price that will yield the maximum revenue;
    (d) determine demand (in units) for the product if we charge a price that maximizes revenue;
    (e) graph the revenue function.

### Business and Economcis

15. (Demand and profit) In a survey of customers, the marketing department of our company has collected data that is summarized in the table below.

| PRICE (p) | $50 | $75 | $100 |
|---|---|---|---|
| QUANTITY DEMANDED (x) | 2000 | 1600 | 1200 |

In this table price is measured in dollars and quantity demanded in units of our product.
(a) Determine whether this data determines $x$ as a linear function of $p$.
(b) Find and graph the demand function $x = f(p)$.
(c) Find the revenue as a function of price.
(d) Given a cost function $C(x) = 20x + 800$, find the profit as a function of price.

(e) Determine the price we should charge to maximize profit. What is the maximum profit? What is the quantity demanded at the optimal (profit-maximizing) price?
(f) Graph the profit function.

16. (Maximum profit) In Section 1.3 of this chapter, we encountered the profit function
$P = f(p) = -4p^2 + 2300p - 190000$, where $p$ was the price per unit in dollars and $P$ was the profit in dollars.
(a) Determine the profit-maximizing price and also find the maximum profit.
(b) Graph the profit function.

17. (Maximum profit) Referring to the perfume example of this section, use the old demand schedule in Table 1.5 and the cost function $C(x) = 50x + 600$ to determine whether our previous price of $200 per ounce was the profit-maximizing price. *Hint:* You must determine the demand function, the revenue function, and the profit function in that order. Also, remember that our production capacity is 20 ounces per week.

18. (Recouping lost profits) Referring to the perfume company example of this section, what can we do (besides pricing optimally for our demand schedule at $162.50 per ounce) in order to try to regain more of our lost profit?

19. (Minimum cost) Given the cost function $C(x) = 0.001x^2 - 4x + 8000$, where $x$ is the number of units produced per week and $C(x)$ is the weekly cost in dollars of producing $x$ units, find the weekly production level that minimizes cost. What is the minimum cost? Graph the cost function $C(x)$.

20. (Motion) A rock is thrown directly upward with an initial velocity of 64 feet per second and is released at a height of 7 feet. The height $s$ of the rock at any time $t$ is given by the quadratic function

$$s = -16t^2 + 64t + 7$$

(a) Find the maximum height attained by the rock.
(b) After how many seconds does the rock hit the ground?

21. (Advertising and profit) If a company spends $x$ thousand dollars on advertising a certain product per week, the weekly profit from this product will be

$P(x) = -\frac{11}{75}x^2 + \frac{88}{25}x - 4$ thousand dollars

How much should the company spend per week on advertising this product in order to maximize weekly profit? What is the maximum weekly profit from the product?

22. (Maximum revenue) The revenue of a company is given by the function

$R(x) = -0.125x^2 + 300x$ dollars

where $x$ is the number of units sold. Find the maximum revenue.

**Business and Economics**

23. (Cost, revenue, and profit) A company that manufactures power saws has fixed costs of $2000 per month and variable costs (for labor and materials) of $30 per saw. The manufacturer sells the saws to a retail store at a price that depends upon the number of saws purchased. If $x$ saws are purchased, the retailer receives a discount of $x/50$ percent from the nominal price of $65 per saw. Thus, for example, if 50 saws are purchased, the discount is 1 percent; if 75 saws are purchased, the discount is 1.5 percent; and so forth. The manufacturer can supply a maximum of 1000 saws per month to the retailer.
(a) Find the monthly cost function $C(x)$ of the manufacturer.

(b) Show that the price per saw that the retailer pays is $p = 65 - 0.013x$ when he purchases $x$ saws.
(c) Find the monthly revenue function $R(x)$ of the manufacturer.
(d) Find the monthly profit function $P(x)$ of the manufacturer.
(e) Find the number of saws that the manufacturer should supply to the retailer to maximize monthly profit. What is the maximum profit?
(f) If the production capabilities of the manufacturer are increased so that he can supply a maximum of 1500 saws per month to the retailer but other conditions remain the same, how many saws should the manufacturer supply to the retailer to maximize profit?

24. (Construction of a dog kennel) The owner of a dog kennel wishes to use 1200 feet of fencing to enclose and partition a rectangular area in which the size of each of the partitioned areas is equal, as illustrated in Figure 1.49. Find the length of the sides $x$ and $y$ that maximize the total area enclosed while using all of the available fencing.

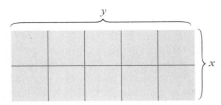

**FIGURE 1.49**

25. (Manufacturing) In a certain production process, a piece of wire 36 inches long must be bent into the shape of a rectangle. Find the dimensions of the rectangle that will maximize the area.

## OPTIONAL
## GRAPHING CALCULATOR/ COMPUTER EXERCISES

26. Graph the function $s = -16t^2 + 64t + 7$ of Exercise 20 and verify graphically that the maximum height of the rock is the one you obtained algebraically in Exercise 20.

27. Graph the profit function $P(x) = -\frac{11}{75}x^2 + \frac{88}{25}x - 4$ of Exercise 21 and verify graphically that the maximum profit is the one you obtained algebraically in Exercise 21.

28. Graph the revenue function $R(x) = -0.125x^2 + 300x$ of Exercise 22 and verify graphically that the maximum revenue is the one you obtained algebraically in Exercise 22.

29. Graph the profit function $P(x)$ that you derived in part (d) of Exercise 23 for $0 \le x \le 1500$ and verify graphically that the maximum profit is the one you obtained algebraically in Exercise 23.

## 1.6  CHAPTER ONE REVIEW

**KEY TERMS**

function
functional notation
dependent variable
independent variable

variable cost
graph
vertical line test
break-even point

demand function
quadratic function
extrapolation
line of symmetry

| | | |
|---|---|---|
| domain | x-intercept | vertex |
| range | y-intercept | profit-maximizing price |
| revenue function | marginal profit | composition of functions |
| cost function | delta notation | linear function |
| profit function | slope | straight-line depreciation |
| fixed cost | demand schedule | |

**KEY FORMULAS AND RESULTS**

$f(x) = mx + b$ (linear function)

$m = \dfrac{\Delta y}{\Delta x} = \dfrac{y_2 - y_1}{x_2 - x_1}$ (slope)

### Equations of a Straight Line

$y = mx + b$ (slope–intercept form)

$y - y_1 = m(x - x_1)$ (point–slope form)

$y - y_1 = \dfrac{y_2 - y_1}{x_2 - x_1}(x - x_1)$ (two-point form)

$Ax + By = C$ (general form)

Parallel lines $\quad m_1 = m_2$

Perpendicular lines $\quad m_1 = -\dfrac{1}{m_2}$

### Miscellaneous Formulas

$f(x) = ax^2 + bx + c$ (quadratic function)

$x = \dfrac{-b \pm \sqrt{b^2 - 4ac}}{2a}$ (quadratic formula)

$V = \left(-\dfrac{b}{2a}, f\left(-\dfrac{b}{2a}\right)\right)$ (coordinates of the vertex of a parabola)

profit = revenue − cost

revenue = (price)(quantity)

**CHAPTER 1 SHORT ANSWER REVIEW QUESTIONS**

1. In the formula $R = 60x + 200$, the independent variable is ____ and the dependent variable is ____.
2. The domain of $f(x) = 2x/(x + 3)$ is all real numbers x *except* x = ____.
3. Given the cost function $C(x) = 25x + 500$, where C is in dollars and x is the number of units produced, the cost of producing 45 units is ____.
4. Given the function $f(x) = 2x^2 + 3x - 1$, then $f(0) =$ ____, $f(-5) =$ ____, and $f(10) =$ ____.
5. If your fixed cost is $1200 per month and your variable cost is $25 per unit produced, find the linear function $C(x)$ that expresses your monthly cost in terms the number x of units produced. ____

6. True or False:
   (a) The function $f(x) = 25.9x - 17.6$ is a linear function. _____
   (b) The graph $y = 7 - 3x - x^2$ is a parabola that opens upward. _____
   (c) A negative profit corresponds to a loss. _____
   (d) Perpendicular lines have equal slopes. _____
   (e) All profit functions are linear functions. _____
7. If $f(x) = 2/x$ and $g(x) = 2x + 3$, then (a) $f(2) + g(2) =$ _____;
   (b) $f(1) \cdot g(1) =$ _____; (c) $f(g(0)) =$ _____; (d) $g(f(\tfrac{1}{2})) =$ _____.
8. Given the linear profit function $P(x) = 200x - 1800$, where $P(x)$ is profit in dollars corresponding to a sale of $x$ units, the break-even point is: (a) 0 units; (b) 5 units; (c) 9 units; (d) 200 units. _____
9. Given that two points on a certain straight line are $(-1, -2)$ and $(2, -3)$, the slope of the line is _____.
10. The line that passes through $(1, 2)$ and is perpendicular to $2x + 3y = 7$ has slope (a) $-\tfrac{2}{3}$; (b) $\tfrac{1}{3}$; (c) $\tfrac{2}{3}$; (d) $\tfrac{3}{2}$.
11. The $y$-coordinate of the vertex of the parabola with formula $y = 12 - 12x + 2x^2$ is (a) $-6$; (b) 3; (c) $\tfrac{13}{2}$; (d) 1.
12. Given the demand function $x = f(p) = 200 - 4p$, where $p$ is the price per unit in dollars and $x$ is the number of units demanded per month at price $p$, the monthly demand when the price is $10 per unit is _____.
13. Using straight-line depreciation, the value in dollars of a certain asset after $t$ years is given by $V(t) = 100000 - 5000t$. We are depreciating this asset over a period of _____ years.
14. If a certain graph passes through the points $(3, 0)$, $(-5, 2)$, and $(3, 2)$, then (a) this graph represents a function; (b) this graph does not represent a function; (c) this graph may or may not represent a function.
15. The slope of a certain line is $-2$. If $x$ increases by ten units, then (a) $y$ increases by 20 units; (b) $y$ increases by 10 units; (c) $y$ decreases by 20 units; (d) $y$ may either increase or decrease.

## CHAPTER 1 REVIEW EXERCISES

*Exercises marked with an asterisk (*) constitute a Chapter Test.*

1. If $f(x) = 3x - 5$, find (a) $f(0)$; (b) $f(-5)$; (c) $f(a)$.
2. If $f(x) = 4x^2 - 7x + 5$, find (a) $f(-3)$; (b) $f(10)$; (c) $f(x + h)$.
*3. If $g(x) = \sqrt{1 - 2x}$, find (a) $g(\tfrac{1}{2})$; (b) $g(-5)$; (c) $g(x + h)$.
4. Find the domain and range of $f(x) = 450x$.
*5. Find the domain and range of $f(x) = 3 - x^2$.
6. Find the domain of $f(p) = 1/p$.
7. Find the domain of $f(x) = \sqrt{x}/(x - 3)$.
8. (Demand function) Given a demand function $p = D(x) = 2000 - 4x$, find and interpret $D(100)$ and $D(200)$.
9. (Fixed and variable costs) If the fixed cost of producing your product is $3000 per week and the variable cost is $30 per unit, find the linear cost function $C(x)$ for your product where $x$ is the number of units produced per week.

10. Sketch the graph of $y = \sqrt{x}$.
11. Sketch the graph of $y = \sqrt[3]{x}$.
12. Sketch the graph of $f(x) = x/(x + 2)$.
*13. Sketch the graph of $f(x) = 1/(x^2 - 1)$.
14. (Cost function) Sketch the graph of the cost function $C(x) = 30x + 3000$ for $x \geq 0$.
15. (Demand function) Sketch the graph of the demand function $p = D(x) = 800 - 4x$ for $x \geq 0$.
16. (Volume discounts) For orders of fewer than 100 parts, your supply company charges $1.00 per part. For orders of at least 100 parts, your cost per part is $0.95.
    (a) Determine the amount $C(x)$ that you must pay to purchase $x$ parts, where $x \geq 0$.
    (b) Graph the function $C(x)$.
*17. (Cost and revenue) If your cost function is $C(x) = 20x + 1000$ and your revenue function is $R(x) = 40x$, where $x$ is units and $C(x)$ and $R(x)$ are measured in dollars,
    (a) determine your profit function $P(x)$;
    (b) find your profit on the sale of 100 units;
    (c) determine your break-even point.
18. If $f(x) = 3x - 5$ and $g(x) = 2x - x^2$, find (a) $f(x) + g(x)$; (b) $f(x) - g(x)$; (c) $f(x) \cdot g(x)$; (d) $f(x)/g(x)$.
*19. Given $f(x) = \sqrt{x - 2}$ and $g(x) = x^2$, find (a) $f(6) + g(6)$; (b) $f(2) - g(2)$; (c) $f(3) \cdot g(3)$; (d) $f(5)/g(5)$.
20. Given $f(x) = \sqrt{2 - 3x}$ and $g(x) = 5x$, find (a) $f(g(x))$; (b) $g(f(x))$.
*21. If $f(x) = x^2 + 2$ and $g(x) = \sqrt{x - 5}$, find and simplify $(f \circ g)(x)$ and find the domain of $f \circ g$. What is $(f \circ g)(6)$?
22. If $f(x) = 2x/(3x - 2)$ and $g(x) = x^2 + 3$, find and simplify $(f \circ g)(x)$ and find the domain of $f \circ g$. What is $(f \circ g)(10)$?

Business and Economics

*23. (Maximum profit) Given a profit function $P(x) = -x^2 + 900x - 20000$ and a demand function $x = q(p) = -5p + 1500$, where $x$ is units sold and $p$ is selling price in dollars per unit,
    (a) express profit as a function of price per unit.
    (b) If we charge $100 per unit, what is our profit?
    (c) Determine the price we should charge to maximize profit. What is our maximum profit? What number of units will be demanded at the profit-maximizing price?

Business and Economics

24. (Break-even point) If your profit function is linear, your marginal profit is $20 per unit, and your fixed costs are $1000 per week, (a) find your (weekly) profit function; (b) find your break-even point.
25. A horizontal line and a vertical line intersect at (200, 500).
    (a) Find an equation of the horizontal line.
    (b) Find an equation of the vertical line.
26. Identify each of the following functions as linear, quadratic, or neither linear nor quadratic. (a) $f(x) = 500x - 20000$; (b) $f(x) = (17x^2 + 12x)/3$; (c) $g(x) = \sqrt{x^2 + 1}$; (d) $f(x) = \frac{1}{2}x - x^2 + 13$; (e) $f(p) = -p^3 + 7p^2 + 30p - 5$
27. Find an equation of the line with slope $-3$ and that passes through the origin.
28. Find an equation of the line with slope 1/2 and that passes through the point $(-2, -1)$.

*29. Find an equation of the line with slope $-1/5$ and that passes through the point (5, 400).

30. Find an equation of the line that passes through the points (3, 600) and (5, 100).

31. Sketch the graph of $f(x) = -4x + 12$.

32. Sketch the graph of $f(x) = \frac{1}{4}x + 8$.

33. Find an equation of the line that is parallel to $C(x) = 30x + 500$ and that passes through the point (0, 700).

34. Find an equation of the line that passes through the point $(-1, 3)$ and that is perpendicular to the line with equation $2x - 5y = 20$.

*35. (Demand function) If you charge $100 per unit, demand will be 5000 units. If you reduce your price to $75 per unit, demand will increase to 7500 units. Assuming that the demand function is linear, (a) find the demand function $x = f(p)$, where $x$ is the quantity demanded at price $p$; (b) find the demand if you charge $85 per unit.

36. (Depreciation) You use a car exclusively for business purposes and decide to use straight-line depreciation to devaluate it over a period of six years for income tax purposes. If you bought the car for $12,000, determine an equation for $V(t)$, the value of the car $t$ years after purchase for $0 \le t \le 6$.

*37. Find the $x$-intercepts, the $y$-intercept, and the coordinates of the vertex of the parabola represented by $f(x) = 3x^2 - 7x + 2$. Sketch the graph of the parabola.

38. Find the $x$-intercepts, the $y$-intercept, and the coordinates of the vertex of the parabola represented by $f(x) = 6 - 5x - x^2$. Sketch the graph of the parabola.

39. (Demand function) Given the linear demand function $x = 1000 - 20p$, where $p$ is the price in dollars and $x$ is the quantity demanded in units, (a) find the revenue as a function of price; (b) determine the price that will yield the maximum revenue, and find the maximum revenue; (c) find the demand (in units) for this product if we charge a price that maximizes revenue.

40. (Cost function) Given the quadratic cost function $C(x) = \frac{1}{200}x^2 - 3x + 2000$, find the production level that minimizes monthly cost and find that minimum cost if $x$ is the number of units produced per month and $C(x)$ is the monthly cost in dollars.

## OPTIONAL
## GRAPHING CALCULATOR/COMPUTER EXERCISES

Graph each of the following functions using a graphing calculator or a microcomputer.

41. $f(x) = x^3 - 7x^2 + 3x + 5$

42. $g(x) = 2x^4 - 4x^3 + x^2 - 10$

43. $f(x) = \dfrac{2\sqrt{x-3}}{x^2 + 5}$

44. $y = \dfrac{10x^2 - 100}{8x^2 - x + 1}$

45. $f(x) = x^5 - 12x^3 + 7x + 1$

46. $y = \dfrac{x^3 + 1}{x^2 + 2x + 1}$

# 2 LIMITS AND AN INTRODUCTION TO THE DERIVATIVE

## CHAPTER CONTENTS

2.1 An Introduction to Limits
2.2 Limits at Infinity and Infinite Limits—Asymptotes
2.3 Continuity
2.4 The Tangent Line to a Curve
2.5 The Derivative of a Function
2.6 The Derivative as a Rate of Change
2.7 Chapter 2 Review

# CHAPTER 2 LIMITS AND AN INTRODUCTION TO THE DERIVATIVE

This chapter begins the study of calculus by considering the concept of a limit. There are two main branches of calculus: differential calculus and integral calculus. As we will see in this chapter, the idea of a limit is essential in defining the derivative of a function, and the derivative is the most important concept in differential calculus. In Chapter 6 we will see that limits are also important in discussing a key concept of integral calculus—the definite integral.

## 2.1 AN INTRODUCTION TO LIMITS

### Definition of a Limit

Let us consider the function $y = f(x) = 2x$ and ask the following question:

*If we let x take on values that become closer and closer to 3, do the corresponding values of y get closer and closer to any single number L?*

As you can see from Table 2.1 and Table 2.2, as $x$ takes on values that become closer to 3, whether from the left (for values of $x$ less than 3) or from the right (for values of $x$ greater than 3), the corresponding values of $y$ get closer to 6.

| $x$ | 2.0 | 2.5 | 2.9 | 2.99 | 2.999 | 2.9999 |
|---|---|---|---|---|---|---|
| $y = f(x) = 2x$ | 4.0 | 5.0 | 5.8 | 5.98 | 5.998 | 5.9998 |

TABLE 2.1  The value of $x$ approaches 3 from the left.

| $x$ | 4.0 | 3.5 | 3.1 | 3.01 | 3.001 | 3.0001 |
|---|---|---|---|---|---|---|
| $y = f(x) = 2x$ | 8.0 | 7.0 | 6.2 | 6.02 | 6.002 | 6.0002 |

TABLE 2.2  The value of $x$ approaches 3 from the right.

In symbols, we write

$$\lim_{x \to 3} (2x) = 6$$

which is read, "the limit of $2x$ as $x$ approaches 3 is equal to 6."

You can see from Figure 2.1 that as the $x$-values get closer to 3 (as indicated by the arrows), the corresponding $y$-values get closer to 6.

You may have already discovered that the limit we just considered can be computed in the following way, without having to resort to tables of values:

$$\lim_{x \to 3} (2x) = 2(3) = 6$$

This limit was a very easy one. Not all limits can be computed by this short-cut method. Before we look at some more difficult examples, let us first consider the following informal definition of a limit:

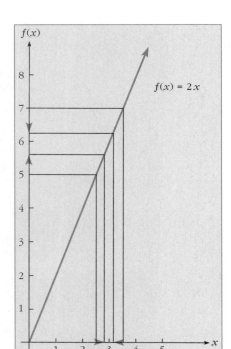

FIGURE 2.1  The graph of $f(x) = 2x$, illustrating that $\lim_{x \to 3} f(x) = 6$.

## 2.1 AN INTRODUCTION TO LIMITS

**DEFINITION**

> The limit of $f(x)$ as $x$ approaches $a$ is equal to $L$ (in symbols, $\lim_{x \to a} f(x) = L$) if, as we let $x$ take on values that get closer and closer to $a$, but not equal to $a$, the corresponding values of $f(x)$ get closer and closer to the number $L$.

Notice the fact that in this definition, the value of $\lim_{x \to a} f(x)$ does not depend upon the value of the function *at a* but only on the values of the function for $x$ "close to but not equal to $a$." In fact, a limit may exist even if $f(a)$ is not defined. As an illustration of this, consider the functions $g(x)$ and $h(x)$ defined by

$$g(x) = \begin{cases} 2x & \text{if } x \neq 3 \\ 8 & \text{if } x = 3 \end{cases} \quad \text{and} \quad h(x) = \begin{cases} 2x & \text{if } x \neq 3 \\ \text{undefined} & \text{if } x = 3 \end{cases}$$

The graph of $g(x)$ in Figure 2.2 is the same as the graph of the straight line $y = 2x$ *except* when $x = 3$. Since $g(3) = 8$, instead of the point $(3, 6)$ appearing on the graph (as happens with $y = 2x$), the point $(3, 8)$ is on the graph of $g$. This is the reason for the "hole" in the graph of $g(x)$ when $x = 3$. The graph of $h(x)$ in Figure 2.3 is the same as the graph of the straight line $y = 2x$ except $h(x)$ has a "hole" in the graph above $x = 3$. This is so because there is *no y*-value assigned to $x = 3$ by the function $h$.

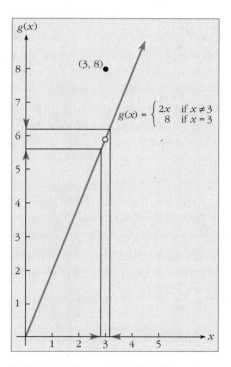

FIGURE 2.2 The graph of $g(x)$, illustrating that $\lim_{x \to 3} g(x) = 6$.

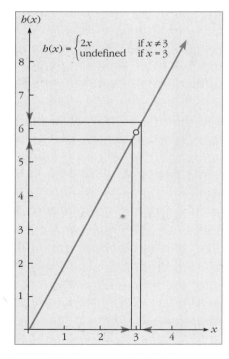

FIGURE 2.3 The graph of $h(x)$, illustrating that $\lim_{x \to 3} h(x) = 6$.

Now let us try to find $\lim_{x \to 3} g(x)$. Remember that we should be asking ourselves the following question: "As $x$ takes on values closer and closer to 3, but not equal to 3, do the corresponding values of $g(x)$ get closer and closer to any number $L$?" The answer is that $L = 6$, but if you do not see this, here are some suggestions:

i. Replace $f(x)$ in Table 2.1 and Table 2.2 by $g(x)$. The numbers in these tables stay the same. So as $x$ approaches 3, the value of $g(x)$ approaches 6.
ii. Look at the graph in Figure 2.2 again and notice that as the $x$-values approach 3 along the $x$-axis, the corresponding $y$-values approach 6 along the $y$-axis.

If you understand why $\lim_{x \to 3} g(x) = 6$, you should also see that $\lim_{x \to 3} h(x) = 6$.
We conclude that

$$\lim_{x \to 3} f(x) = \lim_{x \to 3} g(x) = \lim_{x \to 3} h(x) = 6$$

even though the functions $f$, $g$, and $h$ are all *different* when $x = 3$. That is, $f(3) = 6$, $g(3) = 8$, and $h(3)$ is undefined. The reason why these limits are all equal is that $f$, $g$, and $h$ are the same for all $x \neq 3$ and, in finding a limit, the value of the function *at* the limit point (3 in this case) does not determine the value of the limit. What determines the value of $\lim_{x \to a} f(x)$ is the value of $f(x)$ at points *close to but not equal to a*.

---

EXAMPLE 1

Let

$$f(x) = \begin{cases} 3x & \text{if } x \leq 1 \\ 3x + 1 & \text{if } x > 1 \end{cases}$$

Compute $\lim_{x \to 1} f(x)$.

Solution

To find this limit, first note that $f(x)$ is a multipart function with a definition that changes at $x = 1$. In Table 2.3 we choose values of $x$ that approach 1 from the left and compute the corresponding functional values using the formula $3x$. In Table 2.4 we choose values of $x$ that approach 1 from the right and compute the corresponding functional value using the formula $3x + 1$.

| $x$ | 0 | 0.5 | 0.9 | 0.99 | 0.999 | 0.9999 | $\to 1$ |
|---|---|---|---|---|---|---|---|
| $f(x)$ | 0 | 1.5 | 2.7 | 2.97 | 2.997 | 2.9997 | $\to 3$ |

TABLE 2.3 The value of $x$ approaches 1 from the left.

| $x$ | 2 | 1.5 | 1.1 | 1.01 | 1.001 | 1.0001 | $\to 1$ |
|---|---|---|---|---|---|---|---|
| $f(x)$ | 7 | 5.5 | 4.3 | 4.03 | 4.003 | 4.0003 | $\to 4$ |

TABLE 2.4 The value of $x$ approaches 1 from the right.

Looking at Table 2.3, as $x$ approaches 1 from the left, $f(x)$ apparently approaches 3. Upon inspecting Table 2.4, however, as $x$ approaches 1 from the

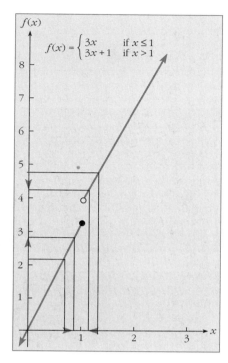

**FIGURE 2.4** The graph of $f(x)$ in Example 1.

right, $f(x)$ approaches 4. Therefore, there is no *single* number $L$ that $f(x)$ approaches as $x$ approaches 1. We say that $\lim_{x \to 1} f(x)$ does not exist. Figure 2.4 makes it clear that as $x$ approaches 1 from the left (see the arrow on the $x$-axis to the left of 1), the corresponding values of $y = f(x)$ approach 3 (see the arrow on the $y$-axis below 3). We also see from the figure that as $x$ approaches 1 from the right, $f(x)$ approaches 4. ∎

### Left-Hand and Right-Hand Limits

In Example 1, even though $\lim_{x \to 1} f(x)$ does not exist, we can say that the *left-hand limit* of $f(x)$ as $x$ approaches 1 does exist and is equal to 3. In symbols:

$$\lim_{x \to 1^-} f(x) = 3$$

The symbol $1^-$ in this limit means that $x$ is restricted to approach 1 through values that are always less than 1; that is, $x$ approaches 1 from the left, as in Table 2.3. In a similar way, we can say that the *right-hand limit* of $f(x)$ as $x$ approaches 1 exists and is equal to 4. In symbols:

$$\lim_{x \to 1^+} f(x) = 4$$

The symbol $1^+$ in this limit means that $x$ is required to approach 1 through values that are always greater than 1; that is, $x$ approaches 1 from the right, as in Table 2.4. The limit $\lim_{x \to 1^-} f(x)$ is called a "left-hand limit" and the limit $\lim_{x \to 1^+} f(x)$ is called a "right-hand limit." A left- or right-hand limit is sometimes referred to as a *one-sided limit*. We have seen that one-sided limits may exist at a number $a$, even though the limit may not exist at $a$. In this situation, $\lim_{x \to 1} f(x)$ does not exist because $\lim_{x \to 1^-} f(x) \neq \lim_{x \to 1^+} f(x)$. That is, the limit fails to exist in this case because $f(x)$ approaches two *different* values as $x$ approaches 1 from the left and from the right. Thus $f(x)$ does not approach any *single number* as $x$ approaches 1. In general, we can state the following rule:

---

Let $L$ be a real number. Then

$$\lim_{x \to a} f(x) = L$$

if and only if

$$\lim_{x \to a^-} f(x) = L \quad \text{and} \quad \lim_{x \to a^+} f(x) = L$$

---

**EXAMPLE 2**

Compute $\lim_{x \to 2} (x^2 - 4)/(x - 2)$.

**Solution**

We will compute the limit in two different ways. First, let us construct tables of values: see Table 2.5 and Table 2.6.

| $x$ | 1 | 1.5 | 1.9 | 1.99 | 1.999 | 1.9999 | →2⁻ |
|---|---|---|---|---|---|---|---|
| $f(x) = \dfrac{x^2 - 4}{x - 2}$ | 3 | 3.5 | 3.9 | 3.99 | 3.999 | 3.9999 | →4 |

**TABLE 2.5** The value of $x$ approaches 2 from the left.

| $x$ | 3 | 2.5 | 2.1 | 2.01 | 2.001 | 2.0001 | →2⁺ |
|---|---|---|---|---|---|---|---|
| $f(x) = \dfrac{x^2 - 4}{x - 2}$ | 5 | 4.5 | 4.1 | 4.01 | 4.001 | 4.0001 | →4 |

**TABLE 2.6** The value of $x$ approaches 2 from the right.

Table 2.5 makes it clear that $\lim_{x \to 2^-} (x^2 - 4)/(x - 2) = 4$, and Table 2.6 indicates that $\lim_{x \to 2^+} (x^2 - 4)/(x - 2) = 4$. Since both the left- and the right-hand limits are equal to 4, we conclude that

$$\lim_{x \to 2} \frac{x^2 - 4}{x - 2} = 4$$

Here is another way to compute the limit:

$$\lim_{x \to 2} \frac{x^2 - 4}{x - 2} = \lim_{x \to 2} \frac{(x + 2)(x - 2)}{(x - 2)} = \lim_{x \to 2} (x + 2) = 2 + 2 = 4$$

There is a subtle point about the second method (which we call the "algebraic method") of computing the limit of Example 2. The point is that the functions $f(x) = (x^2 - 4)/(x - 2)$ and $g(x) = x + 2$ are not the same. The number 2 is in the domain of $g(x)$, since $g(2) = 2 + 2 = 4$, but the number 2 is *not* in

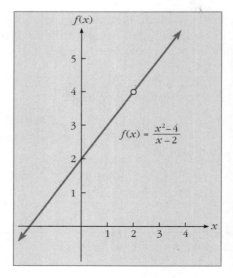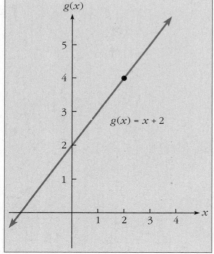

**FIGURE 2.5** $\lim_{x \to 2} f(x) = \lim_{x \to 2} g(x) = 4$

the domain of $f(x)$, since $f(2) = (2^2 - 4)/(2 - 2) = 0/0$, and $0/0$ is indeterminate. However, if you look carefully at Example 2 again, you see the statement

$$\lim_{x \to 2} \underbrace{\frac{x^2 - 4}{x - 2}}_{f(x)} = \lim_{x \to 2} \underbrace{(x + 2)}_{g(x)}$$

How can we say that these two limits are equal when we have just shown that $f(x)$ and $g(x)$ are not equal? The answer is that $f(x)$ and $g(x)$ are the same *except* when $x = 2$, and we are taking the limit as $x$ approaches 2. Since, in finding a limit, the value of the function at the limit point does not determine the value of the limit, we may say that $\lim_{x \to 2} f(x) = \lim_{x \to 2} g(x)$. The functions $f(x)$ and $g(x)$ are sketched in Figure 2.5.

*Note: In the next example you should participate in finding the solution by filling in the blanks. Correct answers are given after the example.*

---

**EXAMPLE 3 (Participative)**

Find $\lim_{x \to 0} (x^3 + 3x)/x$ by two different methods.

**Solution**

(a) *Method I—Tables of Values.* To compute this limit using tables of values, first complete the tables below.

| $x$ | −1 | −0.5 | −0.1 | −0.01 | −0.001 |
|---|---|---|---|---|---|
| $f(x) = \frac{x^3 + 3x}{x}$ | 4 | 3.25 | 3.01 | 3.0001 | |

TABLE 2.7

| $x$ | 1 | 0.5 | 0.1 | 0.01 | 0.001 |
|---|---|---|---|---|---|
| $f(x) = \frac{x^3 + 3x}{x}$ | 4 | 3.25 | 3.01 | 3.0001 | |

TABLE 2.8

Table 2.7 indicates that $\lim_{x \to 0^-} (x^3 + 3x)/x = \underline{3}$. Table 2.8 indicates that $\lim_{x \to 0^+} (x^3 + 3x)/x = \underline{3}$. Since the left-hand limit and the right-hand limit are both equal to 3, we conclude that

$$\lim_{x \to 0} \frac{x^3 + 3x}{x} = \underline{3}$$

(b) *Method II—Algebraic Method.* Since $x^3 + 3x$ may be factored as $x(\underline{x^2+3})$, we find that

$$\lim_{x \to 0} \frac{x^3 + 3x}{x} = \lim_{x \to 0} \frac{x(x^2 + 3)}{x} = \lim_{x \to 0} \underline{x^2+3} = \underline{3}$$

## Properties of Limits

Many limits can be computed without having to construct tables of values; instead we can use the properties of limits. In many other cases, although these properties may not apply directly, the given limit can be converted into one where these properties may be used. Here are the important properties of limits:

1. If $c$ is any constant, then $\lim_{x \to a} c = c$.
2. If $n$ is any positive integer, then $\lim_{x \to a} x^n = a^n$.
3. If $\lim_{x \to a} f(x)$ exists and if $c$ is any constant, then $\lim_{x \to a} [cf(x)] = c \lim_{x \to a} f(x)$.
4. If $\lim_{x \to a} f(x)$ and $\lim_{x \to a} g(x)$ both exist, then $\lim_{x \to a} [f(x) \pm g(x)] = \lim_{x \to a} f(x) \pm \lim_{x \to a} g(x)$.
5. If $\lim_{x \to a} f(x)$ and $\lim_{x \to a} g(x)$ both exist, then $\lim_{x \to a} [f(x) \cdot g(x)] = \lim_{x \to a} f(x) \cdot \lim_{x \to a} g(x)$.
6. If $\lim_{x \to a} f(x)$ exists and if $\lim_{x \to a} g(x)$ exists and is not equal to zero, then

$$\lim_{x \to a} \left[\frac{f(x)}{g(x)}\right] = \frac{\lim_{x \to a} f(x)}{\lim_{x \to a} g(x)}$$

7. If $\lim_{x \to a} f(x)$ exists and if $n$ is a positive integer, then $\lim_{x \to a} \sqrt[n]{f(x)} = \sqrt[n]{\lim_{x \to a} f(x)}$, provided that $\lim_{x \to a} f(x)$ is positive if $n$ is even.

### Examples of the Properties of Limits

**EXAMPLE 4**

$\lim_{x \to 2} (5x + 7) = \lim_{x \to 2} (5x) + \lim_{x \to 2} 7$    Property 4

$= 5 \lim_{x \to 2} x + 7$    Properties 3 and 1

$= 5(2) + 7 = 17$    Property 2

---

**SOLUTIONS TO EXAMPLE 3**

(a) 4, 3.25, 3.01, 3.0001, 3.000001, 4, 3.25, 3.01, 3.0001, 3.000001, 3, 3, 3
(b) $x^2 + 3$, $x^2 + 3$, $0^2 + 3 = 3$

**EXAMPLE 5**

$$\lim_{x \to -1} (x^2 - 7x + 3) = \lim_{x \to -1} (x^2) - \lim_{x \to -1} (7x) + \lim_{x \to -1} 3 \quad \text{Property 4}$$

$$= (-1)^2 - 7 \lim_{x \to -1} (x) + 3 \quad \text{Properties 2, 3, and 1}$$

$$= 1 - 7(-1) + 3 = 11 \quad \text{Property 2} \quad \blacksquare$$

**EXAMPLE 6**

$$\lim_{x \to 0} [(1 - x^3)(2x^2 + 3)] = \lim_{x \to 0} (1 - x^3) \lim_{x \to 0} (2x^2 + 3) \quad \text{Property 5}$$

$$= \left(\lim_{x \to 0} 1 - \lim_{x \to 0} x^3\right)\left(\lim_{x \to 0} 2x^2 + \lim_{x \to 0} 3\right) \quad \text{Property 4}$$

$$= (1 - 0^3)\left(2 \lim_{x \to 0} x^2 + 3\right) \quad \text{Properties 1, 2, and 3}$$

$$= (1)[2(0^2) + 3] = 3 \quad \text{Property 2} \quad \blacksquare$$

**EXAMPLE 7**

$$\lim_{x \to 3} \frac{5}{x^2 - 7} = \frac{\lim_{x \to 3} 5}{\lim_{x \to 3} (x^2 - 7)} \quad \text{Property 6}$$

$$= \frac{5}{\lim_{x \to 3} x^2 - \lim_{x \to 3} 7} \quad \text{Properties 1 and 4}$$

$$= \frac{5}{3^2 - 7} = \frac{5}{2} \quad \text{Properties 2 and 1} \quad \blacksquare$$

**EXAMPLE 8**

$$\lim_{x \to 5} \sqrt[3]{-2 - x^2} = \sqrt[3]{\lim_{x \to 5} (-2 - x^2)} \quad \text{Property 7}$$

$$= \sqrt[3]{\lim_{x \to 5} (-2) - \lim_{x \to 5} (x^2)} \quad \text{Property 4}$$

$$= \sqrt[3]{-2 - (5)^2} \quad \text{Properties 1 and 2}$$

$$= \sqrt[3]{-27} = -3 \quad \blacksquare$$

*Note:* In the next example, you should participate in finding the solution by filling in the blanks. Correct answers are given after the example.

**EXAMPLE 9 (Participative)**

Let's compute $\lim_{x \to 3} (x^2 - 2)/\sqrt{x + 5}$ using Properties 1–7.

**Solution**

Consider:

$$\lim_{x \to 3} \frac{x^2 - 2}{\sqrt{x + 5}} = \frac{\lim_{x \to 3} (x^2 - 2)}{\lim_{x \to 3} \sqrt{x + 5}} \quad \text{Property 6}$$

74  CHAPTER 2  LIMITS AND AN INTRODUCTION TO THE DERIVATIVE

$$= \frac{\lim_{x \to 3} x^2 - \lim_{x \to 3} 2}{\sqrt{\lim_{x \to 3} (x+5)}}$$  Properties 4 and 7

$$= \frac{3^2 - 2}{\sqrt{\lim_{x \to 3} (x) + \lim_{x \to 3} (5)}}$$  Properties 2, 1, and 4

$$= \frac{7}{\sqrt{3+5}}$$  Properties 2 and 1

$$= 7/\sqrt{8} = \frac{7\sqrt{2}}{4}$$ ∎

Properties 1–7 also apply to one-sided limits, provided all the appropriate one-sided limits exist.

---

**EXAMPLE 10**

$$\lim_{x \to 3^+} (x^2 + x - 2) = \lim_{x \to 3^+} (x^2) + \lim_{x \to 3^+} (x) - \lim_{x \to 3^+} (2)$$  Property 4

$$= (3)^2 + 3 - 2$$  Properties 2 and 1

$$= 10$$ ∎

Properties 1–7 provide formal justification of the "evaluation method" of computing a limit, in which $\lim_{x \to a} f(x)$ is computed by simply computing $f(a)$. For example, a quick way to compute $\lim_{x \to 1} \sqrt{5x^2 + 3}/(7x - 2)$ is simply to evaluate the function $f(x) = \sqrt{5x^2 + 3}/(7x - 2)$ at $x = 1$, obtaining $f(1)$. That is,

$$\lim_{x \to 1} \frac{\sqrt{5x^2 + 3}}{7x - 2} = \frac{\sqrt{5(1)^2 + 3}}{7(1) - 2} = \frac{\sqrt{8}}{5}$$

Formal justification of this procedure can be accomplished using Properties 1–7. However, it is *not* true that $\lim_{x \to a} f(x)$ is *always* equal to $f(a)$. For example, consider the multipart function of Example 1:

$$f(x) = \begin{cases} 3x & \text{if } x \leq 1 \\ 3x + 1 & \text{if } x > 1 \end{cases}$$

We found that $\lim_{x \to 1} f(x)$ does not exist, even though $f(1) = 3(1) = 3$. We will now consider some limits that can be computed with the aid of Properties 1–7 only after some algebraic manipulations are performed on the functions in question.

---

**SOLUTIONS TO EXAMPLE 9**   $x^2 - 2, x + 5, 3^2 - 2, 3 + 5, 7/\sqrt{8} = 7\sqrt{2}/4$

## EXAMPLE 11

Compute $\lim_{x \to 3} (x^2 - 9)/(x - 3)$.

**Solution**

First, note that Property 6 does *not* apply to this limit, since $\lim_{x \to 3} (x - 3) = 3 - 3 = 0$. The evaluation method would yield 0/0 if applied to this problem. The form 0/0 is called an "indeterminate form" and is usually an indication that some algebra must be performed before the limit can be evaluated by Properties 1–7. We write:

$$\lim_{x \to 3} \frac{x^2 - 9}{x - 3} = \lim_{x \to 3} \frac{(x + 3)(x - 3)}{(x - 3)}$$

$$= \lim_{x \to 3} (x + 3) \qquad \text{(Properties 1–7 may be used now)}$$

$$= \lim_{x \to 3} (x) + \lim_{x \to 3} (3) \qquad \text{Property 4}$$

$$= 3 + 3 = 6 \qquad \text{Properties 2 and 1}$$

## EXAMPLE 12

Compute $\lim_{x \to 0} (\sqrt{x + 1} - 1)/x$.

**Solution**

In this problem, it is not as obvious as in Example 11 what algebra must be performed before we can use Properties 1–7 or the evaluation method. Let us rationalize the numerator by multiplying numerator and denominator of the given expression by $\sqrt{x + 1} + 1$. We obtain

$$\lim_{x \to 0} \frac{\sqrt{x + 1} - 1}{x} = \lim_{x \to 0} \frac{\sqrt{x + 1} - 1}{x} \cdot \frac{\sqrt{x + 1} + 1}{\sqrt{x + 1} + 1}$$

$$= \lim_{x \to 0} \frac{x + 1 - 1}{x(\sqrt{x + 1} + 1)}$$

$$= \lim_{x \to 0} \frac{x}{x(\sqrt{x + 1} + 1)}$$

$$= \lim_{x \to 0} \frac{1}{\sqrt{x + 1} + 1} \qquad \text{(Properties 1–7 may now be used)}$$

$$= \frac{1}{\sqrt{0 + 1} + 1} = \frac{1}{2}$$

*Note:* In the next example, you should participate in finding the solution by filling in the blanks. Correct answers are given after the example.

## EXAMPLE 13 (Participative)

Let's compute $\lim_{x \to -3} (2x^2 + x - 15)/(x^2 + 2x - 3)$.

Solution

First, note that Property 6 cannot be used on this limit, since the limit of the denominator is equal to zero. That is, $\lim_{x \to -3} (x^2 + 2x - 3) =$ _____. However, note that we may factor the numerator and denominator of the given expression. In fact, $2x^2 + x - 15 = (2x - 5)(\underline{\phantom{xxx}})$ and $x^2 + 2x - 3 = (x - 1)(\underline{\phantom{xxx}})$. So we write

$$\lim_{x \to -3} \frac{2x^2 - x + 15}{x^2 + 2x - 3} = \lim_{x \to -3} \frac{(2x - 5)(x + 3)}{(x - 1)(x + 3)}$$

$$= \lim_{x \to -3} \underline{\phantom{xxxx}} \quad \text{(Properties 1–7 may now be used)}$$

$$= \underline{\phantom{xxxxxxxx}}$$

## SECTION 2.1
**SHORT ANSWER EXERCISES**

1. In computing $\lim_{x \to 5} (7x + 2)$, we would let $x$ become closer and closer to _____ and see if $7x + 2$ gets closer and closer to any single number $L$. What is $\lim_{x \to 5} (7x + 2)$? _____

2. (True or False) When computing $\lim_{x \to 3} f(x)$ by the table-of-values method, it is enough to construct a table of values with $x$-values 2, 2.5, 2.9, 2.99, 2.999 and 2.9999. _____

3. To compute $\lim_{x \to 7} 10$, we ask ourselves what number 10 approaches as $x$ approaches 7. The answer is $\lim_{x \to 7} 10 =$ _____, since 10 stays fixed as $x$ approaches 7.

4. If $\lim_{x \to 1^-} f(x) = 12$ and $\lim_{x \to 1^+} f(x) = 12$, what is $\lim_{x \to 1} f(x)$? _____

5. If $\lim_{x \to 0^-} f(x) = 5$ and $\lim_{x \to 0^+} f(x) = 7$, then $\lim_{x \to 0} f(x)$ (a) is equal to 5; (b) is equal to 7; (c) is equal to 5 and 7; (d) does not exist.

6. (True or False) $\lim_{x \to 0} x/x = 0$ _____

7. (True or False) Properties 1–7 can be used on every limit. _____

8. When the evaluation method is used to attempt to find a limit and the form 0/0 arises: (a) Properties 1–7 apply directly to the limit; (b) the limit does not exist; (c) some algebra must be performed on this limit before we can use Properties 1–7.

9. Evaluate $\lim_{x \to 5} (7)$. _____

10. Evaluate $\lim_{x \to 3} x^3$. _____

11. Evaluate $\lim_{x \to 1} x(x - 1)/(x - 1)$. _____

12. (True or False) Properties 1–7 do not apply to one-sided limits. _____

---

**SOLUTIONS TO EXAMPLE 13**

$(-3)^2 + 2(-3) - 3 = 0$, $x + 3$, $x + 3$, $(2x - 5)/(x - 1)$, $[2(-3) - 5]/(-3 - 1) = \frac{11}{4}$

## SECTION 2.1 EXERCISES

*In Exercises 1–11, compute the limits by constructing tables of values.*

1. $\lim_{x \to 4} (5x)$
2. $\lim_{x \to 2} (7x + 1)$
3. $\lim_{x \to 0} (1 - 3x)$
4. $\lim_{x \to -1} (2x - 5)$
5. $\lim_{x \to 3} (x^2 - x)$
6. $\lim_{x \to 0} (7 - x^3)$
7. $\lim_{x \to -2} (1 - 2x^2)$
8. $\lim_{x \to 1} f(x)$, where $f(x) = \begin{cases} 2x & \text{if } x \leq 1 \\ 2x + 3 & \text{if } x > 1 \end{cases}$
9. $\lim_{x \to 0} f(x)$, where $f(x) = \begin{cases} -x & \text{if } x < 0 \\ -x + 2 & \text{if } x \leq 0 \end{cases}$
10. $\lim_{x \to -1} f(x)$, where $f(x) = \begin{cases} x + 1 & \text{if } x \leq -1 \\ x^2 - x & \text{if } x > -1 \end{cases}$
11. $\lim_{x \to -2} f(x)$, where $f(x) = \begin{cases} 2x & \text{if } x < -2 \\ x^2 - 8 & \text{if } x \geq -2 \end{cases}$

*In Exercises 12–20, compute the limit by two different methods.*

12. $\lim_{x \to 5} \dfrac{x^2 - 25}{x - 5}$
13. $\lim_{x \to 1} \dfrac{x^2 - 2x + 1}{x - 1}$
14. $\lim_{x \to 0} \dfrac{2x^4 + 7x^2}{x^2}$
15. $\lim_{x \to 0} \dfrac{x^3 - 2x^2}{2x}$
16. $\lim_{x \to -1} \dfrac{3x^2 + 4x + 2}{x + 1}$
17. $\lim_{x \to -1/2} \dfrac{2x^2 - x - 1}{2x + 1}$
18. $\lim_{x \to -3} \dfrac{2x^2 + 3x - 9}{x^2 + 2x - 3}$
19. $\lim_{x \to 4} \dfrac{5x^2 - 19x - 4}{x^2 - 16}$
20. $\lim_{x \to 1/3} \dfrac{6x^2 + 7x - 3}{3x^2 - x}$

*In Exercises 21–25, compute each of the following limits: $\lim_{x \to a^-} f(x)$, $\lim_{x \to a^+} f(x)$, and $\lim_{x \to a} f(x)$ for the given value of $a$, if it exists. Draw a graph of each function and interpret your results graphically.*

21. $f(x) = \begin{cases} 4x & \text{if } x \leq 2 \\ 4x + 2 & \text{if } x > 2 \end{cases}$; $a = 2$
22. $f(x) = \begin{cases} -3x & \text{if } x < 1 \\ 1 + 3x & \text{if } x \geq 1 \end{cases}$; $a = 1$
23. $f(x) = \begin{cases} 5 & \text{if } x \leq 0 \\ 5 - x & \text{if } x > 0 \end{cases}$; $a = 0$
24. $f(x) = \begin{cases} -2 & \text{if } x < -3 \\ x + 1 & \text{if } x \geq -3 \end{cases}$; $a = -3$
25. $f(x) = \begin{cases} x^2 & \text{if } x \leq 0 \\ -x^2 & \text{if } x > 0 \end{cases}$; $a = 0$

26. Given the graph of the function $f(x)$ shown in Figure 2.6, compute the following:

   (a) $\lim_{x \to 4^-} f(x)$
   (b) $\lim_{x \to 4^+} f(x)$
   (c) $\lim_{x \to 4} f(x)$

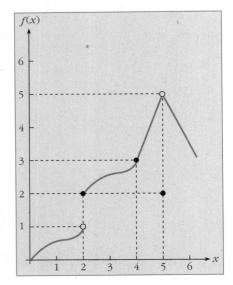

FIGURE 2.6

(d) $f(4)$  (e) $\lim_{x \to 2^-} f(x)$  (f) $\lim_{x \to 2^+} f(x)$
(g) $\lim_{x \to 2} f(x)$  (h) $f(2)$  (i) $\lim_{x \to 5^-} f(x)$
(j) $\lim_{x \to 5^+} f(x)$  (k) $\lim_{x \to 5} f(x)$  (l) $f(5)$

*In Exercises 27–55, evaluate the limits if they exist.*

27. $\lim_{x \to 2} (2x + 3)$

28. $\lim_{x \to -5} (4 - 6x)$

29. $\lim_{x \to -1} (x^2 - 1)$

30. $\lim_{x \to 3^+} (2x^2 + x)$

31. $\lim_{x \to 3^-} (2x^2 - x + 4)$

32. $\lim_{x \to -4} (5 - x - 3x^2)$

33. $\lim_{x \to 2} [(x - 5)(2x + 1)]$

34. $\lim_{x \to 0} [(2x - 7)(3x + 5)]$

35. $\lim_{x \to 4} \dfrac{x^2}{2x + 1}$

36. $\lim_{x \to -2} \dfrac{3x + 1}{3x - 4}$

37. $\lim_{x \to -3} \dfrac{7x^2 + 3x}{2x^2 + 5}$

38. $\lim_{x \to 6^-} \dfrac{x^2 - 36}{x - 6}$

39. $\lim_{x \to -7^+} \dfrac{x^2 - 49}{x + 7}$

40. $\lim_{x \to 1} \sqrt{2x^2 + 7}$

41. $\lim_{x \to 3} \sqrt[3]{3x^2 + 3}$

42. $\lim_{x \to -1} \sqrt[3]{x}$

43. $\lim_{x \to -2} \sqrt{x^2 + 5}$

44. $\lim_{x \to 0} \dfrac{(2x + 3)^2 - 9}{x}$

45. $\lim_{x \to 0} \dfrac{(x - 1)^3 + 1}{3x}$

46. $\lim_{x \to 2} \dfrac{x - 2}{x^2 - 4}$

47. $\lim_{x \to -3} \dfrac{x + 3}{x^2 - 9}$

48. $\lim_{x \to -2} \dfrac{x^2 + x - 2}{x^2 + 5x + 6}$

49. $\lim_{x \to 1^+} \dfrac{2x^2 - x - 1}{3x^2 - 2x - 1}$

50. $\lim_{x \to 2^-} \dfrac{3x^2 - 4x - 4}{3x^2 - 8x + 4}$

51. $\lim_{x \to 1} \dfrac{3x^2 - 7x + 1}{5x^2 + 2x - 4}$

52. $\lim_{x \to 0} \dfrac{\sqrt{x + 16} - 4}{x}$

53. $\lim_{x \to 0} \dfrac{\sqrt{2x + 4} - 2}{x}$

54. $\lim_{x \to 0} \dfrac{\sqrt{x + 5} - \sqrt{5}}{x}$

55. $\lim_{x \to 0} \dfrac{\sqrt{x^2 + 3} - \sqrt{3}}{x}$

*In Exercises 56–62,* $\lim_{x \to 2} f(x) = 3$ *and* $\lim_{x \to 2} g(x) = -5$. *Evaluate the given limits.*

56. $\lim_{x \to 2} [2f(x) + 3g(x)]$

57. $\lim_{x \to 2} [5f(x) - 4g(x)]$

58. $\lim_{x \to 2} [f(x) \cdot g(x)]$

59. $\lim_{x \to 2} \dfrac{7f(x) + 1}{2g(x) + 3}$

60. $\lim_{x \to 2} \dfrac{3g(x)}{2f(x) + 1}$

61. $\lim_{x \to 2} \sqrt{4f(x) + g(x)}$

62. $\lim_{x \to 2} \dfrac{\sqrt[3]{f(x)}}{\sqrt[3]{g(x) + 1}}$

## 2.2 LIMITS AT INFINITY AND INFINITE LIMITS—ASYMPTOTES

### An Example of a Limit at Infinity

If a certain bank gives eight percent annual interest compounded quarterly (four times a year), consider the amounts $A$ of money you would have after several quarters if you invested $10,000 initially:

After 1 quarter: $A = 10000\left(1 + \frac{0.08}{4}\right) = 10000(1.02) = \$10200$

After 2 quarters: $A = 10200\left(1 + \frac{0.08}{4}\right) = 10200(1.02) = 10000(1.02)^2$
$= \$10404$

After 3 quarters: $A = 10404\left(1 + \frac{0.08}{4}\right) = 10404(1.02) = 10000(1.02)^3$
$= \$10612.08$

After 4 quarters: $A = 10000(1.02)^4 = \$10824.32$.

Apparently, the general formula for finding the amount of money accumulated after $q$ quarters at a yearly interest rate of 8% is

$$A = 10000\left(1 + \frac{0.08}{4}\right)^q = 10000(1.02)^q$$

Since there are four quarters in one year, if $t$ is time in years then $q = 4t$; hence

$$A(t) = 10000\left(1 + \frac{0.08}{4}\right)^{4t} = 10000(1.02)^{4t}, \quad \text{where } A \text{ is a function of } t$$

For example, after five years, the amount accumulated is

$$A(5) = 10000(1.02)^{4(5)} = 10000(1.02)^{20} = \$14859.47$$

Now let's consider what happens if we invest $10,000 for a period of one year at an annual interest rate of 8% compounded $x$ times per year. If $A(x)$ is the amount accumulated after one year, then

$$A(x) = 10000\left(1 + \frac{0.08}{x}\right)^x$$

In Table 2.9, the amounts $A(x)$ are computed (using a calculator) for various numbers of compounding times, $x$, per year.

| $x$ | 1 | 2 | 4 | 12 | 365 | 1000 | 100000 | 1000000 | 10000000 |
|---|---|---|---|---|---|---|---|---|---|
| $A(x)$ | 10800.00 | 10816.00 | 10824.32 | 10830.00 | 10832.78 | 10832.84 | 10832.87 | 10832.87 | 10832.87 |

TABLE 2.9

Apparently as $x$ increases, $A(x)$ also increases. However, it appears that $A(x)$ approaches the limiting value of 10832.87 as $x$ gets very large. Since we have rounded our numbers to two decimal places, this limiting value is only approximate. In symbols:

$$\lim_{x \to \infty} A(x) \approx 10832.87$$

where the notation $x \to \infty$ is read "$x$ approaches infinity." The statement $\lim_{x \to \infty} A(x) \approx 10832.87$ means that as $x$ takes on values that are larger and larger, the corresponding values of $A(x)$ get closer and closer to 10832.87 (approximately). "Letting $x$ take on values that are larger and larger" is sometimes referred to as "letting $x$ increase without bound."

**Limits at Infinity**

**DEFINITION**

> Let $L$ be a real number. Then the limit of $f(x)$ as $x$ approaches infinity is equal to $L$ (in symbols, $\lim_{x \to \infty} f(x) = L$) if, as we let $x$ take on values that become larger and larger, the corresponding values of $f(x)$ get closer and closer to the number $L$.

The limit $\lim_{x \to \infty} f(x)$ is called a "limit at infinity" because we are attempting to find the limit of the function $f(x)$ as $x$ "approaches infinity"; thus we are computing the limit "at infinity." (Of course, no matter how large we let $x$ become, it cannot really approach infinity, since any real number $x$ will always be infinitely distant from infinity! Nonetheless, we still use the terminology "$x$ approaches infinity.") A second type of limit at infinity is $\lim_{x \to -\infty} f(x)$. This notation is read, "the limit of $f(x)$ as $x$ approaches negative infinity." In this second type of limit, we are attempting to determine whether $f(x)$ approaches a limiting value as $x$ decreases without bound through negative values.

**EXAMPLE 1**

Find (a) $\lim_{x \to \infty} (1/x)$ and (b) $\lim_{x \to -\infty} (1/x)$.

**Solution**

(a) To find this limit, we construct a table of values, choosing the $x$-values so that they become larger and larger, and we observe whether the corresponding values of $1/x$ approach any limiting number $L$.

| $x$ | 1 | 10 | 100 | 1000 | 10000 | 100000 | $\to \infty$ |
|---|---|---|---|---|---|---|---|
| $f(x) = 1/x$ | 1 | 0.1 | 0.01 | 0.001 | 0.0001 | 0.00001 | $\to 0$ |

**TABLE 2.10**

From Table 2.10 we see that as $x$ increases without bound, $1/x$ approaches zero, so that

$$\lim_{x \to \infty} \frac{1}{x} = 0$$

## 2.2 LIMITS AT INFINITY AND INFINITE LIMITS—ASYMPTOTES

(b) We construct Table 2.11 by choosing $x$-values that "approach minus infinity," and we observe that the corresponding values of $1/x$ approach a limiting value of zero. We conclude that

$$\lim_{x \to -\infty} \frac{1}{x} = 0$$

| $x$ | $-1$ | $-10$ | $-100$ | $-1000$ | $-10000$ | $-100000$ | $\to -\infty$ |
|---|---|---|---|---|---|---|---|
| $f(x) = 1/x$ | $-1$ | $-0.1$ | $-0.01$ | $-0.001$ | $-0.0001$ | $-0.00001$ | $\to 0$ |

TABLE 2.11

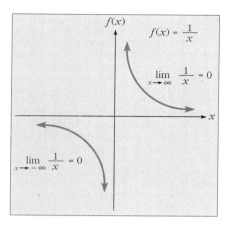

FIGURE 2.7 The graph of $f(x) = 1/x$ illustrates that $\lim_{x \to \pm\infty} (1/x) = 0$.

Figure 2.7 shows that as $x$ becomes larger and larger positively, the graph of $f(x) = 1/x$ approaches the $x$-axis; that is, $y = 1/x$ approaches zero as $x$ becomes very large. Also, as $x$ becomes larger and larger negatively, the graph of $f(x)$ again approaches the $x$-axis; that is, $y = 1/x$ approaches zero as $x$ decreases without bound through negative values.

---

### EXAMPLE 2 (Participative)

Let's compute $\lim_{x \to \infty} (2/x^2)$.

### Solution

First complete Table 2.12.

| $x$ | 1 | 10 | 100 | 1,000 | 10,000 |
|---|---|---|---|---|---|
| $f(x) = 2/x^2$ | 2 | 0.02 | 0.0002 | 0.000002 | 0.00000002 |

TABLE 2.12

We conclude that $\lim_{x \to \infty} (2/x^2) = \underline{\phantom{0}}$.

---

### SOLUTIONS TO EXAMPLE 2

| $x$ | 1 | 10 | 100 | 1,000 | 10,000 |
|---|---|---|---|---|---|
| $f(x) = 2/x^2$ | 2 | 0.02 | 0.0002 | 0.000002 | 0.00000002 |

TABLE 2.13

$$\lim_{x \to \infty} (2/x^2) = 0$$

In general, we may state the following rules:

> i. If $a$ is any constant and $p > 0$, then $\lim_{x \to \infty} (a/x^p) = 0$.
>
> ii. Also, if we restrict $p$ to being a positive number such that $x^p$ is defined for all $x < 0$, then $\lim_{x \to -\infty} (a/x^p) = 0$.

In the second result, $p$ might be 2 or $\frac{1}{3}$, but $p$ could not be $\frac{1}{2}$, for example, since $x^{1/2}$ is not defined for $x < 0$. Thus $\lim_{x \to -\infty} (1/x^{1/2})$ does not exist.

We can use these properties to compute certain limits at infinity without having to construct tables of values. Before we attempt the next example, we comment that Properties 1–7 of Section 2.1 also apply to limits at infinity.

**EXAMPLE 3**

Compute the following limits:

(a) $\lim_{x \to \infty} \dfrac{5}{x^3}$  (b) $\lim_{x \to -\infty} \left[1 - \dfrac{4}{\sqrt[3]{x}}\right]$  (c) $\lim_{x \to \infty} \left[\dfrac{5 - x - x^2}{x^3}\right]$

**Solution**

(a) By Property i with $a = 5$ and $p = 3$, we have $\lim_{x \to \infty} (5/x^3) = 0$.

(b) $\lim_{x \to -\infty} \left[1 - \dfrac{4}{\sqrt[3]{x}}\right] = \lim_{x \to -\infty} 1 - \lim_{x \to -\infty} \left(\dfrac{4}{\sqrt[3]{x}}\right)$  Property 4 of Section 2.1

$= 1 - \lim_{x \to -\infty} \left(\dfrac{4}{x^{1/3}}\right)$  Property 1 of Section 2.1

$= 1 - 0 = 1$  Property ii with $a = 4$ and $p = \frac{1}{3}$

(c) $\lim_{x \to \infty} \left[\dfrac{5 - x - x^2}{x^3}\right] = \lim_{x \to \infty} \left[\dfrac{5}{x^3} - \dfrac{x}{x^3} - \dfrac{x^2}{x^3}\right] = \lim_{x \to \infty} \left[\dfrac{5}{x^3} - \dfrac{1}{x^2} - \dfrac{1}{x}\right]$

$= \lim_{x \to \infty} \left(\dfrac{5}{x^3}\right) - \lim_{x \to \infty} \left(\dfrac{1}{x^2}\right) - \lim_{x \to \infty} \left(\dfrac{1}{x}\right)$  Property 4 of Section 2.1

$= 0 - 0 - 0 = 0$  Property i

### Rational Functions

A *rational function* is defined to be a quotient of two polynomials. For example, each of the following functions is a rational function:

$$f(x) = \dfrac{x^2 - 7x + 3}{x^3 + 5x + 4} \qquad g(x) = \dfrac{7x^3 + 5}{2x^2 - 3x + 1}$$

These functions are not rational:

$$f(x) = \frac{\sqrt{x+1}}{x^3 + 5} \qquad (\sqrt{x+1} \text{ is } not \text{ a polynomial})$$

$$g(x) = \frac{x^2 + 3x + 1}{\sqrt[3]{x} + 7x + 5} \qquad (\sqrt[3]{x} + 7x + 5 \text{ is } not \text{ a polynomial})$$

### Limits of Rational Functions at Infinity

We will now consider finding limits of rational functions at infinity. When finding such limits, we follow the general procedure of dividing numerator and denominator of the function by the highest power of the variable that appears in the function as a first step in computing the limit. Then we use Properties 1–7 of Section 2.1 and Properties i and ii of this section to complete the problem.

**EXAMPLE 4**

Compute $\lim_{x \to \infty} (5 - x^2)/(7 - x - x^3)$.

**Solution**

In this problem, the highest power of $x$ that appears in the rational function is $x^3$, so we begin by dividing numerator and denominator by $x^3$.

$$\lim_{x \to \infty} \frac{5 - x^2}{7 - x - x^3} = \lim_{x \to \infty} \frac{\dfrac{5 - x^2}{x^3}}{\dfrac{7 - x - x^3}{x^3}}$$

$$= \lim_{x \to \infty} \frac{\dfrac{5}{x^3} - \dfrac{x^2}{x^3}}{\dfrac{7}{x^3} - \dfrac{x}{x^3} - \dfrac{x^3}{x^3}} = \lim_{x \to \infty} \frac{\dfrac{5}{x^3} - \dfrac{1}{x}}{\dfrac{7}{x^3} - \dfrac{1}{x^2} - 1}$$

$$= \frac{\lim_{x \to \infty}\left(\dfrac{5}{x^3} - \dfrac{1}{x}\right)}{\lim_{x \to \infty}\left(\dfrac{7}{x^3} - \dfrac{1}{x^2} - 1\right)} \qquad \text{Property 6 of Section 2.1}$$

$$= \frac{\lim_{x \to \infty}\left(\dfrac{5}{x^3}\right) - \lim_{x \to \infty}\left(\dfrac{1}{x}\right)}{\lim_{x \to \infty}\left(\dfrac{7}{x^3}\right) - \lim_{x \to \infty}\left(\dfrac{1}{x^2}\right) - \lim_{x \to \infty}(-1)} \qquad \text{Property 4 of Section 2.1}$$

$$= \frac{0 - 0}{0 - 0 - (-1)} = 0 \qquad \text{Property 1 of Section 2.1 and Property ii.} \blacksquare$$

**EXAMPLE 5 (Participative)**

Let's find $\lim_{x \to -\infty} (3x + 1)/(5x - 10)$.

# 84 CHAPTER 2 LIMITS AND AN INTRODUCTION TO THE DERIVATIVE

**Solution**

We begin by noting that the highest power of $x$ in this rational function is $\underline{x}$, so we divide numerator and denominator by $x$.

$$\lim_{x\to-\infty} \frac{3x+1}{5x-10} = \lim_{x\to-\infty} \frac{\frac{3x+1}{x}}{\frac{5x-10}{x}} = \lim_{x\to-\infty} \frac{\frac{3x}{x}+\frac{1}{x}}{\frac{5x}{x}-\frac{10}{x}}$$

$$= \lim_{x\to-\infty} \frac{3+\frac{1}{x}}{5-\frac{10}{x}} = \frac{\lim_{x\to-\infty}\left(3+\frac{1}{x}\right)}{\lim_{x\to-\infty}\left(5-\frac{10}{x}\right)}$$

$$= \frac{\lim_{x\to-\infty} 3 + \lim_{x\to-\infty}\frac{1}{x}}{\lim_{x\to-\infty} 5 - \lim_{x\to-\infty}\left(\frac{10}{x}\right)}$$

$$= \frac{3+0}{5-0} = \frac{3}{5}$$

A graph of $f(x) = (3x+1)/(5x-10)$ appears in Figure 2.8. By replacing each $-\infty$ in this example by $\infty$, it can be seen that $\lim_{x\to\infty}(3x+1)/(5x-10)$ is also equal to $\frac{3}{5}$.

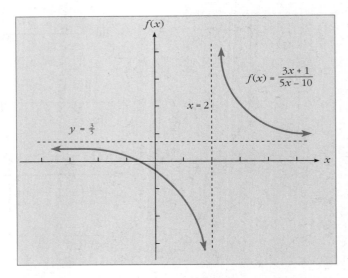

**FIGURE 2.8** The horizontal line $y = \frac{3}{5}$ is a horizontal asymptote of the graph of $f(x)$.

---

**SOLUTIONS TO EXAMPLE 5**   $x, x, 10/x, 5, 5-(10/x), (10/x), 5-0, \frac{3}{5}$

### Horizontal Asymptotes

An important feature of the graph in Figure 2.8 is that as $x \to \infty$, the graph of $f(x)$ approaches the horizontal line $y = \frac{3}{5}$. This is a consequence of the fact that $\lim_{x \to \infty} (3x + 1)/(5x - 10) = \frac{3}{5}$, as was shown in Example 5. The horizontal line $y = \frac{3}{5}$ is called a *horizontal asymptote* of the graph of $f$.

**DEFINITION**

> The horizontal line $y = a$ is a *horizontal asymptote* of the graph of $f$ if and only if either
> $$\lim_{x \to \infty} f(x) = a$$
> or
> $$\lim_{x \to -\infty} f(x) = a$$
> where $a$ is some real number.

In Figure 2.9 some horizontal asymptotes are illustrated graphically. The function in Figure 2.9(a) has $y = -1$ as a horizontal asymptote, and the function in Figure 2.9(b) has the lines $y = 0$ (that is, the $x$-axis) and $y = 2$ as horizontal asymptotes.

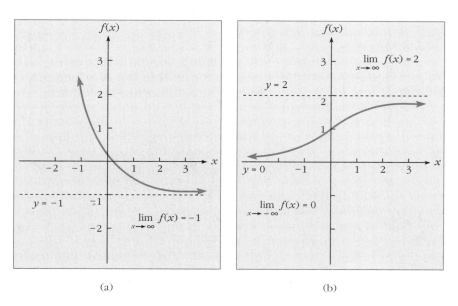

FIGURE 2.9 The graph in (a) has $y = -1$ as a horizontal asymptote. The graph in (b) has $y = 0$ and $y = 2$ as horizontal asymptotes.

### Infinite Limits

One reason why a limit $\lim_{x \to a} f(x)$ may not exist is that $f(x)$ may become unbounded (positively or negatively) as $x$ approaches $a$. This type of limit is called an "infinite limit."

**EXAMPLE 6**

Examine $\lim_{x \to 0} (1/x^2)$.

**Solution**

To attempt to determine this limit, we construct Table 2.14 and Table 2.15.

| $x$ | $-1$ | $-0.1$ | $-0.01$ | $-0.001$ | $\to 0^-$ |
|---|---|---|---|---|---|
| $f(x) = 1/x^2$ | 1 | 100 | 10,000 | 1,000,000 | $\to \infty$ |

**TABLE 2.14** The value of $x$ approaches zero from the left.

| $x$ | 1 | 0.1 | 0.01 | 0.001 | $\to 0^+$ |
|---|---|---|---|---|---|
| $f(x) = 1/x^2$ | 1 | 100 | 10,000 | 1,000,000 | $\to \infty$ |

**TABLE 2.15** The value of $x$ approaches zero from the right.

Observe that as $x$ approaches 0, the values of $f(x) = 1/x^2$ become larger and larger, so $1/x^2$ does not approach any real number. We conclude that $\lim_{x \to 0} 1/x^2$ does not exist. Tables 2.14 and 2.15 show, however, that $f(x) = 1/x^2$ becomes indefinitely large as $x \to 0$. We write this as $\lim_{x \to 0} (1/x^2) = \infty$. Since $\infty$ is *not* a real number, we are not saying that this limit exists. The statement $\lim_{x \to 0} (1/x^2) = \infty$ means that this limit does not exist *because* $1/x^2$ becomes indefinitely large as $x$ approaches 0. The function $f(x) = 1/x^2$ is graphed in Figure 2.10. Note that as $x$ approaches 0 from either side, the corresponding values of $f(x)$ increase indefinitely. ∎

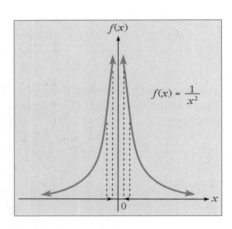

**FIGURE 2.10** The graph of $f(x) = 1/x^2$ illustrates that $\lim_{x \to 0} (1/x^2) = \infty$.

### Vertical Asymptotes

In Example 6, the vertical line $x = 0$ (that is, the $y$-axis) is called a *vertical asymptote* of the graph $f(x) = 1/x^2$.

**DEFINITION**

The vertical line $x = a$ is a *vertical asymptote* of the graph of $f$ if and only if any of the following statements is true:

(a) $\lim_{x \to a^+} f(x) = \infty$    (b) $\lim_{x \to a^+} f(x) = -\infty$

(c) $\lim_{x \to a^-} f(x) = \infty$    (d) $\lim_{x \to a^-} f(x) = -\infty$

See Figure 2.11.

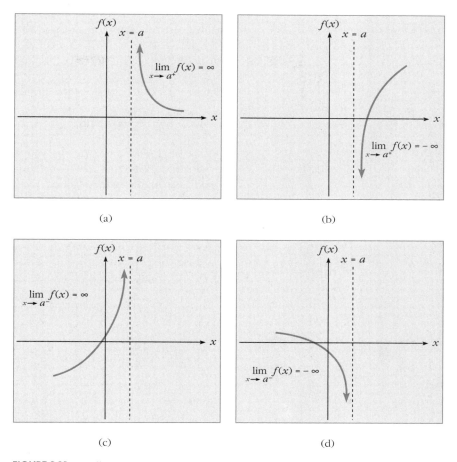

**FIGURE 2.11** An illustration of vertical asymptotes.

Typically, if a rational function $f(x)/g(x)$ is in lowest terms, vertical asymptotes occur at values of $x$ for which the denominator $g(x)$ is zero.

In finding infinite limits, the following properties are useful.

---

If $\lim_{x \to a} f(x) = N \neq 0$ and if $\lim_{x \to a} g(x) = 0$, then

I. $\lim_{x \to a} \dfrac{f(x)}{g(x)} = \infty$ if $N > 0$ and $g(x)$ approaches 0 through positive values;

II. $\lim_{x \to a} \dfrac{f(x)}{g(x)} = -\infty$ if $N > 0$ and $g(x)$ approaches 0 through negative values;

III. $\lim_{x \to a} \dfrac{f(x)}{g(x)} = -\infty$ if $N < 0$ and $g(x)$ approaches 0 through positive values;

IV. $\lim_{x \to a} \dfrac{f(x)}{g(x)} = \infty$ if $N < 0$ and $g(x)$ approaches 0 through negative values.

EXAMPLE 7

Solution

Find the vertical asymptotes of $F(x) = 2x/(x^2 + x - 2)$.

First, note that $F(x)$ is in lowest terms, and the denominator factors as $x^2 + x - 2 = (x - 1)(x + 2)$, so our "candidates" for vertical asymptotes are $x = 1$ and $x = -2$. Let us consider the limits:

(a) $\lim\limits_{x \to 1^-} \dfrac{2x}{x^2 + x - 2} = \lim\limits_{x \to 1^-} \dfrac{2x}{(x - 1)(x + 2)}$

As $x \to 1^-$, we have $2x \to 2(1) = 2$. As $x \to 1^-$, $x$ will be less than but "close to" 1, so that $x - 1 < 0$ and $x + 2 > 0$. Therefore $(x - 1)(x + 2) \to 0$ through *negative* values. By Property II, $\lim\limits_{x \to 1^-} F(x) = -\infty$.

(b) $\lim\limits_{x \to 1^+} \dfrac{2x}{x^2 + x - 2} = \lim\limits_{x \to 1^+} \dfrac{2x}{(x - 1)(x + 2)}$

As $x \to 1^+$, we have $2x \to 2(1) = 2$. As $x \to 1^+$, $x$ will be greater than but "close to" 1, so that $x - 1 > 0$ and $x + 2 > 0$. Therefore $(x - 1)(x + 2) \to 0$ through *positive* values. By Property I, $\lim\limits_{x \to 1^+} F(x) = \infty$.

Note that (a) and (b) show that the vertical line $x = 1$ is a vertical asymptote of $F(x)$.

(c) $\lim\limits_{x \to -2^-} \dfrac{2x}{x^2 + x - 2} = \lim\limits_{x \to -2^-} \dfrac{2x}{(x - 1)(x + 2)}$

As $x \to -2^-$, we have $2x \to 2(-2) = -4$. As $x \to -2^-$, $x$ will be less than but "close to" $-2$, so that $x - 1 < 0$ and $x + 2 < 0$. Therefore $(x - 1)(x + 2) \to 0$ through *positive* values. By Property III, $\lim\limits_{x \to -2^-} F(x) = -\infty$.

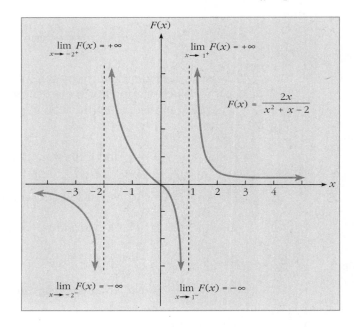

FIGURE 2.12   The function $F(x)$ has vertical asymptotes $x = -2$ and $x = 1$.

## 2.2 LIMITS AT INFINITY AND INFINITE LIMITS—ASYMPTOTES

(d) $\lim_{x \to -2^+} \dfrac{2x}{x^2 + x - 2} = \lim_{x \to -2^+} \dfrac{2x}{(x-1)(x+2)}$

As $x \to -2^+$, we have $2x \to 2(-2) = -4$. As $x \to -2^+$, $x$ will be greater than but "close to" $-2$, so that $x - 1 < 0$ and $x + 2 > 0$. Therefore $(x-1)(x+2) \to 0$ through *negative* values. By Property IV, $\lim_{x \to -2^+} F(x) = \infty$.

Note that (c) and (d) show that the vertical line $x = -2$ is a vertical asymptote. A graph of $F(x)$ appears in Figure 2.12.

### EXAMPLE 8 (Participative)

**Solution**

Let's find $\lim_{x \to 3} 2x/(x - 3)$.

For this we compute as follows.

(a) First we compute $\lim_{x \to 3^-} 2x/(x - 3)$. As $x \to 3^-$, we have $2x \to \underline{6}$. As $x \to 3^-$, the quantity $x - 3 \to 0$ through $\underline{NEGATIVE}$ values. We conclude that $\lim_{x \to 3^-} \dfrac{2x}{x - 3} = \underline{-\infty}$.

(b) Next we compute $\lim_{x \to 3^+} 2x/(x - 3)$. As $x \to 3^+$, we have $2x \to \underline{6}$. As $x \to 3^+$, the quantity $x - 3 \to 0$ through $\underline{positive}$ values. We conclude that $\lim_{x \to 3^+} \dfrac{2x}{x - 3} = \underline{\infty}$.

Note that (a) and (b) show that the vertical line $x = 3$ is a vertical asymptote of $y = 2x/(x - 3)$. Since the left-hand limit at 3 is $-\infty$ and the right-hand limit at 3 is $\infty$, we say that $\lim_{x \to 3} 2x/(x - 3)$ does not exist. Specifically it is *neither* $\infty$ nor $-\infty$.

### EXAMPLE 9

In the problem of ordering and holding inventory, the ordering cost $C_0$ is the cost of placing an order (for labor, record keeping, etc.), regardless of the number of units that are ordered. If we require $d$ units per year in the production of our product and we order in batches of $q$ units during the year, then we must make $d/q = n$ orders per year, and our annual ordering cost is

$$A = C_0 n = \dfrac{C_0 d}{q}$$

Given that $C_0 = \$10$ per order and $d = 5000$ units per year,

(a) find the annual ordering cost function $A(q)$ for $q \geq 0$;

---

**SOLUTIONS TO EXAMPLE 8**   (a) 6, negative, $-\infty$; (b) 6, positive, $\infty$

(b) determine any horizontal or vertical asymptotes of this function; and
(c) graph the function.

**Solution**

(a) Substituting $C_0 = 10$ and $d = 5000$ into the given formula produces

$$A(q) = \frac{(10)(5000)}{q} = \frac{50000}{q}, \quad q \geq 0$$

(b) Since $\lim_{q \to 0^+} 50000/q = +\infty$, the vertical line $q = 0$ (that is, the $y$-axis) is a vertical asymptote. Since $\lim_{q \to \infty} 50000/q = 0$, the horizontal line $y = 0$ (that is, the $q$-axis) is a horizontal asymptote.

(c)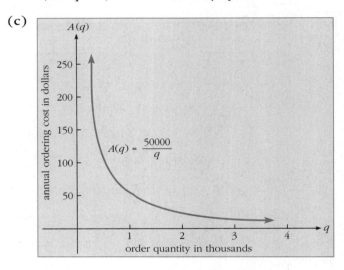

**FIGURE 2.13** The annual ordering cost function $A(q)$.

### Optional Graphing Calculator/Computer Material

Before reading this material, read Exercise 48 of this section to learn about the "short-cut" method for finding limits at infinity.

A graphing calculator or computer graphing software can help you to find certain infinite limits quickly and give you a good visual representation of the function for which you are finding the limits. As an example, consider

$$f(x) = \frac{x^2 + 7x - 3}{2x^2 - 7x - 15}$$

Even though a graphing calculator or computer graphing software could be used to find the horizontal asymptotes of a function, typically these are difficult to read exactly from a graph, so it is probably better to use the short-cut method of Exercise 48:

$$\lim_{x \to \infty} \frac{x^2 + 7x - 3}{2x^2 - 7x - 15} = \lim_{x \to \infty} \frac{x^2}{2x^2} = \lim_{x \to \infty} \frac{1}{2} = \frac{1}{2}$$

Similarly $\lim_{x \to -\infty} f(x) = \frac{1}{2}$. We conclude that the horizontal line $y = \frac{1}{2}$ is a horizontal asymptote of $f(x)$. Now,

$$f(x) = \frac{x^2 + 7x - 3}{(x - 5)(2x + 3)}$$

Candidates for vertical asymptotes are $x = 5$ and $x = -\frac{3}{2}$. By using a graphing calculator or computer graphing software, we obtain the graph in Figure 2.14. From this graph we can immediately observe that, for example,

$$\lim_{x \to -3/2^-} f(x) = -\infty \qquad \lim_{x \to -3/2^+} f(x) = +\infty$$
$$\lim_{x \to 5^-} f(x) = -\infty \qquad \lim_{x \to 5^+} f(x) = +\infty$$

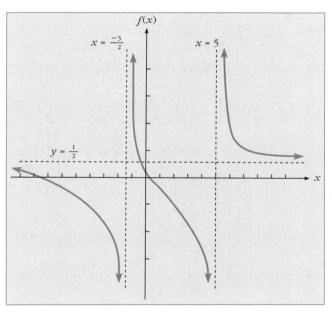

**FIGURE 2.14** The graph of $f(x) = \dfrac{x^2 + 7x - 3}{(x - 5)(2x + 3)}$, showing vertical and horizontal asymptotes.

## SECTION 2.2
**SHORT ANSWER EXERCISES**

1. The limit $\lim_{x \to \infty} (1/x^3)$ is (a) a limit at infinity; (b) an infinite limit; (c) a limit that does not exist.

2. The limit $\lim_{x \to 0} (1/x^4)$ is (a) a limit at infinity; (b) an infinite limit; (c) a limit that does not exist; (d) all of the preceding.

3. (True or False) A horizontal asymptote of $f(x)$ is a horizontal line that the graph of $f(x)$ approaches as $x \to \infty$ or $x \to -\infty$. _____

4. (True or False) If $\lim_{x \to 5} f(x) = -\infty$, then this limit exists. _____

5. In computing a limit at infinity of a rational function $F(x)$, we divide numerator and denominator by (a) the highest power of $x$ in the numerator; (b) the highest power of $x$ in the denominator; (c) the highest power of $x$ in both the numerator and the denominator.

6. Given that $\lim_{x \to 2} f(x) = \infty$ and $\lim_{x \to \infty} f(x) = 7$, we know that the vertical line $x =$ _____ is a vertical asymptote of $f(x)$ and the horizontal line $y =$ _____ is a horizontal asymptote of $f(x)$.

7. $\lim_{x \to -\infty} \left( \dfrac{100}{x} - 300 \right) =$ _____.

8. Which of the following functions are rational functions?

   (a) $f(x) = \dfrac{x}{x-1}$   (b) $g(x) = \dfrac{\sqrt{x}}{x^2 - 1}$   (c) $h(x) = \dfrac{2x^4 - 7}{3x + 5}$   Answer: _____

9. (True or False) Any function can have at most two horizontal asymptotes. _____

10. (True or False) Any function can have at most two vertical asymptotes. _____

11. (True or False) Every polynomial function has both horizontal and vertical asymptotes. _____

12. The vertical asymptote of the graph of $y = 1/(x + 3)$ is $x =$ _____.

13. The horizontal asymptote of the graph of $y = 2/x^3$ is $y =$ _____.

## SECTION 2.2
### EXERCISES

In Exercises 1–13, find the limits at infinity when they exist.

1. $\lim_{x \to \infty} \dfrac{2}{x}$

2. $\lim_{x \to -\infty} \dfrac{7}{x^2}$

3. $\lim_{x \to -\infty} \left( \dfrac{1}{x} + 3 \right)$

4. $\lim_{x \to \infty} \left( 10 - \dfrac{4}{x} + \dfrac{5}{x^2} \right)$

5. $\lim_{x \to \infty} \dfrac{5x + 1}{2x - 3}$

6. $\lim_{x \to -\infty} \dfrac{1 - 7x}{4x + 5}$

7. $\lim_{x \to \infty} \dfrac{3}{5x + 8}$

8. $\lim_{x \to \infty} \dfrac{-10}{6 - 3x}$

9. $\lim_{x \to -\infty} \dfrac{4x^2 + 3x - 7}{2x^2 - x + 2}$

10. $\lim_{x \to \infty} \dfrac{5x^2 + 7x}{2x^2 + 9}$

11. $\lim_{x \to -\infty} \dfrac{10x^3 + 7x^2 + 3x + 5}{4x^4 + 3x^2 + 2}$

12. $\lim_{x \to \infty} \dfrac{x^3 + 7x + 3}{x^2 - 4x + 2}$

13. $\lim_{x \to \infty} \dfrac{2x^2 + 3x - 5}{7x - 4}$

In Exercises 14–27, select the appropriate classification for the following limits that do not exist: (a) *limits that are* $\infty$; (b) *limits that are* $-\infty$; (c) *other limits that do not exist.*

14. $\lim_{x \to 0} \dfrac{5}{x^2}$

15. $\lim_{x \to 1} \dfrac{-1}{(x - 1)^2}$

16. $\lim_{x \to -2} \dfrac{4}{x + 2}$

17. $\lim_{x \to 3^+} \dfrac{x}{x - 3}$

18. $\lim_{x \to -1^+} \dfrac{x-1}{x+1}$

19. $\lim_{x \to 3} \dfrac{2}{(x-3)^3}$

20. $\lim_{x \to 4^+} \dfrac{6}{x^2-16}$

21. $\lim_{x \to -2^-} \dfrac{4x}{x^2+5x+6}$

22. $\lim_{x \to 5^+} \dfrac{x+3}{x^2-3x-10}$

23. $\lim_{x \to 0} \dfrac{4}{x^3-x^2}$

24. $\lim_{x \to 5} \dfrac{x+5}{x^2-25}$

25. $\lim_{x \to 1^-} \dfrac{x^2+2x+7}{x^2+6x-7}$

26. $\lim_{x \to 4} \dfrac{10x+5}{x^3-8x^2+16x}$

27. $\lim_{x \to 2^+} \dfrac{(x+3)(x-2)}{x(x-2)^2(x-3)}$

*In Exercises 28–39, find the horizontal and vertical asymptotes of the functions. Also, graph the indicated functions.*

28. $f(x) = 5/x$: graph

29. $f(x) = 4/x^2$: graph

30. $f(x) = \dfrac{2x}{x-1}$: graph

31. $g(x) = \dfrac{-7x}{2x-4}$: graph

32. $f(x) = \dfrac{6}{5-x}$: graph

33. $f(x) = \dfrac{-5}{6-x}$: graph

34. $g(x) = \dfrac{10}{x^2-1}$: graph

35. $f(x) = \dfrac{7x}{x^2-36}$

36. $f(x) = \dfrac{2x+2}{x^2-2x+1}$

37. $g(x) = \dfrac{x^3+3x^2+5}{x^2+2x-15}$

38. $f(x) = \dfrac{7-x-x^2}{x^2+5x+6}$

39. $g(x) = \dfrac{3x-2x^2}{x^2+2x+1}$

40. **(Revenue)** Given the revenue function $R(x) = 5x/(x+1)$, $x \geq 0$, where $x$ is the number of units produced and $R(x)$ is measured in thousands of dollars, find the value that $R(x)$ approaches as production increases without bound.

41. **(Compound interest)** Construct a table of values similar to Table 2.9 of this section, showing the amounts $A(x)$ that $50,000 will grow to in one year at 10% interest compounded $x$ times per year. What is $\lim_{x \to \infty} A(x)$? This is the amount that $50,000 grows to in one year at 10% compounded "continuously," as the banks say.

**Business and Economics**

42. **(Ordering costs)** Suppose, in producing one of your products, your company needs 20,000 of a certain type of component per year. If each order of these parts has an ordering cost of $15, and if you order in equal batches of $q$ units during the year,
    (a) find the annual ordering cost function $A(q)$ for $q \geq 0$;
    (b) determine the annual ordering cost if you order $q = 1000$ parts per order;
    (c) determine any horizontal or vertical asymptotes of $A(q)$;
    (d) graph $A(q)$ for $q \geq 0$.

**Business and Economics**

43. **(Average cost)** Given a cost function $C(x)$, the average cost of producing $x$ units is defined to be $\overline{C}(x) = C(x)/x$.
    (a) Given the cost function $C(x) = 50x + 600$, $x \geq 0$, find the average cost function $\overline{C}(x)$ for $x > 0$.

(b) Find and interpret $\overline{C}(0.01)$, $\overline{C}(1)$, $\overline{C}(10)$, and $\overline{C}(10,000)$.
(c) Determine any horizontal or vertical asymptotes of $\overline{C}(x)$.
(d) Graph the function $\overline{C}(x)$ for $x > 0$.
(e) Interpret any horizontal or vertical asymptotes of your graph in terms of the average cost of producing certain levels of output.

44. (Average cost) The average cost $\overline{C}(x)$ of producing $x$ units of a certain product is

$$\overline{C}(x) = \frac{x^2 + 5x + 6}{x^2 - 1}, \quad \text{if } x > 2$$

(a) Find the average cost of producing 1000 units of this product.
(b) If we increase production without bound, what does the average cost per unit approach?

45. (Employee training) After $t$ hours of special training, an average new employee working on an assembly line in a certain factory can assemble

$$N(t) = \frac{500t}{t + 50}$$

parts per day. Find $\lim_{t \to \infty} N(t)$ and interpret your result.

**Social Sciences**

46. (Psychology) In a psychology experiment, the time $T$ (in seconds) required for a subject to complete a certain task depends upon the number $r$ of repetitions of the task that the subject has completed, according to the function

$$T(r) = \frac{40r + 140}{4r + 7}$$

Find $\lim_{r \to \infty} T(r)$ and interpret your result.

47. (Sociology) The number $n$ of thousands of people in a town who have heard a certain rumor $t$ days after the rumor begins is given by

$$N(t) = \frac{100t + 29}{4t + 10}$$

Find $\lim_{t \to \infty} N(t)$ and interpret your result.

48. There is a "short-cut" method for finding limits at infinity (and hence horizontal asymptotes) of rational functions. Let $F(x) = f(x)/g(x)$ be a rational function in lowest terms. Note that this implies that $f(x)$ and $g(x)$ are polynomials with no common factors between them. If the term involving the highest power of $x$ in $f(x)$ is $ax^n$ and if the term involving the highest power of $x$ in $g(x)$ is $bx^m$, then

$$\lim_{x \to \infty} \frac{f(x)}{g(x)} = \lim_{x \to \infty} \frac{ax^n}{bx^m}$$

where an equivalent statement may be made if we replace $\infty$ by $-\infty$. For example,

$$\lim_{x \to \infty} \frac{7x^2 + 3x + 5}{2x^2 - 7x - 10} = \lim_{x \to \infty} \frac{7x^2}{2x^2} = \lim_{x \to \infty} \frac{7}{2} = \frac{7}{2}$$

Redo Exercises 5–13 of this section using this short-cut method.

OPTIONAL

GRAPHING CALCULATOR/
COMPUTER EXERCISES

*For each of the following functions, (a) find the horizontal asymptotes using the shortcut method; (b) determine candidates for vertical asymptotes by finding all values that make the denominator equal to zero; (c) graph the function and determine all right- and left-hand limits at each of the vertical asymptote candidates you found in part (b).*

**49.** $f(x) = \dfrac{2x^2 + 5x + 3}{x^2 + 4x - 21}$

**50.** $f(x) = \dfrac{7x^2 - 10x + 5}{3x^2 - 17x + 10}$

**51.** $f(x) = \dfrac{x^3 + 3x^2 + 12x - 5}{x^2 - 1}$

**52.** $f(x) = \dfrac{2x - 12}{5x^2 - 22x - 8}$

**53.** $g(x) = \dfrac{5 - 2x - x^2}{x^2 + 3x + 1}$

**54.** $g(x) = \dfrac{3 + 7x - x^2}{x^3 + 5x^2 + 3x}$

## 2.3 CONTINUITY

### Continuity at a Point

The use of much of the analysis of calculus requires that the graphs of functions have a property called "continuity." Roughly speaking, the graph of $y = f(x)$ is continuous at $x = a$ if the graph is "connected" there; that is, if there are no breaks, gaps, or disruptions at $x = a$. In Figure 2.15, five graphs are shown. Only the graph shown in Figure 2.15(e) is continuous at the point $x = a$. For each of the other graphs in this figure, we have to lift our pen or pencil from the paper when we get to $x = a$ in order to sketch the graph. The graph in Figure 2.15(e), on the other hand, can be drawn with one continuous motion of the pen or pencil as we move through the area around $x = a$.

You should now have an intuitive idea of what it means for a function to be continuous at a number $a$. To obtain a more formal mathematical definition of continuity, we should note that the notion of continuity is related to the concept of the limit. Inspecting Figure 2.15 again, note the following:

In Figure 2.15(a), $\lim\limits_{x \to a} f(x) = n$ and $f(a)$ is not defined. This gives us a graph with a hole in it at $x = a$, so it is not continuous at $x = a$.

In Figure 2.15(b), $\lim\limits_{x \to a} f(x) = n$ but $n \neq f(a)$. This gives us a graph with a hole in it and a single isolated dot above $x = a$. This graph is not continuous at $x = a$.

In Figure 2.15(c), $\lim\limits_{x \to a} f(x)$ does not exist, since $\lim\limits_{x \to a^-} f(x) = f(a)$ but $\lim\limits_{x \to a^+} f(x) = n \neq f(a)$, so the left- and right-hand limits are different at $a$. This produces a gap at $x = a$ and the graph is once again discontinuous at $a$.

In Figure 2.15(d), $\lim\limits_{x \to a} f(x)$ does not exist, since $\lim\limits_{x \to a^-} f(x) = -\infty$ and $\lim\limits_{x \to a^+} f(x) = +\infty$. The vertical line $x = a$ is a vertical asymptote of the graph, and the graph is not continuous at $a$.

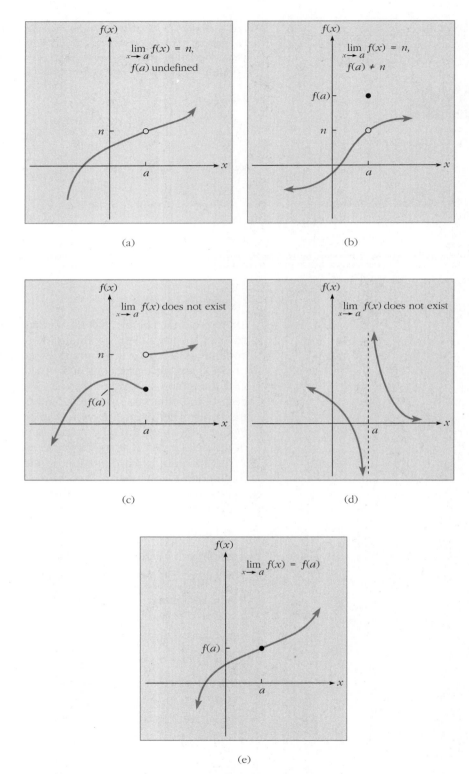

**FIGURE 2.15** An illustration of continuity (or discontinuity) of a graph at a point.

In Figure 2.15(e), $\lim_{x \to a} f(x) = f(a)$, and $f(x)$ is continuous at $x = a$.

It seems that in Figure 2.15(e) we have found the necessary and sufficient condition for continuity.

**DEFINITION: CONTINUITY AT A POINT**

A function $f$ is said to be continuous at a point $a$ if *all* of the following are true:

i. $f(a)$ is defined;
ii. $\lim_{x \to a} f(x)$ exists;
iii. $\lim_{x \to a} f(x) = f(a)$.

**Examples of Continuity at a Point**

EXAMPLE 1

Show that $f(x) = x^2$ is continuous at $a = 2$.

Solution

To solve this problem we note that

i. $f(2) = 2^2 = 4$, so that $f(2)$ is defined;
ii. $\lim_{x \to 2} x^2 = 2^2 = 4$, by Property 2 of Section 2.1, so that $\lim_{x \to 2} f(x)$ exists;
iii. $\lim_{x \to 2} f(x) = 4 = f(2)$.

We conclude that $f(x) = x^2$ is continuous at $a = 2$. ∎

Observe that in Example 1, there was nothing special about the value $a = 2$. That is, for any real number $a$, it is easily shown that $f(x) = x^2$ is continuous at $a$. The graph of $f(x) = x^2$ is drawn in Figure 2.16. More generally, it can be shown that

Any polynomial function is continuous at every real number.

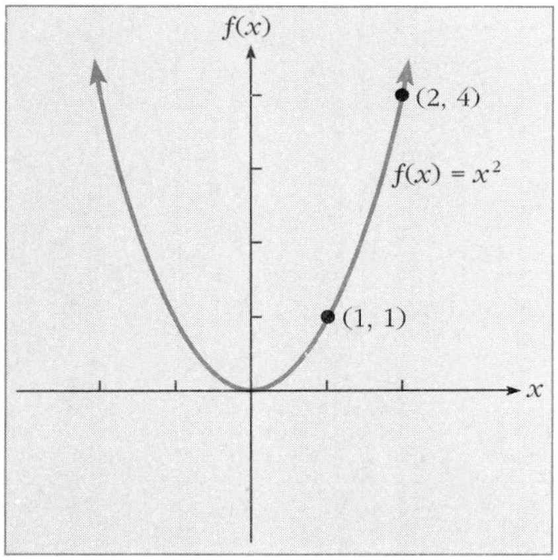

FIGURE 2.16 The function $f(x) = x^2$ is continuous at $a = 2$.

For example, the function $f(x) = 7x^3 - 5x^2 + 3x - 2$ is continuous at every real number $a$, since it is a polynomial function.

EXAMPLE 2

Determine the points at which $f(x) = (x^2 + 3)/(x - 1)$ is continuous.

Solution

If $a$ is any real number $a \neq 1$, then each of the following is true.

i. $f(a) = \dfrac{a^2 + 3}{a - 1}$

ii. $\lim_{x \to a} f(x) = \lim_{x \to a} \dfrac{x^2 + 3}{x - 1} = \dfrac{a^2 + 3}{a - 1}$

iii. $\lim_{x \to a} f(x) = \dfrac{a^2 + 3}{a - 1} = f(a)$

We conclude that $f(x)$ is continuous at every real number $a \neq 1$. If $a = 1$, then $f(1) = 4/0$, which is undefined, so $f(x)$ is not continuous at $a = 1$. The graph of $f(x)$ is sketched in Figure 2.17. Note that the vertical line $x = 1$ is a vertical asymptote of $f(x)$.

By extending the results of Example 2 to the class of rational functions, we can see that *a rational function is continuous except at points where the denominator is zero*. For example, the function

$$f(x) = \dfrac{3x^3 - 7x^2 + 2x + 1}{2x^2 - 5x - 3}$$

is continuous at all points except those where $2x^2 - 5x - 3 = 0$, or where $(2x + 1)(x - 3) = 0$. That occurs at $x = -\tfrac{1}{2}$ and at $x = 3$, so $f(x)$ is continuous at all real numbers except $x = -\tfrac{1}{2}$ and $x = 3$.

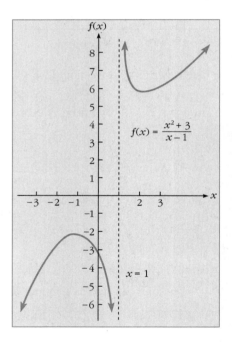

**FIGURE 2.17** The function $f(x)$ is continuous at every real number $a$ except $a = 1$.

---

EXAMPLE 3     A janitorial service that your company uses charges $50 to sweep an area of less than 10,000 square feet, $75 for an area greater than or equal to 10,000 square feet but less than 20,000 square feet, and $100 for an area greater than or equal to 20,000 square feet but less than or equal to 30,000 square feet. The janitorial service does not have the manpower to take on jobs of greater than 30,000 square feet.

(a) Find the function $F(x)$ that describes the amount charged by the service in terms of $x$, the number of thousands of square feet of area in the job.
(b) Graph $F(x)$.
(c) Discuss the continuity of $F(x)$ for the points included in $0 < x < 30$.

**Solution**

(a) $F(x) = \begin{cases} 50 & \text{if } 0 < x < 10 \\ 75 & \text{if } 10 \leq x < 20 \\ 100 & \text{if } 20 \leq x \leq 30 \end{cases}$

(b)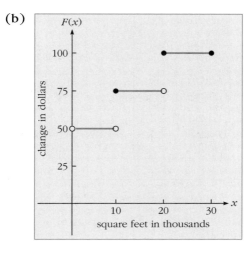

FIGURE 2.18 The function $F(x)$ is continuous at all points $x$ in the interval $0 < x < 30$ except $x = 10$ and $x = 20$.

(c) The graph of $F(x)$ for $0 < x < 30$ is continuous except at $x = 10$ and $x = 20$, where gaps occur. At $x = 10$,

$$\lim_{x \to 10^-} F(x) = \lim_{x \to 10^-} 50 = 50$$

$$\lim_{x \to 10^+} F(x) = \lim_{x \to 10^+} 75 = 75$$

So $\lim_{x \to 10} F(x)$ does not exist and $F(x)$ is not continuous at $x = 10$. At $x = 20$,

$$\lim_{x \to 20^-} F(x) = \lim_{x \to 20^-} 75 = 75$$

$$\lim_{x \to 20^+} F(x) = \lim_{x \to 20^+} 100 = 100$$

Therefore, $\lim_{x \to 20} F(x)$ does not exist and $F(x)$ is not continuous at $x = 20$.

**EXAMPLE 4**

Let $f(x) = (x^2 - x)/(x - 1)$. Determine if $f(x)$ is continuous at $x = 1$ and sketch the graph of $f(x)$.

**Solution**

Since $f(1) = 0/0$, we see that $f(1)$ is not defined. Thus $f(x)$ is not continuous at $x = 1$. Furthermore,

$$f(x) = \frac{x^2 - x}{x - 1} = \frac{x(x - 1)}{(x - 1)} = \begin{cases} x & \text{if } x \neq 1 \\ \text{undefined} & \text{if } x = 1 \end{cases}$$

The graph of $f(x)$ is just the straight line $y = x$ with a hole in it when $x = 1$. See Figure 2.19.

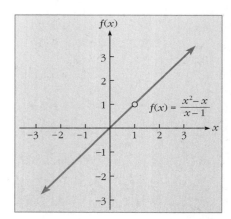

**FIGURE 2.19** The function $f(x)$ is continuous at all real numbers except $x = 1$.

### EXAMPLE 5

Let

$$f(x) = \begin{cases} \dfrac{x^2 - x}{x - 1} & \text{if } x \neq 1 \\ 2 & \text{if } x = 1 \end{cases}$$

Determine if $f(x)$ is continuous at $x = 1$ and sketch the graph of $f(x)$.

**Solution**

Following our three-step procedure to determine continuity at a point,

i. $f(1) = 2$, so $f$ is defined at 1.

ii. $\lim\limits_{x \to 1} f(x) = \lim\limits_{x \to 1} \dfrac{x^2 - x}{x - 1} = \lim\limits_{x \to 1} \dfrac{x(x - 1)}{x - 1} = \lim\limits_{x \to 1} x = 1$

iii. Since $1 \neq 2$, we see that $\lim\limits_{x \to 1} f(x) \neq f(1)$.

We find that $f(x)$ is not continuous at $x = 1$. The graph is sketched in Figure 2.20.

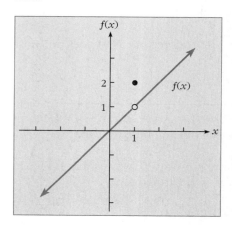

**FIGURE 2.20** The function $f(x)$ of Example 5 is not continuous at $x = 1$.

## EXAMPLE 6 (Participative)

(a) Let's determine the points at which the function $f(x) = (x^2 - 3x + 1)/(x^2 - 16)$ is continuous.

(b) Let
$$g(x) = \begin{cases} 2x + 1 & \text{if } x \leq 2 \\ 2x^2 - 3 & \text{if } x > 2 \end{cases}$$

Let's determine if $g(x)$ is continuous at the point $a = 2$ and sketch the graph of $g(x)$.

### Solution

(a) First, note that $f(x)$ is a ratio of two polynomials, so it is a _____ function. Hence $f(x)$ is continuous *except* at points where $x^2 - 16 =$ _____. We conclude that $f(x)$ is continuous at all real numbers *except* $x =$ _____ and $x =$ _____.

(b) Using our three-step method,

i. $g(2) =$ _____, so $g(2)$ is defined.

ii. To find $\lim_{x \to 2} g(x)$, if it exists, we must use left-hand and right-hand limits, since the definition of $f(x)$ changes at $x = 2$. We obtain

$$\lim_{x \to 2^-} g(x) = \lim_{x \to 2^-} (2x + 1) = \underline{\hspace{1cm}}$$

$$\lim_{x \to 2^+} g(x) = \lim_{x \to 2^+} (\underline{\hspace{1cm}}) = \underline{\hspace{1cm}}$$

Since $\lim_{x \to 2^-} g(x) = \lim_{x \to 2^+} g(x) = 5$, we conclude that $\lim_{x \to 2} g(x) =$ _____.

iii. Since $\lim_{x \to 2} g(x) = g(2)$, we find that _____.

Since $g(x) = 2x + 1$ if $x \leq 2$, the graph of $g(x)$ for $x \leq 2$ is the straight line $y = 2x + 1$. Since $g(x) = 2x^2 - 3$ for $x > 2$, the graph of $g(x)$ for $x > 2$ is the parabola $y = 2x^2 - 3$. Sketch the graph of $g(x)$ on the coordinate axis provided here.

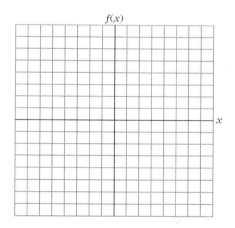

**FIGURE 2.21**

---

**SOLUTIONS TO EXAMPLE 6** (a) rational, 0, −4, 4; (b) 5, 5, $2x^2 - 3$, 5, 5, $g(x)$ is continuous at $x = 2$.

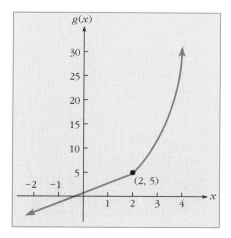

**FIGURE 2.22** The graph of $g(x)$ in Example 6.

## EXAMPLE 7

Let $f(x) = \sqrt{x-1}$. Determine if $f(x)$ is continuous at $x = 1$.

### Solution

i. $f(1) = \sqrt{1-1} = \sqrt{0} = 0$

ii. To find $\lim_{x \to 1} f(x)$, if it exists, first note that the domain of $f(x)$ is all real numbers $x$ such that $x - 1 \geq 0$, or $x \geq 1$. To find $\lim_{x \to 1} f(x)$, we must be able to "approach" 1 both from the left and from the right. Since for this function $a = 1$ is the left-hand endpoint of the domain $[1, \infty)$, we cannot approach 1 from the left, so $\lim_{x \to 1^-} f(x)$ does not exist and so $\lim_{x \to 1} f(x)$ does not exist. See Figure 2.23.

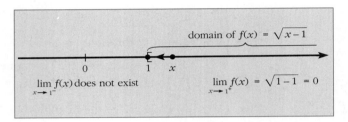

**FIGURE 2.23** The domain of $f(x) = \sqrt{x-1}$ and one-sided limits at $x = 1$.

Therefore $f(x)$ is not continuous at the point $x = 1$. However, $f(x)$ is continuous at all points $x > 1$, since if $a$ is a number greater than 1, then

$$\lim_{x \to a} f(x) = \sqrt{x-1} = \sqrt{a-1} = f(a)$$

See Figure 2.24 for a sketch of $f(x) = \sqrt{x-1}$.

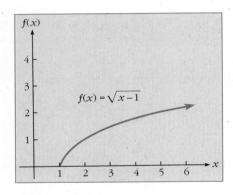

**FIGURE 2.24** The function $f(x) = \sqrt{x-1}$ is not continuous at the point $x = 1$ but is continuous at all points $x > 1$.

In general, if the domain of $f(x)$ is $[a, b]$, then $f$ will not be continuous at either of the points $a$ or $b$. Can you explain why? See Exercise 47.

### Continuity on Intervals

We now consider the notion of continuity on *intervals*, as opposed to continuity at points.

### DEFINITION: CONTINUITY ON INTERVALS

i. A function $f(x)$ is continuous on the open interval $(a, b)$ if $f(x)$ is continuous at all points in $(a, b)$.

ii. A function $f(x)$ is continuous on the closed interval $[a, b]$ if $f(x)$ is continuous on $(a, b)$ and if $\lim_{x \to a^+} f(x) = f(a)$ and $\lim_{x \to b^-} f(x) = f(b)$.

iii. A function $f(x)$ is continuous on the half-open interval $[a, b)$ if $f(x)$ is continuous on $(a, b)$ and if $\lim_{x \to a^+} f(x) = f(a)$.

iv. A function $f(x)$ is continuous on the half-open interval $(a, b]$ if $f(x)$ is continuous on $(a, b)$ and if $\lim_{x \to b^-} f(x) = f(b)$.

## EXAMPLE 8

Determine the largest interval on which $f(x) = \sqrt{x-1}$ is continuous.

### Solution

As we showed in Example 6, the function $f(x)$ is continuous at all points $x > 1$, so that $f(x)$ is continuous on the interval $(1, \infty)$. Also,

$$\lim_{x \to 1^+} f(x) = \lim_{x \to 1^+} \sqrt{x-1} = \sqrt{1-1} = 0 = f(0)$$

so that $f(x)$ is continuous on the interval $[1, \infty)$.

It is important to note from the results of Examples 7 and 8 that $f(x) = \sqrt{x-1}$ is *not* continuous at the *point* $x = 1$ even though $f(x)$ is continuous on the *interval* $[1, \infty)$, which includes 1.

## SECTION 2.3 SHORT ANSWER EXERCISES

1. A graph of a certain function has the vertical line $x = 3$ as a vertical asymptote. Is the function continuous at the point $x = 3$? _____

2. (True or False) The function $f(x) = x^3 - 7x - 5$ is continuous at all real numbers. _____

3. The function $g(x) = (x - 1)/(x + 3)$ is continuous except at the point $x =$ _____.

4. If $f(-1) = 5$, then (a) $f(x)$ is continuous at $a = -1$; (b) $f(x)$ is continuous at $a = 5$; (c) $f(x)$ is not continuous at $a = -1$; (d) $f(x)$ may or may not be continuous at $a = -1$.

5. If $f(4) = 7$ and $\lim_{x \to 4} f(x) = 8$, then (a) $f(x)$ is continuous at $a = 4$; (b) $f(x)$ is not continuous at $a = 4$; (c) $f(x)$ may or may not be continuous at $a = 4$.

6. (True or False) The function $f(x) = \sqrt{x}$ is continuous at the point $x = 0$. _____

7. The demand function $q = -0.16p + 44$, $150 \leq p \leq 250$, is continuous on the open interval (150, _____). Is this function continuous on the closed interval [150, 250]? _____

8. Given the function with the following graph, on what intervals is $f(x)$ continuous? _____

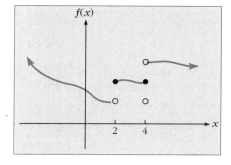

FIGURE 2.25

9. A function $f(x)$ is not continuous at the point $x = 3$ because $f(3)$ is not defined. Then (a) $f(x)$ is not continuous at $x = 3$; (b) $\lim_{x \to 3} f(x)$ may or may not exist; (c) we cannot include 3 in any interval of continuity; (d) all of the above.

10. Which of the following functions are continuous at the point $x = 5$?

$$f(x) = 12x^4 - x^2 + 1 \qquad g(x) = \frac{x^2 - 7x + 3}{x^2 - 2x - 15} \qquad h(x) = \sqrt{x + 5}$$

(a) $f(x)$ and $g(x)$      (b) $f(x)$ and $h(x)$      (c) $g(x)$ and $h(x)$
(d) only $f(x)$      (e) only $h(x)$

## SECTION 2.3
## EXERCISES

*In Exercises 1–20, determine whether the given function is continuous at the indicated points. When required, sketch the graph.*

1. $f(x) = 3x - 5$ at $x = -2, x = 0, x = 2$: sketch
2. $f(x) = 2x^2 + 3x + 1$ at $x = -1, x = 0, x = 3$: sketch
3. $f(x) = \dfrac{x^2 + 2x}{x}$ at $x = -1, x = 0, x = 1$: sketch
4. $f(x) = \dfrac{x^2 - 3x}{x - 3}$ at $x = -1, x = 0, x = 3$: sketch
5. $f(x) = \dfrac{x + 2}{x - 2}$ at $x = -2, x = 0, x = 2$: sketch
6. $f(x) = \dfrac{2x - 1}{x + 1}$ at $x = -2, x = -1, x = 0$: sketch
7. $f(x) = \begin{cases} x + 1 & \text{if } x < 2 \\ 2x - 1 & \text{if } x \geq 2 \end{cases}$ at $x = 2$: sketch
8. $f(x) = \begin{cases} 1 - 2x & \text{if } x \leq 1 \\ 5 & \text{if } x > 1 \end{cases}$ at $x = 1$: sketch
9. $f(x) = \begin{cases} x^2 & \text{if } x \leq 0 \\ -x^2 & \text{if } x > 0 \end{cases}$ at $x = 0$: sketch
10. $f(x) = \begin{cases} 3x - 1 & \text{if } x < -1 \\ x^2 - 5 & \text{if } x \geq -1 \end{cases}$ at $x = -1$: sketch
11. $f(x) = \begin{cases} \dfrac{x^2 - x - 2}{x - 2} & \text{if } x \neq 2 \\ 3 & \text{if } x = 2 \end{cases}$ at $x = 2$: sketch
12. $f(x) = \begin{cases} \dfrac{6x^2 + x - 1}{2x + 1} & \text{if } x \neq -\tfrac{1}{2} \\ -\tfrac{5}{2} & \text{if } x = -\tfrac{1}{2} \end{cases}$ at $x = -\tfrac{1}{2}$: sketch
13. $f(x) = \begin{cases} \dfrac{x^2 - 49}{x - 7} & \text{if } x \neq 7 \\ 12 & \text{if } x = 7 \end{cases}$ at $x = 7$: sketch
14. $f(x) = \begin{cases} \dfrac{x^2 + 6x + 9}{x + 3} & \text{if } x \neq -3 \\ 5 & \text{if } x = -3 \end{cases}$ at $x = -3$: sketch
15. $f(x) = \sqrt{x - 3}$ at $x = 0, x = 3, x = 4$: sketch
16. $f(x) = \sqrt{2 + x}$ at $x = -3, x = -2, x = -1$: sketch

17. $f(x) = \dfrac{x^3 - 7x + 3}{x^3 + 2x^2 + x}$ at $x = -2, x = -1, x = 0$.

18. $f(x) = \begin{cases} \sqrt{x^2 + 3} & \text{if } x \le 1 \\ \sqrt[3]{x} & \text{if } x > 1 \end{cases}$ at $x = -1, x = 1, x = 2$.

19. $f(x) = \begin{cases} \sqrt{2x^2 + 5} & \text{if } x < 0 \\ \sqrt{3x^2 + 5} & \text{if } x \ge 0 \end{cases}$ at $x = -1, x = 0, x = 1$.

20. $f(x) = \begin{cases} x & \text{if } x < 0 \\ -1 & \text{if } x = 0 \\ 2x & \text{if } x > 0 \end{cases}$ at $x = -1, x = 0, x = 1$; sketch

*Exercises 21–40: For each of Exercises 1–20, determine the largest interval or set of intervals on which each function is continuous.*

41. (Delivery charges) A private delivery service charges $5.00 to deliver a package of less than ten pounds to a point less than 20 miles from the delivery center, $15.00 to deliver a package of less than ten pounds to a point at least 20 but less than 50 miles from the delivery center, and $25.00 to deliver a package of less than ten pounds to a point at least 50 miles but less than 100 miles from the delivery center. The delivery service does not deliver any package to a point 100 or more miles from the delivery center.
    (a) Find the function $C(x)$ that represents the cost of delivering a package of less than ten pounds to a point $x$ miles from the delivery center.
    (b) Sketch the graph of $C(x)$.
    (c) Determine the points at which $C(x)$ is continuous.
    (d) Determine the intervals on which $C(x)$ is continuous.

42. (Transportation) A local taxi service charges $2.00 for the first mile and $0.50 for each additional mile or fraction thereof. Assuming that the taxi will not travel beyond ten miles,
    (a) find the function $C(x)$ that represents the cost of a trip of $x$ miles;
    (b) sketch the graph of $C(x)$;
    (c) determine the points at which $C(x)$ is continuous;
    (d) determine the intervals on which $C(x)$ is continuous.

**Business and Economics**

43. (Supply costs) A chemical supply company has the following pricing policy regarding one of its chemicals. If 100 or less pounds of the chemical are ordered, the price per pound is $10. If more than 100 pounds are ordered, the price per pound is $10 per pound for each pound less than 100 and $9.50 per pound for every pound over 100.
    (a) Find the function $C(x)$ that represents the cost of purchasing $x$ pounds of the chemical from this supplier.
    (b) Sketch the graph of $C(x)$.
    (c) Determine the points at which $C(x)$ is continuous.
    (d) Determine the intervals on which $C(x)$ is continuous.

44. (Revenue) Given the revenue function

$$R(x) = \begin{cases} 10x & \text{if } 0 \le x < 100, \\ 9.75x + \sqrt{x} & \text{if } x \ge 100, \end{cases}$$

    (a) sketch the graph of $R(x)$;
    (b) determine the points at which $R(x)$ is continuous;
    (c) determine the intervals on which $R(x)$ is continuous.

**Business and Economics**

**45.** (Profit) If you sell 10,000 or fewer units of your product, your profit is $5.00 per unit. If you sell more than 10,000 units, your profit is $5.00 per unit on the first 10,000 units and $4.50 per unit on every unit over 10,000.
(a) Find the profit function $P(x)$, where $x$ is the number of units sold.
(b) Sketch the graph of $P(x)$.
(c) What can you conclude about the continuity of $P(x)$?

**Social Sciences**

**46.** (Salary of public officials) The average yearly salary (in thousands of dollars) of public officials in a certain region over a period of three years is given by

$$S(t) = \begin{cases} 24 & 0 \le t < 1 \\ 25.2 & 1 \le t < 2 \\ 26.4 & 2 \le t < 3 \end{cases}$$

(a) Sketch the graph of $S(t)$.
(b) What can you conclude about the continuity of $S(t)$?

**47.** Suppose the domain of $f(x)$ is $[a, b]$.
(a) Show that $f(x)$ is not continuous at the point $x = a$.
(b) Show that $f(x)$ is not continuous at the point $x = b$.

## 2.4 THE TANGENT LINE TO A CURVE

### An Intuitive Notion of Tangent Lines

In this section we will discuss the key concept of the derivative of a function and how this concept is related to the tangent line to a curve at a point. You probably already have an intuitive concept of what "tangent" means from your high school geometry. For example, a tangent line to a circle at a point $P$ is the line that passes through $P$ and no other point on the circle. See Figure 2.26. Now inspect Figure 2.27, where tangent lines have been drawn to the graph of a function $y = f(x)$ at points $P$, $Q$, and $R$. By inspecting the tangent line $t_Q$ at point $Q$ and the tangent line $t_R$ at point $R$, we see that when we generalize the concept of a tangent line to an arbitrary graph, we have to give up the idea of

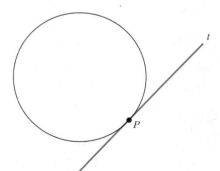

FIGURE 2.26 The line $t$ is tangent to the circle at point $P$.

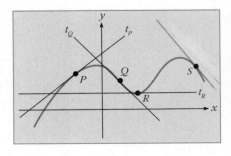

FIGURE 2.27 Tangent lines to a curve at points $P$, $Q$, and $R$.

the line intersecting the curve at only one point. The point $S$ in Figure 2.27 is provided so that you might try your own hand at drawing a tangent line. Try it!

### An Example

Before we continue with our discussion of tangent lines, let us see how such a concept might be useful in an applied setting. Suppose, for example, that your company has determined that their monthly profit in thousands of dollars from the sale of $x$ thousand items, for $1 \leq x \leq 5$, is given by the function

$$P(x) = x^3 - 16x^2 + 65x - 50, \quad 1 \leq x \leq 5$$

An approximate graph of $P(x)$ is drawn in Figure 2.28. Note the points $P$, $Q$, $R$, $S$, and $T$ on this graph, where tangent lines have been drawn to the profit curve. A tangent line gives us a measure of how rapidly the curve is rising or falling at a point. For this profit function, the curve seems to be rising more rapidly at $P$ than at $Q$, as is evidenced by the fact that the tangent line at $P$ is steeper (and hence has a larger positive slope) than the tangent line at $Q$. Correspondingly, it appears that the profit curve is falling more sharply at point $T$ than at point $S$. Observe that the slope of the tangent line at $T$ is negative and larger in absolute value than the slope of the tangent line at $S$. Observe also that at point $R$ (where the graph seems to attain a maximum), the tangent line is apparently horizontal and so has zero slope. Summarizing, these tangent lines give us an indication of

(a) how rapidly profit is increasing or decreasing at a given level of sales, $x$;
(b) what level of sales, $x$, will maximize profit.

Thus the tangent line can be used to extract some very useful information from this profit function. It would be useful to have a *formula* that gives the slope of the tangent line. Then we could, for example, determine exactly how rapidly profit is changing at a given level of sales and exactly what level of sales will maximize profit. In order to obtain such a formula, we will have to formalize some of our vague ideas about tangent lines to curves. We will return to this example in Section 2.5, after we have developed the necessary theory.

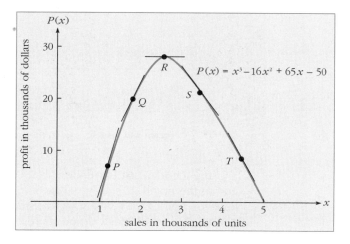

FIGURE 2.28 A profit function and some tangent lines.

## Slope of the Tangent Line

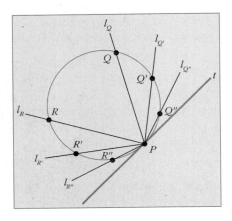

FIGURE 2.29 Secant and tangent lines to a circle.

As a first step in obtaining a formula for the slope of the tangent line to the graph of a function $y = f(x)$ at a point $P$, let us revisit the case of the tangent line to a circle. Another way to view the tangent line to a circle at a point $P$ is as a limit of secant lines. To understand what is meant by this statement, take a look at Figure 2.29. The lines $l_Q$, $l_{Q'}$, and $l_{Q''}$ in this diagram are not tangent lines. Lines such as these that pass through two points on the circle are referred to as *secant lines*. The tangent line at point $P$ is the line $t$ in this diagram. What is the relationship between these secant lines and the tangent line? Note that if we consider the point $Q$ on this graph as "movable," then as $Q$ slides down the circle in a clockwise direction and becomes, successively, $Q'$, $Q''$, etc., the secant lines thorugh $PQ$, $PQ'$, and $PQ''$ also rotate in the clockwise direction, their slopes becoming more and more similar to the slope of the tangent line $t$ until, in the limit, as $Q$ approaches $P$, the secant line "becomes" the tangent line.

A similar argument applies if we approach $P$ in the counterclockwise direction via the sequence of points $R$, $R'$, $R''$, etc. These secant lines again rotate into the position of the tangent line $t$. This idea generalizes fairly easily to the graph of an arbitrary function $y = f(x)$; see Figure 2.30 for an example. For this graph we see that if we fix the point $P$ and let $Q$ slide along the graph and approach $P$ becoming, successively, $Q'$, $Q''$, etc., the corresponding secant lines $l_Q$, $l_{Q'}$, $l_{Q''}$, etc., rotate into the position of the tangent line. Similar comments apply to the sequence of secant lines $l_R$, $l_{R'}$, $l_{R''}$, etc. It is in this sense that the tangent line may be viewed as a limit of secant lines.

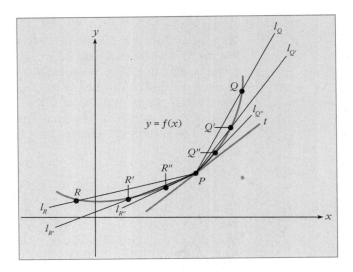

FIGURE 2.30 Secant and tangent lines to the graph of $y = f(x)$.

Using the idea just developed, let us attempt to find the slope of the tangent line to the graph of $f(x) = x^2$ at the point $(1, 1)$. Refer to Figure 2.31. We let $P$ be the point $(1, 1)$, which we consider fixed. Recall from Chapter 1 that the symbol $\Delta x$ is read "delta $x$" and represents the difference in $x$ or the increment in $x$. The "movable" point $Q$ has coordinates $(1 + \Delta x, 1 + \Delta y)$, where $\Delta x$ and

### 2.4 THE TANGENT LINE TO A CURVE

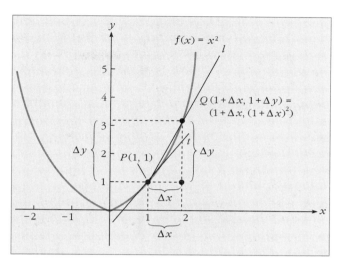

**FIGURE 2.31** The tangent line to the graph of $y = x^2$ at the point $(1, 1)$.

$\Delta y$ are the differences in the $x$ and $y$ coordinates between $P$ and $Q$. In Figure 2.31, $\Delta x$ and $\Delta y$ are assumed to be positive. In general, however, $\Delta x$ and $\Delta y$ may also be negative numbers, for example, if we took $Q$ to be below $P$ on the graph of the parabola, say, $Q = (0, 0)$. The slope of the secant line $l$ is

$$m_{sec} = \frac{(1 + \Delta y) - 1}{(1 + \Delta x) - 1} = \frac{\Delta y}{\Delta x},$$

as expected. Now, letting $Q \to P$ along the curve corresponds to letting $\Delta x \to 0$, and we obtain

$$m_{tan} = \lim_{\Delta x \to 0} \frac{\Delta y}{\Delta x}$$

The symbol normally used for this limit, introduced by the mathematician Gottfried Leibniz (1646–1716), is $dy/dx$, and is called the *Leibniz notation*. That is,

$$m_{tan} = \lim_{\Delta x \to 0} \frac{\Delta y}{\Delta x} = \frac{dy}{dx}$$

This is a completely general result that applies to finding $m_{tan}$ (if it exists) at any point on the graph of a function. To compute $m_{tan}$ for our particular problem, we observe that the $y$-coordinate of the point $Q$ must be the square of the $x$-coordinate, since this point lies on the graph of $y = x^2$. That is, $Q$ can be described as having coordinates $(x + \Delta x, (x + \Delta x)^2)$, so that the slope of the secant line is

$$m_{sec} = \frac{(1 + \Delta x)^2 - 1}{1 + \Delta x - 1} = \frac{1 + 2\Delta x + (\Delta x)^2 - 1}{\Delta x}$$

$$= \frac{2\Delta x + (\Delta x)^2}{\Delta x} = \frac{\Delta x(2 + \Delta x)}{\Delta x} = 2 + \Delta x, \quad \text{if } \Delta x \neq 0$$

Thus, for example, if the $x$-coordinate $1 + \Delta x$ of the point $Q$ is equal to 3, then $\Delta x = 2$, and $m_{sec} = 2 + 2 = 4$. As another example, if the $x$-coordinate of the point $Q$ is equal to $\frac{1}{2}$, then $1 + \Delta x = \frac{1}{2}$, so $\Delta x = -\frac{1}{2}$ and $m_{sec} = 2 - \frac{1}{2} = \frac{3}{2}$. Some other sample values of $\Delta x$ and $m_{sec}$ are shown in Tables 2.16 and 2.17.

| $\Delta x$ | 1 | 0.1 | 0.01 | 0.001 | 0.0001 | $\to 0^+$ |
|---|---|---|---|---|---|---|
| $m_{sec} = 2 + \Delta x$ | 3 | 2.1 | 2.01 | 2.001 | 2.0001 | $\to 2$ |

**TABLE 2.16**

| $\Delta x$ | $-1$ | $-0.1$ | $-0.01$ | $-0.001$ | $-0.0001$ | $\to 0^-$ |
|---|---|---|---|---|---|---|
| $m_{sec} = 2 + \Delta x$ | 1 | 1.9 | 1.99 | 1.999 | 1.9999 | $\to 2$ |

**TABLE 2.17**

Recall that the objective is to let $\Delta x$ approach 0, since this is equivalent to allowing $Q$ approach $P$ along the graph. The results of Tables 2.16 and 2.17 suggest that $m_{sec} \to 2$ as $\Delta x \to 0$, and indeed,

$$m_{tan} = \lim_{\Delta x \to 0} m_{sec} = \lim_{\Delta x \to 0} (2 + \Delta x) = 2 + 0 = 2$$

We conclude that the slope of the tangent line to the graph of $y = x^2$ at the point $(1, 1)$ is equal to 2. Using the point–slope form of the equation of a straight line,

$$y - y_1 = m(x - x_1)$$

we can now easily find an equation of the tangent line

$$y - 1 = 2(x - 1)$$
$$y = 2x - 1$$

We will now develop a general formula for finding the slope of the tangent line to the graph of $y = f(x)$ at the point $(x, y) = (x, f(x))$. For this we will simply generalize the methods used in the example we just worked. Refer to Figure 2.32. To obtain the Leibniz form, consider:

$$m_{sec} = \frac{y + \Delta y - y}{x + \Delta x - x} = \frac{\Delta y}{\Delta x}$$

and

$$m_{tan} = \lim_{\Delta x \to 0} m_{sec} = \lim_{\Delta x \to 0} \frac{\Delta y}{\Delta x} = \frac{dy}{dx}$$

Notice the little triangle in Figure 2.32 formed by the line segments $PR$ (which is equal to $\Delta x$), $RQ$ (which is equal to $\Delta y$), and $PQ$ (which lies along the secant line $l$). As $\Delta x \to 0$, this triangle becomes smaller and smaller and the secant line rotates toward the position of the tangent line $t$. Eventually, as $\Delta x$ and $\Delta y$ become infinitesimally small, the secant line "becomes" the tangent line.

### 2.4 THE TANGENT LINE TO A CURVE 111

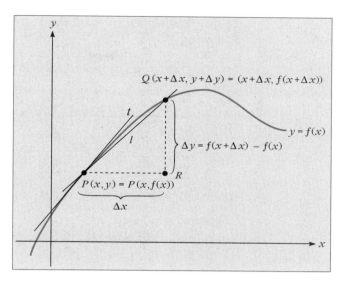

**FIGURE 2.32** The line $t$ is tangent to the graph of $y = f(x)$ at the point $P$.

Although the Leibniz form for $m_{\tan}$ is a nice theoretical result, it may not be immediately clear how this form can be used in a particular problem. Observe that $Q$ may also be described as having coordinates $(x + \Delta x, f(x + \Delta x))$, and so

$$m_{\sec} = \frac{f(x + \Delta x) - f(x)}{x + \Delta x - x} = \frac{f(x + \Delta x) - f(x)}{\Delta x}$$

so that

$$m_{\tan} = \lim_{\Delta x \to 0} m_{\sec} = \lim_{\Delta x \to 0} \frac{f(x + \Delta x) - f(x)}{\Delta x}$$

In comparing this with the Leibniz form, observe that

$$\Delta y = f(x + \Delta x) - f(x)$$

To compute $m_{\tan}$ (when it exists) from the formula, we typically find $m_{\sec} = (f(x + \Delta x) - f(x))/\Delta x$ first and then simplify this expression to remove the $\Delta x$ from the denominator. After simplifying $m_{\sec}$, we then compute $m_{\tan} = \lim_{\Delta x \to 0} m_{\sec}$.

---

**EXAMPLE 1**     Find the slope of the tangent line to the graph of $y = x^2$ at the point $(2, 4)$.

## Solution

In this problem, $(x, y) = (x, f(x)) = (2, 4)$, so $x = 2$ and $f(2) = 4$. We find that

$$m_{sec} = \frac{f(2 + \Delta x) - f(2)}{\Delta x} = \frac{(2 + \Delta x)^2 - 2^2}{\Delta x}$$

$$= \frac{4 + 4\Delta x + (\Delta x)^2 - 4}{\Delta x}$$

$$= \frac{4\Delta x + (\Delta x)^2}{\Delta x} = \frac{\Delta x(4 + \Delta x)}{\Delta x} = 4 + \Delta x, \quad \text{if } \Delta x \neq 0$$

Therefore, $m_{tan} = \lim_{\Delta x \to 0} m_{sec} = \lim_{\Delta x \to 0} (4 + \Delta x) = 4 + 0 = 4$. This tangent line is illustrated in Figure 2.33.

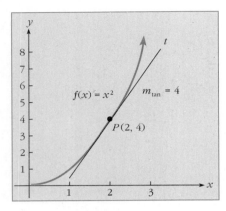

**FIGURE 2.33** The line $t$ is tangent to the graph of $y = x^2$ at the point $(2, 4)$.

### EXAMPLE 2

Find an equation of the tangent line to the graph of $f(x) = x^3 + 1$ at the point $(0, 1)$.

## Solution

In this case, $x = 0$, and $f(0) = 1$, so that

$$m_{sec} = \frac{f(0 + \Delta x) - f(0)}{\Delta x} = \frac{f(\Delta x) - 1}{\Delta x}$$

$$= \frac{(\Delta x)^3 + 1 - 1}{\Delta x} = \frac{(\Delta x)^3}{\Delta x} = (\Delta x)^2, \quad \text{if } \Delta x \neq 0$$

Thus, $m_{tan} = \lim_{\Delta x \to 0} m_{sec} = \lim_{\Delta x \to 0} (\Delta x)^2 = 0$. Since the point $(0, 1)$ is on the tangent line, the point–slope form of the equation of a straight line gives us

$$y - 1 = 0(x - 0)$$
$$y = 1$$

as the equation of the tangent line to the graph of $f(x) = x^3 + 1$ at the point $(0, 1)$. A sketch is drawn in Figure 2.34.

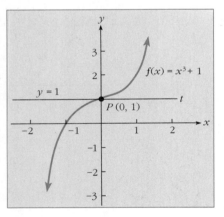

**FIGURE 2.34** The line $y = 1$ is tangent to the graph of $y = x^3 + 1$ at the point $(0, 1)$.

### EXAMPLE 3 (Participative)

Let's find the slope of the tangent line to the graph of $f(x) = 2x^2 + 4x$ at the point $(-3, 6)$. Also, let's sketch the graphs of $f(x)$ and the tangent line.

## Solution

(a) We begin by noting that $x = -3$ and $f(-3) = $ \_\_\_\_\_, so that

$$m_{sec} = \frac{f(-3 + \Delta x) - f(-3)}{\Delta x}$$

$$= \frac{[2(-3 + \Delta x)^2 + 4(\underline{\quad})] - 6}{\Delta x}$$

$$= \frac{[2(\underline{\quad}) - 12 + 4\Delta x] - 6}{\Delta x}$$

$$= \frac{18 - 12\Delta x + 2(\Delta x)^2 - 12 + 4\Delta x - 6}{\Delta x}$$

$$= \frac{2(\Delta x)^2 - \underline{\quad}}{\Delta x}$$

$$= \frac{\Delta x(\underline{\quad})}{\Delta x}$$

$$= \underline{\quad}, \quad \text{if } \Delta x \neq 0$$

(b) To find $m_{tan}$, we compute $\lim_{\Delta x \to 0} m_{sec}$. For this problem

$$m_{tan} = \lim_{\Delta x \to 0} (\underline{\quad}) = \underline{\quad}$$

(c) To sketch the graphs, note that $f(x) = 2x^2 + 4x$ is a parabola. The x-intercepts are $x = -2$ and $x = \underline{\quad}$. The y-intercept is $y = \underline{\quad}$. The x-coordinate of the vertex is $x = -b/(2a) = \underline{\quad}$, and the y-coordinate of the vertex is $f(-1) = \underline{\quad}$. For the tangent line, we are given that one point on the line is $(-3, 6)$. Since the slope is equal to $-8$, an increase in $x$ of one unit produces a corresponding decrease in $y$ of 8 units. So the point $(-3 + 1, 6 - \underline{\quad}) = \underline{\quad}$ is another point on the tangent line. Sketch the graphs on the coordinate axis provided here.

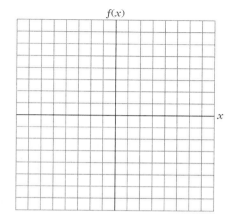

FIGURE 2.35

---

**SOLUTIONS TO EXAMPLE 3**

(a) $6, -3 + \Delta x, 9 - 6\Delta x + (\Delta x)^2, 8\Delta x, 2\Delta x - 8, 2\Delta x - 8$;  (b) $2\Delta x - 8, -8$;
(c) $0, 0, -1, -2, 8, (-2, -2)$

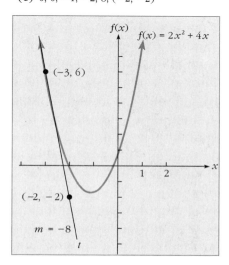

FIGURE 2.36  Line $t$ is tangent to the graph of $f(x) = 2x^2 + 4x$ at the point $(-3, 6)$.

## EXAMPLE 4

Find the slope of the tangent line to the graph of $y = 1/x$ at the point where $x = 2$.

**Solution**

To begin, we have $x = 2$ and $f(2) = \frac{1}{2}$, so

$$m_{\sec} = \frac{f(2 + \Delta x) - f(2)}{\Delta x}$$

$$= \frac{\frac{1}{2 + \Delta x} - \frac{1}{2}}{\Delta x}$$

$$= \frac{\frac{2 - (2 + \Delta x)}{2(2 + \Delta x)}}{\Delta x}$$

$$= \frac{\frac{-\Delta x}{2(2 + \Delta x)}}{\Delta x}$$

$$= -\frac{\Delta x}{2(2 + \Delta x)} \cdot \frac{1}{\Delta x}$$

$$= -\frac{1}{2(2 + \Delta x)}, \quad \text{if } \Delta x \neq 0$$

So

$$m_{\tan} = \lim_{\Delta x \to 0} m_{\sec} = \lim_{\Delta x \to 0} \left(-\frac{1}{2(2 + \Delta x)}\right)$$

$$= -\frac{1}{2(2 + 0)} = -\frac{1}{4}$$

See Figure 2.37 for the graph.

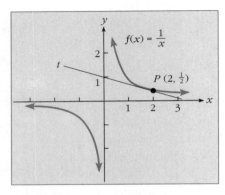

**FIGURE 2.37** The line $t$ is tangent to the graph of $y = 1/x$ at the point $\left(2, \frac{1}{2}\right)$.

## EXAMPLE 5

Find the slope of the tangent line to the graph of $f(x) = \sqrt{x}$ at the point $(3, \sqrt{3})$.

**Solution**

We have $x = 3$ and $f(3) = \sqrt{3}$, so that

$$m_{\sec} = \frac{f(3 + \Delta x) - f(3)}{\Delta x} = \frac{\sqrt{3 + \Delta x} - \sqrt{3}}{\Delta x}$$

To find $m_{\tan}$, we must evaluate

$$\lim_{\Delta x \to 0} \frac{\sqrt{3 + \Delta x} - \sqrt{3}}{\Delta x}$$

This limit can be evaluated if we first rationalize the numerator of the expression for $m_{\sec}$. Consider:

$$m_{\sec} = \frac{\sqrt{3 + \Delta x} - \sqrt{3}}{\Delta x} = \frac{\sqrt{3 + \Delta x} - \sqrt{3}}{\Delta x} \cdot \frac{\sqrt{3 + \Delta x} + \sqrt{3}}{\sqrt{3 + \Delta x} + \sqrt{3}}$$

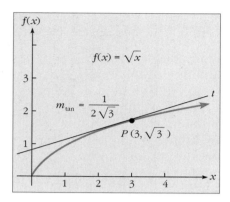

FIGURE 2.38 The line $t$ is tangent to the graph of $f(x) = \sqrt{x}$ at the point $(3, \sqrt{3})$.

$$= \frac{3 + \Delta x - 3}{\Delta x(\sqrt{3 + \Delta x} + \sqrt{3})}$$

$$= \frac{\Delta x}{\Delta x(\sqrt{3 + \Delta x} + \sqrt{3})}$$

$$= \frac{1}{\sqrt{3 + \Delta x} + \sqrt{3}}, \quad \text{if } \Delta x \neq 0$$

Now $m_{\tan}$ is easier to evaluate:

$$m_{\tan} = \lim_{\Delta x \to 0} m_{\sec} = \lim_{\Delta x \to 0} \frac{1}{\sqrt{3 + \Delta x} + \sqrt{3}}$$

$$= \frac{1}{\sqrt{3 + 0} + \sqrt{3}} = \frac{1}{2\sqrt{3}} = \frac{\sqrt{3}}{6}$$

See Figure 2.38 for a graph.

The next example shows that a tangent line need not exist.

## EXAMPLE 6

Show that $f(x) = |x|$ has no tangent line at the point $(0, 0)$.

### Solution

To begin, $x = 0$ and $f(0) = 0$, so that

$$m_{\sec} = \frac{f(0 + \Delta x) - f(0)}{\Delta x} = \frac{f(\Delta x) - 0}{\Delta x} = \frac{|\Delta x|}{\Delta x}$$

Now, viewing $m_{\sec}$ as a function of $\Delta x$,

$$\frac{|\Delta x|}{\Delta x} = \begin{cases} -1 & \text{if } \Delta x < 0 \\ \text{undefined} & \text{if } \Delta x = 0 \\ 1 & \text{if } \Delta x > 0 \end{cases}$$

A graph of $m_{\sec} = |\Delta x|/\Delta x$ is drawn in Figure 2.39.

Consider $m_{\tan} = \lim_{\Delta x \to 0} m_{\sec} = \lim_{\Delta x \to 0} |\Delta x|/\Delta x$. Now,

$$\lim_{\Delta x \to 0^+} \frac{|\Delta x|}{\Delta x} = \lim_{\Delta x \to 0^+} 1 = 1$$

and

$$\lim_{\Delta x \to 0^-} \frac{|\Delta x|}{\Delta x} = \lim_{\Delta x \to 0^-} (-1) = -1$$

We conclude that $m_{\tan} = \lim_{\Delta x \to 0} |\Delta x|/\Delta x$ does not exist. The graph of $f(x) = |x|$ is drawn in Figure 2.40.

The reason that $m_{\tan}$ does not exist for this function at $x = 0$ is that if we fix $P(0, 0)$ and choose $Q$ to the right of $P$, then as we let $Q$ slide down the graph toward $P$, all these secant lines have slope 1, whereas if we choose $R$ to the left of $P$ and let $R$ slide down the graph toward $P$, all of these secant lines have

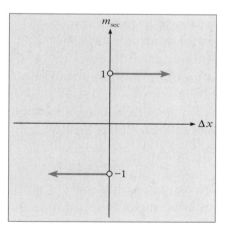

FIGURE 2.39 The graph of $m_{\sec} = |\Delta x|/\Delta x$ as a function of $\Delta x$.

slope −1. Since we obtain a different limiting value for the secant line slopes (namely, 1 and −1) from either side of P, there is no unique tangent line at P.

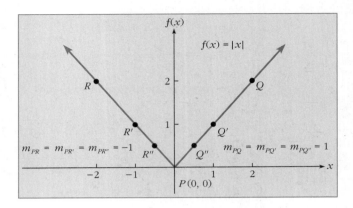

**FIGURE 2.40** No tangent line exists to the graph of $f(x) = |x|$ at the point (0, 0).

## SECTION 2.4
### SHORT ANSWER EXERCISES

1. (True or False) Any tangent line to an arbitrary function $y = f(x)$ at a point $P(x, f(x))$ intersects the graph only once: at the point P. _____
2. (True or False) The tangent line to the graph of a function $y = f(x)$ at a point P always exists. _____
3. Given a profit function $y = P(x)$ that is continuous and smooth, the maximum profit will occur at a point $(x, P(x))$ such that the tangent line at that point is _____.
4. The tangent line to the graph of $y = f(x)$ at a point P may be viewed as a limit of _____ lines.
5. If the slope of the secant line joining $P(x, y)$ and $Q(x + \Delta x, y + \Delta y)$ is equal to $3 - 7\Delta x + (\Delta x)^2$ if $\Delta x \neq 0$, then the slope of the tangent line is equal to _____.
6. A tangent line to the graph of $y = f(x)$ at the point P (a) can always be found; (b) has slope $\Delta y/\Delta x$; (c) is equal to $\lim_{\Delta x \to 0} P$; (d) may not exist.
7. If an equation of the tangent line to the graph of $y = f(x)$ at the point P is $y = 3$, then (a) no tangent line exists; (b) the maximum value of $f(x)$ is equal to 3; (c) the tangent line is horizontal; (d) none of the above.
8. Given that $f(x) = 2x + 1$, then $f(x + \Delta x)$ is equal to (a) $2\Delta x + 1$; (b) $2x + 2\Delta x$; (c) $2x + 1$; (d) $2x + 2\Delta x + 1$.
9. If, at a certain point P on the graph of $y = f(x)$ we have $\lim_{\Delta x \to 0^-} (\Delta y/\Delta x) = -2$ and $\lim_{\Delta x \to 0^+} (\Delta y/\Delta x) = -2$, then (a) the slope of the secant line is equal to −2; (b) the slope of the tangent line is equal to −2; (c) the secant and tangent lines are identical; (d) none of the above.
10. (True or False) The slope of the tangent line to the graph of $y = |x|$ at the point (2, 2) is equal to 1. _____

## SECTION 2.4
### EXERCISES

In Exercises 1–8, find a general expression for the slope of the secant line joining the given point P and an arbitrary point Q on the graph of the function. Evaluate $m_{sec}$ for $\Delta x = 1$, $\Delta x = 0.1$, and $\Delta x = 0.01$. Also, evaluate $m_{sec}$ for $\Delta x = -1$, $\Delta x = -0.1$, and $\Delta x = -0.01$. Sketch the graph of $f(x)$, the secant lines corresponding to $\Delta x = 1$ and $\Delta x = -1$, and the tangent line at P.

1. $f(x) = x^2$, $P(-1, 1)$
2. $f(x) = x^2$, $P(-2, 4)$
3. $f(x) = x^2 - 2x$, $P(1, -1)$
4. $f(x) = x^2 + 2x + 1$, $P(0, 1)$
5. $f(x) = x^3 + 1$, $P(1, 2)$
6. $f(x) = x^3 - 2$, $P(-1, -3)$
7. $f(x) = 1/(x + 2)$, $P(0, \frac{1}{2})$
8. $f(x) = |x - 1|$, $P(-1, 2)$

In Exercises 9–23, find the slope of the tangent line to the graph of the function at the indicated point P, if it exists. Where indicated, sketch the curve and the tangent line at P.

9. $f(x) = x^2 - 1$, $P(-1, 0)$: sketch
10. $f(x) = x^2 + 3x$, $P(2, 10)$: sketch
11. $f(x) = x^2 + 4x + 4$, $P(0, 4)$: sketch
12. $f(x) = x^2 + x - 2$, $P(-2, 0)$: sketch
13. $f(x) = x^3 + x$, $P(1, 2)$
14. $f(x) = 2x^3 - x^2$, $P(2, 12)$
15. $f(x) = 1/(x-1)$, $P(3, \frac{1}{2})$: sketch
16. $f(x) = 1/(2x + 1)$, $P(-1, -1)$: sketch
17. $g(x) = \sqrt{x}$, $P(4, 2)$
18. $g(x) = \sqrt{x - 2}$, $P(3, 1)$
19. $g(x) = x + (1/x)$, $P(-1, -2)$
20. $g(x) = (3x^2 + 2x)/x$, $P(-2, -4)$
21. $f(x) = |x|$, $P(2, 2)$
22. $f(x) = |x - 3|$, $P(3, 0)$
23. $f(x) = |2x - 8|$, $P(4, 0)$

In Exercises 24–33, find an equation of the tangent line to the graph of each of the functions at the indicated point P, if it exists.

24. $f(x) = x^2 + 3x + 4$, $P(1, 8)$
25. $f(x) = -x^2 - x + 3$, $P(0, 3)$
26. $g(x) = 2x^3 - x^2 + 1$, $P(-1, -2)$
27. $g(x) = \sqrt{x + 1}$, $P(0, 1)$
28. $f(x) = 2x - 3$, $P(1, -1)$
29. $f(x) = 3x + 9$, $P(-2, 3)$
30. $f(x) = \sqrt{x}$, $P(0, 0)$
31. $f(x) = \sqrt{x + 1}$, $P(-1, 0)$
32. $f(x) = 1/(3x + 2)$, $P(0, \frac{1}{2})$
33. $f(x) = 1/(4x - 6)$, $P(1, -\frac{1}{2})$

**Business and Economics**

34. (Profit) Given the profit function $P(x) = -x^2 + 140x - 4000$, where $x$ is the number of units produced per month and $P(x)$ is profit in dollars per month,
    (a) Sketch the graph of $P(x)$;
    (b) Find the slope of the tangent line to the graph of $P(x)$ at the point where $x = 50$. Sketch this tangent line. Is it profitable to increase production beyond 50 units?
    (c) Find the slope of the tangent line to the graph of $P(x)$ at the point where $x = 80$. Sketch this tangent line. Is it profitable to increase production beyond 80 units?
    (d) Find the slope of the tangent line to the graph of $P(x)$ at the point where $x = 70$. Sketch this tangent line. What can you conclude about the profit-maximizing production level?
    (e) Find the maximum profit.

**Business and Economics**

35. (Minimum average cost) The average cost (in thousands of dollars) of producing $x$ thousand units of a certain product is $\overline{C}(x) = x^2 - 3x + 9$.
    (a) Sketch the graph of $\overline{C}(x)$.

(b) Find the slope of the tangent line to the graph of $\overline{C}(x)$ at the point where $x = 1$. Sketch this tangent line. Will the average cost increase or decrease if we increase production beyond 1000 units?
(c) Find the slope of the tangent line to the graph of $\overline{C}(x)$ at the point where $x = 3$. Sketch this tangent line. Will the average cost increase or decrease if we increase production beyond 3000 units?
(d) Find the slope of the tangent line to the graph of $\overline{C}(x)$ when the level of production is 1500 units. Sketch this tangent line. What can you conclude about the production level at which average cost is minimized?
(e) Find the minimum average cost.

Life Sciences

36. (Population) It is estimated that the population (in thousands) of a certain city in $t$ years will be

$$P(t) = t^2 + 4t + 1, \quad \text{for } 0 \leq t \leq 5$$

(a) Graph the function $P(t)$.
(b) Find the slope of the tangent line when $t = 1$ and $t = 4$. Sketch each of these tangent lines.
(c) Is the population of this city growing faster when $t = 1$ or when $t = 4$?

OPTIONAL

GRAPHING CALCULATOR/
COMPUTER EXERCISE

37. (Profit) Given the profit function $P(x) = x^3 - 19x^2 + 104x - 140$, where $P(x)$ is the monthly profit in thousands of dollars from the sale of $x$ thousand items, $2 \leq x \leq 7$,
(a) graph the function $P(x)$;
(b) find the slope of the tangent line when $x = 3$ and sketch this tangent line. Is it profitable to increase production beyond $x = 3$?
(c) Estimate the level of sales at which profit is maximized.
(d) For the level of sales you estimated in part (c), compute the slope of the tangent line. This slope should be close to zero if your estimate in part (c) was good.
(e) Estimate the maximum profit.

## 2.5 THE DERIVATIVE OF A FUNCTION

### Definition of the Derivative and Tangent Lines

We begin this section by considering the definition of the derivative of a function $y = f(x)$.

**DEFINITION**

Given a function $y = f(x)$, the *derivative* of $f(x)$, denoted by the symbol $f'(x)$ [read "$f$ prime of $x$"] is another function of $x$ defined by

$$f'(x) = \lim_{\Delta x \to 0} \frac{f(x + \Delta x) - f(x)}{\Delta x}$$

The formula that defines the derivative should look familiar to you, since it is the same formula that we used to define the slope of the tangent line to the graph of $y = f(x)$ at a point $P(x, f(x))$. In general, the derivative of a function $y = f(x)$ is another function designated by $f'(x)$ that gives us a general formula for the slope of the tangent line to the graph of $y = f(x)$ at the point $P(x, f(x))$. In other words, $f'(x)$ is the slope of the tangent line to the graph of $y = f(x)$ at the point $P(x, f(x))$.

**EXAMPLE 1**

(a) Find the derivative of $f(x) = x^2$.
(b) Find the slope of the tangent lines to the graph of $f(x) = x^2$ at the points where $x = -3$, $x = 1$, and $x = 3$.
(c) Sketch the graph of $f(x)$ and draw the tangent lines to the graphs at the points indicated in part (b).

**Solution**

(a) First, note that

$$f(x + \Delta x) = (x + \Delta x)^2 = x^2 + 2x\Delta x + (\Delta x)^2$$

Second, we have

$$f(x + \Delta x) - f(x) = x^2 + 2x\Delta x + (\Delta x)^2 - x^2 = 2x\Delta x + (\Delta x)^2$$

Third, we compute

$$\frac{f(x + \Delta x) - f(x)}{\Delta x} = \frac{2x\Delta x + (\Delta x)^2}{\Delta x} = \frac{\Delta x(2x + \Delta x)}{\Delta x}$$

$$= 2x + \Delta x, \quad \text{if } \Delta x \neq 0$$

Fourth, we find the derivative:

$$f'(x) = \lim_{\Delta x \to 0} \frac{f(x + \Delta x) - f(x)}{\Delta x} = \lim_{\Delta x \to 0} (2x + \Delta x) = 2x + 0 = 2x$$

We conclude that if $f(x) = x^2$, then $f'(x) = 2x$. That is, the derivative of $x^2$ is equal to $2x$.

(b) Since, in general, $f'(x) = m_{\tan}$, we have for this problem $f'(x) = m_{\tan} = 2x$, so that

$f'(-3) = 2(-3) = -6$ is the slope of the tangent line at the point where $x = -3$;

$f'(1) = 2(1) = 2$ is the slope of the tangent line at the point where $x = 1$;

$f'(3) = 2(3) = 6$ is the slope of the tangent line at the point where $x = 3$.

(c) The graph of $f(x) = x^2$ and the appropriate tangent line are sketched in Figure 2.41. ∎

In Example 1(a), we illustrated the *four-step method* of computing the derivative of a function $y = f(x)$ by using the definition. This four-step method works as follows.

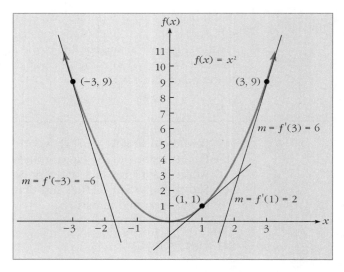

**FIGURE 2.41** Some tangent lines to the graph of $f(x) = x^2$.

> Step 1  Compute $f(x + \Delta x)$ by replacing all $x$'s in the formula for $f(x)$ by $x + \Delta x$.
>
> Step 2  Find $f(x + \Delta x) - f(x)$ simply by subtracting the formula for $f(x)$ from the quantity $f(x + \Delta x)$, which you found in Step 1.
>
> Step 3  Find and simplify $(f(x + \Delta x) - f(x))/\Delta x$ by dividing the quantity $f(x + \Delta x) - f(x)$, which you found in Step 2, by $\Delta x$.
>
> Step 4  Compute the limit $\lim_{\Delta x \to 0} (f(x + \Delta x) - f(x))/\Delta x$, if it exists. This limit is, of course, the derivative $f'(x)$.

The process of finding a derivative is called *differentiation*. Thus, the four-step method gives us a way to *differentiate* a function $y = f(x)$.

Let's look at some more examples of finding derivatives.

**EXAMPLE 2**

(a) Find the derivative of $f(x) = x^2 + 2x - 3$.
(b) Find equations of the tangent lines to the graph of $f(x)$ at the points $(-2, -3)$ and $(2, 5)$.
(c) Sketch the graph of $f(x)$ and the tangent lines found in part (b).

Solution

(a) We are given $f(x) = x^2 + 2x - 3$. We will follow the four-step method to find $f'(x)$.

Step 1  $f(x + \Delta x) = (x + \Delta x)^2 + 2(x + \Delta x) - 3$
$= x^2 + 2x\Delta x + (\Delta x)^2 + 2x + 2\Delta x - 3$

**Step 2**

$$f(x + \Delta x) - f(x) = x^2 + 2x\Delta x + (\Delta x)^2 + 2x + 2\Delta x - 3 - (x^2 + 2x - 3)$$

Note the parentheses!

$$= x^2 + 2x\Delta x + (\Delta x)^2 + 2x + 2\Delta x - 3 - x^2 - 2x + 3$$
$$= 2x\Delta x + (\Delta x)^2 + 2\Delta x$$

**Step 3**
$$\frac{f(x + \Delta x) - f(x)}{\Delta x} = \frac{2x\Delta x + (\Delta x)^2 + 2\Delta x}{\Delta x}$$
$$= \frac{\Delta x(2x + \Delta x + 2)}{\Delta x}$$
$$= 2x + \Delta x + 2, \quad \text{if } \Delta x \neq 0$$

**Step 4**
$$\lim_{\Delta x \to 0} \frac{f(x+\Delta x) - f(x)}{\Delta x} = \lim_{\Delta x \to 0} (2x + \Delta x + 2) = 2x + 2$$

Therefore, $f'(x) = 2x + 2$.

**(b)** The slope of the tangent line to the graph of $f(x)$ at the point $(-2, -3)$ is equal to

$$f'(-2) = 2(-2) + 2 = -2$$

Using the point–slope form of the equation of a straight line, we find that

$$y - (-3) = -2[x - (-2)]$$

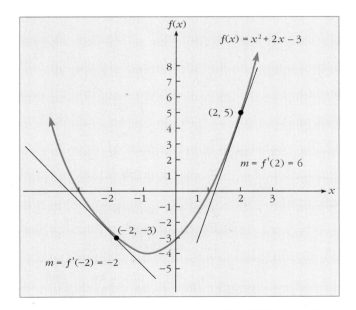

**FIGURE 2.42** Some tangent lines to the graph of $f(x) = x^2 + 2x - 3$.

This reduces to $2x + y = -7$, which is an equation of the tangent line to the graph of $f(x)$ at the point $(-2, -3)$. The slope of the tangent line to the graph of $f(x)$ at the point $(2, 5)$ is equal to

$$f'(2) = 2(2) + 2 = 6$$

Again using the point–slope form of the equation of a straight line, we obtain

$$y - 5 = 6(x-2)$$
$$6x - y = 7$$

This is an equation of the tangent line to the graph of $f(x)$ at the point $(2, 5)$.

(c) A sketch of $f(x)$ and the tangent lines appear in Figure 2.42 on the previous page.

---

### EXAMPLE 3 (Participative)

(a) Find the derivative of $f(x) = x^3$.
(b) Find and interpret $f'(-1)$ and $f'(2)$.
(c) Sketch the graph of $f(x)$ and the tangent lines at the points $(-1, -1)$ and $(2, 8)$.

**Solution**

(a) We use the four-step method with $f(x) = x^3$.

Step 1 $f(x + \Delta x) = (x + \Delta x)^3 = $ _____

Step 2 $f(x + \Delta x) - f(x) = x^3 + 3x^2\Delta x + 3x(\Delta x)^2 + (\Delta x)^3 - ($ \_\_\_\_\_$)$

Step 3 $\dfrac{f(x + \Delta x) - f(x)}{\Delta x} = \dfrac{3x^2\Delta x + 3x(\Delta x)^2 + (\Delta x)^3}{\Delta x}$

$= \dfrac{\Delta x(\underline{\phantom{xxxxxxxxxxxxxxxxxx}})}{\Delta x}$

$=$ _____, if $\Delta x \neq 0$

Step 4 $\lim\limits_{\Delta x \to 0} \dfrac{f(x + \Delta x) - f(x)}{\Delta x} = \lim\limits_{\Delta x \to 0} [\underline{\phantom{xxxxxxxxxxxxx}}]$

$=$ _____

Thus $f'(x) = $ _____.

(b) $f'(-1) = 3(\underline{\phantom{xx}})^2 = $ \_\_\_\_\_, which represents the slope of the tangent line to the graph of $f(x) = x^3$ at the point $(-1, -1)$.

$f'(2) = $ \_\_\_\_\_, which is the slope of the tangent line to the graph of $f(x) = x^3$ at the point _____.

(c) Sketch the graph of $f(x) = x^3$ and the two tangent lines on the coordinate axes provided here. ∎

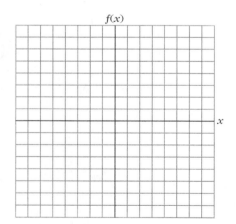

FIGURE 2.43

As we saw in Section 2.4, an alternative notation for the slope of the tangent line is the Leibniz notation, $dy/dx$. Since the slope of the tangent line at $(x, f(x))$ is just the derivative $f'(x)$, we have

$$f'(x) = \dfrac{dy}{dx}$$

Therefore the symbol $dy/dx$ also represents the derivative of $y = f(x)$. To evaluate $dy/dx$ at a point $x = a$ in the domain of the derivative function, we use the symbol

$$\left.\frac{dy}{dx}\right|_{x=a}$$

As an example, we already know (see Example 1 of this section) that the derivative of $x^2$ is equal to $2x$. Therefore if $y = x^2$, then

$$\frac{dy}{dx} = 2x$$

and, for example,

$$\left.\frac{dy}{dx}\right|_{x=-3} = 2(-3) = -6$$

$$\left.\frac{dy}{dx}\right|_{x=1} = 2(1) = 2$$

$$\left.\frac{dy}{dx}\right|_{x=3} = 2(3) = 6$$

**SOLUTIONS TO EXAMPLE 3**

(a) $x^3 + 3x^2\Delta x + 3x(\Delta x)^2 + (\Delta x)^3$, $x^3$, $3x^2 + 3x\Delta x + (\Delta x)^2$, $3x^2 + 3x\Delta x + (\Delta x)^2$, $3x^2 + 3x\Delta x + (\Delta x)^2$, $3x^2$, $3x^2$;
(b) $-1, 3, 3(2)^2 = 12, (2, 8)$;
(c) The function and appropriate tangent lines are graphed in Figure 2.44.

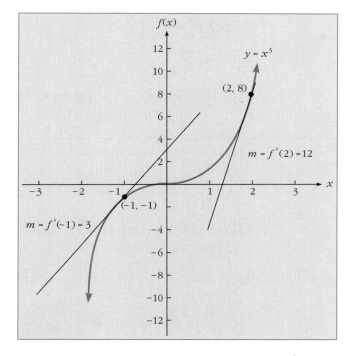

**FIGURE 2.44** Some tangent lines to the graph of $f(x) = x^3$.

**EXAMPLE 4**

(a) Find the derivative $dy/dx$ if $y = 1/x$.
(b) Find and interpret $\dfrac{dy}{dx}\bigg|_{x=-1}$ and $\dfrac{dy}{dx}\bigg|_{x=3}$.
(c) Sketch the graph of $y = 1/x$ and the tangent lines at the points $(-1, -1)$ and $(3, \tfrac{1}{3})$.

**Solution**

(a) Step 1 $f(x + \Delta x) = \dfrac{1}{x + \Delta x}$

Step 2 $f(x + \Delta x) - f(x) = \dfrac{1}{x + \Delta x} - \dfrac{1}{x}$

$= \dfrac{x - (x + \Delta x)}{x(x + \Delta x)} = \dfrac{-\Delta x}{x(x + \Delta x)}$

Step 3 $\dfrac{f(x + \Delta x) - f(x)}{\Delta x} = \dfrac{\dfrac{-\Delta x}{x(x + \Delta x)}}{\Delta x}$

$= \dfrac{-\Delta x}{x(x + \Delta x)} \cdot \dfrac{1}{\Delta x} = \dfrac{-1}{x(x + \Delta x)}, \quad \text{if } \Delta x \neq 0$

Step 4 $\lim\limits_{\Delta x \to 0} \dfrac{f(x + \Delta x) - f(x)}{\Delta x} = \lim\limits_{\Delta x \to 0} \dfrac{-1}{x(x + \Delta x)} = \dfrac{-1}{x^2}$

We conclude that $dy/dx = -1/x^2$.

(b) $\dfrac{dy}{dx}\bigg|_{x=-1} = -\dfrac{1}{(-1)^2} = -1$ is the slope of the tangent line to the graph of $y = 1/x$ at the point $(-1, -1)$.

$\dfrac{dy}{dx}\bigg|_{x=3} = -\dfrac{1}{3^2} = -\dfrac{1}{9}$ is the slope of the tangent line to the graph of $y = 1/x$ at the point $(3, \tfrac{1}{3})$.

(c) The sketch of $y = 1/x$ and the tangent line are drawn in Figure 2.45. ∎

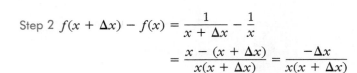

**FIGURE 2.45** Some tangent lines to the graph of $y = 1/x$.

---

**EXAMPLE 5**

(a) Find the derivative of $f(x) = \sqrt{x}$.
(b) Find $f'(1)$ and $f'(4)$.
(c) Sketch the graph of $f(x)$ and the tangent lines at the points $(1, 1)$ and $(4, 2)$.

**Solution**

(a) Step 1 $f(x + \Delta x) = \sqrt{x + \Delta x}$

Step 2 $f(x + \Delta x) - f(x) = \sqrt{x + \Delta x} - \sqrt{x}$

Step 3 $\dfrac{f(x + \Delta x) - f(x)}{\Delta x} = \dfrac{\sqrt{x + \Delta x} - \sqrt{x}}{\Delta x}$

Step 4 $\lim\limits_{\Delta x \to 0} \dfrac{f(x + \Delta x) - f(x)}{\Delta x} = \lim\limits_{\Delta x \to 0} \dfrac{\sqrt{x + \Delta x} - \sqrt{x}}{\Delta x}$

In order to evaluate this limit, we rationalize the numerator as follows:

$$\frac{\sqrt{x+\Delta x}-\sqrt{x}}{\Delta x} = \frac{\sqrt{x+\Delta x}-\sqrt{x}}{\Delta x} \cdot \frac{\sqrt{x+\Delta x}+\sqrt{x}}{\sqrt{x+\Delta x}+\sqrt{x}}$$

$$= \frac{x+\Delta x - x}{\Delta x(\sqrt{x+\Delta x}+\sqrt{x})}$$

$$= \frac{\Delta x}{\Delta x(\sqrt{x+\Delta x}+\sqrt{x})}$$

$$= \frac{1}{\sqrt{x+\Delta x}+\sqrt{x}}, \quad \text{if } \Delta x \neq 0$$

Thus, $f'(x) = \lim_{\Delta x \to 0} \dfrac{1}{\sqrt{x+\Delta x}+\sqrt{x}} = \dfrac{1}{\sqrt{x+0}+\sqrt{x}} = \dfrac{1}{2\sqrt{x}}$.

**(b)** $f'(1) = 1/(2\sqrt{1}) = 1/2;\ \ f'(4) = 1/(2\sqrt{4}) = 1/4$

**(c)** The graph of $f(x)$ and the tangent lines are sketched in Figure 2.46.

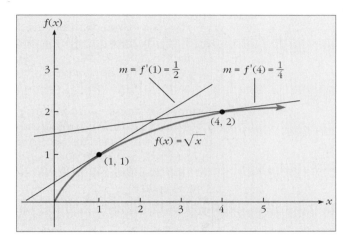

**FIGURE 2.46** Some tangent lines to the graph of $f(x) = \sqrt{x}$.

### Differentiability

Observe that, from the results of Example 4, if $f(x) = \sqrt{x}$, then $f'(x) = 1/(2\sqrt{x})$. Note that $f(0) = \sqrt{0} = 0$, but $f'(0) = 1/(2\sqrt{0}) = 1/0$ is undefined. Therefore, the derivative may not exist at a point $a$ even though $a$ is a number in the domain of $f$.

**DEFINITION**

> If $a$ is a number in the domain of $f$ and if $f'(a)$ exists, then we say that $f$ is *differentiable at the point a*. If $f$ is differentiable at all points in the open interval $(c, d)$ we say that $f$ is *differentiable on the interval $(c, d)$*. If $f$ is differentiable on $(-\infty, \infty)$, we say that $f$ is *differentiable everywhere*.

EXAMPLE 6

(a) If $f(x) = x^2$, then $f'(x) = 2x$ (see Example 1) and since the domain of $f'(x)$ is all real numbers, $f(x)$ is differentiable everywhere.
(b) If $f(x) = 1/x$, then $f'(x) = -1/x^2$ (see Example 4). Since the domain of $f'(x)$ is all nonzero real numbers, $f(x)$ is differentiable at all points $a$ where $a \neq 0$. Alternatively, $f(x)$ is differentiable on the intervals $(-\infty, 0)$ and $(0, \infty)$.
(c) If $f(x) = \sqrt{x}$, then $f'(x) = 1/(2\sqrt{x})$ (see Example 5). Since the domain of $f'(x)$ is all real numbers $x > 0$, then $f(x)$ is differentiable at all points $a$ where $a > 0$. Alternatively, $f(x)$ is differentiable on the interval $(0, \infty)$. ■

In general, the domain of $f'(x)$ is included in (and so may be equal to) the domain of $f(x)$, and the domain of $f'(x)$ is the set of all points at which $f(x)$ is differentiable. There are several ways that a function $y = f(x)$ may fail to be differentiable at a point $a$ in the domain of $f$. Some possibilities are these:

i. $f(x)$ may have a vertical tangent line at $x = a$;
ii. $f(x)$ may have a "cusp" or "sharp point" at $x = a$, and so $f(x)$ will have no tangent line at $x = a$;
iii. $f(x)$ may not be continuous at $x = a$, and so $f(x)$ will have no tangent line at $x = a$.

Examples of these situations are provided in Figure 2.47(a)–(c). In each case $f'(0)$ fails to exist. We have already investigated (see Example 6 in Section 2.4) the reason why $f'(0)$ fails to exist for $f(x) = |x|$. Since $f(x) = |x|$ has no tangent line at $(0, 0)$ and since $f'(0)$ represents the slope of the tangent line at $x = 0$, clearly $f'(0)$ does not exist. We now consider the cases represented by Figure 2.47(a) and (c).

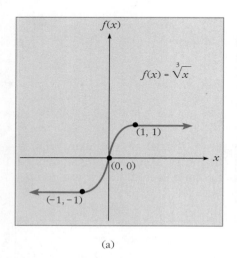
(a)

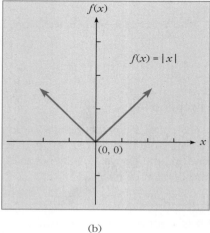
(b)

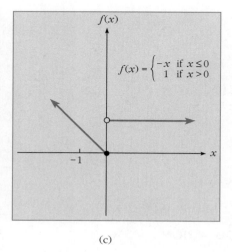
(c)

**FIGURE 2.47** Functions that are not differentiable at $(0, 0)$: (a) $f(x)$ has a vertical tangent line at $x = 0$; (b) $f(x)$ has a cusp at $x = 0$; (c) $f(x)$ is not continuous at $x = 0$.

## 2.5 THE DERIVATIVE OF A FUNCTION

**EXAMPLE 7**

Show that if $f(x) = \sqrt[3]{x}$, then $f(x)$ has a vertical tangent line at $x = 0$.

**Solution**

We use the four-step method with $x = 0$ to attempt to compute $f'(0)$.

Step 1 $f(0 + \Delta x) = f(\Delta x) = \sqrt[3]{\Delta x}$

Step 2 $f(0 + \Delta x) - f(0) = \sqrt[3]{\Delta x} - \sqrt[3]{0} = \sqrt[3]{\Delta x} = (\Delta x)^{1/3}$

Step 3 $\dfrac{f(0 + \Delta x) - f(0)}{\Delta x} = \dfrac{(\Delta x)^{1/3}}{\Delta x} = \dfrac{1}{(\Delta x)^{2/3}} = \dfrac{1}{\sqrt[3]{(\Delta x)^2}}$

Step 4 $f'(0) = \lim\limits_{\Delta x \to 0} \dfrac{f(0 + \Delta x) - f(0)}{\Delta x} = \lim\limits_{\Delta x \to 0} \dfrac{1}{\sqrt[3]{(\Delta x)^2}}$

Now, $\lim\limits_{\Delta x \to 0} 1 = 1 > 0$ and $\lim\limits_{\Delta x \to 0} \sqrt[3]{(\Delta x)^2} = 0$. Furthermore, $\sqrt[3]{(\Delta x)^2}$ approaches zero through positive values as $\Delta x \to 0$. By Property I of Section 2.2, we have

$$f'(0) = \lim\limits_{\Delta x \to 0} \dfrac{1}{\sqrt[3]{(\Delta x)^2}} = \infty$$

We conclude that the tangent line to the graph of $f(x) = \sqrt[3]{x}$ at the point $x = 0$ is vertical, and that $f(x)$ is not differentiable at $x = 0$. ■

In general, if we obtain a limit of either $\infty$ or $-\infty$ when attempting to find $f'(a)$, then the tangent line to the graph of $f(x)$ at $x = a$ is vertical and $f$ is not differentiable at $x = a$. Also, if we obtain a limit that does not exist (and is not equal to $\infty$ or $-\infty$) when attempting to find $f'(a)$, then there is no tangent line at $x = a$ and $f$ is not differentiable at $x = a$. An example of the latter case is $f(x) = |x|$ at $x = 0$. In Example 6 of Section 2.4, we showed that there is no tangent line at $x = 0$. Another example is considered next.

**EXAMPLE 8 (Participative)**

Let's show that if

$$f(x) = \begin{cases} -x & \text{if } x \leq 0 \\ 1 & \text{if } x > 0 \end{cases}$$

then $f'(0)$ does not exist, so that $f(x)$ is not differentiable at $x = 0$.

**Solution**

Let's use the four-step method with $x = 0$ to attempt to compute $f'(0)$.

Step 1 $f(0 + \Delta x) = f(\Delta x)$. Since $f(x)$ is a multipart function defined by different formulas for $x \leq 0$ and $x > 0$, and since $\Delta x$ may be positive or negative, we will have to consider two cases in this problem.

Case 1 $\Delta x > 0$: then $f(\Delta x) = $ _____.

Case 2 $\Delta x \leq 0$: then $f(\Delta x) = $ _____.

Step 2 Case 1 $\Delta x > 0$: then $f(0 + \Delta x) - f(0) = $ _____.

Case 2 $\Delta x \leq 0$: then $f(0 + \Delta x) - f(0) = $ _____.

Step 3 Case 1 $\Delta x > 0$: then $\dfrac{f(0 + \Delta x) - f(0)}{\Delta x} = $ _____.

Case 2 $\Delta x \leq 0$: then $\dfrac{f(0 + \Delta x) - f(0)}{\Delta x} = $ _____.

Step 4 Note that $f'(0) = \lim\limits_{\Delta x \to 0} \dfrac{f(0 + \Delta x) - f(0)}{\Delta x}$ if this limit exists. We will now show that the limit does not exist. Consider:

Case 1 $\Delta x > 0$: then $\lim\limits_{\Delta x \to 0^+} \dfrac{f(0 + \Delta x) - f(0)}{\Delta x} = \lim\limits_{\Delta x \to 0^+}$ _____ $=$ _____.

Case 2 $\Delta x \leq 0$: then $\lim\limits_{\Delta x \to 0^-} \dfrac{f(0 + \Delta x) - f(0)}{\Delta x} = \lim\limits_{\Delta x \to 0^-}$ _____ $=$ _____.

Since the right- and left-hand limits at zero are different, we conclude that

$$f'(0) = \lim_{\Delta x \to 0} \dfrac{f(0 + \Delta x) - f(0)}{\Delta x}$$

does not exist, so $f(x)$ is not differentiable at the point $x = 0$. Further, since this limit is not equal to $+\infty$ or $-\infty$, no well-defined tangent line exists at $x = 0$. ∎

### Alternate Definition of the Derivative

There is an alternate definition of the derivative of $f(x)$ at a point $x = a$ in the domain of $f(x)$. To understand this alternate definition, look at Figure 2.48. In this figure, we fix the point $P(a, f(a))$ and choose another point $Q(x, f(x))$ on the graph $y = f(x)$. The slope of the secant line joining $P$ and $Q$ is then

$$m_{\text{sec}} = \dfrac{f(x) - f(a)}{x - a}$$

This slope is sometimes called a *difference quotient*, since it is a quotient of two differences. Now, as we let $Q$ slide along the curve toward $P$, we obtain the tangent line $t$ that has slope $f'(a)$. But letting $Q$ approach $P$ along the curve is equivalent to letting $x$ approach $a$ along the $x$-axis. Therefore,

$$f'(a) = m_{\text{tan}} = \lim_{x \to a} \dfrac{f(x) - f(a)}{x - a}$$

---

**SOLUTIONS TO EXAMPLE 8**

Step 1 $1, -\Delta x$;
Step 2 $1, -\Delta x$;
Step 3 $1/\Delta x,\ -\Delta x/\Delta x = -1$ if $\Delta x \neq 0$;
Step 4 $1/\Delta x,\ \infty,\ -1,\ -1$

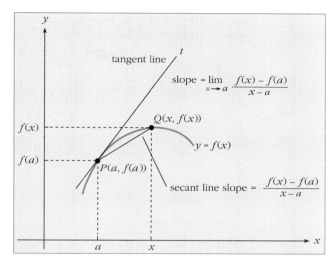

**FIGURE 2.48** The alternate definition of $f'(x)$.

EXAMPLE 9

Use the alternate definition of the derivative to compute $f'(x)$ if $f(x) = x^2$.

Solution

We compute:

$$f'(a) = \lim_{x \to a} \frac{f(x) - f(a)}{x - a} = \lim_{x \to a} \frac{x^2 - a^2}{x - a}$$

$$= \lim_{x \to a} \frac{(x + a)(x - a)}{(x - a)} = \lim_{x \to a} (x + a) = a + a = 2a$$

Thus since $f'(a) = 2a$, we conclude that $f'(x) = 2x$ by replacing $a$ by $x$ in the formula for this derivative. ∎

### Differentiability and Continuity

We can use the alternate definition of the derivative to prove the following important relationship between differentiability and continuity at a point $a$.

THEOREM

If a function $f$ is differentiable at a point $a$, then $f$ is continuous at $a$.

Proof

Assuming that $f$ is differentiable at $a$ (that is, assuming that $f'(a)$ exists), we want to show that $f$ is continuous at $a$ (that is, we want to show that $\lim_{x \to a} f(x) = f(a)$). To accomplish this task, we will use the following little "trick":

$$f(x) = \left[ \frac{f(x) - f(a)}{x - a} \right](x - a) + f(a) \qquad \text{if } x \neq a$$

To see that this is true, try multiplying out the right-hand side of the preceding equation and simplifying the result. We continue the proof by taking limits of both sides of the equation as $x$ approaches $a$, and then using some limit properties from Section 2.1.

$$\begin{aligned}
\lim_{x \to a} f(x) &= \lim_{x \to a} \left\{ \left[ \frac{f(x) - f(a)}{x - a} \right] (x - a) + f(a) \right\} \\
&= \lim_{x \to a} \left\{ \left[ \frac{f(x) - f(a)}{x - a} \right] (x - a) \right\} + \lim_{x \to a} f(a) \quad \text{Property 4} \\
&= \lim_{x \to a} \frac{f(x) - f(a)}{x - a} \cdot \lim_{x \to a} (x - a) + f(a) \quad \begin{array}{l}\text{Properties 5 and 1}\\ \text{Note that } f(a) \text{ is}\\ \text{a constant.}\end{array} \\
&= f'(a) \cdot (a - a) + f(a) \\
&= f'(a) \cdot 0 + f(a) = f(a)
\end{aligned}$$

This completes the proof of the theorem. ∎

We should observe that the converse of the above theorem is false. That is, if $f$ is continuous at $a$, then $f$ need not be differentiable at $a$. As an example, recall that $f(x) = |x|$ is continuous at $x = 0$ since $\lim_{x \to 0} |x| = |0| = 0 = f(0)$ (see Figure 2.47(b)), but it is not differentiable at $x = 0$ since the graph has a sharp point there. The relationship between differentiability and continuity at a point $a$ is illustrated in Figure 2.49.

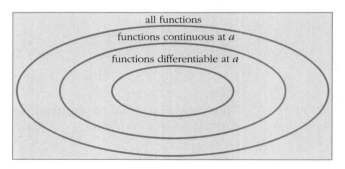

FIGURE 2.49  Differentiability and continuity at a point $a$.

### SECTION 2.5
### SHORT ANSWER EXERCISES

1. (True or False) For a certain function, $f'(4) = 10$. Therefore the slope of the tangent line to the graph of $y = f(x)$ at the point $(4, f(4))$ is equal to 10. _____
2. If $f(x) = 2x^4$, it can be shown that $f'(x) = 8x^3$. Then $f'(-1) =$ _____, and the slope of the tangent line to the graph of $f(x)$ at the point $(-1, 2)$ is equal to _____.
3. If $f(x) = 2 - x^2$, then $f(x + \Delta x) =$ _____.
4. Given that $(f(x + \Delta x) - f(x))/\Delta x = 4x + 2\Delta x$ if $\Delta x \neq 0$, then $f'(x) =$ _____.
5. The slope of the secant line joining $(x, y)$ and $(x + \Delta x, y + \Delta y)$ is equal to $6x + 3\Delta x$ if $\Delta x \neq 0$ for a certain function $y = f(x)$. Therefore the derivative of this function is equal to _____.
6. If $\dfrac{dy}{dx} = 12x - 5$, then $\left. \dfrac{dy}{dx} \right|_{x=2} =$ _____.

7. (True or False) The symbols $f'(x)$ and $dy/dx$ are two different ways to express the derivative of $y = f(x)$. _____

8. If $f(x) = 1/(x - 1)$, then (a) $f'(1)$ does not exist, since $x = 1$ is not in the domain of $f(x)$; (b) $f(x)$ is not continuous at $x = 1$ so it is not differentiable at $x = 1$; (c) both (a) and (b) are true.

9. It can be shown that if $f(x) = \sqrt[3]{x}$, then $f'(x) = 1/(3\sqrt[3]{x^2})$. We may conclude that (a) $f(x)$ is differentiable everywhere; (b) the domain of $f'(x)$ is all real numbers; (c) $f(x)$ is not differentiable at $x = 0$; (d) none of the above.

10. A certain function $y = f(x)$ is differentiable at the point $x = 3$. We may conclude that (a) $f(x)$ is continuous at $x = 3$; (b) $f(x)$ is not continuous at $x = 3$; (c) no tangent line exists at $x = 3$; (d) a vertical tangent line exists at $x = 3$.

11. A certain function $y = g(x)$ is continuous at the point $x = -2$. We may conclude that (a) $g(x)$ is differentiable at $x = -2$; (b) $g(x)$ is not differentiable at $x = -2$; (c) no tangent line exists at $x = -2$; (d) none of the above.

## SECTION 2.5
## EXERCISES

In Exercises 1–11, use the definition of the derivative to find $f'(x)$. Find the slope of the tangent lines to the graph of $y = f(x)$ at the indicated points. Where indicated, sketch the graph of $f(x)$ and draw the tangent lines to the graph at the given points.

1. $f(x) = x^2 + 1$, at $x = -2, x = 0, x = 2$: sketch
2. $f(x) = 3 - x^2$, at $x = -2, x = 1, x = 3$: sketch
3. $f(x) = 2x^2 + x$, at $x = -1, x = 0, x = 2$: sketch
4. $f(x) = 4x - 6x^2$, at $x = -3, x = 1, x = 2$: sketch
5. $f(x) = x^2 - 6x + 5$, at $x = 0, x = 3, x = 5$: sketch
6. $f(x) = x^2 - 2x - 8$, at $x = -1, x = 2, x = 5$: sketch
7. $f(x) = x^3 + 2$, at $x = -1, x = 0, x = 1$
8. $f(x) = 2x^3 - 4$, at $x = -1, x = 0, x = 1$
9. $f(x) = 3/x$, at $x = -1, x = 3$
10. $f(x) = -1/x$, at $x = -2, x = 2$
11. $f(x) = 2\sqrt{x}$, at $x = 1, x = 4$

In Exercises 12–17, find $dy/dx$.

12. $y = 3x^2 + 4x - 2$
13. $y = 7 - x - 3x^2$
14. $y = 2x^3 - 3x$
15. $y = x^3 + 3x^2 - x$
16. $y = 2/(x + 1)$
17. $y = 3/(x - 1)$

In Exercises 18–23, determine the points at which $f(x)$ is not differentiable by (a) drawing a sketch of $f(x)$ and (b) attempting to compute $f'(a)$ at points that look "suspicious" on your graph.

18. $f(x) = -\sqrt[3]{x}$
19. $f(x) = |x - 2|$
20. $f(x) = 3|x + 1|$
21. $f(x) = x^{2/3}$
22. $f(x) = \begin{cases} x & \text{if } x \leq 1 \\ 2 & \text{if } x > 1 \end{cases}$
23. $f(x) = \begin{cases} x^2 & \text{if } x < 2 \\ x & \text{if } x \geq 2 \end{cases}$

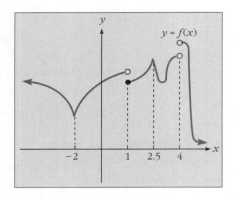

**FIGURE 2.50**

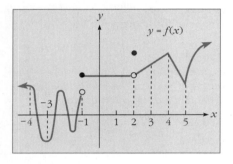

**FIGURE 2.51**

In Exercises 24–29, use the alternate definition of the derivative to compute $f'(x)$.

24. $f(x) = 5x^2 - 4$
25. $f(x) = 2 - 2x - 3x^2$
26. $f(x) = 2/x$
27. $f(x) = 3/x^2$
28. $f(x) = x^3$ (Hint: $x^3 - a^3 = (x - a)(x^2 + ax + a^2)$)
29. $f(x) = 3/(x + 4)$

30. Referring to Figure 2.50, determine where $f(x)$ is (a) not continuous; (b) not differentiable; (c) both continuous and differentiable.

31. Referring to Figure 2.51, determine where $f(x)$ is: (a) not continuous and (b) not differentiable.

32. (State lottery) The percentage of registered voters who favor the institution of a state lottery in a certain county of Georgia is given by

$$P(t) = 53 + 0.3t + 0.1t^2, \quad 0 \leq t \leq 3$$

where $t = 0$ corresponds to the beginning of 1988 and $t$ is time measured in years.
(a) Find the percentage of voters who favored the lottery at the beginning of 1991.
(b) Find $P'(t)$ using the definition of the derivative.
(c) Find the slope of the tangent line when $t = 1$ and $t = 2$.

33. (Smoking at a university) At a large university, the number of students who smoke is given by

$$N(t) = 5000 - 25t - 9t^2, \quad 0 \leq t \leq 4$$

where $t = 0$ corresponds to the beginning of 1987 and $t$ is time measured in years.
(a) Find the number of students who smoked at the beginning of 1990.
(b) Find $N'(t)$ using the definition of the derivative.
(c) Find the slope of the tangent line when $t = 2$ and $t = 3$.

## 2.6 THE DERIVATIVE AS A RATE OF CHANGE

Let's review what we have learned so far about the derivative. First, in Section 2.4 we saw that the slope of the tangent line to the graph of $y = f(x)$ at the point $P(x, f(x))$ is given by a limit of secant line slopes when this limit exists:

$$m_{\text{tan}} = \lim_{\Delta x \to 0} m_{\text{sec}} = \lim_{\Delta x \to 0} \frac{f(x + \Delta x) - f(x)}{\Delta x}$$

Next, we defined this limit of the difference quotient $[f(x + \Delta x) - f(x)]/\Delta x$ to be the derivative of $f(x)$, denoted by the symbol $f'(x)$ or by the symbol $dy/dx$ (Section 2.5). In this section we consider another important interpretation of the derivative $f'(a)$, namely, "the instantaneous rate of change of $y$ with respect to $x$ at the value $x = a$." Our goal in this section is to understand what is meant by this statement.

### Average Velocity

One of the easiest rates of change to understand is the rate of change of distance with respect to time, otherwise known as the *velocity*. As an example, suppose you travel in your car for three hours and suppose that the distance $s$ traveled in miles as a function of time $t$ in hours is given by

$$s = f(t) = 10t^2 + 6t, \qquad 0 \le t \le 3$$

In Table 2.18 we list the distance traveled for various values of time $t$.

| $t$ | 0 | 1 | 2 | 3 |
|---|---|---|---|---|
| $s = 10t^2 + 6t$ | 0 | 16 | 52 | 108 |

**TABLE 2.18**

The *average velocity* between two times $t$ and $t + \Delta t$ is the average rate of change of distance $s$ with respect to time $t$ over the time interval $\Delta t$. For the example in Table 2.18,

The average velocity between $t = 0$ and $t = 1$ is equal to

$$\frac{\text{change in distance between } t = 0 \text{ and } t = 1}{\text{change in time between } t = 0 \text{ and } t = 1} = \frac{f(1) - f(0)}{1 - 0} = \frac{\Delta s}{\Delta t}$$

$$= \frac{16 - 0}{1 - 0} = \frac{16 \text{ miles}}{\text{hour}}$$

The average velocity between $t = 1$ and $t = 2$ is equal to

$$\frac{\text{change in distance between } t = 1 \text{ and } t = 2}{\text{change in time between } t = 1 \text{ and } t = 2} = \frac{f(2) - f(1)}{2 - 1} = \frac{\Delta s}{\Delta t}$$

$$= \frac{52 - 16}{2 - 1} = \frac{36 \text{ miles}}{\text{hour}}$$

The average velocity between $t = 2$ and $t = 3$ is equal to

$$\frac{\text{change in distance between } t = 2 \text{ and } t = 3}{\text{change in time between } t = 2 \text{ and } t = 3} = \frac{f(3) - f(2)}{3 - 2} = \frac{\Delta s}{\Delta t}$$

$$= \frac{108 - 52}{3 - 2} = \frac{56 \text{ miles}}{\text{hour}}$$

In general, the average velocity between times $t$ and $t + \Delta t$ is equal to

$$\frac{\text{change in distance between } t \text{ and } t + \Delta t}{\text{change in time between } t \text{ and } t + \Delta t} = \frac{f(t + \Delta t) - f(t)}{(t + \Delta t) - t}$$

$$= \frac{f(t + \Delta t) - f(t)}{\Delta t} = \frac{\Delta s}{\Delta t}$$

The difference quotient $[f(t + \Delta t) - f(t)]/\Delta t = \Delta s/\Delta t$ should be familiar to you by now. You already know that this quotient represents the slope of the secant line joining the points $P(t, f(t))$ and $Q(t + \Delta t, f(t + \Delta t))$ on the graph of $s = f(t)$. We conclude that the average velocity between times $t$ and

$t + \Delta t$ is simply the slope of the secant line joining $P$ and $Q$. In Figure 2.52, the graph of $s = f(t) = 10t^2 + 6t$ is drawn for $0 \leq t \leq 3$, and the average velocity between $t = 1$ and $t = 2$ is shown to be numerically equal to the slope of the corresponding secant line, which is $\Delta s/\Delta t = \frac{36}{1} = 36$.

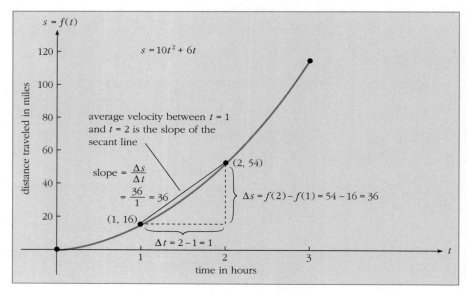

FIGURE 2.52  Average velocity between $t = 1$ and $t = 2$.

### Average Rate of Change

Generalizing to an arbitrary function $y = f(x)$, we may say the following:

> The average rate of change in $y$ with respect to $x$ between $x$ and $x + \Delta x$ is equal to the slope of the secant line joining the points $P(x, f(x))$ and $Q(x + \Delta x, f(x + \Delta x))$, which is
> 
> $$m_{\text{sec}} = \frac{\Delta y}{\Delta x} = \frac{f(x + \Delta x) - f(x)}{\Delta x}$$

See Figure 2.53 for a graphical illustration of this idea.

---

EXAMPLE 1

Let $f(x) = x^2$.

(a) Find the average rate of change in $y$ with respect to $x$ between $x = 0$ and $x = 2$.

(b) Illustrate the result of part (a) graphically by drawing the graph of $f(x)$ and the appropriate secant line.

2.6 THE DERIVATIVE AS A RATE OF CHANGE    135

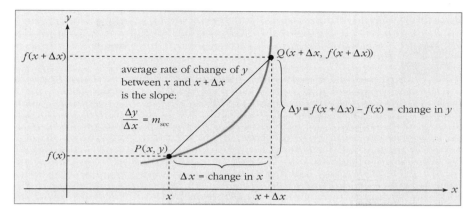

**FIGURE 2.53**  Average rate of change of $y$ with respect to $x$ between $x$ and $x + \Delta x$.

Solution

(a) The average rate of change of $y$ with respect to $x$ between $x$ and $x + \Delta x$ is given by

$$\frac{\Delta y}{\Delta x} = \frac{f(x + \Delta x) - f(x)}{\Delta x}$$

In this problem, we are given $f(x) = x^2$, as well as $x = 0$ and $x + \Delta x = 2$, so $0 + \Delta x = 2$ and $\Delta x = 2$. We find that the average rate of change of $y$ with respect to $x$ between $x = 0$ and $x = 2$ is given by

$$\frac{f(0 + 2) - f(0)}{2} = \frac{2^2 - 0^2}{2} = \frac{4}{2} = \frac{2 \text{ units of } y}{1 \text{ unit of } x}$$

Thus as $x$ changes from 0 to 2, the average rate of change of $y$ for this function is 2 units of $y$ per unit of $x$.

(b) The required graph is sketched in Figure 2.54.

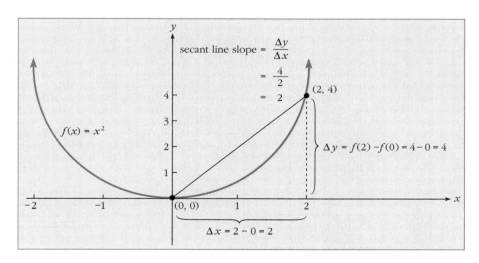

**FIGURE 2.54**  The average rate of change of $y = x^2$ between $x = 0$ and $x = 2$.

# 136 CHAPTER 2 LIMITS AND AN INTRODUCTION TO THE DERIVATIVE

EXAMPLE 2 (Participative)

Solution

If $f(x) = \sqrt{x}$, let's find the average rate of change of $y$ with respect to $x$ between $x = 1$ and $x = 4$ and illustrate the result graphically.

First we recall that the average rate of change of $y$ with respect to $x$ between $x$ and $x + \Delta x$ is

$$\frac{\Delta y}{\Delta x} = \frac{f(x + \Delta x) - f(x)}{\Delta x}$$

In this case, $x = $ \_\_\_\_ and $x + \Delta x = $ \_\_\_\_, so $\Delta x = $ \_\_\_\_. Thus the average rate of change of $y$ with respect to $x$ between $x = 1$ and $x = 4$ is equal to

$$\frac{f(1 + 3) - f(1)}{3} = \frac{\sqrt{4} - \underline{\phantom{xx}}}{3} = \underline{\phantom{xxxxx}}$$

So as $x$ changes from 1 to 4, the average rate of change of $y$ for this function is one unit of $y$ per \_\_\_\_ units of $x$. Sketch the graph of $f(x)$ and the appropriate secant line on the coordinate axes provided here.

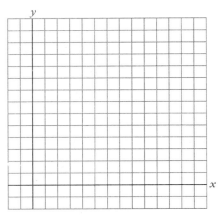

FIGURE 2.55

---

SOLUTIONS TO EXAMPLE 2

1, 4, 3, $\sqrt{1}$, (1 unit of $y$)/(3 units of $x$), 3

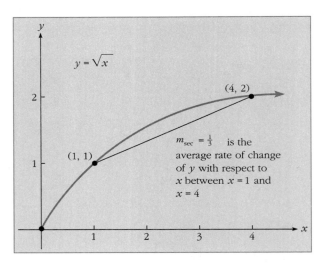

FIGURE 2.56 The average rate of change of $y = \sqrt{x}$ between $x = 1$ and $x = 4$.

## EXAMPLE 3

The following table provides the Real Gross National Product (GNP) in the U.S. (in billions of 1972 dollars) for selected years from 1970 to 1980.

| Year | 1970 | 1974 | 1976 | 1980 |
|------|------|------|------|------|
| GNP  | 1090 | 1250 | 1300 | 1500 |

(a) Find the average rate of change of GNP with respect to time between 1970 and 1974.
(b) Find the average rate of change of GNP with respect to time between 1974 and 1980.

**Solution**

(a) $\dfrac{\text{GNP in 1974} - \text{GNP in 1970}}{1974 - 1970} = \dfrac{1250 - 1090}{4}$

$= \dfrac{40 \text{ billions of 1972 dollars}}{1 \text{ year}}$

That is, between 1970 and 1974, GNP increased on average by 40 billions of 1972 dollars per year.

(b) $\dfrac{\text{GNP in 1980} - \text{GNP in 1974}}{1980 - 1974} = \dfrac{1500 - 1250}{6}$

$\approx \dfrac{41.67 \text{ billions of 1972 dollars}}{1 \text{ year}}$ ■

## EXAMPLE 4

It is estimated that the population in a certain small town for the next six years will be

$$N(t) = 100 - 2t + t^2, \quad 0 \leq t \leq 6$$

where $t$ is time measured in years.

(a) Find the average rate of change of the population of this town over the next two years, that is, between $t = 0$ and $t = 2$.
(b) Find the average rate of change of the population of this town between $t = 4$ and $t = 6$.
(c) Sketch the graph of $N(t)$ and the appropriate secant lines with slopes computed in parts (a) and (b).
(d) Would you say that the rate of growth of the population of this town is increasing or decreasing?

**Solution**

(a) The average rate of change of the population between $t = 0$ and $t = 2$ is given by the difference quotient:

$$\dfrac{N(2) - N(0)}{2 - 0} = \dfrac{[100 - 2(2) + 2^2] - [100 - 2(0) + 0^2]}{2} = \dfrac{0 \text{ people}}{1 \text{ year}}$$

That is, over the next two years, the town's population will neither increase nor decrease on the average.

(b) The average rate of change of the population between $t = 4$ and $t = 6$ is also given by the difference quotient:

$$\frac{N(6) - N(4)}{6 - 4} = \frac{[100 - 2(6) + 6^2] - [100 - 2(4) + 4^2]}{2}$$

$$= \frac{124 - 108}{2} = \frac{8 \text{ people}}{1 \text{ year}}$$

So between years 4 and 6, the town's population will increase on the average by eight people per year.

(c) A graph of the function $N(t)$ and the appropriate secant lines are sketched in Figure 2.57.

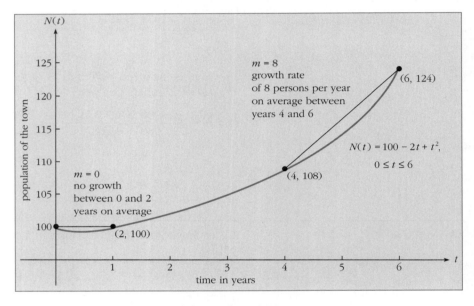

FIGURE 2.57 Some average rates of change of population.

(d) It appears that the average rate of growth (in persons per year) of the population of this town is increasing. To verify this, we construct Table 2.19, which shows the average rate of growth between years $t$ and $t + 1$ for the relevant range of $t$ values. This table makes it clear that the average growth rate is increasing.

| Years | $t=0$ to $t=1$ | $t=1$ to $t=2$ | $t=2$ to $t=3$ | $t=3$ to $t=4$ | $t=4$ to $t=5$ | $t=5$ to $t=6$ |
|---|---|---|---|---|---|---|
| Average Growth Rate | −1 | 1 | 3 | 5 | 7 | 9 |

TABLE 2.19

### Instantaneous Velocity

Let us now return to the example considered at the beginning of this section, in which we considered distance as a function of time during a car trip, where

$$s = f(t) = 10t^2 + 6t, \quad 0 \le t \le 3$$

Recall that distance $s$ in this formula is measured in miles and time $t$ in hours. We computed several average velocities for this situation. For example, the average velocity between times $t = 2$ and $t = 3$ was computed as

$$\frac{f(3) - f(2)}{3 - 2} = \frac{56 \text{ miles}}{\text{hour}}$$

We now pose an interesting question. What is the *instantaneous* velocity precisely at time $t = 2$? That is, if you were riding in this car and glanced at the speedometer precisely 2 hours after this trip began, what would the reading on the speedometer be at that precise instant? To answer this question, let us compute the average velocity over time intervals 2 to $2 + \Delta t$, where $|\Delta t|$ is small. By computing such average velocities over smaller and smaller time intervals we will approach what we intuitively feel is the *instantaneous* velocity at $t = 2$. As an example, if we let $\Delta t = 0.1$, then the average velocity between $t = 2$ and $t = 2.01$ is equal to

$$\frac{f(2.01) - f(2)}{2.01 - 2} = \frac{[10(2.01)^2 + 6(2.01)] - [10(2)^2 + 6(2)]}{0.01}$$

$$= \frac{52.461 - 52}{0.01} = 46.1 \text{ miles/hour}$$

Tables 2.20 and 2.21 show the average velocity for various values of $\Delta t$. In Table 2.20 we let $\Delta t$ approach zero through positive values, and in Table 2.21 we let $\Delta t$ approach zero through negative values. In each case, the average velocity figures seem to approach a limit of 46 miles/hour: this is the instantaneous velocity of the car at $t = 2$.

| $\Delta t$ | 0.1 | 0.01 | 0.001 | 0.0001 | $\to 0^+$ |
|---|---|---|---|---|---|
| Average Velocity | 47 | 46.1 | 46.01 | 46.001 | $\to 46$ |

**TABLE 2.20**

| $\Delta t$ | $-0.1$ | $-0.01$ | $-0.001$ | $-0.0001$ | $\to 0^-$ |
|---|---|---|---|---|---|
| Average Velocity | 45 | 45.9 | 45.99 | 45.999 | $\to 46$ |

**TABLE 2.21**

In general, the average velocity between times $t = 2$ and $t = 2 + \Delta t$ is given by

$$\frac{f(2 + \Delta t) - f(2)}{\Delta t}$$

In Tables 2.20 and 2.21 we computed this difference quotient for smaller and smaller values of $\Delta t$; that is, we let $\Delta t \to 0$, and the limit was 46. Therefore,

$$46 = \lim_{\Delta t \to 0} \frac{f(2 + \Delta t) - f(2)}{\Delta t}$$

But the limit of this difference quotient is just the *derivative* of $f(t)$ evaluated at $t = 2$. That is,

$$f'(2) = \lim_{\Delta t \to 0} \frac{f(2 + \Delta t) - f(2)}{\Delta t} = 46$$

We conclude that the instantaneous rate of change of distance with respect to time (that is, the instantaneous velocity) when $t = 2$ is equal to $f'(2)$. Generalizing, we find that

$$f'(t) = \lim_{\Delta t \to 0} \frac{f(t + \Delta t) - f(t)}{\Delta t}$$

is the instantaneous rate of change of distance with respect to time (that is, the instantaneous velocity) at time $t$. Since $f'(t)$ also represents the slope of the tangent line to the graph of $f(t)$ at the point $P(t, f(t))$, we see that the slope of the tangent line at $P(t, f(t))$ represents the instantaneous rate of change of distance with respect to time (that is, the instantaneous velocity) at time $t$.

---

**EXAMPLE 5**

(a) Find the instantaneous velocity at times $t = 1$ and $t = 2.5$ if $s = f(t) = 10t^2 + 6t$, $0 \leq t \leq 3$.
(b) Sketch the graph of $f(t)$ and draw the tangent line that has slope equal to the instantaneous velocity at times $t = 1$ and $t = 2.5$.

**Solution**

(a) We compute the derivative $f'(t)$ of $f(t) = 10t^2 + 6t$.

Step 1 $f(t + \Delta t) = 10(t + \Delta t)^2 + 6(t + \Delta t)$
$= 10t^2 + 20t\Delta t + 10(\Delta t)^2 + 6t + 6\Delta t$

Step 2 $f(t + \Delta t) - f(t) = 10t^2 + 20t\Delta t + 10(\Delta t)^2 + 6t$
$\qquad\qquad + 6\Delta t - (10t^2 + 6t)$
$= 20t\Delta t + 10(\Delta t)^2 + 6\Delta t$
$= \Delta t(20t + 10\Delta t + 6)$

Step 3 $\dfrac{f(t + \Delta t) - f(t)}{\Delta t} = \dfrac{\Delta t(20t + 10\Delta t + 6)}{\Delta t}$
$= 20t + 10\Delta t + 6, \quad$ if $\Delta t \neq 0$

Step 4 $\displaystyle\lim_{\Delta t \to 0} \frac{f(t + \Delta t) - f(t)}{\Delta t} = \lim_{\Delta t \to 0} (20t + 10\Delta t + 6)$
$= 20t + 10(0) + 6 = 20t + 6$

Therefore $f'(t) = 20t + 6$, and so the instantaneous velocity at time $t = 1$ is equal to

$f'(1) = 20(1) + 6 = 26$ miles/hour

Also, the instantaneous velocity at time $t = 2.5$ is equal to

$$f'(2.5) = 20(2.5) + 6 = 56 \text{ miles/hour}$$

(b) Figure 2.58 displays the graph of $f(t)$ and the appropriate tangent lines, with slopes that represent the instantaneous rates of change of distance with respect to time (that is, the instantaneous velocities).

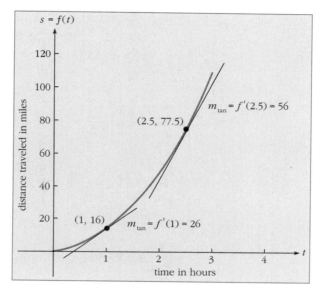

**FIGURE 2.58** Instantaneous velocity at $t = 1$ and at $t = 2.5$.

### Instantaneous Rate of Change

We may make the following generalization to an arbitrary function $y = f(x)$:

> The instantaneous rate of change of $y$ with respect to $x$ at $x = a$ is equal to the derivative $f'(a) = \dfrac{dy}{dx}\bigg|_{x=a}$ and also represents the slope of the tangent line to the graph of $y = f(x)$ at the point $P(a, f(a))$.

The Leibniz notation $dy/dx$ is especially suggestive of the rate of change of $y$ with respect to $x$.

EXAMPLE 6 (Participative)

The cost function of a certain company is $C(x) = 4x^2 + 1200$, where $x$ is measured in hundreds of units produced, $x \geq 0$.

(a) Let's find the rate at which cost is increasing when production is 100 units and when production is 500 units. These rates are called the *marginal costs* when production is 100 units and 500 units respectively.

# 142 CHAPTER 2 LIMITS AND AN INTRODUCTION TO THE DERIVATIVE

**Solution**

(b) Let's draw the cost function and the tangent lines with slopes representing the rates mentioned in part (a).

(a) We find the derivative $C'(x)$ by the four-step procedure and then evaluate the derivative at $x = 1$ and $x = 5$ to obtain the appropriate rates of change.

Step 1 $C(x + \Delta x) = 4(\underline{\qquad})^2 + 1200$
$= 4x^2 + 8x\Delta x + 4(\Delta x)^2 + 1200$

Step 2 $C(x + \Delta x) - C(x) = 4x^2 + 8x\Delta x + 4(\Delta x)^2$
$+ 1200 - (\underline{\qquad})$
$= 8x\Delta x + 4(\Delta x)^2$

Step 3 $\dfrac{C(x + \Delta x) - C(x)}{\Delta x} = \dfrac{\Delta x(8x + 4\Delta x)}{\Delta x} = \underline{\qquad},$
if $\Delta x \neq 0$

Step 4 $C'(x) = \lim\limits_{\Delta x \to 0} \dfrac{C(x + \Delta x) - C(x)}{\Delta x} = \lim\limits_{\Delta x \to 0} (\underline{\qquad}) = \underline{\qquad}$

Thus, $\dfrac{dC}{dx} = C'(x) = \underline{\qquad}$, and so the rate at which cost is increasing when production is 100 units (that is, when $x = 1$) is $\dfrac{dC}{dx}\bigg|_{x=1} = C'(1) = \underline{\qquad}$. Also, the rate at which cost is increasing when production is 500 units is $\dfrac{dC}{dx}\bigg|_{x=5} = C'(5) = \underline{\qquad}$.

(b) Draw the cost function and the tangent lines on the coordinate axes provided here. ∎

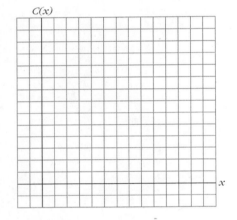

**FIGURE 2.59**

We are now ready to answer the questions posed at the beginning of Section 2.4. Recall that we assumed that your company's monthly profit from the sale of $x$ items, where $x$ is measured in thousands, is given by

$$P(x) = x^3 - 16x^2 + 65x - 50, \quad 1 \leq x \leq 5$$

where $P(x)$ is in thousands of dollars. See Figure 2.28 in Section 2.4 for a graph of $P(x)$. We now know that the tangent lines drawn to the graph of $P(x)$ at the various points represent the (instantaneous) rates of change of profit with respect to units sold. This rate of change $P'(x)$ is called the *marginal profit*.

---

**SOLUTIONS TO EXAMPLE 6**

Step 1 $x + \Delta x$;

Step 2 $4x^2 + 1200$;

Step 3 $8x + 4\Delta x$;

Step 4 $8x + 4\Delta x$, $8x$, $8x$, 8 dollars/100 units, 40 dollars/100 units

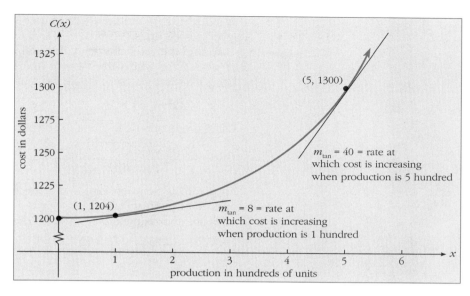

**FIGURE 2.60** Rates of change of $C(x) = 4x^2 + 1200$.

---

**EXAMPLE 7**

Find $P'(x)$, evaluate $P'(1)$ and $P'(4)$, and find the level of sales that maximizes profit for the profit function just given.

**Solution**

Step 1 $P(x + \Delta x) = (x + \Delta x)^3 - 16(x + \Delta x)^2 + 65(x + \Delta x) - 50$
$= x^3 + 3x^2\Delta x + 3x(\Delta x)^2 + (\Delta x)^3 - 16x^2 - 32x\Delta x$
$\phantom{=} - 16(\Delta x)^2 + 65x + 65\Delta x - 50$

Step 2 $P(x + \Delta x) - P(x) = x^3 + 3x^2\Delta x + 3x(\Delta x)^2 + (\Delta x)^3 - 16x^2 - 32x\Delta x$
$\phantom{=} - 16(\Delta x)^2 + 65x + 65\Delta x - 50 - (x^3 - 16x^2 + 65x - 50)$
$= 3x^2\Delta x + 3x(\Delta x)^2 + (\Delta x)^3 - 32x\Delta x$
$\phantom{=} - 16(\Delta x)^2 + 65\Delta x$
$= \Delta x[3x^2 + 3x\Delta x + (\Delta x)^2 - 32x - 16\Delta x + 65]$

Step 3 $\dfrac{P(x + \Delta x) - P(x)}{\Delta x} = \dfrac{\Delta x[3x^2 + 3x\Delta x + (\Delta x)^2 - 32x - 16\Delta x + 65]}{\Delta x}$
$= 3x^2 + 3x\Delta x + (\Delta x)^2 - 32x - 16\Delta x + 65,$
$\phantom{=}$ if $\Delta x \ne 0$

Step 4 $P'(x) = \lim\limits_{\Delta x \to 0} \dfrac{P(x + \Delta x) - P(x)}{\Delta x}$
$= \lim\limits_{\Delta x \to 0} [3x^2 + 3x\Delta x + (\Delta x)^2 - 32x - 16\Delta x + 65]$
$= 3x^2 - 32x + 65$

So, since $P'(x) = 3x^2 - 32x + 65$, we find that $P'(1) = 3(1)^2 - 32(1) + 65 = 36$; thus that profit is increasing at a rate of 36 thousand dollars per 1000 units of sales when the production level is 1000 units. Also, $P'(4) = 3(4)^2 - 32(4) + 65 = -15$, so that profit is decreasing (because of the negative sign) at the rate of 15 thousand dollars per 1000 units of sales when the production level is 4000 units.

To find the level of sales $x$ that will maximize profit, we must find the level of sales $x$ for which the tangent line is horizontal (see Figure 2.28). Since a horizontal line has zero slope, we must find $x$ so that $P'(x) = 0$, or

$$3x^2 - 32x + 65 = 0$$

Using the quadratic formula to solve this equation yields

$$x = \frac{32 \pm \sqrt{(-32)^2 - 4(3)(65)}}{2(3)} = \frac{32 \pm \sqrt{244}}{6}$$

Approximate solutions are $x \approx 2.73$ and $x \approx 7.94$. Since 7.94 is outside the interval of interest (which is $1 \leq x \leq 5$), we find that profit is maximum when sales are approximately 2730 units. The maximum profit is approximately

$$P(2.73) = (2.73)^3 - 16(2.73)^2 + 65(2.73) - 50 = 28.55 \text{ thousands of dollars} \quad \blacksquare$$

## SECTION 2.6
### SHORT ANSWER EXERCISES

1. The secant line drawn through the points $(2, 5)$ and $(6, 17)$ on the graph of $y = f(x)$ has a slope of _____. This slope represents the _____ rate of change of $y$ with respect to $x$ between $x = 2$ and $x = 6$.

2. If distance as a function of time is given by $d = f(t) = 2t^2 + t$, where $d$ is in feet and $t$ is in seconds, $0 \leq t \leq 6$, then $f(1) =$ _____ and $f(5) =$ _____, so the average velocity between times $t = 1$ second and $t = 5$ seconds is equal to _____ ft/sec.

3. (True or False) The slope of the secant line yields an average rate of change. _____

4. If the derivative of the profit function of your company is $P'(x) = -2x + 10$, where $x$ is measured in hundreds of units and $P(x)$ is measured in thousands of dollars, then $P'(2) =$ _____ thousands of dollars/hundred units and represents the rate at which profit is increasing when production is equal to _____ units.

5. The slope of the tangent line to the graph of $y = f(x)$ at the point $(5, 100)$ is equal to 15. We can conclude that (a) the slope of the secant line joining $(5, 100)$ and $(6, 120)$ is less than 15; (b) the average rate of change of $y$ with respect to $x$ between $x = 5$ and $x = 6$ is equal to 15 units of $y$/unit of $x$; (c) the instantaneous rate of change of $y$ with respect to $x$ at $x = 5$ is equal to 15 units of $y$/unit of $x$.

6. If the tangent line to an average cost function $\overline{C}(x)$ at a certain point is positive, then (a) an increase in production of 100 units will produce an increase in average cost; (b) a "small" increase in production will produce an increase in average cost; (c) average cost per unit will eventually start to decrease.

7. A certain function $y = f(x)$ has derivative $f'(3) = 10$. We can therefore say that (a) the instantaneous rate of change of $y$ when $x = 3$ is 10 units of $y$/unit of $x$; (b) the slope of the tangent line when $x = 3$ is equal to 10; (c) a "small" increase in $x$ will produce an increase in $y$; (d) all of the above.

8. The average rate of change of $y = f(x)$ between $x = 1$ and $x = 3$ is equal to 8 units of $y$ per unit of $x$. Therefore if $f(1) = 5$, then $f(3) = $ _____.
9. (True or False) The slope of the tangent line yields an instantaneous rate of change. _____
10. If the derivative of the profit function of your company is $P'(x) = -2x + 10$, the tangent line to the profit function $P(x)$ is horizontal when $x = $ _____.

## SECTION 2.6 EXERCISES

*In Exercises 1–10, find the average rate of change of $y$ with respect to $x$ between the two indicated x-values. Draw a sketch of $f(x)$ and the appropriate secant line with a slope that represents this average rate of change where indicated.*

1. $f(x) = x^2$, from $x = 2$ to $x = 4$: sketch
2. $f(x) = -x^2$, from $x = -1$ to $x = 2$: sketch
3. $f(x) = 3x + 1$, from $x = -2$ to $x = 3$: sketch
4. $f(x) = 1 - x$, from $x = 1$ to $x = 5$: sketch
5. $f(x) = 5$, from $x = 0$ to $x = 2$: sketch
6. $f(x) = 2x^2 + 4x$, from $x = -1$ to $x = 3$: sketch
7. $f(x) = 6x - x^2$, from $x = 0$ to $x = 2$
8. $f(x) = x^3$, from $x = -1$ to $x = 1$
9. $f(x) = 1 - x^3$, from $x = 1$ to $x = 2$
10. $f(x) = -2x^3$, from $x = -1$ to $x = 2$

*In Exercises 11–21, find the instantaneous rate of change of $y$ with respect to $x$ at the indicated point. Draw a sketch of $f(x)$ and the appropriate tangent line with slope representing this instantaneous rate of change.*

11. $f(x) = x^2 + 2$, at $x = 1$
12. $f(x) = x^2 + 1$, at $x = -2$
13. $f(x) = \frac{1}{3}x^3$, at $x = 3$
14. $f(x) = 1 - 2x^2$, at $x = 0$
15. $f(x) = 2 - 3x$, at $x = 4$
16. $f(x) = -4$, at $x = 2$
17. $f(x) = x^2 + x - 2$, at $x = 0$
18. $f(x) = 3 + 2x - x^2$, at $x = 3$
19. $f(x) = 2x - x^2$, at $x = 4$
20. $f(x) = \sqrt{x}$, at $x = 9$
21. $f(x) = 1/x$, at $x = -2$

22. (Velocity) The distance traveled by a certain bike rider as a function of time is given by $s = f(t) = t^2 + 3t$ for $0 \le t \le 4$, where $s$ is measured in miles and $t$ is measured in hours.
    (a) Find the average velocity of the rider between times $t = 1$ and $t = 2$ and also between $t = 2$ and $t = 4$.
    (b) Find the instantaneous velocity of the rider at time $t = 2$.
    (c) Sketch the function $f(t)$ and the appropriate secant and tangent lines with slopes representing the velocities of parts (a) and (b).

23. (Falling object) When an object falls freely from rest (like, for example, a rock dropped from the top of a building), if we neglect air resistance, the distance (in feet) traveled by the object after $t$ seconds is given by $s = f(t) = 16t^2$.
    (a) Find the average velocity of the object between times $t = 3$ seconds and $t = 5$ seconds.
    (b) Find the instantaneous velocity of the object after 3 seconds.
    (c) Sketch $f(t)$ and the secant and tangent lines with slopes that represent the velocities of parts (a) and (b).

24. **(Falling object)** If an object is dropped from a height of 500 feet, its distance *from the ground* after $t$ seconds (neglecting air resistance) is given by $s = f(t) = 500 - 16t^2$.
    (a) Find the average rate of change of distance with respect to time between $t = 2$ seconds and $t = 4$ seconds. Why is this rate of change negative?
    (b) Find the instantaneous rate of change of distance with respect to time when $t = 2$ seconds.
    (c) Sketch $f(t)$ and the secant and tangent lines with slopes representing the rates of change of parts (a) and (b).

25. **(Marginal cost)** A certain company has the weekly cost function $C(x) = 0.02x^2 + 1500$, where cost is measured in dollars and $x$ is the number of units produced.
    (a) Find the average rate of change of cost with respect to production between $x = 100$ and $x = 200$ units.
    (b) Find the rate at which cost is increasing when production is 100 units. This is the marginal cost when production is 100 units.
    (c) Draw the cost function $C(x)$ and the secant and tangent lines with slopes that represent the rates of change in parts (a) and (b).

**Business and Economics**

26. **(Demand)** The demand for one of the products a company manufactures is $d = f(p) = 1200 - 2p^2$, where $d$ is the number of units demanded when the price is $p$ dollars per unit.
    (a) Find the average rate of change of demand with respect to price between $p = \$15$ and $p = \$20$.
    (b) Find the (instantaneous) rate of change of demand with respect to price with $p = \$15$.
    (c) Sketch the graph of this demand function and the secant and tangent lines with slopes representing the rates of change in parts (a) and (b).

**Business and Economics**

27. **(Mortgage rates)** The average monthly mortgage rates for a period of six months at a certain bank are tabulated below.

| Month | 1 | 2 | 3 | 4 | 5 | 6 |
|---|---|---|---|---|---|---|
| Mortgage Rate (%) | 10.2 | 10.4 | 10.6 | 10.5 | 10.8 | 11.0 |

    (a) Find the average rate of change in mortgage rates at this bank between $t = 1$ month and $t = 6$ months
    (b) Find the average rate of change in mortgage rates at this bank between $t = 3$ months and $t = 4$ months

28. **(Revenue)** The weekly revenue of your company for the next eight weeks is estimated to be

$$R(t) = t^3 - 4t^2 - 4t + 50, \quad 0 \le t \le 8$$

where revenue is measured in hundreds of dollars and $t$ is time measured in weeks.
    (a) Find the average rate of change of revenue with respect to time between $t = 1$ and $t = 3$ and also between $t = 4$ and $t = 5$.
    (b) At what rate is revenue changing when $t = 2$? Is revenue increasing or decreasing? At what rate is revenue changing when $t = 5$? Is revenue increasing or decreasing?

**Social Sciences**

**29.** (Education) After $x$ hours of classroom instruction in a certain class, the average student with no previous trigonometry background can correctly answer $N(x) = x^2 + x$ questions on a standardized trigonometry test with 100 questions, where $0 \leq x \leq 8$.
  (a) Find the average rate of change of the number of questions answered correctly with respect to hours of instruction between $x = 2$ and $x = 4$.
  (b) Find the instantaneous rate of change of the number of questions answered correctly with respect to hours of instruction when $x = 4$.
  (c) Graph the function $N(x)$ and the secant and tangent lines with slopes representing the rates of change in parts (a) and (b).

**30.** (Marginal profit) Your company has profit function

$$P(x) = -x^2 + 10x + 40, \quad 0 \leq x \leq 15$$

where $P(x)$ is weekly profit in hundreds of dollars from the sale of $x$ hundred units of your product.
  (a) Find the average rate of change of profit with respect to the number of units sold between $x = 1$ and $x = 3$ and also between $x = 6$ and $x = 8$.
  (b) Determine how fast profit is increasing or decreasing (the marginal profit) when $x = 3$, $x = 5$, and $x = 7$.
  (c) At what level of sales is the rate of change of profit equal to zero?
  (d) Sketch the graph of $P(x)$ and determine the level of sales that produces the maximum profit. What is the maximum profit derived from this level of sales?

**Life Sciences**

**31.** (Liquor consumption) The approximate per capita consumption of hard liquor (in gallons of ethanol per adult) in the United States between 1981 and 1986, according to the U.S. Centers for Disease Control, is tabulated as follows.

| Year | 1981 | 1982 | 1983 | 1984 | 1985 | 1986 |
|---|---|---|---|---|---|---|
| Consumption | 1.2 | 1.1 | 0.9 | 0.7 | 0.5 | 0.4 |

Find the average rate of change of per capita consumption (a) between 1981 and 1986; (b) between 1982 and 1984.

**OPTIONAL**

**GRAPHING CALCULATOR/ COMPUTER EXERCISES**

**32.** (Profit) The monthly profit in hundreds of dollars from the sale of $x$ thousand units of your product is

$$P(x) = x^3 - 20x^2 + 40x + 40, \quad 0.5 \leq x \leq 3$$

  (a) Sketch the graph of $P(x)$ and use your graph to estimate the level of sales that produces the maximum profit. Estimate the maximum profit.
  (b) Find the derivative $P'(x)$.
  (c) Determine the level of sales $x$ for which $P'(x)$ is equal to zero. Find the profit derived from this level of sales.
  (d) Compare the results you obtained in parts (a) and (c).

**33.** (Revenue) Sketch the graph of the revenue function

$$R(t) = t^3 - 4t^2 - 4t + 50, \quad 0 \leq t \leq 8$$

which was discussed in Exercise 28. Include in your sketch the appropriate secant and tangent lines that have slopes representing the rates of change discussed in parts (a) and (b) of Exercise 28.

## 2.7 CHAPTER TWO REVIEW

**KEY TERMS**

- limit
- right-hand limit
- left-hand limit
- one-sided limit
- properties of limits
- evaluation method
- limits at infinity
- infinite limit
- tangent line
- secant line
- derivative
- four-step method
- polynomial
- rational function
- horizontal asymptote
- vertical asymptote
- ordering cost
- continuity at a point
- continuity on an interval
- three-step method
- differentiable
- difference quotient
- instantaneous rate of change
- average rate of change

**KEY FORMULAS AND RESULTS**

$\lim_{x \to a} f(x) = L$ if and only if $\lim_{x \to a^-} f(x) = L$ and $\lim_{x \to a^+} f(x) = L$

$\lim_{x \to a} c = c$

$\lim_{x \to a} x^n = a^n$

Provided $\lim_{x \to a} f(x)$ and $\lim_{x \to a} g(x)$ exist:

$\lim_{x \to a} [cf(x)] = c \lim_{x \to a} f(x)$

$\lim_{x \to a} [f(x) \pm g(x)] = \lim_{x \to a} f(x) \pm \lim_{x \to a} g(x)$

$\lim_{x \to a} [f(x) \cdot g(x)] = \lim_{x \to a} f(x) \cdot \lim_{x \to a} g(x)$

$\lim_{x \to a} \left[ \dfrac{f(x)}{g(x)} \right] = \dfrac{\lim_{x \to a} f(x)}{\lim_{x \to a} g(x)}$, if $\lim_{x \to a} g(x) \neq 0$

$\lim_{x \to a} \sqrt[n]{f(x)} = \sqrt[n]{\lim_{x \to a} f(x)}$, where $\lim_{x \to a} f(x)$ is positive if $n$ is even

$\lim_{x \to \infty} \dfrac{a}{x^p} = 0$  if $a$ is any constant and $p > 0$

$\lim_{x \to -\infty} \dfrac{a}{x^p} = 0$  if $p$ is positive and $x^p$ is defined for all $x < 0$

If $\lim_{x \to a} f(x) = N$:

If $N > 0$, then $\lim_{x \to a} \dfrac{f(x)}{g(x)} = \begin{cases} \infty & \text{if } g(x) \to 0 \text{ through positive values} \\ -\infty & \text{if } g(x) \to 0 \text{ through negative values} \end{cases}$

If $N < 0$, then $\lim_{x \to a} \dfrac{f(x)}{g(x)} = \begin{cases} -\infty & \text{if } g(x) \to 0 \text{ through positive values} \\ \infty & \text{if } g(x) \to 0 \text{ through negative values} \end{cases}$

**Continuity at a Point**

$\lim_{x \to a} f(x) = f(a)$

### Continuity on an Interval

i. A function $f$ is continuous on $(a, b)$ if it is continuous at all points in $(a, b)$.
ii. A function $f$ is continuous on $[a, b]$ if it is continuous on $(a, b)$ and one-sided limits from within are equal to functional values at the endpoints.
iii. A function $f$ is continuous on $[a, b)$ if it is continuous on $(a, b)$ and if $\lim_{x \to a^+} f(x) = f(a)$.
iv. A function $f$ is continuous on $(a, b]$ if it continuous on $(a, b)$ and if $\lim_{x \to b^-} f(x) = f(b)$.

### Slope of a Secant Line

$$m_{\text{sec}} = \frac{f(x + \Delta x) - f(x)}{\Delta x} = \frac{\Delta y}{\Delta x}$$

### Slope of a Tangent Line

$$m_{\tan} = \lim_{\Delta x \to 0} m_{\text{sec}} = \lim_{\Delta x \to 0} \frac{f(x + \Delta x) - f(x)}{\Delta x}$$

$$= \lim_{\Delta x \to 0} \frac{\Delta y}{\Delta x} = \frac{dy}{dx}$$

### Definition of the Derivative

$$f'(x) = \lim_{\Delta x \to 0} \frac{f(x + \Delta x) - f(x)}{\Delta x}$$

### Alternate Definition of the Derivative

$$f'(a) = \lim_{x \to a} \frac{f(x) - f(a)}{x - a}$$

### Differentiability

1. A function $f$ is differentiable at a point $a$ if $f'(a)$ exists.
2. A function $f$ is differentiable on the interval $(c, d)$ if $f'(x)$ exists for all $x$ in $(c, d)$.
3. If $f'(x)$ exists for all $x$ in $(-\infty, \infty)$, then $f(x)$ is differentiable everywhere.

### Relation between Differentiability and Continuity

If $f$ is differentiable at a point $a$, then $f$ is continuous at $a$. However, the converse is false. That is, if $f$ is continuous at $a$, then $f$ need not be differentiable at $a$.

## CHAPTER 2 SHORT ANSWER REVIEW QUESTIONS

1. Given that $\lim_{x \to 1^+} f(x) = 5$ and $f(1) = 5$, then (a) $f(x)$ is continuous at $x = 1$; (b) $f(x)$ is not continuous at $x = 1$; (c) $f(x)$ may or may not be continuous at $x = 1$.

2. Symbolically, $\lim_{x \to -1} (2x + 5) = 3$ indicates that as $x$ approaches $-1$, the function $2x + 5$ approaches _____.

3. (True or False) If $\lim_{x \to 3} f(x)$ does not exist, then $f(3)$ is not defined. _____

4. If $\lim_{x \to -2} f(x) = 18$, what must $f(-2)$ be equal to in order for $f$ to be continuous at $x = -2$? _____

5. When we attempt to evaluate $\lim_{x \to 0} (x^3 - 3x)/x$ by the evaluation method, we obtain 0/0. This indicates that (a) the limit does not exist; (b) the limit is equal to zero; (c) more work is required to determine whether or not this limit exists.

6. $\lim_{x \to 0} (x^3 - 3x)/x =$ _____.

7. $\lim_{x \to 0^+} \sqrt{x}$ (a) does not exist since $\sqrt{0}$ is not defined; (b) does not exist since 0 is an endpoint of the domain of $f(x) = \sqrt{x}$; (c) is equal to zero.

8. $\lim_{x \to 0^-} \sqrt{x}$ (a) does not exist since $\sqrt{0}$ is not defined; (b) does not exist since it makes no sense for $x$ to approach zero from the left for this function; (c) is equal to zero.

9. $\lim_{x \to 0} \sqrt{x}$ (a) does not exist; (b) is equal to zero; (c) is beyond the scope of this course.

10. When evaluating a certain limit by the evaluation method, we obtain 12/0. We conclude that (a) more work is required to determine whether or not this limit exists; (b) the limit is equal to $\infty$; (c) the limit is equal to $-\infty$; (d) the limit does not exist.

11. Given a cost function $C(x) = \frac{1}{500}x^2 - 2x + 1500$, the average cost function $\overline{C}(x)$ is _____.

12. The polynomial function $f(x) = x^4 - 7x^3 + x - 5$ is continuous at all real numbers or on the interval _____.

13. (True or False) The function $f(x) = (x^3 - x)/(x - 1)$ is continuous at the point $x = 1$. _____

14. If $\lim_{x \to 2} f(x) = 7$ and $f(2) = 7$, then $f(x)$ is continuous at the point $x =$ _____.

15. (True or False) If we define

$$f(x) = \begin{cases} \dfrac{x}{x} & \text{if } x \neq 0 \\ 1 & \text{if } x = 0 \end{cases}$$

then $f(x)$ is continuous at the point $x = 0$. _____

16. (True or False) The slope of the tangent line to the graph of $y = f(x)$ at the point $P(x, f(x))$ is $f'(x)$ if $f'(x)$ exists. _____

17. The slope of the secant line joining $P(x, f(x))$ and $Q(x + \Delta x, f(x + \Delta x))$ is equal to

(a) $\dfrac{dy}{dx}$  (b) $\dfrac{\Delta y}{\Delta x}$  (c) $f(x + \Delta x) - f(x)$  (d) $f'(x)$

18. If $f(x) = 5$, then $f(x + \Delta x) - f(x) =$ _____.

19. The function $f(x) = x^2$ (a) is differentiable at the point $x = 3$; (b) is differentiable on the interval $(0, 3)$; (c) is differentiable everywhere; (d) all of the above.

20. If $\dfrac{dy}{dx} = 5x + 3$, then $\dfrac{dy}{dx}\bigg|_{x=2} = $ _____.

21. The points $(1, -3)$ and $(4, -9)$ lie on the graph of $y = f(x)$. The average rate of change of $y$ with respect to $x$ between $x = 1$ and $x = 4$ is equal to _____.

22. The rate of change of distance with respect to time is usually called the _____.

23. If $y = f(x)$, then the instantaneous rate of change of $y$ with respect to $x$ at $x = a$ is given by

   (a) $\dfrac{\Delta y}{\Delta x}$   (b) $f(x + \Delta x) - f(x)$   (c) $f'(a)$   (d) $dy$

## CHAPTER 2 REVIEW EXERCISES

*Exercises marked with an asterisk (*) constitute a Chapter Test.*

*Find the following limits. If the limit does not exist and is equal to either $\infty$ or $-\infty$, be sure to specify this fact.*

1. $\lim\limits_{x \to 0} (5x^3 - 7x + 1)$

2. $\lim\limits_{x \to 1^-} (7x + x^2)$

3. $\lim\limits_{x \to 2} \dfrac{x^3 + 2x}{x + 2}$

4. $\lim\limits_{x \to 3} \dfrac{4x^2 - 3}{x + 3}$

*5. $\lim\limits_{x \to -3} f(x)$, where $f(x) = \begin{cases} 12x & \text{if } x < -3 \\ -36 & \text{if } x \geq -3 \end{cases}$

6. $\lim\limits_{x \to 4} f(x)$, where $f(x) = \begin{cases} x^2 + 2x & \text{if } x \leq 4 \\ 4x + 2 & \text{if } x > 4 \end{cases}$

7. $\lim\limits_{x \to 0} \dfrac{x^3 + 3x^2 - x}{x^2 - x}$

8. $\lim\limits_{x \to -2} \dfrac{x^3 + 2x^2}{3x^2 + 8x + 4}$

*9. $\lim\limits_{x \to 1^+} \dfrac{x^2 - 2x + 1}{x^3 - x}$

10. $\lim\limits_{x \to 1^-} \sqrt{5 - x}$

11. $\lim\limits_{x \to 2^+} \dfrac{4x + 5}{x - 2}$

12. $\lim\limits_{x \to 2^-} \dfrac{x^2 - 4x + 4}{x - 2}$

13. $\lim\limits_{x \to -4} \sqrt{x^2 + 5}$

14. $\lim\limits_{x \to -4} \sqrt{2x + 3}$

15. $\lim\limits_{x \to 5} \dfrac{x^2 - 25}{x^2 - 2x - 15}$

16. $\lim\limits_{x \to 0} \dfrac{\sqrt{x + 4} - 2}{x}$

*17. $\lim\limits_{x \to 0^+} \dfrac{\sqrt{x + 36} - 6}{x}$

18. $\lim\limits_{x \to \infty} \left(12 - \dfrac{1}{x^2}\right)$

19. $\lim\limits_{x \to -\infty} \left(\dfrac{1}{x} - \dfrac{2}{x^2} + 3\right)$

20. $\lim\limits_{x \to \infty} \dfrac{7x^2 - 4x + 5}{2x^2 + 3x - 1}$

*21. $\lim\limits_{x \to \infty} \dfrac{10 - x - x^3}{4 - 2x^2 + x^3}$

22. $\lim\limits_{x \to -\infty} \dfrac{10x^2 + 50x + 40}{2x^3 + x - 5}$

*23. $\lim\limits_{x \to -\infty} \dfrac{7x^3 + 12x^2 + 3}{x^4 - x^2}$

24. $\lim\limits_{x \to \infty} \dfrac{12x^2 - 3}{10x + 5}$

25. $\lim\limits_{x \to \infty} \dfrac{3 - 7x - x^2}{4x - 5}$

26. $\lim\limits_{x \to 4^-} \dfrac{10}{x - 4}$

27. $\lim_{x \to 2^+} \dfrac{12x^2}{2 - x}$

28. $\lim_{x \to 2^-} \dfrac{3}{(x - 2)^3}$

*29. $\lim_{x \to -1^+} \dfrac{x^2 + 3x - 4}{7x^2 + 10x + 3}$

30. $\lim_{x \to 0^-} \dfrac{4x^3 - x}{x^3 + x^2}$

31. $\lim_{x \to -3} \dfrac{2x - 4}{x^2 + 10x + 21}$

32. $\lim_{x \to -6} \dfrac{x^3 + 2x - 3}{x^2 + 4x - 12}$

33. $\lim_{x \to 5^+} \dfrac{\sqrt{x-5}}{x - 5}$

34. $\lim_{x \to -4} \dfrac{x^2 - x - 20}{(x + 3)(x + 4)^2 (x - 4)}$

*Find the horizontal and vertical asymptotes of the following functions.*

35. $f(x) = \dfrac{10x}{x^2 - 4}$

36. $f(x) = \dfrac{2x - 6}{6 - 3x}$

*37. $g(x) = \dfrac{x^2 + 3x - 5}{x^2 + 2x + 1}$

38. $g(x) = \dfrac{5 - x^3}{7 - x^2}$

39. $f(x) = \dfrac{\sqrt{2x}}{x^2 - 3x - 18}$

40. $g(x) = \dfrac{x^2 + 4x + 3}{x(x + 1)(x - 1)(x + 3)}$

*Determine whether the given function is continuous at the indicated points. Also, determine the largest interval or set of intervals on which each function is continuous. Where indicated, sketch the graph.*

41. $f(x) = \dfrac{x - 1}{x^2 - x}$ at $x = 1$: sketch

42. $g(x) = \dfrac{x - 1}{x + 1}$ at $x = -1$: sketch

*43. $f(x) = \begin{cases} \dfrac{x^2 - x - 6}{x + 2} & \text{if } x \neq -2 \\ -5 & \text{if } x = -2 \end{cases}$ at $x = -2$: sketch

44. $f(x) = \begin{cases} 10x + 1 & \text{if } x < 3 \\ 31 & \text{if } x \geq 3 \end{cases}$ at $x = 3$: sketch

45. $f(x) = \begin{cases} x & \text{if } x \leq -1 \\ 2 + x & \text{if } x > -1 \end{cases}$ at $x = -1$: sketch

46. $f(x) = \sqrt{2x + 4}$ at $x = -2$: sketch

*47. $g(x) = \sqrt{10 - 5x}$ at $x = 2$: sketch

48. $f(x) = \begin{cases} x^3 + 7x - 5 & \text{if } x < 0 \\ \sqrt{2x^2 + 12x} & \text{if } x \geq 0 \end{cases}$ at $x = 0$

49. $f(x) = \begin{cases} -1 & \text{if } x < 0 \\ 0 & \text{if } x = 0 \\ 1 & \text{if } x > 0 \end{cases}$ at $x = 0$: sketch

*In Exercises 50–54, find the slope of the secant line joining the given points P and Q and the slope of the tangent line to $y = f(x)$ at point P.*

50. $f(x) = 2x^2 - 8$, $P(1, -6)$, $Q(2, 0)$

*51. $f(x) = x^2 + 2x - 1$, $P(0, -1)$, $Q(2, 7)$

52. $f(x) = \dfrac{2}{2x + 1}$, $P(-1, -2)$, $Q\left(3, \tfrac{2}{7}\right)$

53. $f(x) = \sqrt{2x + 4}$, $P(0, 2)$, $Q(6, 4)$

**54.** $f(x) = \dfrac{3}{2 - 5x}$, $P(1, -1)$, $Q\left(4, -\tfrac{1}{6}\right)$

*In Exercises 55–61, use the definition of the derivative to find $dy/dx$.*

**55.** $y = 4x - 9$

**56.** $y = 12 - x$

**\*57.** $y = 3x^2 + x$

**58.** $y = 2x - 5x^2$

**59.** $y = 2x^3 - 2x$

**60.** $y = \dfrac{1}{1 - x}$

**61.** $y = \sqrt{x + 1}$

*In Exercises 62–66, determine whether $f(x)$ is differentiable at the indicated value of $x$.*

**62.** $f(x) = \sqrt{x^3}$ at $x = 0$

**63.** $f(x) = 2|2x - 6|$ at $x = 3$

**64.** $f(x) = \sqrt[3]{x}$ at $x = 0$

**65.** $f(x) = \begin{cases} 2x & \text{if } x < 0 \\ x^2 & \text{if } x \geq 0 \end{cases}$ at $x = 0$

**66.** $f(x) = \begin{cases} -x^3 & \text{if } x \leq 1 \\ x^3 + 1 & \text{if } x > 1 \end{cases}$ at $x = 1$

*In Exercises 67–71, find the average rate of change of $y$ with respect to $x$ between the two indicated $x$-values. Also, find the instantaneous rate of change of $y$ with respect to $x$ at the first $x$-value given.*

**\*67.** $f(x) = x^2 - 5x$ from $x = 1$ to $x = 5$

**68.** $f(x) = x^3 + 2x - 5$ from $x = 0$ to $x = 2$

**69.** $f(x) = -x^2 + 7x + 2$ from $x = 1$ to $x = 1.1$

**70.** $f(x) = \sqrt{x} + x$ from $x = 4$ to $x = 4.01$ (Round your answer to two decimal places)

**71.** $f(x) = \dfrac{4}{x + 3}$ from $x = 0$ to $x = 0.01$

**72.** (Average cost) A company's average cost function is

$$\overline{C}(x) = \dfrac{10x + 50}{x + 2}$$

where $x$ is the number of units produced and $\overline{C}(x)$ is the average cost per unit in dollars. Find $\lim\limits_{x \to \infty} \overline{C}(x)$ and interpret your result.

**\*73.** (Ordering cost) Given the annual ordering cost function $A(Q) = 20000/Q$, $Q > 0$, where $A(Q)$ is measured in dollars and $Q$ is the order quantity,
(a) determine any vertical or horizontal asymptotes of $A(Q)$;
(b) graph the function $A(Q)$.

**74.** (Salesman's salary) A salesman makes a base salary of $200 per week and 10 percent of all sales over $300.
(a) Find the function $M(x)$ that is the saleman's weekly pay in terms of dollar sales $x$.
(b) Graph the function $M(x)$.
(c) Determine if $M(x)$ is continuous at $x = 300$.

**\*75.** (Landscaper's salary) A landscaper charges $10 per hour for a daily job requiring six hours or less and $12.50 for each additional hour beyond six hours.
(a) Find the function $S(x)$ that represents the landscaper's daily salary in terms of the number of hours worked, $x$.
(b) Graph the function $S(x)$.
(c) Determine if $S(x)$ is continuous at $x = 6$.

**Business and Economics**

**76.** (Manufacturing) Your company estimates that it will require

$$H(x) = \frac{400x^2 + 30x + 50}{x^2}$$

hours to build the $x$th unit of a certain product, $x \geq 1$.
(a) Find and interpret $H(1)$, $H(10)$, $H(100)$, and $H(1000)$.
(b) Find $\lim_{x \to \infty} H(x)$ and interpret your result.

**77.** (Velocity) The distance in feet traveled by a certain object after $t$ seconds is given by

$$s = f(t) = \frac{25t^2}{10t^2 + 4}, \qquad 0 \leq t \leq 5$$

If the object travels in a straight line,
(a) find the average velocity of the object between $t = 0$ and $t = 2$;
(b) find the (instantaneous) velocity at $t = 2$.

**78.** (Profit) Given the monthly profit function

$$P(x) = -\frac{x^2}{10} + 120x - 6000$$

where $P(x)$ is the profit in dollars and $x$ is the monthly number of units produced and sold,
(a) find the slope of the tangent line to the graph of $P(x)$ at the point where $x = 200$ and where $x = 400$;
(b) find the value of $x$ that makes the slope of the tangent line equal to zero.

**Social Sciences**

**79.** (Education) After $t$ hours of study, the number of new vocabulary words that a certain student can learn is given by

$$N(t) = 45\sqrt[3]{t^2}, \qquad 0 \leq t \leq 10$$

Find the average rate at which the student is learning between $t = 0$ and $t = 2$; between $t = 8$ and $t = 10$.

**Life Sciences**

**80.** (Liquor consumption) In a certain county of Georgia, the per capita consumption of hard liquor (in gallons of ethanol per person, per year) has been decreasing according to

$$C(t) = 0.8 - 0.05t - 0.01t^2, \qquad 0 \leq t \leq 4$$

where $t = 0$ corresponds to 1987.
(a) Find the average rate at which consumption decreased between 1987 and 1990.
(b) Find the rate at which consumption was decreasing in 1989.

# 3 THE DERIVATIVE

## CHAPTER CONTENTS

3.1 Basic Differentiation Formulas
3.2 The Chain Rule
3.3 Higher Derivatives
3.4 Implicit Differentiation
3.5 Related Rates
3.6 Differentials
3.7 Chapter 3 Review

## 3.1 BASIC DIFFERENTIATION FORMULAS

**Differentiation Rules**

As we have seen in Chapter 2, the derivative of a function is a very useful concept. However, computing the derivative of a function directly from the definition can be long and tedious. Fortunately, a number of rules and formulas exist that allow us to find the derivatives of many functions with a minimum of effort. Each of these differentiation formulas can be proved by using the definition of the derivative stated in Section 2.5:

$$f'(x) = \lim_{\Delta x \to 0} \frac{f(x + \Delta x) - f(x)}{\Delta x}$$

Of the rules that follow, in this text we will prove only Rule 1. The proofs of the remaining rules can be found in more advanced calculus texts.

Rule 1: Constant Rule

> If $f(x) = c$, where $c$ is a constant, then $f'(x) = 0$.

This rule states that the derivative of any constant is equal to zero. This makes sense, since if a function is constant it does not change, so the derivative (which measures the *rate* of change of the function) must be equal to zero.

Proof

$$f'(x) = \lim_{\Delta x \to 0} \frac{f(x + \Delta x) - f(x)}{\Delta x} = \lim_{\Delta x \to 0} \frac{c - c}{\Delta x} = \lim_{\Delta x \to 0} 0 = 0 \quad \blacksquare$$

**EXAMPLE 1**

If $f(x) = 7$, then $f'(x) = 0$. To illustrate this example graphically, note that the graph of $f(x) = 7$ is a horizontal line passing through the point $(0, 7)$. The slope of the tangent line at any point along this horizontal line is equal to zero, since any tangent line has equation $y = 7$. $\quad \blacksquare$

Rule 2: Power Rule

> If $f(x) = x^n$, where $n$ is any real number, then $f'(x) = nx^{n-1}$.

This rule states that the derivative of $x$ raised to a power $n$ is equal to $n$ multiplied by $x$ raised to the power $n - 1$.

**EXAMPLE 2**

For the function $f(x) = x^3$, we have $n = 3$. Using the power rule, we find that $f'(x) = 3x^{3-1} = 3x^2$. $\quad \blacksquare$

When using the power rule, we will often have to convert functions to the exponential form $x^n$ using the rules

$$\frac{1}{x^n} = x^{-n} \quad \text{and} \quad \sqrt[q]{x^p} = x^{p/q}$$

## 3.1 BASIC DIFFERENTIATION FORMULAS

**EXAMPLE 3**

If $f(x) = 1/x^2$, then to find $f'(x)$ we write $f(x) = x^{-2}$, so that, using the power rule with $n = -2$,

$$f'(x) = -2x^{-2-1} = -2x^{-3} = -2/x^3$$

**EXAMPLE 4**

If $y = \sqrt{x^3}$, then to find $dy/dx$ we write $y = x^{3/2}$, so that, using the power rule with $n = \frac{3}{2}$,

$$\frac{dy}{dx} = \frac{3}{2}x^{3/2-1} = \frac{3}{2}x^{1/2}$$

If we write this expression in radical form we have

$$\frac{dy}{dx} = \frac{3}{2}\sqrt{x}.$$

*Note:* In the next example you should participate in finding the solution by filling in the blanks. Correct answers are given after the example.

**EXAMPLE 5 (Participative)**

Let's find the derivatives of (a) $f(x) = 1/\sqrt[3]{x^4}$ and (b) $f(x) = x$.

(a) Writing $f(x) = 1/\sqrt[3]{x^4}$ in the exponential form $x^n$, we have

$$f(x) = \frac{1}{\sqrt[3]{x^4}} = \frac{1}{x^{4/3}} = \underline{\qquad}$$

To find $f'(x)$, we use the power rule with $n = \underline{\qquad}$ to obtain

$$f'(x) = -\frac{4}{3}x \underline{\qquad} = -\frac{4}{3}x^{-7/3}$$

Rewriting the expression for $f'(x)$ with a positive exponent gives us

$$f'(x) = \underline{\qquad}$$

If we write this expression in radical form we have

$$f'(x) = \underline{\qquad}$$

(b) To find the derivative of $f(x) = x = x^1$ we use the power rule with $n = \underline{\qquad}$ to obtain

$$f'(x) = 1x^{1-1} = \underline{\qquad}$$

The next rule states that the derivative of a constant times a function is the constant times the derivative of the function.

**Rule 3: Constant Multiple Rule**

> If $f(x) = c \cdot g(x)$, where $c$ is a constant and $g(x)$ is differentiable, then
> $$f'(x) = c \cdot g'(x)$$

**SOLUTIONS TO EXAMPLE 5**

(a) $x^{-4/3}$, $-\frac{4}{3}$, $-\frac{4}{3} - 1$, $-4/(3x^{7/3})$, $-4/(3\sqrt[3]{x^7})$; (b) 1, 1

**EXAMPLE 6**  If $f(x) = 6x^3$, then to find $f'(x)$, we use the constant multiple rule with $c = 6$ and $g(x) = x^3$ to obtain

$$f'(x) = 6(3x^2) = 18x^2$$

(Note that we used the power rule to find $g'(x) = 3x^2$.) ∎

**EXAMPLE 7**  If $f(x) = 3/x^4$, then to find $f'(x)$ we first write $f(x) = 3x^{-4}$. Next we use the constant multiple rule with $c = 3$ and $g(x) = 1/x^4 = x^{-4}$ to obtain

$$f'(x) = 3(-4x^{-5}) = -12x^{-5} = \frac{-12}{x^5}$$

(Note that we used the power rule to find $g'(x) = -4x^{-5}$.) ∎

**EXAMPLE 8**  If $f(x) = 3/x^4$, then to find $f'(2)$, notice that from Example 7 we already know that $f'(x) = -12/x^5$, so that

$$f'(2) = \frac{-12}{2^5} = \frac{-12}{32} = -\frac{3}{8}$$

∎

Rule 4 states that the derivative of the sum or difference of two individually differentiable functions is the sum or difference of the derivatives of the individual functions.

Rule 4: Sum or Difference Rule

> If $f(x) = g(x) \pm h(x)$, where $g(x)$ and $h(x)$ are differentiable, then
> 
> $$f'(x) = g'(x) \pm h'(x)$$

**EXAMPLE 9**  If $f(x) = 3x^2 + 5x$, then to find $f'(x)$ we use the sum rule, with $g(x) = 3x^2$ and $h(x) = 5x$, to obtain

$$f'(x) = 3(2x) + 5(1) \quad \text{(Using the power rule and the constant multiple rule)}$$

$$= 6x + 5$$

∎

The sum or difference rule can be generalized to apply to a sum or difference of more than two functions. As an example, consider the following.

**EXAMPLE 10**  If $f(x) = (2/x) - 4x^2 + 3\sqrt{x}$, then to find $f'(x)$ we first write each term of $f(x)$ in exponential form, $x^n$. Thus,

$$f(x) = 2x^{-1} - 4x^2 + 3x^{1/2}$$

We now use the sum or difference rule extended to three functions to find that

$$f'(x) = 2(-1x^{-2}) - 4(2x) + 3\left(\frac{1}{2}x^{-1/2}\right) \quad \text{(Using the power rule and the constant multiple rule)}$$
$$= -\frac{2}{x^2} - 8x + \frac{3}{2\sqrt{x}}$$

## EXAMPLE 11 (Participative)

Let's find the derivative of $f(x) = 5x^3 - \sqrt[3]{x} + (4/x^3)$. First we write each term of $f(x)$ in exponential form $x^n$:

$$f(x) = 5x^3 - x^{1/3} + \underline{\qquad}$$

Next, we use the sum or difference rule extended to three functions to find that

$$f'(x) = 5(\underline{\qquad}) - \tfrac{1}{3}x^{-2/3} + 4(\underline{\qquad}) \quad \text{(Using the power rule and the constant multiple rule)}$$

Rewriting the expression with positive exponents, we get

$$f'(x) = 15x^2 - \tfrac{1}{3}(\underline{\qquad}) - 12(\underline{\qquad})$$

Finally, writing $x^{2/3}$ in radical form, we obtain

$$f'(x) = \underline{\qquad\qquad\qquad}$$

## EXAMPLE 12

A certain company makes a weekly profit in dollars of

$$P(x) = -1000 - 0.5x^2 + 0.1x^3$$

on the sale of $x$ items, $0 \le x \le 50$. To find the rate of change in weekly profit for 20 items, we take the derivative of $P(x)$, using Rules 1–4:

$$P'(x) = -x + 0.3x^2$$

This represents the rate of change in weekly profit for $x$ items, where $0 \le x \le 50$. Economists call $P'(x)$ the *marginal profit*. Thus for $x = 20$ items,

$$P'(20) = \frac{100 \text{ dollars}}{\text{item}}$$

Therefore when the sales level is 20 items, profit is increasing at the rate of 100 dollars per item. Economists say that the *marginal profit* is $100 per item when the sales level is 20 items.

The next rule states that the derivative of a product of two individually differentiable functions is equal to the first function times the derivative of the second function plus the derivative of the first function times the second function.

---

**SOLUTIONS TO EXAMPLE 11**

$$4x^{-3},\ 3x^2,\ -3x^{-4},\ 1/x^{2/3},\ 1/x^4,\ 15x^2 - \frac{1}{3\sqrt[3]{x^2}} - \frac{12}{x^4}$$

**Rule 5: Product Rule**

> If $f(x) = g(x) \cdot h(x)$, where $g(x)$ and $h(x)$ are differentiable, then
> $$f'(x) = g(x) \cdot h'(x) + g'(x) \cdot h(x)$$

**EXAMPLE 13**

To find the derivative of $f(x) = (4 - 3x^2)(x + 2x^3)$, we may use the product rule with $g(x) = (4 - 3x^2)$ and $h(x) = (x + 2x^3)$ to obtain

$$f'(x) = \underset{g(x)}{(4 - 3x^2)}\underset{h'(x)}{(1 + 6x^2)} + \underset{g'(x)}{(-6x)}\underset{h(x)}{(x + 2x^3)}$$

$$= 4 + 21x^2 - 18x^4 - 6x^2 - 12x^4$$
$$= 4 + 15x^2 - 30x^4$$ ∎

**EXAMPLE 14**

An alternate way to find the derivative of the function $f(x) = (4 - 3x^2)(x + 2x^3)$ of Example 13 is to expand the function first and then find the derivative. That is,

$$f(x) = 4x + 5x^3 - 6x^5$$
$$f'(x) = 4 + 15x^2 - 30x^4$$

Note that if we expand the product before we find the derivative, we do *not* use the product rule. ∎

Rule 6 states that the derivative of the quotient of two individually differentiable functions is equal to the denominator times the derivative of the numerator minus the numerator times the derivative of the denominator, all divided by the denominator squared.

**Rule 6: Quotient Rule**

> If $f(x) = g(x)/h(x)$, where $g(x)$ and $h(x)$ are both differentiable, then
> $$f'(x) = \frac{h(x)g'(x) - g(x)h'(x)}{[h(x)]^2}$$
> at all points $x$ for which $h(x) \neq 0$.

**EXAMPLE 15**

To find the derivative of $f(x) = (3x - 4)/(2x^2 + 1)$, we use the quotient rule with $g(x) = 3x - 4$ and $h(x) = 2x^2 + 1$ to obtain

$$f'(x) = \frac{\overset{h(x)}{(2x^2 + 1)}\overset{g'(x)}{(3)} - \overset{g(x)}{(3x - 4)}\overset{h'(x)}{(4x)}}{\underset{[h(x)]^2}{(2x^2 + 1)^2}}$$

$$= \frac{6x^2 + 3 - (12x^2 - 16x)}{(2x^2 + 1)^2}$$

$$= \frac{-6x^2 + 16x + 3}{(2x^2 + 1)^2}$$

∎

---

**EXAMPLE 16 (Participative)**

Let's find the derivative of

(a) $f(x) = (2x - 5)(5 - 7x)$ using the product rule, and

(b) $f(x) = \dfrac{4x}{2x - 7}$ using the quotient rule.

**Solution**

(a) For this we use the product rule with $g(x) = 2x - 5$ and $h(x) = 5 - 7x$ to obtain

$$f'(x) = (2x - 5)(\underline{\phantom{xx}}) + (\underline{\phantom{xx}})(5 - 7x)$$
$$\phantom{f'(x) = }\;\uparrow\phantom{xxxxx}\uparrow\phantom{xxxxxx}\uparrow\phantom{xxxxx}\uparrow$$
$$\phantom{f'(x) = }\;g(x)\phantom{xx}h'(x)\phantom{xxxx}g'(x)\phantom{xx}h(x)$$

Expanding these products and collecting similar terms, we get

$f'(x) = \underline{\phantom{xxxxxxxx}}$

(b) Let's use the quotient rule with $g(x) = 4x$ and $h(x) = \underline{\phantom{xxxx}}$ to obtain

$$\phantom{f'(x) = }\;h(x)\phantom{xx}g'(x)\phantom{xxxx}g(x)\phantom{xx}h'(x)$$
$$\phantom{f'(x) = }\;\downarrow\phantom{xxxxx}\downarrow\phantom{xxxxxx}\downarrow\phantom{xxxxx}\downarrow$$
$$f'(x) = \frac{(2x - 7)(\underline{\phantom{xx}}) - (4x)(\underline{\phantom{xx}})}{(2x - 7)^2}$$
$$\phantom{f'(x) = xxxxxxxxxxx}\uparrow$$
$$\phantom{f'(x) = xxxxxxxxx}[h(x)]^2$$

Expanding the numerator of $f'(x)$ and simplifying, we get

$f'(x) = \underline{\phantom{xxxxxxxx}}$

∎

---

**EXAMPLE 17**

Let $f(x) = x/(x^2 + 1)$. To find an equation of the tangent line to the graph of $f(x)$ at the point $\left(2, \tfrac{2}{5}\right)$, we first find the derivative $f'(x)$, which represents the slope of the tangent line to the graph of $f(x)$ at the point $P(x, f(x))$. Using the quotient rule, we have

$$f'(x) = \frac{(x^2 + 1)(1) - x(2x)}{(x^2 + 1)^2}$$

$$= \frac{1 - x^2}{(x^2 + 1)^2}$$

So the slope of the tangent line to the graph of $f(x) = x/(x^2 + 1)$ at $\left(2, \tfrac{2}{5}\right)$ is equal to

$$f'(2) = \frac{1 - 2^2}{(2^2 + 1)^2} = -\frac{3}{25}$$

---

**SOLUTIONS TO EXAMPLE 16**

(a) $-7, 2, 45 - 28x$; (b) $2x - 7, 4, 2, -28/(2x - 7)^2$

Since the point $\left(2, \frac{2}{5}\right)$ lies on the tangent line, we can use the point–slope form of the equation of a straight line:

$$y - y_1 = m(x - x_1)$$

$$y - \frac{2}{5} = -\frac{3}{25}(x - 2)$$

Multiplying both sides of this equation by 25 to clear the fractions, we get

$$25y - 10 = -3(x - 2)$$

Simplifying, we obtain

$$3x + 25y = 16$$

as the equation of the tangent line. ∎

**EXAMPLE 18**

Suppose that the demand $x$ (in units) for a product as a function of the price $p$ (in dollars) is given by

$$x = f(p) = \frac{7000 - 700p}{2p^2}$$

To find how fast demand is changing when the price is \$5 per unit, we compute $\left.\frac{dx}{dp}\right|_{p=5}$. First, we find the derivative $dx/dp$ by using the quotient rule to obtain

$$\frac{dx}{dp} = \frac{2p^2(-700) - (7000 - 700p)(4p)}{[2p^2]^2}$$

Simplifying, we obtain

$$\frac{dx}{dp} = \frac{-1400p^2 - (28000p - 2800p^2)}{4p^4}$$

$$= \frac{1400p^2 - 28000p}{4p^4}$$

$$= \frac{4p(350p - 7000)}{4p^4}$$

$$= \frac{350p - 7000}{p^3}, \quad \text{if } p \neq 0$$

Now substituting $p = 5$ into this formula, we get

$$\left.\frac{dx}{dp}\right|_{p=5} = \frac{350(5) - 7000}{5^3} = \frac{-42 \text{ units}}{\text{dollar}}$$

That is, when the price per unit is \$5, demand is decreasing at the rate of 42 units for every dollar increase in price. ∎

**EXAMPLE 19**

The number of unemployment claims (in thousands) filed per week in a certain county for the next several weeks is expected to be

$$U(t) = \frac{250t}{5t + 6}, \quad 1 \leq t \leq 7$$

where $t$ is time measured in weeks. Find the rate at which the number of claims is changing when $t = 5$.

Solution

Using the quotient rule to find $dU/dt$, we get

$$\frac{dU}{dt} = \frac{(5t + 6)(250) - (250t)(5)}{(5t + 6)^2} = \frac{1250t + 1500 - 1250t}{(5t + 6)^2} = \frac{1500}{(5t + 6)^2}$$

Therefore

$$\left.\frac{dU}{dt}\right|_{t=5} = \frac{1500}{[5(5) + 6]^2} \approx 1.561 \, \frac{\text{thousands of claims}}{\text{week}}$$

Thus when $t = 5$, the number of unemployment claims filed will be increasing at the rate of approximately 1561 claims per week. ∎

### Notations for the Derivative

We end this section with a word about the different notations for the derivative of a function. We have already seen that if $y = f(x)$, then two notations for the derivative of $f(x)$ are $f'(x)$ and $dy/dx$. In addition to these notations, the following notations for the derivative of $y = f(x)$ are also common:

$$y' \qquad \frac{df}{dx} \text{ or } \frac{d}{dx}f(x) \qquad D_x y \text{ or } D_x[f(x)]$$

↑ the "prime" notation

↑ the "Leibniz" or "differential" notation

↑ the "operator" notation

As a specific instance of the usage of the various notations, the result of Example 2, in which we found the derivative of $y = f(x) = x^3$, can be written

$$y' = 3x^2 \qquad \frac{df}{dx} = 3x^2 \qquad \frac{d}{dx}(x^3) = 3x^2 \qquad D_x y = 3x^2 \qquad D_x[x^3] = 3x^2$$

## SECTION 3.1
### SHORT ANSWER EXERCISES

1. Given $f(x) = -12$, then $f'(x) =$ _____.
2. If $y = x^4$ then $dy/dx =$ _____.
3. In order to find the derivative of $f(x) = 1/x^3$, we first write $f(x)$ in the exponential form $x^n$, which is _____.
4. Write the radical expression $\sqrt[3]{x^5}$ using a fractional exponent. _____
5. Using the power rule, we find that $D_x[x^5]$ is equal to
   (a) $5x^5$; (b) $5x^4$; (c) $4x^5$; (d) $4x^4$.
6. (True or False) The sum or difference rule can be used to help find the derivative of $f(x) = 4x - 7$. _____
7. If $f(x) = 7x^5$, then $f'(x) = 35x^4$ and $f'(-1) =$ _____.
8. (True or False) $\frac{d}{dx}(x^{1/2}) = \frac{1}{2}x^{1/2}$ _____
9. The derivative of a certain function is $dy/dx = 10x$. One possible function $y$ is (a) $4x^2 - 3$; (b) $10x^3$; (c) $5x^2 - 40$; (d) $\frac{1}{4}x^4 + 3$.

10. If $F(x) = u(x) \cdot v(x)$ and if $u(x)$ and $v(x)$ are differentiable, then by the product rule, $F'(x) = $ _____.

11. If $F(x) = u(x)/v(x)$ and if $u(x)$ and $v(x)$ are differentiable, then by the quotient rule, $F'(x) = $ _____.

12. If $y = f(x)$, which of the following symbols does *not* represent the derivative of $y$ with respect to $x$?

 (a) $\dfrac{df}{dx}$; (b) $D_x[f(x)]$; (c) $y'$; (d) $\dfrac{\Delta y}{\Delta x}$

## SECTION 3.1
## EXERCISES

In Exercises 1–46, differentiate the given function.

1. $f(x) = -19$
2. $g(x) = 28.6$
3. $f(x) = x^6$
4. $f(x) = x^7$
5. $g(x) = x^{-4}$
6. $g(x) = x^{-6}$
7. $f(t) = 3t^3$
8. $g(t) = 8t^9$
9. $f(t) = \frac{1}{2}t^2$
10. $g(x) = \frac{4}{5}x^5$
11. $y = 2x + 4$
12. $y = 12 - 7x$
13. $f(x) = \sqrt[3]{x}$
14. $f(x) = \sqrt[3]{x^5}$
15. $f(t) = 1/t^4$
16. $g(t) = 3/t^5$
17. $f(x) = 2/\sqrt{x}$
18. $g(x) = -4/\sqrt[3]{x}$
19. $f(x) = x^2 - 2x$
20. $y = 3x^2 + 4x$
21. $y = 2x^2 - 5x + 3$
22. $f(x) = 12x^2 + 6x - 4$
23. $f(t) = 4 - 3t - 6t^2$
24. $g(t) = 6t^3 - 7t^2$
25. $f(p) = -p^2 + 100$
26. $f(p) = -3p^2 + 400$
27. $f(x) = 3x^4 - 4x^3 + 2x^2 - 10$
28. $f(x) = 15 - x - x^2 + 3x^3 - 6x^4$
29. $y = 2x - \sqrt[3]{x^4} + \dfrac{1}{x}$
30. $y = 17x^3 + \dfrac{1}{\sqrt[4]{x^5}} - \dfrac{3}{x^4}$
31. $f(x) = \dfrac{4}{x^2} - 5\sqrt{x} + 2\sqrt[3]{x^2}$
32. $y = \sqrt[5]{x^2} - \dfrac{10}{x^5} + 5x^4$
33. $f(x) = (2x + 1)(x + 3)$
34. $f(x) = (3 - 2x)(5x + 4)$
35. $f(x) = (4x^2 + 5)(2x^2 - 3)$
36. $f(x) = (10x^2 + x)(4x^2 - 3)$
37. $y = (4 - x - x^2)(x^2 + 12)$
38. $y = (2x^2 + 3x + 1)(3x^2 - 7x + 2)$
39. $f(x) = 2/(x + 1)$
40. $g(x) = -3/(2 - x)$
41. $f(x) = 2x/(x + 2)$
42. $g(x) = (2x - 4)/(x + 3)$
43. $y = \dfrac{x^2}{x^3 + 1}$
44. $y = \dfrac{4x - x^2}{x^2 + 1}$
45. $f(x) = \dfrac{x^2 + x - 1}{x^2 + 3x + 4}$
46. $f(x) = \dfrac{2x^2 - 3x + 4}{3x^2 + x + 7}$

### Business and Economics

47. (Marginal profit) The weekly profit (in hundreds of dollars) from producing and selling $x$ hundred units of a certain product is $P(x) = 75x/(x^2 + 2)$.
 (a) Find the rate of change of profit (the marginal profit) for a sales level of 50 units.
 (b) Is it profitable to increase sales beyond 50 units?

## Business and Economics

**48.** (Demand) If the demand $x$ (in units) as a function of price $p$ (in dollars) is given by

$$x = f(p) = \frac{4000 - 400p}{p^4}, \quad 0 < p \leq 10$$

determine how fast demand is decreasing when the price is $4 per unit.

**49.** (Marginal cost) The weekly cost in dollars of producing $x$ hundred units of our product is

$$C(x) = \frac{x^2}{10x + 5} + 500$$

Determine how fast cost is increasing (the marginal cost) when the level of production is 300 units.

**50.** (Falling object) A stone is dropped from a bridge that is 300 feet high. If we neglect air resistance, the height of the stone above the ground $t$ seconds after it is released is given by $f(t) = 300 - 16t^2$. Find the velocity of the stone four seconds after it is released.

**51.** (Retention of knowledge) The percentage of original knowledge retained $t$ months after taking the final exam in a certain course is

$$P(t) = 100 - \frac{100.5t^2}{t^2 + 1}, \quad 0 \leq t \leq 14$$

Find the rate at which the percentage of knowledge is decreasing when $t = 1$ month.

**52.** (Voting) The percentage of voters who would vote for a particular candidate for mayor $t$ days after the beginning of an advertising campaign promoting the candidate is

$$P(t) = 40 + \frac{20t}{t + 80}, \quad 0 \leq t \leq 14$$

Find the rate at which the percentage of voters is changing when $t = 7$ days.

## Social Sciences

**53.** (Divorce rate) In a certain county, the percentage of marriages that end in divorce within $t$ years is given by $P(t) = 60t/(t + 5)$. Find the rate at which the percentage of divorces is changing when $t = 5$ years.

## OPTIONAL

### GRAPHING CALCULATOR/ COMPUTER EXERCISES

**54.** Graph the profit function

$$P(x) = -1000 - 0.5x^2 + 0.1x^3, \quad 0 \leq x \leq 50$$

of Example 12 and sketch the tangent line to this graph when $x = 20$. Use your graph to estimate the sales level that produces the maximum profit. Estimate the maximum profit.

**55.** Graph the demand function $x = f(p) = (7000 - 700p)/(2p^2)$ of Example 18 and sketch the tangent line to this graph when $p = 5$.

**56.** Graph the profit function $P(x) = 75x/(x^2 + 2)$ of Exercise 47 and sketch the tangent line to this graph when $x = 0.50$. Use your graph to estimate the sales level that produces the maximum profit. What is the maximum profit, approximately?

**57.** Sketch the graph of the demand function

$$x = f(p) = \frac{4000 - 400p}{p^4}, \quad 0 < p \leq 10$$

of Exercise 48. Sketch the tangent line to this graph when $p = 4$.

**58.** Sketch the function

$$P(t) = 100 - \frac{100.5t^2}{t^2 + 1}, \quad 0 \leq t \leq 14$$

of Exercise 51. Sketch the tangent line to this graph when $t = 1$.

## 3.2 THE CHAIN RULE

### Composition of Functions

Suppose that the monthly revenue from the sale of $x$ units of one of our company's products is given by

$$R(x) = \sqrt{3x^2 + x}, \quad x \geq 0$$

In order to determine the rate of change of revenue with respect to the number of units sold, we need to find $R'(x) = dR/dx$. However, none of the differentiation formulas included in the previous section applies to this situation. In addition, computing $R'(x)$ by using the definition of the derivative is quite difficult. Therefore, we would like to have a formula for differentiating such a function. Note that the function $R(x)$ can be written as the composition of functions:

$$R(x) = (f \circ g)(x) = f(g(x))$$

where $g(x) = 3x^2 + x$ (which is the function "inside" the square root) and $f(x) = \sqrt{x}$ (which is the "outside" function in $R(x) = \sqrt{3x^2 + x}$). This is true because

$$f(g(x)) = f(3x^2 + x) = \sqrt{3x^2 + x}$$

### The Chain Rule

If we had a rule that would allow us to find the derivative of a composition of functions $(f \circ g)(x)$ in terms of the derivative of the individual functions $f$ and $g$, then finding $R'(x)$ would be relatively easy. Such a rule is given here; it is called the *chain rule*.

**Rule 6: Chain Rule**

If $f(x)$ and $g(x)$ are differentiable functions and if $F(x) = f(g(x))$, then

$$F'(x) = f'(g(x)) \cdot g'(x)$$

The chain rule states that the derivative of the composition of two functions $f \circ g$ is the derivative of the first function, $f$, evaluated at the second function, $g$, multiplied by the derivative of $g$. The proof of the chain rule is beyond the scope of this book. In this rule, the composition $f(g(x))$ may be likened to a "chain" of functions linked together. The chain rule can be generalized to longer "chains" of functions, such as $f(g(h(x)))$, but we will not consider such generalizations here.

To apply the chain rule to the revenue function $R(x) = \sqrt{3x^2 + x}$, we have already noted that $R(x) = f(g(x))$, where $g(x) = 3x^2 + x$ and $f(x) = \sqrt{x} = x^{1/2}$. Also note that $f'(x) = \frac{1}{2}x^{-1/2}$ and $g'(x) = 6x + 1$. Thus,

$$R'(x) = f'(g(x))g'(x) = f'(3x^2 + x)[6x + 1]$$
$$= \frac{1}{2}(3x^2 + x)^{-1/2}[6x + 1] = \frac{6x + 1}{2\sqrt{3x^2 + x}}$$

The steps in applying the chain rule to the composition $F(x) = f(g(x))$ are as follows:

---

Step 1  Identify the "inside" function $g(x)$.
Step 2  Identify the "outside" function $f(x)$.
Step 3  Find $f'(x)$ and $g'(x)$.
Step 4  Evaluate $f'(g(x))$ by replacing every $x$ in the formula for $f'(x)$ by $g(x)$.
Step 5  Complete the task by forming the product:
$$F'(x) = f'(g(x)) \cdot g'(x)$$

---

EXAMPLE 1

Find $F'(x)$ if $F(x) = (2x + 1)^4$.

Solution

To find $F'(x)$ using the chain rule, note that $F(x) = f(g(x))$ where

$g(x) = 2x + 1$   (The inside function in $(2x + 1)^4$)
$f(x) = x^4$   (The outside function in $(2x + 1)^4$)

Then we have $f'(x) = 4x^3$ and $g'(x) = 2$, so that, by the chain rule,

$$F'(x) = f'(g(x))g'(x) = f'(2x + 1)(2)$$
$$= 4(2x + 1)^3(2) = 8(2x + 1)^3$$

This problem can also be solved by expanding $(2x + 1)^4$ to obtain

$$F(x) = 16x^4 + 32x^3 + 24x^2 + 8x + 1$$

Then

$$F'(x) = 64x^3 + 96x^2 + 48x + 8$$

Note that we had a choice of methods available in Example 1 for finding $F'(x)$. That is, we were able to find $F'(x)$ either by using the chain rule or by expanding

$F(x)$ to obtain a polynomial and then taking the derivative of the polynomial. You should note, however, that

(a) no such choice of methods is available when finding derivatives involving radical expressions such as the derivative of $R(x) = \sqrt{3x^2 + x}$ or the derivative of $F(x) = \sqrt[3]{12x - 10}$, since neither of these expressions can be simplified or "multiplied out" readily;

(b) although the derivative of $F(x) = (x^2 + 3x - 5)^{100}$ *could* be found by expanding $F(x)$ first and then differentiating the resulting polynomial, the chain rule provides a much easier way of finding this derivative.

---

### EXAMPLE 2

If $F(x) = (x^2 + 3x - 5)^{100}$, find $dF/dx$.

**Solution**

In preparing to use the chain rule, we identify the inside function as $g(x) = x^2 + 3x - 5$ and the outside function as $f(x) = x^{100}$. Then we have $F(x) = f(g(x))$. Also, $f'(x) = 100x^{99}$ and $g'(x) = 2x + 3$, so that

$$F'(x) = f'(g(x))g'(x) = f'(x^2 + 3x - 5)[2x + 3]$$
$$= 100(x^2 + 3x - 5)^{99}[2x + 3]$$
$$= 100(2x + 3)(x^2 + 3x - 5)^{99}$$

---

### EXAMPLE 3 (Participative)

Let's find the derivative of $F(x) = \sqrt[3]{2x^2 + 1}$ using the chain rule.

**Solution**

To solve this problem, note that the inside function is $g(x) = \underline{2x^2+1}$ and the outside function is $f(x) = \underline{x^{1/3}}$. Then $F(x) = f(g(x))$. Further,

$$g'(x) = \underline{4x} \quad \text{and} \quad f'(x) = \underline{\tfrac{1}{3}x^{-2/3}}$$

By the chain rule,

$$F'(x) = f'(g(x))g'(x) = f'(2x^2 + 1)g'(x) = \underline{\tfrac{1}{3}(2x^2+1)^{-2/3}(4x)}$$

Writing the expression for $F'(x)$ in radical form, we get

$$F'(x) = \underline{4x/3\sqrt[3]{(2x^2+1)^2}}$$

### Alternate Form of the Chain Rule

There is an alternate form of the chain rule, which uses the Leibniz notation. In the composition $F(x) = f(g(x))$, let $y = f(u)$ and $u = g(x)$. Then $y = f(g(x))$, and the chain rule $F'(x) = f'(g(x))g'(x)$ becomes

$$F'(x) = f'(u)\frac{du}{dx}$$

or

---

**SOLUTIONS TO EXAMPLE 3**  $2x^2 + 1$, $\sqrt[3]{x} = x^{1/3}$, $4x$, $\tfrac{1}{3}x^{-2/3}$, $\tfrac{1}{3}(2x^2 + 1)^{-2/3}(4x)$, $4x/(3\sqrt[3]{(2x^2 + 1)^2})$

Alternate Form of the Chain Rule

$$\frac{dy}{dx} = \frac{dy}{du}\frac{du}{dx}$$

Remember that in the alternate form of the chain rule, we are assuming that the composition $y = f(g(x))$ is written $y = f(u)$, where $u = g(x)$. Thus the inside function of the composition is equal to $u$ and the outside function is equal to $y$. A device for remembering the alternate form of the chain rule is that the $du$'s on the right-hand side can be "cancelled" to give $dy/dx$. Let's look at an example of the use of this alternate form.

**EXAMPLE 4**

Let $y = \sqrt[3]{x^2 + 2x + 10}$. Find $dy/dx$.

Solution

To use the alternate form of the chain rule, we let $u$ be equal to the inside function, so $u = x^2 + 2x + 10$. Also, $y = f(u) = \sqrt[3]{u} = u^{1/3}$ is the outside function, written as a function of $u$. Then:

$$\frac{dy}{dx} = \frac{dy}{du} \cdot \frac{du}{dx}$$
$$= \tfrac{1}{3}u^{-2/3} \cdot (2x + 2)$$
$$= \tfrac{1}{3}(x^2 + 2x + 10)^{-2/3}(2x + 2) \quad \text{(Replacing } u \text{ by } x^2 + 2x + 10\text{)}$$
$$= \frac{2x + 2}{3(x^2 + 2x + 10)^{2/3}}$$
$$= \frac{2x + 2}{3\sqrt[3]{(x^2 + 2x + 10)^2}}$$

**EXAMPLE 5 (Participative)**

Solution

Let's use the alternate form of the chain rule to find the derivative of $y = (x^3 - 10x)^6$.

First, we identify the inside function $u = \underline{x^3 - 10x}$. Next, we write the outside function as a function of $u$; that is, $y = f(u) = \underline{u^6}$. Therefore

$$\frac{dy}{dx} = \frac{dy}{du} \cdot \frac{du}{dx}$$
$$= 6u^5 \,(\underline{3x^2 - 10}\,)$$
$$= 6(\underline{x^3 - 10x}\,)^5(3x^2 - 10) \quad \text{(Replacing } u \text{ by } x^3 - 10x\text{)}$$
$$= (18x^2 - 60)\,\underline{(x^3 - 10x)^5}$$

**SOLUTIONS TO EXAMPLE 5**

$x^3 - 10x$, $u^6$, $3x^2 - 10$, $x^3 - 10x$, $(x^3 - 10x)^5$

## The Generalized Power Rule

Recall that the power rule of Section 3.1 stated that if $f(x) = x^n$ then $f'(x) = nx^{n-1}$ for any real number $n$. The chain rule can be used to generalize the power rule and differentiate functions of the form

$$y = F(x) = [g(x)]^n$$

Using the alternate form of the chain rule, we let

$$u = g(x) \quad \text{(the inside function)}$$

and

$$y = f(u) = u^n \quad \text{(the outside function written as a function of } u\text{)}$$

Then

$$\frac{dy}{dx} = \frac{dy}{du} \cdot \frac{du}{dx}$$
$$= nu^{n-1} \cdot g'(x)$$
$$= n[g(x)]^{n-1} \cdot g'(x) \quad \text{(Replacing } u \text{ by } g(x)\text{)}$$

We have just proven the *generalized power rule*.

**Generalized Power Rule**

> If $g(x)$ is a differentiable function and if
> $$F(x) = [g(x)]^n$$
> then
> $$F'(x) = n[g(x)]^{n-1} \cdot g'(x)$$

The generalized power rule tells us how to find the derivative of a function ($g$) raised to a power ($n$). In words, to find the derivative of a function $g$ raised to a power $n$, we multiply the power $n$ by the function $g$ raised to the power $n - 1$ and multiply this result by the derivative of $g$.  <u>WARNING</u>:  The most common error students make in attempting to apply the generalized power rule is forgetting to multiply by the factor $g'(x)$.

---

**EXAMPLE 6**  Let $F(x) = (7 - x - 2x^2)^4$. Find $F'(x)$ using the generalized power rule.

**Solution**  Note that $g(x) = 7 - x - 2x^2$ (which is the function that is raised to the power) and $n = 4$. Therefore,

$$F'(x) = n[g(x)]^{n-1} \cdot g'(x)$$
$$= 4[7 - x - 2x^2]^{4-1} \cdot \underbrace{(-1 - 4x)}_{\text{Don't forget to multiply by } g'(x)!}$$
$$= -4(1 + 4x)(7 - x - 2x^2)^3 \qquad \blacksquare$$

## EXAMPLE 7 (Participative)

**Solution**

Let's determine the derivative of $y = 1/\sqrt[4]{x^2 + 5}$ using the generalized power rule.

We must first write the given function in the exponential form $y = [g(x)]^n$ before we can apply the generalized power rule. With this in mind, we write

$$y = \frac{1}{(x^2 + 5)^{1/4}} = (\underline{x^2+5})^{-1/4}$$

We see now that $g(x) = \underline{x^2+5}$ and $n = \underline{-1/4}$, so that

$$y' = n[g(x)]^{n-1} \cdot g'(x) = -\frac{1}{4}[\underline{x^2+5}]^{(-1/4)-1} \cdot (\underline{2x})$$

Simplifying, we get

$$y' = \underline{-x/(2\sqrt[4]{(x^2+5)^5})}$$

At this point you may be a little confused about when you should use the chain rule and when you should use the generalized power rule to find the derivative of a function. The chain rule is a more general rule than the generalized power rule, since it can be used to find the derivative of *any* composition of functions $f(g(x))$, whereas the generalized power rule only applies to the specific composition $f(g(x))$ when $f(x) = x^n$. Although for the problems in this section you can use either the chain rule or the generalized power rule, we'll see instances later in the book (when we discuss exponential and logarithmic functions) where we must use the chain rule rather than the generalized power rule.

## EXAMPLE 8

**Solution**

Find the derivative of $F(x) = 2x(5 - x^2)^7$.

This problem requires a combination of the product rule and either the chain rule or the generalized power rule. If we write

$$F(x) = \underset{\underset{f(x)}{\uparrow}}{(2x)}\underset{\underset{g(x)}{\uparrow}}{(5 - x^2)^7}$$

then $f(x)$ and $g(x)$ are the factors to be used in applying the product rule. We have

$$F'(x) = \underset{\underset{f(x)}{\uparrow}}{(2x)}\underset{\underset{g'(x)}{\uparrow}}{D_x[(5 - x^2)^7]} + \underset{\underset{f'(x)}{\uparrow}}{D_x(2x)}\underset{\underset{g(x)}{\uparrow}}{(5 - x^2)^7}$$

---

**SOLUTIONS TO EXAMPLE 7**  $x^2 + 5$, $x^2 + 5$, $-\frac{1}{4}$, $x^2 + 5$, $2x$, $-x/(2\sqrt[4]{(x^2 + 5)^5})$

$$= (2x)[(7)(5 - x^2)^6(-2x)] + 2(5 - x^2)^7$$

↑<br>Applying the generalized<br>power rule

$$= -28x^2(5 - x^2)^6 + 2(5 - x^2)^7$$

The above expression is a valid one for $F'(x)$ and the *calculus* part of the problem is done. Now we'll apply a bit of *algebra* to simplify our answer:

$$F'(x) = 2(5 - x^2)^6[-14x^2 + (5 - x^2)^1] \quad \text{(Factoring out the common factor } 2(5 - x^2)^6\text{)}$$

$$= 2(5 - x^2)^6(5 - 15x^2)$$

$$= 10(5 - x^2)^6(1 - 3x^2) \quad \blacksquare$$

---

**EXAMPLE 9**  Find the derivative of $y = \sqrt[3]{\dfrac{x^2}{x^2 + 3}}$.

**Solution**  We write

$$y = \left(\frac{x^2}{x^2 + 3}\right)^{1/3}$$

and use the generalized power rule with

$$g(x) = \frac{x^2}{x^2 + 3} \quad \text{and} \quad n = \frac{1}{3}$$

to obtain

$$y' = \frac{1}{3}\left(\frac{x^2}{x^2 + 3}\right)^{-2/3} D_x\left(\frac{x^2}{x^2 + 3}\right)$$

To complete the problem, we need to find the remaining derivative using the quotient rule:

$$y' = \frac{1}{3}\left(\frac{x^2}{x^2 + 3}\right)^{-2/3} \left[\frac{(x^2 + 3)(2x) - x^2(2x)}{(x^2 + 3)^2}\right]$$

Time for some algebraic simplification: we obtain

$$y' = \frac{1}{3}\left[\frac{x^{-4/3}}{(x^2 + 3)^{-2/3}}\right]\left[\frac{6x}{(x^2 + 3)^2}\right]$$

$$= \frac{2x^{-1/3}}{(x^2 + 3)^{4/3}} = \frac{2}{x^{1/3}(x^2 + 3)^{4/3}}$$

$$= \frac{2}{\sqrt[3]{x}\sqrt[3]{(x^2 + 3)^4}} = \frac{2}{(x^2 + 3)\sqrt[3]{x^3 + 3x}} \quad \blacksquare$$

In the next example we will revisit an example considered in Section 1.3 and use the chain rule to find the derivative of a profit function with respect to price.

---

**EXAMPLE 10**  Given that our company makes a weekly profit of $P(x) = -x^2 + 850x - 40000$ dollars on the sale of $x$ units and that the weekly demand for our product is

$x = q(p) = -2p + 1000$, where $p$ is the price in dollars that we charge per unit, find $dP/dp$ using the chain rule.

**Solution**

We begin by noting that profit can be expressed as the composition of functions, $(P \circ q)(p) = P(q(p))$, so that, by the chain rule,

$$\frac{dP}{dp} = P'(p) = P'(q(p))q'(p)$$

Now, $P'(x) = -2x + 850$ and $q'(p) = -2$, so that

$$\frac{dP}{dp} = P'(-2p + 1000) \cdot (-2)$$
$$= [-2(-2p + 1000) + 850](-2)$$
$$= -8p + 2300$$

As an example, when our price is $150 per unit, our profit is increasing at the rate of

$$\left.\frac{dP}{dp}\right|_{p=150} = 1100 \text{ dollars/dollar increase in price}$$

Thus, setting our price at $150 will not maximize our profit, since the rate of change of this profit function is increasing when $p = \$150$.

We could have also solved this problem by first directly computing profit ($P$) as a function of price ($p$), as we did in Section 1.3. That is,

$$P(q(p)) = P(-2p + 1000)$$
$$= -(-2p + 1000)^2 + 850(-2p + 1000) - 40000$$
$$= -4p^2 + 2300p - 190000$$

and so $dP/dp = -8p + 2300$. A graph of $y = P(q(p))$ is drawn in Figure 3.1. Note that the maximum profit is attained when the tangent line to the graph is

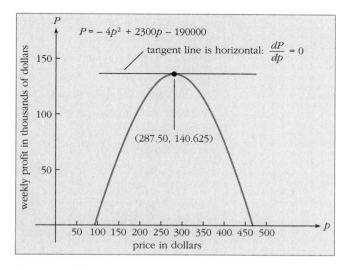

**FIGURE 3.1** The maximum profit occurs at the price level where the tangent line is horizontal.

horizontal; that is, when $dP/dp = -8p + 2300 = 0$, or when $p = \$287.50$. The maximum weekly profit is $p(287.50) = \$140,625$. ∎

**EXAMPLE 11**

One of the jobs on a certain assembly line is fastening bolts to each unit with a pneumatic wrench. Before a worker is put on the assembly line, he goes through a training program. The average newly hired employee can fasten approximately

$$B(t) = \sqrt{1.5t^2 + 25}$$

bolts per minute after $t$ hours of training, where $0 \le t \le 15$. Find the rate at which an average worker is learning after five hours of training.

**Solution**

We write $B(t) = (1.5t^2 + 25)^{1/2}$ and use the generalized power rule to compute $B'(t) = dB/dt$.

$$\frac{dB}{dt} = B'(t) = \frac{1}{2}(1.5t^2 + 25)^{-1/2}(3t)$$

$$= \frac{3t}{2\sqrt{1.5t^2 + 25}}$$

Thus

$$\left.\frac{dB}{dt}\right|_{t=5} = \frac{3(5)}{2\sqrt{1.5(5)^2 + 25}} \approx \frac{0.95 \text{ bolts}}{\text{hour of training}}$$ ∎

**SECTION 3.2**

**SHORT ANSWER EXERCISES**

1. To use the chain rule to differentiate $F(x) = (x^3 + x)^5$, we write $F(x)$ in the form $f(g(x))$, where $g(x) = $ _____ and $f(x) = $ _____.
2. For the function $F(x) = \sqrt{x^2 - 3x}$, the inside function is $g(x) = $ _____ and the outside function is $f(x) = $ _____.
3. If we apply the generalized power rule to find the derivative of $y = (2x + 1)^{10}$, we obtain **(a)** $10(2x + 1)^9$; **(b)** $20(2x + 1)^{10}$; **(c)** $20(2x + 1)^9$; **(d)** $10(2x + 1)^{10}$.
4. (True or False) The generalized power rule applies to a larger class of functions than the chain rule. _____
5. (True or False) The generalized power rule may be used to find the derivative of a function of the form $F(x) = [g(x)]^n$ but can be used only if $n$ is a positive whole number. _____
6. In using the alternate form of the chain rule to find the derivative of $y = \sqrt{4x + 5}$, we choose as the inner function $u = 4x + 5$ and as the outer function $y = f(u) = $ _____.
7. (True or False) Theoretically, we could find the derivative of $y = (2x - 3)^{250}$ by multiplying out the expression $(2x - 3)^{250}$ and then differentiating the resulting polynomial. _____
8. The chain rule allows us to differentiate the _____ of two functions $f$ and $g$.

9. In the Leibniz form (or the alternate form) of the chain rule, $y = f(u)$ and $u = g(x)$, and $dy/dx = $ _____.

10. In order to use the generalized power rule to find the derivative of $F(x) = 1/\sqrt{(3x + 1)^3}$, we must first write $F(x)$ in exponential form. This form is
(a) $(3x + 1)^{3/2}$; (b) $(3x + 1)^{2/3}$; (c) $(3x + 1)^{-2/3}$; (d) $(3x + 1)^{-3/2}$.

## SECTION 3.2
## EXERCISES

In Exercises 1–10, write the given function as a composition of functions $f(g(x))$ by identifying two appropriate functions $g(x)$ and $f(x)$. Note that there may be more than one way to accomplish this for certain exercises.

1. $F(x) = (2x - 5)^6$
2. $G(x) = (7x^2 + 3)^5$
3. $y = 1/(5 - x)^4$
4. $y = -10/(17x^2 + 4x)^7$
5. $F(x) = \sqrt{x^2 + 1}$
6. $G(x) = \sqrt[3]{3x^2 + 7x + 2}$
7. $y = \sqrt[5]{(x^2 + 4x)^3}$
8. $y = 1/\sqrt{3x^2 - x}$
9. $G(x) = -5/\sqrt[3]{(x^2 + 10)^2}$
10. $y = -12/\sqrt[5]{(10 - 2x - x^3)^4}$

11–20. Find the derivative of each of the functions given in Exercises 1–10.

Compute the derivative of each of the functions in Exercises 21–42.

21. $F(x) = (2x^2 - 5)^{15}$
22. $F(x) = (6 - 7x^2)^{10}$
23. $y = (3x^2 + x)^{-2}$
24. $y = (6 - 11x)^{-7}$
25. $F(x) = 1/(x^2 + 2x)^3$
26. $G(x) = 12/(2x - 5)^6$
27. $y = \sqrt{3x^2 + 5}$
28. $F(x) = \sqrt[3]{x^2 - 7x + 3}$
29. $G(x) = 4/\sqrt{3x^2 + 6}$
30. $y = -3/\sqrt{x^2 + 2x}$
31. $y = -2/\sqrt[3]{x^3 + 12x}$
32. $y = 3x(x^2 + 10)^6$
33. $y = -x^2(2x - 5)^{10}$
34. $F(x) = x\sqrt{x^2 + 1}$
35. $G(x) = x^2 \sqrt[3]{3x + 5}$
36. $y = 5x^3 \sqrt{10x^2 + 7x}$
37. $y = -7x^2 \sqrt[3]{x^3 + 6}$
38. $y = \left(\dfrac{2}{3 - x^2}\right)^4$
39. $y = \left(\dfrac{3}{5 - 2x}\right)^5$
40. $F(x) = \sqrt{\dfrac{2x}{2x^2 + 3}}$
41. $G(x) = \sqrt[3]{\dfrac{x + 1}{2x + 3}}$
42. $y = \sqrt[3]{\dfrac{3x^2}{2x^2 + 5}}$

43. Find $F'(1)$ in Exercises 21 and 22.
44. Find $y'(-1)$ in Exercises 30 and 31.
45. Find $y'(0)$ in Exercises 38 and 39.

## Business and Economics

46. (Marginal revenue) If the monthly revenue in hundreds of dollars from the sale of $x$ units of a certain product is
$$R(x) = \sqrt[3]{0.01x^3 + 5x}, \quad x \geq 0$$
(a) find $dR/dx$; (b) find and interpret $\left.\dfrac{dR}{dx}\right|_{x=15}$. This is the marginal revenue when $x = 15$.

**47. (Profit)** If your company makes a monthly profit of $P(x) = -x^2 + 140x - 4000$ dollars on the sale of $x$ units of its product, and if monthly demand for the product is $x = q(p) = 140 - p$, where $p$ is the price per unit in dollars,
(a) find $dP/dp$ using the chain rule;
(b) find $dP/dp$ by substituting the expression given for $x$ into the formula for profit and differentiating the resulting expression;
(c) find the rate at which profit is changing when the price per unit is $50.

**48. (Learning rate)** After $t$ days of training with a certain word-processing package, an average student can process $W(t) = \sqrt{36 + 0.1t^3}$ words per minute where $0 \le t \le 30$. Find the rate at which an average student is learning after 15 hours of training.

**49. (Compound interest)** If a principal amount of $10,000 is invested in a savings account with an annual interest rate $r\%$ compounded quarterly, the amount of money in the account after five years is

$$A(r) = 10000\left(1 + \frac{r}{400}\right)^{20}$$

(a) Find $dA/dr$.
(b) Find $\left.\dfrac{dA}{dr}\right|_{r=6}$ and $\left.\dfrac{dA}{dr}\right|_{r=8}$.

**50. (Salary)** A car salesperson named Debby is paid a base salary of $150 per week and receives an average commission of $100 on every car she sells. The number of cars $x$ that she sells depends on the number of hours per week $h$ that she spends in the showroom according to the function

$$x = \sqrt{0.1h^2 + h}, \qquad 0 \le h \le 60$$

Determine the rate of change of Debby's weekly salary when $h = 40$.

### Life Sciences

**51. (Spread of disease)** The number of persons infected by a certain contagious disease $t$ days after it is initially detected is equal to

$$N = f(t) = \sqrt[3]{10t^5 - 0.5t^3}, \qquad 0 \le t \le 42$$

Find the rate at which the disease is spreading 21 days after it is initially detected.

**52. (Population density)** The population density (in thousands of people per square mile) $x$ miles from the center of a city is given by

$$P(x) = \sqrt{10000 - 250(x + 1)^4} \qquad 0 \le x \le 1.2$$

Find the rate at which the population density is changing one-half mile from the center of the city.

### Social Sciences

**53. (Anti-smoking campaign)** The percentage of adults in a certain community who are aware of a local anti-smoking campaign $t$ days after its inception is given by

$$P(t) = 100\left(1 - \sqrt{\frac{t^2 + 1}{t^3 + 1}}\right)$$

Find the rate at which the percentage is changing four days after the inception of the campaign.

OPTIONAL

GRAPHING CALCULATOR/
COMPUTER EXERCISES

**54.** Graph the function $P(p)$ you obtained in Exercise 47(b) and sketch the tangent line to the graph when $p = \$50$.

**55.** Sketch the graph of the function $W(t)$ discussed in Exercise 48 and draw the tangent line to this graph when $t = 15$.

**56.** Sketch the graph of the function that defines the weekly salary of the car salesperson described in Exercise 50 as a function of the number of hours per week she spends in the showroom. Sketch the tangent line to this graph at the point when $h = 40$.

**57.** Sketch the graph of the function $f(t)$ described in Exercise 51. Sketch the tangent line to this graph when $t = 21$.

## 3.3 HIGHER DERIVATIVES

### The Idea of Higher Derivatives

Since the derivative of a function $y = f(x)$ is just another function, $f'(x)$, we can find the derivative of $f'(x)$ (if it exists). We call this function the *second derivative* of $f(x)$ and designate it by the symbol $f''(x)$. For example, if

$$y = f(x) = x^2 + 4x - 5$$

then

$$y' = f'(x) = 2x + 4 \quad \text{(first derivative)}$$

and

$$y'' = f''(x) = 2 \quad \text{(second derivative)}$$

We can continue with this process by finding the derivative of $f''(x)$ to obtain the third derivative of $f(x)$, denoted by $f'''(x)$; and so forth. For our example function above,

$$y''' = f'''(x) = 0 \quad \text{(third derivative)}$$
$$y'''' = f''''(x) = 0 \quad \text{(fourth derivative)}$$

etc.

For this particular example, it is clear that if we continue to find successive derivatives, all derivatives beyond the second one will be zero. The second, third, fourth, etc., derivatives are referred to as *higher derivatives* or *higher-order derivatives*. The order of the first derivative is 1, the order of the second derivative is 2, and so forth.

There are no new differentiation formulas for finding higher derivatives. We simply use the differentiation formulas we have already learned in Sections 3.1 and 3.2 and apply them repeatedly to find, in turn, $f'$, $f''$, $f'''$, etc.

### Notation for Higher Derivatives

One of the more difficult things to master about higher derivatives is the various notations. For example, the second derivative of a function $y = f(x)$ can be written in any of the following ways:

$$y'' \text{ or } f''(x) \qquad \frac{d^2y}{dx^2} \text{ or } \frac{d^2f}{dx^2} \qquad D_x^2 y \text{ or } D_x^2 f$$

$\qquad\uparrow \qquad\qquad\qquad \uparrow \qquad\qquad\qquad \uparrow$

the "prime" notation $\qquad$ the "Leibniz" or "differential" notation $\qquad$ the "operator" notation

Similar notations are used for the third, fourth, and other higher-order derivatives. When we want to *evaluate* the second derivative of $y = f(x)$ at a point $x = a$ in its domain, we write, in the various notations, respectively,

$$y''(a) \qquad f''(a) \qquad \frac{d^2y}{dx^2}\bigg|_{x=a} \qquad \frac{d^2f}{dx^2}\bigg|_{x=a} \qquad D_x^2 y\big|_{x=a} \qquad D_x^2 f\big|_{x=a}$$

In the "prime" notation, when the number of primes is larger than three, we usually replace the primes by the order of the derivative written as a superscript in parentheses. For example, the fifth derivative of $y = f(x)$ is written

$$y^{(5)} \quad \text{or} \quad f^{(5)}(x)$$

rather than $y''''' $ or $f'''''(x)$. It is important to include the parentheses around the order of the derivative, since, for example, $f^5(x)$ means $[f(x)]^5$; that is, $f(x)$ raised to the fifth power, *not* the fifth derivative of $f(x)$.

---

**EXAMPLE 1**

If $y = 2x^3 + \dfrac{1}{x}$, find $\dfrac{d^3y}{dx^3}$ and $\dfrac{d^3y}{dx^3}\bigg|_{x=1}$

**Solution**

Note that $y = 2x^3 + x^{-1}$, so that

$$\frac{dy}{dx} = 6x^2 - x^{-2} \qquad \text{(first derivative)}$$

$$\frac{d^2y}{dx^2} = 12x + 2x^{-3} \qquad \text{(second derivative)}$$

$$\frac{d^3y}{dx^3} = 12 - 6x^{-4} \qquad \text{(third derivative)}$$

Written with a positive exponent, we have

$$\frac{d^3y}{dx^3} = 12 - \frac{6}{x^4}$$

and so

$$\frac{d^3y}{dx^3}\bigg|_{x=1} = 12 - \frac{6}{(1)^4} = 6 \qquad \blacksquare$$

---

**EXAMPLE 2**

Let $f(x) = 1/(2x + 1)^2$. Find $f^{(4)}(x)$ and $f^{(4)}(0)$.

**Solution**

We begin by observing that $f(x) = (2x + 1)^{-2}$, so that, using the generalized power rule, we get

$$f'(x) = -2(2x+1)^{-3}(2) = -4(2x+1)^{-3}$$
$$f''(x) = -4(-3)(2x+1)^{-4}(2) = 24(2x+1)^{-4}$$
$$f'''(x) = 24(-4)(2x+1)^{-5}(2) = -192(2x+1)^{-5}$$
$$f^{(4)}(x) = -192(-5)(2x+1)^{-6}(2) = 1920(2x+1)^{-6}$$

Writing this fourth derivative with a positive exponent, we get

$$f^{(4)}(x) = \frac{1920}{(2x+1)^6}$$

and so

$$f^{(4)}(0) = \frac{1920}{(2 \cdot 0 + 1)^6} = 1920$$

■

## EXAMPLE 3 (Participative)

If $f(x) = 2\sqrt[3]{x} - 7x^3$, let's find $\dfrac{d^2f}{dx^2}$ and $\dfrac{d^2f}{dx^2}\bigg|_{x=1}$

### Solution

We begin by observing that $f(x)$ can be written $f(x) = 2x^{1/3} - 7x^3$, so that

$$\frac{df}{dx} = \frac{2}{3}x^{-2/3} - \underline{\hspace{1cm}}$$

$$\frac{d^2f}{dx^2} = \underline{\hspace{2cm}} \qquad \text{and} \qquad \frac{d^2f}{dx^2}\bigg|_{x=1} = \underline{\hspace{2cm}}$$

■

### Applications of the Second Derivative

Since the second derivative of a function $y = f(x)$ can be written

$$\frac{d}{dx}\left(\frac{dy}{dx}\right)$$

and since the derivative $dy/dx$ gives the (instantaneous) rate of change of $f(x)$, the second derivative gives the "rate of change of the rate of change" of $f(x)$, or the *second-order* rate of change of $f(x)$. We consider several examples of the use of the second derivative.

Suppose that on this evening's newscast you hear the announcer say, "Economists predict that during the next year, home mortgage rates will be increasing, but at a decreasing rate." This statement says something about derivatives. That is, the fact that mortgage rates will be increasing over the next year means that the rate of change of mortgage rates will be positive. In other words, if $M = f(t)$, where $M$ represents mortgage rates as a function of time $t$, the derivative of the mortgage rate function will be positive; that is, $dM/dt > 0$. Furthermore, the fact that mortgage rates will be increasing at a *decreasing* rate indicates that the rate of change of the rate of increase will be negative. In other words, the

**SOLUTIONS TO EXAMPLE 3**    $21x^2, -\frac{4}{9}x^{-5/3} - 42x, -\frac{382}{9}$

derivative of the derivative of the mortgage rate function will be negative, or $d^2M/dt^2 < 0$, since the second derivative is simply the derivative of the first derivative.

To take a specific example, suppose that mortgage rates over the next 12 months are given by

$$M = f(t) = 0.5\sqrt{t} + 8.5, \quad 0 \le t \le 12$$

where $M$ is the mortgage rate and $t$ is time measured in months, with $t = 0$ corresponding to midnight at the end of the present month. For example,

$$f(1) = 0.5\sqrt{1} + 8.5 = 9.0$$

so the mortgage rate for next month will be 9.0%. As another example,

$$f(12) = 0.5\sqrt{12} + 8.5 \approx 10.23$$

so the mortgage rate 12 months (or 1 year) from now will be 10.23%. A graph of $M = f(t)$ is sketched in Figure 3.2. From the graph it is easy to see that the

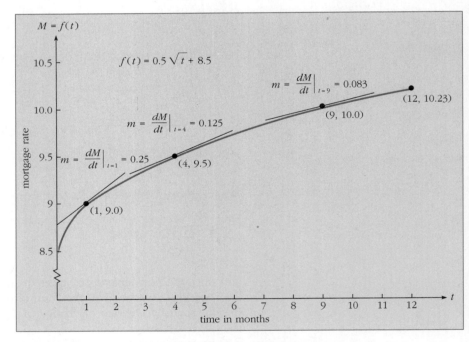

FIGURE 3.2 The slopes of the tangent lines to the graph of $f(t)$ are decreasing.

mortgage rate will be increasing over the next 12 months, since the graph of $f(t)$ is rising as we travel from left to right. If we compute

$$\frac{dM}{dt} = \frac{1}{4\sqrt{t}}, \quad 0 < t \le 12$$

then clearly $dM/dt > 0$ and, for example,

$$\left.\frac{dM}{dt}\right|_{t=1} = 0.25, \quad \left.\frac{dM}{dt}\right|_{t=4} = 0.125, \quad \left.\frac{dM}{dt}\right|_{t=9} \approx 0.083$$

In other words, $\left.\frac{dM}{dt}\right|_{t=1} = 0.25$ indicates that at the end of one month, mortgage rates will be increasing at a rate of 0.25% per month; $\left.\frac{dM}{dt}\right|_{t=4} = 0.125$ indicates that at the end of four months, mortgage rates will be increasing at a rate of 0.125% per month; $\left.\frac{dM}{dt}\right|_{t=9} \approx 0.083$ indicates that at the end of nine months, mortgage rates will be increasing at a rate of 0.083% per month. Note that the values of these derivatives (which represent the rate of change of $M$ with respect to $t$) are getting smaller, indicating that the rate of change of $dM/dt$ is negative; that is, $d^2M/dt^2 < 0$. Thus, although each of these numbers is positive (which indicates that mortgage rates are increasing), the pattern seems to be that they are getting smaller; that is, mortgage rates are increasing at a decreasing rate. To verify this, observe that the second derivative of $f(t)$ is equal to

$$\frac{d^2M}{dt^2} = -\frac{1}{8\sqrt{t^3}}, \quad 0 < t \le 12$$

which verifies that $d^2M/dt^2 < 0$.

As another example of the use of a higher derivative, suppose that distance $s$ traveled by a car in miles as a function of time $t$ in hours is given by

$$s = f(t) = 12t^2 + 7t, \quad 0 \le t \le 3$$

We have used $s$ rather than $d$ to indicate distance here since the symbol $d$ is easily confused with the $d$ that appears in the notation for the derivative. Recall that $ds/dt = v(t)$ measures the instantaneous rate of change of distance $(s)$ with respect to time $(t)$ and yields the instantaneous velocity. For this example,

$$v(t) = \frac{ds}{dt} = 24t + 7, \quad 0 \le t \le 3$$

Now, consider taking another derivative with respect to $t$. That is, consider

$$\frac{dv}{dt} = \frac{d}{dt}\left(\frac{ds}{dt}\right) = \frac{d^2s}{dt^2} = 24, \quad 0 \le t \le 3$$

We obtain the second derivative of distance $(s)$ with respect to time $(t)$. This represents the instantaneous rate of change of velocity $(v)$ with respect to time $(t)$ as given by $dv/dt$. The rate of change of velocity with respect to time is called the *acceleration*. We will denote the instantaneous acceleration by the symbol $a(t)$, so that in general we have

$$a(t) = \frac{dv}{dt} = \frac{d^2s}{dt^2}$$

where $s$ is the distance traveled by an object in $t$ time units, $v$ is the instantaneous velocity of the object, and $a(t)$ is the instantaneous acceleration of the object at time $t$. For our specific example, $a(t) = 24$, which means that at any time $t$, where $0 \le t \le 3$, the acceleration of the car is 24 miles per hour per hour, or 24 miles/hour$^2$.

### EXAMPLE 4

The population of a certain town $t$ years in the future is projected to be approximately

$$P(t) = 5000 + 4\sqrt{t^5}, \quad 0 \le t \le 10$$

(a) Determine the rate of change of the population at $t = 1$, $t = 4$, and $t = 9$.
(b) Determine if the population is increasing at an increasing rate or at a decreasing rate.

**Solution**

(a) We write

$$P(t) = 5000 + 4t^{5/2}$$

so that

$$\frac{dP}{dt} = 10t^{3/2} = 10t\sqrt{t}, \quad 0 \le t \le 10$$

Thus we have

$$\left.\frac{dP}{dt}\right|_{t=1} = 10(1)\sqrt{1} = 10 \text{ people/year}$$

$$\left.\frac{dP}{dt}\right|_{t=4} = 10(4)\sqrt{4} = 80 \text{ people/year}$$

$$\left.\frac{dP}{dt}\right|_{t=9} = 10(9)\sqrt{9} = 270 \text{ people/year}$$

(b) From the increasing pattern of the numbers representing the first derivative in part (a), we conjecture that the population is increasing at an increasing rate. To verify this, we compute the second derivative:

$$\frac{d^2P}{dt^2} = 15t^{1/2} = 15\sqrt{t}, \quad 0 \le t \le 10$$

Since $d^2P/dt^2 > 0$, the first derivative is increasing, so the population of this town is increasing at an increasing rate. ∎

### EXAMPLE 5 (Participative)

The distance $s$ in feet traveled by a certain object is given by

$$s = f(t) = \frac{t^3 + 2t}{3}, \quad 0 \le t \le 15$$

where $t$ is time measured in seconds. Let's find (a) the velocity of the object at $t = 10$ and (b) the acceleration of the object at $t = 10$.

**Solution**

(a) We begin by writing $f(t)$ in the form

$$s = f(t) = \frac{t^3}{3} + \frac{2t}{3} = \frac{1}{3}t^3 + \frac{2}{3}t, \quad 0 \le t \le 15$$

Thus,

$$v(t) = \frac{ds}{dt} = \underline{t^2 + \tfrac{2}{3}}$$

so that

$$v(10) = \left.\frac{ds}{dt}\right|_{t=10} = \underline{100\,\tfrac{2}{3}} \text{ ft/sec}$$

**(b)** Since $v(t) = t^2 + \tfrac{2}{3}$ by part (a), we find that the acceleration at any time $t$, $0 \le t \le 15$, is given by

$$a(t) = \frac{dv}{dt} = \underline{2t}$$

so that

$$a(10) = \left.\frac{dv}{dt}\right|_{t=10} = \underline{20} \text{ ft/sec}^2$$

## SECTION 3.3
**SHORT ANSWER EXERCISES**

1. If $f(x) = x^{3/2}$, then $f'(x) = \tfrac{3}{2}x^{1/2}$, so $f'(0) = $ _____. Also, $f''(x) = 3/(4\sqrt{x})$, so $f''(0)$ is _____.
2. (True or False) If $f'(a)$ exists, then $f''(a)$ exists. _____
3. If $y = x^3$, then $D_x^2 y = D_x^2(x^3) = D_x[D_x(x^3)] = D_x(3x^2) = $ _____.
4. If $y = x^3$, then $D_x^2 y|_{x=5} = $ _____.
5. Which of the following expressions is *not* a correct way to write the fourth derivative of $y = f(x)$?

   **(a)** $y^{(4)}$; **(b)** $f^4(x)$; **(c)** $\dfrac{d^4 y}{dx^4}$; **(d)** $D_x^4 y$

6. Which of the following expressions is *not* a correct way of denoting the second derivative of $s = f(t)$?

   **(a)** $s''$; **(b)** $f''(t)$; **(c)** $\dfrac{ds^2}{dt^2}$; **(d)** $D_t^2 f$

7. (True or False) Given $y = f(x)$, if we find enough successive derivatives, then eventually $f^{(n)}(x)$ will be equal to zero. _____
8. Given a formula $M = f(t)$ that gives mortgage rates on new homes for the next year, the statement, "Mortgage rates are declining at an increasing rate" can be written in derivative notation as follows.

   **(a)** $\dfrac{dM}{dt} < 0$ and $\dfrac{d^2 M}{dt^2} < 0$      **(b)** $\dfrac{dM}{dt} < 0$ and $\dfrac{d^2 M}{dt^2} > 0$

   **(c)** $\dfrac{dM}{dt} > 0$ and $\dfrac{d^2 M}{dt^2} > 0$      **(d)** $\dfrac{dM}{dt} > 0$ and $\dfrac{d^2 M}{dt^2} < 0$

9. The order of the derivative $f'''(x)$ is equal to _____.
10. Acceleration is the first derivative of _____ or the second derivative of _____ with respect to time.

---

**SOLUTIONS TO EXAMPLE 5**      (a) $t^2 + \tfrac{2}{3}$, $100\tfrac{2}{3}$; (b) $2t$, 20

## SECTION 3.3
**EXERCISES**

In Exercises 1–5, find $f''(x)$ and $f''(2)$.

1. $f(x) = x^3 - 4x^2$
2. $f(x) = (1/x) + \sqrt{x}$
3. $f(x) = \sqrt[3]{x} + 5x^2$
4. $f(x) = 3/(x+1)^2$
5. $f(x) = x/(2x - 1)$

In Exercises 6–10, find $\dfrac{d^3y}{dx^3}$ and $\dfrac{d^3y}{dx^3}\bigg|_{x=-1}$.

6. $y = 5x^3$
7. $y = 2/x$
8. $y = \sqrt[3]{x}$
9. $y = -1/(x-1)^3$
10. $y = 10 - x - 4x^2$

In Exercises 11–15, find $D_x^2 y$ and $D_x^2 y\big|_{x=4}$.

11. $y = 10 + 3x + 2x^3$
12. $y = \sqrt{x}$
13. $y = 2(x - 5)^4$
14. $y = (2x + 1)^3 + (x + 1)^2 + 6$
15. $y = -2/(3x + 2)^3$

In Exercises 16–20, find $f^{(4)}(x)$.

16. $f(x) = 100x + 6$
17. $f(x) = 3/x^2$
18. $f(x) = (4x + 3)^3$
19. $f(x) = 3/\sqrt{x}$
20. $f(x) = (x^4 + 4)/2$

In Exercises 21–24, let $s = f(t)$ represent the distance in feet traveled by an object as a function of time $t$ in seconds. Find the instantaneous velocity at $t = 2$ seconds and the instantaneous acceleration at $t = 2$ seconds.

21. $s = f(t) = 5t^2 - t, \quad 0 \le t \le 10$
22. $s = f(t) = t^3 + 0.5t^2 - t, \quad 0 \le t \le 4$
23. $s = f(t) = 16t^2$ (the distance traveled by a freely falling object)
24. $s = f(t) = 5\sqrt{t} + t^2, \quad 0 \le t \le 7$

25. (Falling object) An object is thrown upward with a velocity of 200 feet per second. The height of the object above the ground after $t$ seconds is given by the formula $s = f(t) = 200t - 16t^2$.
    (a) Find the velocity and the acceleration after five seconds.
    (b) What is the maximum height attained by the object?
    (c) When will this object strike the ground?
    (d) Sketch the graph of $s = f(t)$.

**Life Sciences**

26. (Population) The population of a certain town during the next ten years is projected to be
    $$P(t) = 25000 + 5t^3 + 6t\sqrt{t}, \quad 0 \le t \le 10$$
    (a) What is the current population of this town?
    (b) Find the rate of change of the population when $t = 1$, $t = 4$, and $t = 9$.
    (c) Determine whether the population is increasing at an increasing rate or at a decreasing rate.

27. (Mortgage rates) Mortgage rates for the next six months are projected to be
    $$M = f(t) = 8.0 + \tfrac{1}{8}t\sqrt{t}, \quad 0 \le t \le 6$$
    (a) Find and interpret $f'(1), f'(3),$ and $f'(5)$.

(b) Find $f''(t)$.
(c) Will mortgage rates increase at an increasing rate or at a decreasing rate over the next six months?

**28.** (Profit)  The daily profit in dollars of a certain company in terms of the number $x$ of units sold is

$$P = f(x) = 10x\sqrt{x} - 0.1x - 100$$

(a) Find and interpret $f'(10)$, $f'(100)$, and $f'(1000)$.
(b) Find $f''(x)$.
(c) Is profit increasing at an increasing rate or at a decreasing rate?

**29.** (Standardized testing)  The number of grammar questions answered correctly by an average student on a certain standardized test after $t$ hours of instruction is given by

$$N(t) = 15 + 5\sqrt{t} + 3\sqrt[3]{t^2}, \qquad 0 \le t \le 25$$

(a) Find and interpret $\left.\dfrac{dN}{dt}\right|_{t=1}$ and $\left.\dfrac{dN}{dt}\right|_{t=16}$.
(b) Find $d^2N/dt^2$.
(c) Is learning increasing at an increasing rate or at a decreasing rate?

**30.** Determine which of the functions graphed in Figure 3.3 is increasing at an increasing rate, which is increasing at a decreasing rate, which is decreasing at an increasing rate, and which is decreasing at a decreasing rate.

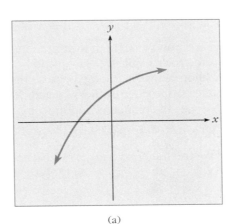
(a)

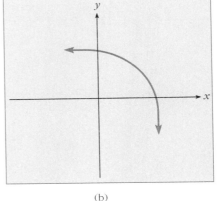
(b)

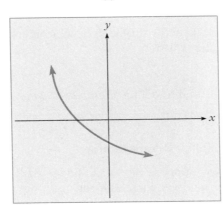
(c)

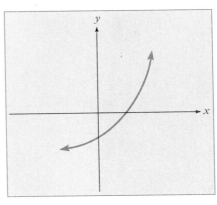
(d)

**FIGURE 3.3**

## OPTIONAL

### GRAPHING CALCULATOR/ COMPUTER EXERCISES

31. Graph the function $s = f(t) = t^3 + 0.5t^2 - t$, $0 \le t \le 4$, of Exercise 22.
32. Graph the function $s = f(t) = 5\sqrt{t} + t^2$, $0 \le t \le 7$, of Exercise 24.
33. Graph the function $P(t) = 25000 + 5t^3 + 6t\sqrt{t}$ of Exercise 26.
34. Graph the function $M = f(t) = 8.0 + \frac{1}{8}t\sqrt{t}$ of Exercise 27. From the shape of this graph, what can you say about the sign of $d^2M/dt^2$?
35. Graph the profit function $P = f(x) = 10x\sqrt{x} - 0.1x - 100$ of Exercise 28. From the shape of your graph, what can you say about the sign of $d^2P/dx^2$?
36. Graph the function $N(t) = 15 + 5\sqrt{t} + 3\sqrt[3]{t^2}$, $0 \le t \le 25$, of Exercise 29. Inspecting your graph, is $dN/dt$ positive or is it negative? Is $d^2N/dt^2$ positive or negative?

## 3.4 IMPLICIT DIFFERENTIATION

### Implicit Relationships between Variables

Suppose that your company sells both computers and modems. Computers and modems are complementary goods, since they are used together. Suppose that the relationship between monthly sales of computers ($x$) and modems ($y$) is given (approximately) by

$$30y = 5x^3 - 18yx^2, \quad 0 \le x \le 30$$

where $x$ and $y$ are both measured in hundreds of units. This relationship is not solved for $y$ explicitly as a function of $x$. We call such a relationship an *implicit* relation between $x$ and $y$ because the relationship between $x$ and $y$ is implied rather than being explicitly stated. Another way to write the relation between the sales of computers and modems is

$$30y - 5x^3 + 18yx^2 = 0, \quad 0 \le x \le 30$$

If we use the notation $F(x, y)$ to denote a function of *two* variables, then, letting $F(x, y) = 30y - 5x^3 + 18yx^2$, we can state the implicit relationship between $x$ and $y$ as

$$F(x, y) = 0$$

---

In general, if $G(x, y)$ is an arbitrary function of two variables $x$ and $y$, then the equation

$$G(x, y) = 0$$

is said to define an *implicit relationship* between $x$ and $y$.

---

### Implicit Differentiation

Now suppose we want to find $dy/dx$, which is the rate at which sales of modems change with increasing sales of computers. Since the relationship is not explicitly solved for $y$ as a function of $x$, we have a small problem. This example is a bit difficult, so we will defer its solution until later in this section. First, let's consider a relatively easier problem that involves finding $dy/dx$ when the relationship between $x$ and $y$ is implicit.

**EXAMPLE 1**

Let $3y = 2x + 7$. Find $dy/dx$ using two different methods.

**Solution**

**Method 1** Solve the equation explicitly for $y$ as a function of $x$ first and then find $dy/dx$.

$$3y = 2x + 7 \quad \text{(Implicit relationship between } x \text{ and } y\text{)}$$

$$y = \frac{2}{3}x + \frac{7}{3} \quad (y \text{ as an explicit function of } x)$$

$$\frac{dy}{dx} = \frac{2}{3}$$

**Method 2** Differentiate both sides of the given implicit relationship first with respect to $x$ and then solve the resulting equation for $dy/dx$.

$$\frac{d}{dx}(3y) = \frac{d}{dx}(2x + 7) \quad \text{(Differentiating the implicit relationship)}$$

$$3\frac{dy}{dx} = 2$$

$$\frac{dy}{dx} = \frac{2}{3} \quad \left(\text{Solving for } \frac{dy}{dx}\right) \quad \blacksquare$$

Method 2 is called *the method of implicit differentiation*, since we begin by differentiating the implicit relationship between $x$ and $y$.

In Example 1, there does not seem to be any advantage in using one of the methods rather than the other. In this particular problem, both methods require about the same amount of work. However, it will soon become clear to you that method 1 is not always convenient (or even possible) to use. In most cases, when you are required to find $dy/dx$ from an implicit relationship of the form $F(x, y) = 0$, method 2 (that is, implicit differentiation) is the method of choice.

The notation $\left.\frac{dy}{dx}\right|_{\substack{x=a \\ y=b}}$, which is used in the next example, means that we are to evaluate the derivative $dy/dx$ by replacing each $x$ in the formula for $dy/dx$ by $a$ and each $y$ by $b$.

**EXAMPLE 2**

If $y^2 + 3y = x$, find $\left.\frac{dy}{dx}\right|_{\substack{x=4 \\ y=1}}$ by two different methods.

**Solution**

**Method 1** In this case it is possible to find $y$ explicitly as a function of $x$ if we first write the equation as $y^2 + 3y - x = 0$. This equation is just a quadratic equation in $y$, so that we may use the quadratic formula (with $a = 1$, $b = 3$, and $c = -x$) to obtain

$$y = \frac{-3 \pm \sqrt{9 - 4(1)(-x)}}{2(1)} = \frac{-3 \pm \sqrt{9 + 4x}}{2}$$

$$= -\tfrac{3}{2} \pm \tfrac{1}{2}(9 + 4x)^{1/2}$$

Note that in solving for $y$ explicitly we do not get a function of $x$. For example, if $x = 0$, then

$$y = -\tfrac{3}{2} \pm \tfrac{1}{2}(9)^{1/2} = -\tfrac{3}{2} \pm \tfrac{3}{2} = 0 \text{ or } -3$$

So for this value of $x$, there are *two* corresponding values of $y$. This shows that $y$ is not a function of $x$. A graph of the relation between $x$ and $y$ is sketched in Figure 3.4. By inspecting this figure, it is clear that this relationship graphs as a parabola and can be viewed as comprised of *two* functions:

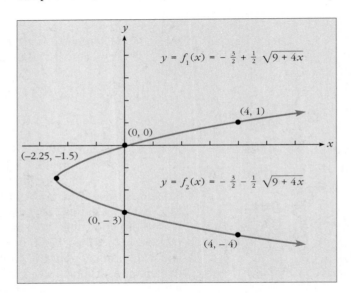

**FIGURE 3.4** Two implicit functions $f_1$ and $f_2$ defined by $y^2 + 3y = x$.

$$y = f_1(x) = -\tfrac{3}{2} + \tfrac{1}{2}\sqrt{9 + 4x} \quad \text{(The *upper branch* of the parabola)}$$

and

$$y = f_2(x) = -\tfrac{3}{2} - \tfrac{1}{2}\sqrt{9 + 4x} \quad \text{(The *lower branch* of the parabola)}$$

Since we are required to find $\left.\dfrac{dy}{dx}\right|_{\substack{x=4 \\ y=1}}$ and since the point $(4, 1)$ lies on the graph of $y = f_1(x) = -\tfrac{3}{2} + \tfrac{1}{2}(9 + 4x)^{1/2}$ (see Figure 3.4), we will use this function and differentiate with respect to $x$ to obtain

$$\frac{dy}{dx} = \tfrac{1}{4}(9 + 4x)^{-1/2}(4) = \frac{1}{\sqrt{9 + 4x}}$$

so that

$$\left.\frac{dy}{dx}\right|_{\substack{x=4\\y=1}} = \frac{1}{\sqrt{9 + 4(4)}} = \frac{1}{5}$$

Note that if we had been required to find, for example, $\left.\frac{dy}{dx}\right|_{\substack{x=4\\y=-4}}$, we would have had to differentiate the formula for $f_2(x)$ and substitute $x = 4, y = -4$. This is because the point $(4, -4)$ lies on the lower branch of the parabola and the corresponding function is $y = f_2(x) = -\tfrac{3}{2} - \tfrac{1}{2}(9 + 4x)^{1/2}$.

**Method 2** Another (and perhaps simpler) way to find $\left.\frac{dy}{dx}\right|_{\substack{x=4\\y=1}}$ is to use the method of implicit differentiation. We begin by differentiating both sides of the original implicit relationship with respect to $x$:

$$\frac{d}{dx}(y^2 + 3y) = \frac{d}{dx}(x)$$

$$\frac{d}{dx}y^2 + 3\frac{dy}{dx} = 1 \qquad (3\text{-}1)$$

Now, the whole idea of implicit differentiation is to be able to find $dy/dx$ without actually solving the given equation for $y$. It suffices to assume that the given implicit relationship *can* be solved for $y$ to yield one or more differentiable functions of $x$. Thus, in order to find $\frac{d}{dx}y^2$, we assume that we can solve the original equation for $y$ to obtain $y = f(x)$. However, note that *we don't actually solve for y!* Then, using the generalized power rule, we get

$$\frac{d}{dx}(y^2) = \frac{d}{dx}[f(x)]^2 = 2[f(x)]^1 f'(x) = 2y\frac{dy}{dx}$$

Substituting this back into Equation (3-1), we get

$$2y\frac{dy}{dx} + 3\frac{dy}{dx} = 1$$

$$(2y + 3)\frac{dy}{dx} = 1 \qquad \text{(Factoring)}$$

$$\frac{dy}{dx} = \frac{1}{2y + 3} \qquad \left(\text{Solving for } \frac{dy}{dx}\right)$$

We conclude that

$$\left.\frac{dy}{dx}\right|_{\substack{x=4\\y=1}} = \frac{1}{2(1) + 3} = \frac{1}{5}$$

■

From our experience in Example 2, we begin to see the usefulness of the implicit differentiation method. Indeed, if the original implicit relationship in Example 2 had been, say, $y^4 + 3y = x$, it would have been considerably more difficult to solve for $y$. Implicit differentiation is certainly the most reasonable way to find $dy/dx$ in such an instance.

We are now in a position to solve the problem posed at the beginning of this section. Recall that the implicit relationship between $x$ (monthly sales of computers) and $y$ (monthly sales of modems) is given by

$$30y = 5x^3 - 18yx^2, \quad 0 \leq x \leq 30$$

where $x$ and $y$ are measured in hundreds of units. Although this equation can be solved for $y$ in terms of $x$, we will use the method of implicit differentiation to find $dy/dx$. We begin by differentiating both sides of the equation with respect to $x$:

$$\frac{d}{dx}(30y) = \frac{d}{dx}(5x^3 - 18yx^2)$$

$$30\frac{dy}{dx} = \frac{d}{dx}(5x^3) - 18\frac{d}{dx}(yx^2) \qquad (3\text{-}2)$$

Now, $\frac{d}{dx}(5x^3) = 15x^2$. To find $\frac{d}{dx}(yx^2)$, however, we must assume that we have solved the original implicit relationship for $y$ to obtain $y = f(x)$. Then

$$yx^2 = [f(x)]x^2 = x^2 \cdot f(x)$$

So to find $\frac{d}{dx}(yx^2) = \frac{d}{dx}(x^2y)$, we use the product rule.

$$\frac{d}{dx}(yx^2) = \frac{d}{dx}(x^2 \cdot f(x)) = x^2 f'(x) + 2xf(x) = x^2 \frac{dy}{dx} + 2xy$$

Substituting these results into Equation (3-2), we have

$$30\frac{dy}{dx} = 15x^2 - 18\left(x^2 \frac{dy}{dx} + 2xy\right)$$

Our next task is to solve this equation for $dy/dx$.

$$30\frac{dy}{dx} = 15x^2 - 18x^2 \frac{dy}{dx} - 36xy$$

$$18x^2 \frac{dy}{dx} + 30\frac{dy}{dx} = 15x^2 - 36xy$$

$$(18x^2 + 30)\frac{dy}{dx} = 15x^2 - 36xy \qquad \text{(Factoring)}$$

$$\frac{dy}{dx} = \frac{15x^2 - 36xy}{18x^2 + 30} = \frac{5x^2 - 12xy}{6x^2 + 10}$$

Note that our expression for $dy/dx$ contains both the variable $x$ and the variable $y$. This will be the case in many implicit differentiation problems when the original implicit relationship involves $x$ and $y$, and it represents no real hardship. To find $dy/dx$ at a particular point, we simply substitute *both* the $x$ and $y$ values at that point into our expression for $dy/dx$. For example, to find the rate at which sales of modems are changing when $x = 15$ hundred (1500) computers are sold per month, we must find the value of $y$ when $x = 15$. Using the original implicit relationship, we find that when $x = 15$,

$$30y = 5(15)^3 - 18y(15)^2 = 16875 - 4050y$$
$$4080y = 16875$$
$$y = 4.14 \quad \text{(approximately)}$$

So if 1500 computers are sold during the month, about 4.14 hundred or 414 modems will be sold during that month. Now, to find the required rate of change, we evaluate $dy/dx$ when $x = 15$ and $y = 4.14$ to obtain

$$\left.\frac{dy}{dx}\right|_{\substack{x=15 \\ y=4.14}} = \frac{5(15)^2 - 12(15)(4.14)}{6(15)^2 + 10} \approx 0.279$$

That is, when computer sales are 15 hundred, sales of modems are increasing at the rate of 0.279 hundred modems/hundred computers, or about 28 modems per 100 computers.

## Steps for Finding $\frac{dy}{dx}$ by Implicit Differentiation

We can now summarize the steps for finding $dy/dx$ from an implicit relationship between $x$ and $y$.

Step 1  Differentiate both sides of the given equation with respect to $x$. Remember to treat $y$ as a function of $x$. This means, for example, that $\frac{d}{dx}(y^3) = 3y^2 \, dy/dx$, and

$$\frac{d}{dx}(x^2 y^4) = x^2\left(4y^3 \frac{dy}{dx}\right) + 2xy^4$$

Step 2  Rearrange the equation derived in Step 1 so that all terms involving $dy/dx$ are on one side of the equation and all other terms are on the other side of the equation.

Step 3  Factor $dy/dx$ out as a common factor from the $dy/dx$ side of the equation.

Step 4  Solve the equation for $dy/dx$.

Step 5  If it is required to evaluate $dy/dx$ at some particular value of $x$, say, $x = a$, and if the expression for $dy/dx$ found in Step 4 involves both $x$ and $y$, then first find the value of $y$ that corresponds to $x = a$ by substituting $x = a$ into the *original* implicit relation and then solving the resulting equation for $y$. If this corresponding $y$-value is, say, $y = b$, then complete the evaluation of $dy/dx$ by substituting $x = a$ and $y = b$ into the expression for $dy/dx$.

**EXAMPLE 3**  Find $\left.\frac{dy}{dx}\right|_{x=1}$ if $x - \frac{y}{2x} = 3y$.

**Solution**

Step 1 We differentiate both sides:

$$\frac{d}{dx}\left(x - \frac{y}{2x}\right) = \frac{d}{dx}(3y)$$

$$\frac{d}{dx}(x) - \frac{d}{dx}\left(\frac{y}{2x}\right) = 3\frac{dy}{dx}$$

$$1 - \frac{2x\left(\frac{dy}{dx}\right) - y(2)}{(2x)^2} = 3\frac{dy}{dx} \quad \text{(Quotient rule)}$$

$$1 - \frac{2x\frac{dy}{dx} - 2y}{4x^2} = 3\frac{dy}{dx}$$

Step 2 To rearrange the equation, we begin by multiplying both sides by $4x^2$ to clear it of fractions:

$$4x^2 - \left(2x\frac{dy}{dx} - 2y\right) = 12x^2\frac{dy}{dx}$$

$$4x^2 - 2x\frac{dy}{dx} + 2y = 12x^2\frac{dy}{dx}$$

$$4x^2 + 2y = 12x^2\frac{dy}{dx} + 2x\frac{dy}{dx}$$

Step 3 We factor $dy/dx$ out as a common factor:

$$4x^2 + 2y = (12x^2 + 2x)\frac{dy}{dx}$$

Step 4 We solve for $dy/dx$:

$$\frac{dy}{dx} = \frac{4x^2 + 2y}{12x^2 + 2x} = \frac{2x^2 + y}{6x^2 + x}$$

Step 5 Here we are required to evaluate $dy/dx$ when $x = 1$. To find the corresponding value of $y$, we substitute $x = 1$ into the original implicit relation:

$$1 - \frac{y}{2} = 3y$$

Solving this equation for $y$ yields $y = \frac{2}{7}$. Finally, we compute

$$\left.\frac{dy}{dx}\right|_{\substack{x=1 \\ y=2/7}} = \frac{2(1)^2 + \frac{2}{7}}{6(1)^2 + 1} = \frac{16}{49}$$

---

**EXAMPLE 4 (Participative)**

Let's find an equation of the tangent line to the graph of $2x^3y + y^2 = 3$ at the point $(1, 1)$.

**Solution**

We will first find $dy/dx$ and then evaluate when $x = 1$ and $y = 1$ to obtain the slope of the tangent line to the graph at $(1, 1)$.

Step 1
$$\frac{d}{dx}(2x^3y + y^2) = \frac{d}{dx}(3)$$

$$2\frac{d}{dx}(x^3y) + \frac{d}{dx}(y^2) = \underline{0}$$

$$2\left(x^3\frac{dy}{dx} + 3x^2y\right) + \underline{2y\frac{dy}{dx}} = 0$$

↑ product rule   ↑ generalized power rule

$$2x^3\frac{dy}{dx} + 6x^2y + 2y\frac{dy}{dx} = 0$$

Step 2  $2x^3\frac{dy}{dx} + 2y\frac{dy}{dx} = \underline{-6x^2y}$

Step 3  $(\underline{2x^3 + 2y})\frac{dy}{dx} = -6x^2y$

Step 4  $\frac{dy}{dx} = \underline{\dfrac{-3x^2y}{x^3 + y}}$

Step 5  $\left.\dfrac{dy}{dx}\right|_{\substack{x=1 \\ y=1}} = \underline{-\dfrac{3}{2}}$

We conclude that the slope of the tangent line to the graph of $2x^3y + y^2 = 3$ at the point (1, 1) is equal to ____. To find an equation of the tangent line, use the point–slope form of the equation of a straight line:

$$y - y_1 = m(x - x_1)$$

$$y - 1 = \underline{\hspace{3cm}}$$

Rearranging this equation into the general form of a straight line gives us

_____

■

---

**EXAMPLE 5**

Suppose that the weekly demand $x$ for our product, if we offer it at price $p$, is given by

$$10x = 20000 - 10p - p^2$$

where $x$ is measured in units and $p$ in dollars.

(a) Find the number of units of our product that can be sold per week if we set the price per unit at $50.

(b) Find $\left.\dfrac{dp}{dx}\right|_{x=1700}$

---

**SOLUTIONS TO EXAMPLE 4**

$0$, $2y\dfrac{dy}{dx}$, $-6x^2y$, $2x^3 + 2y$, $\dfrac{-3x^2y}{x^3 + y}$, $-\dfrac{3}{2}$, $-\dfrac{3}{2}$, $-\dfrac{3}{2}(x - 1)$, $3x + 2y = 5$

**Solution**

(a) Substituting $p = \$50$ into the demand equation gives

$$10x = 20000 - 10(50) - (50)^2 = 17000$$
$$x = 1700$$

Thus, if we set our price at $50 per unit, weekly demand will be 1700 units of the product.

(b) To find $dp/dx$, we use implicit differentiation. In this problem, the role of $y$ is played by $p$.

Step 1  $\dfrac{d}{dx}(10x) = \dfrac{d}{dx}(20000 - 10p - p^2)$

Step 2  $10 = -10\dfrac{dp}{dx} - 2p\dfrac{dp}{dx}$

Step 3  $10 = (-10 - 2p)\dfrac{dp}{dx}$

Step 4  $\dfrac{dp}{dx} = \dfrac{10}{-10 - 2p} = \dfrac{-5}{p + 5}$

Step 5  We already know from part (a) that if $x = 1700$ then the corresponding value of $p$ is $p = 50$, so that

$$\left.\dfrac{dp}{dx}\right|_{\substack{x=1700 \\ p=50}} = \dfrac{-5}{50 + 5} = -\dfrac{1}{11} \approx -0.091$$

Thus when 1700 units are demanded weekly, the price will decrease at the rate of about 9.1 cents per unit of additional demand. ∎

## SECTION 3.4
### SHORT ANSWER EXERCISES

1. (True or False) The method of implicit differentiation requires us to solve the implicit relationship $G(x, y) = 0$ for $y$ in terms of $x$. _____

2. (True or False) The equation $y = 7x^2 + 3x - 5$ defines $y$ explicitly as a function of $x$. _____

3. (True or False) If $x + 2y = 7$, then differentiating both sides with respect to $x$ yields $1 + 2\dfrac{dy}{dx} = 7$. _____

4. If $y$ is a function of $x$, then $\dfrac{d}{dx}(5y^3) = $ _____.

5. If $y$ is a function of $x$, then $\dfrac{d}{dx}(7xy) = $ _____.

6. If $\dfrac{dy}{dx} = -\dfrac{1}{\sqrt{9 + 4x}}$, then $\left.\dfrac{dy}{dx}\right|_{\substack{x=4 \\ y=-4}} = $ _____.

7. If $\dfrac{dy}{dx} = \dfrac{-3x^2y}{x^3 + y}$, then $\left.\dfrac{dy}{dx}\right|_{\substack{x=1 \\ y=-3}} = $ _____.

8. The method of implicit differentiation (a) must always be used to find $dy/dx$ if $F(x, y) = 0$; (b) is always the way to find $dy/dx$ if $y = f(x)$; (c) can be used to find $dy/dx$ if $F(x, y) = 0$; (d) all of the above.

9. If $F(x, y) = 0$, then (a) we can always solve for $y$ to find $y = f(x)$; (b) $y$ may not be a function of $x$; (c) implicit differentiation may not be used to find $dy/dx$; (d) if $x = 0$ and $y = 0$, then $F = 0$.

10. If $y$ is a function of $x$ then $\dfrac{d}{dx}(y^5 + xy)$ is equal to

(a) $5y^4 \dfrac{dy}{dx} + \dfrac{dy}{dx}$ (b) $5y^4 \dfrac{dy}{dx} + xy$ (c) $5y^4 \dfrac{dy}{dx} + x \dfrac{dy}{dx} + y$

(d) $5y^5 \dfrac{dy}{dx} + x \dfrac{dy}{dx} + y$

## SECTION 3.4
## EXERCISES

In Exercises 1–7, find $dy/dx$ using two different methods.

1. $7y = 14x - 21$
2. $3x - 81 = -6y$
3. $3x - 4y = 15$
4. $3y = 2x^2 + 2xy$
5. $5x^2y = 4 + x^2 - 5y$
6. $y^2 + 2y = x$
7. $2y^2 - y = 2x$

In Exercises 8–18, use implicit differentiation to find $dy/dx$, and evaluate $dy/dx$ at the given point.

8. $y^2 + 5y = 2x$, at $(-1, -1)$
9. $2x^2 + y - 3 = 0$, at $(1, 1)$
10. $xy + y = 4$, at $(3, 1)$
11. $y^3 + y = x + 1$, at $(-1, 0)$
12. $10y = 5x - 10xy$, at $(-2, 1)$
13. $x^3y^2 + 7xy = 2x - 7$, at $\left(2, -\dfrac{3}{2}\right)$
14. $2xy^3 - xy^2 = 24$, at $(2, 2)$
15. $\dfrac{x}{y} + \dfrac{y}{x} = -\dfrac{10}{3}$, at $(-1, 3)$
16. $(2y + 1)/x - 2xy = x^2$, at $\left(2, -\dfrac{7}{6}\right)$
17. $2\sqrt{x} + \sqrt{y} = y$, at $(1, 4)$
18. $\dfrac{3}{\sqrt{y}} + 2xy = y + \dfrac{3}{2}$, at $\left(\dfrac{1}{2}, 4\right)$

In Exercises 19–25, find the equation(s) of the tangent line(s) to the graphs of the given equations at the point with given x-coordinate.

19. $xy + y = x$, at $x = 1$
20. $x^3 + 2y = 2x$, at $x = -2$
21. $\dfrac{5y}{x^2} + x = 10 + y$, at $x = 2$
22. $5x^2y + 3xy = 2x^3 - 4x + 1$, at $x = -1$
23. $x^2 + y^2 = 25$, at $x = 3$
24. $xy^2 + 6xy + y + 15 = 0$, at $x = 2$
25. $y^2 + xy + 4x = -4y$, at $x = 3$

**Business and Economics**

26. (Complementary goods) Cigars and lighters are complementary goods in the sense that they are purchased and used together. The daily sales of cigars ($x$) and lighters ($y$) in a small store are related (approximately) by the equation

$$25xy - 10x^2 = -4y, \quad 0 \le x \le 50$$

(a) Suppose 20 cigars are sold on a particular day. Approximately how many lighters are sold on that day?
(b) When cigar sales are 30 per day, at what rate is the sale of lighters changing?

**Business and Economics**

27. (Substitute goods) Butter and margarine are substitute goods in the sense that one is purchased and consumed in place of the other. Suppose weekly sales of butter,

$x$ (in hundreds of pounds), and margarine, $y$ (in hundreds of pounds), at a local supermarket are related by the equation

$$12x^2y - 20x = 3, \qquad 0 \le x \le 15$$

(a) How many pounds of margarine are sold when weekly sales of butter are one (hundred)? Six (hundred)?

(b) When weekly sales of butter are 150 pounds, at what rate is the sale of margarine changing?

*In Exercises 28–31, weekly demand of a product is given by x (measured in units) and price per unit in dollars is given by p. Find dp/dx, evaluate it at the given value, and interpret your result.*

28. $15x = 12000 - 20p - 2p^2$, at $x = 750$
29. $2x = \sqrt{5000 - 2p^2}$, at $x = 35$
30. $x = -p^3 + 20p^2 + 5p + 3$, $13.50 \le p \le 20$, at $p = 15$
31. $0.01x = 2p/(p^2 + 5)$, $p \ge 4$, at $x = 35$

OPTIONAL

GRAPHING CALCULATOR/
COMPUTER EXERCISES

32. The equation given in Exercise 13 was $x^3y^2 + 7xy = 2x - 7$.
    (a) Solve the given equation for $y$ in terms of $x$ by using the quadratic formula.
    (b) Sketch each of the functions you found in part (a) on the same coordinate axes to produce a graph of the equation.
    (c) Draw the tangent line to the graph at the point $(2, -\frac{3}{2})$ and compare your (approximate) slope with that given by the solution of Exercise 13.

33. The equation given in Exercise 26 was $25xy - 10x^2 = -4y$, $0 \le x \le 50$.
    (a) Solve the equation for $y$ in terms of $x$.
    (b) Sketch the graph of the equation.
    (c) Sketch the tangent line to the graph at the point where $x = 30$.

34. The equation given in Exercise 27 was $12x^2y - 20x = 3$, $0 \le x \le 15$.
    (a) Solve the equation for $y$ in terms of $x$.
    (b) Sketch the graph of the equation.
    (c) Sketch the tangent line to the graph at the point where $x = 1.5$.

35–37. For each of the equations given in Exercises 28, 29, and 31,
    (a) solve the equation for $p$ in terms of $x$ obtaining $p = f(x)$;
    (b) sketch the graph of $p = f(x)$;
    (c) sketch the tangent line to the graph at the indicated value of $x$.

## 3.5 RELATED RATES

A newly hired car salesman makes a base salary of $200 per week and a commission of $150 on each car he sells during the week. His weekly salary $S$ (in dollars) is given by the formula

$$S = 200 + 150x$$

Suppose further that, as the weeks go by, his sales performance improves at the constant rate of two cars per week for the first five weeks of his employment. During the first week of his employment he sells only one car, during the second week he sells three cars, and so forth up to the fifth week. Then $x$ (the number of cars sold per week) depends on time $t$ according to the functional relationship

$$x = f(t) = 2t + 1, \quad 1 \le t \le 5$$

For example, if $t = 4$, then $x = f(4) = 9$. That is, the salesman sells nine cars during the fourth week of employment. We now pose the following question: At what rate is the salesman's salary increasing during the first five weeks of his employment? This is a fairly simple problem, and we have several methods of solution available.

Method I  Find $S$ as a function of $t$ and then find $dS/dt$.

We require the rate of change of salary $S$ with respect to time $t$. That is, we want $dS/dt$. Since $S = 200 + 150x$ and $x = 2t + 1$, we have

$$S = 200 + 150(2t + 1)$$
$$= 350 + 300t, \quad 1 \le t \le 5.$$

This last equation expresses $S$ as a function of time $t$. Differentiating with respect to $t$, we find that

$$\frac{dS}{dt} = \frac{300 \text{ dollars}}{\text{week}}$$

That is, the salesman's salary is increasing at the rate of $300 per week.

Method II  Reason the problem out logically.

Since for each week that passes, the salesman sells two more cars than the previous week, for each passing week he earns an additional $2(\$150) = \$300$ of salary. So the weekly rate at which his salary is increasing is $300 per week.

Method III  Differentiate the given equation $S = 200 + 150x$ implicitly with respect to $t$ and use the given fact that $dx/dt = 2$.

We have $S = 200 + 150x$, where $x$ and $S$ change from week to week. Thus $S$ and $x$ are functions of time $t$. That is, we may think of $S$ and $x$ as $S(t)$ and $x(t)$. Since both $S$ and $x$ are functions of $t$, we may differentiate each with respect to $t$ to obtain $dS/dt$ and $dx/dt$, respectively. Differentiating both sides of $S = 200 + 150x$ with respect to $t$, we get

$$\frac{d}{dt}(S) = \frac{d}{dt}(200 + 150x)$$

$$\frac{dS}{dt} = 150 \frac{dx}{dt}$$

This last equation relates the rates of change of $S$ and $x$ in that it says that the rate of change of $S$ is equal to 150 times the rate of change of $x$. Since we are given the fact that the salesman's performance improves at the rate of two cars per week, we have $dx/dt = 2$, and so

$$\frac{dS}{dt} = 150(2) = \frac{300 \text{ dollars}}{\text{week}}$$

The problem which we have just solved is called a *related rates problem*. For most problems of this type, Method II, the method of reasoning the problem out logically, is considerably more difficult than it was for this problem and is usually not used. The difference between the Method I solution and the Method III solution is that in Method I, we first obtained an explicit formula for $S$ in terms of $t$ and then differentiated this explicit formula to find $dS/dt$, whereas in Method III we used the relationship $S = 200 + 150x$ in which the time variable $t$ did not appear explicitly and then differentiated this implicit relation with respect to $t$ and used the given rate of change $(dx/dt = 2)$ to obtain $dS/dt$. Since it is usually not easy (or even possible) to obtain an explicit formula for the variable of interest in terms of the time variable $t$, we will use the technique of Method III for most related rates problems. In using Method III, we obtain a relationship between the rates of change (that is, the derivatives with respect to time $t$) of the variables of interest; hence the name related rates problem. In the previous example, Method III yielded the following relationship between the rates of change of $S$ and $x$:

$$\frac{dS}{dt} = 150 \frac{dx}{dt}$$

In general, if two or more variables vary with time (that is, are functions of time) and if we can obtain a relationship relating the variables (that is, an equation) that is valid for all times $t$, then by differentiating both sides of this relationship we obtain a relationship between the rates of change (that is, the derivatives with respect to time) of the variables. In a related rates problem we are always concerned with how fast a variable is changing. In other words, we are concerned with the rate of change of a variable with respect to time. Let's look at some more examples.

### EXAMPLE 1

Suppose that the weekly demand for our product is given by $x = f(p) = 500 - 0.5p$. Suppose further that because of heavy competition, the price of the product is decreasing at the rate of \$2 per week. If our current price per unit is \$250, how fast is the demand changing?

Solution

Since price is changing from week to week, we know that price is a function of time, so we can denote it by $p(t)$. Also, since the weekly demand is related to price via the equation $x = 500 - 0.5p$, and since price $p$ is changing, demand $x$ is also changing with time. Thus quantity demanded is also a function of time, which we denote by $x(t)$. We conclude that for any $t$,

$$x(t) = 500 - 0.5p(t)$$

In this problem we *could* find $p(t)$ and $x(t)$ explicitly and use Method I (see Exercise 17), but we choose to use Method III. We now differentiate implicitly both sides of the relationship between $x$ and $p$ with respect to time $t$ to obtain

$$\frac{d}{dt}[x(t)] = \frac{d}{dt}[500 - 0.5p(t)]$$

$$\frac{dx}{dt} = -0.5 \frac{dp}{dt}$$

So the relationship between the rates of change of $x$ and $p$ is that the rate of change of $x$ is equal to $-0.5$ times the rate of change of $p$. Now, since price is decreasing at the rate of $2 per week, we have $dp/dt = -\$2/\text{week}$. Note that since $p$ is decreasing, its rate of change is negative. Substituting this value into the relationship between the rates,

$$\frac{dp}{dt} = -0.5(-2) = 1 \; \frac{\text{unit}}{\text{week}}$$

We conclude that when our price is $250 per unit and price is decreasing at $2 per week, demand is increasing at a rate of 1 unit per week. ∎

## EXAMPLE 2

Coffee is being poured into a cylindrical coffee mug at the rate of 100 cubic centimeters (cm³) per second. If the mug is 9 centimeters (cm) high and has a radius of 4 cm, how fast is the level of coffee in the mug increasing when it is half full?

### Solution

In this case, it is helpful to draw a diagram of the situation: see Figure 3.5. Recall that the volume of a cylinder is

$$V = \pi r^2 h$$

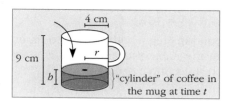

FIGURE 3.5 A cylindrical coffee mug.

where $r$ is the radius, $h$ is the height, and $V$ is the volume. After we have begun pouring the coffee into the mug, the coffee forms a "cylinder" of coffee in the mug. As we continue to pour, the volume of this cylinder of coffee increases— that is, its volume is a function of time. Let $V$, $r$, and $h$ denote, respectively, the volume, radius, and height of the cylinder of coffee at any time $t$. Note that since $V$ is increasing with time, we may denote it by $V(t)$, and since $h$ is increasing with time, we may denote it as $h(t)$. The radius $r$ of this cylinder of coffee is *not* changing, however, since it is always equal to the radius of the mug, which is 4 cm. We conclude that the volume of the cylinder of coffee in the mug is

$$V(t) = \pi(4)^2 h(t) = 16\pi h(t)$$

Differentiating this relationship with respect to $t$, we obtain

$$\frac{d}{dt}[V(t)] = \frac{d}{dt}[16\pi h(t)]$$

$$\frac{dV}{dt} = 16\pi \frac{dh}{dt}$$

We are given that $dV/dt = 100$ cm³/sec and we are required to find $dh/dt$ when $h = \frac{1}{2}(9) = 4.5$ cm. Solving the relationship between the rates of change $V$ and $h$ for $dh/dt$, we get

$$\frac{dh}{dt} = \frac{\frac{dV}{dt}}{16\pi}$$

so that

$$\left.\frac{dh}{dt}\right|_{h=4.5} = \frac{100}{16\pi} \approx 1.99 \text{ cm/sec}$$

So when the cup is half full, the level of coffee in the mug is rising at the rate of 1.99 cm/sec. Note that this result does not depend on the fact that the cup is half full. That is, $\frac{dh}{dt} = \frac{1}{16\pi} \cdot \frac{dV}{dt}$, so the rate at which the height is increasing does not depend on the height $h$. In other words, we never substituted $h = 4.5$ anywhere into this formula. Thus, the rate of change at which the height is increasing is a constant equal to 1.99 cm/sec for *any* value of height $h$. ∎

Before continuing, let's outline some guidelines for solving related rates problems.

**GUIDELINES FOR SOLVING RELATED RATES PROBLEMS**

Step 1 Identify the variables of interest in the problem and name them. Example: $V$ for volume, $r$ for radius, etc.

Step 2 Make a labeled sketch of the situation if it is appropriate and helpful.

Step 3 Note which variables are changing with time (that is, which are functions of time) and which are fixed or constant. Example: In Example 2, $V$ and $h$ were changing with time and $r = 4$ was fixed.

Step 4 Find a relationship or relationships involving the variables of interest that is valid at any time $t$. Example: $V(t) = 16\pi h(t)$.

Step 5 Differentiate the relationship(s) found in Step 4 with respect to time $t$ to obtain a relationship between the rates of change (that is, the derivative) of the variables of interest. Example:
$$\frac{d}{dt}[V(t)] = \frac{d}{dt}[16\pi h(t)] = 16\pi \frac{dh}{dt}.$$

Step 6 Solve the relationship found in Step 5 for the rate of change that is required. Example: $\frac{dh}{dt} = \frac{1}{16\pi} \cdot \frac{dV}{dt}$.

Step 7 Substitute the given rate(s) of change into the formula found in Step 6 to complete the problem. It may be necessary to find the value(s) of certain of the variables at a certain point in time using the relationship(s) found in Step 4. Example: $\frac{dh}{dt} = \frac{100}{16\pi} \approx 1.99$ cm/sec.

---

**EXAMPLE 3 (Participative)**

Two bumble bees take off from the same flower at the same time. The first flies directly north at 20 ft/sec and the other flies directly east at 25 ft/sec. Let's find the rate at which the distance between them is increasing six seconds after they take off.

**Solution**

Steps 1 and 2  Let's make a sketch of this situation and label the variables of interest. (See Figure 3.6.)

Let $y$ = the distance of the first ("northern") bee from the flower after $t$ seconds;

$x$ = _____ ;

$z$ = the distance between the two bees after $t$ seconds.

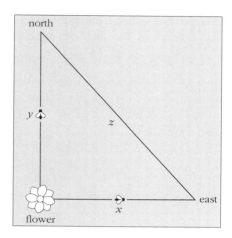

**FIGURE 3.6** The position of two bumble bees.

**Step 3** The variables involved here are $x$, $y$, and $z$. Which of these variables are changing with time? _____ Which of these variables are fixed? _____

**Step 4** A relationship between $x(t)$, $y(t)$, and $z(t)$ is given by the Pythagorean theorem. Thus we have

$$[x(t)]^2 + [y(t)]^2 = \text{_____}$$

**Step 5** When we differentiate both sides of the relationship found in Step 4, we must be careful to use the generalized power rule to differentiate a quantity like $[x(t)]^2$. That is, for example,

$$\frac{d}{dt}[x(t)]^2 = 2[x(t)]x'(t) = 2x(t)\frac{dx}{dt}$$

Differentiating the relationship found in Step 4, we obtain

$$\frac{d}{dt}[x(t)]^2 + \frac{d}{dt}[y(t)]^2 = \frac{d}{dt}[z(t)]^2$$

$$2x(t)\frac{dx}{dt} + 2y(t)\frac{dy}{dt} = \text{_____}$$

**Step 6** The rate of change we want in this problem is $dz/dt$. Solving for $dz/dt$ in the relationship we found in Step 5, we get

$$\frac{dz}{dt} = \text{_____}$$

**Step 7** The rates of change we are given are $\frac{dy}{dt} = 20$ ft/sec and $\frac{dx}{dt} = $ _____. To complete the problem we must substitute these rates into the formula for $dz/dt$ we found in Step 6. However, note also that we are required to find $\left.\frac{dz}{dt}\right|_{t=6}$ and the formula for $dz/dt$ involves $x(t)$, $y(t)$, and $z(t)$ as well as $dx/dt$ and $dy/dt$. Thus we need to find the values of $x$, $y$, and $z$ when $t = 6$. That is, we need to compute $x(6)$, $y(6)$, and $z(6)$. To do this, we use the relationship in Step 4 and the formula

distance = (speed)(time)

Now,

$y(6)$ = the distance traveled by the first ("northern") bee in 6 seconds
     = (speed of the first bee)(time)
     = (20 ft/sec)(6 sec) = 120 ft

$x(6)$ = the distance traveled by the second ("eastern") bee in 6 seconds
     = (speed of the second bee)(time)
     = (_____)(_____) = _____

To find $z(6)$, the distance between the two bees after 6 seconds, we use the relationship

$$[x(t)]^2 + [y(t)]^2 = [z(t)]^2$$

evaluated at time $t = 6$. We obtain

$$(150)^2 + (\underline{\phantom{xxx}})^2 = [z(6)]^2$$

so that $z(6) = \underline{\phantom{xxxx}}$. Finally, we compute

$$\left.\frac{dz}{dt}\right|_{t=6} = \underline{\phantom{xxxxx}}$$

■

### EXAMPLE 4

The output $z$ (in units) of one of our company's products depends on the two inputs, labor and materials. If we spend $x$ dollars on labor per week and $y$ dollars on materials per week, our company will produce

$$z = 0.001xy + 0.1x + 0.04y$$

units of this product per week. At present we are spending $2000 per week for labor and $5000 for materials. Due to a slump in the market for our product, however, we will be decreasing our labor costs at a rate of $75 per week and our materials cost at the rate of $150 per week over the next few weeks. Find the rate at which production will be changing after three weeks.

### Solution

**Steps 1 and 2** The variables of interest are $z$, $x$, and $y$; a sketch is not appropriate for this problem.

**Step 3** All the variables $z$, $x$, and $y$ are changing over time, so we designate them by $z(t)$, $x(t)$, and $y(t)$, respectively.

**Step 4** The relationship involving the variables is given to us:

$$z(t) = 0.001x(t)y(t) + 0.1x(t) + 0.04y(t)$$

**Step 5** Differentiating the relationship of Step 4 with respect to $t$, we obtain

$$\frac{dz}{dt} = 0.001\frac{d}{dt}[x(t)y(t)] + 0.1\frac{dx}{dt} + 0.04\frac{dy}{dt}$$

### SOLUTIONS TO EXAMPLE 3

**Steps 1 and 2** $x$ = the distance of the second ("eastern") bee from the flower after $t$ seconds.

**Step 3** The variables $x$, $y$, and $z$ are all changing with time. None of these variables is fixed.

**Step 4** $[z(t)]^2$

**Step 5** $2z(t)\dfrac{dz}{dt}$

**Step 6** $\dfrac{x(t)\dfrac{dx}{dt} + y(t)\dfrac{dy}{dt}}{z(t)}$

**Step 7** 25 ft/sec, 25 ft/sec, 6 sec, 150 ft, 120 ft, $\sqrt{36900} \approx 192$ ft, 32 ft/sec

$$\frac{dz}{dt} = 0.001[x(t)\frac{dy}{dt} + \underset{\underset{\text{product rule}}{\uparrow}}{\frac{dx}{dt}y(t)}] + 0.1\frac{dx}{dt} + 0.04\frac{dy}{dt}$$

Step 6  The relationship obtained in Step 5 is already solved for the required rate of change $dz/dt$.

Step 7  The given rates of change in this problem are $dx/dt = -\$75/\text{week}$ and $dy/dt = -\$150/\text{week}$. After three weeks our expenditures on labor will be

$$x(3) = 2000 - 3(75) = \$1775/\text{week}$$

and our expenditures on materials will be

$$y(3) = 5000 - 3(150) = \$4550/\text{week}$$

We conclude that

$$\frac{dz}{dt}\bigg|_{t=3} = 0.001[1775(-150) + (-75)(4550)] + 0.1(-75)$$

$$+ 0.04(-150) = -621 \frac{\text{units}}{\text{week}}$$

That is, after three weeks, when expenditures for labor are $1775 per week and expenditures for materials are $4550 per week, production will be decreasing at a rate of 621 units per week.  ■

### EXAMPLE 5

The number of arrests made per month of individuals purchasing illegal drugs in a certain community depends upon the number of duty hours $x$ worked by narcotics officers that month and also on the total number of illegal drug purchases $y$ made in the community that month according to the relationship

$$N = 0.0001y\sqrt[3]{x^2}, \quad \text{for } 0 \le y \le 1000 \text{ and } 0 \le x \le 5000$$

Presently, narcotics officers work a total of 2400 hours monthly and it is estimated that 5000 drug purchases are made per month. If the number of hours worked by narcotics officers is increasing at the rate of 150 hours per month and the number of drug purchases is increasing at the rate of 200 per month, find the rate at which the number of arrests will be changing after two months.

### Solution

Steps 1–3  The variables of interest $N$, $y$, and $x$—all of which change over time—will be designated by $N(t)$, $y(t)$, and $x(t)$, respectively.

Step 4  The given relationship involving the variables is

$$N(t) = 0.0001y(t)[x(t)]^{2/3}$$

Step 5  Differentiating the relationship of Step 4 with respect to $t$ gives

$$\frac{dN}{dt} = 0.0001y(t)\left\{\frac{2}{3}[x(t)]^{-1/3}\frac{dx}{dt}\right\} + 0.0001\frac{dy}{dt}[x(t)]^{2/3}$$

$$\frac{dN}{dt} = 0.0001[x(t)]^{2/3}\left\{\frac{2}{3}\frac{y(t)}{x(t)}\frac{dx}{dt} + \frac{dy}{dt}\right\}$$

Steps 6 and 7 The relationship in Step 5 is already solved for the required rate of change $dN/dt$. The given rates of change are

$$\frac{dy}{dt} = \frac{200 \text{ purchases}}{\text{month}}, \quad \frac{dx}{dt} = \frac{150 \text{ hours}}{\text{month}}$$

After two months, the number of drug purchases will be $y(2) = 5000 + 2(200) = 5400$ and the number of hours worked by narcotics officers will be $x(2) = 2400 + 2(150) = 2700$, so

$$\left.\frac{dN}{dt}\right|_{t=2} = 0.0001(2700)^{2/3} \left\{\frac{2}{3} \cdot \frac{5400}{2700} \cdot 150 + 200\right\} \approx 7.8 \frac{\text{arrests}}{\text{month}}$$

That is, after two months, when the number of drug purchases is 5400 per month and the number of duty hours of narcotics officers is 2700 hours per month, arrests will be increasing at a rate of about eight arrests per month. ■

## SECTION 3.5
### SHORT ANSWER EXERCISES

1. The perimeter of a square is four times the length of its side: $P = 4s$. If the side of a square is increasing at the rate of one foot per minute, what is the rate at which the perimeter is increasing? _____
2. Given $C(x) = 6x + 500$, if $dx/dt = 15$ then $dC/dt =$ _____.
3. (True or False) In a related rates problem, we may sometimes determine all the variables as explicit functions of time $t$. _____
4. (True or False) In a related rates problem, we usually differentiate both sides of the relationship with respect to $x$. _____
5. If $P$ is profit in dollars and $t$ is time in days, then $dP/dt = -20$ means that (a) profit is increasing at the rate of $20 per day; (b) profit is decreasing at the rate of $20 per day; (c) twenty days ago our profit was zero; (d) none of the above.
6. If two variables $z$ and $x$ are related by the equations $2z = x^2$ and if $z$ and $x$ are both functions of time $t$, then (a) $\frac{dz}{dt} = 2\frac{dx}{dt}$; (b) $\frac{dz}{dt} = \frac{dx}{dt}$; (c) $\frac{dz}{dt} = 2x(t)\frac{dx}{dt}$; (d) $2\frac{dz}{dt} = 2x\frac{dx}{dt}$.
7. If our revenue $R$ is increasing with time $t$, then (a) $\frac{dR}{dt} > 0$; (b) $\frac{dR}{dt} < 0$; (c) $\frac{dR}{dt} = 0$; (d) none of the preceding
8. Given the relationship $\frac{dA}{dt} = 2\pi r \frac{dr}{dt}$, find $\frac{dA}{dt}$ if $r = 3$ and $\frac{dr}{dt} = \frac{1}{2}$. _____
9. The statement, "$h$ changes twice as fast as $w$" can be written $dh/dt =$ _____.
10. (True or False) If $A = s^2$, then $A$ changes twice as fast as $s$ when $s = 2$. _____

## SECTION 3.5
### EXERCISES

1. (Perimeter) The formula for the perimeter of a square is $P = 4s$, where $P$ is the perimeter and $s$ is the length of one side. If the length of the side of a certain square is decreasing at the rate of 2 ft/min, determine the rate at which the perimeter is changing when the length of the side is 50 ft.

2. **(Area)** The formula for the area of a square is $A = s^2$, where $A$ is the area and $s$ is the length of one side. If the length of the side of a certain square is increasing at the rate of 3 in./sec, determine the rate at which the area is changing when the length of the side is 12 inches.

3. **(Manufacturing)** A worker in a shoe factory earns $1.75 for every dozen shoes he works on. If he is improving his work performance at the rate of three dozen shoes per day, at what rate is his daily salary increasing?

4. **(Demand)** If daily demand for a product is

$$x = f(p) = 100 - 0.1p^2$$

where $x$ is the number of units demanded at price $p$ (in dollars), and if price is increasing at the rate of 50 cents per day,
(a) determine the rate of change in quantity demanded at a price of $70 per unit;
(b) determine the rate of change in revenue when the price is $70 per unit.

5. **(Cost and revenue)** Given the monthly cost function $C(x) = 600 + 50x$ and the revenue function $R(x) = 200x$, where cost and revenue are measured in dollars and $x$ is the number of units produced and sold per month, if production is increasing at the rate of 20 units per month, find the rate of change in cost, revenue, and profit.

6. **(Demand and cost)** If monthly demand $x$ (in units) for our product is given by

$$x = -0.16p + 44, \quad 150 \le p \le 250$$

and if the monthly cost in dollars of producing $x$ units of the product is $C(x) = 50x + 600$, then if price is increasing at the rate of $5 per month, find
(a) the rate at which revenue is changing when the price is $200 per unit, and
(b) the rate at which profit is changing when the price is $200 per unit.

7. **(Cost in a bakery)** The weekly cost in dollars of baking $x$ cakes in a small bakery is

$$C(x) = \tfrac{1}{500}x^2 - 2x + 1500$$

If production at the bakery is increasing by 20 cakes per week, find the rate of change of cost when the production level is (a) 300 cakes; (b) 500 cakes; (c) 700 cakes.

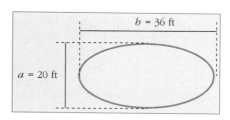

**FIGURE 3.7** An elliptical swimming pool.

8. **(Draining a pool)** An elliptical swimming pool has a uniform depth of five feet and is being drained at the rate of 100 cubic feet per minute. Find the rate at which the depth $h$ is changing if the pool has the dimensions indicated in Figure 3.7. (Hint: The volume of the water in the pool is $V = \tfrac{1}{4}\pi abh$.)

9. **(Area of a ripple)** A boy throws a stone into a lake. The radius of the circular ripple caused by the stone increases at the rate of 2.5 ft/sec. How fast is the area enclosed by the ripple increasing when the radius is 8 ft?

10. **(Volume of a wine glass)** A wine glass has the shape of an inverted cone with a height of 5 cm and a diameter of 4 cm: see Figure 3.8. If wine is poured into the glass at the rate of 25 cm³/sec, how fast is the level of wine in the glass increasing when the height of the wine is 2 cm? (Hint: The volume of a cone is $V = \tfrac{1}{3}\pi r^2 h$.)

11. **(Sliding ladder)** A 20-ft ladder is leaning against a wall. If someone pulls the bottom of the ladder horizontally away from the wall at a rate of 3 ft/sec, find the rate at which the top of the ladder is sliding down the wall when the bottom is 9 ft from the bottom of the wall.

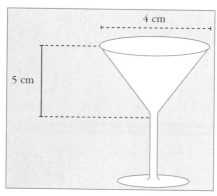

**FIGURE 3.8** A wine glass shaped like an inverted cone.

12. **(Length of a shadow)** A man who is 6 ft tall is walking away from an 18-ft lamppost at a rate of 4 ft/sec.

(a) Find the rate at which the length of his shadow is changing.
(b) Find the rate at which the tip of his shadow is moving away from the lamppost.

13. (Distance) Two planes are traveling at the same altitude at right angles to one another. If one of the planes is traveling at 300 mph and 30 minutes ago passed a point $P$, and if the other plane is traveling at 350 mph, is 200 miles from point $P$, and traveling directly toward it, how fast is the distance between the planes changing at that precise instant?

14. (Profit and price) The monthly profit our company makes in dollars is related to the number of units produced and sold according to the formula $P(x) = -x^2 + 900x - 20,000$. The number of units sold depends upon price as given by the monthly demand function $x = -5p + 1500$. If price is increasing at the rate of $5 per month per unit, determine the rate at which profit is changing when price is $50 per unit.

**Business and Economics**

15. (Production function) The weekly output $z$ at a certain factory depends on the amounts spent on labor and materials. If $x$ dollars is spent on labor per week and $y$ dollars on materials per week, this factory will produce $z = 100\sqrt{x}\sqrt[3]{y^2}$ units of product per week. Find the rate at which production is changing when $2000 is being spent for labor per week and $1000 for materials per week while hiring additional workers whose salaries amount to $300 per week and buying additional materials at the rate of $100 per week.

16. (Teacher salaries) The number $y$ of new teachers that a certain county can employ depends on the annual starting salary $x$ (in thousands) according to

$$y = -0.025x^3 + 3.75x^2 + 1.25x, \qquad 10 \le x \le 70$$

If annual starting salaries are increasing at the rate of $3000/yr, find the rate at which new teachers can be hired if the starting salary is $20,000/yr.

17. In Example 1 of this section, find explicit formulas for $p(t)$ and $x(t)$ and complete the solution of the problem using the technique of method I.

**Social Sciences**

18. (Social workers) The number $N$ of social workers working in a certain community at any time depends on the total population $p$ of the community and the average social class $s$ of the residents of the community according to $N = 0.0005p/\sqrt[3]{(s+1)^4}$. Social class is measured on a scale of 0 to 4. The present population of the community is 500,000 and the present social class index of the residents is 1.5. If the population of the community increases at the rate of 2000 residents per month and the social class index of the residents increases at the rate of 0.05 units per month, determine the rate at which the number of social workers will be changing after four months.

## 3.6 DIFFERENTIALS

Imagine that you are at a meeting with the top management of your company. One purpose of the meeting is to project next month's profits from the sales of your most important product, and you are in charge of this presentation. You know that next month's profits can be predicted from the equation

$$P(x) = -\tfrac{1}{2}x^2 + 25x, \qquad 0 \le x \le 25$$

where $x$ is sales in thousands of units and $P(x)$ is profit in thousands of dollars. When you enter the meeting, the marketing department projects sales next month to be 20,000 units and you compute $P(20) = -\frac{1}{2}(20)^2 + 25(20) = 300$ (that is, $300,000) as the projected profit. During your presentation, however, you are interrupted by a phone call from the marketing department. It seems that because of a new advertising campaign by your major competitor, projected sales for your company for next month must be downgraded to 19,000 units. You make this announcement to management. They are disappointed, of course, and immediately ask you what the effect will be on profits. To give a very quick answer to this question, you could use the theory of approximation by differentials, which we will discuss in this section.

To begin, let us suppose that $y = f(x)$ and that $x$ changes from $x$ to $x + \Delta x$, where $\Delta x$ can be either positive or negative and is usually considered to be a small number. Now a change from $x$ to $x + \Delta x$ produces a change in $y$, since $y$ depends upon $x$, that is, $y$ is a function of $x$. The *exact* change in $y$ produced by a change from $x$ to $\Delta x$ is denoted by $\Delta y$ and can be computed from the formula

$$\Delta y = f(x + \Delta x) - f(x)$$

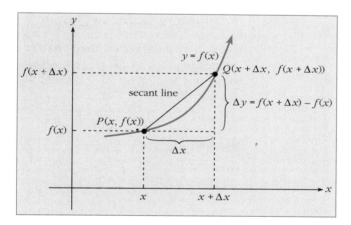

FIGURE 3.9 The definitions of $\Delta x$ and $\Delta y$.

See Figure 3.9 for a graphical illustration when $\Delta x$ is positive. In words, the formula simply states, "To find the actual change in $y = f(x)$ if $x$ changes to $x + \Delta x$, compute the $y$-coordinate at $x + \Delta x$ (that is, $f(x + \Delta x)$) and subtract from this the $y$-coordinate at $x$ (that is, $f(x)$)." For example, if $y = f(x) = \frac{1}{2}x^2$ and if we arbitrarily choose $x = 2$, then $f(2) = \frac{1}{2}(2)^2 = 2$. See Figure 3.10 where the point $P$ has coordinates $(2, 2)$. If we choose $\Delta x$ to be 1, then $f(x + \Delta x) = f(2 + 1) = f(3) = \frac{1}{2}(3)^2 = \frac{9}{2}$, and so the point $Q$ has coordinates $\left(3, \frac{9}{2}\right)$. In addition, we may compute the corresponding change $\Delta y$ in $y$, to be

$$\begin{aligned}\Delta y &= f(x + \Delta x) - f(x) \\ &= f(2 + 1) - f(2) = f(3) - f(2) \\ &= \tfrac{1}{2}(3)^2 - \tfrac{1}{2}(2)^2 = \tfrac{9}{2} - 2 = \tfrac{5}{2}\end{aligned}$$

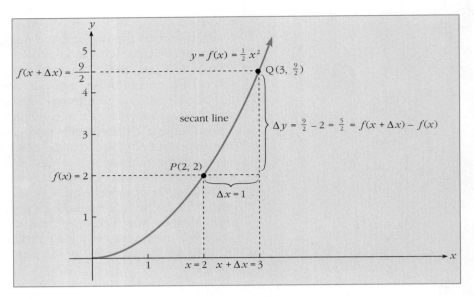

**FIGURE 3.10** An illustration of $\Delta x$ and $\Delta y$ for $f(x) = \frac{1}{2}x^2$.

Note that as $x$ changes from $x = 2$ to $x = 3$ (so that the change $\Delta x$ in $x$ is equal to 1), the corresponding change in $y$ along the curve $y = \frac{1}{2}x^2$ is from $y = 2$ to $y = \frac{9}{2}$, so that the change in $y$ is $\Delta y = \frac{9}{2} - 2 = \frac{5}{2}$.

Referring back to the general setting of Figure 3.9, recall that the slope of the secant line joining $P$ and $Q$ is given by

$$m_{\text{sec}} = \frac{f(x + \Delta x) - f(x)}{\Delta x} = \frac{\Delta y}{\Delta x}$$

Alternatively, we can write

$$\Delta y = (m_{\text{sec}})\Delta x$$

**DEFINITION**

> The *differential* of $x$, denoted by $dx$, is defined by
>
> $$dx = \Delta x, \quad \text{where } \Delta x \text{ is any real number}$$

### The Relationship between Differentials and the Leibniz Notation

We have defined $dx$, the differential of $x$, to be the familiar $\Delta x$, "delta" $x$, which is the change in $x$. Now, we have already used the symbols $dx$ and $dy$ in the Leibniz notation for the derivative, where $dy/dx = f'(x)$. We will now give $dx$ and $dy$ individual meanings consistent with this notation. Refer to Figure 3.11, in which we have drawn the tangent line $t$ to the graph of $y = f(x)$ at the point $P$ with slope $m_{\text{tan}} = f'(x)$. We can see that the slope of this tangent line can also be given by the ratio

$$\frac{dy}{dx} = m_{\text{tan}} \tag{3-3}$$

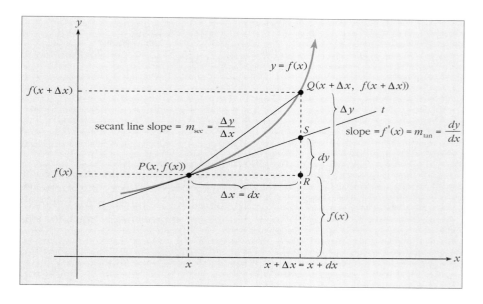

**FIGURE 3.11** The definitions of *dx* and *dy*.

where the symbol $dy$ is used to denote the directed distance from $R$ to $S$ in Figure 3.11. In this diagram, this directed distance is positive, since to travel from $R$ to $S$ we must travel upward, that is, in the direction of increasingly positive $y$-values. From the relationship in Equation (3-3), we conclude that

$$dy = (m_{\tan}) \, dx = f'(x) \, dx$$

We can now state the definition of the differential of $y$.

**DEFINITION**

The differential of $y$, denoted by $dy$, is defined by
$$dy = f'(x) \, dx$$

### The Tangent Line Approximation

We can now give individual interpretations to the symbols $dx$ and $dy$ in the Leibniz notation. In fact, $dx$ is simply the directed distance from $P$ to $R$ in Figure 3.11 and $dy$ is the directed distance from $R$ to $S$. These directed distances are defined so that their ratio, $dy/dx$, is the slope of the tangent line $m_{\tan}$ or $f'(x)$.

As you can see from Figure 3.11, the quantities $\Delta y$ and $dy$ are *not* necessarily equal to one another. However, as we learned in Section 2.4, if $\Delta x$ is small or, equivalently, if point $Q$ is close to point $P$, then the slope of the secant line joining $P$ and $Q$ closely approximates the slope of the tangent line at $P$. In symbols,

$$m_{\sec} \approx m_{\tan} \quad \text{(if } \Delta x \text{ is small)}$$

so that

$$\frac{\Delta y}{\Delta x} \approx \frac{dy}{dx} = \frac{dy}{\Delta x} \quad \text{(since } dx = \Delta x \text{ by definition)}$$

and so

$$\Delta y \approx dy \quad \text{(if } \Delta x \text{ is small)}$$

For a graphical interpretation, see Figure 3.12. Notice that as $Q$ "slides" down the curve $y = f(x)$, becoming, successively, $Q'$ and $Q''$ (that is, as $\Delta x$ approaches 0), the difference between $\Delta y$ and $dy$ becomes smaller and smaller, so that $\Delta y$ and $dy$ are approximately equal for small values of $\Delta x$.

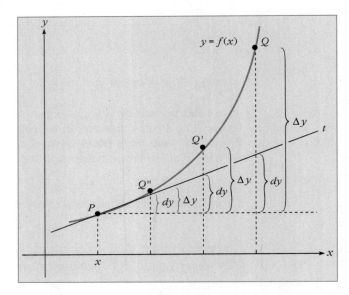

**FIGURE 3.12** The tangent line approximation.

Many of the concepts of this section can be illustrated in an amusing way. In Figure 3.11, consider a tiny ant walking along the graph of $f(x)$. Then $\Delta y$ is the actual change in height from the $x$-axis that the ant experiences if it walks along the curve from the point $P(x, f(x))$ to the point $Q(x + \Delta x, f(x + \Delta x))$. In fact the ant would experience this *same* change in height if it were to walk along the secant line joining $P$ and $Q$ in this diagram. An interpretation of $dy$ can be obtained by having our ant crawl along the *tangent line* (rather than along the curve $y = f(x)$ or the secant line), starting above $x$ (at the point $P$) and stopping above $x + dx$ (at the point $S$). The change in height that this ant experiences with respect to the $x$-axis is $dy$, the differential of $y$. In "ant terms," the approximation $\Delta y \approx dy$ may be summarized by saying that if the ant's walk is quite short, it will experience about the same change in height whether he walks along the curve $y = f(x)$ or along the tangent line to this curve at $P$.

**EXAMPLE 1** If $f(x) = \frac{1}{2}x^2$, compare $\Delta y$ and $dy$ **(a)** when $x = 2$ and $\Delta x = 1$ and **(b)** when $x = 2$ and $\Delta x = 0.2$.

**Solution**

**(a)** First of all, $\Delta y = f(x + \Delta x) - f(x)$, so when $x = 2$ and $\Delta x = 1$, we obtain

$$\Delta y = f(3) - f(2) = \frac{1}{2}(3)^2 - \frac{1}{2}(2)^2 = \frac{5}{2} = 2.5$$

Next, $dy = f'(x)\, dx = x\, dx$, since $f(x) = \frac{1}{2}x^2$. Also, when $x = 2$ and $\Delta x = dx = 1$, we have $dy = 2(1) = 2$. Comparing $\Delta y$ and $dy$, we see that $|\Delta y - dy| = 2.5 - 2 = 0.5$ is the absolute difference between them when $x = 2$ and $\Delta x = 1$.

**(b)** When $x = 2$ and $\Delta x = 0.2$,

$$\Delta y = f(2.2) - f(2) = \frac{1}{2}(2.2)^2 - \frac{1}{2}(2)^2 = 0.42$$

Also, when $x = 2$ and $\Delta x = 0.2$, we get $dy = 2(0.2) = 0.4$, so that the absolute difference between $\Delta y$ and $dy$ is

$$|\Delta y - dy| = 0.42 - 0.4 = 0.02$$

which is smaller than the difference obtained in part (a). The results of Example 1 are illustrated graphically in Figure 3.13.

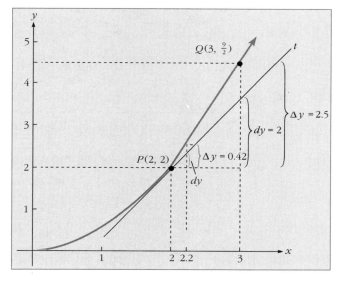

**FIGURE 3.13** A comparison of $\Delta y$ and $dy$ for $f(x) = \frac{1}{2}x^2$

To find the differential of $y$ if $y = f(x)$, we use the formula

$$dy = f'(x)\, dx$$

which says, "Find the derivative and multiply by $dx$." Thus, in finding $dy$ we may use the rules of differentiation developed in Sections 3.1 and 3.2 to find $f'(x)$ and then multiply this derivative by $dx$.

**EXAMPLE 2** Find $dy$ if (a) $f(x) = 3x^3 - 7x^2$; (b) $f(x) = \sqrt{x^2 + 2x}$.

**Solution**
(a) $dy = f'(x)\, dx = (9x^2 - 14x)\, dx$
(b) In this case, $f(x) = \sqrt{x^2 + 2x} = (x^2 + 2x)^{1/2}$, so, using the generalized power rule to find $f'(x)$, we have

$$dy = f'(x)\, dx = \frac{1}{2}(x^2 + 2x)^{-1/2}(2x + 2)\, dx = \frac{x+1}{\sqrt{x^2 + 2x}}\, dx \qquad \blacksquare$$

The approximation $\Delta y \approx dy$ if $\Delta x$ is small can be written in a slightly different form, namely,

$$\Delta y \approx dy$$
$$f(x + \Delta x) - f(x) \approx f'(x)\, dx$$

**THE TANGENT LINE APPROXIMATION**

$$f(x + \Delta x) \approx f(x) + f'(x)\, dx \qquad \text{(if } \Delta x \text{ is small)}$$

The preceding formula is sometimes called the *tangent line approximation* to $f(x + \Delta x)$. The reason is that $f(x) + f'(x)\, dx = f(x) + dy$ represents the $y$-coordinate of the point on the tangent line with $x$-coordinate $x + \Delta x$. See Figure 3.11 where the quantity $f(x) + dy$ is the $y$-coordinate of the point $S$. On the other hand, $f(x + \Delta x)$ is just the $y$-coordinate of the point $Q$ in Figure 3.11 that lies on the curve $y = f(x)$. Thus the approximation

$$f(x + \Delta x) \approx f(x) + f'(x)\, dx$$

says that the $y$-coordinate of $Q$ is approximately equal to the $y$-coordinate of $S$.

**EXAMPLE 3** Approximate $\sqrt[3]{66}$ using differentials.

**Solution** We will use the tangent line approximation

$$f(x + \Delta x) \approx f(x) + f'(x)\, dx$$

Let us choose $f(x) = \sqrt[3]{x}$, since we are required to approximate $\sqrt[3]{66}$. We now need to choose a value for $x$ and a value for $\Delta x$. Since we'll have to calculate $f(x) = \sqrt[3]{x}$ in our tangent line approximation, we'll choose for $x$ a number close to 66 with a cube root that is easily found. Since $\sqrt[3]{64} = 4$, let's choose $x = 64$. Then $x + \Delta x = 64 + \Delta x = 66$, hence $\Delta x = dx = 2$. Note also that $f'(x) = \frac{1}{3}x^{-2/3} = 1/(3\sqrt[3]{x^2})$, so that $f(66) \approx f(64) + f'(64)\, dx$, or

$$\sqrt[3]{66} \approx \sqrt[3]{64} + \frac{1}{3\sqrt[3]{(64)^2}} \quad (2)$$

$$= 4 + \frac{2}{3(\sqrt[3]{64})^2} = 4 + \frac{2}{3(4)^2}$$

$$= 4 + \frac{1}{24} \qquad \left(\text{Here } f(64) = 4 \text{ and } dy\big|_{\substack{x=64 \\ dx=2}} = \frac{1}{24}\right)$$

$$\approx 4.0417$$

To four decimal places, a calculator gives $\sqrt[3]{66} = 4.0412$, so that our approximation by differentials is quite good. A graphical interpretation is provided in Figure 3.14. From this figure, it is clear why our tangent line approximation overestimates the true value of $\sqrt[3]{66}$ slightly.

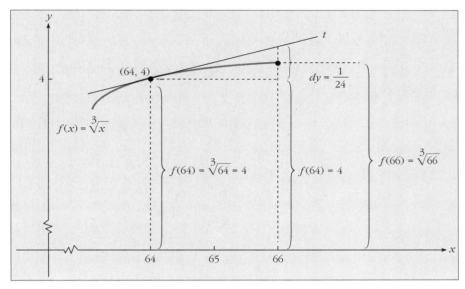

**FIGURE 3.14** Approximating $\sqrt[3]{66}$ using differentials.

**EXAMPLE 4**

If $P(x) = -\frac{1}{2}x^2 + 25x$ for $0 \leq x \leq 25$, where $x$ is sales in thousands of units and $P(x)$ is profit in thousands of dollars, use differentials to approximate the change in profit if sales change from 20,000 units to 19,000 units.

**Solution**

This is the problem we considered at the beginning of this section. The actual change in profits, $\Delta P$, can be approximated by $dP$, where

$$dP = P'(x)\,dx = (-x + 25)\,dx$$

so that

$$dP\Big|_{\substack{x=20 \\ dx=-1}} = (-20 + 25)(-1) = -5$$

Thus our monthly profit will decrease by about \$5000 when sales change from 20,000 units to 19,000 units.

**EXAMPLE 5 (Participative)**

The edge of a cube is measured to be 4.0 inches. If the error in this measurement is $\pm 0.03$ inches, let us use differentials to estimate the maximum error in the volume of the cube.

## Solution

Recall that the volume of a cube is given by $V = e^3$, where $e$ is the length of the edge. We compute:

$$dV = \underline{\phantom{xxxxx}}$$

To approximate the maximum error in the volume, we find

$$dV\Big|_{\substack{e=4.0 \\ de=\pm 0.03}} = 3(\underline{\phantom{xx}})^2(\pm 0.03) = \underline{\phantom{xxxxx}}$$

## SECTION 3.6
### SHORT ANSWER EXERCISES

1. If $dy = (2x - 1)\,dx$, then $dy\big|_{\substack{x=2 \\ dx=0.1}} = \underline{\phantom{xx}}$.

2. If $y = 3x^2$, then $dy = \underline{\phantom{xx}}$.

3. If an ant crawls along the curve $y = f(x)$ from the point $P(x, f(x))$ to the point $Q(x + \Delta x, f(x + \Delta x))$, it experiences a change in height with respect to the $x$-axis equal to **(a)** $dy$; **(b)** $f'(x)\,dx$; **(c)** $\dfrac{dy}{dx}$; **(d)** $\Delta y$.

4. The *actual* change in the value of $y = f(x)$ when $x$ changes to $x + \Delta x$ is equal to **(a)** $dy$; **(b)** $f'(x)\,dx$; **(c)** $m_{\text{tan}}$; **(d)** $\Delta y$.

5. The *approximate* change in the value of $y = f(x)$ when $x$ changes to $x + \Delta x$ is equal to **(a)** $dy$; **(b)** $m_{\text{sec}}$; **(c)** $m_{\text{tan}}$; **(d)** $\Delta y$.

6. If $f(x) = 3x$, where $x = 1$ and $\Delta x = 0.1$, then $\Delta y = \underline{\phantom{xx}}$.

7. (True or False) An ant walking from $P$ to $Q$ along the secant line joining $P(x, f(x))$ and $Q(x + \Delta x, f(x + \Delta x))$ experiences the same change in height relative to the $x$-axis as it would if it walked along the curve $y = f(x)$ from $P$ to $Q$. $\underline{\phantom{xx}}$

8. (True or False) For every function $y = f(x)$, the quantity $\Delta y$ is larger than $dy$. $\underline{\phantom{xx}}$

9. In approximating $\sqrt{24}$ using differentials, we would choose $f(x) = \sqrt{x}$, $x = \underline{\phantom{xx}}$, and $\Delta x = dx = \underline{\phantom{xx}}$.

10. (True or False) If an ant walks from the point $P(x, f(x))$ to the point with $x$-coordinate $x + dx$ along the tangent line to $y = f(x)$ at $P$, the change in height it experiences is equal to $dy$. $\underline{\phantom{xx}}$

## SECTION 3.6
### EXERCISES

*In Exercises 1–7, find $dy$, the differential of $y$.*

1. $f(x) = 5x - 4$
2. $f(x) = 2x^2 - 3$
3. $f(x) = 3x^2 + x$
4. $f(x) = x/(2x + 1)$
5. $f(x) = \sqrt{5x^2 + 21}$
6. $f(x) = (4 - 3x)^3$
7. $f(x) = (2/x) - \sqrt{x} + 4$

---

**SOLUTIONS TO EXAMPLE 5**     $3e^2\,de$, $4.0$, $\pm 1.44$ in.$^3$

*In Exercises 8–12, find $\Delta y$ and $dy$ for the given values of $x$ and $\Delta x$. Make a sketch showing $\Delta y$ and $dy$.*

**8.** $f(x) = x^2$ for $x = 1$, $\Delta x = 0.5$  
**9.** $f(x) = x^3$ for $x = 2$, $\Delta x = 1$  
**10.** $f(x) = \sqrt{x}$ for $x = 9$, $\Delta x = -1$  
**11.** $f(x) = 2x + 1$ for $x = 3$, $\Delta x = -1$  
**12.** $f(x) = 1/x$ for $x = 2$, $\Delta x = 0.5$

*In Exercises 13–18, use the tangent line approximation to approximate the given expression.*

**13.** $\sqrt{37}$  **14.** $\sqrt{48}$  **15.** $\sqrt{98}$  **16.** $\sqrt[3]{27.4}$  **17.** $\sqrt[4]{17}$  **18.** $\sqrt[5]{31.5}$

**19.** (Area) The side of a square is measured to be $7.34 \pm 0.03$ in. Estimate the maximum error in the area.

**20.** (Volume) The radius of a sphere is measured to be $6.06 \pm 0.04$ in. Estimate the maximum error in the volume. (Hint: $V = \frac{4}{3}\pi r^3$, where $r$ is the radius of the sphere.)

**21.** (Material requirement) If a sphere of radius 12 in. is to be spray-painted so that the thickness of the paint is 0.02 in., approximate the volume of paint required.

**22.** (Cost) If the monthly cost in dollars of producing $x$ units of our product is

$$C(x) = 0.5x^2 + 600$$

(a) find the cost of producing 5000 units per month;  
(b) use differentials to approximate the cost of producing the 5001st unit.

**23.** (Revenue) Given the weekly revenue function

$$R(x) = 50x - 0.04x^2$$

where $R(x)$ is the revenue in hundreds of dollars generated from the sale of $x$ units per week, approximate the revenue derived from the 401st unit, using differentials.

Business and Economics

**24.** (Profit) Given the monthly profit function

$$P(x) = -0.05x^2 + 30x - 500$$

where $P(x)$ is the profit in hundreds of dollars generated from the sale of $x$ units per month, use differentials to approximate the change in profit if sales change from 1500 to 1400 units.

Business and Economics

**25.** (Demand) The monthly demand for a product is given by

$$x = f(p) = 5000 - 0.05p^2$$

where $x$ is demand in units and $p$ is price in dollars. Use differentials to find the approximate effect on demand if price decreases from $200 per unit to $195 per unit.

Life Sciences

**26.** (Population) The population of a certain region in $t$ years is estimated at

$$P(t) = 1000 + 10t^2 - 6\sqrt{t}, \quad t \geq 0$$

Find the approximate change in population between (a) $t = 4$ and $t = 4.2$ years; (b) $t = 9$ and $t = 9.1$ years.

**Social Sciences**

**27.** (College tuition) A small private college estimates the cost $y$ in dollars per quarter-hour of taking a course will be

$$y = 35 + 5\sqrt{x}, \quad 0 \le x \le 12$$

$x$ quarters from now. Use differentials to approximate the change in cost per quarter-hour between quarters nine and ten.

**OPTIONAL**

**GRAPHING CALCULATOR/ COMPUTER EXERCISES**

**28–30.** For the functions of Exercises 3–5, (a) find $dy|_{x=0 \atop \Delta x=1}$; (b) find $\Delta y|_{x=0 \atop \Delta x=1}$; (c) sketch the graph of $f(x)$ and label $dy$ and $\Delta y$ on your sketch.

**31.** For the function of Exercise 26, sketch $P(t)$ and identify $dP|_{t=4 \atop \Delta t=0.2}$ and $dP|_{t=9 \atop \Delta t=0.1}$ on your sketch.

## 3.7 CHAPTER THREE REVIEW

**KEY TERMS**

differentiation formula
constant rule
power rule
constant multiple rule
sum or difference rule
product rule
quotient rule
chain rule
inside function

outside function
generalized power rule
higher derivatives
implicit differentiation
related rates
differential of $x$
differential of $y$
tangent line approximation

**KEY FORMULAS AND RESULTS**

**Constant Rule**

$$D_x[c] = 0$$

**Power Rule**

$$D_x[x^n] = nx^{n-1}$$

**Constant Multiple Rule**

$$D_x[c \cdot g(x)] = cD_x[g(x)]$$

**Sum or Difference Rule**

$$D_x[g(x) \pm h(x)] = D_x[g(x)] \pm D_x[h(x)]$$

**Product Rule**

$$D_x[g(x) \cdot h(x)] = g(x) \cdot D_x[h(x)] + D_x[g(x)] \cdot h(x)$$

**Quotient Rule**

$$D_x\left[\frac{g(x)}{h(x)}\right] = \frac{h(x)D_x[g(x)] - g(x)D_x[h(x)]}{[h(x)]^2}$$

**Chain Rule**

$$D_x[f(g(x))] = f'(g(x)) \cdot g'(x)$$

**Chain Rule (alternate form)**

If $y = f(u)$ and $u = g(x)$, then $\dfrac{dy}{dx} = \dfrac{dy}{du} \cdot \dfrac{du}{dx}$.

**Generalized Power Rule**

$$D_x[g(x)]^n = n[g(x)]^{n-1}g'(x)$$

**Differential of x**

$$dx = \Delta x$$

**Differential of y**

$$dy = f'(x)\,dx$$

**Tangent Line Approximation**

$$f(x + \Delta x) \approx f(x) + f'(x)\,dx \quad \text{if } \Delta x \text{ is small}$$

$$\Delta y \approx dy \quad \text{if } \Delta x \text{ is small}$$

## CHAPTER 3 SHORT ANSWER REVIEW QUESTIONS

1. The derivative of any constant is equal to _____.
2. If $f(x) = \sqrt{x}$, then $f'(4) =$ _____.
3. Using the quotient rule, $\dfrac{d}{dx}\left[\dfrac{x}{3x+2}\right] =$ _____.
4. (True or False) $D_x(5x^3) = D_x(5)D_x(x^3) = 0 \cdot (3x^2) = 0.$ _____
5. If $f(x) = (5x + 1)^6$, then $f'(x)$ is equal to (a) $6(5x + 1)^5$; (b) $30(5x + 1)^6$; (c) $6(5x)^5$; (d) none of the preceding.
6. If we apply the chain rule to find the derivative of $y = (2x + 1)^{1/2}$, we obtain $dy/dx =$ _____.
7. (True or False) The chain rule and the generalized power rule are two different names for the same rule. _____
8. When we use the term "higher-order derivative," we mean a derivative of an order larger than _____.
9. The second derivative of $y = f(x)$ gives the rate of change of (a) $y$; (b) $dy$; (c) $dy/dx$; (d) $dx$.
10. If the velocity of an object is given by $v(t) = 32t + 20$, then the acceleration is equal to _____.

11. (True or False) If $y$ is a function of $x$, then $\frac{d}{dx}(y^4) = 4y^3$. _____

12. If $x$ is weekly demand in units and $p$ is price in dollars, and if $\frac{dp}{dx}\big|_{x=200} = -0.25$, then price is decreasing at a rate of $0.25 per unit of additional demand beyond _____ units.

13. If $\frac{dy}{dx} = \frac{7xy}{3x + 2y}$, then $\frac{dy}{dx}\big|_{\substack{x=1 \\ y=-1}} =$ _____.

14. (True or False) If $y = f(x) = x$, then $dy = dx$. _____

15. If $y = f(x) = x^3$ and $x$ changes from 10 to 10.5, then the actual change in $y$ is (a) $dy|_{\substack{x=10 \\ dx=0.5}}$; (b) $f(10.5) - f(10)$; (c) $dx|_{\substack{x=10 \\ dy=0.5}}$; (d) none of the preceding.

16. If $dA = 2\pi r \, dr$, then if $r = 4$ and $dr = 0.1$, it follows that $dA =$ _____.

17. In a related rates problem, (a) all variables vary with time; (b) some variables may not vary with time; (c) time is not a factor; (d) none of the above.

18. If $R(x) = 200x$ and $dx/dt = 2$, then $dR/dt =$ _____.

19. (True or False) To solve a related rates problem, we usually find a relationship between the rates of change of the variables of interest. _____

## CHAPTER 3 REVIEW EXERCISES

The problems marked with an asterisk (*) constitute a Chapter Test.

In Exercises 1–25 find the derivative of the given function.

1. $f(x) = 3x^2 + 2x - 5$
2. $f(x) = 12 - x^2 - 3x^3$
*3. $g(x) = 18x^4 - 4x^3$
4. $f(x) = (1/\sqrt{x}) + (\sqrt{x}/3)$
5. $f(t) = 2\sqrt{t^3} - (1/\sqrt[3]{t})$
6. $g(t) = (4/t^6) - 7t^3$
*7. $f(x) = 3x^4 - 5\sqrt[3]{x}$
8. $f(x) = (x^2 + 2)(3x^2 - 2)$
9. $g(x) = (x + 5)(6x - 11)$
10. $h(x) = (4x - x^3)(2x + x^2)$
*11. $g(t) = (4t^2 + 7t - 5)(2t^2 + t - 6)$
12. $f(x) = 5/(2x + 7)$
*13. $g(x) = \frac{2x}{4x + 1}$
14. $f(t) = \frac{3t^2}{4t^2 - t}$
15. $g(t) = \frac{2t^2 - 5}{5 - t - 3t^2}$
16. $f(x) = \frac{x^2 - 3x + 4}{2x^2 - 5x - 1}$
*17. $f(x) = (2 - 3x)^4$
18. $f(x) = (x^2 - 2x + 5)^3$
19. $g(t) = 1/(2t^2 + t)^5$
20. $y = \sqrt{5x^2 + 13} - \sqrt{x}$
*21. $y = \sqrt[3]{7x + 10} + \sqrt[3]{x}$
22. $F(x) = x^3\sqrt{5 + x^2}$
23. $f(x) = \sqrt[3]{\frac{x^2 - 3x}{4x + 5}}$
24. $f(x) = \sqrt{\frac{2x + 1}{3x + 4}}$
25. $G(x) = -10x\sqrt[5]{x^3 + x^2}$

In Exercises 26–31, find $\frac{d^2y}{dx^2}$ and $\frac{d^2y}{dx^2}\big|_{x=2}$.

26. $y = x^3 - 4x^2 + 3x + 2$
27. $y = 3\sqrt[5]{x^2} + 5$
28. $y = \frac{4}{x} + \frac{x}{4}$
*29. $y = \frac{-5}{(2x - 5)^4}$

**30.** $y = \sqrt{6x + 1} - 2\sqrt[3]{1 - 5x}$  
**31.** $y = 1/(5 - 7x) - \sqrt[5]{3x - 14}$

*In Exercises 32–39, use implicit differentiation to find $dy/dx$.*

**32.** $y^3 + y^2 = 4x$  
**33.** $3x^2 - y = 2y^4$  
**34.** $xy - 5x = 10y$  
**\*35.** $x^3 - 2xy = y^4$  
**36.** $x^2y - y^2x = y$  
**37.** $(x^2/y) + 3y^2 = 12x^4$  
**38.** $(3x/y^2) - 3y = 4x^2y$  
**39.** $\sqrt{y} + 3\sqrt{x} = 1/y$

*In Exercises 40–45, assume that $x$ and $y$ are differentiable functions of time $t$. Find $dx/dt$ in terms of $x$, $y$, and $dy/dt$.*

**40.** $x - 5y = 12$  
**\*41.** $x^2 + y^2 = 81$  
**42.** $xy = 25$  
**43.** $25\sqrt{x}\,\sqrt[3]{y} = x$  
**44.** $(x/y) - 10x^2 = y^2$  
**45.** $(1/\sqrt{x}) - (4/\sqrt{y}) = x^3$

*In Exercises 46–51, find $\Delta y$ and $dy$ for the given values of $x$ and $dx$.*

**46.** $y = x^3 - 2x^2 + 4$ for $x = 2$, $dx = 0.5$  
**47.** $y = (1/\sqrt{x}) + x - 1$ for $x = 1$, $dx = -0.1$  
**48.** $y = x^2/(x^2 - 1)$ for $x = 0$, $dx = 0.2$  
**\*49.** $y = 4\sqrt{3x + 10} - x^2$ for $x = -1$, $dx = -0.2$  
**50.** $y = (x^2 + 3x - 1)^3 - 7x$ for $x = -2$, $dx = 0.3$  
**51.** $y = \dfrac{4}{x + 1} - \dfrac{6}{(x - 1)^5}$ for $x = 2$, $dx = -0.5$

**52.** (Learning) After $t$ hours of study, the number of new vocabulary words that a student can learn is given by

$$N(t) = 45\sqrt[3]{t^2}, \quad 0 \le t \le 10$$

At what rate is the number of new words changing when $t = 2$? When $t = 8$?

**\*53.** (Advertising and sales) If we spend $x$ thousand dollars a month advertising one of our company's products, then monthly sales $y$ in thousands of units will be

$$y = \frac{50x}{x + 5} + 8, \quad x \ge 0$$

(a) Use the quotient rule to find the slope of the tangent line to the graph of this function at any point.

(b) Find $\left.\dfrac{dy}{dx}\right|_{x=1}$; $\left.\dfrac{dy}{dx}\right|_{x=10}$; $\left.\dfrac{dy}{dx}\right|_{x=25}$

(c) Interpret the results of part (b) as rates of change of sales with respect to advertising.

**54.** (Profit) The weekly profit your company makes on the sale of $x$ hundred units of one of your products is

$$P(x) = 0.02x^3 - 0.4x - 1500, \quad 0 \le x \le 80$$

Find the rate of change of profit for a sales level of 5000 units.

**55.** (Compound interest) If $5000 is placed in a bank account that pays a rate of $r$ percent per year compounded weekly, the amount $A$ in the account after three years is $A(r) = 5000[1 + (r/5200)]^{156}$. Find $dA/dr$ and evaluate $dA/dr$ when $r = 5$ percent and when $r = 10$ percent.

**Social Sciences**

**56.** (Unemployment) The number of unemployed people in a certain city is increasing according to

$$N(t) = \sqrt[3]{100 - 2t + t^2}, \qquad 0 \le t \le 24$$

where $t$ is measured in months and $N(t)$ is measured in thousands. Find and interpret $N'(12)$.

**Business and Economics**

***57.** (Cost) The monthly cost in hundreds of dollars of producing $x$ thousand units of a certain product is

$$C(x) = x^3 - 15x^2 + 75x + 175, \qquad 0 \le x \le 30$$

(a) When the production level is 3000 units, are costs increasing at an increasing rate or at a decreasing rate?
(b) When the production level is 7000 units, are costs increasing at an increasing rate or at a decreasing rate?
(c) At what level of production does cost change from increasing at a decreasing rate to increasing at an increasing rate?

**58.** (Primitive culture) In a certain primitive culture, the population $P$ of a region and the cultivation index $C$ (which ranges from 0 to 100) are related by $PC = 100C - 40P + 15$. Use implicit differentiation to find $dP/dC$. Interpret $\left.\dfrac{dP}{dC}\right|_{C=70}$.

***59.** (Volume) An ice "cube" that measures 1.5 in. by 1 in. by 1 in. starts melting so that length, width, and height are each decreasing at the rate of 0.1 in./min. At what rate is the volume changing at the instant it begins to melt?

**60.** (Material requirement) A cube that measures 3 in. on each side is to be coated with plastic that is $\frac{1}{16}$ in. thick. Use differentials to approximate the volume of plastic needed.

**OPTIONAL**

**GRAPHING CALCULATOR/ COMPUTER EXERCISES**

**61.** Sketch the graph of the profit function

$$P(x) = 0.02x^3 - 0.4x - 1500, \qquad 0 \le x \le 80$$

of Exercise 54, draw the tangent line to the graph at the point where $x = 50$, and use your sketch to find an approximation of the maximum profit.

**62.** Sketch the graph of the function

$$N(t) = \sqrt[3]{100 - 2t + t^2}, \qquad 0 \le t \le 24$$

of Exercise 56. From the shape of this curve, determine whether the number of unemployed people is increasing at an increasing rate or increasing at a decreasing rate at $t = 5$.

**63.** Sketch the graph of the cost function

$$C(x) = x^3 - 15x^2 + 75x + 175, \qquad 0 \le x \le 30$$

of Exercise 57. Estimate the production level at which cost changes from increasing at a decreasing rate to increasing at an increasing rate and compare your estimate with the result you got in part (c) of Exercise 57.

# 4 APPLICATIONS OF THE DERIVATIVE

## CHAPTER CONTENTS

4.1 Increasing and Decreasing Functions

4.2 Finding Maxima and Minima: The First Derivative Test

4.3 Concavity, Points of Inflection, and the Second Derivative Test

4.4 Absolute Maxima and Minima

4.5 Curve Sketching

4.6 Optimization Problems

4.7 Business and Economics Applications

4.8 Chapter 4 Review

## 4.1 INCREASING AND DECREASING FUNCTIONS

Suppose the monthly profit your company makes from the sale of $x$ thousand units of one of its products is given by

$$P = f(x) = -x^3 + 20x^2 + 40x - 100$$

where $P$ is measured in hundreds of dollars. It would be very valuable to know over which production levels the profit is increasing and over which production levels it is decreasing. You will actually answer this question in Example 5.

### Straight Lines: Increasing and Decreasing

Let us begin our discussion of increasing and decreasing functions by looking at the graphs of some straight lines. In Figure 4.1, three straight lines are sketched. The line in Figure 4.1(a) has positive slope $\left(m = \frac{1}{2}\right)$ and always is rising as we move along it from left to right. We say that $f(x) = \frac{1}{2}x + 1$ is *increasing* on the interval $(-\infty, \infty)$. The line in Figure 4.1(b) has negative slope $\left(m = -\frac{1}{3}\right)$ and is always falling as we move along it from left to right. The function $g(x) = -\frac{1}{3}x + 1$ is *decreasing* on the interval $(-\infty, \infty)$. Finally, the line in Figure 4.1(c) has zero slope ($m = 0$) and neither rises nor falls as we move along it from left to right. We say that $h(x) = 3$ is neither increasing nor decreasing on the interval $(-\infty, \infty)$.

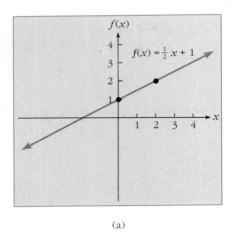
(a)

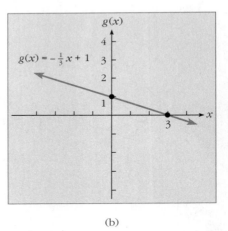
(b)

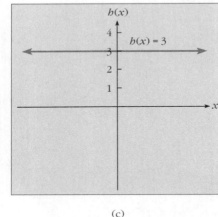
(c)

**FIGURE 4.1** Straight lines: (a) increasing, (b) decreasing, and (c) neither increasing nor decreasing.

### Arbitrary Functions: Increasing and Decreasing

Of course, straight lines cannot change their direction, so that any function of the form $f(x) = mx + b$ is increasing on $(-\infty, \infty)$ if $m > 0$, decreasing on $(-\infty, \infty)$ if $m < 0$, and constant (that is, neither increasing nor decreasing) on $(-\infty, \infty)$ if $m = 0$. However, the graph of an arbitrary function $y = f(x)$ can certainly change from increasing to decreasing and vice versa. For example, referring to the graph sketched in Figure 4.2, the function $f(x)$ is increasing on the interval $(-\infty, -1)$, decreasing on the interval $(-1, 3)$, and increasing on the

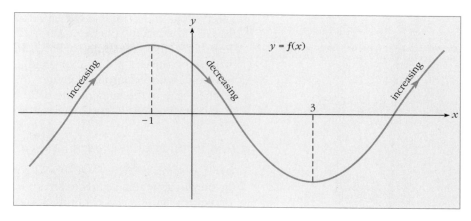

FIGURE 4.2 A function that is increasing on $(-\infty, -1)$ and $(3, \infty)$ and decreasing on $(-1, 3)$.

interval $(3, \infty)$. Typically, we specify intervals on which a function is increasing or decreasing as open intervals in the domain of the function.

To continue our discussion of increasing and decreasing functions, let us consider the graph of the function

$$f(x) = x^2 - x - 6$$

We sketched this parabola in Example 2 of Section 1.5. This graph is reproduced in Figure 4.3. As you can see by inspecting this graph, as we move along the curve from left to right, we first go "downhill" (that is, $y$ decreases) until we get to the point $P(\frac{1}{2}, -\frac{25}{4})$, which is the vertex of the parabola. Moving beyond the vertex from left to right along the curve, we go "uphill" (that is, $y$ increases). We say that $f(x)$ is decreasing on the interval $\left(-\infty, \frac{1}{2}\right)$ and is increasing on the

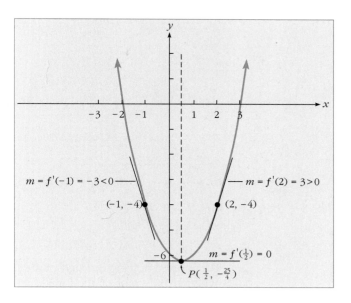

FIGURE 4.3 The graph of $f(x) = x^2 - x - 6$.

interval $\left(\frac{1}{2}, \infty\right)$. Notice that we have also drawn some tangent lines on the graph. In fact, if we draw a tangent line to the graph at any point with $x$-coordinate less than $\frac{1}{2}$, the tangent line will be falling—that is, the slope of the tangent line will be negative. On the other hand, if we draw a tangent line to the graph at any point with $x$-coordinate greater than $\frac{1}{2}$, the tangent line will be rising—that is, the slope of the tangent line will be positive. At the smooth "turning point" $P$, the tangent line is horizontal and so has a slope of zero.

### The Relation of Intervals of Increase and Decrease to the Derivative

An inspection of Figure 4.4 should convince you that the observations we just made will generalize easily to an arbitrary function $y = f(x)$. The curve we have drawn in Figure 4.4 is continuous (no breaks or holes, roughly speaking) and differentiable (smooth, with no sharp corners). This function is increasing on the interval $(-\infty, a)$ so, for any point $P$ having an $x$-coordinate $c < a$, the derivative $f'(c)$ is positive and the tangent line drawn at $P$ has a positive slope. The function is decreasing on the interval $(a, b)$ so, for any point $Q$ having an $x$-coordinate $d$ where $a < d < b$, the derivative $f'(d)$ is negative and the tangent line drawn at $Q$ has a negative slope. Finally, this function is increasing on the interval $(b, \infty)$ so, for any point $R$ having an $x$-coordinate $e > b$, the derivative $f'(e)$ is positive and the tangent line drawn at $R$ has a positive slope. To summarize:

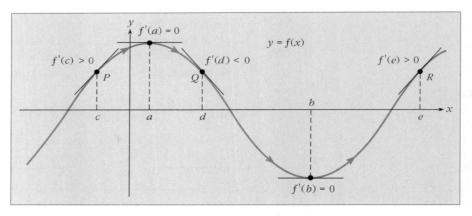

**FIGURE 4.4** Intervals of increase and decrease and the derivative.

---

If a function $f$ is differentiable at all points in an open interval $I$, then

i. $f(x)$ is increasing on $I$ if $f'(x) > 0$ for all numbers $x$ in $I$;
ii. $f(x)$ is decreasing on $I$ if $f'(x) < 0$ for all numbers $x$ in $I$.

---

Therefore, knowing the sign of the first derivative $f'(x)$ on an interval $I$ tells us whether the graph of the function $f(x)$ is increasing or decreasing over that interval. In fact, we say that $f$ is *increasing at a point* $x = a$ in the domain of $f$ if $f'(a) > 0$, and we say that $f$ is *decreasing at a point* $x = a$ in the domain if $f'(a) < 0$.

If a function $f(x)$ is increasing on an interval $I$, then $I$ is called an *interval of increase* of $f(x)$, and if $f(x)$ is decreasing on $I$, then $I$ is called an *interval of decrease* of $f(x)$.

EXAMPLE 1

For a continuous and differentiable function $y = f(x)$, suppose that $f'(x) < 0$ if $x < -3$, that $f'(x) > 0$ if $-3 < x < 2$, and that $f'(x) < 0$ if $x > 2$.

(a) Determine the intervals on which $f(x)$ is increasing and those on which it is decreasing.
(b) Sketch one possible graph $f(x)$.

Solution

(a) Since $f'(x) < 0$ if $x < -3$, we know that $f(x)$ is decreasing on the interval $(-\infty, -3)$. Since $f'(x) > 0$ if $-3 < x < 2$, we know that $f(x)$ is increasing on the interval $(-3, 2)$. Finally, $f'(x) < 0$ if $x > 2$, so $f(x)$ is decreasing on $(2, \infty)$.
(b) One possible graph of $f(x)$ is sketched in Figure 4.5.

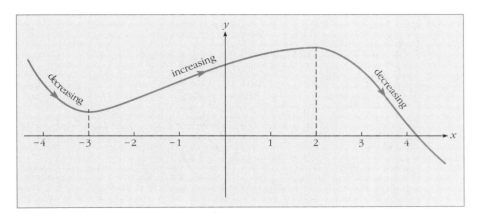

**FIGURE 4.5** A possible graph of the function described in Example 1.

Although any possible graph of $f(x)$ must be "connected" and "smooth" (since $f(x)$ is continuous and differentiable), other graphs are possible. Try drawing a different one.

### Critical Numbers

If we are given the graph of a function, it is an easy matter to determine (approximately) the intervals on which it is increasing and those on which it is decreasing. In general, however, we will usually be interested in finding intervals on which the function increases or decreases given only the formula for $f(x)$. Indeed, the general idea will be to analyze the formula for $f(x)$ and its derivatives to help us sketch the graph of $f(x)$. In determining the intervals on which a function $y = f(x)$ increases or decreases, it is important to decide where the function can *change* from increasing to decreasing and vice versa. Inspecting Figure 4.4 again, we see that $f(x)$ changes from increasing to decreasing when $x = a$, which is where $f'(a) = 0$. Also, the function in Figure 4.4 changes from decreasing to increasing at $x = b$, when $f'(b) = 0$. So apparently a function

$f(x)$ may change from increasing to decreasing or vice versa at a number $c$ in the domain of $f$ having the property that $f'(c) = 0$. Such a number is critical in determining the behavior of the graph of $f(x)$ and it is, in fact, called a *critical number*, or a *first-order critical number*, since it is related to the first derivative. Inspecting Figure 4.6(a) should convince you that a function can also change from increasing to decreasing or vice versa at a number $c$ in the domain of $f$ having the property that $f'(c)$ does not exist. This type of number is also called a critical number.

**DEFINITION: FIRST-ORDER CRITICAL NUMBER**

If $c$ is a number in the domain of $f(x)$ and if $f'(c) = 0$ or $f'(c)$ does not exist, then $c$ is a critical number of $f(x)$.

Finally, in Figure 4.6(b) we see that $f(x)$ can change from increasing to decreasing or vice versa at a number where $f(x)$ is undefined or discontinuous—such as at a vertical asymptote of $f(x)$.

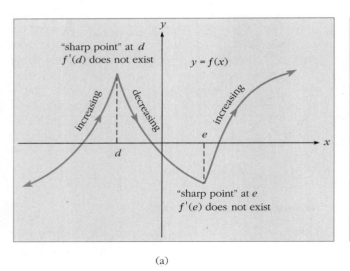
(a)

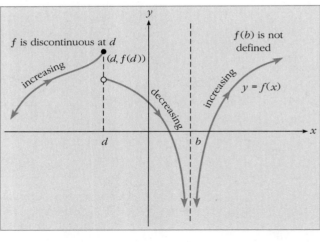
(b)

**FIGURE 4.6** A function $f(x)$ can change from increasing to decreasing or vice versa at a number where (a) $f'$ does not exist, or (b) where $f$ is discontinuous.

Summarizing, we have the following.

**NUMBERS WHERE f(x) CAN CHANGE FROM INCREASING TO DECREASING OR VICE VERSA**

1. Any numbers $c$ in the domain of $f$ such that $f'(c) = 0$.
2. Any numbers $c$ in the domain of $f$ such that $f'(c)$ does not exist.

} First-Order Critical Numbers

3. Any numbers $c$ at which $f(c)$ is undefined or discontinuous.

### The Intermediate Value Theorem and Intervals of Increase and Decrease

In order to develop a procedure for determining when a function is increasing and when it is decreasing, we will use a result called the *intermediate value theorem*. This theorem states, "If $G(x)$ is continuous on $[a, b]$ and if $k$ is any number between $G(a)$ and $G(b)$, then there exists at least one number $x_0$ in $[a, b]$ such that $G(x_0) = k$." More specifically, we need a corollary of this theorem that can be obtained by taking $k = 0$ in the intermediate value theorem. This corollary states that if a function is continuous and never zero on an interval, then the function cannot change signs within the interval. More formally, we have the following.

**COROLLARY OF THE INTERMEDIATE VALUE THEOREM**

> If $G(x)$ is continuous on an interval $I$ and if $G(x) \neq 0$ for all $x$ in $I$, then either $G(x) > 0$ for all $x$ in $I$ or $G(x) < 0$ for all $x$ in $I$.

Intuitively, this corollary is quite appealing. If a graph is connected (continuous), it cannot pass from being above the $x$-axis to being below the $x$-axis (or vice versa) without passing *through* the $x$-axis. So if a continuous graph does not pass through the $x$-axis (that is, $G(x) \neq 0$), then it must either stay above the $x$-axis ($G(x) > 0$) or stay below the $x$-axis ($G(x) < 0$). Although a formal proof of these results is beyond the scope of this text, drawing some graphs (see, for example, Figure 4.7) makes the corollary plausible. The graphs in Figures 4.7(a) and (b) satisfy the corollary, and the graph in Figure 4.7(c) shows that in order for a continuous function $G(x)$ to change from being positive to being negative, it must pass through the intermediate value—zero—at some point $x_0$.

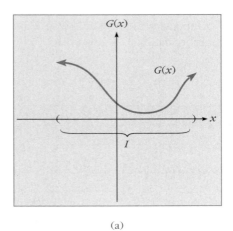
(a)

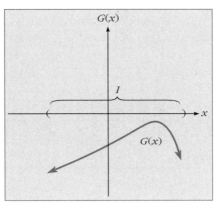
(b)

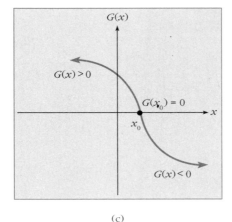
(c)

**FIGURE 4.7** Some illustrations of the corollary of the intermediate value theorem.

## EXAMPLE 2

The function $G(x) = 3x - 6$ is continuous and never zero on the interval $(2, \infty)$. Determine the sign of $G(x)$ in this interval.

### Solution

By the corollary to the intermediate value theorem, $G(x)$ cannot change its sign within the interval $(2, \infty)$. Therefore it is enough to determine the sign of $G(x)$ for a *single number x* within this interval. Using $x = 3$ as a test value, $G(3) = 3(3) - 6 = 3 > 0$, hence we conclude that $G(x) > 0$ for all $x$ in $(2, \infty)$ because it is positive for one such value ($x = 3$) and cannot change its sign within the interval. Note that any other number greater than 2 (for example, 10 or $\sqrt{503}$) could have been used as our test value in this example. ∎

## EXAMPLE 3

Determine the intervals on which $f(x) = \frac{3}{2}x^2 - 6x$ is increasing and those on which it is decreasing.

### Solution

First, observe that $f'(x) = 3x - 6$. We can find critical numbers of $f(x)$ for which $f'(x) = 0$ by solving the equation $f'(x) = 0$; that is $3x - 6 = 0$, so $x = 2$. Since $f'(x)$ exists for all real numbers, there are no other critical numbers. Also, there are no numbers at which $f(x)$ is undefined, since the domain of $f(x)$ is all real numbers. We conclude that there is only one number where $f(x)$ can change from increasing to decreasing or vice versa, and that is at the critical number $x = 2$. We can now use the corollary to the intermediate value theorem with $G(x) = f'(x) = 3x - 6$ to determine the sign of $f'(x)$ in each of the two intervals $(-\infty, 2)$ and $(2, \infty)$, since $G = f'$ is continuous and nonzero in each of the intervals. In the interval $(-\infty, 2)$, let us choose $x = 0$ as a test value and compute $G(0) = f'(0) = 3(0) - 6 = -6 < 0$. Therefore $f'(x) < 0$ for all $x$ in the interval $(-\infty, 2)$ so $f(x)$ is decreasing on that interval. In the interval $(2, \infty)$, let us choose $x = 4$ as a test value to test the sign of $G(x) = f'(x)$. We find that $G(4) = f'(4) = 3(4) - 6 = 6 > 0$, so that $f'(x) > 0$ for all $x$ in the interval $(2, \infty)$, and so $f(x)$ is increasing on the interval $(2, \infty)$.

These results are summarized in the accompanying chart (which is sometimes called a *sign diagram* of $f'$), where the abbreviation **CN** is used for "critical number."

| $x$ | 0 | 2 (CN) | 4 |
|---|---|---|---|
| $f'$ | $-----$ | 0 | $+++++$ |
| $f$ | decreasing | | increasing |

∎

Our procedure for determining the intervals on which $f(x)$ is increasing and those on which $f(x)$ is decreasing can be summarized in the following steps.

Step 1 Find the numbers where $f(x)$ can change from increasing to decreasing or vice versa: critical numbers and numbers at which $f(x)$ is undefined or discontinuous.

Step 2 The numbers (if any) found in Step 1 will divide the $x$-axis into a number of open intervals. Choose one test value in each of these intervals.

Step 3 Evaluate $f'(x)$ at each of the test values chosen in Step 2 and note the sign of each evaluation.

Step 4 The function $f(x)$ is increasing on the intervals where the evaluated derivative tests positive in Step 3 and $f(x)$ is decreasing on the intervals where the evaluated derivative tests negative in Step 3.

### EXAMPLE 4

Determine the intervals on which $f(x) = x^4 - \frac{4}{3}x^3 - 12x^2 + 1$ is increasing and those on which it is decreasing.

### Solution

First observe that $f'(x) = 4x^3 - 4x^2 - 24x$ and that the only places where $f(x)$ can change from increasing to decreasing or vice versa are at those numbers $x$ where $f'(x) = 0$. This is because $f'(x)$ and $f(x)$ are defined for all real numbers $x$. Accordingly, we find the critical numbers by solving the equation $4x^3 - 4x^2 - 24x = 0$:

$$4x^3 - 4x^2 - 24x = 0$$
$$(4x)(x^2 - x - 6) = 0$$
$$(4x)(x - 3)(x + 2) = 0$$
$$x = -2, \quad x = 0, \quad x = 3$$

In this case there are three critical numbers of $f(x)$. To complete the solution we construct the following chart, using $-3$, $-1$, $1$, and $4$ as our test values in the intervals $(-\infty, -2)$, $(-2, 0)$, $(0, 3)$, and $(3, \infty)$, respectively.

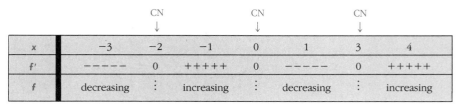

Our conclusions are that $f(x)$ is decreasing on $(-\infty, -2)$, increasing on $(-2, 0)$, decreasing on $(0, 3)$, and increasing on $(3, \infty)$. With a small amount of additional work, we can obtain a good graph of $f(x)$. We compute: $f(-2) = -\frac{61}{3}$, $f(0) = 1$, and $f(3) = -62$. We now have the values of $f(x)$ at each of the three critical points. Using this information together with the intervals of increase and decrease and the fact that $f(x)$ is continuous and differentiable, we can roughly sketch the graph in Figure 4.8.

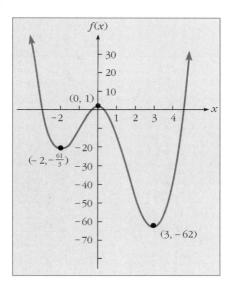

**FIGURE 4.8** The graph of $f(x) = x^4 - \frac{4}{3}x^3 - 12x^2 + 1$.

## EXAMPLE 5 (Participative)

Let us determine the intervals on which the monthly profit function

$$P(x) = -x^3 + 20x^2 + 40x - 100$$

is increasing and those on which it is decreasing. For this function, $x$ is thousands of units produced and sold and $P$ is profit in hundreds of dollars.

**Solution**

We begin by finding $P'(x) = $ _____. Since $P(x)$ and $P'(x)$ exist for all real numbers, $P(x)$ can change sign only at critical numbers $x$ such that $P'(x) = 0$. Accordingly, we find the critical numbers of $P(x)$ by solving the equation

$$-3x^2 + 40x + 40 = 0$$

The solution of this quadratic equation can be found by using the quadratic formula $x = (-b \pm \sqrt{b^2 - 4ac})/(2a)$ with $a = -3$, $b = $ _____, and $c = $ _____. We find that

$$x = \frac{-40 \pm \sqrt{(40)^2 - 4(-3)(40)}}{2(-3)} = \frac{-40 \pm \sqrt{2080}}{-6}$$

This gives approximate solutions $x = -0.935$ and $x = $ _____. Since we are interested only in values of $x$ that are nonnegative, let us choose as test values $x = 1$ and $x = 20$ in each of the intervals $(0, 14.268)$ and $(14.268, \infty)$ for use in our sign diagram of $P'$.

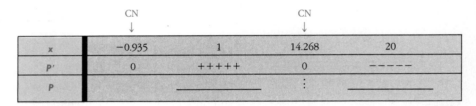

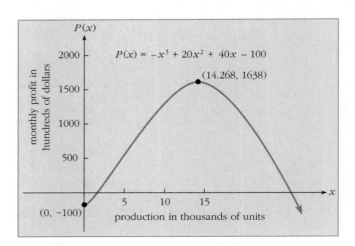

**FIGURE 4.9** The graph of $P(x) = -x^3 + 20x^2 + 40x - 100$.

Our conclusion is that $P(x)$ is increasing on the interval $(0, 14.268)$ and decreasing on the interval _____. Thus our profit will increase up to a production and sales level of about 14,268 units. If we produce more than this number of units, our profit will decrease.

We can graph the profit function by observing that $P(0) = -100$ and $P(14.268) \approx 1637.62$. Using these facts and the intervals of increase and decrease just found, we obtain the sketch of Figure 4.9.

### EXAMPLE 6

Determine the intervals on which $f(x) = 5x^{2/5}$ is increasing and those on which it is decreasing.

### Solution

Note that $f'(x) = 2x^{-3/5} = 2/\sqrt[5]{x^3}$. Now, $f'(x)$ is never zero since for a fraction $P/Q$ to be equal to zero we must have $P = 0$, and the numerator of $f'(x)$ is equal to 2. Also, $f(x) = 5x^{2/5} = 5\sqrt[5]{x^3}$ is defined for all $x$, so if $f(x)$ is to have a critical number, it must be at a point at which $f'(x)$ is undefined. However, when $x = 0$, notice that $f'(0) = 2/\sqrt[5]{0^2}$, which is undefined, since we are attempting to divide by zero. To continue our analysis of the graph of $f(x)$, we construct the following chart, where the abbreviation **ND** stands for "not defined."

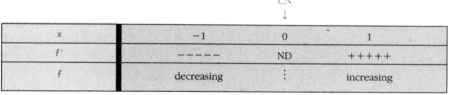

| $x$ | | $-1$ | | $0$ | | $1$ |
|---|---|---|---|---|---|---|
| $f'$ | | $-----$ | | ND | | $+++++$ |
| $f$ | | decreasing | | | | increasing |

We find that $f(x)$ is decreasing on the interval $(-\infty, 0)$ and increasing on the interval $(0, \infty)$. Notice also that as $x \to 0$, we have $|f'(x)| = |2/\sqrt[5]{x^3}| \to \infty$. This indicates that the tangent line to the graph of $f(x)$ at the origin is vertical, since as $x$ approaches 0, the absolute value of its slope increases without bound. See Figure 4.10 for a graph of $f(x)$.

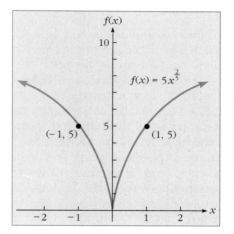

**FIGURE 4.10** A function with a vertical tangent line at (0, 0).

### EXAMPLE 7

Determine the intervals on which $f(t) = 2/(t-1)^2$ is increasing and those on which it is decreasing.

### Solution

We begin by observing that $f(t) = 2(t-1)^{-2}$, so that $f'(t) = -4(t-1)^{-3}(1) = -4/(t-1)^3$. The function $f(t)$ has no critical numbers, since $f'(t)$ is never zero and, even though $f'(1)$ is undefined, 1 is not in the domain of $f(t)$. Thus the only place that the graph of $f(t)$ can change from increasing to decreasing or vice versa is at $t = 1$, where $f(t)$ is undefined. Note that the vertical line $t = 1$ is also a vertical asymptote of the graph of $f(t)$. Next, we construct the sign diagram, where we again use the abbreviation ND for "not defined."

---

**SOLUTIONS TO EXAMPLE 5**   $-3x^2 + 40x + 40, 40, 40, 14.268$, increasing, decreasing, $(14.268, \infty)$

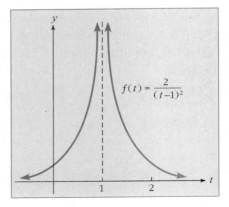

**FIGURE 4.11** A function that changes from increasing to decreasing at a number at which $f(t)$ is not defined.

|  | CN ↓ |  |  |
|---|---|---|---|
| $t$ | 0 | 1 | 2 |
| $f'$ | +++++ | ND | ----- |
| $f$ | increasing | ND | decreasing |

Thus $f(t)$ is increasing on $(-\infty, 1)$ and decreasing on $(1, \infty)$. Figure 4.11 is a sketch of the graph of $f(t)$. Observe that since $\lim_{t \to \pm\infty} f(t) = 0$, the horizontal line $y = 0$ (that is, the $t$-axis) is a horizontal asymptote of the graph of $f(t)$. ∎

### Optional Graphing Calculator/Computer Material

A graphing calculator or graphing software for a computer can be a valuable tool for finding the intervals on which a function is increasing and/or decreasing. Of course, the intervals found by these methods will usually not be exact. Even so, the use of a graphing calculator or the appropriate software can help to strengthen your understanding of the material in this section. Indeed, when you combine the calculus you have learned in this section with the use of a graphing calculator or microcomputer, you have some powerful tools for finding the intervals on which a function is increasing and those on which it is decreasing. As examples of the use of a graphing calculator or microcomputer, sketch the graph of $f(x) = x^4 - \frac{4}{3}x^3 - 12x^2 + 1$ and try to read off the intervals where $f(x)$ increases and decreases from your sketch. Check your work against the exact answers we found by using calculus in Example 4 of this section. Try the same exercise using $P(x) = -x^3 + 20x^2 + 40x - 100$. We considered $P(x)$ in Example 5. The exercises in this section contain some more graphing problems for you to do using a graphing calculator or microcomputer. Some of these exercises would be quite difficult to solve using only the knowledge of calculus we have at this point.

## SECTION 4.1
### SHORT ANSWER EXERCISES

1. For the function $f(x) = x^3 + x$, we have $f'(x) = 3x^2 + 1$, so that $f'(x) > 0$ for all numbers $x$. We conclude that $f(x)$ is _____ on the interval $(-\infty, \infty)$.

2. The graph of a certain function $g(x)$ is increasing at $x = -2$ and decreasing at $x = 0$. Thus (a) $g'(-2) < 0$ and $g'(0) < 0$; (b) $g'(-2) > 0$ and $g'(0) > 0$; (c) $g'(-2) < 0$ and $g'(0) > 0$; (d) $g'(-2) > 0$ and $g'(0) < 0$.

3. A function $f(x)$ can change from increasing to decreasing or vice versa at (a) a critical number $c$ for which $f'(c) = 0$; (b) a critical number $c$ for which $f'(c)$ is undefined; (c) a number at which $f(x)$ is undefined; (d) all of the above.

4. (True or False) The function $f(x) = 7 - 5x$ has no critical numbers. _____

5. (True or False) The function $f(x) = 1/x$ has no critical numbers. _____

6. An ant walking from left to right along the graph of the continuous function $G(x)$ marches from quadrant I into quadrant IV. We can be certain that (a) the graph of $G(x)$ is decreasing on some interval; (b) the ant has crossed the $x$-axis; (c) there is a point $x_0$ in the domain of $f$ such that $f(x_0) = 0$; (d) all of the above.

7. (True or False) If, for a certain function, $f'(2) = 0$, we are sure that $f(x)$ changes from increasing to decreasing or vice versa when $x = 2$. _____

8. (True or False) A function $f(x)$ may change from increasing to decreasing or vice versa as $x$ passes through a vertical asymptote of $f(x)$. _____

9. When we use test values to determine the sign of $f'(x)$ in the various intervals determined by the critical points of $f(x)$, what theorem are we using?
_____

10. A function $y = f(x)$ has critical numbers at $x = -2$ and $x = 5$ and is undefined at $x = 0$. What are the intervals in which our test values must be chosen when we are finding the intervals on which $f$ is increasing and decreasing?
_____

## SECTION 4.1 EXERCISES

In Exercises 1–24, determine the intervals on which each function is increasing or decreasing.

1. $f(x) = 2x - 5$
2. $g(x) = 10 - 5x$
3. $y = x^2 + x - 2$
4. $y = 6x^2 + x - 1$
5. $f(x) = 2 + x - 3x^2$
6. $y = 9x - x^3$
7. $g(x) = x^3 - x^2 - x + 3$
8. $g(x) = \frac{2}{3}x^3 - \frac{1}{2}x^2 - 3x - 2$
9. $f(x) = 2x^3 + 3x^2 - x - 1$
10. $f(x) = 5 + 3x - x^2 - x^3$
11. $f(x) = x^4 + 2$
12. $f(x) = x^4 - 2x^3$
13. $y = 4x^2 - x^4$
14. $f(x) = \frac{1}{4}x^4 + \frac{1}{3}x^3 - x^2 - 3$
15. $y = \frac{1}{2}x^4 - \frac{5}{3}x^3 + \frac{3}{2}x^2 - 1$
16. $g(x) = x^{2/3}$
17. $f(x) = -3x^{2/5}$
18. $y = 3\sqrt[3]{x}$
19. $y = 2/\sqrt[3]{x}$
20. $f(x) = 1/x^2$
21. $f(x) = \dfrac{4}{x - 2}$
22. $f(x) = \dfrac{2x + 1}{x + 3}$
23. $f(x) = 2x - (1/x)$
24. $f(x) = \sqrt{x} - x$

25. (Efficiency of a worker) The efficiency $E$ (measured on a scale from 0 to 1) of a certain worker on an assembly line during a three-hour work period is given by

$$E(t) = \frac{1}{1.2 + (t - 1.2)^2}, \quad 0 \le t \le 3$$

Find the intervals on which the worker's efficiency is increasing and those on which it is decreasing.

26. (Cost and demand) The monthly cost function of our company is $C(x) = 500 + 40x$, where $x$ is the number of units produced per month and $C(x)$ is the cost of production in dollars. Also, suppose we can sell $x$ units per month if our price is $p$ dollars, where $p = 1000 - 2x$.
(a) Find the revenue function and the intervals on which it increases and decreases.
(b) Find the profit function and the intervals on which it increases and decreases.

**27.** (Average cost)  Given the average cost function

$$\overline{C}(x) = \frac{x^3 + 1000}{2x}, \quad x > 0$$

where $\overline{C}(x)$ is the average weekly cost of producing $x$ thousand units of a certain product, determine the intervals on which $\overline{C}(x)$ is decreasing and those on which $\overline{C}(x)$ is increasing.

**28.** (Profit)  The monthly profit in dollars on the sale of $x$ hundred units of a certain product is $P(x) = -2000 + 6x^2 - 0.02x^3$. Find the intervals on which $P(x)$ is increasing and those on which it is decreasing.

Business and Economics

**29.** (Average cost)  The manufacturing department of your company finds that its fixed costs are $800 per day, that there is a unit production cost of $5 per unit produced, and that maintenance and repairs cost $x^2/10000$ dollars per day when $x$ is the production level.
 (a) Find an expression for the total daily cost $C(x)$ of producing $x$ units.
 (b) Find an expression for $\overline{C}(x) = C(x)/x$, the average daily cost per unit.
 (c) Find the intervals on which $\overline{C}(x)$ is increasing and those on which it is decreasing.

**30.** (Marginal cost)  If the monthly cost of producing $x$ thousand units is $C(x) = \frac{1}{3}x^3 - 5x^2 + 28x + 30$, where $C(x)$ is measured in hundreds of dollars, find the intervals on which the marginal cost $dC/dx$ is increasing and those on which it is decreasing.

**31.** (Advertising and sales)  Daily sales in units of our product $t$ days after we begin a new advertising campaign are projected to be

$$S(t) = \tfrac{1}{3}t^3 - 21t^2 + 392t + 500, \quad 0 \le t \le 42$$

Find the intervals on which sales are increasing and those on which they are decreasing.

Social Sciences

**32.** (Crime rate)  In a certain small town, the average number of crimes $y$ reported per day depends upon the day of the year $x$, where $1 \le x \le 365$, according to

$$y = -0.0003x^2 + 0.1098x + 10$$

Determine the intervals on which the average number of crimes reported per day are increasing and those on which they are decreasing.

---

OPTIONAL

**GRAPHING CALCULATOR/ COMPUTER EXERCISES**

*Graph each of the following functions and determine approximately the intervals on which each function is increasing or decreasing. Where indicated, use calculus to find the exact intervals and compare with the answer you got using your graphing calculator or computer.*

**33.** $f(x) = x^4 - 3x^3 + 7x^2 - 10x + 3$
**34.** $f(x) = -5 - 4x - 3x^2 + x^3 + x^4$
**35.** $f(x) = 5x^3 - 2x^2 + 3x - 4$: check using calculus
**36.** $f(x) = -x^3 + 2x^2 - 7x + 6$: check using calculus
**37.** $f(x) = x^5 - 4x^4 + 3x^3 - 7x^2 + 2x + 1$

## 4.2 FINDING MAXIMA AND MINIMA: THE FIRST DERIVATIVE TEST

### Relative Extrema

In the last section we discovered how to determine the intervals on which a function is increasing and those on which it is decreasing. Intervals of increase and decrease of a function are closely related to what are called relative maxima and relative minima.

Informally, a *relative maximum* point of the graph of a function $y = f(x)$ is a point on the graph where the function changes from increasing to decreasing. At the relative maximum point, the function is neither increasing nor decreasing. Relative maxima are the hilltop or peaks of the graph. For example, in Figure 4.12, the points $P(a, f(a))$ and $R(c, f(c))$ are relative maximum points of the graph of $y = f(x)$. A *relative minimum* point of the graph of $y = f(x)$ is a point on the graph where the function changes from decreasing to increasing. Relative minima are the valleys of the graph. In Figure 4.12, the relative minimum point of the graph is $Q(b, f(b))$.

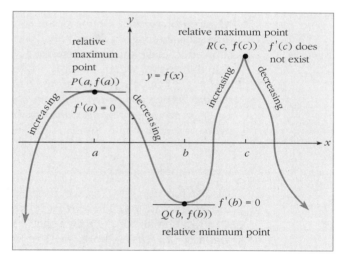

**FIGURE 4.12** Relative maximum and relative minimum points of a function.

The point $P(a, f(a))$ in Figure 4.12 is a *relative* maximum (as opposed to an *absolute* maximum, which we will discuss in a later section of this book) because it is the highest point on the graph relative to immediately neighboring points. That is, if we localize our viewpoint to a small region of the graph near $P$, then the point $P$ is the highest point on the graph in this local neighborhood. For this reason a relative maximum is sometimes referred to as a *local maximum*. Similar comments hold for the relative (or local) minimum point $Q(b, f(b))$, which is the lowest point on the graph relative to immediately neighboring points. Notice that a relative maximum point need not be the highest point on the entire graph (see the point P in Figure 4.12, for example). It only must be the highest point in a local vicinity around itself. Together, relative maxima and minima are referred to as *relative extrema*. A formal definition follows.

### DEFINITION: RELATIVE EXTREMA

i. A point $(c, f(c))$ is called a *relative maximum point of the graph of* $y = f(x)$ if there is an open interval $I$ about $c$ such that $f(x) \leq f(c)$ for all numbers $x$ in the interval.

ii. A point $(c, f(c))$ is called a *relative minimum point of the graph of* $y = f(x)$ if there is an open interval $I$ about $c$ such that $f(x) \geq f(c)$ for all numbers $x$ in the interval.

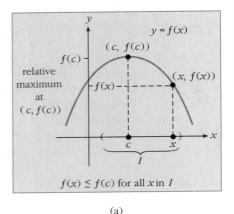

Figure 4.13 illustrates the definition of relative extrema. In Figure 4.13(a), if you pick any $x$ in the open interval $I$ that contains $c$, then $f(x) \leq f(c)$; that is, the point $(x, f(x))$ is a lower point on the graph than the point $(c, f(c))$, which is the relative maximum point. In Figure 4.13(b), choosing any $x$ in the interval $I$ shown will produce $f(x) \geq f(c)$; that is, the point $(x, f(x))$ is a higher point on the graph than the relative minimum point $(c, f(c))$. Try a different $x$ in $I$ for yourself and verify this. You should also observe that the intervals chosen in Figure 4.13 are not unique. Many other intervals containing $c$ also have the required properties. By the definition of relative extrema, we need only find *one* such interval $I$ to verify that we have a relative extremum, and we are allowed to make this open interval $I$ as small as we wish.

In this text we will distinguish between a relative extreme *point* on the graph of $y = f(x)$ and a relative extreme *value* of $y = f(x)$. *If the point* $(c, f(c))$ *is a relative maximum* **point** *of the graph of* $y = f(x)$, *we say that the function* $y = f(x)$ *has a relative maximum* **value** $f(c)$, *and that this relative maximum value occurs at* $x = c$. Similar comments hold for relative minimum points.

### Locating Relative Extrema

Given a function $y = f(x)$, where can relative extrema of $f(x)$ occur? By inspecting Figure 4.12, you will see that relative extrema of $f(x)$ can occur at numbers $k$ in the domain of $f(x)$ such that $f'(k) = 0$ (that is, the tangent line to the graph at $(k, f(k))$ is horizontal) or where $f'(k)$ does not exist (because the tangent line to the graph at $(k, f(k))$ either is vertical or does not exist there). But these numbers $k$ are just the critical numbers of $y = f(x)$.

If $(c, f(c))$ is a relative extreme point of $y = f(x)$, then $c$ is a critical number of $f(x)$.

FIGURE 4.13 An illustration of the definition of relative extrema.

### EXAMPLE 1

Find the values of $x$ where relative extrema of $f(x)$ can occur if $f(x) = x^2 - x$.

Solution

Since relative extrema can occur only at critical numbers, we first find $f'(x) = 2x - 1$ and equate $f'(x)$ to zero:

$$2x - 1 = 0, \quad \text{or} \quad x = \frac{1}{2}$$

Thus, *if* $f(x)$ has a relative extremum, that extremum must occur at $x = \frac{1}{2}$. If we sketch the graph of $f(x)$, we see that a relative minimum does actually occur at $x = \frac{1}{2}$, and that relative minimum value is $f(\frac{1}{2}) = (\frac{1}{2})^2 - \frac{1}{2} = -\frac{1}{4}$. See Figure 4.14.

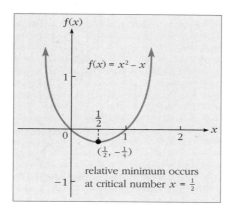

**FIGURE 4.14** The graph of $f(x) = x^2 - x$.

---

EXAMPLE 2

Find the values of $x$ where relative extreme values of $f(x)$ can occur if $f(x) = x^3$.

Solution

We know that relative extreme values can occur only at critical numbers, so we compute $f'(x) = 3x^2$. To find the critical numbers of $f(x)$ we set $f'(x)$ equal to zero to obtain $3x^2 = 0$, so $x = 0$. We conclude that *if* $f(x)$ has a relative extreme value, then that value must occur at $x = 0$. However, if we sketch the graph of $f(x)$, we find that no relative extremum occurs when $x = 0$. See Figure 4.15.

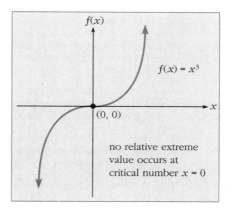

**FIGURE 4.15** The graph of $f(x) = x^3$.

As Example 2 shows, the critical numbers of $y = f(x)$ give us all possible *candidates* of numbers where relative extreme values of $f(x)$ can exist. However, *not every candidate critical number need result in a relative extreme value of $f(x)$*! As examples of the range of possibilities, see Figure 4.16 and Figure 4.17.

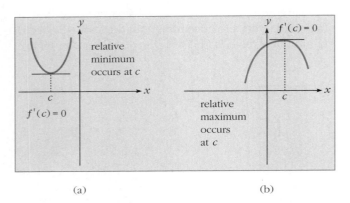
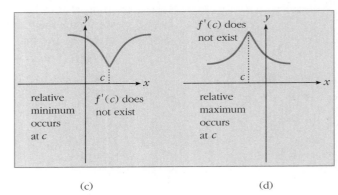

**FIGURE 4.16** Relative extrema occur at the critical number $x = c$.

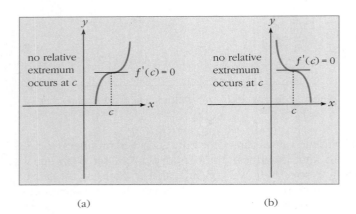
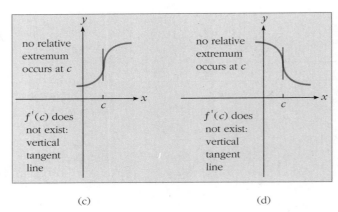

**FIGURE 4.17** No relative extrema occur at the critical number $x = c$.

### First Derivative Test

Once we have found all the critical numbers of $y = f(x)$, we need a test to determine which of these yield relative maxima, which yield relative minima, and which yield no relative extrema. One such test is called the *first derivative test*. The basis for the first derivative test is that at a relative maximum, the function $f(x)$ changes from increasing to decreasing, whereas at a relative minimum, the function $f(x)$ changes from decreasing to increasing: see Figure 4.18. Thus *if c is a critical number at which a relative maximum of $f(x)$ occurs, then $f'(x)$ changes from positive to negative at c, whereas if c is a critical number at which a relative minimum of $f(x)$ occurs, then $f'(x)$ changes from negative to positive at c*. Therefore if $f'(x)$ does not change its sign at $c$ (see the graphs in Figure 4.17), then no relative extremum occurs at $c$. Formally, it looks like this:

## 4.2 FINDING MAXIMA AND MINIMA: THE FIRST DERIVATIVE TEST

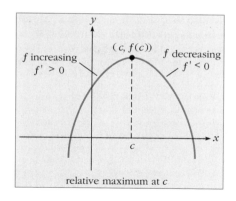
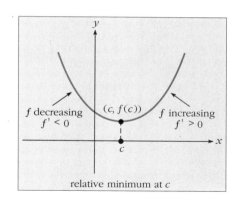

**FIGURE 4.18** The relation between relative extrema and the sign of $f'$.

**FIRST DERIVATIVE TEST**

Let $f$ be continuous on an open interval $I$ and let $c$ be any number in $I$. Then

i. If $f'$ changes from positive to negative at $c$, then $f(c)$ is a relative maximum value of $f(x)$;

ii. If $f'$ changes from negative to positive at $c$, then $f(c)$ is a relative minimum value of $f(x)$.

---

**EXAMPLE 3**

Find any relative extreme values of $f(x) = x^3 - 3x^2 - 9x + 14$.

**Solution**

To begin, we find $f'(x) = 3x^2 - 6x - 9$. To find the critical numbers of $f(x)$, we set $f'(x)$ equal to zero:

$$3x^2 - 6x - 9 = 0$$
$$3(x^2 - 2x - 3) = 0$$
$$3(x - 3)(x + 1) = 0$$
$$x = -1, \quad x = 3 \quad \text{(critical numbers)}$$

We now construct a chart showing the sign of $f'$ on the various intervals, as we did in the last section.

| $x$ | $-2$ | $-1$ | $0$ | $3$ | $5$ |
|---|---|---|---|---|---|
| $f'$ | +++++ | 0 | ----- | 0 | +++++ |
| $f$ | increasing | rel max | decreasing | rel min | increasing |

From this diagram, we see that the sign of $f'$ changes from positive to negative at the critical number $-1$, so that a relative maximum occurs at $x = -1$. The relative maximum value of $f(x)$ at $x = -1$ is

$$f(-1) = (-1)^3 - 3(-1)^2 - 9(-1) + 14 = 19$$

Also, since the sign of $f'$ changes from negative to positive at the critical number 3, a relative minimum occurs at $x = 3$. The relative minimum value of $f(x)$ at $x = 3$ is

$$f(3) = (3)^3 - 3(3)^2 - 9(3) + 14 = -13$$

See Figure 4.19 for a sketch of $f(x)$.

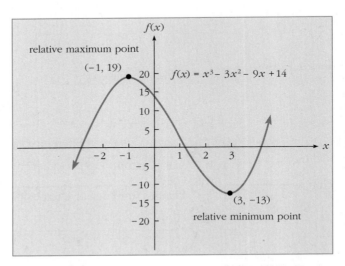

**FIGURE 4.19** The graph of $f(x) = x^3 - 3x^2 - 9x + 14$.

## EXAMPLE 4

Find any local extrema of $f(x) = (2x^2 + 1)/(x - 1)$.

### Solution

Our first task is to find $f'(x)$. For this, we use the quotient rule to obtain

$$f'(x) = \frac{(x - 1)\frac{d}{dx}(2x^2 + 1) - (2x^2 + 1)\frac{d}{dx}(x - 1)}{(x - 1)^2}$$

$$= \frac{(x - 1)(4x) - (2x^2 + 1)(1)}{(x - 1)^2}$$

$$= \frac{2x^2 - 4x - 1}{(x - 1)^2}$$

Critical numbers for which $f'(x) = 0$ can be found by setting the numerator of $f'(x)$ equal to zero:

$$2x^2 - 4x - 1 = 0$$

Using the quadratic formula, we get

$$x = \frac{-(-4) \pm \sqrt{(-4)^2 - 4(2)(-1)}}{2(2)} = \frac{4 \pm \sqrt{24}}{4}$$

Since $\sqrt{24}$ can be simplified to $2\sqrt{6}$, the two critical points are $x = 1 \pm \frac{1}{2}\sqrt{6}$ or, approximately, $x \approx -0.255$ and $x \approx 2.225$. There are no other critical points because, although $f'(x)$ is not defined at $x = 1$, the number $x = 1$ is not in the domain of $f(x)$. We observe that a possible sign change in $f'$ can occur at $x = 1$, however, since $f(1)$ is undefined. This possible sign change in $f'$ is important in determining intervals of increase and decrease of $f(x)$, as we discovered in Section 4.1. But—and this is important—a relative extremum *cannot* occur at $x = 1$ since $x = 1$ is not a critical number. In fact, the vertical line $x = 1$ is a vertical asymptote of $f(x)$. We will, nonetheless, include $x = 1$ in our chart, since a possible change in the sign of $f'$ is important behavior that we should observe to graph the function properly. In the chart, the abbreviation **VA** stands for "vertical asymptote."

| x | −2 | −0.225 | 0 | 1 | 2 | 2.225 | 3 |
|---|---|---|---|---|---|---|---|
| f' | +++++ | 0 | ----- | ND | ----- | 0 | +++++ |
| f | increasing | rel max | decreasing | VA | decreasing | rel min | increasing |

CN ↓ above −0.225; CN ↓ above 2.225

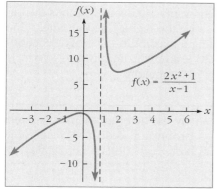

**FIGURE 4.20** The graph of $f(x) = (2x^2 + 1)/(x - 1)$.

From the chart we see that a relative maximum occurs at $x \approx -0.225$ and the relative maximum value is (approximately)

$$f(-0.225) = \frac{2(-0.225)^2 + 1}{-0.225 - 1} \approx -0.90$$

Also, a relative minimum of $f(x)$ occurs at $x \approx 2.225$ and the relative minimum value is (approximately)

$$f(2.225) = \frac{2(2.225)^2 + 1}{2.225 - 1} \approx 8.90$$

The graph is sketched in Figure 4.20.  ∎

---

EXAMPLE 5 (Participative)

Solution

Let's find the relative extrema of $f(x) = \sqrt[3]{(x-1)^2} + 1$.

To begin, we write $f(x) = (x - 1)^{2/3} + 1$ and compute

$f'(x) = $ _____

The simplified form of $f'(x)$ is $f'(x) = 2/(3\sqrt[3]{x - 1})$. To find critical numbers, we must determine for what values $f'(x)$ is equal to zero and for what values $f'(x)$ is _____. Note that there are no critical numbers for which $f'(x) = 0$, since the numerator of $f'(x)$ is equal to 2 and so can never be equal to zero. However, since the denominator of $f'(x)$ is $3\sqrt[3]{x - 1}$, this denominator is equal to zero (and hence $f'$ is undefined) when $x = $ \_\_\_\_\_. Therefore $x = 1$ is a critical number. Complete the given chart.

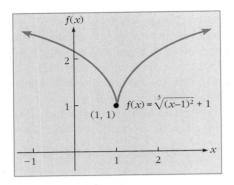

**FIGURE 4.21** A function with a vertical tangent line at $(1, 1)$.

| $x$ | | 0 | 1 | 2 |
|---|---|---|---|---|
| $f'$ | | —— | ND | —— |
| $f$ | | decreasing | rel min | increasing |

CN ↓ (above the column for $x = 1$)

We find that a relative minimum of $f(x)$ occurs at the critical number $x = 1$; the relative minimum value of $f(x)$ there is $f(1) = $ ____. See Figure 4.21 for a sketch of $f(x)$. Note that the tangent line to $f(x)$ at the point $(1, 1)$ is vertical. ∎

The procedure of finding maxima and minima of functions is of great value in solving applied problems in the areas of business and social science. For example, a company may desire to find the level of output that will generate the maximum profit, or a social worker might want to know the periods during which the unemployment rate is expected to be the highest over the next few years. Each of these problems involves finding a maximum value of a function. We will conclude this section with an applied example from business in which we are required to find the minimum value of a function.

## EXAMPLE 6

The manufacturing department of your company has determined that its fixed costs are \$1500 per day, that there is a unit production cost of \$30 per unit produced, and that maintenance and repairs cost $\frac{1}{5000}x^2$ dollars per day when the production level is $x$ units per day.

(a) Find an expression for the total daily cost $C(x)$ of producing $x$ units.
(b) Find an expression for the average daily cost per unit.
(c) Find the level of production necessary to minimize the average daily cost per unit and find the minimum average daily cost per unit.

Solution

(a) Fixed costs are \$1500 per day, and if $x$ units are produced per day, then production costs are $30x$ dollars, so when we add the maintenance cost of $\frac{1}{5000}x^2$ dollars per day, we have a total daily cost (in dollars) of

$$C(x) = 1500 + 30x + \frac{x^2}{5000}, \quad x \geq 0$$

when $x$ units are produced daily.

(b) Since average cost is total cost divided by the production level, we have

$$\overline{C}(x) = \frac{C(x)}{x} = \frac{1500}{x} + 30 + \frac{x}{5000}, \quad x > 0$$

as the average daily cost per unit when $x$ units are produced per day.

---

**SOLUTIONS TO EXAMPLE 5**  $\frac{2}{3}(x - 1)^{-1/3}$, undefined, 1, −, +, 1

(c) First note that

$$\overline{C}'(x) = -\frac{1500}{x^2} + \frac{1}{5000}$$

To find critical numbers, we must solve the equation $\overline{C}'(x) = 0$. We have

$$-\frac{1500}{x^2} + \frac{1}{5000} = 0$$

$$-7{,}500{,}000 + x^2 = 0 \quad \text{(Multiply by } 5000x^2\text{;}$$
$$\qquad\qquad\qquad\qquad \text{we may do so since } x \neq 0\text{)}$$

$$x^2 = 7{,}500{,}000$$

$$x \approx 2738.6 \quad \text{(Critical number)}$$

Note that although $\overline{C}'(0)$ does not exist, $x = 0$ is *not* a critical number since $\overline{C}(0)$ is not defined. Remember that a critical number of a function must be a number in the domain of the function. In fact, by looking at the formula for $\overline{C}(x)$ you can easily see that the line $x = 0$ is a vertical asymptote of $\overline{C}(x)$. We now construct our chart noting that since $x > 0$ we need not take any test values less than zero.

| $x$ | | 2737 | 2738.6 (CN) | 2740 |
|---|---|---|---|---|
| $\overline{C}'$ | | $-----$ | 0 | $+++++$ |
| $\overline{C}$ | | decreasing | rel min | increasing |

We find that a relative minimum value of $\overline{C}(x)$ occurs when $x \approx 2738.6$ and that that relative minimum value is approximately

$$\overline{C}(2738.6) = \frac{1500}{2738.6} + 30 + \frac{2738.6}{5000} \approx 31.1$$

Since $\overline{C}(x)$ is decreasing for all $x < 2738.6$ and increasing for all $x > 2738.6$, the point (2738.6, 31.1) is not only a relative minimum point of the graph of $\overline{C}(x)$ but also an absolute minimum point. That is, it is *the* lowest point on the graph: see Figure 4.22. So to minimize average daily cost per unit, we need to manufacture about 2738.6 units. At this level of production, we will attain the minimum average daily cost of about $31.10 per unit. ∎

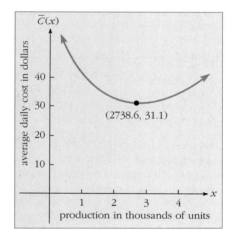

**FIGURE 4.22** The graph of $\overline{C}(x) = \frac{1500}{x} + 30 + \frac{x}{5000}$.

## SECTION 4.2
### SHORT ANSWER EXERCISES

1. (True or False) Every critical number corresponds to either a relative maximum or a relative minimum. _____

2. (True or False) A relative maximum point $P$ of a graph is always the highest point on the graph. _____

3. (True or False) Relative extrema of a function $y = f(x)$ can occur only at critical numbers of $f(x)$. _____

**4.** For a certain function $f(x)$ that has the set of all real numbers for its domain, $f'(x) = x^3 - 4x^2$. Then **(a)** $x = 0$ is a critical point of $f(x)$; **(b)** the point $(0, 0)$ is a relative extremum of $f(x)$; **(c)** $f(x)$ may have critical numbers $c$ for which $f'(c)$ is undefined; **(d)** $f(x)$ has no relative extreme values.

**5.** If $x = 2$ is a critical number of $f(x)$, then a relative minimum occurs at $x = 2$ if **(a)** $f'$ does not change its sign at 2; **(b)** $f'$ changes from positive to negative at 2; **(c)** $f'$ changes from negative to positive at 2; **(d)** none of the above.

**6.** A function $y = f(x)$, with a domain of all the real numbers, has critical numbers $x = -2$, $x = 0$, and $x = 3$. In constructing the critical number chart, appropriate test values are: **(a)** $-3, 0, 1$, and 4; **(b)** $-2, -1, 1$, and 2; **(c)** $-3, -1, 1$, and 4; **(d)** any of the above.

**7.** For the function $f(x) = \sqrt[3]{x}$, **(a)** $x = 0$ is a critical point; **(b)** $f'(0)$ does not exist; **(c)** $f(x)$ has no relative extrema; **(d)** all of the above.

**8.** The function $f(x) = (2x - 3)/(x + 1)$ cannot have a relative extremum at $x = -1$ because $-1$ is not in the _____ of $f(x)$.

**9.** If a profit function $P(x)$ is increasing on the interval $(0, 2000)$ and decreasing on the interval $(2000, \infty)$, then the maximum profit occurs at $x = $ _____.

**10.** (True or False) Every function $y = f(x)$ has at least one relative extremum. _____

## SECTION 4.2
### EXERCISES

In Exercises 1–20, find the relative extrema of the given function and sketch its graph.

**1.** $f(x) = 2x^2 - 4x + 1$
**2.** $f(x) = 5x^2 - 7x + 6$
**3.** $f(x) = 2 - 3x - x^2$
**4.** $g(x) = 5 - 4x - 2x^2$
**5.** $f(x) = 2x^3 - 21x^2 + 36x + 3$
**6.** $f(x) = 2x^3 - 15x^2 + 36x - 4$
**7.** $g(x) = \frac{1}{3}x^3 + x^2 + x - 2$
**8.** $g(x) = \frac{1}{3}x^3 - 2x^2 + 4x + 3$
**9.** $f(x) = 10 + 60x - 9x^2 - 2x^3$
**10.** $f(x) = 12 - 4x + 7x^2 - 3x^3$
**11.** $f(x) = x^3 + 3x^2 + 3x + 1$
**12.** $y = x^4 + x^3 + x^2$
**13.** $y = 6x^2 - 4x^4$
**14.** $f(x) = 3 - 2x^2 + x^3 - 3x^4$
**15.** $f(x) = 9x^4 + 4x^3 - 4x^2 - 5$
**16.** $f(x) = 10 + 15x^2 - 14x^3 - 9x^4$
**17.** $g(x) = x^5 - 80x + 8$
**18.** $g(x) = 2x^6 + 10x^5 - x^4 + 4$
**19.** $f(x) = \dfrac{x}{x - 3}$
**20.** $f(x) = \dfrac{3x}{4x + 1}$

In Exercises 21–26, find the relative extrema of the function.

**21.** $f(x) = \dfrac{x^2}{x - 2}$
**22.** $f(x) = \dfrac{2x - x^2}{x + 2}$
**23.** $f(x) = \sqrt[3]{x^2}$
**24.** $f(x) = \sqrt[5]{(x - 2)^2} + 3$
**25.** $g(x) = x + \sqrt[3]{x}$
**26.** $g(x) = 1 - 4x + 2\sqrt[5]{x}$

### Business and Economics

**27.** (Minimum average cost) A certain manufacturer has fixed costs of $2500 per day, a unit production cost of $12 per unit, and a maintenance and repair cost of $x^2/10000$ dollars per day when the production level is $x$ units per day.
**(a)** Find an expression for the total daily cost $C(x)$ of producing $x$ units.

That is, the cost of producing the 4001st unit is only about $0.40. Thus, as we increase production, each unit will cost less to produce than the preceding one. This happy state of affairs ends, however, at production level c on the graph of Figure 4.23. Beyond that production level, the graph of C(x) begins to turn upward and cost begins to increase at an increasing rate. Thus, beyond production level c, the cost of producing each additional unit will be more than the cost of producing each current unit. The reason for this may be that to produce units beyond output c, we may be required to pay workers overtime wages, to lease additional equipment, and so forth. Imagine a factory that is set up to produce 5000 units of product a month trying to produce 100,000 units a month! The tremendous overcrowding of men and machinery in our small factory would lead to inefficiency and waste, and this would be reflected in a higher and higher cost of producing each additional unit. For our cost function, the cost of producing the 100,001st unit is

$$C(100.001) - C(100) \approx 9.03 \text{ hundreds of dollars}$$

So the cost of producing the 100,001st unit would be about $903. The production level c in Figure 4.23 is called the *point of diminishing returns*. In Example 4 we will find the cost efficient production level c.

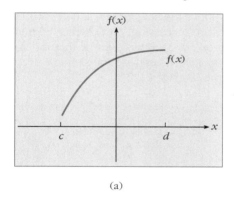
(a)

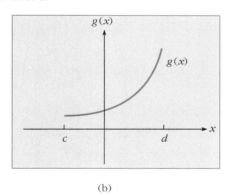
(b)

**FIGURE 4.24** Increasing functions with different concavity: (a) $f(x)$ is increasing and concave down on $(c, d)$; (b) $g(x)$ is increasing and concave up on $(c, d)$.

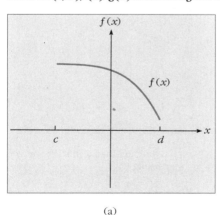
(a)

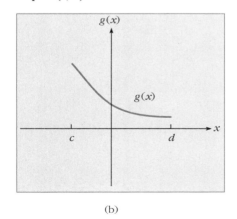
(b)

**FIGURE 4.25** Decreasing functions with different concavity: (a) $f(x)$ is decreasing and concave down on $(c, d)$; (b) $g(x)$ is decreasing and concave up on $(c, d)$.

## Concavity

Consider the graphs sketched in Figure 4.24 and Figure 4.25. Both of the graphs in Figure 4.24 are rising on the interval $(c, d)$. That is, both of the functions in Figure 4.24 are increasing on $(c, d)$. However, the shapes of these graphs are certainly quite different. The graph in Figure 4.24(a) is turning down or bending down as we move from left to right, and the graph in Figure 4.24(b) is turning up or bending up as we move from left to right. We say that the curve in Figure 4.24(a) is *concave down* on the interval $(c, d)$ and that the curve in Figure 4.24(b) is *concave up* on $(c, d)$. Inspecting the graphs in Figure 4.25, we see that both curves are falling on the interval $(c, d)$. In other words, both of the functions sketched in Figure 4.25 are decreasing on the interval $(c, d)$. The curve in Figure 4.25(a) is bending down as we move from left to right, and we say that this curve is *concave down* on the interval $(c, d)$. The curve in Figure 4.25(b) is bending up as we move from left to right, and we say that this curve is *concave up* on the interval $(c, d)$.

In Figure 4.26, the curve $f(x) = \frac{1}{4}x^2 + \frac{1}{2}$, which is concave up on the interval $(0, 2)$, has been drawn, along with some tangent lines to the graph. Note that $f'(x) = \frac{1}{2}x$. As you can see, the slope or steepness of these tangent lines is increasing as we move from left to right. You should also observe that the curve lies above the tangent line at each of these points. Since the slopes of the tangent lines are increasing, $f'(x)$, which represents these slopes, is an increasing function on $(0, 2)$. But we know that the derivative of an increasing function must be positive. Therefore the derivative of $f'$, that is, $(f')' = f''$ must be positive on the interval $(0, 2)$.

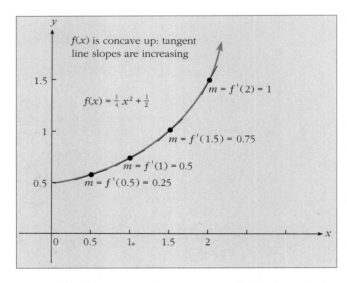

**FIGURE 4.26** A function that is concave up on $(0, 2)$: tangent line slopes are increasing.

In Figure 4.27, the curve $f(x) = 2\sqrt{x}$, which is concave down on the interval $(0, 4)$, has been sketched, along with some tangent lines to the graph. Note that $f'(x) = 1/\sqrt{x}$. In this case the slope of the tangent lines is decreasing as we

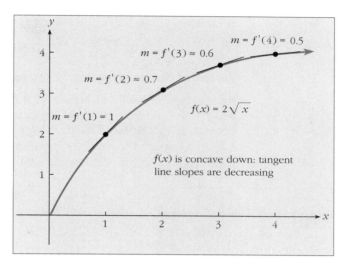

**FIGURE 4.27** A function that is concave down on $(0, 4)$: tangent line slopes are decreasing.

move from left to right. Also notice that the graph lies below the tangent line at each of these points. Since the slopes of the tangent lines, as represented by $f'(x)$, are decreasing on $(0, 4)$, we know that $f'$ must be a decreasing function on $(0, 4)$, so that $(f')' = f''$ must be negative on the interval $(0, 4)$.

We have in Figures 4.26 and 4.27 the motivation for the following definition.

**DEFINITION: CONCAVITY ON AN INTERVAL**

A function $y = f(x)$ is *concave up on an interval* $(c, d)$ included in its domain if the graph of $y = f(x)$ lies above its tangent line at each point in $(c, d)$. The function is *concave down on an interval* $(c, d)$ if the graph of $y = f(x)$ lies below its tangent line at each point in $(c, d)$.

The next result relates concavity on an interval and the sign of the second derivative.

**THE RELATION OF CONCAVITY TO THE SECOND DERIVATIVE**

If $f''(x) > 0$ for all $x$ in the interval $(c, d)$ then $f$ is concave up on this interval. If $f''(x) < 0$ for all $x$ in the interval $(c, d)$ then $f$ is concave down on this interval.

Do not confuse concavity with increasing and decreasing. A function can be concave up and increasing (Figure 4.24(b)) or concave up and decreasing (Figure 4.25(b)). Also a function can be concave down and increasing (Figure 4.24(a)) or concave down and decreasing (Figure 4.25(a)). Remember that whether a graph is increasing or decreasing is governed by the *first* derivative. A positive first derivative indicates that the function is increasing and a negative first derivative indicates that the function is decreasing. Concavity is governed by the *second* derivative. A positive second derivative indicates that the function is concave up and a negative second derivative indicates that the function is concave down.

In Section 3.3 on the second derivative, we discussed how $f''(x)$ represents the rate of change of the first derivative, since $f''(x) = \frac{d}{dx}(f'(x))$. For example, the function sketched in Figure 4.24(a) has $f'(x) > 0$ (since it is increasing) and $f''(x) < 0$ (since it is concave down). This function is increasing, but at a decreasing rate. It has a similar shape to the graph of mortgage rates (Figure 3.2), which we considered in Section 3.3. Table 4.1 summarizes some information concerning the graphs of Figures 4.24 and 4.25.

| GRAPH RESEMBLES | | SIGN OF $f'$ | $f$ IS | SIGN OF $f''$ | $f$ IS | $f$ IS |
|---|---|---|---|---|---|---|
| Fig. 4.24(a) | ⌒ | positive | increasing | negative | concave down | increasing at a decreasing rate |
| Fig. 4.24(b) | ⌣ | positive | increasing | positive | concave up | increasing at an increasing rate |
| Fig. 4.25(a) | ⌐ | negative | decreasing | negative | concave down | decreasing at an increasing rate |
| Fig. 4.25(b) | ⌡ | negative | decreasing | positive | concave up | decreasing at a decreasing rate |

**TABLE 4.1**

Given a function $y = f(x)$, it is important to be able to find the intervals on which $f(x)$ is concave up and those on which it is concave down. The key to this is to determine where the concavity of a function changes from up to down or vice versa. Since the concavity changes whenever the sign of $f''$ changes, we must determine where the sign of the second derivative changes from positive to negative or vice versa. Just as in our discussion of increasing and decreasing functions and changes in the sign of the first derivative, we can conclude that there are three types of numbers where $f''$ can change its sign. These are numbers $c$ in the domain of $f$ where $f''(c) = 0$ or $f''(c)$ does not exist, sometimes referred to as *second-order critical numbers* because they deal with the second-order derivative of $f(x)$, and numbers $c$ at which $f(c)$ is undefined or discontinuous, such as vertical asymptotes.

**NUMBERS WHERE f(x) CAN CHANGE FROM CONCAVE UP TO CONCAVE DOWN OR VICE VERSA**

1. Numbers $c$ in the domain of $f$ such that $f''(c) = 0$ ⎫ second-order
2. Numbers $c$ in the domain of $f$ such that $f''(c)$ does not exist ⎬ critical numbers
3. Numbers $c$ at which $f(c)$ is undefined or discontinuous.

**EXAMPLE 1**

Determine the numbers at which $f(x) = x^4 - 2x^3$ can change concavity.

**Solution**

Note that $f'(x) = 4x^3 - 6x^2$ and $f''(x) = 12x^2 - 12x$. We find that the only possible changes in the concavity of $f(x)$ can occur when $f''(x) = 0$, since $f''$ is defined everywhere and $f$ is continuous everywhere. We continue as follows:

$$12x^2 - 12x = 0$$
$$12x(x - 1) = 0$$
$$x = 0, \quad x = 1$$

Thus, *if f(x) changes concavity, it must do so when $x = 0$ or when $x = 1$, and $x = 0$ and $x = 1$ are the second-order critical numbers of $f(x)$.*

**EXAMPLE 2**

Determine the numbers at which $f(x) = \sqrt[3]{x} + x^2$ can change concavity.

**Solution**

We have $f'(x) = \frac{1}{3}x^{-2/3} + 2x$, and $f''(x) = -\frac{2}{9}x^{-5/3} + 2 = (-2/(9\sqrt[3]{x^5})) + 2$. Observe that since $f''(0)$ does not exist and 0 is in the domain of $f$, the number $x = 0$ is a second-order critical number of $f(x)$. Other second-order critical numbers can be found by equating $f''$ to zero:

$$-\frac{2}{9\sqrt[3]{x^5}} + 2 = 0$$

$$-\frac{1}{9\sqrt[3]{x^5}} + 1 = 0$$

$$-1 + 9\sqrt[3]{x^5} = 0 \quad \text{(Multiplying both sides by } 9\sqrt[3]{x^5}\text{)}$$

$$9\sqrt[3]{x^5} = 1$$

$$\sqrt[3]{x^5} = \frac{1}{9}$$

$$x^{5/3} = \frac{1}{9}$$

$$(x^{5/3})^{3/5} = \left(\frac{1}{9}\right)^{3/5} \quad \left(\text{Raising both sides to the } \tfrac{3}{5} \text{ power}\right)$$

$$x = \left(\frac{1}{9}\right)^{3/5} \approx 0.26758$$

Finally, since $f(x)$ is continuous everywhere, we conclude that *if $f(x)$ changes concavity, it must do so when $x = 0$ or $x = \left(\frac{1}{9}\right)^{3/5}$ and these are the second-order critical numbers of $f(x)$.*

It should be emphasized that the numbers at which $f(x)$ can change concavity are only *candidates* at which a change may occur. To determine where the concavity of a function actually changes, we can use the method of test values developed in Section 4.1 in connection with finding intervals of increase or decrease. An example should help to clarify this procedure.

**EXAMPLE 3**

Find the intervals on which $f(x) = x^4 - 2x^3$ is concave up and those on which it is concave down.

**Solution**

We already know from Example 1 that $f''(x) = 12x^2 - 12x$ and that the concavity of $f(x)$ can change only when $x = 0$ or $x = 1$. We must now determine the sign of $f''$ in each of the three intervals $(-\infty, 0)$, $(0, 1)$, and $(1, \infty)$. In the interval $(-\infty, 0)$, let us choose the test value $x = -1$ and compute $f''(-1) = 12(-1)^2 - 12(-1) = 24 > 0$. Thus, since $f''$ is continuous on $(-\infty, 0)$, we conclude by the corollary to the intermediate value theorem that $f''(x) > 0$ for all $x$ in the interval $(-\infty, 0)$. Therefore $f(x)$ is concave up on the interval

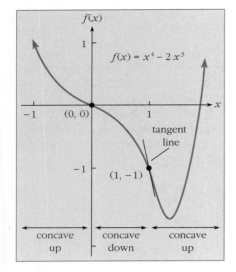

**FIGURE 4.28** The graph of $f(x) = x^4 - 2x^3$.

$(-\infty, 0)$. In the interval $(0, 1)$ let's choose $x = \frac{1}{2}$ as a test value and compute $f''(\frac{1}{2}) = 12(\frac{1}{2})^2 - 12(\frac{1}{2}) = -3 < 0$. Thus $f(x)$ is concave down on the interval $(0, 1)$. In the interval $(1, \infty)$, we choose $x = 2$ as a test value and compute $f''(2) = 12(2)^2 - 12(2) = 24 > 0$. Thus $f(x)$ is concave up on the interval $(1, \infty)$. We summarize these results in the accompanying sign chart for $f''$, where **SOCN** stands for "second-order critical number."

| $x$ | | $-1$ | 0 | $\frac{1}{2}$ | 1 | 2 |
|---|---|---|---|---|---|---|
| $f''$ | | $+++++$ | 0 | $-----$ | 0 | $+++++$ |
| $f$ | | concave up | : | concave down | : | concave up |

SOCN ↓ above 0 (at 0) and SOCN ↓ above 1.

The graph of $f(x)$ is sketched in Figure 4.28 for reference. ∎

## Points of Inflection

In Example 3, we saw that the concavity of $f(x) = x^4 - 2x^3$ changed from concave up to concave down at the point $(0, 0)$ and from concave down to concave up at the point $(1, -1)$. Points on the graph of a function where the concavity changes are called *points of inflection* of $f(x)$, provided that the function has a well-defined tangent line there.

**POINT OF INFLECTION**

> If $f(x)$ has a tangent line (possibly vertical) at $x = c$ and if $f(x)$ changes concavity at $x = c$, then the point $(c, f(c))$ is called a *point of inflection of the graph of $f(x)$* or simply a *point of inflection of $f(x)$*.

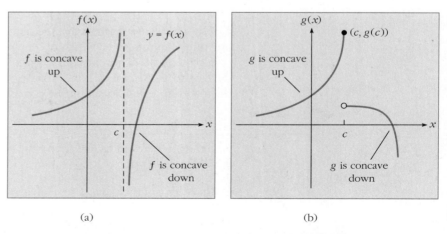

**FIGURE 4.29** Concavity changes occur at $c$ but no point of inflection occurs at $x = c$.

A graph may have a change of concavity at a number $c$ and yet a point of inflection may not occur at $x = c$. One reason for this is that $f(x)$ may not be continuous (and hence not differentiable) at $x = c$: see Figure 4.29 for examples. In Figure 4.29(a), the graph of $f(x)$ has the line $x = c$ as a vertical asymptote so, because $f(c)$ does not exist, $f'(c)$ does not exist and there is no tangent line at $x = c$. In Figure 4.29(b), the graph of $g(x)$ is discontinuous at $x = c$, so that $g'(c)$ does not exist (remember that not continuous implies not differentiable). Therefore no tangent line exists at $x = c$. Notice that at an inflection point (for example, at the point $(1, -1)$ in Figure 4.28), the tangent line will lie above the graph of the function on the side that is concave down and below the graph of the function on the side that is concave up.

EXAMPLE 4 (Participative)

Let's determine the intervals on which $f(x) = \sqrt[3]{x} + x^2$ is concave up and those on which it is concave down and find any points of inflection of $f(x)$.

Solution

Using the results of Example 2, we already know that $f''(x) = (-2/(9\sqrt[3]{x^5})) + 2$ and that the second-order critical numbers are $x =$ ____ and $x =$ ______. Now we can complete the following diagram.

| $x$ | $-1$ | $0$ (SOCN ↓) | $0.1$ | $0.27$ (SOCN ↓) | $1$ |
|---|---|---|---|---|---|
| $f''$ | $+++++$ | ND | $-----$ | $0$ | $+++++$ |
| $f$ | concave ______ | pt of inf | concave ______ | pt of inf | concave ______ |

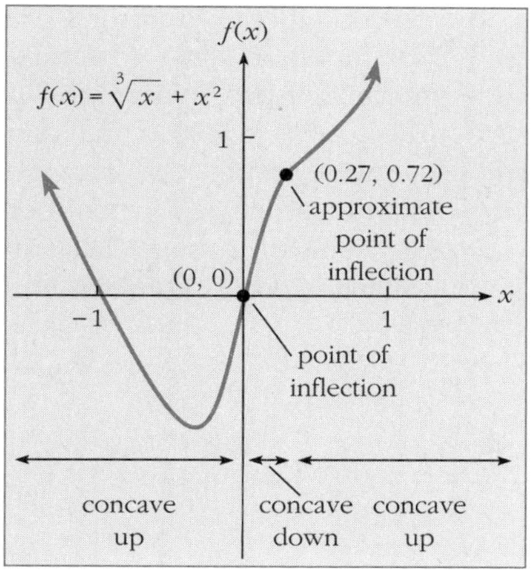


**FIGURE 4.30** The graph of $f(x) = \sqrt[3]{x} + x^2$.

Changes of concavity occur at $x = 0$ and $x = \left(\frac{1}{9}\right)^{3/5} \approx 0.26758$. Because $f'(x) = (1/(3\sqrt[3]{x^2})) + 2x$, we know that $f'(0)$ does not exist. But since

$$\lim_{x\to 0} f'(x) = \lim_{x\to 0}\left[\frac{1}{3\sqrt[3]{x^2}} + 2x\right] = \infty,$$ the tangent line at $x = 0$ is

____________. Clearly $f'\left[\left(\frac{1}{9}\right)^{3/5}\right]$ exists and is finite, so a (nonvertical) tangent line exists when $x = \left(\frac{1}{9}\right)^{3/5}$ We conclude that $f(x)$ has points of inflection at $(0, f(0)) = (0, 0)$ and at ______________. A graph of $f(x)$ is sketched in Figure 4.30 for reference. ■

EXAMPLE 5

The monthly cost $C(x)$ in hundreds of dollars of producing $x$ thousand units of a certain product is

$$C(x) = \frac{1}{3}x^3 - 5x^2 + 28x + 30 \qquad x \geq 0$$

Find the intervals on which $C(x)$ is concave up and those on which it is concave down and find the inflection point $I$.

**SOLUTIONS TO EXAMPLE 4**

$0, \left(\frac{1}{9}\right)^{3/5} \approx 0.26758$, up, down, up, vertical, $(0.27, 0.72)$ approximately

Solution   This is the problem with which this section began. We compute:

$$C'(x) = x^2 - 10x + 28$$
$$C''(x) = 2x - 10$$

To find the numbers at which a change of concavity can occur, we solve the equation $C''(x) = 0$ for $x$ to obtain $2x - 10 = 0$ or $x = 5$.

Now we can construct the sign chart of $C''$.

|   |   | SOCN ↓ |   |
|---|---|---|---|
| x | 2 | 5 | 8 |
| C″ | − − − − − | 0 | + + + + + |
| C | concave down | pt of inf | concave up |

We conclude that the graph of $C(x)$ is concave down on the interval $(0, 5)$ and concave up on the interval $(5, \infty)$, so that the point of inflection $I$ is $(5, C(5)) \approx (5, 86.67)$. Based upon the discussion at the beginning of this section, cost is increasing at a decreasing rate from 0 units to 5000 units, so in this interval, each unit costs less to produce than the preceding one. In this interval we are making more and more efficient use of our inputs of labor, materials, etc., with each successive unit of output. The cost efficient production level is $c = 5$ (thousand) units. As we increase production beyond 5000, unit cost increases at an increasing rate and each unit costs more to produce than the preceding one. In this interval we are making less and less efficient use of our inputs of labor, materials, etc., with each successive unit of output. The production level $c = 5$ (thousand) units is sometimes called the *point of diminishing returns*. ∎

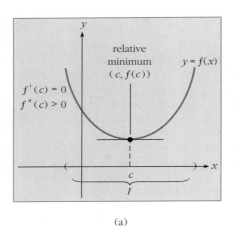

(a)

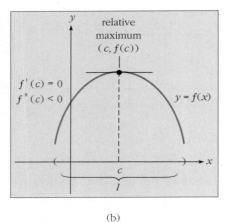

(b)

**FIGURE 4.31** Graphical illustrations of the second derivative test.

### The Second Derivative Test

We will now investigate how the second derivative can be used to determine whether a relative maximum or a relative minimum exists at a (first-order) critical number $c$. Let's suppose that $c$ is a critical number of $f(x)$ and that $f'(c) = 0$. Now, if $f''(c) > 0$ and if $f''$ is continuous near $c$, then there is an open interval $I$ around $c$ such that $f''(x) > 0$ for all $x$ in $I$. Therefore $f(x)$ is concave up on the open interval $I$ that contains $c$. We conclude (see Figure 4.31(a)) that a relative minimum occurs at $c$. Similarly, if we assume that $f''(c) < 0$, we can show that $f(x)$ is concave down on an open interval containing $c$ and conclude that a relative maximum occurs at $c$. See Figure 4.31(b). Finally, if $f'(c) = 0$ and $f''(c) = 0$, then either a relative maximum or minimum may occur at $c$, or no relative extremum at all may occur at $c$. See Figure 4.32 for an illustration of this for three different functions at $c = 0$. You should verify these results for yourself.

We can now state the *second derivative test* for identifying relative extrema of a function.

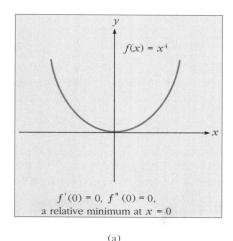
$f'(0) = 0$, $f''(0) = 0$,
a relative minimum at $x = 0$

(a)

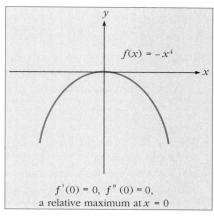
$f'(0) = 0$, $f''(0) = 0$,
a relative maximum at $x = 0$

(b)

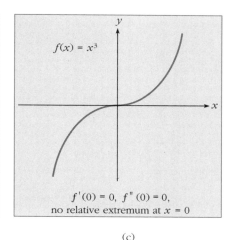
$f'(0) = 0$, $f''(0) = 0$,
no relative extremum at $x = 0$

(c)

**FIGURE 4.32** Functions for which $f'(0) = 0$ and $f''(0) = 0$.

## THE SECOND DERIVATIVE TEST

Let $c$ be a critical number of $f(x)$ such that $f'(c) = 0$. Then if $f''$ is continuous on an open interval containing $c$,

i. When $f''(c) > 0$ a relative minimum occurs at $c$;
ii. When $f''(c) < 0$ a relative maximum occurs at $c$;
iii. When $f''(c) = 0$ the test fails and a relative maximum, a relative minimum, or no relative extremum may occur at $c$. Use the first derivative test.

### EXAMPLE 6

Use the second derivative test to find the relative extreme points of the graph of $f(x) = 3x^4 + 8x^3 - 18x^2 + 1$.

### Solution

To begin, we compute $f'(x) = 12x^3 + 24x^2 - 36x$. The first-order critical numbers of $f(x)$ are found by solving the equation

$$12x^3 + 24x^2 - 36x = 0$$
$$12x(x^2 + 2x - 3) = 0$$
$$12x(x + 3)(x - 1) = 0$$
$$x = -3, \quad x = 0, \quad x = 1$$

There are three critical numbers of $f(x)$, since $f'(-3) = 0$, $f'(0) = 0$, and $f'(1) = 0$. We will now test each of these critical numbers in turn, using the second derivative test to determine whether a relative maximum or a relative minimum occurs there. First, note that $f''(x) = 36x^2 + 48x - 36$. Then:

$$x = -3: \quad f''(-3) = 36(-3)^2 + 48(-3) - 36 = 144 > 0.$$

So the point $(-3, f(-3)) = (-3, -134)$ is a relative minimum point.

$$x = 0: \quad f''(0) = 36(0)^2 + 48(0) - 36 = -36 < 0.$$

So the point $(0, f(0)) = (0, 1)$ is a relative maximum point.

$x = 1$:   $f''(1) = 36(1)^2 + 48(1) - 36 = 48 > 0$.

So the point $(1, f(1)) = (1, -6)$ is a relative minimum point. ∎

The second derivative test is a useful alternative to the first derivative test for finding maxima and minima of a function. However, the second derivative test does have a few limitations. If $c$ is a first-order critical number of $y = f(x)$, then

(a) When $c$ is a critical number such that $f'(c)$ does not exist, the test cannot be used;
(b) When $f'(c) = 0$ and $f''(c) = 0$, the test fails. You can conclude nothing about the behavior of $f(x)$ at the critical number $c$, and you must use the first derivative test to determine whether a relative maximum, a relative minimum, or neither occurs at $x = c$.
(c) If $f''(x)$ is difficult or tedious to find and concavity or inflection points of $f(x)$ are not of interest, you are probably better off using the first derivative test.

In Example 7 we will see an example of the limitations of the second derivative test.

### EXAMPLE 7 (Participative)

Solution

Let's find the relative extrema of $f(x) = 3x^4 - 16x^3 + 24x^2 - 3$, using the second derivative test if possible.

We begin by finding $f'(x) = $ _____. Setting $f'(x)$ equal to zero, we get

$$12x^3 - 48x^2 + 48x = 0$$
$$12x(x^2 - 4x + 4) = 0$$
$$12x(\underline{\phantom{xx}})^2 = 0$$

The critical numbers are $x = 0$ and $x = $ ____. To use the second derivative test, we compute $f''(x) = 36x^2 - 96x + 48$. For $x = 0$,

$$f''(0) = 36(0)^2 - 96(0) + 48 = 48 > 0$$

So $(0, f(0)) = (0, -3)$ is a relative _____. For $x = 2$,

$$f''(2) = 36(2)^2 - 96(2) + 48 = 0$$

and the second derivative test fails. We use the first derivative test and construct the chart below.

|  |  | CN ↓ |  | CN ↓ |  |
|---|---|---|---|---|---|
| $x$ |  | 0 | 1 | 2 | 3 |
| $f'$ |  | 0 | +++++ | 0 | _____ |
| $f$ |  | rel min |  _____ | ⋮ | _____ |

We conclude that neither a relative maximum nor a relative minimum occurs at $x = 2$. In this problem, note that we took care to put the critical number $x = 0$ on our chart in order to be able to determine the appropriate intervals in which to choose test values. This was done even though we were not applying the first derivative test to the critical number $x = 0$. ∎

## SECTION 4.3
### SHORT ANSWER EXERCISES

1. Given that $f''(x) = x^2 + 1$, since $f''(x) > 0$ for all $x$, the graph of $f(x)$ is concave _____.

2. (True or False) A graph can be concave up and decreasing. _____

3. As we move along the graph of a certain function $f(x)$, the slopes of the tangent lines are decreasing. We can conclude that (a) $f(x)$ is decreasing at a decreasing rate; (b) $f(x)$ has a point of inflection; (c) the graph of $f(x)$ is concave down; (d) the graph of $f(x)$ is concave up.

4. The graph of a function $y = f(x)$ lies above its tangent line at each point in the interval (2, 5). We can conclude that (a) $f(x)$ is increasing on (2, 5); (b) $f(x)$ is increasing at an increasing rate on (2, 5); (c) $f(x)$ is concave up on (2, 5); (d) all of the above.

5. The graph of $y = x^2$ on the interval $(0, \infty)$ is (a) increasing; (b) increasing at an increasing rate; (c) concave up; (d) all of the above.

6. (True or False) The second derivative test can be applied to critical numbers $c$ if $f'(c) = 0$ or if $f'(c)$ does not exist. _____

7. For a certain function, $f''(1) = 0$, $f''(2)$ does not exist, and the line $x = 3$ is a vertical asymptote of the graph of $f(x)$. Then (a) the graph of $f(x)$ changes concavity at $x = 1$ and at $x = 2$; (b) the graph of $f(x)$ has a point of inflection at $x = 1$; (c) the concavity of $f(x)$ can change at $x = 1$, $x = 2$, or $x = 3$; (d) all of the above.

8. If $f''(x) = 2x - 10$, then the second-order critical number of $f(x)$ is $x = $ _____.

9. (True or False) A point of inflection $(c, f(c))$ of the graph of $f(x)$ can occur at a number $c$ having the property that $f''(c)$ does not exist. _____

10. For a weekly cost function, if $C''(x) = 5x - 50000$ and if $C(x)$ has no relative extreme values, then the point of diminishing returns is $x = $ _____.

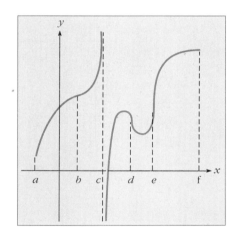

FIGURE 4.33

## SECTION 4.3
### EXERCISES

1. For the function with graph sketched in Figure 4.33, find (a) the intervals on which $f(x)$ is concave up; (b) the intervals on which $f(x)$ is concave down; (c) the $x$-coordinates of any points of inflection.

*In Exercises 2–24, determine the intervals on which the graph of each given function is concave up and those on which it is concave down. Also, find any points of inflection for each graph.*

2. $f(x) = x^2 + 3x - 5$
3. $f(x) = 7 - x - 4x^2$
4. $f(x) = 2x^3 - 5x + 1$
5. $f(x) = -x^3 - 7x + 3$
6. $f(x) = 2x^3 - 3x^2 - 12x + 4$
7. $f(x) = 4x^3 + 15x^2 - 18x - 5$

**SOLUTIONS TO EXAMPLE 7**   $12x^3 - 48x^2 + 48x$, $x - 2$, 2, minimum, +, increasing, increasing

8. $f(x) = 9x^3 + 6x^2 + 2x - 1$
9. $f(x) = (2x - 3)^3$
10. $f(x) = 2x^4 - 4x^2$
11. $f(x) = 6x^2 - x^4$
12. $f(x) = 3x^4 + 16x^3 + 18x^2 - 3$
13. $f(x) = \frac{1}{4}x^4 - \frac{4}{3}x^3 + 2x + 5$
14. $f(x) = 6 + x^2 - 2x^3 - x^4$
15. $f(x) = \sqrt[3]{x} + x$
16. $f(x) = \sqrt[3]{x} - 2x - 3x^2$
17. $f(x) = 10 - 5x - \sqrt{x^2 + 1}$
18. $f(x) = (x - 5)^{1/3}$
19. $f(x) = \sqrt[3]{(2x + 5)^2}$
20. $f(x) = x + \frac{2}{x}$
21. $f(x) = \frac{2}{x + 3}$
22. $f(x) = \frac{-3}{x - 1}$
23. $f(x) = \frac{-3x}{x + 1}$
24. $f(x) = \frac{2x}{x - 5}$

*In Exercises 25–43, use the second derivative test to determine the relative extrema of the given function. If the second derivative test fails, use the first derivative test.*

25. $f(x) = 10 - 5x - x^2$
26. $f(x) = 7x^2 - 20x + 30$
27. $f(x) = 4x^3 - 17x^2 - 6x + 7$
28. $f(x) = 4 + 20x - 17x^2 - 4x^3$
29. $f(x) = x^3 - x^2 - x - 1$
30. $f(x) = x^3 + 3x^2 + 3x + 1$
31. $f(x) = 4x^3 + 20x^2 + 25x$
32. $f(x) = 2x^5 - 10x^3$
33. $f(x) = x^4 - 4x^2$
34. $f(x) = x^4 + 8x^3 + 18x^2 - 3$
35. $f(x) = 6 - 4x^2 - 4x^3 - x^4$
36. $f(x) = 2x^4 + 3x^3 + x^2 - 5$
37. $f(x) = \sqrt[3]{x} - x$
38. $f(x) = \sqrt[5]{x^2} + 2x - 1$
39. $f(x) = 2x - \frac{3}{x}$
40. $f(x) = \frac{x^2 + 4}{x}$
41. $f(x) = \frac{-x}{2x - 1}$
42. $f(x) = \frac{x}{x^2 + 1}$
43. $f(x) = \frac{x^2}{x^2 + 1}$

### Social Sciences

44. **(Crime rate)** A sociologist has determined that the number $N$ of crimes committed per month in a certain city depends on the unemployment rate $r$ in that city according to the function

    $N(r) = r^3 - 24r^2 + 403.6r + 2000, \quad 0 \leq r \leq 30$

    (a) Find the intervals on which $N(r)$ is concave up and those on which it is concave down and locate any points of inflection.
    (b) Determine the interval on which $N(r)$ is increasing at an increasing rate.

45. **(Cost and demand)** The monthly cost function of a certain company is $C(x) = 1000 + 10x$, where $x$ is the number of units produced per month and $C(x)$ is the cost of production in dollars. The company can sell $x$ units per month if the price per unit is $p$ dollars, where $p = -0.0001x^2 + 500$.
    (a) Find the revenue function and the intervals on which it is concave up and those on which it is concave down. Are there any inflection points?
    (b) Find the profit function and the intervals on which it is concave up and those on which it is concave down. Are there any inflection points?

**Business and Economics**

**46.** (Average cost) Given the average cost function $\overline{C}(x) = (2x^3 + 5000)/(5x)$, where $\overline{C}(x)$ is the average weekly cost of producing $x$ thousand units of a certain product, determine the intervals on which $\overline{C}(x)$ is concave up, and those on which it is concave down. Also, find any points of inflection.

**47.** (Diminishing returns) If the monthly cost in hundreds of dollars of producing $x$ thousand units of one of your company's products is $C(x) = x^3 - 30x^2 + 350x + 60$, find the intervals on which $C(x)$ is concave up and those on which it is concave down. What is the inflection point? What is the point of diminishing returns?

**48.** (Advertising and sales) If daily sales in units of our product $t$ days after we begin a new advertising campaign are projected to be

$$S(t) = \tfrac{1}{3}t^3 - 21t^2 + 392t + 500, \quad 0 \le t \le 42$$

(a) find the intervals on which $S(t)$ is concave up and those on which it is concave down, and locate any points of inflection;
(b) determine the intervals in which sales are (i) increasing at a decreasing rate; (ii) decreasing at an increasing rate; (iii) increasing at an increasing rate.

**49.** (Diminishing returns) The marketing department of our company has determined that monthly sales $S(x)$, in units, of one of our products is related to the previous month's advertising expenses $x$ (in thousands of dollars) according to the formula

$$S(x) = -5x^3 + 250x^2 - 1500x + 5000, \quad 4 \le x \le 29$$

(a) Find the intervals on which $S(x)$ is concave up and those on which it is concave down. Find any points of inflection of $S(x)$.
(b) Determine the interval on which $S(x)$ is increasing at an increasing rate and the interval on which $S(x)$ is increasing at a decreasing rate.
(c) What is the point of diminishing returns to advertising expenditures on this product?

**50.** (Research and development) Research and development funds invested in a product typically have a "lag" effect on product demand. That is, it takes some time for the research and development to result in product improvements that in turn result, finally, in increased sales. Suppose it has been determined by our company that this lag effect on a certain product is one year and that quarterly sales $D(x)$ of this product in units is related to research and development expenditures $x$ a year ago, where $x$ is measured in thousands of dollars, according to the equation

$$D(x) = -3x^3 + 150x^2 - 1500x + 9000, \quad 6 \le x \le 25$$

(a) Find the intervals on which $D(x)$ is concave up and those on which it is concave down. Find any points of inflection of $D(x)$.
(b) Determine the interval on which $D(x)$ is increasing at an increasing rate and the interval on which $D(x)$ is increasing at a decreasing rate.
(c) What is the point of diminishing returns to research and development expenditures on this product?

**51.** (Learning) During a study period of 50 minutes, a certain student can learn $N(t)$ new vocabulary words, where

$$N(t) = 0.1923t^2 - 0.002564t^3, \quad 0 \le t \le 50$$

(a) Find the intervals on which $N(t)$ is concave up and those on which it is concave down. Find any points of inflection of $N(t)$.
(b) Determine the interval on which $N(t)$ is increasing at an increasing rate and the interval on which $N(t)$ is increasing at a decreasing rate.
(c) What is the point of diminishing returns to study time invested for this student?

OPTIONAL
**GRAPHING CALCULATOR/
COMPUTER EXERCISES**

In Exercises 52–55, approximate the intervals on which the given function is concave up and those on which it is concave down, and also approximate the x-coordinate of any points of inflection. One suggested way to do this is to graph both $f(x)$ and $f''(x)$ on the same coordinate system. The intervals where the graph of $f''(x)$ is below the x-axis correspond to intervals on which the graph of $f(x)$ is concave down, since on these intervals $f''(x) < 0$. The intervals where the graph of $f''(x)$ is above the x-axis corresponds to intervals on which the graph of $f(x)$ is concave up, since on these intervals $f''(x) > 0$. The points at which the graph of $f''(x)$ crosses the x-axis, changing from negative to positive or vice versa, give the x-coordinates of the points of inflection of $f(x)$.

52. $f(x) = 2x^5 + 10x^4 + 7x^3 + 30x^2 + x + 2$
53. $f(x) = -x^5 + 2x^4 - x^3 + x^2 - x + 1$
54. $f(x) = x^6 - 2x^5 - x^4 + 3x^3 - x^2 + x + 5$
55. $f(x) = \sqrt[3]{x} + x^5 - 2x^4 - x^3 + x - 1$

## 4.4 ABSOLUTE MAXIMA AND MINIMA

### Absolute Extrema

Many applied problems involve finding the maximum or minimum values of a function over an interval. For example, if our monthly profit in dollars on the sale of $x$ hundred units of a certain product is

$$P(x) = -2500 + 2.5x^2 - 0.005x^3$$

but our production capacity for this product is only 30,000 units (or 300 hundred units) per month, how many units should we produce to maximize monthly profit? In this type of problem we are trying to find the *absolute* maximum of $P(x)$ over the interval [0, 300]. This problem is solved in Example 2.

To get a better idea of the difference between absolute extrema and relative extrema, look at Figures 4.34 and 4.35. An *absolute maximum point on a graph* is the highest point on the graph within the interval of interest. For example, in Figure 4.34 the absolute maximum occurs when $x = c$, since $(c, f(c))$ is the highest point on the graph within the interval of interest, $I = (-\infty, \infty)$. Notice that the graph of Figure 4.34 has no absolute minimum point because $f(x) \to -\infty$ as $x \to \infty$. In other words, for large values of $x$, the corresponding values of $f(x)$ are negative and also large in absolute value, so the graph of $f(x)$ falls to lower and lower points as $x$ increases without bound.

An *absolute minimum point on a graph* is the lowest point on the graph within the interval of interest. For example, in Figure 4.35 the absolute minimum occurs when $x = c$, since $(c, f(c))$ is the lowest point on the graph within the interval of interest, $I = (-\infty, \infty)$. The graph of Figure 4.35 has no absolute maximum point because $f(x) \to \infty$ as $x \to \infty$. In other words, for large values of $x$, the corresponding values of $f(x)$ are large and become larger without

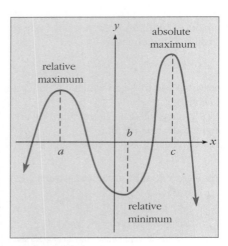

**FIGURE 4.34** This graph has no absolute minimum on $(-\infty, \infty)$.

## 4.4 ABSOLUTE MAXIMA AND MINIMA

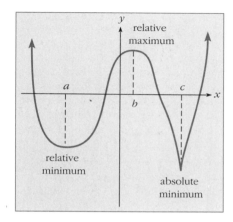

**FIGURE 4.35** This graph has no absolute maximum on $(-\infty, \infty)$.

bound as $x$ increases, so the graph rises to higher and higher points as $x$ increases without bound.

Notice that absolute extrema need not exist for a function on a given interval; Figure 4.36 illustrates several instances where this is the case. The graph of Figure 4.36(a) has no absolute maximum on $(-\infty, \infty)$ because $f(x) \to \infty$ as $x \to 1^-$, and it has no absolute minimum on $(-\infty, \infty)$ because $f(x) \to -\infty$ as $x \to 1^+$. Note that the line $x = 1$ is a vertical asymptote of the graph of $f(x)$ in this case. The graph of Figure 4.36(b) does not attain an absolute maximum in $(-1, 1)$. Although it might appear at first glance that the absolute maximum value is $f(1) = 1$, observe carefully that $x = 1$ is *not* included in $(-1, 1)$, and so no absolute maximum value exists. Similarly, the graph of Figure 4.36(b) has no absolute minimum point since the point $(-1, -1)$ is *not* included on the graph. The graph of Figure 4.36(c) has no absolute maximum point because $f(x) \to \frac{1}{2}$ as $x \to \infty$, but $f(x) < \frac{1}{2}$ for all values of $x$, so no absolute maximum is attained for any number $x$ in $(-\infty, \infty)$. Also, the graph has no absolute minimum for, although the graph gets closer and closer to the horizontal line $y = -\frac{1}{2}$ as $x \to -\infty$, there is no value in $x$ in $(-\infty, \infty)$ such that $f(x) = -\frac{1}{2}$.

The graphs in Figure 4.36 motivate the following definition.

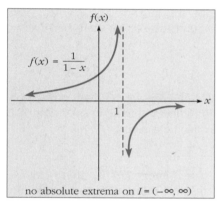
no absolute extrema on $I = (-\infty, \infty)$

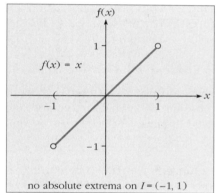
no absolute extrema on $I = (-1, 1)$

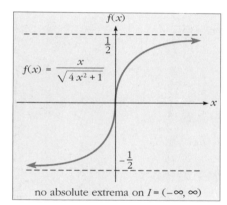
no absolute extrema on $I = (-\infty, \infty)$

**FIGURE 4.36** Functions for which absolute extrema do not exist.

**ABSOLUTE EXTREMA ON AN INTERVAL**

> The point $(c, f(c))$ is an absolute maximum point of the graph of $y = f(x)$ on an interval $I$ if $c$ is in $I$ and $f(c) \geq f(x)$ for all $x$ in $I$. The point $(c, f(c))$ is an absolute minimum point of the graph of $y = f(x)$ on an interval $I$ if $c$ is in $I$ and $f(c) \leq f(x)$ for all $x$ in $I$.

Figure 4.37 graphically illustrates the definition of absolute extrema. The interval $I$ may be open, half-open, or closed. Also note that in Figure 4.37(a), no absolute minimum is attained on the open interval $I = (a, b)$, whereas in Figure 4.37(b), the function $f(x)$ attains an absolute maximum at the endpoint $a$ of the closed interval $I = [a, b]$. This suggests that unless at least one endpoint is included in an interval, an absolute extremum cannot occur at an endpoint. This is true since in the definition of absolute extrema the number $c$ at which

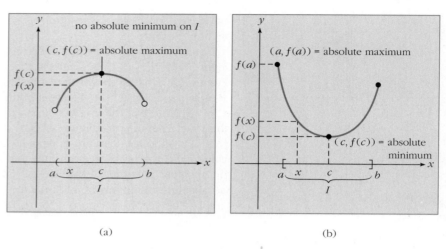

(a)  (b)

**FIGURE 4.37** Illustrations of the definition of absolute extrema.

an extremum occurs must be *in* the interval $I$. It should be emphasized that whether or not absolute extrema exist for a given function $f(x)$ on an interval $I$ depends both upon the function and upon the interval. For example, if in Figure 4.37(b) we choose $I$ to be the *open* interval $(a, b)$, then $f(x)$ has no absolute maximum on $I$.

### The Extreme Value Theorem

Given a function $y = f(x)$ and an interval $I$, when can we be assured that both an absolute maximum and an absolute minimum exist? The answer, which is proved in more advanced texts, is found in our next result.

**EXTREME VALUE THEOREM**

If a function $y = f(x)$ is *continuous* on a *closed* interval $[a, b]$, then $f$ attains both an absolute maximum value and an absolute minimum value on $[a, b]$.

The key words in the extreme value theorem are "continuous" and "closed." As you can see in Figure 4.38, a function that is not continuous on a closed interval $I$ need not have absolute extrema on $I$. Also, in Figure 4.36(b) we saw an example of a continuous function defined on an open interval that does not have absolute extrema.

### Location of Absolute Extrema

Given a function $y = f(x)$ and an interval $I$, where can the absolute extrema of $f$ on $I$ occur? To answer this question, inspect the four graphs of Figure 4.39, all of which depict continuous functions on closed intervals. As you can see, an absolute extremum must occur either at an endpoint of the interval or at a number where a relative extremum occurs. Since relative extrema can occur only at critical numbers of $f$, an absolute extremum of $f(x)$ on $I$ must occur either at an endpoint of $I$ or at a critical number $c$ in $I$. This suggests the following procedure for finding the absolute extrema of a continuous function on a closed interval.

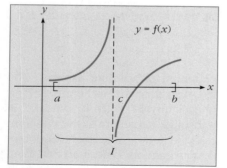

**FIGURE 4.38** A function to which the extreme value theorem does not apply.

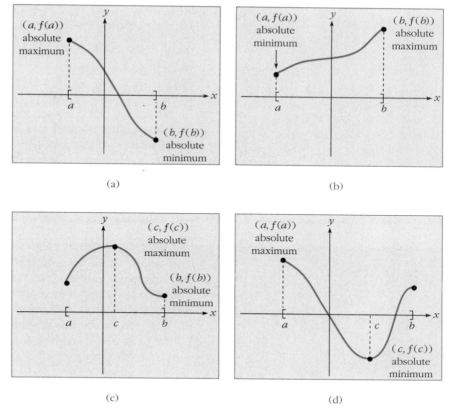

FIGURE 4.39 Occurrences of absolute extrema.

**FINDING ABSOLUTE EXTREMA OF A CONTINUOUS FUNCTION $y = f(x)$ ON A CLOSED INTERVAL $[a, b]$**

Step 1  Be certain that the interval is closed and that $f$ is continuous on the closed interval. Note that $f$ need not be continuous outside the interval $[a, b]$.

Step 2  Find all of the critical numbers of $f$ and discard any that are not in the interval of interest, $[a, b]$.

Step 3  Evaluate $f$ at the endpoints $a$ and $b$ of the interval and at all of the critical numbers that lie in $[a, b]$.

Step 4  The largest value found in Step 3 is the absolute maximum value of $f$ on $[a, b]$, and the smallest value found in Step 3 is the absolute minimum value of $f$ on $[a, b]$.

EXAMPLE 1

Find the absolute extrema of $f(x) = 4x^3 - 9x^2 - 30x + 30$ on the interval $[0, 3]$.

Solution

Step 1  The interval $[0, 3]$ is closed and the polynomial function $f(x) = 4x^3 - 9x^2 - 30x + 30$ is continuous everywhere and so is continuous on $[0, 3]$.

**Step 2** We compute $f'(x) = 12x^2 - 18x - 30$. To find the critical numbers of $f(x)$, we equate $f'(x)$ to zero and solve for $x$:

$$12x^2 - 18x - 30 = 0$$
$$6(2x^2 - 3x - 5) = 0$$
$$6(2x - 5)(x + 1) = 0$$
$$x = \tfrac{5}{2} \quad \text{or} \quad x = -1$$

Since the critical number $x = -1$ does not lie in the interval $[0, 3]$, it is not of interest in this problem and we discard it. We retain the critical number $x = \tfrac{5}{2}$, since it lies in $[0, 3]$.

**Step 3** Observe that

$$f(0) = 4(0)^3 - 9(0)^2 - 30(0) + 30 = 30$$
$$f(3) = 4(3)^3 - 9(3)^2 - 30(3) + 30 = -33$$
$$f(\tfrac{5}{2}) = 4(\tfrac{5}{2})^3 - 9(\tfrac{5}{2})^2 - 30(\tfrac{5}{2}) + 30 = -\tfrac{155}{4} = -38.75$$

**Step 4** The largest of the three values found in Step 3 is $f(0) = 30$, so the absolute maximum value of $f(x)$ on $[0, 3]$ is 30 and this absolute maximum value occurs at the endpoint $x = 0$. The smallest value found in Step 3 is $f(\tfrac{5}{2}) = -38.75$, so the absolute minimum value of $f(x)$ on $[0, 3]$ is $-38.75$ and this absolute minimum value occurs at the critical number $x = \tfrac{5}{2}$. The graph of $f(x)$ on the interval $[0, 3]$ is shown in Figure 4.40 for reference.

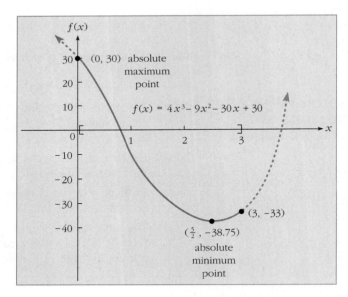

**FIGURE 4.40** The graph of $f(x) = 4x^3 - 9x^2 - 30x + 30$.

---

**EXAMPLE 2 (Participative)**

If $x$ is measured in hundreds of units and $P(x)$ is the monthly profit in dollars, let us find the production and sales level $x$ that will maximize profit if production capacity is limited to 30,000 units and $P(x) = -2500 + 2.5x^2 - 0.005x^3$.

This is the problem with which this section began. We are asked to find the number $x$ in the interval $[0, 300]$ at which the absolute maximum of $P(x)$ occurs.

Solution

Step 1  The interval $[0, 300]$ is closed and $P(x)$ is continuous on $[0, 300]$, so an absolute maximum (and an absolute minimum) exists by the _____ _____ theorem.

Step 2  To find the critical numbers, we compute $P'(x) = $ _____ and equate this to zero to obtain

$$5x - 0.015x^2 = 0$$
$$x(\text{_____}) = 0$$
$$x = 0 \quad \text{or} \quad 5 - 0.015x = 0$$
$$x = 0 \quad \text{or} \quad x = \text{_____}$$

Since the critical number $x = 333.\overline{3}$ is not in the interval $[0, 300]$, we discard it.

Step 3  We compute

$$P(0) = -2500 + 2.5(0)^2 - 0.005(0)^3 = -2500$$
$$P(300) = \text{_____}$$

Step 4  The largest value found in Step 3 is _____ and this is the absolute maximum value of $P(x)$ on $[0, 300]$. In other words, the maximum monthly profit is \$87,500. The production level that produces the maximum profit is $x = $ _____ units. See the graph of $P(x)$ in Figure 4.41. ∎

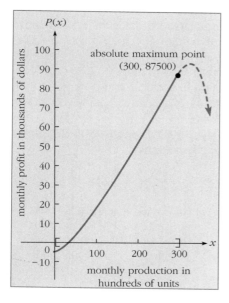

**FIGURE 4.41**  The graph of $P(x) = -2500 + 2.5x^2 - 0.005x^3$.

## Absolute Extrema When the Interval Is Not Closed

In some situations we are interested in finding absolute extrema in cases where the interval may not be closed and may be unbounded. Since the extreme value theorem does not apply to such a situation, we cannot be sure that absolute extrema exist. Our next example considers the case of a continuous function on the unbounded interval $(0, \infty)$.

EXAMPLE 3

Find any absolute extrema of $f(x) = (2x + 1)/(x^2 + 1)$ on $(0, \infty)$.

Solution

To solve this problem, we must modify our usual four-step procedure that we have been using for continuous functions on closed intervals.

Step 1  Note that $f(x)$ is continuous on $(0, \infty)$. However, the interval $(0, \infty)$ is *not* closed, so we cannot use the extreme value theorem to guarantee the existence of absolute extrema. Absolute extrema may or may not exist on $(0, \infty)$.

**SOLUTIONS TO EXAMPLE 2**  extreme value, $5x - 0.015x^2$, $5 - 0.015x$, $333\frac{1}{3}$, 87500, 87500, 300

**Step 2** To find the critical numbers of $f(x)$, we compute

$$f'(x) = \frac{(x^2 + 1)(2) - (2x + 1)(2x)}{(x^2 + 1)^2} \quad \text{(Quotient rule)}$$

$$= \frac{2 - 2x - 2x^2}{(x^2 + 1)^2}$$

The denominator of $f'(x)$ is never equal to zero, since $x^2 + 1$ is positive for all $x$, so there are no critical numbers $c$ of $f(x)$ for which $f'(c)$ does not exist. To find the critical numbers $c$ of $f(x)$ having $f'(c) = 0$, we equate the numerator of $f'(x)$ equal to zero to get $2 - 2x - 2x^2 = 0$. Using the quadratic formula, we obtain the critical numbers

$$x = \frac{-(-2) \pm \sqrt{(-2)^2 - 4(-2)(2)}}{2(-2)} = \frac{2 \pm \sqrt{20}}{-4} = \frac{1 \pm \sqrt{5}}{-2}$$

Using a calculator, we obtain the approximations

$$x = \frac{1 + \sqrt{5}}{-2} \approx -1.618 \quad \text{and} \quad x = \frac{1 - \sqrt{5}}{-2} \approx 0.618$$

Since $x \approx -1.618$ is not in the interval $(0, \infty)$, we discard it.

**Step 3** We compute

$$f\left(\frac{1 - \sqrt{5}}{-2}\right) \approx \frac{2(0.618) + 1}{(0.618)^2 + 1} = 1.618$$

Since this is the only critical point in $(0, \infty)$ and since $f(x)$ is continuous on $(0, \infty)$, this is the only candidate for an absolute extreme value. We now want to determine the behavior of $f(x)$ near the endpoints of the interval $(0, \infty)$. For this we must compute the limits:

$$\lim_{x \to 0^+} f(x) = \lim_{x \to 0^+} \frac{2x + 1}{x^2 + 1} = \frac{2(0) + 1}{0^2 + 1} = 1$$

$$\lim_{x \to \infty} f(x) = \lim_{x \to \infty} \frac{2x + 1}{x^2 + 1}$$

$$= \lim_{x \to \infty} \frac{\frac{2x + 1}{x^2}}{\frac{x^2 + 1}{x^2}} \quad \text{(Dividing numerator and denominator by } x^2\text{)}$$

$$= \lim_{x \to \infty} \frac{\frac{2}{x} + \frac{1}{x^2}}{1 + \frac{1}{x^2}} = \frac{0}{1} = 0$$

**Step 4** Since neither of the limits at the endpoints is larger than 1.618, the absolute maximum value of $f(x)$ on $(0, \infty)$ is approximately 1.618. No absolute minimum value exists since neither 0 nor $\infty$ is included in the open interval $(0, \infty)$. A sketch of $f(x)$ is drawn in Figure 4.42. ∎

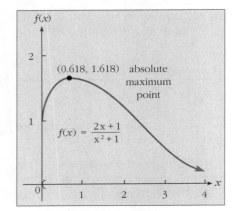

**FIGURE 4.42** The graph of $f(x) = (2x + 1)/(x^2 + 1)$.

### The Second Derivative Test for Absolute Extrema

Note that in Example 3, since the interval on which we were seeking absolute extrema was not closed, we had to modify Step 3 of our procedure somewhat to find limits at endpoints that were not included in the interval. In general, this modified procedure works well for continuous functions defined on an interval $I$ whether or not $I$ is closed. An alternate procedure can sometimes be used, however, in the case that only *one* critical number of $f(x)$ lies interior to (inside of ) $I$. Consider the functions sketched in Figure 4.43.

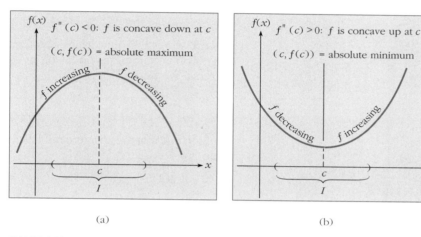

**FIGURE 4.43** Illustration of the second derivative test for absolute extrema.

In both of Figures 4.43(a) and 4.43(b), we are assuming that $f(x)$ is continuous on an interval $I$, that $c$ is the only critical number of $f(x)$ that lies interior to $I$, and that $f'(c) = 0$. Note that the interval $I$ can be open, closed, half-open, or unbounded. In Figure 4.43(a), since $f''(c) < 0$, the graph of $f(x)$ is concave down at $c$. Because $c$ is the only critical number of $f(x)$, the graph of $f$ must be increasing to the left of $c$ and decreasing to the right of $c$ if we restrict our attention to $I$ and if $c$ is interior to $I$. A function $f$ cannot change from increasing to decreasing or vice versa at a point in $I$ other than $c$, since $f(x)$ is continuous and there are no other critical points of $f(x)$ in $I$. We conclude that $(c, f(c))$ is the absolute maximum point of the graph of $f(x)$ on $I$. By inspecting Figure 4.43(b), you can also see that if $f''(c) > 0$, then $f$ is concave up at $c$, and a similar argument shows that $(c, f(c))$ is the absolute minimum point of the graph of $f(x)$ on $I$. Formally, we have the second derivative test for absolute extrema.

**SECOND DERIVATIVE TEST FOR ABSOLUTE EXTREMA**

If $f(x)$ is continuous on an interval $I$, if $c$ is the ONLY critical number interior to $I$, and if $f'(c) = 0$, then

i. if $f''(c) > 0$, the absolute minimum value of $f(x)$ on $I$ occurs at $x = c$;
ii. if $f''(c) < 0$, the absolute maximum value of $f(x)$ on $I$ occurs at $x = c$;
iii. if $f''(c) = 0$, the test fails.

**EXAMPLE 4** Find the absolute minimum value of $f(x) = 3x^4 - 4x^3 - 12x^2 + 10$ on the interval $(0, 20)$.

**Solution** To begin, we compute $f'(x) = 12x^3 - 12x^2 - 24x$ and find the critical numbers of $f(x)$ by solving the equation:

$$12x^3 - 12x^2 - 24x = 0$$
$$12x(x^2 - x - 2) = 0$$
$$12x(x - 2)(x + 1) = 0$$
$$x = 0, \quad x = 2, \quad x = -1$$

Since $x = 2$ is the only critical number that lies in the interval $I = (0, 20)$ and since $f(x)$ is continuous on $I$, we can use the second derivative test for absolute extrema. We find

$$f''(x) = 36x^2 - 24x - 24$$
$$f''(2) = 36(2)^2 - 24(2) - 24 = 72 > 0$$

We conclude that the absolute minimum occurs at $x = 2$. The absolute minimum value is

$$f(2) = 3(2)^4 - 4(2)^3 - 12(2)^2 + 10 = -22 \qquad \blacksquare$$

**EXAMPLE 5** Weekly profit from one of our company's products is related to the previous week's advertising expense $x$ in hundreds of dollars according to the formula

$$P(x) = -x^3 + 43x^2 - 75x + 750, \quad x \geq 5$$

Find the advertising expense $x$ that will yield the maximum profit; find the maximum profit.

**Solution** We need to find the absolute maximum of $P(x)$ on $I = [5, \infty)$. First we must compute $P'(x)$ and solve the equation $P'(x) = 0$ for the critical numbers $x$:

$$P'(x) = -3x^2 + 86x - 75 = 0$$

Using the quadratic formula, we get

$$x = \frac{-86 \pm \sqrt{(86)^2 - 4(-3)(-75)}}{2(-3)} = \frac{-86 \pm \sqrt{6496}}{-6}$$

Approximate critical numbers are $x \approx 0.9004$ and $x \approx 27.7663$. Since only one of these critical numbers is interior to the interval $[5, \infty)$, we can apply the second derivative test for absolute extrema. We obtain

$$P''(x) = -6x + 86$$
$$P''(27.7663) = -6(27.7663) + 86 = -80.5978 < 0$$

Thus the weekly advertising expense that yields the (absolute) maximum profit is $27.7663 hundred dollars, or $2776.63. At this level of advertising, the weekly profit is approximately

$$P(27.7663) = -(27.7663)^3 + 43(27.7663)^2 - 75(27.7663) + 750 \approx \$10412.21 \qquad \blacksquare$$

## EXAMPLE 6

The efficiency $E$ (measured on a scale from 0 to 1) of an employee working on an assembly line during a four-hour shift is given by

$$E(t) = \frac{36}{20t^2 - 60t + 85} \qquad 0 \le t \le 4$$

Find the time during the shift when the employee is working at maximum efficiency and find that maximum efficiency.

**Solution**

We are required to find the absolute maximum of $E(t)$ on $[0, 4]$. Writing $E(t) = 36(20t^2 - 60t + 85)^{-1}$, we compute

$$E'(t) = 36(-1)(20t^2 - 60t + 85)^{-2}(40t - 60) = \frac{-36(40t - 60)}{(20t^2 - 60t + 85)^2}$$

The critical numbers for which $E'(t) = 0$ are found by solving the equation $40t - 60 = 0$, so that $t = 1.5$.

You should verify that there are no critical numbers for which $E'(t)$ does not exist, since the equation $(20t^2 - 60t + 85)^2 = 0$ or, equivalently, $20t^2 - 60t + 85 = 0$ has no real solutions. Although the second derivative test for absolute extrema can be used in this problem, we choose not to use it because $E''(t)$ is not easily found. Rather, since $[0, 4]$ is a closed interval on which $E(t)$ is continuous, we compute

$$E(0) = \frac{36}{20(0)^2 - 60(0) + 85} \approx 0.42$$

$$E(1.5) = \frac{36}{20(1.5)^2 - 60(1.5) + 85} = 0.90$$

$$E(4) = \frac{36}{20(4)^2 - 60(4) + 85} \approx 0.22$$

We conclude that the employee is working at a maximum efficiency of 0.90 one and one-half hours after the shift begins. ∎

---

### SECTION 4.4
**SHORT ANSWER EXERCISES**

1. (True or False) The absolute maximum value of a function on an interval $I$ may also be a relative maximum value of $f(x)$. _____
2. (True or False) The function $f(x) = 5x^4 - 7x^3 + 3x^2 + 2x + 3$ has both an absolute maximum value and an absolute minimum value on $[0, 10]$. _____
3. The absolute maximum value of $f(x) = x^2$ on $[1, 2]$ is equal to _____.
4. (True or False) Given an interval $I$ and a function $f(x)$, we can always find both an absolute maximum value and an absolute minimum value of $f(x)$ on $I$. _____
5. A function has the line $x = 2$ as a vertial asymptote and is continuous otherwise. Then **(a)** $f(x)$ has an absolute maximum value on $[3, 6]$; **(b)** $f(x)$ has an absolute maximum value and an absolute minimum value on $[0, 4]$; **(c)** $f(x)$ has an absolute maximum value and an absolute minimum value on $(-\infty, \infty)$; **(d)** all of the above.

6. The function $f(x)$ is continuous on $[0, 20]$ and the critical numbers of $f(x)$ are $-3, 0,$ and $5$. Also, $f(-3) = 40, f(0) = 10, f(5) = 0,$ and $f(20) = 8$. The absolute maximum value of $f(x)$ on $[0, 20]$ is _____.

7. The function $f(x)$ is continuous in $I = (0, \infty)$ and the critical numbers of $f(x)$ are $-2$ and $10$, where $f''(-2) = -10$ and $f''(10) = 7$. Also, $\lim_{x \to 0^+} f(x) = 4$ and $\lim_{x \to \infty} f(x) = \infty$. Then (a) the absolute minimum value of $f(x)$ on $I$ is $-10$; (b) the absolute maximum value of $f(x)$ on $I$ is $7$; (c) the absolute minimum value of $f(x)$ on $I$ occurs at $x = 10$; (d) all of the above.

8. The function $f(x)$ is continuous on $I = (1, 5)$ and the critical numbers of $f(x)$ are $0$ and $6$. Then (a) an absolute extremum of $f(x)$ on $I$ may occur at $x = 1$; (b) an absolute extremum of $f(x)$ on $I$ may occur at $x = 5$; (c) an absolute extremum of $f(x)$ on $I$ may occur at $x = 0$; (d) $f(x)$ has no absolute extrema on $I$.

9. For the function $f(x) = x^3 + \frac{3}{2}x^2$, the derivative is $f'(x) = 3x^2 + 3x = 3x(x + 1)$. On the interval $[-1, 1]$ the absolute minimum value is _____.

10. If $f(x)$ is continuous on $[0, \infty)$, then (a) $f(x)$ has no absolute maximum value on $[0, \infty)$; (b) $f(x)$ has no absolute minimum value on $[0, \infty)$; (c) $f(x)$ has both an absolute maximum value and an absolute minimum value on $[0, \infty)$; (d) none of the above.

## SECTION 4.4
## EXERCISES

In Exercises 1–30, find the absolute extrema of the function on the given interval $I$ if any exist.

1. $f(x) = x^2 + x; I = [-2, 4]$
2. $f(x) = x - 2x^2; I = [-3, 3]$
3. $f(x) = 2x - 5; I = [0, 2]$
4. $f(x) = 3 - 4x; I = [-1, 1]$
5. $f(x) = 2x^2 + x - 4; I = [0, 3]$
6. $f(x) = 3x^2 + x + 5; I = [-2, 3]$
7. $f(x) = 12 - x - 2x^2; I = [-1, 2]$
8. $f(x) = 3 - 3x - 4x^2; I = [-3, 4]$
9. $f(x) = x^3 - x^2; I = [-2, 2]$
10. $f(x) = 2x^2 - 3x^3; I = [-3, 3]$
11. $f(x) = x^3 - 6x^2 + 9x + 5; I = [0, 3]$
12. $f(x) = 2x^3 - 9x^2 - 24x + 20; I = [-1, 5]$
13. $f(x) = 1 + x + x^2 - x^3; I = [0, 5]$
14. $f(x) = 6 + 6x - 3x^2 - 4x^3; I = [-2, 2]$
15. $f(x) = x^2 + 2x - 2; I = (-3, \infty)$
16. $f(x) = 3x^2 - 2x + 1; I = (-1, 1]$
17. $f(x) = x^3 - x^2; I = (-1, 2]$
18. $f(x) = x^3 - 6x^2 + 9x + 5; I = (0, 3)$
19. $f(x) = x^2 + 2x - 5; I = (-\infty, \infty)$
20. $f(x) = 1 - x - 3x^2; I = (-\infty, \infty)$
21. $f(x) = (x - 2)^4; I = [0, 3]$
22. $f(x) = x^4 - x^2; I = [-1, 1]$
23. $f(x) = 3x^4 + 4x^3 - 12x^2 + 8; I = [-3, 3]$
24. $f(x) = 3x^4 - 2x^3 - 9x^2 - 10; I = [-2, 3]$
25. $f(x) = \dfrac{x}{x^2 + 1}; I = (-2, \infty)$
26. $f(x) = \dfrac{x - 1}{2x^2 + 1}; I = (0, \infty)$
27. $f(x) = 3/x^2; I = (0, \infty)$
28. $f(x) = (4/x^3) + x; I = (0, \infty)$
29. $f(x) = \dfrac{x}{x - 1}; I = [0, 2]$
30. $f(x) = \dfrac{2x}{x + 2}; I = [-3, 0]$

**31.** (Maximum profit) The monthly profit in dollars derived from the sale of $x$ hundred units of one of our company's products is $P(x) = -5000 + 3x^2 - 0.006x^3$. Find the sales level that maximizes profit and the maximum profit if we can produce at most 40,000 units per month.

**32.** (Minimum average cost) Given that the monthly cost $C(x)$ in hundreds of dollars of producing $x$ thousand units of one of our company's products is $C(x) = 4x^2 + 30x + 25$, find the production level that produces the absolute minimum value of average cost, $\overline{C}(x) = C(x)/x$.

*Business and Economics*

**33.** (Cost and demand) Given the monthly cost function $C(x) = 1000 + 25x$, and the fact that the relationship between price $p$ (in dollars) and monthly sales $x$ (in units) is

$$p = 40 - 0.01x, \quad 0 \le x \le 4000$$

(a) find the absolute maximum revenue;
(b) find the absolute maximum profit.

*Social Sciences*

**34.** (Repeat offenders) A police department has determined that the fraction of individuals who will be rearrested for a certain type of crime within $t$ weeks after their release from prison is

$$F(t) = \frac{3t^2}{4t^2 + 750}$$

Find the absolute maximum value of $F(t)$ for the first year after release from prison.

OPTIONAL

GRAPHING CALCULATOR/
COMPUTER EXERCISES

*In Exercises 35–38, determine if $f(x)$ has absolute extrema on the given interval and, if so, approximate the values of $x$ at which the absolute extrema occur.*

**35.** $f(x) = x^4 - 7x^3 + 3x^2 + 4x + 2$; $I = [0, 4]$

**36.** $f(x) = \dfrac{x}{x^2 + 1} + x^4 - x^3$; $I = [-1, 3]$

**37.** $f(x) = \dfrac{x^2}{x^3 + 2x - 1}$; $I = (0, \infty)$

**38.** $f(x) = \sqrt[3]{x} - x + \dfrac{x}{x^3 + 1}$; $I = [0, 1]$

# 4.5 CURVE SKETCHING

In this section we will use the tools developed earlier in this chapter to help us with the task of sketching the graph of a function. Since most of the applied problems we will discuss in this chapter concern polynomial functions or rational functions, we will concentrate mainly on these types of graphs. Toward the end of the section we will discuss sketching other types of graphs.

### Graphing Polynomial Functions

Let's begin by considering the steps necessary to graph a polynomial function.

**272** CHAPTER 4 APPLICATIONS OF THE DERIVATIVE

**GRAPHING A POLYNOMIAL FUNCTION** $y = f(x)$

Step 1 Find the $y$-intercept by computing $f(0)$. If feasible, find the $x$-intercept(s) by solving the equation $f(x) = 0$.

Step 2 Find $f'(x)$ and $f''(x)$.

Step 3 Find the first-order critical numbers of $f(x)$ by solving the equation $f'(x) = 0$. These are numbers at which relative extrema can occur. Use the second derivative test on any critical number $c$: (i) If $f''(c) > 0$, a relative minimum occurs at $c$; (ii) If $f''(c) < 0$, a relative maximum occurs at $c$; (iii) If $f''(c) = 0$, the second derivative test fails: use the first derivative test. For each critical number $c$, find $f(c)$.

Step 4 Find the second-order critical numbers of $f(x)$ by solving the equation $f''(x) = 0$. These are numbers at which points of inflection can occur. Use the second-order critical numbers to divide the $x$-axis into intervals and construct a chart that shows the sign of $f''$ in each of the intervals. Use the chart to determine intervals of concavity and points of inflection.

Step 5 Plot and label the $y$-intercept, the $x$-intercept(s), relative extrema, and points of inflection on the graph. Indicate the intervals where $f$ increases and decreases and the intervals of concavity on your sketch. Remember that a graph must be rising to the left of a relative maximum point and falling to the right. Also a graph must be falling to the left of a relative minimum point and rising to the right. If necessary, plot a few additional points on the graph and complete the sketch, remembering that, since any polynomial is differentiable and continuous, we must draw a smooth connected curve.

---

EXAMPLE 1

Graph the function $f(x) = x^3 - 3x^2$.

Solution

Step 1 To find the $y$-intercept, we compute $f(0) = 0^3 - 3(0)^2 = 0$. To find the $x$-intercept(s) we must solve the equation

$$x^3 - 3x^2 = 0$$
$$x^2(x - 3) = 0$$
$$x = 0, \quad x = 3 \quad \text{($x$-intercepts)}$$

Step 2 $f'(x) = 3x^2 - 6x$ and $f''(x) = 6x - 6$.

Step 3 To find the first-order critical numbers, we solve the equation

$$3x^2 - 6x = 0$$
$$3x(x - 2) = 0$$
$$x = 0, \quad x = 2 \quad \text{(critical numbers)}$$

Using the second derivative test on each of these critical numbers, we get

$$f''(0) = 6(0) - 6 = -6 < 0$$

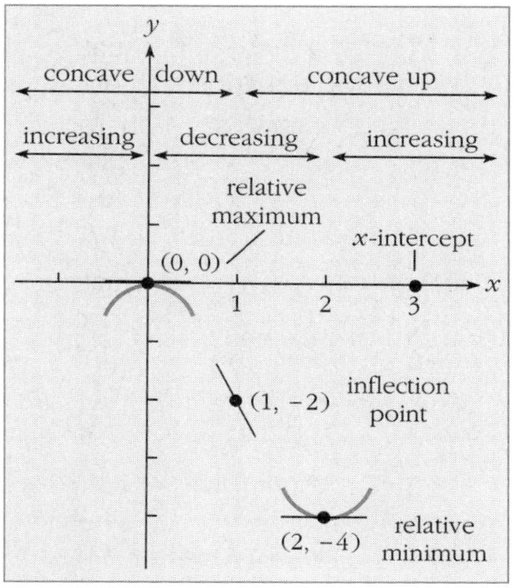

so a relative maximum occurs at $x = 0$. Note that $f(0) = 0^3 - 3(0)^2 = 0$ is the relative maximum value at $x = 0$.

$$f''(2) = 6(2) - 6 = 6 > 0$$

so a relative minimum occurs at $x = 2$. Note that $f(2) = 2^3 - 3(2)^2 = -4$ is the relative minimum value at $x = 2$.

Step 4  To find the second-order critical numbers, we solve the equation

$$6x - 6 = 0$$
$$6x = 6$$
$$x = 1 \quad \text{(second-order critical number)}$$

We now construct the chart.

| $x$ | | 0 | 1 | 2 |
|---|---|---|---|---|
| $f''$ | | ----- | 0 | +++++ |
| $f$ | | concave down | pt of inf | concave up |

SOCN ↓ (at $x = 1$)

We conclude that $f(x)$ is concave down on $(-\infty, 1)$ and concave up on $(1, \infty)$. Since the concavity changes at $x = 1$, the point $(1, f(1)) = (1, -2)$ is a point of inflection of the graph. Note that $f'(1) = 3(1)^2 - 6(1) = -3$ is the slope of the tangent line at the inflection point.

Step 5  Summarizing, we have the following:

$y$-intercept: 0

$x$-intercepts: 0, 3

Relative maximum point: $(0, 0)$

Relative minimum point: $(2, -4)$

Inflection point: $(1, -2)$

Next we plot each of these points on our graph and label them. We show this intermediate plot in Figure 4.44, where we have also sketched the horizontal tangent lines and the tangent line at the inflection point, and we have indicated intervals where $f$ increases and decreases, and intervals of concavity in this sketch.

The final sketch can now be completed; it is shown in Figure 4.45. ∎

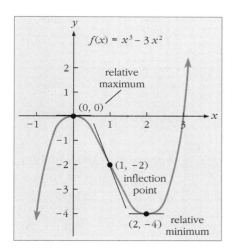

FIGURE 4.44  A preliminary plot of $f(x) = x^3 - 3x^2$.

FIGURE 4.45  The final sketch of $f(x) = x^3 - 3x^2$.

EXAMPLE 2   Graph the function $f(x) = 3x^4 + 16x^3 + 24x^2 - 10$.

Solution   Step 1  The $y$-intercept is $f(0) = 3(0)^4 + 16(0)^3 + 24(0)^2 - 10 = -10$. To find the $x$-intercept(s), we have to solve the equation

$$3x^4 + 16x^3 + 24x^2 - 10 = 0$$

But it is not feasible for us to solve this equation in sketching this graph. If information about the $x$-intercept(s) were absolutely required in this problem, we could use a graphing calculator or a computer to obtain approximate roots of this equation.

Step 2  $f'(x) = 12x^3 + 48x^2 + 48x$ and $f''(x) = 36x^2 + 96x + 48$.

Step 3  To find first-order critical numbers, we solve the equation

$$12x^3 + 48x^2 + 48x = 0$$
$$12x(x^2 + 4x + 4) = 0$$
$$12x(x + 2)^2 = 0$$
$$x = 0, \quad x = -2 \quad \text{(Critical numbers)}$$

Using the second derivative test on the critical number $x = 0$, we find that

$$f''(0) = 36(0)^2 + 96(0) + 48 = 48 > 0$$

so a relative minimum occurs at $x = 0$. Also, $f(0) = 3(0)^4 + 16(0)^3 + 24(0)^2 - 10 = -10$ is the relative minimum value at $x = 0$. When we attempt to use the second-derivative test on the critical number $x = -2$, however,

$$f''(-2) = 36(-2)^2 + 96(-2) + 48 = 0$$

and the test fails. We proceed to use the first derivative test on $x = -2$ by constructing the following chart.

|  |  | CN ↓ |  | CN ↓ |
|---|---|---|---|---|
| $x$ | $-3$ | $-2$ | $-1$ | $0$ |
| $f'$ | $-----$ | $0$ | $-----$ | $0$ |
| $f$ | decreasing | no extremum | decreasing |  |

We conclude that no relative extremum occurs at $x = -2$, since $f(x)$ is decreasing on $(-\infty, -2)$ and on $(-2, 0)$.

Step 4  To find the second-order critical numbers, we solve the equation

$$36x^2 + 96x + 48 = 0$$
$$12(3x^2 + 8x + 4) = 0$$
$$12(x + 2)(3x + 2) = 0$$
$$x = -2, \quad x = -\tfrac{2}{3} \quad \text{(Second-order critical numbers)}$$

Next, we construct the chart.

|  |  | SOCN ↓ |  | SOCN ↓ |  |
|---|---|---|---|---|---|
| $x$ | $-3$ | $-2$ | $-1$ | $-\tfrac{2}{3}$ | $0$ |
| $f''$ | $+++++$ | $0$ | $-----$ | $0$ | $+++++$ |
| $f$ | concave up | pt of inf | concave down | pt of inf | concave up |

Note that $f(x)$ is concave up on the interval $(-\infty, -2)$, concave down on the interval $\left(-2, -\frac{2}{3}\right)$, and concave up on the interval $\left(-\frac{2}{3}, \infty\right)$. Since the concavity changes at $x = -2$ and at $x = -\frac{2}{3}$, both of the points $(-2, f(-2)) = (-2, 6)$ and $\left(-\frac{2}{3}, f\left(-\frac{2}{3}\right)\right) = \left(-\frac{2}{3}, -\frac{94}{27}\right)$ are points of inflection. Also note that $f'(-2) = 12(-2)^3 + 48(-2)^2 + 48(-2) = 0$ and $f'\left(-\frac{2}{3}\right) = 12\left(-\frac{2}{3}\right)^3 + 48\left(-\frac{2}{3}\right)^2 + 48\left(-\frac{2}{3}\right) = -14\frac{2}{9}$, which gives the slope of the tangent line at each of the inflection points.

Step 5 Summarizing, we have the following:

$y$-intercept: $-10$

Relative minimum point: $(0, -10)$

Inflection points: $(-2, 6)$ and $\left(-\frac{2}{3}, -\frac{94}{27}\right)$

We now make a preliminary sketch in Figure 4.46.

In this case, before drawing the final sketch, it is helpful to compute a few more points on the graph. This is done in Table 4.2. The final sketch is drawn in Figure 4.47.

| $x$ | $f(x)$ |
|---|---|
| $-1$ | $1$ |
| $1$ | $33$ |
| $2$ | $262$ |

**TABLE 4.2**

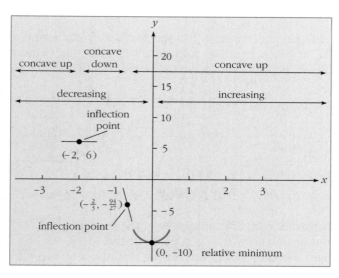

**FIGURE 4.46** A preliminary plot of $f(x) = 3x^4 + 16x^3 + 24x^2 - 10$.

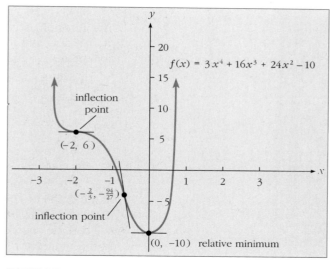

**FIGURE 4.47** The final sketch of $f(x) = 3x^4 + 16x^3 + 24x^2 - 10$.

### Asymptotes of Rational Functions

Recall that a rational function is a quotient of two polynomials. That is, the function $f(x)$ is a rational function if and only if $f(x) = p(x)/q(x)$, where $p(x)$ and $q(x)$ are polynomials. When we graph a rational function, it is important to determine any vertical or horizontal asymptotes. We discussed asymptotes in general in Section 2.2. Here we will mention an efficient specific procedure for finding asymptotes of rational functions.

**VERTICAL ASYMPTOTES OF RATIONAL FUNCTIONS**

If $f(x) = p(x)/q(x)$, where $p(x)$ and $q(x)$ are polynomials, and if $q(a) = 0$ but $p(a) \neq 0$, then the vertical line $x = a$ is a vertical asymptote of the graph of $f(x)$.

**HORIZONTAL ASYMPTOTES OF RATIONAL FUNCTIONS**

If $f(x) = p(x)/q(x)$, where $p(x)$ and $q(x)$ are polynomials, then

i. when the degree of $p(x)$ is larger than the degree of $q(x)$, then $f(x)$ has no horizontal asymptote;
ii. when the degree of $q(x)$ is larger than the degree of $p(x)$, then the horizontal asymptote is $y = 0$ (the $x$-axis);
iii. when the degrees of $p(x)$ and $q(x)$ are equal, then the horizontal asymptote is $y = a/b$, where $a$ is the coefficient of the highest power of $x$ in the numerator of $f(x)$ and $b$ is the coefficient of the highest power of $x$ in the denominator of $f(x)$.

**EXAMPLE 3**

Find the vertical and horizontal asymptotes of $f(x) = 2x^2/(3x^2 - 5x - 2)$.

Solution

The function is a rational function, since it has the form $p(x)/q(x)$, where $p(x) = 2x^2$ and $q(x) = 3x^2 - 5x - 2$ are polynomials. To find the vertical asymptotes, we equate the denominator to zero and solve the resulting equation:

$$3x^2 - 5x - 2 = 0$$
$$(3x + 1)(x - 2) = 0$$
$$3x + 1 = 0 \quad \text{or} \quad x - 2 = 0$$
$$x = -\tfrac{1}{3} \quad \text{or} \quad x = 2.$$

So, since $q\left(-\tfrac{1}{3}\right) = 0$ and $q(2) = 0$ but $p\left(-\tfrac{1}{3}\right) = 2\left(-\tfrac{1}{3}\right)^2 \neq 0$ and $p(2) = 2(2)^2 \neq 0$, we conclude that the lines $x = -\tfrac{1}{3}$ and $x = 2$ are vertical asymptotes of the graph of $f(x)$.

For the horizontal asymptote, note that the degree of the numerator of $f(x)$ is 2 and the degree of the denominator is also 2. So, since the coefficient of $x^2$ in the numerator is 2 and the coefficient of $x^2$ in the denominator is 3, the horizontal asymptote is the line $y = 2/3$. ∎

**EXAMPLE 4 (Participative)**

Let's find the vertical and horizontal asymptotes of $f(x) = (x^2 + 2x + 1)/(x^3 - x)$.

Solution

First we note that the function is rational, since it has the form $p(x)/q(x)$, where $p(x) = x^2 + 2x + 1$ and $q(x) = $ _____. To find vertical asymptotes, we solve the equation

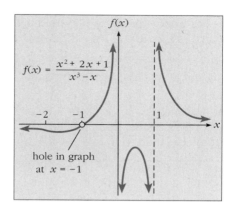

FIGURE 4.48 The graph of $f(x) = (x^2 + 2x + 1)/(x^3 - x)$.

$$\underline{\quad\quad} = 0$$
$$x(x^2 - 1) = 0$$
$$x(x + 1)(x - 1) = 0$$
$$x = \underline{\quad}, \quad x = \underline{\quad}, \quad x = \underline{\quad}$$

Note that $p(0) \neq 0$ and $p(1) \neq 0$, so the lines $x = 0$ and $x = 1$ are vertical asymptotes. However, $p(-1) = \underline{\quad}$, so the line $x = -1$ is *not* a vertical asymptote. The behavior of $f(x)$ at $x = -1$ is a hole or gap in the graph: see Figure 4.48.

To find the horizontal asymptote, we observe that the degree of the numerator is 2 and the degree of the denominator is $\underline{\quad}$. Since the degree of the denominator is larger than that of the numerator, the horizontal asymptote is the line $y = \underline{\quad}$. For reference, the graph of $f(x)$ is drawn in Figure 4.48.

### Graphing Rational Functions

Now let's look at the steps that are necessary to graph a rational function. These steps include those we have already considered in graphing a polynomial function and, additionally, in determining asymptotes.

**GRAPHING A RATIONAL FUNCTION** $f(x) = p(x)/q(x)$

Step 1 Find the $y$-intercept by computing $f(0)$. If feasible, find the $x$-intercept(s) by finding the number(s) $x$ such that $p(x) = 0$ but $q(x) \neq 0$.

Step 2 Find all vertical and horizontal asymptotes of $f(x)$.

Step 3 Find $f'(x)$ and $f''(x)$.

Step 4 Find the numbers at which $f'$ can change its sign, which consist of:

i. first-order critical numbers $c$ such that $f'(c) = 0$—these can be found by setting the numerator of $f'(x)$ equal to zero if $f'(x)$ is written in the form $N(x)/D(x)$;

ii. numbers $c$ such that $f'(c)$ does not exist—these can be found by setting the denominator of $f'(x)$ equal to zero and solving the resulting equation.

You may use the second derivative test on critical numbers found in (i) above, but you should use the first derivative test if you also found any numbers in (ii). In this text we will usually use the first derivative test on the critical numbers.

Step 5 Find the numbers at which $f''$ can change its sign, which consist of:

i. second-order critical numbers $c$ such that $f''(c) = 0$—these can be found by setting the numerator of $f''$ equal to zero if $f''$ is written as a quotient of two polynomials;

**SOLUTIONS TO EXAMPLE 4**  $x^3 - x, x^3 - x, 0, -1, 1, 0, 3, 0$

ii. numbers $c$ such that $f''(c)$ does not exist—these can be found by setting the denominator of $f''$ equal to zero and solving the resulting equation.

Determine the intervals of concavity and points of inflection by drawing a sign chart of the second derivative.

**Step 6** Plot and label all of the important points found in Steps 1–5. Draw in the vertical and horizontal asymptotes as dashed lines. If necessary, plot a few additional points on the graph and then complete the sketch.

---

**EXAMPLE 5**       Graph the function $f(x) = (x^3 - 2)/(2x)$.

**Solution**        First note that the function is rational, since it is of the form $f(x) = p(x)/q(x)$, where $p(x) = x^3 - 2$ and $q(x) = 2x$ are polynomials.

**Step 1** The $y$-intercept is $f(0) = (0^3 - 2)/(2(0))$, which is undefined, so the graph of $f(x)$ never crosses the $y$-axis. To find the $x$-intercepts, we must solve the equation

$$\frac{x^3 - 2}{2x} = 0$$

$$x^3 - 2 = 0$$

$$x = \sqrt[3]{2} \approx 1.26$$

**Step 2** The vertical asymptote can be found by solving the equation $2x = 0$, and is the vertical line $x = 0$, or the $y$-axis. Since the degree of the numerator of $f(x)$ is greater than the degree of the denominator, there is no horizontal asymptote. Note, however, that

$$\lim_{x \to \infty} \left( \frac{x^3 - 2}{2x} \right) = \lim_{x \to \infty} \left( \frac{1}{2}x^2 - \frac{1}{x} \right) = +\infty$$

and

$$\lim_{x \to -\infty} \left( \frac{x^3 - 2}{2x} \right) = \lim_{x \to -\infty} \left( \frac{1}{2}x^2 - \frac{1}{x} \right) = +\infty$$

which tells us how the "ends" or "tails" of the graph behave.

**Step 3** $f'(x) = \dfrac{(2x)(3x^2) - (x^3 - 2)(2)}{(2x)^2} = \dfrac{4x^3 + 4}{4x^2} = \dfrac{x^3 + 1}{x^2}$

$f''(x) = \dfrac{x^2(3x^2) - (x^3 + 1)(2x)}{(x^2)^2} = \dfrac{x^3 - 2}{x^3}$

**Step 4** In this step we find the first-order critical numbers. Recall that $f'(x) = (x^3 + 1)/x^2$.

i. To find first-order critical numbers for which $f'(x) = 0$, we set the numerator of $f'$ equal to zero to obtain

$$x^3 + 1 = 0$$
$$x^3 = -1$$
$$x = \sqrt[3]{-1} = -1$$

ii. To find the numbers for which $f'(x)$ is undefined, we set the denominator of $f'$ equal to zero to obtain $x^2 = 0$ or $x = 0$. However, since $x = 0$ is *not* in the domain of $f(x)$, a relative extremum cannot occur at $x = 0$, and $x = 0$ is not a critical number.

Note that since $f(0)$ is undefined, a possible sign change of $f'$ may occur at $x = 0$ and we include $x = 0$ in the accompanying sign chart.

| $x$ | $-2$ | $-1$ | $-\frac{1}{2}$ | $0$ | $1$ |
|---|---|---|---|---|---|
| $f'$ | $-----$ | $0$ | $+++++$ | ND | $+++++$ |
| $f$ | decreasing | rel min | increasing | VA | increasing |

CN ↓ (above the $-1$ column)

We find that $f(x)$ is decreasing on $(-\infty, -1)$, increasing on $(-1, 0)$, and increasing on $(0, \infty)$; therefore $(-1, f(-1)) = \left(-1, \frac{3}{2}\right)$ is a relative minimum point.

**Step 5** In this step we find the second-order critical numbers of $f$. Recall that $f''(x) = (x^3 - 2)/x^3$.

i. We solve the equation

$$\frac{x^3 - 2}{x^3} = 0$$

$$x^3 - 2 = 0 \quad \text{(Setting the numerator equal to zero)}$$

to obtain $x^3 = 2$ or $x = \sqrt[3]{2} \approx 1.26$.

ii. To determine where $f''$ does not exist, we set the denominator equal to zero to obtain $x^3 = 0$ or $x = 0$. However, $x = 0$ is not in the domain of $f$, so $x = 0$ is not a second-order critical number.

Since $f(0)$ is not defined, a possible sign change of $f''$ may occur at $x = 0$, so we include $x = 0$ in our chart.

| $x$ | $-1$ | $0$ | $1$ | $1.26$ | $3$ |
|---|---|---|---|---|---|
| $f''$ | $+++++$ | ND | $-----$ | $0$ | $+++++$ |
| $f$ | concave up | VA | concave down | pt of inf | concave up |

SOCN ↓ (above the $1.26$ column)

So $f(x)$ is concave up on $(-\infty, 0)$ and $(1.26, \infty)$ and concave down on $(0, 1.26)$. The point $(\sqrt[3]{2}, f(\sqrt[3]{2})) = (1.26, 0)$ is a point of inflection.

Observe that $f'(\sqrt[3]{2}) \approx 1.89$ is the slope of the tangent line at the inflection point.

Step 6 Summarizing, we have

$y$-intercept: none
$x$-intercept: $\sqrt[3]{2} \approx 1.26$
Vertical asymptote: $x = 0$
Horizontal asymptote: none; however, $\lim\limits_{x \to \pm\infty} f(x) = +\infty$
Relative minimum point: $\left(-1, \frac{3}{2}\right)$
Inflection point: $(1.26, 0)$

We make a preliminary sketch in Figure 4.49. Some additional points on the graph are computed in Table 4.3. The completed sketch is drawn in Figure 4.50.

**TABLE 4.3**

| $x$ | $f(x)$ |
|---|---|
| $-2$ | 2.5 |
| 1 | $-0.5$ |
| 2 | 1.5 |

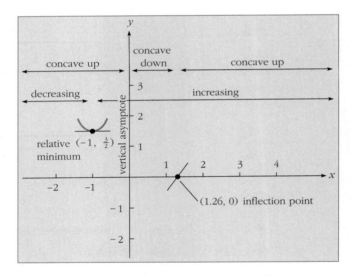

FIGURE 4.49 A preliminary plot of $f(x) = (x^3 - 2)/(2x)$.

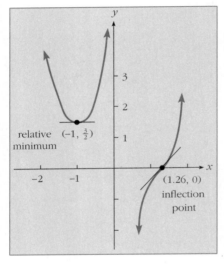

FIGURE 4.50 The final sketch of $f(x) = (x^3 - 2)/(2x)$.

### Graphing Arbitrary Functions

To sketch the graph of an arbitrary function $y = f(x)$, we will use the following general procedure.

**GRAPHING AN ARBITRARY FUNCTION $y = f(x)$**

Step 1 Find the $y$-intercept by computing $f(0)$. If feasible, find the $x$-intercept(s) by solving the equation $f(x) = 0$.
Step 2 Find any asymptotes of $f(x)$.
Step 3 Find the domain of $f(x)$.
Step 4 Find $f'(x)$ and $f''(x)$.

Step 5  Find the first-order critical numbers of $f(x)$ by using $f'(x)$. Determine the relative extrema of $f(x)$ and the intervals where $f$ is increasing or decreasing by using either the first derivative test or the second derivative test.

Step 6  Find the second-order critical numbers of $f(x)$ by using $f''(x)$. Determine the points of inflection of $f(x)$ and the intervals of concavity by constructing a sign chart.

Step 7  Plot and label all of the important points found in Steps 1–6. Draw in any asymptotes as dashed lines. If necessary, plot a few additional points on the graph and then complete the sketch.

---

EXAMPLE 6

Sketch the graph of $f(x) = \sqrt{x} - x$.

Solution

Step 1  $f(0) = \sqrt{0} - 0 = 0$ is the $y$-intercept. To find the $x$-intercepts, we must solve the equation

$$\sqrt{x} - x = 0$$
$$\sqrt{x} = x$$
$$x = x^2 \quad \text{(Squaring both sides)}$$
$$x^2 - x = 0$$
$$x(x - 1) = 0$$
$$x = 0, \quad x = 1$$

In solving this equation, we squared both sides. This can lead to extraneous solutions. However, both $x = 0$ and $x = 1$ check in the original equation $\sqrt{x} - x = 0$ and so both are $x$-intercepts.

Step 2  For any real number $a > 0$, we have $\lim\limits_{x \to a} (\sqrt{x} + x) = \sqrt{a} + a$. Since this limit is neither $\infty$ nor $-\infty$, the number $x = a$ is not a vertical asymptote for $a > 0$. Also, $\lim\limits_{x \to 0^+} (\sqrt{x} + x) = 0$, and so $x = 0$ is not a vertical asymptote. For horizontal asymptotes, we consider the limit

$$\lim_{x \to \infty} (\sqrt{x} - x) = \lim_{x \to \infty} \frac{\sqrt{x} - x}{1} \cdot \frac{\sqrt{x} + x}{\sqrt{x} + x}$$

$$= \lim_{x \to \infty} \frac{x - x^2}{x^{1/2} + x} = \lim_{x \to \infty} \frac{\frac{x - x^2}{x^2}}{\frac{x^{1/2} + x}{x^2}}$$

$$= \lim_{x \to \infty} \frac{\frac{1}{x} - 1}{\frac{1}{x^{3/2}} + \frac{1}{x}}$$

Since the numerator approaches $-1$ and the denominator approaches zero through positive values, this limit is equal to $-\infty$ by Property III of Section 2.2. That is,

$$\lim_{x \to \infty} (\sqrt{x} - x) = -\infty$$

In addition, $\lim_{x \to -\infty} (\sqrt{x} - x)$ does not exist, since $\sqrt{x}$ is defined only if $x \geq 0$. Therefore $f(x)$ has no horizontal asymptotes.

Step 3 Since $\sqrt{x}$ is defined only if $x \geq 0$, the domain of $f(x) = \sqrt{x} - x$ is all nonnegative real numbers.

Step 4 $f'(x) = \frac{1}{2}x^{-1/2} - 1 = (1/(2\sqrt{x})) - 1$
$f''(x) = -\frac{1}{4}x^{-3/2} = -1/(4\sqrt{x^3}) = -1/(4x\sqrt{x})$

Step 5 In this step we use $f'(x)$ to find the first-order critical numbers. First, we solve the equation

$$\frac{1}{2\sqrt{x}} - 1 = 0$$

$$2\sqrt{x} = 1$$

$$\sqrt{x} = \frac{1}{2}$$

$$x = \frac{1}{4} \quad \text{(First-order critical number)}$$

Also, note that $f'(0)$ is undefined and $x = 0$ is in the domain of $f(x)$. Therefore $x = 0$ is also a first-order critical number of $f(x)$. However, because the domain of $f(x)$ is $\{x \mid x \geq 0\}$, a relative extremum cannot occur at $x = 0$, an endpoint of the domain of definition of $f(x)$. Now we calculate that

$$\lim_{x \to 0^+} f'(x) = \lim_{x \to 0^+} \left(\frac{1}{2\sqrt{x}} - 1\right) = \infty$$

Hence there is a vertical tangent line at $x = 0$. We now construct a sign chart of $f'(x)$ for $x \geq 0$.

| | | CN $\downarrow$ | | CN $\downarrow$ | |
|---|---|---|---|---|---|
| $x$ | | 0 | $\frac{1}{16}$ | $\frac{1}{4}$ | 1 |
| $f'$ | | ND | +++++ | 0 | ----- |
| $f$ | | | increasing | rel max | decreasing |

Observe that $f(x)$ is increasing on $\left(0, \frac{1}{4}\right)$ and decreasing on $\left(\frac{1}{4}, \infty\right)$, so that a relative maximum point is $\left(\frac{1}{4}, f\left(\frac{1}{4}\right)\right) = \left(\frac{1}{4}, \frac{1}{4}\right)$.

Step 6 Recall that $f''(x) = -1/(4x\sqrt{x})$. The equation $f''(x) = 0$ yields $-1/(4x\sqrt{x}) = 0$, which has no solution because the numerator of the fraction is never equal to zero. Also, $f''(0)$ is undefined, so $x = 0$ is a second-order critical number of $f(x)$. However, since $f(x)$ is defined only for $x \geq 0$, the concavity cannot change at $x = 0$. Our sign chart for $f''(x)$ for $x \geq 0$ is as follows.

|  | | SOCN↓ | |
|---|---|---|---|
| $x$ |  | 0 | 1 |
| $f''$ |  | ND | ----- |
| $f$ |  |  | concave down |

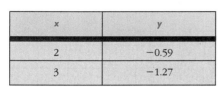

TABLE 4.4

We conclude that $f(x)$ is concave down on $(0, \infty)$ and that there are no points of inflection.

Step 7  Summarizing, we have the following:

y-intercept:  0

x-intercepts:  0 and 1

Domain:  $\{x \mid x \geq 0\}$

Asymptotes:  None; however, $\lim_{x \to \infty} f(x) = -\infty$

Relative maximum point:  $\left(\frac{1}{4}, \frac{1}{4}\right)$

Inflection point:  None; the function is concave down on $(0, \infty)$

In this example we omit the preliminary sketch. In Table 4.4 we have computed two additional points on the graph. The completed sketch is shown in Figure 4.51. ∎

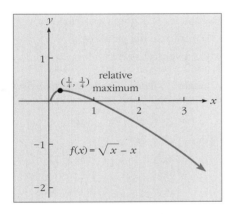

**FIGURE 4.51**  The final sketch of $f(x) = \sqrt{x} - x$.

## SECTION 4.5
### SHORT ANSWER EXERCISES

1. The y-intercept of the polynomial function $y = x^4 - 7x^3 + x - 6$ is $y =$ _____.

2. (True or False) The graph of a polynomial function has neither vertical nor horizontal asymptotes. _____

3. (True or False) The graph of a polynomial function is always smooth and connected. _____

4. (True or False) The graph of a function always has a y-intercept and an x-intercept. _____

5. The horizontal asymptote of $f(x) = \dfrac{2x^2 - 5x + 3}{x^2 + x - 7}$ is $y =$ _____.

6. The rational function $f(x) = x/(x^2 + 1)$ has no vertical asymptote, because $x^2 + 1$ is never equal to _____.

7. Given $f(x) = x/(x - 1)$, we can compute that $f'(x) = -1/(x - 1)^2$. Does $f(x)$ have any critical points? _____

8. Given that $\lim_{x \to \infty} f(x) = \infty$ and $\lim_{x \to -\infty} f(x) = -\infty$, which of the following graphs are possible graphs of $f(x)$?

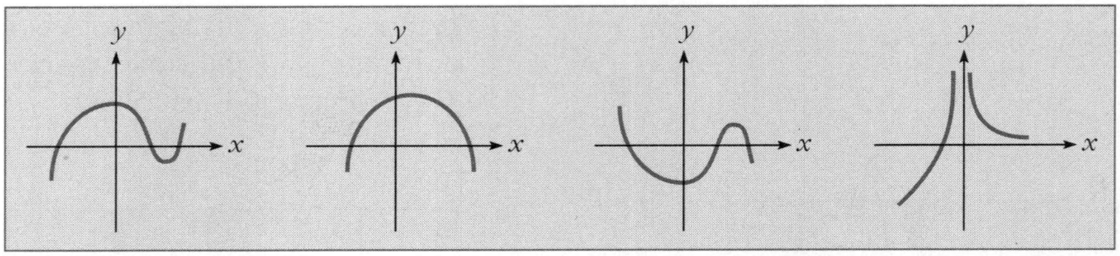

**FIGURE 4.52**     (a)     (b)     (c)     (d)

9. Refer to the graphs in Exercise 8 for this problem. Among these four graphs, how many of them can represent the graph of a polynomial function?
   (a) 1; (b) 2; (c) 3; (d) 4

10. Refer to the graph in Exercise 8(d) for this problem. This graph (a) has no $x$-intercept; (b) has a point of inflection; (c) could be the graph of a polynomial function; (d) has the line $x = 0$ as a vertical asymptote.

## SECTION 4.5
### EXERCISES

In Exercises 1–35, sketch the graph of the given function. If feasible and applicable, find intercepts, asymptotes, relative extrema, intervals of concavity, and points of inflection.

1. $f(x) = x^3 - x$
2. $f(x) = 1 - 2x - 2x^3$
3. $f(x) = 2x^2 - 1 - x^3$
4. $f(x) = x^3 + 2x^2 - 10$
5. $f(x) = 2x^3 + 3x^2 - 6x + 5$
6. $f(x) = 4x^3 + 15x^2 - 72x - 12$
7. $f(x) = 3 - x + 2x^2 - x^3$
8. $f(x) = 3x^3 - 5x^2 + 7x - 8$
9. $f(x) = x^4 - 4x$
10. $f(x) = 2x^4 - 12x - 6$
11. $f(x) = x^4 - 8x^2 - 2$
12. $f(x) = 6 - 10x^2 - 5x^4$
13. $f(x) = x^4 - 4x^3 - 1$
14. $f(x) = 6x^4 - 20x^3 - 18x^2 + 5$
15. $f(x) = 9x^4 - 52x^3 + 24x^2 - 6$
16. $f(x) = x^5 - 10x^2$
17. $f(x) = 40x - x^5$
18. $f(x) = x^5 - 10x^4 + 6$
19. $f(x) = 1/(x - 1)$
20. $f(x) = x/(x - 2)$
21. $f(x) = 2x/(x + 3)$
22. $f(x) = (x - 1)/(x - 3)$
23. $f(x) = (x^2 - 4)/x^2$
24. $f(x) = (x^3 + 2)/(3x)$
25. $f(x) = x/(x - 3)^2$
26. $f(x) = (x - 1)/(x + 2)$
27. $f(x) = (x - 2)/(x + 3)$
28. $f(x) = 1/(x^2 + 1)$
29. $f(x) = x/(x^2 + 1)$
30. $f(x) = \sqrt[3]{x} - x$
31. $f(x) = 2x - 4\sqrt{x}$
32. $f(x) = \sqrt[3]{(x - 1)^2}$
33. $f(x) = \sqrt{(x + 1)^3}$
34. $f(x) = \sqrt[3]{x^2} - \frac{1}{2}x$
35. $f(x) = x\sqrt{x} - 4x - 1$

36. (Profit) If the monthly profit in dollars on the sale of $x$ hundred units of one of our company's products is

$$P(x) = -5000 + 2x + 3x^2 - 0.009x^3, \quad x \geq 0$$

sketch the graph of $P(x)$.

### Business and Economics

37. (Demand and cost) If the demand equation for one of our company's products is

$$p = 5000 - 0.04x^2, \quad 0 \leq x \leq 100$$

where $p$ is the price per unit in dollars and $x$ is the monthly demand in units, and the monthly cost of producing $x$ units of this product is

$$C(x) = 250 + 1200x, \quad 0 \le x \le 100$$

(a) graph the revenue function $R(x)$ for this product;
(b) graph the profit function $P(x)$ for this product.

**38.** (Cost) If the monthly cost in hundreds of dollars of producing $x$ thousand units of one of our company's products is

$$C(x) = 2x^3 - 45x^2 + 400x + 90, \quad x \ge 0$$

sketch the graph of $C(x)$.

**39.** (Average and marginal cost) For the cost function of Exercise 38, (a) sketch the average cost function $\overline{C}(x) = C(x)/x$ (*Hint:* The critical number of $\overline{C}(x)$ is approximately $x = 11.4$); (b) sketch the graph of $C'(x) = 6x^2 - 90x + 400$ on the same coordinate axes as the average cost function. The function $C'(x)$ is called the marginal cost. (c) Observe that the average cost curve and the marginal cost curve intersect at the output level $x$ for which average cost is a minimum.

## Social Sciences

**40.** (Poverty level) The proportion of people below the poverty level in a certain city in $t$ years is projected to be

$$P(t) = -\frac{t+3}{5t+20} + \frac{3}{10}, \quad 0 \le t \le 15$$

Sketch the graph of $P(t)$.

## OPTIONAL
## GRAPHING CALCULATOR/ COMPUTER EXERCISES

*Even with the powerful tools of calculus that we have developed in this chapter, there are still many functions to which simple calculus methods are difficult to apply. For example, if $f(x) = x^5 - 5x^4 - x^3 + 2x^2 - 5x + 10$ then:*
(a) *to find x-intercepts we must solve the (difficult) equation*

$$x^5 - 5x^4 - x^3 + 2x^2 - 5x + 10 = 0$$

(b) *Since $f'(x) = 5x^4 - 20x^3 - 3x^2 + 4x - 5$, finding the first-order critical numbers at which relative extrema can occur involves solving the (difficult) equation*

$$5x^4 - 20x^3 - 3x^2 + 4x - 5 = 0$$

(c) *Since $f''(x) = 20x^3 - 60x^2 - 6x + 4$, finding the second-order critical numbers to help us determine intervals of concavity and points of inflection involves solving the (difficult) equation*

$$20x^3 - 60x^2 - 6x + 4 = 0$$

*It is evident that a graphing problem like this one is best handled using a graphing calculator or a graphing utility on a computer.*

*In Exercises 41–47, sketch the given function. Approximate the x-intercepts to the nearest tenth of a unit and estimate the x-coordinate of any relative extrema or points of inflection to the nearest tenth of a unit.*

**41.** $f(x) = x^5 - 5x^4 - x^3 + 2x^2 - 5x + 10$
**42.** $f(x) = x^4 - 7x^3 + 3x^2 - 5x - 10$
**43.** $f(x) = x^4 - 1.4x^3 - 20.11x^2 - 18.424x - 3.7632$

44. $f(x) = 2x^5 - x^4 - 10x^3 - x^2 + 5x + 5$

45. $f(x) = \dfrac{x^2 + 7x + 2}{x^2 + 3x - 4}$ (You need not estimate the inflection point.)

46. $f(x) = \dfrac{x^2 + \sqrt{x} - 1}{x^2 + x + 1}$  47. $f(x) = \sqrt[3]{x} - \dfrac{10}{x^2} + \dfrac{x^2}{10}$

## 4.6 OPTIMIZATION PROBLEMS

To *optimize* a quantity means to find the absolute maximum or the absolute minimum value of the quantity on an interval. For example, we might want to optimize the weekly profit of our company, given that we can produce only 20,000 units of product per week. This means that we want to find the absolute maximum value of the profit function $P(x)$ on the interval $0 \leq x \leq 20{,}000$. As another example, if we manufacture plastic boxes with a square base and a volume of 256 cubic inches (in$^3$), we might want to optimize the cost of materials used in producing such a box. This means that we would want to find the absolute minimum value of the cost function subject to the condition that the volume is 256 in$^3$. The interval on which this cost function is to be optimized requires some further analysis. We will solve this problem in Example 2.

In optimization problems, some work is usually required to find a formula for the quantity to be optimized, as well as the required interval. Once we have found the formula and the interval, we can complete the problem using the methods for finding absolute extrema developed in Section 4.4. Let's look at some examples.

EXAMPLE 1

Our small perfume company has demand and cost equations

$$p = 275 - 6.25x, \quad 0 \leq x \leq 20$$
$$C(x) = 50x + 600$$

Here $x$ is the number of ounces of perfume manufactured and sold per week and $p$ and $C(x)$ are measured in dollars.

(a) Find the maximum weekly revenue.
(b) Find the maximum weekly profit, the profit-optimizing production level, and the profit-optimizing price.

Solution

(a) revenue = (price)(quantity) = $(275 - 6.25x)(x)$

$$R(x) = 275x - 6.25x^2, \quad 0 \leq x \leq 20$$

To optimize revenue on the given interval, note that $R(x)$ is continuous and $[0, 20]$ is a closed interval, so we can use the methods of Section 4.4 to obtain the absolute maximum value of $R(x)$. To find the critical numbers, we compute $R'(x) = 275 - 12.5x$ and solve the equation

$$275 - 12.5x = 0$$
$$x = 22$$

Since the critical number does not lie in the interval [0, 20], the absolute maximum value must occur at one of the endpoints. We compute:

$$R(0) = 275(0) - 6.25(0)^2 = 0$$
$$R(20) = 275(20) - 6.25(20)^2 = 3000$$

So the maximum weekly revenue is $3000 when we manufacture and sell 20 ounces of perfume per week.

(b) profit = revenue − cost

$$P(x) = 275x - 6.25x^2 - (50x + 600), \quad 0 \le x \le 20$$
$$P(x) = -6.25x^2 + 225x - 600, \quad 0 \le x \le 20$$

Since $P(x)$ is continuous on the closed interval [0, 20], we know that the absolute maximum value occurs either at an endpoint of [0, 20] or at a critical number within this interval. We compute $P'(x) = -12.5x + 225$. To find the critical numbers, we solve the equation $-12.5x + 225 = 0$ to obtain $x = 18$. In this case the critical number lies in the interval [0, 20], so to find the absolute maximum value of $P(x)$ we compute

$$P(0) = -6.25(0)^2 + 225(0) - 600 = -600$$
$$P(18) = -6.25(18)^2 + 225(18) - 600 = 1425$$
$$P(20) = -6.25(20)^2 + 225(20) - 600 = 1400$$

We conclude that the maximum weekly profit is $1425, which is obtained when the production level is 18 ounces per week. To obtain the profit optimizing price, we compute $p = 275 - 6.25(18) = \$162.50$.

You should compare the results of part (b) of this example with those we obtained in Section 1.5 for our perfume company example, which used the same demand and cost functions. A sketch of the revenue and cost functions is shown in Figure 4.53. Note that in this figure, when $R(x) > C(x)$, profit is positive since profit = revenue − cost. Thus profit is the distance between the revenue

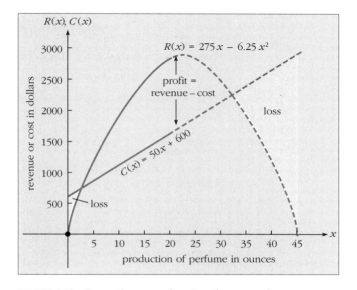

**FIGURE 4.53** Cost and revenue functions for our perfume company.

and cost curves, and for $0 \leq x \leq 20$, this distance is a maximum when $x = 18$, which is the profit optimizing output. When $R(x) < C(x)$, a loss is incurred. ■

EXAMPLE 2

Your company manufactures topless plastic boxes with a square base and a volume of 256 in$^3$. Find the dimensions of the box which would use the smallest amount of plastic. If the cost of plastic is 6¢/ft$^2$, what is the minimum cost of the plastic to be used in the construction of each box?

Solution

We must find the minimum surface area of a box with base $x$ inches by $x$ inches, height $y$ inches, and volume 256 in$^3$. See Figure 4.54. Let $A$ be the total surface area of the five faces. Then

$$A = \text{(area of 4 sides)} + \text{(area of bottom)}$$
$$= 4xy + x^2$$

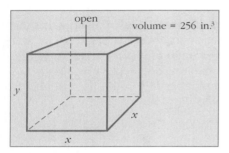

**FIGURE 4.54** A topless box with a square base.

The volume of the box is $V = \text{(area of base)(height)} = x^2 y$. Since we require the volume to be 256 in$^3$, we have $256 = x^2 y$, so that $y = 256/x^2$ and the formula for the total surface area becomes

$$A = 4x\left(\frac{256}{x^2}\right) + x^2 = \frac{1024}{x} + x^2, \quad x > 0$$

To optimize $A$ on the interval $(0, \infty)$, we compute

$$\frac{dA}{dx} = -1024 x^{-2} + 2x = \frac{-1024}{x^2} + 2x$$

We find the critical numbers by solving the equation

$$\frac{-1024}{x^2} + 2x = 0$$
$$-1024 + 2x^3 = 0 \quad \text{(Multiplying by } x^2\text{)}$$
$$x^3 = 512$$
$$x = 8 \quad \text{(Critical number)}$$

Since $x = 8$ is the only critical number in $(0, \infty)$, we apply the second derivative test for absolute extrema:

$$\frac{d^2 A}{dx^2} = \frac{2048}{x^3} + 2$$

and, since $\left.\dfrac{d^2 A}{dx^2}\right|_{x=8} > 0$, we conclude that the absolute minimum value of $A$ occurs at $x = 8$. Given that $x^2 y = 256$, when $x = 8$ we obtain $y = 4$, so the optimal size for the box is 8 in. by 8 in. by 4 in. For such a box, the surface area is $A = 4(8 \times 4) + 8^2 = 192$ in$^2$. Now, one square foot contains 144 in$^2$, so the surface area is $192 \div 144 = 1.33$ ft$^2$. Therefore the plastic to be used in the construction of the box has an optimal cost of

$$C = (1.33)(0.06) = \$0.08$$

For reference, the graph of $A$ for $x > 0$ is sketched in Figure 4.55. ■

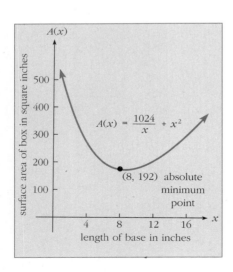

**FIGURE 4.55** The surface area of a topless plastic box.

Optimization problems involve translating the given facts into a mathematical model of the situation. Although each problem typically involves its own unique aspects, some general guidelines for solving optimization problems are presented here.

**GUIDELINES FOR SOLVING OPTIMIZATION PROBLEMS**

Step 1  Read the problem carefully and decide what quantity is to be optimized (maximized or minimized).

Step 2  Choose variables to represent the quantity to be optimized and the other quantities involved in the problem.

Step 3  Draw a figure if you feel it will be helpful.

Step 4  Write an equation for the variable to be optimized in terms of the other variables in the problem. The quantity to be optimized will be the dependent variable. Any other variables are independent variables. If the equation contains only one independent variable, skip to Step 6.

Step 5  Write other equations relating the independent variables and use these to eliminate all but one of the independent variables in the equation found in Step 4. You should now have the quantity to be optimized as a function of just one independent variable.

Step 6  Identify the appropriate interval for the independent variable.

Step 7  Apply the methods of Section 4.4 to find the absolute maximum or the absolute minimum of the function found in Step 5 on the interval found in Step 6.

Step 8  Make certain that you have answered the questions that were asked in the original problem.

**EXAMPLE 3**

A manufacturer produces a certain type of calculator at a cost of $5 each. The current price of this calculator is $15 and sales of the calculator have been averaging 5000 calculators a week. The manufacturer is planning a price increase and from surveys conducted by the marketing department, he knows that for each $0.50 increase in price, weekly demand will decrease by 250 calculators. If the manufacturer wants to optimize profit, what price should he charge per calculator?

Solution

In this problem, we must find the price that optimizes profit. Let $x$ be the number of $0.50 increases in price and $P(x)$ the weekly profit if there are $x$ increases in price. Then

$15 + 0.50x$ is the new price per calculator

and

$5000 - 250x$ is the number sold per week

Since profit = (number of calculators sold)(profit per calculator) and since

profit per calculator = selling price − cost
= (15 + 0.50x) − 5
= 10 + 0.50x

we obtain

profit = $P(x) = (5000 − 250x)(10 + 0.50x) = 50000 − 125x^2$

We now have the quantity to be optimized (profit) in terms of just one independent variable, $x$. Because there will be a price *increase*, we know that $x \geq 0$. In addition, if there are 20 price increases of \$0.50 each, then no calculators will be sold; therefore any increases beyond $x = 20$ are unreasonable. Thus the appropriate interval for $x$ is $0 \leq x \leq 20$. Our problem is to find the absolute maximum of the continuous function $P(x)$ on the closed interval $[0, 20]$. We compute that $P'(x) = −250x$, so to find the critical numbers of $P(x)$, we solve the equation

$$-250x = 0$$
$$x = 0 \quad \text{(Critical number)}$$

To find the absolute maximum, we compute

$$P(0) = 50000 − 125(0)^2 = 50000$$
$$P(20) = 50000 − 125(20)^2 = 0$$

We conclude that the absolute maximum value of $P(x)$ is 50,000 and this absolute maximum occurs when $x = 0$. To answer the question posed in the problem, $x = 0$ implies we should *not* increase the price at all. Our price per calculator should remain at \$15 each to obtain a maximum profit of \$50,000/week. For reference, the graph of $P(x)$ is sketched in Figure 4.56. ∎

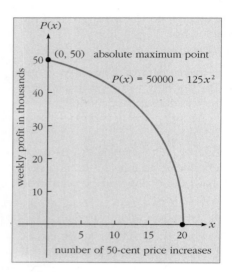

FIGURE 4.56 The profit function for the calculator manufacturer.

---

EXAMPLE 4 (Participative)

Solution

A metal box with a rectangular base and an open top is to be made from a rectangular piece of sheet metal 12 in. by 24 in. by cutting out four equal squares from the corners and folding up the sides. Let us find the dimensions of the box that will yield the largest possible volume.

Refer to Figure 4.57. In this figure, $x$ is the length of the edge of one of the four equal squares to be cut out and discarded from the four corners. The volume of the resulting box is given by

volume = (length)(width)(height)

where length = $24 − 2x$, width = $12 − 2x$, and height = \_\_\_\_\_. We obtain

$$V(x) = (24 − 2x)(12 − 2x)(x)$$
$$= 4x^3 − 72x^2 + 288x$$

To find an appropriate interval for $x$, observe that $x \geq 0$ and that $12 − 2x \geq 0$, so that $x \leq$ \_\_\_\_\_. An appropriate interval on which to optimize (maximize) the continuous function $V$ is therefore the closed interval _____. Next we find

$$V'(x) = \underline{\hspace{2cm}} = 12(x^2 − 12x + 24)$$

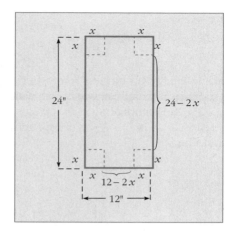

**FIGURE 4.57** Constructing a metal box from a rectangular piece of sheet metal.

To find the critical numbers of $V$, we must solve the equation

$$12(x^2 - 12x + 24) = 0$$

or

$$x^2 - 12x + 24 = 0$$

Using the quadratic formula with $a = 1$, $b = -12$ and $c =$ _____, we get

$$x = \frac{12 \pm \sqrt{(-12)^2 - 4(1)(24)}}{2}$$

This expression simplifies to $x = 6 \pm 2\sqrt{3}$, which gives us two critical numbers. However, since $x = 6 + 2\sqrt{3}$ is not in the interval $[0, 6]$, we discard it. The remaining critical number, _____, is approximately equal to 2.54, which is in $[0, 6]$, so to find the absolute maximum we compute

$$V(0) = 4(0)^3 - 72(0)^2 + 288(0) = 0$$
$$V(6 - 2\sqrt{3}) = 4(6 - 2\sqrt{3})^3 - 72(6 - 2\sqrt{3})^2 + 288(6 - 2\sqrt{3})$$
$$\approx \underline{\hspace{1in}}$$
$$V(6) = 4(6)^3 - 72(6)^2 + 288(6) = 0$$

The absolute maximum value of $V$ is _____, which occurs when $x = 6 - 2\sqrt{3} \approx 2.54$ in. The dimensions of the box of largest possible volume are

$$\text{length} = 24 - 2x = 24 - 2(6 - 2\sqrt{3}) \approx 18.93 \text{ in.}$$
$$\text{width} = 12 - 2x \approx \underline{\hspace{0.5in}} \text{ in.}$$
$$\text{height} = \underline{\hspace{0.5in}} \text{ in.}$$

■

## EXAMPLE 5

A new social service facility is to be built to serve three communities, $A$, $B$, and $C$, that have relative locations as illustrated in Figure 4.58. The planners have decided that the facility should be located along the roadway shown in the figure at a position that minimizes the sum of the population of each community times the square of its distance from the facility. Find the optimal location of the facility if the populations of communities $A$, $B$, and $C$ (in thousands) are 15, 20, and 35, respectively.

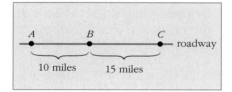

**FIGURE 4.58** The location of three communities along a roadway.

Solution

In Figure 4.59, we have let the roadway be the $x$-axis with the origin at community $A$, and $x$ the position of the facility along the roadway. Then the required criterion is that we minimize the function

---

**SOLUTIONS TO EXAMPLE 4**  $x$, 6, $[0, 6]$, $12x^2 - 144x + 288$, 24, $6 - 2\sqrt{3}$, 332.55, 332.55, 6.93, 2.54

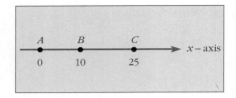

**FIGURE 4.59** The social service facility will be located at position $x$ along the roadway.

$$S(x) = 15(x - 0)^2 + 20(x - 10)^2 + 35(x - 25)^2$$
$$= 15x^2 + 20(x^2 - 20x + 100) + 35(x^2 - 50x + 625)$$
$$= 70x^2 - 2150x + 23875$$

In order to minimize $S(x)$, we require that the facility be located somewhere between $A$ and $C$, in which case the appropriate interval on which we will minimize the continuous function $S(x)$ is $[0, 25]$. We compute $S'(x) = 140x - 2150$. The solution of $S'(x) = 0$, or $140x - 2150 = 0$, is $x = \frac{215}{14} \approx 15.36$, and this is the critical number of $S$. Next, we compute

$$S(0) = 70(0)^2 - 2150(0) + 23875 = 23875$$
$$S\left(\tfrac{215}{14}\right) = 70\left(\tfrac{215}{14}\right)^2 - 2150\left(\tfrac{215}{14}\right) + 23875 \approx 7366.07$$
$$S(25) = 70(25)^2 - 2150(25) + 23875 = 13875$$

Thus the absolute minimum of $S$ occurs when $x = \frac{215}{14} \approx 15.36$, meaning that the new facility should be located along the roadway between $A$ and $C$, approximately 15.36 miles from $A$. ∎

EXAMPLE 6

Two costs of owning and maintaining a piece of machinery in business are the *capital cost*, which is the original purchase price less salvage value, and the *operating cost*, which is the cost associated with owning and maintaining the machinery. For example, if a machine originally costs $50,000 and is sold for salvage for $5000 after several years of use, then its capital cost is $45,000. Operating costs include the cost of insurance on the machine, preventative maintenance, repairs, and so forth. In trying to determine when to replace a piece of machinery, some companies try to minimize the sum of average capital cost and average operating cost.

Suppose our company purchases a piece of machinery for $75,000 to produce component parts for one of our products. The salvage value of the machine in dollars depends on the number of parts produced according to

$$V(x) = 72000 - 0.15x$$

The average operating cost in dollars per part is given by

$$\overline{O}(x) = 0.0000001x + 0.45$$

(a) Determine the number of parts that should be produced before replacing this machine if the replacement decision is determined by minimizing the sum of average capital cost and average operating cost. Assume that only a whole number of parts can be produced.
(b) What is the total cost of owning and operating the machine in producing the optimal number of parts?

Solution

(a) Let $T$ be the sum of average capital cost $\overline{K}$ and average operating cost $\overline{O}$. Note that

$$\overline{K} = \frac{\text{purchase price} - \text{salvage value}}{\text{parts produced}}$$

so that

$$\overline{K}(x) = \frac{75000 - (72000 - 0.15x)}{x}$$

$$= \frac{3000 + 0.15x}{x} = \frac{3000}{x} + 0.15$$

Thus we have

$$T(x) = \overline{K}(x) + \overline{O}(x)$$

$$= \frac{3000}{x} + 0.15 + 0.0000001x + 0.45$$

$$= 0.60 + 0.0000001x + \frac{3000}{x}, \quad x > 0$$

The objective is to find the absolute minimum of $T(x)$ on the interval $(0, \infty)$. We compute that $T'(x) = 0.0000001 - (3000/x^2)$. To find the critical number of $T(x)$, we solve the equation

$$0.0000001 - \frac{3000}{x^2} = 0$$

$$0.0000001x^2 - 3000 = 0 \quad \text{(Multiplying by } x^2\text{)}$$

$$x^2 = \frac{3000}{0.0000001} = 30000000000$$

$$x \approx 173205.08 \quad \text{(Critical number)}$$

Since $T(x)$ is continuous on the interval $(0, \infty)$ and since there is only one critical number in the interval, we apply the second derivative test for absolute extrema. Note that $T''(x) = 6000/x^3$. Since $T''(173205.08) > 0$, the absolute minimum value of $T(x)$ occurs at $x = 173205.08$. The graph of $T(x)$ is decreasing if $x < 173205.08$ and increasing if $x > 173205.08$: see Figure 4.60. Since only a whole number of parts can be produced, the absolute minimum total cost $T$ must occur at either $x = 173,205$ or $x = 173,206$. We compute:

$$T(173205) \approx 0.6346410162$$

$$T(173206) \approx 0.6346410162$$

Since these values are identical to ten decimal places, it makes little practical difference which one we choose. For definiteness, let us choose $x = 173,205$ as the number where the absolute minimum occurs. Hence we should produce 173,205 parts before replacing the machine.

(b) Total capital and operating costs of this machine to produce 173,205 parts at an average of 63.464¢/part is

$$(173205)(0.63464) \approx \$109,923.$$

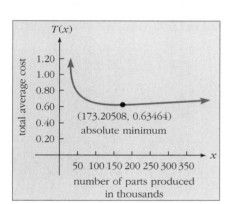

**FIGURE 4.60** The total average cost function for our machine.

EXAMPLE 7  A landowner wants to install a rectangular fence to enclose 1960 square feet. See Figure 4.61. One side of the fence is bordered by a straight river and requires no fencing. Another side is bordered by a straight road and the cost of this fencing is $5 per foot. The cost of fencing the other two sides is $4 per foot. Find the dimensions of the fenced plot of ground that will minimize the cost of fencing and the minimum cost if the distance between the river and the road is 100 feet.

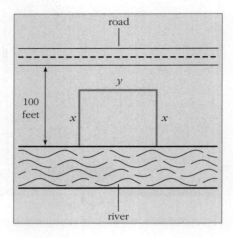

**FIGURE 4.61** The landowner installs a fence.

Solution

Let $C$ be the total cost of the fencing. Then

$$C = 4x + 4x + 5y = 8x + 5y$$

We must optimize (that is, minimize) the cost $C$. However, since $C$ contains two independent variables, we must search for a relationship between $x$ and $y$ that we can use to express $C$ as a function of a single independent variable. The fence must enclose an area of 1960 ft², so we must have

$$\text{area} = xy = 1960$$

so $y = 1960/x$. Substituting for $y$ into our expression for cost, we obtain

$$C(x) = 8x + 5\left(\frac{1960}{x}\right) = 8x + \frac{9800}{x}$$

Since $x$ is a distance, we must have $x > 0$. Now, the distance between the river and the road is 100 feet, so we have $x \leq 100$. Thus we must find the absolute minimum of $C(x)$ on the interval $(0, 100]$. We compute that $C'(x) = 8 - (9800/x^2)$. To find the critical numbers we solve the equation

$$8 - \frac{9800}{x^2} = 0$$
$$8x^2 - 9800 = 0$$
$$x^2 = 1225$$
$$x = \pm 35 \quad \text{(Critical numbers)}$$

But $x$ must be positive, hence we can reject the critical number $-35$. To show that an absolute minimum of $C(x)$ occurs at $x = 35$, we compute $C''(x) = (19600/x^3)$. Since $C''(35) > 0$, we conclude by the second derivative test for absolute extrema that an absolute minimum occurs at $x = 35$. Because $y = 1960/x$, when $x = 35$, we get $y = 1960/35 = 56$. The dimensions that minimize the cost of fencing are 56 ft of fencing parallel to the road and 35 ft of fencing on either side perpendicular to the river. The minimum cost is

$$C(35) = 8(35) + \frac{9800}{35} = \$560$$

EXAMPLE 8

You are the proud owner of a small island, which you have recently purchased at a bargain price and which is 2400 feet offshore from a straight beach. Unfortunately, the island has no source of fresh water. A water-pumping facility is located 6000 feet down the beach (see Figure 4.62). If the cost of laying pipe is $10 per foot along the beach and $20 per foot under water, what is the path along which to run the pipe in order to minimize the cost?

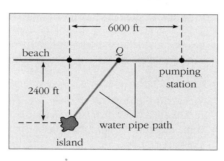

FIGURE 4.62  Pumping water to your island.

Solution

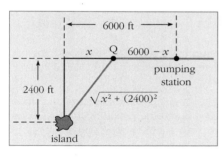

FIGURE 4.63

We begin by drawing and labeling a figure such as Figure 4.63. The number of feet of pipe along the beach is $6000 - x$. Using the Pythagorean theorem, the number of feet of pipe under water is $\sqrt{x^2 + (2400)^2}$. The cost of pipe is $10 per foot along the beach and $20 per foot under water, therefore the total cost is

$$C(x) = 10(6000 - x) + 20\sqrt{x^2 + (2400)^2}$$

Since $x \geq 0$ and $6000 - x \geq 0$, we have $0 \leq x \leq 6000$, so we must find the absolute minimum of $C(x)$ on the closed interval $[0, 6000]$. We compute:

$$C'(x) = -10 + 10[x^2 + (2400)^2]^{-1/2}(2x)$$
$$= -10 + \frac{20x}{\sqrt{x^2 + (2400)^2}}$$

To find any critical numbers, we solve the equation

$$-10 + \frac{20x}{\sqrt{x^2 + (2400)^2}} = 0$$
$$-10\sqrt{x^2 + (2400)^2} + 20x = 0 \quad \text{(Multiplying by } \sqrt{x^2 + (2400)^2}\text{)}$$
$$\sqrt{x^2 + (2400)^2} = 2x$$
$$x^2 + (2400)^2 = 4x^2 \quad \text{(Squaring both sides)}$$
$$3x^2 = (2400)^2$$
$$x = \pm\frac{2400}{\sqrt{3}} \approx \pm 1385.64$$

We reject the negative value of $x$. The critical number that lies in $[0, 6000]$ is $x = 1385.64$. We now compute:

$$C(0) = 10(6000 - 0) + 20\sqrt{0^2 + (2400)^2} = 108000$$
$$C(1385.64) = 10(6000 - 1385.64) + 20\sqrt{(1385.64)^2 + (2400)^2} \approx 101569$$
$$C(6000) = 10(6000 - 6000) + 20\sqrt{(6000)^2 + (2400)^2} \approx 129244$$

Therefore, the minimum cost of laying pipe to carry water from the pumping station is about \$101,569, and this minimum cost can be attained if the point $Q$ at which we begin to lay pipe under water is $6000 - 1385.64 = 4614.36$ ft down the beach from the pumping station. ■

## SECTION 4.6
## EXERCISES

1. **(Perimeter)** The perimeter of a certain rectangle is 400 ft. Find the dimensions of such a rectangle with a maximum area.

2. **(Numbers)** Find two positive numbers with a sum of 20 and a maximum product.

3. **(Demand and cost)** Given the monthly demand and cost functions

$$p = 20000 - 0.50x, \quad 0 \leq x \leq 40000$$
$$C(x) = 5000x + 10,000,000$$

where $x$ is the number of units manufactured and sold per month and $p$ and $C(x)$ are measured in dollars, find **(a)** the maximum monthly revenue; **(b)** the maximum monthly profit, the profit-optimizing production level, and the profit-optimizing price.

4. **(Demand and cost)** Answer the questions asked in Exercise 3 for the monthly demand and cost functions

$$p = 40000 - 0.25x, \quad 0 \leq x \leq 400$$
$$C(x) = 75x + 750$$

5. **(Packaging)** A cardboard box open at the top has a square base and rectangular sides. Find the dimensions of the box with a minimum surface area and a volume of 4 ft$^3$.

6. **(Construction cost)** A metal box with a top has a square base and rectangular sides. If the metal for the top and bottom costs 6¢/in$^2$ and the metal for the four sides costs 3¢/in$^2$, find the dimensions of the box with minimum cost and a volume of 100 in$^3$. What is the cost of the metal for such a box?

7. **(Fencing)** A landowner wants to fence a parcel of land bordered on one side by a straight river. If she has 1000 ft of fencing available, what is the maximum rectangular area she can fence if one side of the rectangle lies along the river and needs no fence?

8. **(Fencing)** If you want to build a small rectangular storage area using the rear of your house as one side, find the dimensions of the largest area that can be enclosed with 15 feet of fencing.

9. **(Fencing)** A store owner wants to enclose a rectangular area of 1200 ft$^2$ with wire fencing. What are the dimensions that will minimize the amount of fencing?

10. **(Production problem)** You have 200,000 pieces of 24-in. wire stored in a warehouse. Each wire is to be cut into two pieces. One cut piece is to be bent to form a circle and the other bent to form a square. Determine where each wire should be cut to **(a)** maximize and **(b)** minimize the sum of the areas of the circle and the square.

**11. (Production problem)** If each wire in Exercise 10 is to be cut into two pieces and each piece is to be bent to form a square (the two squares can be of different dimensions), where should each wire be cut to minimize the sum of the areas of the squares thus formed?

**12. (Retail sales)** A manufacturer produces watches at a cost of $7 each. The current price of each watch is $24 and weekly sales are 1500 watches. For every $1 that the manufacturer increases the price, weekly demand will decrease by 100 watches. What price per watch should the manufacturer charge in order to maximize weekly profit?

## Business and Economics

**13. (Retail sales)** A large retail store is selling a coffee maker for $30 and is contemplating a price change. The cost to the retailer per coffee maker is $18 and monthly sales are 25,000 coffee makers. The retailer estimates that for every $0.50 increase in price, monthly demand will decrease by 500 coffee makers. What price per coffee maker should the manufacturer charge in order to maximize monthly profit?

**14. (Agriculture)** The owner of a peach orchard estimates that if 35 trees are planted per acre, each tree will yield 400 peaches per year. He has also determined that for each tree planted per acre beyond 35 (up to twelve), the yield per tree will be reduced by ten peaches. Find the number of trees that should be planted per acre to maximize the yield. What is the maximum yield of peaches per acre?

**15. (Agriculture)** A pecan grower has determined that if 25 trees are planted per acre, each tree will yield 70 pounds of nuts per year. For each tree beyond 25 that is added (up to fifteen), however, the yield per tree will be reduced by two pounds. Find the number of trees that should be planted per acre to maximize the yield and find the maximum yield per acre.

**16. (Packaging)** A cardboard box with a square base and an open top is to be made from a 12-in.-square piece of cardboard by cutting out four equal squares from the corners and folding up the sides. Find the dimensions of the box that yield the largest possible volume. What is that largest possible volume?

**17. (Postal regulations)** Many post offices will accept a parcel for delivery only if the length plus the girth (the distance around) is no more than 84 in.
  (a) Find the dimensions of a rectangular parcel with a square base and the largest volume that can be mailed.
  (b) Find the dimensions of a cylindrical parcel with the largest volume that can be mailed.
  (c) Find the dimensions of a cubical parcel with the largest volume that can be mailed.

## Business and Economics

**18. (Replacement decision)** Your company purchases a car for $14,000. The salvage value of the car in dollars depends on the number $x$ of miles that the car has been driven, according to $V(x) = 12500 - 0.08x$. The average cost of operating the car in dollars per mile is $\overline{O}(x) = 0.00000035x + 0.18$. If a replacement decision is determined by minimizing the sum of average capital cost and average operating cost, (a) determine the number of miles the car should be driven before replacement; (b) find the total capital and operating costs when the car has been driven the optimal number of miles.

**19. (Replacement decision)** The Nut Company purchases a $100,000 machine to use in the production of bolts. The salvage value of the machine is a function of the number $x$ of bolts produced and is $V(x) = 90000 - 0.025x$. The average operating cost of the machine in dollars per bolt is given by $\overline{O}(x) = 0.00000005x + 0.05$.

(a) Determine the number of bolts that should be produced before replacing this machine if the replacement decision is based on minimizing the sum of average capital cost and average operating cost.

(b) What is the total cost of owning and operating the machine in the production of the optimal number of bolts?

20. (Maximum revenue) A local nautical club plans to rent a large boat from a charter service for an outing. If 15 or fewer persons go on the outing, the cost per person is $75. If more than 15 people go, the price paid by every passenger will be reduced by $1 for every passenger in excess of 15. If the boat can hold no more than 50 people, find the number of people that will maximize the revenue of the charter service.

Life Sciences

21. (Medicine) The concentration of a certain drug in the bloodstream $t$ hours after it is administered to a patient is

$$C(t) = \frac{12t}{4t^2 + 5}, \quad t \geq 0$$

Find the time at which the concentration is maximum.

22. (Packaging cost) A cylindrical can with no top is to be constructed to hold $200\pi$ in$^3$. If the material for the bottom costs 10¢/in$^2$ and the material for the sides costs only 5¢/in$^2$, find the dimensions of the can that will minimize the cost of materials. What is the minimum cost per can?

23. (Gasoline consumption) A car consumes gasoline at the rate of $C(x) = (1/x) + (x/1000)$ gal/mile when it travels at an average speed of $x$ miles per hour on the highway, $x \geq 5$. If gas costs $1.25/gal, find the speed that will minimize the cost of gas for a 200-mile trip. How much will the gasoline for the trip cost if the car travels at the optimal average speed?

24. (Minimum cost) If it costs a limousine service $(100 + x)/200$ dollars/mile for fuel, insurance, repairs, and depreciation to drive a limousine at an average speed of $x$ mph and if the driver must be paid $10/hr, at what speed should the limousine be driven in order to minimize the cost per mile?

25. (Retail sales) The producer of a local brand of fruit pies has estimated that its total weekly sales of pies are

$$S = 2000(0.5x + 0.75y - 0.09xy + 0.2)$$

pies when it spends $x$ dollars for advertising and $y$ dollars for better quality ingredients. If the producer will spend $0.50 per pie on both advertising and improved ingredients, find the amount he should spend per pie on advertising to produce the maximum sales. What are the maximum weekly sales?

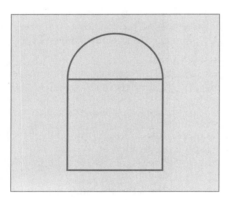

FIGURE 4.64  A Norman window.

26. (Construction problem) A Norman window has the shape of a rectangle with a semicircle on top; see Figure 4.64. Find the dimensions of the window of maximum possible area if the perimeter is 15 ft.

27. (Construction problem) A racetrack in the shape shown in Figure 4.65 must have the sum of the arc lengths of the two semicircles and the perimeter of the rectangle equal to one mile. Find the dimensions of the racetrack that will enclose the maximum area.

28. (Restaurant seating capacity) In planning for the construction of a restaurant, it is estimated that if the seating capacity is 50 to 100 people, weekly profit will be $15 per seat. If seating capacity is more than 100, weekly profit will be reduced by $0.10

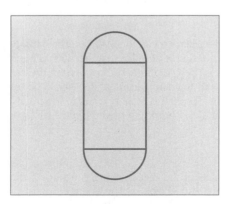

FIGURE 4.65  A racetrack.

per seat for each seat above 100. Find the seating capacity that will optimize weekly profit.

29. (Maximum revenue) A golf club charges annual dues of $500 per member if exactly 700 people join. For each person fewer than 700 who joins, fees of everyone increase $1. For each person more than 700 who joins, fees of everyone decrease by $1. Find the membership number that maximizes the revenue of the club.

30. (Pollution) When a factory discharges particulate matter into the air, the concentration of particles in parts per million is directly proportional to the rate at which the factory discharges the matter and inversely proportional to the square of the distance from the factory. If two factories, $A$ and $B$, are located eight miles apart and if $A$ emits particulate matter five times as fast as $B$, then the concentration of particulate matter at any point between $A$ and $B$ is given by

$$C(x) = \frac{5r}{x^2} + \frac{r}{(8-x)^2}, \quad 0.25 \leq x \leq 7.75$$

where $r$ is a constant of proportionality. See Figure 4.66. Find the distance $x$ from $A$ at which the atmospheric concentration of particulate matter is a minimum.

31. (Publishing) The pages of a book to be published must contain 40 in² of printed matter each. In addition, each page must have $1\frac{1}{2}$-in. margins at the top and the bottom and 1-in. margins on each side. Find the minimum area for each page in the book.

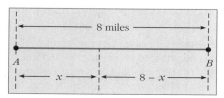

FIGURE 4.66 Polluting factories $A$ and $B$.

32. (Construction) In the construction of rain gutters, long rectangular metal sheets that have a width of 20 in. are formed into gutters by bending up equal lengths on two sides until they form right angles with the center portion, which is not bent. Find the maximum cross-sectional area of the gutters.

33. (Strength of a beam) If the strength $S$ of a beam of a given length is proportional to the width $w$ and the square of the depth $d$ of a cross section, find the dimensions that will make the strongest beam if it is cut from a cylindrical log; see Figure 4.67.

34. (Stiffness of a beam) If the stiffness $T$ of a rectangular beam of a given length is proportional to the width $w$ and the cube of the depth $d$ of a cross section, find the dimensions of the stiffest beam that can be cut from a cylindrical log of diameter 26 in.; see Figure 4.67.

35. (Productivity) During a four-hour shift, an average factory worker in a certain plant can produce

$$P(t) = 10t + 8t^2 - t^3, \quad 0 \leq t \leq 4$$

units of a certain product. If a worker is performing most efficiently when his *rate* of production $dP/dt$ is maximum, find the time during the shift when the worker is performing most efficiently.

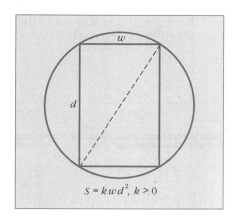

FIGURE 4.67

36. (Falling object) An object that is thrown upward with an initial velocity of 64 ft/sec and is released 7 ft above the ground has a height after $t$ seconds of $h(t) = 7 + 64t - 16t^2$. After how many seconds will the object reach its maximum height? What is the maximum height?

37. (Fencing) A fence is to be built next to a house 50 ft long, and part or all of the entire length of the house is to serve as one side of the enclosed rectangular area. If 200 ft of fencing are available, find the dimensions of the largest enclosed area.

38. (Minimum slope) Find the point on the graph of $f(x) = x^3 - 2x^2 + x + 1$ where the slope of the tangent line is a minimum.

**39.** (Minimum distance) Find the point on the graph of $f(x) = x + 1$ that is closest to the point $(3, 1)$.

**40.** (Minimum cost) The time $T$ in hours needed to perform a certain task is a function of the number of workers $w$ who are working on the task according to the function

$$T(w) = 2 + \frac{3}{w} + \frac{50}{w^2}, \quad w \geq 1$$

If all workers are paid $6.00/hr, find the minimum cost of wages necessary to perform the task.

**41.** (Minimum distance) At 2:00 P.M., an oil tanker is 12 miles north of a pleasure yacht. The tanker is traveling west at 4 mph and the yacht is traveling north at 15 mph. Determine the time when the two craft are nearest one another.

Social Sciences

**42.** (Spread of a rumor) The rate at which a rumor spreads in a population of 20,000 people is equal to $N(x) = kx(20000 - x)$, where $x$ is the number of people who have already heard the rumor and $k$ is a positive constant of proportionality. When the rumor is being spread the fastest, what proportion of the population has already heard it?

**43.** (Minimum cost) An offshore oil platform is located at point $P$ four miles from the nearest point $Q$ on a straight beach. If oil must be pumped to a refinery $R$ ten miles down the beach from $Q$, find the path over which the pipe should be run to minimize the cost if it costs 1.5 times as much to lay pipe in the water as it does on land.

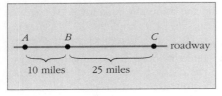

**FIGURE 4.68** The location of three communities along a roadway.

**44.** (Optimal location) A new health care facility is to be constructed to serve three communities, $A$, $B$, and $C$, that have populations (in thousands) of 25, 14, and 30 and relative locations along a roadway as illustrated in Figure 4.68. If the facility is to be located along the roadway at a position that will minimize the sum of the population of each community times the square of its distance from the facility, find the optimal location of the facility.

## 4.7 BUSINESS AND ECONOMICS APPLICATIONS

### Marginal Cost, Revenue, and Profit

We are already familiar with the cost, revenue, and profit functions of a firm or a whole industry. Summarizing,

$$C(x) = \text{the cost of producing } x \text{ units of product per time period}$$

$$R(x) = \text{the revenue generated by the sale of } x \text{ units of product per time period}$$

$$P(x) = \text{the profit resulting from the production and sale of } x \text{ units of product per time period}$$

If we assume, as we always shall in this text, that the number of units produced per time period is equal to the number of units sold in that time period, then

$$P(x) = R(x) - C(x)$$

The *true marginal cost* is the cost of producing one more unit "at the margin" than the current production level. For example, if our production level is $x$ units, then the true marginal cost at $x$ is the cost of producing the $(x + 1)$st unit, which is given by

(total cost of producing $x + 1$ units) − (total cost of producing $x$ units)

This quantity represents the change in cost, $\Delta C$, that results in a change in production from $x$ to $x + 1$, so that true marginal cost at production level $x$ is

**TRUE MARGINAL COST**

$$\Delta C = C(x + 1) - C(x)$$

The true marginal cost is sometimes referred to as the "incremental cost." We may also define *true marginal revenue* as the revenue generated by selling the $(x + 1)$st unit, so that the true marginal revenue at sales level $x$ is

**TRUE MARGINAL REVENUE**

$$\Delta R = R(x + 1) - R(x)$$

*True marginal profit* is derived from producing and selling the $(x + 1)$st unit, so that the true marginal profit at production and sales level $x$ is

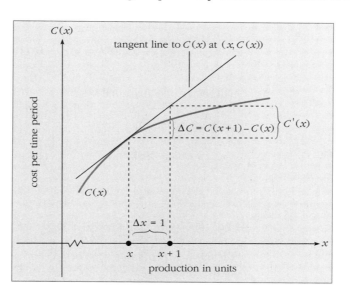

**FIGURE 4.69** A comparison of true marginal cost $\Delta C$ and $C'(x)$.

**TRUE MARGINAL PROFIT**

$$\Delta P = P(x + 1) - P(x)$$

Recall that if $\Delta x = dx$ is small relative to $x$, then $\Delta y \approx dy = f'(x)\, dx$. Choosing $\Delta x = 1$, which is usually a small relative increase in production or sales when we deal with large numbers of units $x$, then, for example,

$$\Delta C \approx dC = C'(x)\, dx = C'(x)(1) = C'(x)$$

We obtain the fact that to a good approximation, the true marginal cost $\Delta C$ is equal to the derivative $C'(x)$ of the total cost function; see Figure 4.69.

Economists usually *define* the marginal cost to be $C'(x)$ with corresponding definitions for marginal revenue and marginal profit. That is,

**DEFINITION OF MARGINAL FUNCTIONS**

Marginal cost at $x$ is $C'(x)$.
Marginal revenue at $x$ is $R'(x)$.
Marginal profit at $x$ is $P'(x)$.

---

**EXAMPLE 1**

Given the monthly cost function $C(x) = 500000 + 80x$ and the weekly demand function $p = 400 - 0.02x$, where $C(x)$ and $p$ are in dollars and $x$ is units, determine marginal cost, revenue, and profit at a production and sales level of 5,000 units; compare each of these with its corresponding true marginal value.

**Solution**

Note that since revenue = (number of units sold)(price per unit), we have

$$R(x) = xp = x(400 - 0.02x) = 400x - 0.02x^2$$

Also,

$$P(x) = R(x) - C(x) = 400x - 0.02x^2 - (500000 + 80x)$$
$$= -0.02x^2 + 320x - 500000$$

Therefore the marginal cost at $x$ is $C'(x) = 80$; the marginal revenue at $x$ is $R'(x) = 400 - 0.04x$; and the marginal profit at $x$ is $P'(x) = -0.04x + 320$. Now we compute:

$C'(5000) = \$80$     (Approximate cost of producing the 5001st unit)

$R'(5000) = 400 - 0.04(5000) = \$200$     (Approximate revenue from selling the 5001st unit)

$P'(5000) = -0.04(5000) + 320 = \$120$     (Approximate profit from the 5001st unit)

Clearly it is profitable to produce and sell the 5001st unit. Let us compare each of the values obtained with their corresponding true marginal values.

$$\Delta C = C(5001) - C(5000)$$
$$= [500000 + 80(5001)] - [500000 + 80(5000)] = \$80$$
$$\Delta R = R(5001) - R(5000)$$
$$= [400(5001) - 0.02(5001)^2] - [400(5000) - 0.02(5000)^2] = \$199.98$$
$$\Delta P = P(5001) - P(5000) = [-0.02(5001)^2 + 320(5001) - 500000]$$
$$- [-0.02(5000)^2 + 320(5000) - 500000]$$
$$= \$119.98$$

Notice that these values agree very closely with the ones we obtained using the derivative definition of marginality and that using the derivative definition results in easier computations. ∎

Recall that if the profit function $P(x)$ is differentiable for $x \geq 0$, then the maximum profit must occur at a critical number of $P(x)$, that is, a number $b$ such that $P'(b) = 0$. This assumes, of course, that the maximum profit does not occur at minimum production level $x = 0$ or at the other endpoint of production, which represents the largest production capability of the firm or industry. Since $P(x) = R(x) - C(x)$, assuming profit is not maximized at an endpoint, profit is maximized at the production level $x$ where

$$P'(x) = R'(x) - C'(x) = 0$$

or when $R'(x) = C'(x)$. We have proven here a familiar fact of elementary economics.

> Profit is maximized at the production level for which marginal revenue is equal to marginal cost.

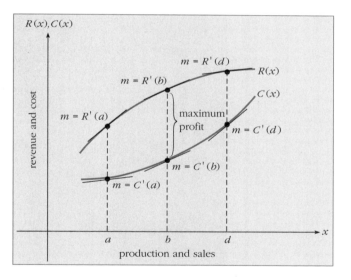

**FIGURE 4.70** Maximum profit occurs at production level $b$, where $R'(b) = C'(b)$.

A graphical interpretation of this fact is presented in Figure 4.70. At production and sales level $a$, we see that $R'(a) > C'(a)$, so that marginal revenue is greater than marginal cost. In other words, the revenue derived from the sale of the $(a + 1)$st unit is greater than the cost of producing the $(a + 1)$st unit. It is, therefore, profitable to produce unit $a + 1$. At production and sales level $b$, we see that $R'(b) = C'(b)$, so the revenue derived from selling the $(b + 1)$st unit is equal to the cost of producing the $(b + 1)$st unit. Therefore the $(b + 1)$st unit contributes nothing to profit. Beyond production and sales level $b$, however, say at $d$, we can see that $R'(d) < C'(d)$, so that the revenue derived from the sale of the $(d + 1)$st unit is less than the cost of producing the $(d + 1)$st unit. It is not profitable, therefore, to produce the $(d + 1)$st unit. We conclude that the profit-maximizing level of production and sales is level $b$, where $R'(b) = C'(b)$, that is, the level at which marginal revenue is equal to marginal cost. Note that when the graph of $R(x)$ lies above the graph of $C(x)$, the maximum profit occurs when the distance between the graphs of $R(x)$ and $C(x)$ is greatest.

**EXAMPLE 2**   For the monthly cost function $C(x) = 500000 + 80x$ and revenue function $R(x) = 400x - 0.02x^2$, determine the production level at which profit is maximized.

**Solution**   These are the cost and revenue functions considered in Example 1. We have $C'(x) = 80$ and $R'(x) = 400 - 0.04x$, so profit is maximized when $400 - 0.04x = 80$, or $x = 8000$ units per month. The graphs of $C(x)$ and $R(x)$ are sketched in Figure 4.71(a). The graph of the profit function $P(x) = -0.02x^2 + 320x - 500000$ is drawn in Figure 4.71(b).

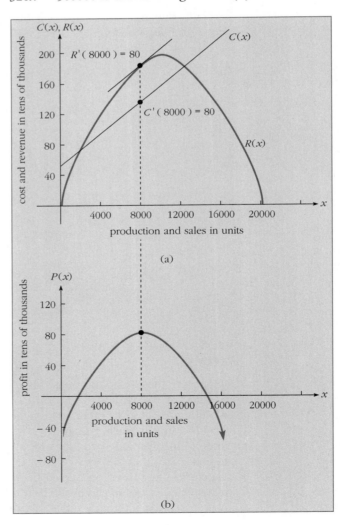

**FIGURE 4.71** Maximum profit occurs at a production level of 8000 units, where marginal cost equals marginal revenue.

## EXAMPLE 3 (Participative)

Solution

If weekly cost in dollars is $C(x) = 200 + 4x + 0.02x^2$ and weekly revenue is $R(x) = 100x - 0.01x^2$, let us determine the production level at which profit is maximized.

We'll solve this problem by equating marginal cost and marginal _____ and solving the resulting equation for $x$. We compute:

$C'(x) =$ _____  (Marginal cost)

$R'(x) =$ _____  (Marginal revenue)

To find the profit-maximizing output, we solve the equation

$$4 + 0.04x = 100 - 0.02x$$
$$0.06x = 96$$
$$x = \underline{\qquad}$$

We conclude that profit is maximized when the weekly production level is _____ units.

### The Relationship between Marginal Cost and Average Cost

Recall that the average cost $\overline{C}(x)$ is defined by $\overline{C}(x) = C(x)/x$. An important relationship between average cost and marginal cost is illustrated by the next example.

## EXAMPLE 4

Given the monthly cost function $C(x) = 900 + \frac{1}{4}x^2$, where $C(x)$ is the cost in dollars of producing $x$ units, (a) find the production level that minimizes average cost, and (b) find the production level at which average cost is equal to marginal cost.

Solution

(a) The average cost

$$\overline{C}(x) = C(x)/x = \frac{900}{x} + \frac{1}{4}x$$

We compute that $\overline{C}'(x) = -(900/x^2) + \frac{1}{4}$. Then to find critical numbers of $\overline{C}(x)$, we solve the equation

$$-\frac{900}{x^2} + \frac{1}{4} = 0$$
$$-3600 + x^2 = 0 \quad \text{(Multiplying by } 4x^2\text{)}$$
$$x^2 = 3600$$
$$x = \pm 60$$

We reject the critical number $x = -60$, since $x > 0$, and conclude that $x = 60$ is the critical number of interest. Now, $\overline{C}''(x) = 1800/x^3$ and we see that $\overline{C}''(60) > 0$, therefore, since $\overline{C}(x)$ is continuous for $x > 0$ and $x = 60$ is the only critical number in $(0, \infty)$, by the second derivative test for

---

**SOLUTIONS TO EXAMPLE 3**  revenue, $4 + 0.04x$, $100 - 0.02x$, $1600$, $1600$

absolute extrema, $\overline{C}(x)$ has an absolute minimum at $x = 60$. The absolute minimum value of the average cost is

$$\overline{C}(60) = \frac{900}{60} + \frac{1}{4}(60) = \$30 \text{ per unit}$$

(b) Marginal cost is $C'(x) = \frac{1}{2}x$ and so average cost is equal to marginal cost when

$$\frac{900}{x} + \frac{1}{4}x = \frac{1}{2}x$$

$$3600 + x^2 = 2x^2 \quad \text{(Multiplying by } 4x\text{)}$$

$$x^2 = 3600$$

$$x = \pm 60$$

Rejecting $x = -60$ since $x > 0$, we find that average cost is equal to marginal cost when the production level is $x = 60$ units. See Figure 4.72 for a graphical illustration.

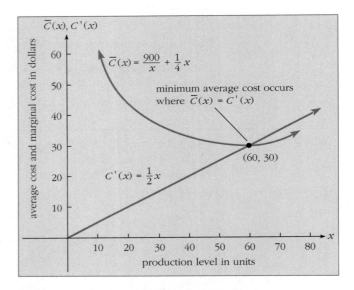

**FIGURE 4.72** Average cost and marginal cost functions.

Example 4 illustrates the general principle that *average cost is a minimum when marginal cost is equal to average cost*. The reason for this is that when marginal cost is less than average cost, it costs less to produce an additional unit than the average cost of the units already being produced. Thus producing an additional unit will decrease the average cost per unit in this case. On the other hand, when marginal cost is greater than average cost, it costs more to produce an additional unit than the average cost of the units already being produced. Thus producing an additional unit will increase the average cost per unit in this case. Finally, if marginal cost is equal to average cost, the cost of producing an additional unit is equal to the average cost of the units already being produced. Thus producing an additional unit will neither increase nor decrease the average cost per unit in this case, and so the minimum average cost occurs at this production level.

### Economic Order Quantity and Economic Lot Size

Suppose a company needs to order $N$ units of a certain item per year and suppose the item is used at a constant rate during the year. If the company orders all $N$ units at once, it will have to pay a large amount in storage costs (for warehouse space, insurance, interest on money needed to purchase the items, etc.). On the other hand, if $N$ is large and the company makes many small orders, it will incur a large amount of expense in reordering costs (for delivery charges, clerical time, etc.). If we let $x$ be the number of units per order, then

total cost = storage cost + reorder cost
$$T(x) = S(x) + R(x)$$

The order size $x$ that minimizes the total cost function $T(x)$ is called the *economic order quantity*.

It is common to diagram the number of units in inventory as in Figure 4.73, where we have assumed that $N = 3000$ and $x = 1000$, so that we place $3000/1000 = 3$ orders/yr. From this diagram you can see that each order of 1000 units is used uniformly throughout the four-month period and that a new order arrives just as the previous order is used up. The average inventory over any period is then $\frac{1}{2}(1000 + 0) = 500$, and this is the average number of units in storage during the year. In general, when the order size is $x$ units, the average number of units in storage is $x/2$.

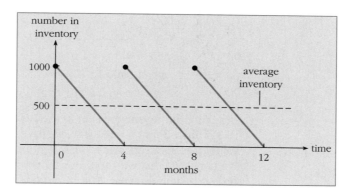

**FIGURE 4.73** Inventory level for a yearly demand of 3000 units ordered three times a year.

To obtain a general expression for the total inventory cost $T(x)$, suppose that it costs $s$ dollars to store one unit for a year. Then

storage cost = (cost of storing one unit)(average number of units in storage)
$$S(x) = s\left(\frac{x}{2}\right)$$

If the fixed cost to reorder is $r$ dollars and there is an additional variable charge of $u$ dollars per unit,

reorder cost = (fixed cost of each order)(number of orders)
    + (variable cost per unit)(number of units)
$$R(x) = r\left(\frac{N}{x}\right) + uN$$

Our objective is to minimize the total inventory cost:

**TOTAL INVENTORY COST**

$$T(x) = \frac{sx}{2} + \frac{rN}{x} + uN, \quad x > 0$$

**EXAMPLE 5** A department store sells a total of 1600 stereos at a fairly constant rate throughout the year. It costs $10 to store one stereo for one year and the cost of placing an order is $20 plus $2 per stereo. Find the number of units that should be ordered each time to minimize yearly inventory cost and find the minimum yearly inventory cost.

**Solution** We have $N = 1600$, $s = \$10$, $r = \$20$, and $u = \$2$, so that

$$T(x) = 10\left(\frac{x}{2}\right) + 20\left(\frac{1600}{x}\right) + 2(1600)$$

$$= 5x + \frac{32000}{x} + 3200, \quad x > 0$$

To find the minimum of $T(x)$, we compute that $T'(x) = 5 - (32000/x^2)$. Critical numbers of $T(x)$ are then found by solving the equation

$$5 - \frac{32000}{x^2} = 0$$

$$5x^2 - 32000 = 0$$

$$x^2 = 6400$$

$$x = \pm 80 \quad \text{(Critical numbers)}$$

Rejecting the negative value, $x = 80$ is the only critical number in the interval $(0, \infty)$. Since $T''(x) = 64000/x^3$, we have $T''(80) > 0$, so that an absolute minimum value of $T(x)$ occurs at $x = 80$. In other words, 80 is the economic order quantity. We conclude that we should order 80 stereos each time, placing $1600/80 = 20$ orders during the year. The minimum yearly inventory cost will be

$$T(80) = 5(80) + \frac{32000}{80} + 3200 = \$4000 \quad \blacksquare$$

A problem that is similar to finding the economic order quantity is that of finding the *economic lot size*. In this problem a company that manufactures $N$ units per year must decide the most economical batch or lot size to produce in order to minimize the sum of manufacturing and storage costs. Thus let $x$ be the number of lots produced per year; then

total cost = manufacturing cost + storage cost

$$T(x) = M(x) + S(x)$$

The number of lots $x$ that minimizes this total cost function $T(x)$ is called the *economic lot size*.

## 4.7 BUSINESS AND ECONOMICS APPLICATIONS

The manufacturing cost for each batch will consist of a fixed cost of $f$ dollars of setting up the production run and a variable cost of $v$ dollars per unit produced. Since each production run will consist of $N/x$ units, if we let $m(x)$ be the manufacturing cost *per batch*,

manufacturing cost/batch = fixed set-up cost
$\qquad\qquad$ + (variable cost/unit)(number of units)

$$m(x) = f + v\left(\frac{N}{x}\right)$$

Since there are $x$ lots produced annually,

total manufacturing cost = (manufacturing cost per batch)(number of batches)

$$M(x) = \left(f + \frac{vN}{x}\right)(x) = fx + vN$$

Turning to the storage cost component of total cost, suppose that $s$ dollars is the cost of storing one unit for one year. Then, since each production run consists of $N/x$ units, the average number of units in storage is $\frac{1}{2}(N/x) = N/2x$, and

storage cost = (cost of storing one unit)(average number of units in storage)

$$S(x) = s\left(\frac{N}{2x}\right)$$

Our objective, then, is to minimize the total cost:

### TOTAL PRODUCTION AND STORAGE COST

$$T(x) = fx + vN + \frac{sN}{2x}, \quad x > 0$$

---

**EXAMPLE 6**  A company expects an annual demand of 10,000 units of one of its products per year. It costs $500 for set-up costs to prepare the factory to run a batch of these units. If storage costs are $10/yr per item and demand for this product is expected to be fairly uniform throughout the year, find the economic lot size.

**Solution**  Note that $N = 10000$, $f = 500$, and $s = 10$. The variable cost $v$ per unit is not given, and so

$$T(x) = 500x + 10000v + \frac{(10)(10000)}{2x}$$
$$= 500x + 10000v + \frac{50000}{x}, \quad x > 0$$

To find the number $x$ for which $T(x)$ is minimum, we compute that $T'(x) = 500 - (50000/x^2)$. (Note that since $v$ is a constant, the derivative of $10000v$ is zero.) To find the critical numbers of $T(x)$, we solve the equation

$$500 - \frac{50000}{x^2} = 0$$
$$500x^2 - 50000 = 0$$
$$x^2 = 100$$
$$x = \pm 10$$

Thus $x = 10$ is the only critical number in the interval $(0, \infty)$. Since $T'''(x) = 100000/x^3$, we find that $T'''(10) > 0$, so that the absolute minimum value of $T(x)$ occurs at $x = 10$, which is the number of lots per year that minimizes production and storage costs. The economic lot size is $10000/10 = 1000$ units per batch. Note that we need not know the variable cost per unit $v$ in order to determine the economic lot size. ∎

### Elasticity of Demand

For ordinary goods, an increase in the price of a product is usually accompanied by a decrease in consumer demand. If a small relative change in price produces a large relative change in demand, then the demand for the product is very sensitive to changes in price. Many items whose purchases can be delayed for a while, or that have cheaper substitutes, fall into this category. For example, for vacations, steak, cars, and designer clothes, a small relative increase in price produces a large relative decrease in demand. For other items, such as salt, laundry detergent, or light bulbs, a small relative change in price will not result in any significant decrease in demand. Demand for these items is not sensitive to changes in price.

One measure of the sensitivity of demand to changes in price is the *average elasticity of demand*, which is the ratio of the relative change in price to the relative change in demand. If the price of a product changes from $p$ to $p + \Delta p$, the *relative change in price* is $\Delta p/p$ and the *percentage change in price* is $100(\Delta p/p)$. If the demand for a product changes from $x$ to $x + \Delta x$, the *relative change in demand* is $\Delta x/x$ and the *percentage change in demand* is $100(\Delta x/x)$.

**AVERAGE ELASTICITY OF DEMAND**

If price changes from $p$ to $p + \Delta p$, the *average elasticity of demand with respect to price* is defined to be

$$\frac{\text{relative change in demand}}{\text{relative change in price}} = \frac{\frac{\Delta x}{x}}{\frac{\Delta p}{p}}$$

**EXAMPLE 7** If the price of a certain car increases from \$10,000 to \$11,000, monthly demand at a dealership will decrease from 50 to 40 cars. Find the average elasticity demand for this change in price.

**Solution** Note that $p = 10000$, $\Delta p = 11000 - 10000 = 1000$, $x = 50$, and $\Delta x = 40 - 50 = -10$. Thus the average elasticity is

$$\frac{\text{relative change in demand}}{\text{relative change in price}} = \frac{\frac{-10}{50}}{\frac{1000}{10000}} = \frac{-0.20}{0.10} = -2$$

In this example, a 10% increase in price results in a 20% decrease in demand, producing an average elasticity of $-2$. ∎

We define the *point elasticity of demand at p* as the limit of the average elasticity as the change in price approaches zero. Using the symbol $E(p)$ for the point elasticity of demand, we have

$$E(p) = \lim_{\Delta p \to 0} \frac{\Delta x / x}{\Delta p / p} = \lim_{\Delta p \to 0} \frac{p}{x} \cdot \frac{\Delta x}{\Delta p}$$

$$= \frac{p}{x} \lim_{\Delta p \to 0} \frac{\Delta x}{\Delta p} \quad \left(\text{Since } \frac{p}{x} \text{ does not depend on } \Delta p\right)$$

Now, if $x = f(p)$ represents demand as a function of price, then $\Delta x = f(p + \Delta p) - f(p)$, and

$$E(p) = \frac{p}{f(p)} \lim_{\Delta p \to 0} \frac{f(p + \Delta p) - f(p)}{\Delta p} = \frac{p}{f(p)} f'(p)$$

For ordinary goods, demand is a decreasing function of price. Therefore $f'(p) \leq 0$ and so $E(p) \leq 0$ for all $p$, since both price $p$ and quantity demanded $f(p)$ are positive. The function $E(p)$ can be viewed as the approximate percentage change in demand corresponding to a 1% increase in price. To see this, let $\Delta p = 0.01p$, which represents a 1% price increase. Then

$$E(p) = \lim_{\Delta p \to 0} \frac{p}{x} \cdot \frac{\Delta x}{\Delta p} \approx \frac{p}{x} \cdot \frac{\Delta x}{0.01p} = 100 \frac{\Delta x}{x} = \text{percent change in demand}$$

↑<br>Since $\Delta p = 0.01p$<br>is the price change

**POINT ELASTICITY OF DEMAND**

If demand is a function of price given by $x = f(p)$, then the *point elasticity of demand* is

$$E(p) = \frac{pf'(p)}{f(p)}$$

Furthermore, $E(p)$ can be interpreted as the approximate percentage change in demand corresponding to a 1% change in price.

If $-1 < E(p) \leq 0$, demand is *inelastic*.

If $E(p) < -1$, demand is *elastic*.

If $E(p) = -1$, demand has *unit elasticity*.

### EXAMPLE 8

Let the monthly demand $x$ in units be given by $x = f(p) = 4000 - 5p$, where $p$ is the price per unit in dollars. Find and interpret $E(p)$ if **(a)** $p = 200$; **(b)** $p = 400$; **(c)** $p = 600$.

**Solution**

First we compute $f'(p) = -5$, so that

$$E(p) = \frac{pf'(p)}{f(p)} = \frac{p(-5)}{4000 - 5p} = \frac{-p}{800 - p}$$

**(a)** $E(200) = \dfrac{-200}{800 - 200} = -\dfrac{1}{3} \approx -0.333$

Therefore, since $-1 < E(200) \leq 0$, demand is inelastic at a price of $200. A change in price of 1% will produce a change in demand of approximately only 0.333% when the price is $200.

**(b)** $E(400) = \dfrac{-400}{800 - 400} = -1$

Therefore, since $E(400) = -1$, demand has unit elasticity at a price of $400. A change in price of 1% will produce a change in demand of approximately 1% when the price is $400.

**(c)** $E(600) = \dfrac{-600}{800 - 600} = -3$

Therefore, since $E(600) < -1$, demand is elastic at a price of $600. A change in price of 1% will produce a change in demand of approximately 3% when the price is $600.

A sketch of the demand function $x = 4000 - 5p$ is drawn in Figure 4.74.

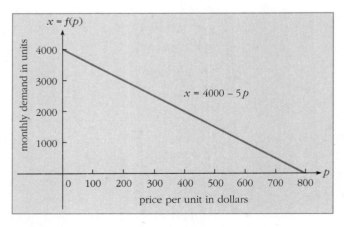

**FIGURE 4.74** The demand function $x = 4000 - 5p$.

### EXAMPLE 9 (Participative)

If weekly demand is given by $x = 10(50 - p)^2$, $0 \leq p \leq 50$, let's find the price at which demand has unit elasticity.

## Solution

First, we compute $E(p) = pf'(p)/f(p)$. Since $f(p) = 10(50 - p)^2$, we calculate that $f'(p) = $ _____ and so

$$E(p) = \frac{p[-20(50 - p)]}{10(50 - p)^2} = \underline{\hspace{1cm}}, \quad 0 \le p < 50$$

To find the price at which demand has unit elasticity, we set $E(p)$ equal to \_\_\_\_ and solve the resulting equation:

$$\frac{-2p}{50 - p} = -1$$

$$-2p = -50 + p \quad \text{(Multiplying by } 50 - p\text{)}$$

$$p = \underline{\hspace{1cm}}$$

## EXAMPLE 10

For the demand function of Example 8, find the interval on which demand is elastic and the interval on which demand is inelastic.

## Solution

We have $x = f(p) = 4000 - 5p$, $0 \le p \le 800$, and from Example 8,

$$E(p) = \frac{-p}{800 - p}, \quad 0 \le p < 800$$

Since demand is inelastic if $E(p) > -1$, demand is inelastic if $E(p) + 1 > 0$. Because demand is elastic if $E(p) < -1$, demand is elastic if $E(p) + 1 < 0$. Lastly, demand has unit elasticity if $E(p) = -1$, so demand has unit elasticity if $E(p) + 1 = 0$. We conclude that a sign change in $E(p) + 1$ signals a change in elasticity. As in our discussion of sign changes of $f'(x)$ earlier in this chapter, a sign change of $E(p) + 1$ can occur only at a number at which $E(p) + 1$ is equal to zero or at a number at which $E(p) + 1$ is undefined or discontinuous. We compute:

$$E(p) + 1 = \frac{-p}{800 - p} + 1 = \frac{800 - 2p}{800 - p}, \quad 0 \le p < 800$$

The quantity $E(p) + 1$ is equal to zero when $800 - 2p = 0$, or $p = 400$. The quantity $E(p) + 1$ is undefined when $800 - p = 0$ or $p = 800$, but $p = 800$ is not in the interval $[0, 800)$. We draw a sign diagram of $E(p) + 1$ as follows.

| | Domain Begins ↓ | | | | Possible Sign Change ↓ | | | Domain Ends ↓ |
|---|---|---|---|---|---|---|---|---|
| p | 0 | | 200 | | 400 | 600 | | 800 |
| E(p) + 1 | + | + | + | + | + 0 − | − | − | − |
| DEMAND | i n e l a s t i c | | | | unit elasticity | e l a s t i c | | |

---

**SOLUTIONS TO EXAMPLE 9**  $-20(50 - p)$, $\dfrac{-2p}{50 - p}$, $-1$, $\$16.67$

We conclude that demand is inelastic for $0 \le p < 400$ and elastic for $400 < p < 800$. When $p = 400$, because $E(p) = -1$, demand has unit elasticity. ∎

There is an important relationship between elasticity and revenue. Recall that

revenue = (price per unit)(number of units sold)

$$R(p) = px = pf(p)$$

so that

$$R'(p) = pf'(p) + f(p) = f(p)\left[\frac{pf'(p)}{f(p)} + 1\right]$$
$$= f(p)[E(p) + 1]$$

Now, since $f(p)$ is positive, we may state the following relationships:

---

$R'(p) > 0$ (revenue is increasing) if and only if $E(p) + 1 > 0$. (demand is inelastic).

$R'(p) < 0$ (revenue is decreasing) if and only if $E(p) + 1 < 0$. (demand is elastic).

$R'(p) = 0$ (revenue is maximum) if and only if $E(p) + 1 = 0$. (demand has unit elasticity).

---

EXAMPLE 11

If the monthly demand function for a product is $x = f(p) = 4000 - 5p$, find the maximum monthly revenue and verify that this maximum occurs at the price for which demand has unit elasticity.

Solution

Note that $R(p) = px = pf(p) = p(4000 - 5p)$, $0 \le p \le 800$, so that

$$R(p) = 4000p - 5p^2, \quad 0 \le p \le 800$$

We compute that $R'(p) = 4000 - 10p$ and find the critical numbers of $R(p)$ by solving the equation $4000 - 10p = 0$ to obtain $p = 400$.

Since $R''(p) = -10$ and $R''(400) = -10 < 0$, the maximum monthly revenue is obtained when $p = 400$, which, by Example 10, is the price for which demand has unit elasticity. A sketch of the demand and revenue curves is drawn in Figure 4.75. ∎

SECTION 4.7

SHORT ANSWER EXERCISES

1. Given the cost function $C(x) = 500 + 0.50x^2$, where $C(x)$ is the cost of producing $x$ units, then the marginal cost function is $C'(x) = x$ and $C'(100) = \$100$ is the approximate cost of producing the _____ unit.
2. When production rises from 100 to 101 units, the true marginal cost for the cost function of Exercise 1 is _____.
3. Given that the marginal revenue is $R'(x) = 200 - 0.4x$ and the marginal cost is $C'(x) = 100$, profit is maximized when production and sales are $x = $ _____ units.

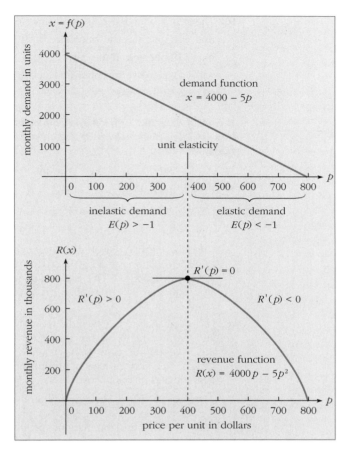

**FIGURE 4.75** Maximum revenue occurs when price per unit is $400, which is the price for which demand has unit elasticity.

4. (True or False) The average and marginal cost curves intersect in the point at which average cost is a minimum. _____

5. (True or False) If marginal cost is less than average cost, then producing another unit will decrease average cost. _____

6. If the cost of storing one unit for one year is $20 and if, on the average, 200 units are in storage during the year, then the yearly storage cost is _____.

7. If the economic lot size for the production of a certain product produced by our company is 500 units per batch, and if the variable cost of producing a unit of this product doubles, then the new economic lot size is (a) 1000 units; (b) 500 units; (c) 250 units; (d) not able to be determined from the information.

8. If we sell 10,000 radios a year and order 500 radios at a time, what is our yearly reorder cost if the fixed cost of processing an order is $25? _____

9. If the price of a container of salt were to double from $0.75 to $1.50, you would expect that (a) demand will be halved; (b) demand will double; (c) demand will decrease only slightly; (d) none of the above.

10. The demand for a certain product is inelastic for a price between $0 and $10 per unit, has unit elasticity at a price of $10 per unit, and is elastic for prices higher than $10 per unit. If the current price is $11 per unit, a price decrease of $0.50 will (a) increase revenue; (b) decrease revenue; (c) leave revenue unaffected; (d) cause unit elasticity.

## SECTION 4.7
## EXERCISES

*In each of Exercises 1–4, determine marginal cost, revenue, and profit at the indicated sales and production level and compare each of these with the corresponding true marginal values.*

1. $C(x) = 200 + 12x + \frac{1}{4}x^2$, $p = 40 - 0.1x$, $x = 100$
2. $C(x) = 1000 + 1500x$, $p = 500 - 0.25x$, $x = 1000$
3. $C(x) = 2000 + 15x + 0.40x^2$, $p = 16900 - x^2$, $x = 50$
4. $C(x) = 5x + 0.01x^2$, $p = 125000 - 0.001x^3$, $x = 200$

5. (Marginal profit) Given the weekly cost function $C(x) = 200 + 15x + 0.1x^2$ and demand function $p = 50 - 0.02x$, (a) find the profit function $P(x)$ and the marginal profit function $P'(x)$; (b) find and interpret the quantity $P'(100)$; (c) find the level of production that maximizes profit by equating marginal revenue and marginal cost.

6. (Marginal analysis) Given the monthly cost function $C(x) = 250 + 150x - 0.02x^2$ and the monthly profit function $P(x) = -0.1x^3 + 0.02x^2 + 850x - 250$, (a) find the revenue function $R(x)$; (b) find and interpret the quantities $R'(50)$ and $C'(50)$; (c) find the level of production that maximizes profit by equating marginal revenue and marginal cost.

7. (Marginal profit) If the monthly profit in dollars from the sale of $x$ television sets is $P(x) = 200x - \frac{1}{2}x^2 - 500$, (a) find the true profit from the sale of the 51st television; (b) find the marginal profit $P'(50)$ and compare with part (a).

8. (Marginal analysis) Given the monthly demand and cost functions $p = 80 - \frac{1}{10}x$ and $C(x) = 30x + 300$, respectively, (a) find the marginal revenue and marginal cost functions; (b) find and interpret $R'(250)$ and $C'(250)$; (c) find the production and sales level that maximizes profit.

9. (Marginal analysis) The weekly cost in dollars of producing $x$ tape recorders is $C(x) = 0.05x^2 + 27x + 500$, and weekly demand is given by $p = 150 - 0.1x$.
   (a) Find the weekly revenue function and the weekly profit function.
   (b) Find the marginal cost and the marginal revenue functions.
   (c) Interpret $C'(400)$, $R'(400)$, $C'(420)$, and $R'(420)$.
   (d) Find the production level that maximizes weekly profit.

10. (Marginal and average cost) If the weekly cost function is $C(x) = 250 + 15x + 0.005x^2$, (a) find the marginal cost; (b) find the average cost; (c) show that the average cost is a minimum when marginal cost is equal to average cost.

11. (Marginal and average cost) Given the monthly cost function $C(x) = 1600 + \frac{1}{9}x^2$, (a) find the production level that minimizes average cost; (b) find the production level at which average cost is equal to marginal cost.

12. (Economic order quantity) A retailer sells about 5000 pairs of women's shoes during the year at a uniform rate. Ordering costs are $40 per order and $0.25 per pair of shoes. The cost of holding a pair of shoes in stock for the year is $4. Find the economic order quantity for this retailer and find the minimum yearly inventory cost.

**13.** (Economic order quantity) The owner of a bar and grill sells about 900 cases of Big Horse Beer per year. The ordering costs for the beer are $9 per order and $0.50 per case. If it costs $2.00 per case to keep the beer in inventory for a year, find the economic order quantity for the owner and find the minimum yearly inventory cost of Big Horse Beer.

**14.** (Economic lot size) The Big Horse Brewery expects an annual demand of 20,000 cases of Big Horse Beer from local taverns. It costs $400 in set-up costs to prepare the brewery to make a batch of cases of beer. Annual storage costs are $1.50 per case and demand is expected to be fairly uniform throughout the year. Find the economic lot size for the brewery and the total of manufacturing and storage costs for the year if they produce in optimal batches and if the variable cost per case is $1.25.

**15.** (Economic lot size) The Goodride Tire Company expects an annual demand of 75,000 radial tires. It costs $6000 to set up the factory to run a batch of tires. The variable cost per tire is $12 and annual storage cost is $4 per tire. Find the economic lot size and the minimum possible total of yearly manufacturing and storage costs.

**16.** (Average elasticity) If the price of a certain computer increases from $2000 to $2500, monthly demand at a retail store will decrease from 40 to 25. Find and interpret the average elasticity of demand for this change in price.

**17.** (Average elasticity) The price of light bulbs at a certain store rises from $1.50 for a box of four to $2.50 for a box of four. Weekly demand falls from 200 boxes to 180 boxes. Find and interpret the average elasticity of demand for this change in price.

**18.** (Average elasticity) The price of pepper at a supermarket drops dramatically from $0.50 per carton to $0.20 per carton. Weekly demand rises only slightly, from 400 cartons per week to 420 cartons per week. Find and interpret the average elasticity of demand for this change in price.

*In each of Exercises 19–22, (a) find $E(p)$ for the given demand function; (b) evaluate $E(p)$ at the given price levels; (c) determine whether demand is inelastic, elastic, or has unit elasticity at the price levels given; (d) interpret your results in terms of the effect on demand of a 1% price increase and a 1% price decrease.*

**19.** $x = 600 - 1.5p$ for (i) $p = \$100$; (ii) $p = \$200$; (iii) $p = \$300$
**20.** $x = 2000 - 0.50p$ for (i) $p = \$100$; (ii) $p = \$2000$; (iii) $p = \$3900$
**21.** $x = 12500 - 5p^2$ for (i) $p = \$20$; (ii) $p = \$30$; (iii) $p = \$40$
**22.** $x = 8000 - 2p^2$ for (i) $p = \$5$; (ii) $p = \$40$; (iii) $p = \$60$

*In Exercises 23–30, determine the intervals on which demand is elastic and those on which it is inelastic.*

**23.** $x = 500 - 0.4p$      **24.** $x = 6000 - 10p$
**25.** $x = 50000 - 5p^2$      **26.** $x = 12000 - 2p^2$
**27.** $x = 15625 - \frac{1}{8}p^3$      **28.** $x = 21952 - \frac{1}{27}p^3$
**29.** $x = 5 + (100/p)$      **30.** $x = 2 + (200/3p)$

*In Exercises 31–36, find the revenue function corresponding to the given demand function, determine the maximum revenue, and verify that the maximum revenue occurs at the price for which demand has unit elasticity.*

**31.** $x = 4000 - 8p$      **32.** $x = 2000 - p$

**33.** $x = 150 - 0.2p^2$　　**34.** $x = 500 - 2p^2$
**35.** $x = 50 - 10\sqrt{p}$　　**36.** $x = 200 - 4\sqrt{p}$

*In many economic texts, the demand equation is written $p = f(x)$, where $p$ is price and $x$ is quantity demanded. In this case it can be shown that the point elasticity of demand is $E(x) = f(x)/(xf'(x))$. In Exercises 37–42, use this formula to find $E(x)$ for the given value of $x$.*

**37.** $p = 100 - 0.2x$, for $x = 50$　　**38.** $p = 40 - 0.1x$, for $x = 200$
**39.** $p = 1000 - 2x^2$, for $x = 15$　　**40.** $p = 2500 - 10x^2$, for $x = 15$
**41.** $p = 50 - 0.2\sqrt{x} - 0.001x^2$, for $x = 25$
**42.** $p = 100 - 0.1\sqrt{x} - 0.003x^2$, for $x = 64$

## OPTIONAL
### GRAPHING CALCULATOR/ COMPUTER EXERCISES

**43.** (Average and marginal cost) The cost of producing $x$ thousand units of a certain product is

$$C(x) = x^3 - 20x^2 + 140x + 50, \quad 0 \le x \le 20$$

Graph the average cost curve and the marginal cost curve on the same coordinate axis and approximate the production level that produces the minimum average cost.

**44.** (Average and marginal cost) Follow the same directions as in Problem 43 for the monthly cost function

$$C(x) = \tfrac{1}{3}x^3 - 5x^2 + 28x + 30, \quad 0 \le x \le 15$$

where $C(x)$ is in hundreds of dollars and $x$ is in thousands of units.

*In Exercises 45 and 46, for the given demand function, (a) determine the approximate price levels for which demand $x$ is nonnegative by sketching the graph of $x = f(p)$; (b) determine the approximate intervals on which demand is elastic and inelastic and the approximate price at which demand has unit elasticity.*

**45.** $x = 1000 - p - 0.02p^3$　　**46.** $x = 500 - 0.1p - 0.01p^2 - 0.001p^3$

## 4.8　CHAPTER FOUR REVIEW

**KEY TERMS**

increasing function
decreasing function
first-order critical number
intermediate value theorem
sign diagram
relative (local) maximum point
relative (local) minimum point
test values
first derivative test
concave up
concave down
second-order critical number
point of inflection

extreme value theorem
second derivative test for absolute extrema
vertical asymptote
horizontal asymptote
optimize
marginal cost
marginal revenue
marginal profit
average cost
economic order quantity
economic lot size
average elasticity of demand

point of diminishing returns
second derivative test
absolute maximum
absolute minimum

point elasticity of demand
elastic demand
inelastic demand
unit elasticity

**KEY FORMULAS AND RESULTS**

**Increasing Function**

$f'(x) > 0$ on $I$

**Decreasing Function**

$f'(x) < 0$ on $I$

**First-Order Critical Number c**

A number $c$ in the domain of $f(x)$ is a first-order critical number of $f$ if $f'(c) = 0$ or $f'(c)$ does not exist.

**Numbers Where f'(x) Can Change Sign**

(a) First-order critical numbers
(b) Numbers at which $f(x)$ is undefined or discontinuous

**Corollary to the Intermediate Value Theorem**

A continuous function that is never equal to zero on $I$ cannot change sign on $I$.

**Relative Extrema at c**

(a) A number $c$ yields a relative maximum value of $f$ if $f(x) \leq f(c)$ for all $x$ in some open interval containing $c$.
(b) A number $c$ yields a relative minimum value of $f$ if $f(x) \geq f(c)$ for all $x$ in some open interval containing $c$.

**Occurrence of a Relative Extreme Point**

A relative extreme point of $f(x)$ can occur only at a critical number of $f$.

**First Derivative Test**

(a) $f(c)$ is a relative maximum if $f'$ changes from positive to negative at $c$.
(b) $f(c)$ is a relative minimum if $f'$ changes from negative to positive at $c$.
(c) There is no relative extremum at $c$ if $f'$ does not change sign at $c$.

**Concavity of a Function**

(a) $f(x)$ is concave up on an interval $I$ if $f''(x) > 0$ for all $x$ in $I$.
(b) $f(x)$ is concave down on an interval $I$ if $f''(x) < 0$ for all $x$ in $I$.

### Second-Order Critical Number c

A number $c$ in the domain of $f(x)$ is a second-order critical number of $f$ if $f''(c) = 0$ or $f''(c)$ does not exist.

### Numbers Where f''(x) Can Change Sign

(a) Second-order critical numbers
(b) Numbers at which $f(x)$ is undefined or discontinuous

### Second Derivative Test

If $c$ is a critical number such that $f'(c) = 0$, then

(a) $f''(c) > 0$ indicates a relative minimum at $c$;
(b) $f''(c) < 0$ indicates a relative maximum at $c$;
(c) $f''(c) = 0$ means the test fails.

### Absolute Extrema on I

(a) A function $f$ has an absolute maximum at $c$ if $f(c) \geq f(x)$ for all $x$ in $I$.
(b) A function $f$ has an absolute minimum at $c$ if $f(c) \leq f(x)$ for all $x$ in $I$.

### Extreme Value Theorem

A continuous function attains both an absolute maximum and an absolute minimum on a closed interval $[a, b]$.

### Numbers Where Absolute Extrema on [a, b] Can Occur

(a) At an endpoint $a$ or $b$
(b) At a critical number in $[a, b]$

### Second Derivative Test for Absolute Extrema

If $c$ is a critical number in an interval $I$ and $f(x)$ is a continuous function on $I$, then

(a) $f''(c) > 0$ indicates an absolute minimum at $c$;
(b) $f''(c) < 0$ indicates an absolute maximum at $c$;
(c) $f''(c) = 0$ means the test fails.

### Vertical Asymptotes of a Rational Function p(x)/q(x)

The vertical line $x = a$ is an asymptote for $p(x)/q(x)$, if $q(a) = 0$ and $p(a) \neq 0$.

### Horizontal Asymptotes of a Rational Function p(x)/q(x)

(a) None if degree $(p) >$ degree $(q)$
(b) The line $y = 0$ if degree $(p) <$ degree $(q)$

(c) The line $y = a/b$ if degree $(p)$ = degree $(q)$, where $a$ and $b$ are coefficients of the highest powers of $x$ in the numerator and denominator

**Marginal Cost**

$C'(x)$, where $C(x)$ is the cost function

**Marginal Revenue**

$R'(x)$, where $R(x)$ is the revenue function

**Marginal Profit**

$P'(x)$, where $P(x)$ is the profit function

$$P(x) = R(x) - C(x)$$

**Maximum Profit**

$R'(x) = C'(x)$

**Minimum Average Cost**

$$C'(x) = \frac{C(x)}{x}$$

**Elasticity of Demand**

$$E(p) = \frac{pf'(p)}{f(p)}$$

**Inelastic Demand**

$-1 < E(p) \leq 0$: revenue is increasing

**Elastic Demand**

$E(p) < -1$: revenue is decreasing

**Unit Elasticity**

$E(p) = -1$: revenue is a maximum

## CHAPTER 4 SHORT ANSWER REVIEW QUESTIONS

1. For the function $f(x) = -x^3$, the derivative is $f'(x) = -3x^2$. We conclude that $f(x)$ is (increasing, decreasing) on the interval $(-\infty, \infty)$.

2. The number $x = -1$ is a first-order critical number of $f(x)$. We know that (a) $f'(-1) = 0$; (b) $f'(-1)$ does not exist; (c) a relative extremum occurs at $x = -1$; (d) none of the above.

3. The function $f(x)$ is continuous on $[0, 5]$; furthermore, $f(0) = 2$ and $f(5) = 7$. We know that (a) $f(x)$ is increasing on $[0, 5]$; (b) $f(x)$ does not change sign on $[0, 5]$; (c) $f(x)$ has no critical points in $[0, 5]$; (d) none of the above.

4. The critical points of a continuous function $f(x)$ are $-2$ and $3$. If $f'(0) = 5$, we conclude that (a) $f(x)$ has a relative minimum at $-2$; (b) $f(x)$ has a relative extremum at $3$; (c) $f(x)$ is increasing on $[-2, 3]$; (d) all of the above.

5. (True or False) The absolute minimum of $f(x)$ on $I$ is also a relative minimum of $f(x)$ on $I$. _____

6. (True or False) If $f'(c)$ does not exist then $f(c)$ does not exist. _____

7. (True or False) If $f(c)$ does not exist then $f'(c)$ does not exist. _____

8. (True or False) Every quadratic function $f(x) = ax^2 + bx + c$, where $a \neq 0$, has a relative extremum. _____

9. The point of diminishing returns of a certain monthly cost function $C(x)$ is $x = 7000$ units. The cost of producing units beyond 7000 is (a) increasing at a decreasing rate; (b) increasing at an increasing rate; (c) decreasing at a decreasing rate; (d) decreasing at an increasing rate.

10. A continuous function $f(x)$ has only one second-order critical number at $x = 3$; furthermore, $f''(1) < 0$ and $f''(4) < 0$. We conclude that (a) a point of inflection occurs at $x = 3$; (b) $f(x)$ is concave down on $(-\infty, \infty)$; (c) $f(x)$ changes concavity at $x = 3$; (d) none of the above.

11. A continuous function $f(x)$ has second derivative $f''(x) = 3x - 6$. A point of inflection of $f(x)$ occurs at $x =$ _____.

12. (True or False) Whenever the concavity of $f(x)$ changes at a number $x = c$, the point $(c, f(c))$ is a point of inflection of $f(x)$. _____

13. A continuous function $f(x)$ has only one relative extremum at $x = -2$. Then (a) an absolute extremum may or may not exist; (b) $f(x)$ has an absolute extremum; (c) $f(x)$ is decreasing on $(-\infty, \infty)$; (d) $f(x)$ changes concavity at $x = -2$.

14. The critical numbers of a continuous function $f(x)$ are $-2$ and $3$. We know that $f(-2) = 15, f(0) = -1, f(3) = -4$, and $f(5) = -7$. The absolute maximum value of $f(x)$ on $[0, 5]$ is _____.

15. The absolute minimum value of $f(x) = 10x - 5$ on $[0, 2]$ is _____.

16. The vertical asymptotes of $f(x) = \dfrac{2x + 1}{x^2 - x}$ are $x = 0$ and $x =$ _____.

17. (True or False) A rational function must have either a horizontal asymptote or a vertical asymptote. _____

18. A monthly profit function $P(x)$ has derivative $P'(x) = 2x^2 + 4$, $x \geq 0$. We can conclude that (a) any level of output $x$ results in a positive profit; (b) the profit function has a relative maximum value; (c) the profit function is increasing on $[0, \infty)$; (d) none of the above.

**19.** In finding two numbers that have a difference of 25 and a minimum product, we have $x - y = 25$ and we must optimize $xy$ or $P(x) = x(\underline{\phantom{xxx}})$.

**20.** If $P(x)$ is profit in dollars from the sale of $x$ units and if $P'(1000) = 20$, then \$20 is the approximate profit derived from the manufacture and sale of the \underline{\phantom{xxx}}st unit.

**21.** Given that $C'(x) = 4$ and $R'(x) = 10 - 0.01x$, where $x$ is units of product, profit is maximized when $x = \underline{\phantom{xxx}}$ units.

**22.** Which of the following four conditions is *not* equivalent to the others?
(a) Revenue is decreasing. (b) Demand is elastic. (c) $E(p) + 1 = 0$
(d) $R'(p) < 0$

**23.** The marginal and average cost curves intersect at the production level that minimizes which cost?
(a) marginal; (b) total; (c) incremental; (d) average

**24.** If it costs \$20 to place an order for radios plus an additional \$1 per radio, if 20 orders are placed, the total yearly reorder cost for 500 radios is \underline{\phantom{xxx}}.

## CHAPTER 4 REVIEW EXERCISES

*Exercises marked with an asterisk (\*) constitute a Chapter Test.*

*In Exercises 1–6, determine the intervals on which each function is increasing and those on which it is decreasing.*

**1.** $f(x) = x^2 - 4x + 6$
**2.** $f(x) = 7 - 10x - x^2$
**\*3.** $g(x) = 2x^3 - 12x^2 - 72x + 40$
**4.** $g(x) = 6 + 15x - 6x^2 - x^3$
**5.** $f(x) = 1/x^3$
**6.** $f(x) = 4/(3x - 9)$

*In Exercises 7–16, find all the relative extrema of the given function.*

**7.** $f(x) = x^3 - x^2$
**8.** $f(x) = 4x^3 - 12x$
**\*9.** $f(x) = 2x^3 - 15x^2 - 84x + 100$
**10.** $f(x) = 4 + 24x + 3x^2 - x^3$
**11.** $g(x) = 4 - x^4$
**12.** $g(x) = x^4 - 4x$
**13.** $f(x) = 3x^4 - 8x^3 - 18x^2 + 4$
**14.** $f(x) = x^5 - 5x - 1$
**15.** $f(x) = \dfrac{2x}{x + 4}$
**16.** $g(x) = \dfrac{2x - 5}{x + 4}$

*In Exercises 17–22, find the intervals on which the graph of the given function is concave up and those on which it is concave down. Find any points of inflection.*

**17.** $f(x) = x^4 - 12x^2$
**18.** $f(x) = x^3 - 2x^2 + 7x - 5$
**19.** $g(x) = 10 - x - x^2 - 4x^3$
**20.** $f(x) = 2x^4 + 3x^3 - x^2 + x - 3$
**\*21.** $f(x) = \sqrt[3]{x} - 2$
**22.** $g(x) = x/(2x - 1)$

*In Exercises 23–27, find the absolute extrema of the given function on the given interval I if any exist.*

**23.** $f(x) = 2x^2 + 5x - 3$ on $I = [0, 1]$
**24.** $f(x) = x^3 - 3x$ on $I = [-2, 2]$
**\*25.** $g(x) = 3x^4 - 4x^3$ on $I = [-1, 2]$
**26.** $f(x) = x^3 - 3x^2 - 5x - 2$ on $I = [1, 5]$
**27.** $f(x) = \dfrac{2 - 3x}{x + 1}$ on $I = [-2, 0]$

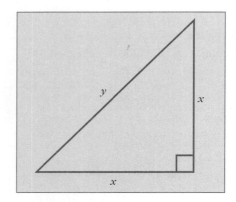

**FIGURE 4.76**

Business and Economics

In Exercises 28–32, sketch the graph of the given function. If feasible and applicable, find intercepts, asymptotes, relative extrema, intervals of concavity, and points of inflection.

**28.** $f(x) = x^4 - 2x^2$

**29.** $f(x) = 2x^3 - 15x^2 + 36x - 6$

**30.** $f(x) = 3x^4 + 6x^3 - 15x^2 - 5$

**\*31.** $f(x) = -2x/(x^2 + 4)$

**32.** $g(x) = \sqrt[3]{x} - \frac{1}{4}x$

**33.** (Maximum profit) The profit in hundreds of dollars on the sale of $x$ hundred units of a product is $P(x) = -x^3 + 12x^2 - 25x$. Find the sales level that optimizes profit and the maximum profit.

**34.** (Fencing) If the triangular area in Figure 4.76 is to be enclosed using 100 feet of fencing, find the dimensions that maximize the enclosed area.

**\*35.** (Marginal analysis) If a firm produces and sells $x$ units of a product, its weekly cost and revenue functions are $C(x) = 300 + 200x$ and $R(x) = 400x - 20x^2$ in dollars. If production and sales are 10 units, find the marginal cost, marginal revenue, and marginal profit.

**36.** (Packaging) One of our company's products is to be packaged in a closed cylindrical container having a volume of $64\pi$ in$^3$. If the surface area of the container is to be a minimum, find the radius and height of the container.

**37.** (Material cost) If the material for the top and bottom of the container in Exercise 36 costs three times as much as the material for the sides, find the dimensions of the $64\pi$-in$^3$ container that will minimize the cost of materials.

**38.** (Bus company costs) It costs a certain bus company $(150 + 2x)/200$ dollars per mile for fuel, insurance, repairs, and depreciation to drive a bus at an average speed of $x$ miles per hour. If the driver must be paid \$12/hr, at what speed should the bus be driven to minimize the cost per mile?

**\*39.** (Parking lot revenue) A downtown parking lot with 400 parking spaces is completely filled every weekday when the parking rate is \$6/day. For each \$0.25 increase in the daily rate, ten fewer spaces are occupied. Find the daily rate that the owner should charge to maximize revenue; find that maximum revenue.

**40.** (Production cost) The monthly cost of producing $x$ units of one of our company's products is $C(x) = 2500 + 8x + 0.05x^2$. Find the production level that minimizes monthly average cost per unit. Graph the average cost function and the marginal cost function on the same axes.

**41.** (Minimum area) The sum of the circumference of a circle and the perimeter of a square is 25 in. What are the dimensions of the circle and square that give a minimum total area?

**42.** (Inventory cost) A store sells a total of 2400 calculators at a constant rate throughout the year. It costs \$3 to store a calculator for one year and the cost of placing an order is \$25 plus \$1 per calculator. Find the number of calculators that should be ordered each time to minimize yearly inventory cost. What is the minimum yearly inventory cost?

**43.** (Number of retirees) It is estimated that the number of retired persons (in hundreds) living in a certain community in Florida in $t$ years will be

$$N(t) = 12 + 8t^2 - 1.4t^3, \quad 0 \le t \le 6$$

(a) What will be the maximum number of retired persons in this community within the next six years?

(b) When will the rate of increase of the number of retired persons be largest?

Social Sciences

44. (Economic lot size) A company expects a uniform annual demand of 25,000 units of one of its products next year. It costs $750 for set-up costs to prepare the factory to run a batch of these units. If storage costs are $6/yr per unit, find the economic lot size.

45. (Arrests of minors) In a large city, the number $N$ of arrests of minors per year depends on the yearly income $m$ (in thousands of dollars) of the minors' parents according to the function

$$N(m) = 0.08m^2 - 8m + 400, \quad 0 \le m \le 75$$

where it is assumed that $m$ is measured to the nearest whole unit. This function is called a *discrete* function because it is defined at only finitely many numbers. Approximate this discrete function by the corresponding continuous function defined on the interval [0, 75] and determine the income level of the minors' parents that yields the minimum number of yearly arrests; calculate that minimum number of arrests.

46. (Sociology) In a study of the relationship between economic satisfaction $E$ and annual yearly income $m$ (in thousands of dollars) of retirees, a sociologist developed the following formula:

$$E(m) = -0.0000415m^3 + 0.004353m^2 + 0.02156m + 0.2, \quad 0 \le m \le 50$$

In this equation, economic satisfaction is measured on a scale from 0 to 5.
(a) Sketch the graph of $E(m)$.
(b) Determine the annual income level at which economic satisfaction changes from increasing at an increasing rate to increasing at a decreasing rate.

*In Exercises 47–50, determine the intervals on which demand is elastic and those on which demand is inelastic. Sketch the graph of each demand function.*

*47. $x = 1000 - 4p$  
49. $x = 3 + (125/2p)$  
48. $x = 18000 - 4p^2$  
50. $x = 10 + (100/\sqrt{p})$

*In Exercises 51–53, find the revenue function corresponding to the given demand function, determine the maximum revenue, and verify that the maximum revenue occurs at the price for which demand has unit elasticity.*

51. $x = 10000 - 20p$  
52. $x = 400 - 2p^2$  
*53. $x = 91125 - p^3$

OPTIONAL

GRAPHING CALCULATOR/
COMPUTER EXERCISES

**Exercises 54–56:** *For each of the Exercises 54–56, use a graphing calculator or a computer to approximate to the nearest tenth of a unit the x-coordinates of any relative extrema or points of inflection of the given function.*

54. $f(x) = x^4 - 7x^3 + 4x^2 - x + 3$
55. $f(x) = 10 - 3x - 4x^2 + x^3 + 3x^4 - x^5$
56. $f(x) = 0.01x^5 - 0.1x^4 + x^2 - x + 2$

# 5 EXPONENTIAL AND LOGARITHMIC FUNCTIONS

## CHAPTER CONTENTS

5.1 Exponential Functions

5.2 Logarithmic Functions

5.3 Derivatives of Exponential and Logarithmic Functions

5.4 Exponential Growth and Decay

5.5 Chapter 5 Review

## 5.1 EXPONENTIAL FUNCTIONS

In Section 2.2 we considered the problem of finding the amount of money $A$ accumulated after $q$ quarters if \$10,000 was invested in a savings account paying 8% interest compounded quarterly. We found that

$$A = f(q) = 10000\left(1 + \frac{0.08}{4}\right)^q = 10000(1.02)^q$$

If we replace $q$ by $x$, we obtain the function

$$f(x) = 10000(1.02)^x$$

In this function the variable $x$ appears as an exponent. We call such a function—one in which the independent variable appears as an exponent or part of an exponent—an *exponential function*. Exponential functions have many applications and can be used to study the growth of money due to interest, the growth of a population, the decay of radioactive material, and the growth of learning. Some examples of exponential functions are

$$f(x) = 5^x$$
$$g(x) = 2^{-x^2}$$
$$h(x) = 8(1.5)^{-2.3x+1}$$

In order to use exponential functions, you may need to brush up on some of the properties of exponents usually studied in algebra. Let's spend a little time reviewing this material.

If $a$ is a positive real number and $n$ and $m$ are positive integers, then we define

1. $a^n = a \cdot a \cdots a$   ($n$ factors)

    Examples:  $2^3 = 2 \cdot 2 \cdot 2 = 8$
    $5^4 = 5 \cdot 5 \cdot 5 \cdot 5 = 625$

2. $a^0 = 1$. This property is valid for any nonzero real number $a$.

    Examples:   $2^0 = 1$
    $(-3.14)^0 = 1$

3. $a^{1/n} = \sqrt[n]{a}$

    Examples:  $2^{1/3} = \sqrt[3]{2}$
    $5^{1/2} = \sqrt{5}$

4. $a^{m/n} = (\sqrt[n]{a})^m$

    Examples:  $9^{3/2} = (\sqrt{9})^3 = 3^3 = 27$
    $16^{3/4} = (\sqrt[4]{16})^3 = 2^3 = 8$

5. $a^{-n} = \dfrac{1}{a^n}$. This property is also valid if $n$ is a rational number.

Examples: $2^{-3} = \dfrac{1}{2^3} = \dfrac{1}{8}$

$\left(\dfrac{1}{2}\right)^{-4} = \dfrac{1}{\left(\frac{1}{2}\right)^4} = \dfrac{1}{\frac{1}{16}} = 16$

The preceding five-part definition allows us to raise any positive number $a$ to a rational power. Let's look at some more examples.

## EXAMPLE 1

Find $3^{-2.5}$ and $2^{1.7}$.

### Solution

$$3^{-2.5} = 3^{-5/2} = \dfrac{1}{3^{5/2}} = \dfrac{1}{(\sqrt{3})^5} \approx 0.06415$$

$$2^{1.7} = 2^{17/10} = \left(\sqrt[10]{2}\right)^{17} \approx 3.2490 \quad \blacksquare$$

We have yet to define what an expression like $2^{\sqrt{3}}$ means. That is, how do we raise a positive number like 2 to an irrational power like $\sqrt{3}$? Although a completely formal treatment of this matter is beyond the intent of this book, in order to get an idea of how this works, we consider the decimal approximation of $\sqrt{3}$:

$$\sqrt{3} \approx 1.732050808$$

Since we know how to find $2^r$ if $r$ is rational, we consider the sequence of computations illustrated in Table 5.1.

| $r$ | 1.7 | 1.73 | 1.7320 | 1.732050 | 1.73205080 | 1.732050808 | $\to \sqrt{3}$ |
|---|---|---|---|---|---|---|---|
| $2^r$ | 3.2490 | 3.3173 | 3.32188 | 3.321995 | 3.32199707 | 3.32199709 | $\to 2^{\sqrt{3}}$ |

**TABLE 5.1**

This table is reminiscent of the ones we constructed when computing limits in Chapter 2. In this case the number $r$ is approaching $\sqrt{3}$ through rational values and the corresponding values of $2^r$ are approaching what we will define to be $2^{\sqrt{3}}$, which, from Table 5.1, appears to be 3.3219971 rounded to seven decimal places. From now on we will assume that for a function such as $f(x) = 2^x$, the domain is all real numbers $x$.

To continue our review of the properties of exponents, we consider the *laws of exponents* summarized as follows.

## LAWS OF EXPONENTS

If $a$ and $b$ are positive real numbers and if $x$ and $y$ are any real numbers, then

1. $a^x a^y = a^{x+y}$   2. $\dfrac{a^x}{a^y} = a^{x-y}$   3. $(a^x)^y = a^{xy}$

4. $(ab)^x = a^x b^x$   5. $\left(\dfrac{a}{b}\right)^x = \dfrac{a^x}{b^x}$

## CHAPTER 5 EXPONENTIAL AND LOGARITHMIC FUNCTIONS

**EXAMPLE 2** Compute (a) $3^2 \cdot 3^5$; (b) $\dfrac{4^{1.5}}{4^{0.5}}$; (c) $(2^3)^4$; (d) $2^{-3}5^{-3}$; (e) $(\sqrt{2}/\sqrt{3})^4$

**Solution**

(a) $3^2 \cdot 3^5 = 3^{2+5} = 3^7 = 2187$
(b) $\dfrac{4^{1.5}}{4^{0.5}} = 4^{1.5-0.5} = 4^1 = 4$
(c) $(2^3)^4 = 2^{3(4)} = 2^{12} = 4096$
(d) $2^{-3}5^{-3} = (2 \cdot 5)^{-3} = 10^{-3} = 1/10^3 = 1/1000$
(e) $(\sqrt{2}/\sqrt{3})^4 = (2^{1/2}/3^{1/2})^4 = 2^2/3^2 = 4/9$ ∎

One simple but useful class of exponential functions is considered in the next definition.

**DEFINITION**

> If $b > 0$ and $b \neq 1$, the function
> $$f(x) = b^x$$
> is an exponential function with **base** $b$.

In defining an exponential function with base $b$, we require that $b > 0$ because if, for example, $b = -2$, then $(-2)^x$ is not real for numbers like $x = \frac{1}{2}, \frac{1}{4}, \frac{3}{4}, \frac{5}{2}$, and so forth. We require that $b \neq 0$ to avoid the indeterminate expression $0^0$. We require that $b \neq 1$ to avoid the trivial function $f(x) = 1^x = 1$, which graphs as a horizontal straight line.

As examples, consider $y = 2^x$, which is an exponential function with base 2; or $y = 5^x$, which is an exponential function with base 5; or $y = \left(\frac{1}{2}\right)^x$, which is an exponential function with base $\frac{1}{2}$.

**EXAMPLE 3** Sketch the graphs of $y = 2^x$, $y = 3^x$, $y = \left(\frac{1}{2}\right)^x$, and $y = \left(\frac{1}{3}\right)^x$.

**Solution** In Table 5.2 we compute functional values of each of these exponential functions for various values of $x$. The graphs are drawn in Figure 5.1. Note that we cannot use the methods of calculus to assist us in graphing, since we do not yet know how to find the derivative of a function like $f(x) = 2^x$. It is *not* equal to $x2^{x-1}$. In the next section we will discuss derivatives of exponential functions.

| $x$ | $2^x$ | $3^x$ | $\left(\frac{1}{2}\right)^x$ | $\left(\frac{1}{3}\right)^x$ |
|---|---|---|---|---|
| $-2$ | $\frac{1}{4}$ | $\frac{1}{9}$ | 4 | 9 |
| $-1$ | $\frac{1}{2}$ | $\frac{1}{3}$ | 2 | 3 |
| 0 | 1 | 1 | 1 | 1 |
| 1 | 2 | 3 | $\frac{1}{2}$ | $\frac{1}{3}$ |
| 2 | 4 | 9 | $\frac{1}{4}$ | $\frac{1}{9}$ |

**TABLE 5.2**

# 5.1 EXPONENTIAL FUNCTIONS

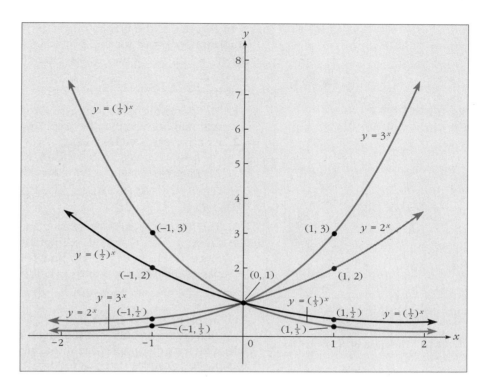

**FIGURE 5.1** The graphs of a few exponential functions.

The graphs of $y = 2^x$ and $y = 3^x$ are representative of the graphs of the class of functions $f(x) = b^x$ for $b > 1$. A function of this type is said to exhibit *exponential growth*. We summarize the properties of this class of functions as follows.

**PROPERTIES OF $f(x) = b^x$ IF $b > 1$**

1. The domain of $f(x)$ is all real numbers and the range is all nonnegative real numbers, so that the graph of $f(x)$ lies above the $x$-axis.
2. As $x$ increases without bound, $f(x)$ increases very rapidly, whereas as $x$ decreases without bound the graph of $f(x)$ approaches the negative $x$-axis asymptotically. In the language of limits

$$\lim_{x \to \infty} f(x) = \infty \quad \text{and} \quad \lim_{x \to -\infty} f(x) = 0$$

3. The function $f(x)$ is increasing and concave up throughout its domain.
4. Since $f(0) = b^0 = 1$, all such graphs pass through the point $(0, 1)$.
5. Since $f(-1) = b^{-1} = 1/b$ and $f(1) = b^1 = b$, the graph of $f(x) = b^x$ passes through the points $(-1, 1/b)$ and $(1, b)$.

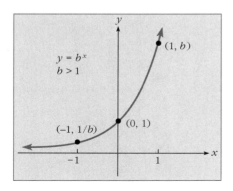

**FIGURE 5.2** The graph of $f(x) = b^x$ for $b > 1$.

The graph of $f(x) = b^x$ for $b > 1$ is illustrated in Figure 5.2. The steepness of the graph depends on the base $b$. The larger the number $b$, the steeper the graph.

In Figure 5.1, the graphs of $y = \left(\frac{1}{2}\right)^x$ and $y = \left(\frac{1}{3}\right)^x$ are representative of the graphs of $f(x) = b^x$ for $0 < b < 1$. A function of this type exhibits *exponential decay* and the properties of this type of function are summarized as follows.

**PROPERTIES OF $f(x) = b^x$ IF $0 < b < 1$**

1. The domain of $f(x)$ is all real numbers and the range is all nonnegative real numbers, so that the graph of $f(x)$ lies above the $x$-axis.
2. As $x$ increases without bound, $f(x)$ decreases very rapidly approaching the positive $x$-axis asymptotically, whereas as $x$ decreases without bound, $f(x)$ increases without bound. In the language of limits

$$\lim_{x \to \infty} f(x) = 0 \quad \text{and} \quad \lim_{x \to -\infty} f(x) = \infty$$

3. The function $f(x)$ is decreasing and concave up throughout its domain.
4. Since $f(0) = b^0 = 1$, all such graphs pass through the point $(0, 1)$.
5. Since $f(-1) = b^{-1} = 1/b$ and $f(1) = b^1 = b$, the graph of $f(x) = b^x$ passes through the points $(-1, 1/b)$ and $(1, b)$.

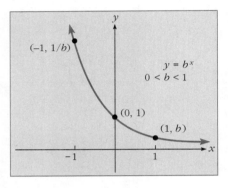

**FIGURE 5.3** The graph of $f(x) = b^x$ for $0 < b < 1$.

The graph of $f(x) = b^x$ for $0 < b < 1$ is illustrated in Figure 5.3.

An interesting property of the graph of $y = b^x$, $b > 0$, is that any horizontal line intersects the graph in at most one point; see Figure 5.4(a). Algebraically, this property can be stated as follows:

If $b > 0$ and $b^n = b^m$, then $n = m$.

More formally, we say that the function $y = b^x$ is *one-to-one*, meaning that any $y$-value corresponds to a single $x$-value. A function that is not one-to-one is $y = f(x) = x^2$. This is true since, for example, $f(-2) = f(2) = 4$, so the value $y = 4$ can be obtained from two *different* $x$-values, $x = -2$ and $x = 2$; see Figure

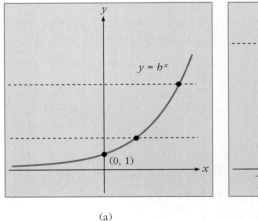

(a)

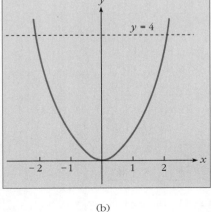

(b)

**FIGURE 5.4** The graph in (a) is one-to-one. The graph in (b) is not one-to-one.

5.4(b). In this figure the horizontal line $y = 4$ intersects the graph of $f(x) = x^2$ in two different points.

**DEFINITION**

> The function $y = f(x)$ is *one-to-one* if $f(n) = f(m)$ implies that $n = m$. Equivalently, $f(x)$ is one-to-one if any horizontal line intersects the graph of $f(x)$ in at most one point.

We can use the one-to-one property of $y = b^x$ to help us solve certain exponential equations—that is, equations in which the unknown appears as an exponent or part of an exponent.

---

**EXAMPLE 4**

Solve the equation $3^{2x} = 243$.

**Solution**

Notice that 243 can be written as a power of 3, namely, $3^5$, so that the given equation can be written $3^{2x} = 3^5$. Now, using the one-to-one property of $y = 3^x$, we conclude that $2x = 5$ and $x = \frac{5}{2}$. ∎

---

**EXAMPLE 5 (Participative)**

Let's solve the equation $4^x = 8^{2x-1}$.

**Solution**

We begin by observing that both 4 and 8 can be written as powers of 2. Thus $4 = 2^2$ and $8 = $ _____. The given equation can then be written

$$(2^2)^x = (2^3)^{2x-1}$$
$$2^{2x} = 2^{6x-3}$$

Using the one-to-one property of $y = 2^x$, we can write $2x = $ _____. Solving for $x$, we obtain $x = $ _____. ∎

A somewhat more general form of exponential function than $y = b^x$ has the representation

$$y = k b^{g(x)}$$

where $k$ is a constant and $g(x)$ is some function of $x$.

---

**EXAMPLE 6**

Graph the function $y = 2(3)^{-x^2}$.

**Solution**

First, note that this function is *not* equal to $6^{-x^2}$. We construct a table of values of $x$ and $y$ in Table 5.3.

---

**SOLUTIONS TO EXAMPLE 5**  $2^3$, $6x - 3$, $\frac{3}{4}$

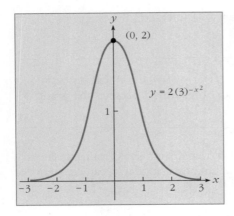

**FIGURE 5.5** The graph of $y = 2(3)^{-x^2}$

| $x$ | $-3$ | $-2$ | $-1$ | $0$ | $1$ | $2$ | $3$ |
|---|---|---|---|---|---|---|---|
| $y = 2(3)^{-x^2}$ | $\frac{2}{19683}$ | $\frac{2}{81}$ | $\frac{2}{3}$ | $2$ | $\frac{2}{3}$ | $\frac{2}{81}$ | $\frac{2}{19683}$ |

**TABLE 5.3**

Since the function can be written $y = 2/3^{x^2}$, we see that $\lim_{x \to \pm\infty} (2/3^{x^2}) = 0$, so that $y = 0$ (the $x$-axis) is a horizontal asymptote of the graph. We sketch the graph in Figure 5.5.

### The Number e

The most important base for an exponential function is undoubtedly the number designated by the letter **e**. This number is irrational and was named by the Swiss mathematician Leonhard Euler (1707–1783). The approximation

$$e \approx 2.718281828$$

is accurate to the nine decimal places shown. Mathematicians have approximated $e$ to thousands of decimal places, although for most practical purposes, the accuracy we have shown is more than sufficient. The exact value of $e$ is defined by the limit

$$e = \lim_{x \to \infty} \left(1 + \frac{1}{x}\right)^x$$

In Table 5.4 we have shown the value of $\left(1 + \frac{1}{x}\right)^x$ for increasingly large values of $x$. These values have been computed using a calculator with a $y^x$ key. The graph of $f(x) = e^x$ is sketched in Figure 5.6, in which the graphs of $y = 2^x$ and $y = 3^x$ are also shown for comparison purposes. Table 5.5, which accompanies the figure, was constructed with the aid of a calculator with an $e^x$ key. Alternatively, a calculator with a $y^x$ key would suffice, since $e^x \approx (2.71828)^x$. If you do not have access to a calculator with the required keys, you can use the table on the front endsheets to compute values for $e^x$ and $e^{-x}$. In Table 5.5, the value of $e^x$ is computed for various values of $x$. As you can see from Figure 5.6, because $2 < e < 3$, the graph of $y = e^x$ lies between the graphs of $y = 2^x$ and $y = 3^x$ (since $2^x < e^x < 3^x$).

Although the base $e$ seems to be rather esoteric, we will see that it occurs naturally in the very practical problem of computing interest that is compounded continuously. It is also a natural choice for use in calculus because, as we will see in the next section, the derivative of $e^x$ is particularly simple. It is not surprising, then, that the function $f(x) = e^x$ is sometimes referred to as the "natural" exponential function or, alternatively, as *the* exponential function.

| $x$ | $\left(1 + \frac{1}{x}\right)^x$ |
|---|---|
| 100 | 2.7048 |
| 10000 | 2.71815 |
| 1000000 | 2.7182805 |
| 100000000 | 2.71828182 |
| $\to \infty$ | $\to e$ |

**TABLE 5.4**

| $x$ | $e^x$ |
|---|---|
| $-2$ | 0.1353 |
| $-1$ | 0.3679 |
| $0$ | 1 |
| $1$ | 2.7183 |
| $2$ | 7.3891 |

**TABLE 5.5**

### Simple Interest

Suppose we invest a sum of money (the principal $P$) for $t$ years at a simple yearly interest rate $r$. Since with simple interest we earn interest only on the principal $P$, the interest earned after the first year is $Pr$ and the amount $A(t)$ in the account after one year is

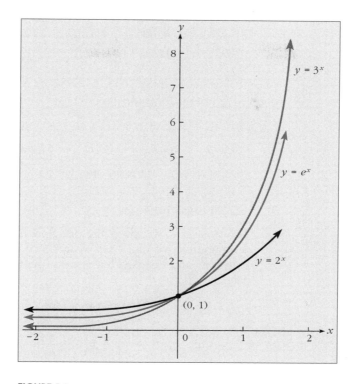

**FIGURE 5.6** A comparison of the graphs of some exponential functions.

$$A(1) = P + Pr$$
$\quad\quad\ \ \uparrow\ \ \ \ \ \uparrow$
principal  interest for
$\quad\quad\quad\ \ $ 1st year

The corresponding amount at the end of the second year is

$$A(2) = (P + Pr) + Pr = P + 2Pr$$
$\quad\quad\quad\ \ \uparrow\quad\quad\quad\ \uparrow$
amount after  interest for
1st year  $\quad\ $ 2nd year

After $t$ years we will have accumulated the amount $P + tPr$. Summarizing, we have

**SIMPLE INTEREST FORMULA**

$$A(t) = P + tPr = P(1 + rt)$$

$P$ = principal
$r$ = annual interest rate
$t$ = time in years
$A$ = amount accumulated after $t$ years

## EXAMPLE 7

If you invest $1000 in a savings account at a simple yearly interest rate of 8%, what will be the amount in the account after 12 years?

**Solution**

We have $P = \$1000$ and $r = 0.08$, so that

$$A(t) = 1000(1 + 0.08t)$$

Since $t = 12$, we obtain

$$A(12) = 1000[1 + 0.08(12)] = \$1960$$

as the amount in the account after 12 years. ∎

### Compound Interest

Now suppose we invest the principal $P$ for $t$ years at an annual interest rate $r$ compounded $k$ times per year. Since compound interest yields "interest on interest," and since the interest rate per period is $r/k$, the amount in the account after

1 period is $P + P\left(\dfrac{r}{k}\right) = P\left(1 + \dfrac{r}{k}\right)$;

2 periods is $P\left(1 + \dfrac{r}{k}\right) + P\left(1 + \dfrac{r}{k}\right)\left(\dfrac{r}{k}\right) = P\left(1 + \dfrac{r}{k}\right)\left(1 + \dfrac{r}{k}\right) = P\left(1 + \dfrac{r}{k}\right)^2$

3 periods is $P\left(1 + \dfrac{r}{k}\right)^2 + P\left(1 + \dfrac{r}{k}\right)^2\left(\dfrac{r}{k}\right) = P\left(1 + \dfrac{r}{k}\right)^2\left(1 + \dfrac{r}{k}\right) = P\left(1 + \dfrac{r}{k}\right)^3$

$\vdots$

$k$ periods or 1 year is $P\left(1 + \dfrac{r}{k}\right)^{k-1} + P\left(1 + \dfrac{r}{k}\right)^{k-1}\left(\dfrac{r}{k}\right)$

$$= P\left(1 + \dfrac{r}{k}\right)^{k-1}\left(1 + \dfrac{r}{k}\right) = P\left(1 + \dfrac{r}{k}\right)^k$$

$\vdots$

$kt$ periods is $P\left(1 + \dfrac{r}{k}\right)^{kt-1} + P\left(1 + \dfrac{r}{k}\right)^{kt-1}\left(\dfrac{r}{k}\right) = P\left(1 + \dfrac{r}{k}\right)^{kt-1}\left(1 + \dfrac{r}{k}\right)$

$$= P\left(1 + \dfrac{r}{k}\right)^{kt}$$

Since there are $kt$ periods in $t$ years, we obtain the following formula.

**COMPOUND INTEREST FORMULA**

$$A(t) = P\left(1 + \dfrac{r}{k}\right)^{kt}$$

$P$ = principal
$r$ = annual interest rate
$k$ = number of compounding periods per year
$t$ = time in years
$A(t)$ = amount accumulated after $t$ years

## EXAMPLE 8

If you invest $1000 in a savings account that pays an annual rate of 8%, find the amount in the account after 12 years if interest is compounded (a) quarterly; (b) monthly; (c) daily.

**Solution**

In all parts of this problem, $P = 1000$, $r = 0.08$, and $t = 12$, so that the compound interest formula becomes

$$A(12) = 1000\left(1 + \frac{0.08}{k}\right)^{12k}$$

(a) Here $k = 4$ and so

$$A(12) = 1000\left(1 + \frac{0.08}{4}\right)^{12(4)} = 1000(1.02)^{48} = \$2587.07$$

(b) In this case $k = 12$ and

$$A(12) = 1000\left(1 + \frac{0.08}{12}\right)^{12(12)} = \$2603.39$$

(c) Take $k = 365$ to obtain

$$A(12) = 1000\left(1 + \frac{0.08}{365}\right)^{365(12)} = \$2611.42$$

## EXAMPLE 9 (Participative)

You have $25,000 to invest for a period of two years. Let's determine whether you can earn more interest by investing the money at 9.75% compounded monthly or at a simple interest rate of 10%.

**Solution**

First let's compute the amount we can have after two years by investing at 9.75% compounded monthly. For this, we use the compound interest formula

$$A(t) = P\left(1 + \frac{r}{k}\right)^{kt}$$

with $P = 25,000$, $r = $ _____, $k = $ _____, and $t = $ _____ to obtain

$$A(2) = 25000\left(1 + \frac{0.0975}{12}\right)^{12(2)} = \underline{\hspace{2cm}}$$

Next we use the simple interest formula to decide how much we can accumulate after two years by investing the money at 10% simple interest. We have $A(t) = P(1 + rt)$ with $P = 25000$, $r = 0.10$, and $t = 2$ to obtain $A(2) = $ _____. We conclude that we can earn $358.84 more interest by investing the money at 9.75% compounded monthly.

**SOLUTIONS TO EXAMPLE 9**   0.0975, 12, 2, $30,358.84, $30,000

## Continuously Compounded Interest

We now consider the question of what the interest formula should be if interest is *compounded continuously*. Essentially this means that the number of compounding periods per year, $k$, approaches infinity. If we let $x = k/r$, then $k = rx$ and as $k \to \infty$, so does $x \to \infty$. Thus we obtain

$$\lim_{k \to \infty} P\left(1 + \frac{r}{k}\right)^{kt} = \lim_{x \to \infty} P\left(1 + \frac{1}{x}\right)^{rxt} = \lim_{x \to \infty} P\left[\left(1 + \frac{1}{x}\right)^x\right]^{rt}$$

$$= P\left(\lim_{x \to \infty} \left[\left(1 + \frac{1}{x}\right)^x\right]\right)^{rt}$$

Now, since $\lim_{x \to \infty} \left(1 + \frac{1}{x}\right)^x = e$, this reduces to

$$\lim_{k \to \infty} P\left(1 + \frac{r}{k}\right)^{kt} = Pe^{rt}$$

as the formula for interest compounded continuously.

**CONTINUOUSLY COMPOUNDED INTEREST FORMULA**

$$A(t) = Pe^{rt}$$

$P$ = principal
$r$ = annual interest rate
$t$ = time in years
$A(t)$ = amount accumulated after $t$ years

---

**EXAMPLE 10**

If you invest $1000 in a savings account that pays an annual rate of 8%, find the amount in the account after 12 years if interest is compounded continuously.

**Solution**

We have $P = \$1000$ and $r = 0.08$, so that $A(t) = 1000e^{0.08t}$. With $t = 12$ years, we get

$$A(12) = 1000e^{0.08(12)} = 1000e^{0.96} \approx \$2611.70.$$ ∎

---

**SECTION 5.1**

**SHORT ANSWER EXERCISES**

1. Which of the following functions is *not* an exponential function?
   (a) $f(x) = (1.6)^x$;  (b) $g(x) = x^{2.1}$;  (c) $h(x) = 7(1.73)^{x^2}$;  (d) $r(x) = \frac{1}{5}e^{-x^2}$
2. Written as a decimal to four places, an approximate value of $e^{2/5}$ is _____.
3. (True or False) The number $(-8)^{1/3}$ is not a real number. _____
4. The function $f(x) = b^x$ is said to exhibit exponential growth if $b >$ _____.
5. The graph of $f(x) = e^x$ and $g(x) = e^{-x}$ intersect at the point _____.
6. By the one-to-one property of $f(x) = b^x$, if $7^{x^2} = 7^x$ then $x^2 =$ _____.
7. (True or False) The simple interest on $5000 invested at 7% for six years is $2100. _____

8. If $10,000 is invested for one year at 9% in an account that gives compound interest, for which number of corresponding periods would the largest amount of interest be earned?
   (a) quarterly;  (b) monthly;  (c) weekly;  (d) daily

9. If you invest $20,000 into an account that gives 7% interest compounded continuously, the formula for the amount in this account after $t$ years is
   $A(t) = $ _____.

10. (True or False) In the compound interest formula, as we increase the number of compounding periods without bound, the yearly interest we can earn also increases without bound. _____

## SECTION 5.1
## EXERCISES

*In Exercises 1–10, simplify the given expressions. Write your answers without fractional or negative exponents but do not use your calculator to approximate the results.*

1. $16^{1/2}$
2. $27^{4/3}$
3. $12/5^{-3}$
4. $(5^2)^3(5^{-1})^2$
5. $7^5/49^2$
6. $6^{-3}7^{-3}$
7. $\dfrac{(4^3)(3^2)}{(2^2)(3^{-1})}$
8. $(5^{2/3})(5^{4/3})$
9. $\dfrac{3^{1.75}}{9^{0.5}}$
10. $\dfrac{(12)(7^3)}{4(49)}$

*In each of Exercises 11–20, approximate the given expression to four decimal places using a calculator.*

11. $5^{4.75}$
12. $7^{2.20}$
13. $(-3)^{7/9}$
14. $10^{-1.29}$
15. $12^{-0.675}$
16. $3^{\sqrt{2}}$
17. $5^{\sqrt{3}}$
18. $12^{\sqrt{5}+1}$
19. $10^{1-\sqrt{7}}$
20. $7^{2\sqrt{3}-2}$

*In each of Exercises 21–30, graph the given exponential function.*

21. $y = 5^x$
22. $y = 10^x$
23. $y = \left(\tfrac{1}{5}\right)^x$
24. $y = \left(\tfrac{1}{10}\right)^x$
25. $y = e^{-x}$
26. $y = e^{-2x}$
27. $y = e^{-x^2}$
28. $y = 5(2)^{-x^2}$
29. $y = 50e^{0.2x}$
30. $y = 50e^{-0.2x}$

31. (Simple interest) If $20,000 is invested at an annual rate of 10% compounded quarterly, find the amount accumulated after five years.

32. (Compound interest) A sum of $50,000 is invested in a money market certificate that yields an annual rate of 12% compounded monthly for a period of two years. How much is the certificate worth at maturity?

33. (Compound interest) If you invest $100,000 in a savings account that pays an annual rate of 7%, find the amount in the account after six years if interest is compounded (a) semiannually; (b) monthly; (c) daily.

34. (Simple interest) Find the simple interest on $5000 invested for seven years at 8% per year.

35. (Simple interest) Find the amount in a savings account after ten years if $5000 is invested at a simple yearly interest rate of 6.5%.

36. (Interest) If $15,000 is invested at 7.25% per year, find the accumulated amount after eight years if interest is compounded (a) daily; (b) continuously.

37. (Interest) If $25,000 is invested at 8.75% per year, find the accumulated amount after five years if interest is compounded (a) semiannually; (b) continuously.

38. (Compound interest) Will you earn more interest by investing $10,000 for three years at 10.25% compounded daily or at 10% compounded semiannually?

39. (Interest) Will you earn more interest by investing $30,000 for five years at 9.50% compounded quarterly or at 9% compounded continuously?

40. (Interest) Will you earn more interest by investing $75,000 for nine years at 9.75% compounded semiannually or by investing this sum for eight years at 9.75% compounded continuously?

*Sometimes the amount $A(t)$ accumulated after $t$ years is called the* future value *of the principal amount $P$, in which case the principal $P$ is called the* present value *of the future amount $A(t)$. We can find the present value by solving the interest formulas of this section for $P$. For example, if we solve $A = Pe^{rt}$ for $P$, we find that the present value of a future amount $A$ compounded continuously at $100r\%$ per year for $t$ years is $P = Ae^{-rt}$.*

41. (Present value) Find the present value of a future amount of $10,000 in five years if interest is compounded continuously and the annual rate is 9%.

42. (Present value) Find the present value of a future amount of $25,000 in four years if interest is compounded continuously and the annual rate is 10.25%.

43. (Present value) How much must you invest today in a savings account that pays an annual rate of 7% compounded quarterly in order to have $50,000 in ten years?

44. (Present value) How much must you invest in a savings account that pays an annual rate of 7.5% compounded daily in order to accumulate $15,000 in five years?

45. (Employee training) The number of components assembled per hour by an average worker on an assembly line after $t$ hours of formal training is

    $$N(t) = 25(1 - e^{-0.1t})$$

    (a) Find the number of components assembled by a worker who has had 20 hours of training.
    (b) Find and interpret $\lim_{t \to \infty} N(t)$.
    (c) Sketch the graph of $N(t)$ for $t \geq 0$.

46. (Sales) The Euler Company projects its monthly sales in dollars will be

    $$S(t) = 20000e^{0.05t}, \quad 0 \leq t \leq 23$$

    where $t$ is time in months and $t = 0$ represents sales for January, 1991.
    (a) Find the projected sales for January 1992 and March 1992.
    (b) Graph the sales function $S(t)$.

47. (Population growth) A mathematical model for the growth of a population (of people, animals, bacteria, etc.) over a short time interval is $P(t) = P_0 e^{rt}$, where $P_0$ is the population at time $t = 0$ (sometimes called the *initial population*), $r$ is the continuous compound growth rate, $t$ is time, and $P(t)$ is the population at time $t$. If the population of a certain city is 2,500,000 and if it continues to grow at a

$$\log_4 1 = 0 \qquad 4^0 = 1$$
$$\log_{15} 15 = 1 \qquad 15^1 = 15$$
$$\log_e \frac{1}{e} = -1 \qquad e^{-1} = \frac{1}{e}$$

**EXAMPLE 1**  Find each of the following: **(a)** $\log_{10} \frac{1}{100}$; **(b)** $\log_2 32$; **(c)** $\log_e e^5$.

**Solution**

**(a)** $\log_{10} \frac{1}{100}$ is the exponent to which the base 10 must be raised in order to produce $\frac{1}{100}$. We conclude that

$$\log_{10} \frac{1}{100} = -2 \quad \text{since} \quad 10^{-2} = \frac{1}{100}$$

**(b)** $\log_2 32$ is the exponent to which the base 2 must be raised in order to produce 32. Thus

$$\log_2 32 = 5 \quad \text{since} \quad 2^5 = 32$$

**(c)** $\log_e e^5$ represents the exponent to which the base $e$ must be raised in order to produce $e^5$. Therefore

$$\log_e e^5 = 5 \quad \text{since} \quad e^5 = e^5$$

Note carefully that we cannot find the logarithm of zero or of a negative number. For example, $\log_2 0 = x$ has equivalent exponential form $2^x = 0$, which has no solution $x$, and $\log_2(-1) = x$ has exponential form $2^x = -1$, which also has no solution $x$.

### Logarithmic Functions and Their Graphs

**DEFINITION OF THE LOGARITHMIC FUNCTION WITH BASE $b$**

If $b > 0$ and $b \neq 1$, then

$$y = \log_b x \quad \text{if and only if} \quad x = b^y$$

The number $b$ is called the *base* of the logarithmic function $y = \log_b x$.

As examples, $y = \log_2 x$ is a logarithmic function with base 2, $y = \log_{10} x$ is a logarithmic function with base 10, and $y = \log_e x$ is a logarithmic function with base $e$.

In most applications, the bases that are commonly used are either 10 or $e$. We call a logarithm with base 10 a *common logarithm* and write $\log x$ rather than $\log_{10} x$. We call a logarithm with base $e$ a *natural logarithm* and write $\ln x$ rather than $\log_e x$. On your calculator, log means $\log_{10}$ and ln means $\log_e$. The table on the front endsheets is a table of natural logarithms.

Since the logarithmic function is closely related to the exponential function by definition, we might expect the graphs of $y = \log_b x$ and $y = b^x$ to be related in some way. In fact, these graphs are simply mirror images of one another across the straight line $y = x$. The next example illustrates this fact for the logarithmic function with base 2.

## EXAMPLE 2

Sketch the graph of $y = \log_2 x$ and $y = 2^x$ on the same coordinate axes.

### Solution

We begin by constructing two tables of values, Table 5.6(a) and Table 5.6(b). By inspecting these tables, you can see that the points in the set $\{(\frac{1}{4}, -2), (\frac{1}{2}, -1), (1, 0), (2, 1), (4, 2)\}$ lie on the graph of $y = \log_2 x$ (or $x = 2^y$), whereas the points in the set $\{(-2, \frac{1}{4}), (-1, \frac{1}{2}), (0, 1), (1, 2), (2, 4)\}$ lie on the graph of $y = 2^x$. Note that corresponding points in each of these two sets can be obtained by interchanging the $x$ and $y$ coordinates. For example the point $(\frac{1}{4}, -2)$ is on the graph of $y = \log_2 x$, while the point $(-2, \frac{1}{4})$ lies on the graph of $y = 2^x$. These points are mirror images of one another across the line $y = x$. The final graph appears in Figure 5.7. Note that if this graph were "folded" along the line $y = x$, the two halves of the graph would match exactly. Alternatively, if we were first to sketch the graph of $y = 2^x$ and then place a mirror along the line $y = x$, we would see the image of the graph of $y = \log_2 x$ in the mirror.

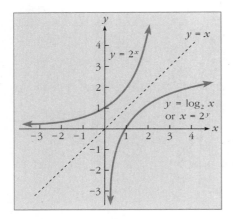

**FIGURE 5.7** The graphs of $y = \log_2 x$ and $y = 2^x$ are mirror images of one another across the line $y = x$.

| $x$ | $y = \log_2 x$ |
|---|---|
| $\frac{1}{4}$ | $-2$ |
| $\frac{1}{2}$ | $-1$ |
| $1$ | $0$ |
| $2$ | $1$ |
| $4$ | $2$ |

**TABLE 5.6(a)**

| $x$ | $y = 2^x$ |
|---|---|
| $-2$ | $\frac{1}{4}$ |
| $-1$ | $\frac{1}{2}$ |
| $0$ | $1$ |
| $1$ | $2$ |
| $2$ | $4$ |

**TABLE 5.6(b)**

When the graphs of $f(x)$ and $g(x)$ are mirror images across the line $y = x$, then whenever the point $(a, b)$ is on the graph of $f(x)$ (so that $f(a) = b$), the point $(b, a)$ is on the graph of $g(x)$ (so that $g(b) = a$). We conclude that

$$g(f(a)) = g(b) = a \qquad \text{for all } a \text{ in the domain of } f(x)$$

and

$$f(g(b)) = f(a) = b \qquad \text{for all } b \text{ in the domain of } g(x)$$

Pairs of functions like these are called *inverse functions*; they "undo" one another. As an example, $f(x) = \log_b x$ and $g(x) = b^x$ are inverse functions, and so

$$g(f(x)) = g(\log_b x) = b^{\log_b x} = x \qquad \text{if } x > 0$$

and

$$f(g(x)) = f(b^x) = \log_b b^x = x \qquad \text{for all } x$$

To take an example with a specific base, let $b = e$, so that $e^{\ln x} = x$ for $x > 0$ and $\ln e^x = x$ for all $x$.

### Properties of Logarithms

There are a number of important properties of logarithms; we summarize these in Table 5.7. The proof of Property III can be found in Example 4. The proofs of the remaining properties are left as exercises for the reader.

**PROPERTIES OF LOGARITHMS**
(Assume $b > 0$, $b \neq 1$, $x > 0$, and $y > 0$)

| Property | Example |
|---|---|
| I. $\log_b 1 = 0$ | $\log_3 1 = 0$ |
| II. $\log_b b = 1$ | $\log 10 = 1$ |
| III. $\log_b xy = \log_b x + \log_b y$ | $\log_2(21) = \log_2(3 \cdot 7)$ |
|  | $\quad\quad\quad\; = \log_2 3 + \log_2 7$ |
| IV. $\log_b \dfrac{x}{y} = \log_b x - \log_b y$ | $\ln\left(\dfrac{5}{7}\right) = \ln 5 - \ln 7$ |
| V. $\log_b x^y = y \log_b x$ | $\ln \sqrt[3]{2} = \ln 2^{1/3} = \frac{1}{3} \ln 2$ |

**TABLE 5.7**

---

**EXAMPLE 3 (Participative)**

Given that $\log_3 5 \approx 1.46497$ and $\log_3 2 \approx 0.63093$ let us find (approximate) values of **(a)** $\log_3 10$; **(b)** $\log_3 \frac{25}{8}$; **(c)** $\log_3 \sqrt{50}$.

**Solution**

**(a)** $\log_3(10) = \log_3(2 \cdot 5) = \log_3 2 + \log_3 5$     (Property III)

$\quad\quad\quad\approx 0.63093 + \underline{\hspace{2cm}}$

$\quad\quad\quad= \underline{\hspace{2cm}}$

**(b)** $\log_3 \frac{25}{8} = \log_3 25 - \log_3 \underline{\hspace{1cm}}$     (Property IV)

$\quad\quad\quad= \log_3 5^2 - \log_3 2^3$     (Property IV)

$\quad\quad\quad= 2 \log_3 5 - \underline{\hspace{2cm}}$

$\quad\quad\quad\approx 2(1.46497) - \underline{\hspace{2cm}}$

$\quad\quad\quad= \underline{\hspace{2cm}}$

**(c)** $\log_3 \sqrt{20} = \log_3(20)^{1/2} = \frac{1}{2}(\underline{\hspace{2cm}})$     (Property V)

$\quad\quad\quad= \frac{1}{2} \log_3(2^2 \cdot 5)$

$\quad\quad\quad= \frac{1}{2}(\log_3 2^2 + \underline{\hspace{2cm}})$

$\quad\quad\quad= \frac{1}{2}(2 \log_3 2 + \log_3 5)$

$\quad\quad\quad\approx \frac{1}{2}[2(\underline{\hspace{2cm}}) + \underline{\hspace{2cm}}]$

$\quad\quad\quad= \underline{\hspace{2cm}}$

---

**SOLUTIONS TO EXAMPLE 3**

(a) 1.46497, 2.09590; (b) 8, 3 $\log_3$ 2, 3(0.63093), 1.03715; (c) $\log_3$ 20, $\log_3$ 5, 0.63093, 1.46497, 1.36342

## EXAMPLE 4

Prove Property III of logarithms: $\log_b xy = \log_b x + \log_b y$ if $b > 0, b \neq 1$, $x > 0$, and $y > 0$.

**Solution**

Let

$$\log_b xy = M \quad \text{so that} \quad b^M = xy$$
$$\log_b x = N \quad \text{so that} \quad b^N = x$$
$$\log_b y = P \quad \text{so that} \quad b^P = y$$

Then

$$b^M = xy = b^N b^P = b^{N+P}$$

where the last equality holds because of Law 1 for exponents in Section 5.1. We conclude by the one-to-one property of the exponential function that $M = N + P$ and so

$$\log_b xy = \log_b x + \log_b y \qquad \blacksquare$$

## EXAMPLE 5

(a) Given that $\log_b x = N$ and $\log_b y = M$, write $\log_b(\sqrt{x}/y^3)$ in terms of $N$ and $M$.

(b) Write $\ln(x + 3) - \ln x + \frac{1}{2}\ln(x^2 + 1)$ as the logarithm of a single quantity.

**Solution**

(a) $\log_b \dfrac{\sqrt{x}}{y^3} = \log_b \dfrac{x^{1/2}}{y^3}$

$\qquad = \log_b x^{1/2} - \log_b y^3 \qquad$ (Property IV)

$\qquad = \frac{1}{2} \log_b x - 3 \log_b y \qquad$ (Property V)

$\qquad = \frac{1}{2} N - 3M$

(b) $\ln(x + 3) - \ln x + \frac{1}{2}\ln(x^2 + 1) = \ln\left(\dfrac{x+3}{x}\right) + \frac{1}{2}\ln(x^2+1) \qquad$ (Property IV)

$\qquad = \ln\left(\dfrac{x+3}{x}\right) + \ln[(x^2+1)^{1/2}] \qquad$ (Property V)

$\qquad = \ln\left[\left(\dfrac{x+3}{x}\right)(x^2+1)^{1/2}\right] \qquad$ (Property III)

$\qquad = \ln\left[\dfrac{(x+3)\sqrt{x^2+1}}{x}\right] \qquad \blacksquare$

### Solving Equations Involving Exponential or Logarithmic Functions

The properties of logarithms are useful when solving equations involving logarithmic or exponential functions. The next example illustrates the techniques involved.

## EXAMPLE 6

Solve each of the equations for $x$.
(a) $\ln(2x) - \ln(x - 10) = 1$
(b) $e^{3x-2} = 10$

Solution  (a) First, we write all logarithms in the equation as the logarithm of a single quantity.

$$\ln(2x) - \ln(x - 10) = 1$$
$$\ln\left[\frac{2x}{x-10}\right] = 1$$

Writing this in exponential form, we get

$$\frac{2x}{x-10} = e^1 = e$$

Solving this equation, we obtain

$$2x = ex - 10e$$
$$(2 - e)x = -10e$$
$$x = \frac{-10e}{2-e} \approx 37.844$$

Observe that the solution $x = 37.844$ checks in the original equation. Solutions to equations involving logarithmic functions *must* be checked to avoid a situation involving the logarithm of zero or a negative number.

(b) To solve this equation for $x$, we begin by applying the natural logarithm function to both sides:

$$e^{3x-2} = 10$$
$$\ln(e^{3x-2}) = \ln 10$$
$$3x - 2 = \ln 10 \quad \text{(Recall that } \ln e^z = z \text{ for } z > 0)$$
$$3x = \ln 10 + 2$$
$$x = \frac{\ln 10 + 2}{3} \approx 1.434$$

*Note:* An alternative way to begin the solution of part (b) is to write the exponential equation $e^{3x-2} = 10$ in logarithmic form, $\log_e 10 = 3x - 2$, or $\ln 10 = 3x - 2$. We can then continue to solve for $x$ as in Example 6. ■

## EXAMPLE 7 (Participative)

Let's solve the equation $e^{0.08t} = 2$ for $t$.

Solution

From the discussion at the beginning of this section, recall that the value of $t$ is the number of years that it takes an investment of $1000 to double when interest of 8% per year is compounded continuously. Writing the exponential equation $e^{0.08t} = 2$ in logarithmic form, we get $\log_e 2 = $ _____ or $\ln 2 = 0.08t$. Therefore $t = $ _____ $\approx 8.664$ years, or about 8 years and 242 days. ■

**SOLUTIONS TO EXAMPLE 7**     $0.08t$, $(\ln 2)/0.08$

**EXAMPLE 8**   After $t$ days of training, an average worker in a certain factory can assemble

$$N(t) = 85(1 - e^{-0.12t})$$

telephones per day. After how many days of training can the average worker assemble 60 telephones?

**Solution**   The average worker will be able to assemble 60 telephones when $N(t) = 60$, so we must solve the equation

$$85(1 - e^{-0.12t}) = 60$$

We proceed as follows:

$$85 - 85e^{-0.12t} = 60$$
$$-85e^{-0.12t} = -25$$
$$e^{-0.12t} = \frac{5}{17}$$
$$\ln e^{-0.12t} = \ln\left(\tfrac{5}{17}\right) \quad \text{(Applying ln to both sides)}$$
$$-0.12t = \ln\left(\tfrac{5}{17}\right) \quad \text{(Using } \ln e^x = x)$$
$$t = \frac{\ln\left(\tfrac{5}{17}\right)}{-0.12} \approx 10.2 \text{ days}$$

Thus the average worker will be able to assemble 60 telephones after approximately 10.2 days of training.   ■

### Logarithms to Bases Other Than e and 10 from a Calculator

A number like $\log_7 28$ cannot be computed directly on most calculators, which can compute only natural logarithms (base $e$) or common logarithms (base 10), using the ln and log keys, respectively. Fortunately, we can derive a simple formula that allows us to compute logarithms to any base by using a calculator or the table on the front endsheets. To begin the derivation, let $y = \log_b x$, so that $b^y = x$. Then, taking the logarithm to the base $k$ of both sides, we get

$$\log_k b^y = \log_k x$$
$$y \log_k b = \log_k x \quad \text{(Property V of logarithms)}$$
$$y = \frac{\log_k x}{\log_k b}$$

To summarize our result,

---

If $b$ and $k$ are positive numbers and neither is equal to 1, then

$$\log_b x = \frac{\log_k x}{\log_k b}$$

---

This formula is called the *change-of-base formula* for logarithms. The formula allows us to express the logarithm of a number $x$ to the base $b$ in terms of the logarithms of $x$ and $b$ to a different base $k$. In particular, if $k = 10$, we obtain the change-of-base formula

$$\log_b x = \frac{\log x}{\log b}$$

whereas if $k = e$, we obtain the change-of-base formula

$$\log_b x = \frac{\ln x}{\ln b}$$

EXAMPLE 9

Approximate $\log_7 28$ to four decimal places using a calculator.

Solution

Using the change-of-base formula with $k = e$,

$$\log_b x = \frac{\ln x}{\ln b}$$

where $b = 7$ and $x = 28$, we get

$$\log_7 28 = \frac{\ln 28}{\ln 7} \approx \frac{3.33220}{1.94591} \approx 1.7124$$

## SECTION 5.2
## SHORT ANSWER EXERCISES

1. The exponent to which the base 3 must be raised to yield 81 is 4. Therefore $\log_3 81 =$ _____.
2. The exponent to which the base 10 must be raised to yield 0.0001 is _____. Therefore $\log 0.0001 = -4$.
3. The logarithmic form $\ln 1 = 0$ is equivalent to the exponential form _____.
4. The functions $f(x) = \log x$ and $g(x) = 10^x$ are inverse functions. Therefore $10^{\log 17} =$ _____.
5. For a certain base $b$, it is known that the point $(c, d)$ is on the graph of $g(x) = b^x$. We can assert that the following point(s) lies on the graph of $f(x) = \log_b x$: **(a)** $(1, 0)$; **(b)** $(d, c)$; **(c)** $(b, 1)$; **(d)** all of the above.
6. (True or False) $\log_5 \frac{21}{10} = \log_5 7 + \log_5 3 - \log_5 2 - 1$. _____
7. (True or False) $\log(3 + 5) = \log 3 + \log 5$. _____
8. If $\ln x = 3$ then $x =$ _____.
9. If $e^x = 10$ then $x =$ _____.
10. If $2^x = e$ then $x =$ _____.

## SECTION 5.2
## EXERCISES

In Exercises 1–6, write the given equations in exponential form.

1. $\log_2 16 = 4$
2. $\log_7 343 = 3$
3. $\log_2 \sqrt{2} = \frac{1}{2}$
4. $\log_6 \frac{1}{36} = -2$
5. $\log \frac{1}{10} = -1$
6. $\ln \sqrt[3]{e} = \frac{1}{3}$

In Exercises 7–12, write the given equations in logarithmic form.

7. $5^4 = 625$
8. $8^3 = 512$
9. $10^{-5} = 0.00001$
10. $e^2 \approx 7.3891$
11. $16^{3/2} = 64$
12. $125^{2/3} = 25$

*In Exercises 13–24, use the definition of the logarithm to evaluate the given expressions.*

13. $\log_9 81$
14. $\log_6 36$
15. $\log_2 32$
16. $\log_3 27$
17. $\log_5 \frac{1}{125}$
18. $\log_{12} \frac{1}{144}$
19. $\log_4 2$
20. $\log_8 \frac{1}{2}$
21. $\log 1 + \log 100$
22. $\log 10 - \log \frac{1}{10}$
23. $\ln 1 - \ln e^2$
24. $\ln(1/e) + \ln e^4$

*In Exercises 25–32, use a calculator to evaluate each expression accurate to four decimal places.*

25. $\log 19$
26. $\log 117$
27. $\log \sqrt{3}$
28. $\log \sqrt{2}$
29. $\ln(0.7)$
30. $\ln(7.25)$
31. $\ln(\sqrt{5} + 1)$
32. $\ln(4 - \sqrt[3]{5})$

*In Exercises 33–38, sketch the graph of each function.*

33. $f(x) = \log x$
34. $f(x) = \log_3 x$
35. $f(x) = \ln x$
36. $f(x) = \log_2(x - 1)$
37. $f(x) = \log_5(2x + 1)$
38. $f(x) = \log(4x - 5)$

*In each of Exercises 39–46, use the properties of logarithms to write the given quantity as a sum, difference, or multiple of logarithms.*

39. $\log \frac{7}{5}$
40. $\ln \frac{2}{5}$
41. $\ln(xy/2)$
42. $\log(x^2 y)$
43. $\log \dfrac{3x}{(x + 1)^2}$
44. $\ln \dfrac{x^3}{\sqrt{x^2 + 1}}$
45. $\ln \sqrt[3]{2x + 1}$
46. $\log[2x(x + 1)^4]$

*In each of Exercises 47–54, write the given expression as the logarithm of a single quantity.*

47. $\ln x + \ln(2x)$
48. $2 \ln x + \ln(x + 1)$
49. $\ln(x - 1) - \ln x + \frac{1}{2} \ln(x + 1)$
50. $-\ln(x + 2) + \ln x - 3 \ln(x^2 + 1)$
51. $4 \ln(x - 1) - 3 \ln(x + 1) + 2 \ln(x^2 + 1)$
52. $\frac{1}{2} \ln x - \frac{1}{2} \ln(x + 7)$
53. $\frac{1}{2} \ln(2x - 5) + \frac{1}{3} \ln x - \frac{1}{3} \ln(2x^2 + 7)$
54. $\frac{1}{2} \ln A - \frac{1}{2} \ln B + \frac{3}{4} \ln C^2$

*In Exercises 55–66, solve the equations for x. Be sure to check your prospective solutions if the equation contains logarithms of the variable.*

55. $e^{2x} = 5$
56. $e^x + 5 = 7$
57. $3^{2x-1} = 12$
58. $10^{4x+3} = 20$
59. $5e^{4x} = 1.75$
60. $e^{x^2+2} = 25$
61. $e^x = 10^{2x+1}$
62. $\ln x - \ln(x - 12) = 1$

63. $\ln(x^2 + x - 1) = 0$ 

64. $\ln(x + 1)^3 = 9$

65. $\ln(3x) - \ln(x + 1) = 2$ 

66. $\ln x + \ln(x + 5) = 0$

67. **(Doubling time)** If we invest $5000 in a savings account that pays 9.5% annual interest compounded continuously, how long will it take for this investment to double?

68. **(Doubling and tripling time)** If we invest a certain amount $P$ into an account that pays 7.5% annual interest compounded continuously, how long will it take our investment to **(a)** double; **(b)** triple?

69. **(Interest rate for doubling)** If money deposited in a certain investment instrument doubles every ten years and if interest is compounded continuously, what is the annual interest rate?

70. **(Interest rate for tripling)** If money deposited in a certain investment instrument triples every 14 years and if interest is compounded continuously, what is the annual interest rate?

71. **(Doubling and tripling time)** If $10,000 is invested in a savings account that pays 8% annual interest compounded quarterly, how long will it take for this investment to **(a)** double; **(b)** triple?

72. **(Doubling and tripling time)** If $25,000 is invested in a savings account that pays 9% annual interest compounded monthly, how long will it take for this investment to **(a)** double; **(b)** triple?

73. **(Cost)** A manufacturer's monthly cost function is given by

$$C(x) = 500 \ln(24x + 3), \quad 0 \le x \le 100$$

where $C(x)$ is the cost in dollars of producing $x$ units of product.
**(a)** Find the cost of producing 50 units per month.
**(b)** Graph this cost function.
**(c)** If 50 units are produced per month, what is the average cost per unit?

74. **(Demand)** The monthly demand for one of our company's products is given by

$$p = \frac{25}{\ln(x + 1.2)} \quad 0 \le x \le 3000$$

where $p$ is the price per unit in dollars.
**(a)** If monthly demand is 1250 units, find the price per unit.
**(b)** Approximately how many units will be sold per month if the price per unit is $3.30?

75. **(Demand and revenue)** For the demand function of Exercise 74, **(a)** find the monthly revenue function $R(x)$; **(b)** find the monthly revenue when 2000 units are demanded per month; **(c)** find the monthly revenue when the price per unit is $3.15.

**Social Sciences**

76. **(Retained information)** A psychologist has determined that the percentage of a certain body of information retained by a specific subject $t$ days after she is exposed to it is given by the function

$$P(t) = \frac{100}{\ln(t + e + 1)}, \quad 0 \le t \le 14$$

**(a)** What percentage of the information is retained by this subject initially (when $t = 0$)?

(b) What percentage of the information is retained by this subject after one week (when $t = 7$)?
(c) After how much time does this subject retain only 40% of the information?

**Business and Economics**

77. (Cost and revenue) The monthly cost and revenue functions from the production and sale of $x$ hundred units of a certain product are

$$C(x) = 0.0001x^2 + 0.1x + 5, \quad 0 \le x \le 1200$$
$$R(x) = \sqrt{x} \ln(x + 5), \quad 0 \le x \le 1200$$

where $R(x)$ and $C(x)$ are measured in hundreds of dollars.
(a) Find the monthly revenue from the sale of 50,000 units of the product.
(b) Find the monthly profit function $P(x)$.
(c) Find the monthly profit derived from the sale of 25,000 units of the product.

78. (Advertising and sales) Market analysis has indicated that monthly sales in units of one of our company's products is related to the amount of money $x$ spent on advertising, in thousands of dollars, according to

$$S(x) = 100 + 10\sqrt{x} \ln(250x + 50), \quad 0 \le x \le 20$$

(a) Find monthly sales if we spend $1000 on advertising.
(b) Find monthly sales if we spend $10,000 on advertising.

79. (Missing children) In a certain city, the percentage of children who are still unaccounted for $t$ months after they are reported missing is given by

$$P(t) = \frac{100}{\ln(\sqrt{t} + e)}, \quad 0 \le t \le 60$$

(a) What percentage of the children are still unaccounted for after one year?
(b) After how many months will 50% of the children be accounted for?

80. (Private schools) The number $N$ of children who are enrolled in private schools in a certain region and whose parents have social class $x$ is $N(x) = 500/\ln(5.2 - x)$, where social class is measured on a scale of 0 to 4 and it is measured to the nearest tenth of a unit. How many children are enrolled by parents having social class $x = 1$? $x = 3.8$?

81. (Earthquakes) Earthquake magnitude is measured using a *Richter scale*, on which the magnitude is expressed $R = \log(I/I_0)$, where $I_0$ is a constant standard intensity used for comparison purposes. For example, if a particular earthquake has magnitude $R = 3$, then $3 = \log(I/I_0)$, so $I/I_0 = 10^3$ and $I = 1000I_0$. Thus this earthquake has magnitude 1000 times greater than the standard comparison earthquake.
(a) An earthquake that occurred in San Francisco in 1906 registered 8.2 on the Richter scale. How many times more intense was this earthquake than the standard intensity?
(b) In 1989, an earthquake that registered 6.9 on the Richter scale struck the San Francisco area. How many times more intense was the 1906 earthquake than the 1989 earthquake?

82. Prove Property IV of logarithms using a technique similar to that used in Example 4.

83. Prove Property V of logarithms using the method illustrated in Example 4 as a guide.

OPTIONAL

GRAPHING CALCULATOR/
COMPUTER EXERCISES

In Exercises 84–88, graph each of the functions.

84. $y = \log x + 0.1x^3 - x^2$
85. $y = x \ln x - 3\sqrt{x}$
86. $y = (\log x)^3 + [\log(x + 1)]^2 - x + 4$
87. $y = x^3 \ln x - \sqrt{x^7} + 2x$
88. $y = x^{5/2} + 3x^2 + x \ln x - e^x - 1$

In Exercises 89–91, approximate the real solutions $x$ in $[0, 10]$ of the given equations to two-decimal-place accuracy.

89. $x^{5/2} + 3x^2 + x \ln x = e^x + 1$ (*Hint:* See Exercise 88.)
90. $x^{4/5}e^x - 2^x x^2 - e^x = (\ln x)^3 - x - 6$
91. $xe^x - x^{10/9}e^{(x-1)^2} + 7(\ln x)^5 + x^2 = -1$

## 5.3 DERIVATIVES OF EXPONENTIAL AND LOGARITHMIC FUNCTIONS

We begin this section by finding the derivatives of the exponential function $f(x) = e^x$ and the natural logarithm function $g(x) = \ln x$. We will also obtain the derivative of $f(x) = b^x$, the exponential function to the base $b$, and $g(x) = \log_b x$, the logarithmic function to the base $b$. Applications of these results will be studied in Section 5.4.

### The Derivative of the Exponential Function $e^x$

Let $f(x) = e^x$. Applying the definition of the derivative,

$$f'(x) = \lim_{\Delta x \to 0} \frac{f(x + \Delta x) - f(x)}{\Delta x}$$

$$= \lim_{\Delta x \to 0} \frac{e^{x+\Delta x} - e^x}{\Delta x}$$

$$= \lim_{\Delta x \to 0} \frac{e^x(e^{\Delta x} - 1)}{\Delta x}$$

$$= \left(\lim_{\Delta x \to 0} e^x\right)\left(\lim_{\Delta x \to 0} \frac{e^{\Delta x} - 1}{\Delta x}\right)$$

$$= e^x \lim_{\Delta x \to 0} \frac{e^{\Delta x} - 1}{\Delta x} \qquad \text{(Since } e^x \text{ is constant with respect to } \Delta x\text{)}$$

To complete the derivation, we must compute $\lim_{\Delta x \to 0} \frac{e^{\Delta x} - 1}{\Delta x}$. Although this limit may be computed by using tables of values (see Exercise 94), we present another method here. For this, recall that

$$e = \lim_{x \to \infty} \left(1 + \frac{1}{x}\right)^x$$

so if we let $z = 1/x$, then $z \to 0$ as $x \to \infty$ and $e = \lim_{z \to 0}(1 + z)^{1/z}$. In this formula $z$ is what we call a "dummy" variable in the sense that it is simply a placeholder and may be replaced by any other variable. If we let $z = \Delta x$ in the formula, we get

$$e = \lim_{\Delta x \to 0} (1 + \Delta x)^{1/\Delta x}$$

Now, if $|\Delta x|$ is small (so that $\Delta x$ is close to zero), then $e \approx (1 + \Delta x)^{1/\Delta x}$, so that $e^{\Delta x} \approx 1 + \Delta x$.

Using this approximation in the formula for $f'(x)$, we have

$$f'(x) = e^x \lim_{\Delta x \to 0} \frac{1 + \Delta x - 1}{\Delta x} = e^x \lim_{\Delta x \to 0} 1 = e^x$$

We conclude that if $f(x) = e^x$, then $f'(x) = e^x$. The exponential function $e^x$ has the unusual (and convenient) property that it is its own derivative! It is this property that makes it a natural choice for many applications.

To generalize our result, let $F(x) = e^{g(x)}$, where $g(x)$ is a differentiable function of $x$. By the chain rule,

$$F'(x) = g'(x)e^{g(x)}$$

In summary, if $g(x)$ is a differentiable function of $x$, then

$$D_x e^x = e^x \tag{1}$$

$$D_x e^{g(x)} = g'(x)e^{g(x)} \tag{2}$$

In words, formula (2) can be stated: "The derivative of $e$ to an exponent $g(x)$ is the derivative of the exponent $g(x)$ times $e$ raised to the original exponent $g(x)$."

### EXAMPLE 1

Find (a) $D_x e^{6x}$; (b) $D_x e^{x^3}$

### Solution

(a) In formula (2), we let $g(x) = 6x$ to obtain

$$D_x e^{6x} = D_x(6x)e^{6x} = 6e^{6x}$$

(b) In formula (2), we let $g(x) = x^3$ to obtain

$$D_x e^{x^3} = D_x(x^3)e^{x^3} = 3x^2 e^{x^3}$$ ∎

### EXAMPLE 2 (Participative)

Let's find $f'(1)$ if $f(x) = xe^{2x^2}$.

## Solution

We write $f(x) = (x)(e^{2x^2})$ and use the product rule to obtain

$$f'(x) = x(\underline{\qquad}) + 1(e^{2x^2})$$
$$= e^{2x^2}(4x^2 + 1)$$

Therefore $f'(1) = \underline{\qquad}$, or, using a calculator to obtain an approximation to four decimal places, $f'(1) \approx \underline{\qquad}$.

## EXAMPLE 3

Find the slope of the tangent line to the graph of $f(x) = 3/(2 + e^{-x})$ at the point where $x = -2$.

## Solution

Using the quotient rule,

$$f'(x) = \frac{(2 + e^{-x})(0) - 3(-e^{-x})}{(2 + e^{-x})^2}$$
$$= \frac{3e^{-x}}{(2 + e^{-x})^2}$$

The required slope, therefore, is

$$f'(-2) = \frac{3e^2}{(2 + e^2)^2} \approx 0.2515$$

## EXAMPLE 4

If the cost to a manufacturer of producing a certain calculator is $4 each in addition to a fixed monthly cost of $500, and if the monthly demand function for these calculators is $x = 5000e^{-0.15p}$, where $x$ is the number of calculators demanded per month if the price is $p$ dollars per calculator, find the price that the manufacturer should charge to optimize profit.

## Solution

The monthly revenue is

revenue = (price per calculator)(number of calculators sold)
$$R(p) = 5000pe^{-0.15p}$$

The monthly cost function is

cost = (cost per calculator)(number of calculators produced) + fixed cost
$$C(p) = 4x + 500 = 20000e^{-0.15p} + 500$$

The monthly profit function is

profit = revenue − cost
$$P(p) = R(p) - C(p)$$
$$= 5000pe^{-0.15p} - 20000e^{-0.15p} - 500$$

To find the price that maximizes profit, we compute:

$$P'(p) = -750pe^{-0.15p} + 5000e^{-0.15p} + 3000e^{-0.15p}$$

---

**SOLUTIONS TO EXAMPLE 2**   $4xe^{2x^2}$, $5e^2$, 36.9453

Equating $P'(p)$ to zero to obtain critical numbers, we get

$$-750pe^{-0.15p} + 8000e^{-0.15p} = 0$$

$$e^{-0.15p}[8000 - 750p] = 0$$

Since $e^{-0.15p}$ cannot be zero for any $p$, we have

$$8000 - 750p = 0$$

$$p = \frac{32}{3} \quad \text{(Critical number)}$$

You may verify that $P'(0) = 8000 > 0$ and $P'(15) \approx -342.55 < 0$, so that by the first derivative test, a maximum of $P(p)$ occurs when $p = \frac{32}{3}$. The manufacturer should charge about $10.67 per calculator to ensure the maximum monthly profit. ■

### The Derivative of the Exponential Function $b^x$

We now consider the problem of finding the derivative of $f(x) = b^x$, where $b > 0$, $b \neq 1$. For this, note that $b = e^{\ln b}$, so that $b^x = (e^{\ln b})^x$. Thus

$$b^x = e^{x \ln b}$$

We conclude from formula (2) that

$$D_x b^x = D_x e^{x \ln b} = D_x(x \ln b) e^{x \ln b}$$
$$= (\ln b) e^{x \ln b} \quad \text{(Note that } \ln b \text{ is a constant here)}$$
$$= (\ln b) b^x$$

To generalize this result, let $g(x)$ be a differentiable function of $x$ and let $F(x) = b^{g(x)}$. Then $b^{g(x)} = e^{g(x) \ln b}$ and, using the chain rule, we get

$$F'(x) = (\ln b) g'(x) b^{g(x)}$$

To summarize, if $g(x)$ is a differentiable function of $x$, then

$$D_x b^x = (\ln b) b^x \qquad (3)$$

$$D_x b^{g(x)} = (\ln b) g'(x) b^{g(x)} \qquad (4)$$

Note that the difference between formula (4) for differentiating $b^{g(x)}$ and formula (2) for differentiating $e^{g(x)}$ is that in formula (4), if the base is not $e$, we must multiply by the additional factor $\ln b$. Of course, if $b = e$, then $\ln e = 1$.

**EXAMPLE 5** Find the derivative of $f(x) = 5^{\sqrt{x} + 2x^2}$.

Solution    We use formula (4) with $g(x) = x^{1/2} + 2x^2$ and $b = 5$ to obtain

$$f'(x) = (\ln 5)\left(\frac{1}{2\sqrt{x}} + 4x\right)5^{\sqrt{x}+2x^2}$$

■

**The Derivative of the Natural Logarithm Function ln x**

To find the derivative of the natural logarithm function, we could apply the definition of the derivative with $f(x) = \ln x$. However, we choose a somewhat simpler method, one that uses the technique of implicit differentiation and the formula for differentiating $e^x$ to develop this formula. For this, we let $y = \ln x$; then $e^y = x$. Differentiating both sides of this equation with respect to $x$, we obtain $D_x e^y = D_x x$ or

$$\frac{dy}{dx} \cdot e^y = 1 \quad \text{(Observe that } g(x) = y \text{ in formula (2))}$$

$$\frac{dy}{dx} = \frac{1}{e^y} = \frac{1}{x}$$

We conclude that $D_x(\ln x) = 1/x$. We can use the chain rule to obtain a more general result. If $g(x)$ is a differentiable function of $x$ and $F(x) = \ln g(x)$, then $F'(x) = g'(x)/g(x)$. Summarizing, if $g(x)$ is a differentiable function of $x$, then

$$D_x(\ln x) = \frac{1}{x} \tag{5}$$

$$D_x[\ln g(x)] = \frac{g'(x)}{g(x)} \quad \text{if } g(x) \neq 0 \tag{6}$$

In words, formula (6) can be stated: "The derivative of $\ln g(x)$ is the derivative of $g$ divided by $g$ itself." Note that, unless $g(x)$ itself contains logarithms, the derivative of $\ln g(x)$ does *not* contain logarithms.

---

EXAMPLE 6    Find the derivatives of **(a)** $f(x) = 5\ln(7x^2 + 10)$; **(b)** $f(x) = e^x \ln x$

Solution    **(a)** We use formula (6) with $g(x) = 7x^2 + 10$ to yield

$$f'(x) = \frac{5D_x(7x^2 + 10)}{7x^2 + 10} = \frac{5(14x)}{7x^2 + 10} = \frac{70x}{7x^2 + 10}$$

**(b)** We use the product rule in conjunction with formulas (1) and (5) to yield

$$f'(x) = e^x D_x(\ln x) + (D_x e^x)\ln x = \frac{e^x}{x} + e^x \ln x$$

■

The next example illustrates that it is useful to rewrite an expression involving logarithms using the Properties I–V *before* attempting to find the derivative of the expression.

## EXAMPLE 7

Find the derivative of $f(x) = \ln \dfrac{\sqrt{2x^2 + 1}}{x^2(2x^3 + 5)^3}$.

**Solution**

If we were to use formula (6) immediately, we would write

$$g(x) = \dfrac{\sqrt{2x^2 + 1}}{x^2(2x^3 + 5)^3}$$

Then we would have to calculate

$$f'(x) = \dfrac{D_x\left[\dfrac{\sqrt{2x^2 + 1}}{x^2(2x^3 + 5)^3}\right]}{\dfrac{\sqrt{2x^2 + 1}}{x^2(2x^3 + 5)^3}}$$

The derivative in the numerator of this function is rather complicated to find, however, so that we will not complete the problem using this method. Instead, let's begin again and rewrite $f(x)$ using the properties of logarithms.

$$\begin{aligned}
f(x) &= \ln \dfrac{\sqrt{2x^2 + 1}}{x^2(2x^3 + 5)^3} \\
&= \ln[(2x^2 + 1)^{1/2}] - \ln x^2 - \ln[(2x^3 + 5)^3] \\
&= \tfrac{1}{2}\ln(2x^2 + 1) - 2\ln x - 3\ln(2x^3 + 5)
\end{aligned}$$

We can now apply formulas (5) and (6) to each of the three natural logarithm expressions on the right-hand side to get

$$\begin{aligned}
f'(x) &= \dfrac{D_x(2x^2 + 1)}{2(2x^2 + 1)} - \dfrac{2}{x} - \dfrac{3D_x(2x^3 + 5)}{2x^3 + 5} \\
&= \dfrac{2x}{2x^2 + 1} - \dfrac{2}{x} - \dfrac{18x^2}{2x^3 + 5}
\end{aligned}$$

∎

---

## EXAMPLE 8 (Participative)

Let's sketch the graph of $f(x) = x \ln x$.

**Solution**

First, we note that the domain of $f(x)$ is all positive numbers. Next, we find the derivatives:

$f'(x) = \underline{\phantom{xxxxxx}}$

$f''(x) = \underline{\phantom{xxx}}$.

Note that $f(x)$ has a first-order critical number when $1 + \ln x = 0$ or $\ln x = -1$. Thus $x = \underline{\phantom{xxx}}$ is a critical number.

Now construct a sign diagram of $f'$, choosing your own test values and remembering that the domain of $f(x)$ is $x > 0$. From this we learn that $f(x)$ is $\underline{\phantom{xxxxxxxxxxx}}$ on $(0, 1/e)$ and $\underline{\phantom{xxxxxxxxxxx}}$ on $(1/e, \infty)$, so

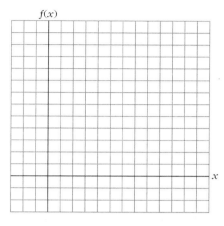

**FIGURE 5.8**

that the point $(1/e, f(1/e)) = (1/e, -1/e)$ is a relative _____ point of the graph of $f(x)$.

Since $f''(x) = 1/x$ and $x > 0$, we know that $f''(x)$ is always positive on the domain of $f$, so that the graph of $f(x)$ is concave _____ on $(0, \infty)$. Sketch the graph of $f(x)$ on the coordinate axes provided here. ∎

### The Derivative of the Logarithmic Function $\log_b x$

Recall from Section 5.2 that

$$\log_b x = \frac{\ln x}{\ln b}$$

Thus we may write

$$D_x(\log_b x) = D_x\left[\frac{\ln x}{\ln b}\right]$$

$$= \frac{1}{\ln b} D_x(\ln x) \quad \text{(Note that } \ln b \text{ is a constant here)}$$

$$= \frac{1}{\ln b} \cdot \frac{1}{x}$$

A more general formula is obtained by using the chain rule and the preceding result:

$$D_x[\log_b g(x)] = \frac{1}{\ln b} \cdot \frac{g'(x)}{g(x)}$$

**SOLUTIONS TO EXAMPLE 8**   $1 + \ln x$, $1/x$, $1/e$, decreasing, increasing, minimum, up

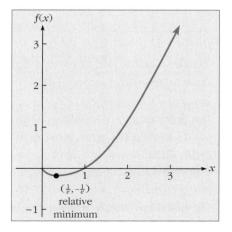

**FIGURE 5.9** The graph of $f(x) = x \ln x$.

Summarizing, if $g(x)$ is a differentiable function of $x$, then

$$D_x(\log_b x) = \frac{1}{\ln b} \cdot \frac{1}{x} \qquad (7)$$

$$D_x[\log_b g(x)] = \frac{1}{\ln b} \cdot \frac{g'(x)}{g(x)} \quad \text{if } g(x) \neq 0 \qquad (8)$$

Note that if we let $b = e$ in formula (8), we obtain formula (6), since $\log_e g(x) = \ln g(x)$ and $\ln e = 1$.

## EXAMPLE 9

Find the derivative of $f(x) = \log_2 \dfrac{x^3 + 3x}{\sqrt{x^2 + 1}}$.

### Solution

We begin by using the properties of logarithms to rewrite $f(x)$ as

$$f(x) = \log_2(x^3 + 3x) - \frac{1}{2} \log_2(x^2 + 1)$$

Using formula (8) on each of the two logarithmic expressions on the right-hand side yields

$$f'(x) = \frac{1}{\ln 2} \cdot \frac{D_x(x^3 + 3x)}{x^3 + 3x} - \frac{1}{2 \ln 2} \cdot \frac{D_x(x^2 + 1)}{x^2 + 1}$$

$$= \left(\frac{1}{\ln 2}\right)\left(\frac{3x^2 + 3}{x^3 + 3x}\right) - \left(\frac{1}{\ln 2}\right)\left(\frac{x}{x^2 + 1}\right) \quad \blacksquare$$

## SECTION 5.3
## SHORT ANSWER EXERCISES

1. (True or False) The derivative of $f(x) = 10e^x$ is equal to itself. _____
2. (True or False) The derivative of $g(x) = e^{10x}$ is equal to itself. _____
3. The slope of the tangent line to the graph of $f(x) = e^x$ at the point $(0, 1)$ is equal to _____.
4. The function $f(x) = 2^x$ has no relative extrema since its derivative is $f'(x) = $ _____, which is (positive, negative) for all $x$.
5. If $f(x) = \ln x$, then $f'(x) = 1/x$ and $f''(x) = $ _____. Since $f''(x)$ is (positive, negative) for all $x > 0$, the graph of $f(x) = \ln x$ is concave _____.
6. (True or False) The derivative of $y = \ln g(x)$ never contains any logarithms. _____
7. The derivative of $y = \ln(\ln x)$ is equal to (a) $1/(\ln x)$; (b) $(2 \ln x)/x$; (c) $1/(x \ln x)$; (d) none of the above.
8. *Before* finding the derivative of $f(x) = \ln\sqrt{x}/[(x + 1)^2(x^2 + 2)^3]$ we should write it as $f(x) = $ _____ using the properties of logarithms.
9. (True or False) The derivative of $\log_2 x$ is $1/x$. _____
10. The derivative of $e^{x^2}$ is (a) $x^2 e^{x^2-1}$; (b) $2xe^{x^2-1}$; (c) $2xe^{x^2}$; (d) none of the above.

## SECTION 5.3
## EXERCISES

In Exercises 1–66, find the derivatives of the given functions.

1. $f(x) = 3e^x$
2. $f(x) = -5e^x$
3. $f(x) = e^{4x}$
4. $f(x) = 2e^{3x}$
5. $f(x) = 7e^{-5x}$
6. $f(x) = 4e^{0.03x}$
7. $f(x) = 6e^{-0.22x}$
8. $f(x) = e^{x^2}$
9. $f(x) = e^{2x-x^2}$
10. $f(x) = e^{2-x^3}$
11. $f(x) = -2e^{\sqrt{x}}$
12. $f(x) = xe^{-x^2}$
13. $f(x) = 3xe^{-4x}$
14. $f(x) = x^2 + x^3 e^{-2x}$
15. $f(x) = \frac{1}{2}e^x + \frac{1}{2}e^{-x}$
16. $f(x) = e^{\sqrt{x+3}} - e^{x^2}$
17. $f(x) = 5^{2x}$
18. $f(x) = 4^{-x^2}$
19. $f(x) = x3^x$
20. $f(x) = 7^{3x^2-7x+1}$
21. $f(x) = 2^x/(5x)$
22. $f(x) = -4x^2 6^{-x^2}$
23. $f(x) = 2^x e^{x^2}$
24. $f(x) = 3^{-x} e^{-x^2}$
25. $f(x) = 1000(1.02)^x$
26. $f(x) = 5000(1.01)^{12x}$
27. $f(x) = 5 \ln x$
28. $f(x) = 6 \ln(x + 1)$
29. $f(x) = \ln(3x + 4)$
30. $f(x) = -2 \ln(2x^2 - 5)$
31. $f(x) = -7 \ln(x^3 + 7x)$
32. $f(x) = (\ln x)^2$
33. $f(x) = \ln x^2$
34. $f(x) = \ln \sqrt{x}$
35. $f(x) = \sqrt{\ln x}$
36. $f(x) = \ln(\ln x)$
37. $f(x) = \ln(5 \ln x)$
38. $f(x) = x^2 \ln x$
39. $f(x) = \ln \sqrt{x^2 + 1}$
40. $f(x) = \sqrt{\ln(x^2 + 1)}$
41. $f(x) = (\ln x)/x$
42. $f(x) = \ln \sqrt{x^3 + 3x}$
43. $f(x) = \ln \sqrt[3]{x^2 + 16}$
44. $f(x) = \ln [3x/(x^2 + 4)]$
45. $f(x) = \ln \dfrac{7x + 3}{2x + 5}$
46. $f(x) = \ln \dfrac{x^2 + 1}{x}$
47. $f(x) = \ln \sqrt{\dfrac{x + 2}{x - 2}}$
48. $f(x) = \ln \sqrt[3]{\dfrac{2x + 1}{x + 3}}$
49. $f(x) = \ln(2x + \sqrt{x^2 + 1})$
50. $f(x) = \ln(x + \sqrt{x - 4})$
51. $f(x) = \ln \dfrac{2x + 3}{\sqrt{x} \sqrt[3]{2x + 1}}$
52. $f(x) = \ln \dfrac{x^2 + 5}{\sqrt{x + 1} \sqrt[3]{x^2 + 9}}$
53. $f(x) = \ln \dfrac{\sqrt{x^2 + 2x}}{(x + 1)^3 (2x - 1)^4}$
54. $f(x) = \ln \dfrac{\sqrt[3]{2x + 3}}{\sqrt{x}(2x - 5)^4}$
55. $f(x) = 6 \log_3 x^2$
56. $f(x) = x \log_7 x$
57. $f(x) = x^2 \log x$
58. $f(x) = \log(\sqrt{x} + 2x^2)$
59. $f(x) = \log_5(x^3 + \sqrt[3]{x})$
60. $f(x) = (\log_5 x)/x^2$
61. $f(x) = \log \dfrac{2x + 4}{x - 5}$
62. $f(x) = \log_8 \sqrt{\dfrac{x^2 - 7}{2x + 7}}$
63. $f(x) = xe^x \ln x$
64. $f(x) = \dfrac{\ln x}{e^x - e^{-x}}$
65. $f(x) = \dfrac{\ln \sqrt{x}}{e^x + e^{-x}}$
66. $f(x) = 2^x \ln(2x + 3)$

*In each of Exercises 67–70, find the slope of the tangent line to the graph of the function at the point with given x-coordinate.*

**67.** $f(x) = x^3 e^{3x}$, $x = 2$

**68.** $f(x) = e^x \ln(x + 5)$, $x = 0$

**69.** $f(x) = 3^x \log_3(4x^2 + 10)$, $x = -3$

**70.** $f(x) = e^{\sqrt[3]{x}} + (\ln x)^3$, $x = 1$

*In Exercises 71–78, sketch the graphs of the given functions. Be sure to find relative extrema, intervals of concavity, and points of inflection in each case.*

**71.** $y = (e^x + e^{-x})/2$

**72.** $y = (e^x - e^{-x})/2$

**73.** $y = xe^x$

**74.** $y = (\ln x)/x$

**75.** $y = e^{-x^2}$

**76.** $y = e^x - x + 1$

**77.** $y = x \log x$

**78.** $y = x5^{x-1}$

**79.** (Marginal cost and average cost) The monthly cost function of a firm is $C(x) = \ln(2x^2 + 15x) + 1200$, where $C(x)$ is the cost in dollars of producing $x$ units.
  (a) Find the marginal cost function.
  (b) Find the average cost function $\bar{C}(x)$.
  (c) Find the marginal average cost function $\bar{C}'(x)$.

**80.** (Marginal analysis) The monthly cost and revenue functions of a certain firm are given by $C(x) = \ln(x^3 + x + 7)$ and $R(x) = \ln(500x^2 + 1)$, where $C(x)$ and $R(x)$ are the cost and revenue (in hundreds of dollars) of producing and selling $x$ hundred units. Find the marginal cost, marginal revenue, and marginal profit functions.

**81.** (Production) If the daily output $y$ of a certain product is equal to $y = 400 - 400e^{-0.2x}$ after $x$ days of production, find the rate of change of daily output after seven days and after 14 days.

**82.** (Depreciation) The value in dollars of a machine $t$ years after it is purchased is equal to $V(t) = 50000e^{-0.3t}$. Find the rate at which the value of this machine is decreasing after seven years.

**83.** (Retail sales) Monthly sales in units of one of our new products is projected to be $S(t) = 5000 - 4500e^{-0.2t}$ for the next two years, where $t$ is time in months and $t = 0$ corresponds to the present. Find the rate of change of sales when $t = 6$ and when $t = 12$.

**84.** (Radioactive decay) The amount of a certain radioactive material present at time $t$ is given by $A(t) = 750e^{-0.05t}$ grams, where $t$ is measured in hours. Find the rate of change of the amount present at (a) $t = 1$; (b) $t = 5$; (c) $t = 10$.

**85.** (Minimum average cost) If the monthly cost in dollars of producing $x$ units of a certain product is

$$C(x) = 1000 + 150x - 200 \ln x, \quad x \geq 1$$

find the level of output that yields the minimum average cost and find the minimum average cost per unit.

**86.** (Product reliability) The proportion of a batch of a new type of light bulb that is still operating after $t$ hours of use is given by $P(t) = e^{-0.0015t}$. Find and interpret $P'(500)$.

**87.** (Information retention) The percentage of information about history that a certain student retains $t$ weeks after learning it is given by

$$P(t) = 88e^{-0.3t} + 12, \quad t \geq 0$$

Find the rate at which the percentage is changing two weeks after the material is learned.

**Social Sciences**

88. (Spread of a rumor) The number $N$ of people who have heard a rumor about a prominent citizen $t$ days after the rumor begins is $N(t) = 1000(1 - 0.95e^{-0.9t})$. Find the rate at which the number of people who have heard the rumor is changing one day after it has begun.

**Life Sciences**

89. (Spread of an epidemic) In a certain school, $t$ days after the beginning of a flu epidemic,

$$N(t) = \frac{600}{1 + 60e^{-0.4t}}$$

students will have had the flu. Find the rate at which the number of infected people is changing three days after the beginning of the epidemic.

90. (Elasticity) Given the monthly demand equation

$$p = 150 - 30 \ln(x + 1), \quad 0 \leq x \leq 145$$

where $p$ is price measured in dollars and $x$ is monthly demand in hundreds of units,
(a) find the price elasticity of demand. Use the formula

$$E(x) = -\frac{p}{x}\frac{dp}{dx}$$

(b) Find and interpret $E(10)$.

**Business and Economics**

91. (Sales and advertising) If monthly sales in units of one of our company's products is related to the amount $x$ spent on advertising, in thousands of dollars, according to the formula

$$S(x) = 200 + 25\sqrt{x} \ln(300x + 20), \quad 0 \leq x \leq 25$$

find the rate at which sales are changing when the monthly advertising level is (a) $5000; (b) $20000.

92. (Maximum profit) Given the monthly cost and revenue functions $C(x) = 5x$ and $R(x) = 750 \ln(x + 1)$, where $C(x)$ is the cost in dollars of producing $x$ units per month and $R(x)$ is the monthly revenue derived from the sale of $x$ units, find (a) the production and sales level at which profit is a maximum, and (b) the maximum monthly profit.

93. (Learning) The exponential function $f(x) = a - be^{-cx}$ is used to describe certain types of learning processes, where $a$, $b$, and $c$ are positive constants. In this equation $f(x)$ measures the degree of learning and $x$ represents the number of positive reinforcements to learning. For example, suppose that after $x$ days of experience on the job, an average worker can assemble $f(x) = 150 - 100e^{-0.20x}$ units per day.
(a) Find the number of units assembled by an average worker after ten days of experience.
(b) Find the rate of change of the number of units assembled when $t = 20$ days.

94. Evaluate $\lim\limits_{\Delta x \to 0} \dfrac{e^{\Delta x} - 1}{\Delta x}$ by constructing tables of values for $\Delta x$ and $\dfrac{e^{\Delta x} - 1}{\Delta x}$.

OPTIONAL

GRAPHING CALCULATOR/
COMPUTER EXERCISES

**95.** For the cost function of Exercise 79, estimate the smallest production level for which the average cost is less than $1 per unit.

**96.** For the cost and revenue functions of Exercise 80, estimate the production and sales levels that maximize profit.

**97.** If the monthly profit from the sale of $x$ hundred units of a certain product is $P(x) = x^2 e^{0.0001x} - x e^{0.018x}$, estimate the sales level $x$ that maximizes monthly profit.

## 5.4 EXPONENTIAL GROWTH AND DECAY

Recall that the formula for interest compounded continuously is

$$A(t) = Pe^{rt}$$

where $P$ is the principal (or initial) amount, $r$ is the annual interest rate, $t$ is time in years, and $A(t)$ is the amount after $t$ years. If we differentiate this formula with respect to time $t$, we obtain

$$A'(t) = Pre^{rt} \quad \text{(Note that } r \text{ is a constant)}$$
$$A'(t) = r[Pe^{rt}]$$
$$\frac{dA}{dt} = rA$$

In words, this last result says that the rate of change of the amount is proportional to the amount present at any time, with constant of proportionality $r$, the annual interest rate.

In general, any quantity that grows (or decays) so that its rate of change at time $t$ is proportional to the amount present at time $t$ satisfies the equation

$$\frac{dA}{dt} = kA, \tag{1}$$

where $k$ is a constant. If $k$ is positive, the amount of the quantity is increasing with time, since the rate of change of $A$ is positive. If $k$ is negative, the amount of the quantity is decreasing with time, since the rate of change of $A$ is negative.

Equation (1) is called a *differential equation*, because it involves the *derivative* of an *unknown function* $A(t)$. We will study differential equations in Chapter 9. For now, we will simply state that any solution to (1) is a function of the form

$$A(t) = Ce^{kt} \tag{2}$$

where $C$ is some constant. It is easy to verify that equation (2) satisfies the differential equation (1) since

$$\frac{dA}{dt} = Cke^{kt} = k[Ce^{kt}] = kA$$

Also, if we know the amount $A_0$ of the substance initially present (at time $t = 0$), then

$$A(0) = Ce^0 = C \cdot 1 = A_0$$

and the solution (2) becomes $A(t) = A_0 e^{kt}$.

In the case $k > 0$, we say that the quantity exhibits *unlimited exponential growth*. If $k < 0$, the quantity exhibits *unlimited exponential decay*. When the meaning is clear, we will omit the word "unlimited" in our discussions. We summarize these results.

> A quantity that grows or decays so that $dA/dt$, its rate of change at time $t$, is proportional to $A(t)$, the amount present at time $t$, satisfies the differential equation
>
> $$\frac{dA}{dt} = kA \qquad (1)$$
>
> where $k$ is a constant designated as the *growth constant* if $k > 0$ and the *decay constant* if $k < 0$. The general solution of equation (1) is the function
>
> $$A(t) = Ce^{kt} \qquad (2)$$
>
> If the initial amount $A(0)$ is equal to $A_0$, then
>
> $$A(t) = A_0 e^{kt} \qquad (3)$$

One example of a quantity that exhibits exponential growth is, of course, money on which interest is compounded continuously. In this case the growth constant is the annual interest rate $r$. Other examples of exponential growth include the uninhibited growth of a population (people, bacteria, and so forth) and the growth in sales of a popular new product during its introductory phase. Examples of exponential decay include the decay of a radioactive substance over time or the decay in sales of a product following a termination of its advertising. A sketch of $A(t) = A_0 e^{kt}$ is drawn in Figure 5.10. You are asked to verify the correctness of this sketch in Exercise 26.

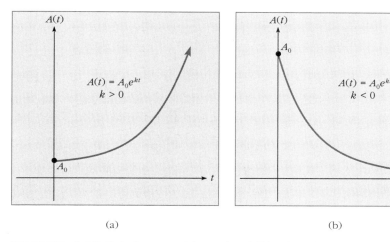

FIGURE 5.10 (a) Unlimited exponential growth and (b) unlimited exponential decay.

Let's look at some examples. First, we consider an example of unlimited exponential growth.

## Exponential Growth (Unlimited)

**EXAMPLE 1**

The population of a large city in the United States is currently 2,500,000 and is growing exponentially at the rate of 2% per year.
(a) Estimate the population of this city in five years.
(b) How long will it take for the population to double to 5,000,000?

**Solution**

(a) The population $A(t)$ of this city in $t$ years is given by $A(t) = A_0 e^{kt}$, where $A_0 = 2,500,000$ and $k = 0.02$, so that

$$A(t) = 2500000 e^{0.02t}$$

In five years, therefore, the population of this city will be

$$A(t) = 2500000 e^{0.02(5)} = 2500000 e^{0.1} \approx 2,762,927$$

(b) Let $T$ be the time required for the population to double. Then $A(T) = 5000000$, so that

$$2500000 e^{0.02T} = 5000000$$
$$e^{0.02T} = 2$$
$$\ln e^{0.02T} = \ln 2 \quad \text{(Applying ln to both sides)}$$
$$0.02T = \ln 2 \quad \text{(Using } \ln e^z = z\text{)}$$
$$T = \frac{\ln 2}{0.02} \approx 34.66 \text{ years}$$

A sketch of $A(t)$ is drawn in Figure 5.11 for reference. In this sketch, the unit of population has been converted to one million. ∎

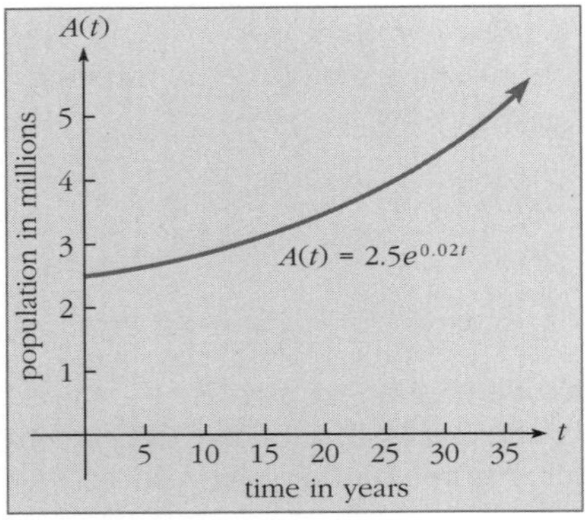

**FIGURE 5.11** The graph of the population function $A(t) = 2.5 e^{0.02t}$.

Part (b) of Example 1 asked us to find the time required for the population to double. In fact, it is not difficult to find the *doubling time* of any quantity that is growing exponentially, and we do so in Example 2.

---

**EXAMPLE 2 (Participative)**

If a quantity is growing exponentially according to $A(t) = A_0 e^{kt}$, let's find the time required for the initial amount $A_0$ of the quantity to double.

**Solution**

Let $T$ be the time required for the initial amount $A_0$ to double. Then $A(T) = 2A_0$, so that

$$A_0 e^{kT} = 2A_0$$
$$e^{kT} = \underline{\phantom{xx}}$$
$$\ln e^{kT} = \ln 2 \quad \text{(Applying ln to both sides)}$$
$$\underline{\phantom{xx}} = \ln 2 \quad \text{(Using } \ln e^z = z\text{)}$$
$$T = \underline{\phantom{xxxx}}$$
∎

---

**SOLUTIONS TO EXAMPLE 2**    2, $kT$, $(\ln 2)/k$

Summarizing the result of Example 2,

> If a quantity is growing exponentially according to $A(t) = A_0 e^{kt}$, where $k > 0$, then the *doubling time* of the quantity is
> $$T_2 = \frac{\ln 2}{k} \qquad (4)$$

More generally, it can be shown (see Exercise 27) that if $N > 1$, the time required for the initial amount $A_0$ to grow to $NA_0$ is

$$T_N = \frac{\ln N}{k} \qquad (5)$$

**EXAMPLE 3** If you invest a sum of money in an account on which interest is compounded continuously,
(a) how long does it take for the principal to double if the annual interest rate is 8.25%?
(b) how long does it take for the principal to quadruple if the annual interest rate is 9.75%?

**Solution** (a) Using formula (4) with $k = 0.0825$, we obtain a doubling time of

$$T_2 = \frac{\ln 2}{0.0825} \approx 8.40 \text{ years}$$

(b) Using formula (5) with $N = 4$ and $k = 0.0975$, we obtain a quadrupling time of

$$T_4 = \frac{\ln 4}{0.0975} \approx 14.22 \text{ years}$$

### Exponential Decay

Let us now consider an example of exponential decay. Specifically, our next example will involve the decay of a radioactive substance. In such an instance, it is important to understand that the *half-life* of a radioactive material is the amount of time required for one-half of the initial amount to decay due to the emission of certain elementary atomic particles.

**EXAMPLE 4** The half-life of radium is approximately 1620 years. If 100 g of radium are stored away today,
(a) how much of this radium will remain after 50 years?
(b) how many years must pass before only 1 g of radium remains?
Assume that radium decays at a rate that is proportional to the quantity present at any time.

**Solution** (a) Let $A(t)$ be the amount of radium present $t$ years from the time the initial amount of 100 g is stored away. Then since, by assumption, $dA/dt = kA$, we have $A(t) = A_0 e^{kt}$, where $A_0 = A(0) = 100$, so that

$$A(t) = 100 e^{kt}$$

To find the decay constant $k$, we use the fact that only 50 g will remain after 1620 years. That is, $A(1620) = 50$. But $A(1620) = 100 e^{1620k}$, so that

$$100 e^{1620k} = 50$$

$$e^{1620k} = \frac{1}{2}$$

$$\ln e^{1620k} = \ln \frac{1}{2}$$

$$1620k = \ln \frac{1}{2}$$

$$k = \frac{\ln \frac{1}{2}}{1620} \approx -0.000428$$

We conclude that

$$A(t) = 100 e^{-0.000428 t}$$

and so

$$A(50) = 100 e^{-0.000428(50)} \approx 97.88 \text{ grams}$$

This is the amount of radium remaining after 50 years.

(b) In order for only 1 g of radium to remain, we require the time $T$ such that $A(T) = 1$. Therefore

$$100 e^{-0.000428 T} = 1$$

$$e^{-0.000428 T} = 0.01$$

$$\ln e^{-0.000428 T} = \ln(0.01)$$

$$-0.000428 T = \ln(0.01)$$

$$T = \frac{\ln(0.01)}{-0.000428} \approx 10760 \text{ years}$$

So approximately 10,760 years must pass before only 1 g of radium is left. A sketch of $A(t)$ is drawn in Figure 5.12 for reference. ∎

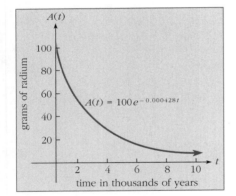

**FIGURE 5.12** The amount of radium present after $t$ years.

Using methods similar to Example 2 (see Exercise 28), it is possible to prove the following summary:

---

If a quantity is decaying exponentially according to $A(t) = A_0 e^{kt}$, $k < 0$, the time required for the initial amount $A_0$ to decay to $A_0/N$, where $N > 1$, is

$$T_{1/N} = \frac{-\ln N}{k} \qquad (6)$$

The half-life of the substance is

$$T_{1/2} = \frac{-\ln 2}{k} \qquad (7)$$

---

## 5.4 EXPONENTIAL GROWTH AND DECAY

**EXAMPLE 5 (Participative)**

Radioactive carbon, $C^{14}$, has a half-life of about 5730 years.
(a) Find the decay constant $k$ for $C^{14}$.
(b) How long will it take 12 g of $C^{14}$ to decay to 4 g?
(c) How long will it take an initial amount $A_0$ of $C^{14}$ to decay to 10% of the initial amount?

**Solution**

(a) Solving formula (7) for $k$, we get $k = $ _____.
Substituting $T_{1/2} = 5730$, we obtain

$$k = -\frac{\ln 2}{5730} \approx \underline{\phantom{xxxxxx}} \quad \text{(3 significant digits)}$$

(b) Since $4 = \frac{1}{3}(12)$, we can use formula (6) with $N = $ \_\_\_\_ to get

$$T_{1/3} = \underline{\phantom{xxxxxxxxxxxxxxxx}} \approx \underline{\phantom{xxxxxx}}$$

(c) Since $10\% A_0 = \frac{1}{10} A_0$, we use formula (6) with $N = $ \_\_\_\_ to get

$$T_{1/10} = \underline{\phantom{xxxxxxxxxxxxxxxx}} \approx \underline{\phantom{xxxxxx}}$$

**EXAMPLE 6**

Living organisms contain two types of carbon, $C^{14}$ and $C^{12}$. The first, $C^{14}$, is radioactive and $C^{12}$ is stable. While an organism is alive, the ratio of $C^{14}$ to $C^{12}$ in an organism is constant and it is approximately equal to the same ratio in the atmosphere. When an organism dies, however, the $C^{14}$ begins to decay exponentially. This fact is used by anthropologists and archaeologists to determine the age of material that was once living. They simply compute the ratio of $C^{14}$ to $C^{12}$ in the material and determine the percentage of the original $C^{14}$ remaining in the material. For this example, if the wooden shaft of an axe is unearthed from an archaeological dig and if 20% of the original amount of $C^{14}$ remains, approximately how old is the axe?

**Solution**

Using the results of Example 5, the amount of $C^{14}$ remaining in the axe $t$ years after cutting down the tree is

$$A(t) = A_0 e^{-0.000121 t}$$

where $A_0$ is the initial amount of $C^{14}$ in the axe. We must determine the time $T$ when $A(T) = 0.20 A_0$:

$$A_0 e^{-0.000121 T} = 0.20 A_0$$
$$e^{-0.000121 T} = 0.20$$
$$\ln(e^{-0.000121 T}) = \ln(0.20)$$
$$-0.000121 T = \ln(0.20)$$
$$T = \frac{\ln(0.20)}{-0.000121} \approx 13{,}301 \text{ years}$$

**SOLUTIONS TO EXAMPLE 5**

(a) $(-\ln 2)/T_{1/2}$, $-0.000121$; (b) 3, $(-\ln 3)/-0.000121$, 9079.44; (c) 10, $(-\ln 10)/-0.000121$, 19029.6

An alternate way to solve this problem is by using formula (6). That is, since $20\% A_0 = \frac{1}{5}A_0$, formula (6) with $N = 5$ gives us

$$T_{1/5} = \frac{-\ln 5}{-0.000121} \approx 13{,}301 \text{ years}$$

### Limited Exponential Growth

There are situations that can be modeled by exponential functions other than the one used for unlimited exponential growth. For example, when a new worker is trained to use a machine, daily output does not increase without bound as the unlimited exponential growth model $A(t) = A_0 e^{kt}$, $k > 0$, would indicate. Rather, it is more realistic to assume that daily output approaches some fixed maximum number $M$. That is, the growth of output is *limited* by $M$. If $A(t)$ is daily output $t$ days after beginning to work on the machine, it is reasonable to assume that the rate of change of output, $dA/dt$, is proportional to the difference between the maximum attainable output $M$ and the current output $A$. That is,

$$\frac{dA}{dt} = k(M - A)$$

It can be shown (see Exercise 36) that a solution of this differential equation is

$$A(t) = M(1 - Ce^{-kt})$$

where $C$ is a constant. We summarize.

> If a quantity $A$ grows so that its rate of change $dA/dt$ is proportional to the difference between a maximum attainable amount $M$ and the amount $A$ at time $t$, then
>
> $$\frac{dA}{dt} = k(M - A), \quad k > 0$$
>
> Such a quantity $A$ is said to exhibit *limited exponential growth* and
>
> $$A(t) = M(1 - Ce^{-kt}) \tag{8}$$
>
> where $C$ is some constant. If $k < 0$, the quantity $A$ exhibits *limited exponential decay*.

A graph of $A(t) = M(1 - Ce^{-kt})$ for $k > 0$ is sketched in Figure 5.13. The details of why this sketch is correct are explored in Exercise 37. This graph is sometimes called a *learning curve* because it is often used by psychologists to model a learning process, where an individual's learning is rapid initially but slows down eventually and approaches a level $M$ beyond which the individual cannot go, no matter how much additional training is given. However, the model of limited exponential growth or decay is used in many other situations as well. For example, it is used by sociologists to model the diffusion of information spreading through a population, so that the curve in Figure 5.13 is also called

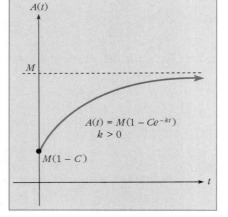

FIGURE 5.13 The graph of the limited exponential growth curve $A(t) = M(1 - Ce^{-kt})$, $k > 0$.

a *diffusion* curve. Other examples of the use of the model of limited exponential growth or decay include the cooling of a warm object or the heating of a cool object to an ambient temperature (that of the surrounding medium), the growth of certain businesses, and the depreciation of certain industrial equipment.

**EXAMPLE 7** A typical new employee with no previous training can assemble about 50 units of a certain product per day and about 100 units after five days of on-the-job training. After many months of on-the-job training, an average employee can assemble about 200 units per day, but additional units beyond 200 per day cannot be produced by the employee. Assuming the model of limited exponential growth applies, (a) find an equation for $A(t)$, the number of units produced by an employee after $t$ days of on-the-job training. (b) How many units can an average employee assemble per day after 14 days of on-the-job training?

**Solution**

(a) In the equation $A(t) = M(1 - Ce^{-kt})$, we have $M = 200$ and so $A(t) = 200(1 - Ce^{-kt})$. Also, $A(0) = 50 = 200(1 - C)$, so that

$$200 - 200C = 50$$
$$C = \tfrac{3}{4}$$

Therefore

$$A(t) = 200\left(1 - \tfrac{3}{4}e^{-kt}\right)$$

To find $k$, we use the fact that $A(5) = 100$. That is,

$$200\left(1 - \tfrac{3}{4}e^{-5k}\right) = 100$$
$$1 - \tfrac{3}{4}e^{-5k} = \tfrac{1}{2}$$
$$e^{-5k} = \tfrac{2}{3}$$
$$-5k = \ln\left(\tfrac{2}{3}\right)$$
$$k = \frac{\ln\left(\tfrac{2}{3}\right)}{-5} \approx 0.08109$$

We conclude that

$$A(t) = 200\left(1 - \tfrac{3}{4}e^{-0.08109t}\right)$$

(b) $A(14) = 200\left(1 - \tfrac{3}{4}e^{-0.08109(14)}\right) \approx 152$ units

So an average employee can assemble about 152 units after 14 days of on-the-job training. ∎

### Logistic Growth

Another type of growth that can be modeled by an exponential function is called *logistic growth*. As an example, consider a culture of bacteria growing in a laboratory experiment. Although in the short run the growth of the bacteria may be modeled well by unlimited exponential growth, it is clear that after some time, this growth will begin to slow down because of lack of food and overcrowding. Indeed, eventually the growth will become very slow as the population approaches the maximum $M$ that the surrounding environment will support. This slow-down in growth is not taken into account by the model of unlimited exponential growth. To correctly model this situation, we should observe that if $A(t)$ is the number of bacteria present at time $t$, then $dA/dt$ (the rate of growth of the bacteria) is proportional not only to the size $A$ of the population, but also to the unused capacity for growth, $M - A$. The required differential equation is therefore

$$\frac{dA}{dt} = kA(M - A), \quad \text{where } k > 0$$

In Exercise 44 you are asked to show that a solution to this differential equation is

$$A(t) = \frac{M}{1 + Ce^{-kMt}}$$

where $C$ is a constant. We summarize:

> If a quantity $A$ grows so that its rate of change $dA/dt$ is proportional to the product of its current size $A$ and the difference between a maximum attainable amount $M$ and the amount $A$ at time $t$, then
>
> $$\frac{dA}{dt} = kA(M - A), \quad k > 0$$
>
> Such a quantity is said to exhibit *logistic growth* and
>
> $$A(t) = \frac{M}{1 + Ce^{-kMt}} \qquad (9)$$
>
> where $C$ is some constant. The number $M$ is called the *carrying capacity* of the growth system.

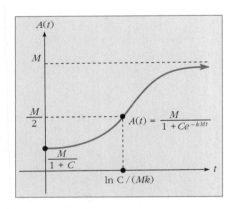

**FIGURE 5.14** The graph of the logistic growth curve $A(t) = M/(1 + Ce^{-kMt})$.

A graph of $A(t) = M/(1 + Ce^{-kMt})$ is sketched in Figure 5.14. You are asked to verify the details of why this S-shaped curve is the correct sketch in Exercise 52. In addition to modeling long-term population growth, logistic growth can be used to model the number of people infected with a disease during an epidemic, the number of people who have heard a certain rumor, and the amount of sales of some new products. Let's look at an example of logistic growth.

**EXAMPLE 8**

The marketing department of a magazine company predicts that subscriptions to a certain new magazine will eventually reach 500,000. When the magazine is introduced, the subscription level is expected to be 10,000 and is expected to rise to 50,000 after four months. Assuming that the growth in subscriptions for this magazine is logistic, (a) find the function $A(t)$ that represents the number of subscribers to this magazine after $t$ months. (b) After how many months will there be a quarter of a million subscribers to the magazine?

**Solution**

(a) We use the logistic growth equation $A(t) = M/(1 + Ce^{-kMt})$ with $M = 500{,}000$. Note that $A(0) = M/(1 + C) = 10000$, so that $500000/(1 + C) = 10000$. Solving for $C$, we get $C = 49$. Also, since $A(4) = 50000$, we get

$$\frac{500000}{1 + 49e^{-k(500000)(4)}} = 50000$$

$$50000 + 2450000e^{-2000000k} = 500000$$

$$e^{-2000000k} = \frac{9}{49}$$

$$-2000000k = \ln\left(\frac{9}{49}\right)$$

It is convenient to solve this equation for $-Mk = -500000k$; to do this, we divide each side by 4, obtaining

$$-Mk = -500000k = \frac{\ln\left(\frac{9}{49}\right)}{4} \approx -0.423649$$

Therefore

$$A(t) = \frac{500000}{1 + 49e^{-0.423649t}}$$

which is the number of subscribers after $t$ months.

(b) We must find the time $T$ for which $A(T) = 250{,}000$. That is, we must find $T$ so that

$$\frac{500000}{1 + 49e^{-0.423649T}} = 250000$$

Solving for $T$, we obtain

$$1 + 49e^{-0.423649T} = 2$$

$$e^{-0.423649T} = \frac{1}{49}$$

$$-0.423649T = \ln\left(\frac{1}{49}\right)$$

$$T = \frac{\ln\left(\frac{1}{49}\right)}{-0.423649} \approx 9.2 \text{ months}$$

Thus there will be a quarter million subscribers to the magazine after about 9.2 months. ∎

## SECTION 5.4
### SHORT ANSWER EXERCISES

1. A quantity $Q$ satisfies the relationship $dQ/dt = 0.05Q$; we conclude that $Q(t) = $ _____.

2. (True or False) If a quantity is growing exponentially with a growth constant 0.10 and if time is measured in years, the amount of time it takes for the quantity to double is 10 years. _____

3. (True or False) If it takes four years for a quantity growing exponentially to double, it will take six years for it to triple. _____

4. (True or False) If the half-life of a certain radioactive substance is 30 years, a quantity of this substance will decay completely in 60 years. _____

5. After a living organism dies, the ratio of $C^{14}$ to $C^{12}$ contained in the organism (a) decreases; (b) increases; (c) remains constant.

6. (True of False) Money growing exponentially because of continuous compounding of interest at a fixed yearly rate $r$ satisfies the equation of limited exponential growth. _____

7. The graph of the function $A(t) = M(1 - Ce^{-kt})$ for $k > 0$ (a) has no horizontal asymptote; (b) is characterized by its typical S-shape; (c) has a point of inflection; (d) none of the above.

8. (True or False) If a quantity $Q$ exhibits limited exponential growth, the amount of the quantity present will never exceed a certain maximum amount $M$. _____

9. The graph of a logistic growth function $A(t)$ (a) has a horizontal asymptote; (b) has one point of inflection; (c) is characterized by its typical S-shape; (d) all of the above.

10. A rumor has begun to spread that a candidate for political office once cheated a poor old widow out of her home. The number of people who have heard this rumor can be described by (a) an exponential growth function; (b) an exponential decay function; (c) a logistic growth function; (d) none of the above.

## SECTION 5.4
## EXERCISES

*In Exercises 1–6, determine the solution $A(t)$ of each differential equation with the given initial condition.*

1. $dA/dt = 0.01A$, where $A(0) = 100$
2. $dA/dt = -0.025A$, where $A(0) = 50$
3. $dA/dt = -0.003(500 - A)$, where $A(0) = 50$
4. $dA/dt = 10 - 0.01A$, where $A(0) = 20$
5. $dA/dt = 0.002A(400 - A)$, where $A(0) = 100$
6. $dA/dt = 2.5A - 0.05A^2$, where $A(0) = 5$

*In Exercises 7–28, assume that the model of* unlimited *exponential growth or decay is valid.*

7. (Asset appreciation) An investment property was purchased in 1989 for $50,000. If the value appreciates at 14% compounded continuously, how much will the property be worth in 2001?

8. (Annual sales) The annual sales of a new company were $75,000 in 1989 and $150,000 in 1990. What are predicted annual sales for 1992?

9. (Continuous compounding) If you invest a sum of money in an account in which interest is compounded continuously, how long does it take for the principal to (a) double if the annual interest rate is 9.5%? (b) triple if the annual interest rate is 8.75%? (c) quadruple if the annual interest rate is 11.25%?

10. (Interest rates) For a certain account in which interest is compounded continuously, find the annual interest rate $r$ if (a) the principal doubles in 7.5 years; (b) the principal triples in 12 years; (c) the principal quadruples in 12.5 years.

11. (Present value) If in five years you need to have $25,000, how much money should you place in an investment today that pays 11.50% interest compounded continuously?

12. (Predicting profit) The annual profits of a "hot" new company were $250,000 in 1989 and $600,000 in 1990. What are predicted annual profits for 1991?

13. (Predicting demand) The demand for one of our company's products is 8000 units per month when the price is $5/unit and 10,000 units per month when the price is $4/unit.
    (a) Find the demand function $x = f(p)$, where $x$ is monthly demand and $p$ is price per unit.
    (b) What will monthly demand be if we raise the price to $6/unit?

14. (Predicting value) A piece of industrial equipment was purchased for $150,000 in 1990. In 1992 its value is $115,000.
    (a) Predict the value of the equipment in 1995.
    (b) When will the value be equal to $20,000?

15. (Population) The population of a certain country is growing at 2.5% per year compounded continuously. If its current population is 25 million, what will its population be in 10 years?

16. (Population) The population of a certain city was 2.5 million in 1988 and 3.0 million in 1990. Estimate the population of this city in 1994.

17. (Advertising and sales) One month after advertising of a certain product was discontinued, sales fell from 20,000 units per month to 18,000 units per month. How many months after the discontinuance of advertising will sales be 5,000 units per month?

18. (Bacteria growth) If the number of bacteria in a culture doubles in two hours, how many hours will it take for the number of bacteria to triple?

19. (Radioactive decay) The half-life of barium 140 is about 13 days. If 50 g of barium 140 are present today, (a) how much of this amount will remain after 20 days? (b) How many days must pass before only 2 g of barium 140 remain?

20. (Radioactive decay) The half-life of flourine 17 is about 66 sec. How long will it take an initial amount of flourine 17 to decay to 1% of the initial amount?

21. (Radioactive decay) If 800 g of a certain radioactive substance is present initially and if 600 g are present 20 years later, how many grams will remain after 100 years?

22. (Safe level of radioactivity) Strontium 90 has a half-life of about 29 years. If a certain area is contaminated with five times the safe level of strontium 90, after how many years will the level be safe, assuming normal radioactive decay?

23. (Carbon dating) A sample of wood found in an excavation has 65% of its original $C^{14}$ remaining. How old is the wood?

24. (Carbon dating) A fossil is found to contain 20% of its original carbon 14. How old is the fossil?

25. (Carbon dating) A specimen that is 25,000 years old contains what percentage of its original carbon 14?

26. Using the technique of calculus, find $A'(t)$ and $A''(t)$ and verify that the sketches drawn in Figure 5.10 are correct for $A(t) = A_0 e^{kt}$.

27. If a quantity grows according to $A(t) = A_0 e^{kt}$, $k > 0$, find the time required for the initial amount of the quantity $A_0$ to grow to $NA_0$, where $N > 0$.

28. If a quantity is decaying according to $A(t) = A_0 e^{kt}$, $k < 0$, find the time required for the initial amount $A_0$ to decay to $A_0/N$, where $N > 1$.

*In Exercises 29–37, assume that the model of limited exponential growth or decay is valid.*

**Business and Economics**

29. (Product awareness) A company begins an advertising campaign for a new product. At the beginning of the campaign, no one in the target market is aware of the product. After five weeks of advertising, 60% of the target market knows of the product. If the target market for this product contains 40 million people, (a) find the formula for $A(t)$, the number of people who are aware of the product after $t$ weeks of advertising. (b) What percentage of the target market will have heard of the product after 12 weeks of advertising?

30. (Learning) Tom Tapp is taking a typing class. Before he begins the class he can type about 15 words per minute. After ten days of class he is able to type 30 words per minute. If the practical limit for the best students ever to complete this course is 95 words per minute, (a) find a formula for $A(t)$, the number of words per minute Tom will be able to type after $t$ days of class. (b) If the class consists of 30 days of instruction, how many words per minute will Tom be able to type when he completes the course?

31. (Employee learning) A new employee with no experience can fasten the casings on about 100 air conditioning units per day. After four weeks of experience he is able to fasten the casings on about 250 air conditioning units per day. If the most experienced workers in the plant can fasten casings at the rate of 350 per day, (a) find the formula for $A(t)$, the number of casings fastened by a typical new worker after $t$ weeks of experience. (b) After how many weeks will an average worker be able to fasten 300 casings per day?

32. (Spread of a rumor) A small university needs a new athletic center. The number of students who have heard that funds have been approved is given by

$$A(t) = 5500(1 - Ce^{-kt})$$

If none of the students knows of the decision before it is announced and if 500 students know of the decision one hour after it is announced, when will 5000 students know of the decision?

33. (Newton's Law of Cooling) A can of beer at 42°F is removed from a cooler at the beach. After five minutes the temperature of the can is 60°F. If the temperature outdoors is 82°F, find the temperature of the beer after ten minutes.

34. (Newton's Law of Cooling) Water that was boiling at 212°F has a temperature of 160°F after standing for four minutes. If the temperature of the room is 75°F, when will the temperature of the water be 100°F? *Hint:* This is an example of limited exponential *decay*, so that $k < 0$ in formula (8).

35. (Depreciation) A piece of industrial equipment is purchased for $75,000 and is valued at $65,000 after one year of use. If its salvage value will never fall below $10,000, what is the value of the equipment after five years of use?

36. Show that the function $A(t) = M(1 - Ce^{-kt})$ is a solution of the differential equation

$$\frac{dA}{dt} = k(M - A)$$

37. Consider the function $A(t) = M(1 - Ce^{-kt})$ for $k > 0$.
    (a) Show that the vertical intercept of the graph is $M(1 - C)$.
    (b) Find $\lim_{t \to \infty} A(t)$.
    (c) Compute $A'(t)$ and show that $A(t)$ is increasing on $(0, \infty)$.
    (d) Compute $A''(t)$ and show that $A(t)$ is concave down on $(0, \infty)$.

*In Exercises 38–52 assume that the model of logistic growth is valid.*

**Life Sciences**

38. (Trout population) One year after the introduction of 1,000 trout into a lake, the population has grown to 2,500 trout. Assuming that the lake can support at most

Social Sciences

20,000 trout and that fishing for trout will be illegal until the lake contains at least 18,000 trout, how many years after the initial introduction will trout-fishing be allowed?

39. **(Bacteria growth)** Bacteria is growing in a medium that can support no more than 15 million bacteria. Initially one million bacteria are present and, two hours later, two million are present.
   (a) Find a formula for $A(t)$, the number of bacteria present after $t$ hours.
   (b) How many bacteria will be present after seven hours?

40. **(Spread of a rumor)** A rumor is spreading through a small town of 15,000 people. Initially only the person who started the rumor knew about it. Five days after the rumor began, 200 people have heard it.
   (a) Find a formula for $A(t)$, the number of people who have heard the rumor $t$ days after it starts.
   (b) After how many days will 5000 people have heard the rumor?

41. **(Spread of an epidemic)** In a certain high school, $t$ days after the beginning of a flu epidemic,

$$A(t) = \frac{500}{1 + 50e^{-0.5t}}$$

   students will have had the flu.
   (a) Find the number of students who will have had the flu after 10 days.
   (b) When will the number of infected students reach 250?

42. **(Spread of a disease)** An infectious disease is spreading through a city of 100,000 people. When the disease is first detected, ten cases are identified. A week later, 300 cases have been identified.
   (a) Find a formula for $A(t)$, the number of people who have been infected after $t$ weeks.
   (b) When will 20% of the city's population have been infected?

43. **(Homeless people)** The number of homeless people in a certain area of a large city was estimated at 3000 in 1989 and 3500 in 1990. If social workers estimate that this area can support at most 10,000 homeless people, predict the number of homeless people in this area in 1995.

44. Show that the logistic function $A(t) = M/(1 + Ce^{-kMt})$ is a solution of the differential equation

$$\frac{dA}{dt} = kA(M - A), \quad k > 0$$

45. Show that the maximum rate of growth of the logistic function

$$A(t) = \frac{M}{1 + Ce^{-kMt}}$$

   occurs at $t = (\ln C)/(Mk)$ when the amount $A(t)$ is equal to $M/2$. Also show that this is the point of inflection of the logistic function.

46–51. For each of the logistic functions developed in Exercises 38–43, determine the time when the rate of growth $dA/dt$ is a maximum.

52. Consider the function $A(t) = M/(1 + Ce^{-kMt})$ for $k > 0$.
   (a) Show that the vertical intercept of the graph is $M/(1 + C)$.
   (b) Find $\lim_{t \to \infty} A(t)$.
   (c) Compute $A'(t)$ and show that $A(t)$ is increasing on $(0, \infty)$.

(d) Compute $A''(t)$ and determine the intervals of concavity and the point of inflection of $A(t)$.

**53.** (Ebbinghaus Model) One model of forgetting that uses an exponential function was suggested by the psychologist Ebbinghaus. If $A(t)$ is the percentage of information retained by an individual $t$ units of time after it is learned, then

$$A(t) = (100 - C)e^{-kt} + C, \quad \text{where } 0 < C < 100 \text{ and } k > 0$$

(a) Sketch the graph of the *forgetting curve* for an individual:

$$A(t) = 70e^{-0.4t} + 30, \quad t \geq 0$$

(b) If $t$ is time measured in weeks, what percentage of originally learned information is retained by this individual after six weeks?

# 5.5 CHAPTER FIVE REVIEW

**KEY TERMS**

exponential function
one-to-one
simple interest
compound interest
continuously compounded interest
future value
present value
logarithmic function
common logarithm
natural logarithm
inverse functions

differential equation
unlimited exponential growth
unlimited exponential decay
growth constant
decay constant
doubling time
half-life
limited exponential growth
limited exponential decay
logistic growth
carrying capacity

**KEY FORMULAS AND RESULTS**

**Definition**

$$a^n = \underbrace{a \cdot a \cdots a}_{n \text{ factors}}$$

$a^0 = 1$
$a^{1/n} = \sqrt[n]{a}$
$a^{m/n} = (\sqrt[n]{a})^m$
$a^{-n} = 1/a^n$

**Laws of Exponents**

1. $a^x a^y = a^{x+y}$
2. $\dfrac{a^x}{a^y} = a^{x-y}$
3. $(a^x)^y = a^{xy}$
4. $(ab)^x = a^x b^x$
5. $\left(\dfrac{a}{b}\right)^x = \dfrac{a^x}{b^x}$

### An Exponential Function with Base b

$f(x) = b^x$

### The Number e

$$e = \lim_{x \to \infty} \left(1 + \frac{1}{x}\right)^x$$

$e \approx 2.718281828$

### One-To-One

A function $f(x)$ is one-to-one if either **(a)** $f(n) = f(m)$ implies $n = m$ or **(b)** any horizontal line intersects the graph of $f(x)$ in at most one point.

### Simple Interest Formula

$A(t) = P(1 + rt)$

where $P$ = principal, $r$ = annual interest rate, $t$ = time in years, and $A$ = amount accumulated after $t$ years

### Compound Interest Formula

$$A(t) = P\left(1 + \frac{r}{k}\right)^{kt}$$

where $k$ = the number of compounding periods per year

### Continuously Compounded Interest Formula

$A(t) = Pe^{rt}$

### Logarithmic Function with Base b

$y = \log_b x$ if and only if $x = b^y$

Natural Logarithm $\quad \log_e y = \ln y$
Common Logarithm $\quad \log_{10} y = \log y$

### ln x and e^x Are Inverse Functions

$e^{\ln x} = x \quad$ for $x > 0$

$\ln e^x = x \quad$ for all $x$

### Properties of Logarithms

I. $\log_b 1 = 0$
II. $\log_b b = 1$
III. $\log_b xy = \log_b x + \log_b y$

IV. $\log_b \dfrac{x}{y} = \log_b x - \log_b y$

V. $\log_b x^y = y \log_b x$

**Change-of-Base Formula for Logarithms**

$$\log_b x = \dfrac{\log_k x}{\log_k b}$$

**Differentiation Formulas**

$D_x e^x = e^x$

$D_x e^{g(x)} = g'(x) e^{g(x)}$

$D_x b^x = (\ln b) b^x$

$D_x b^{g(x)} = (\ln b) g'(x) b^{g(x)}$

$D_x (\ln x) = 1/x$

$D_x [\ln g(x)] = \dfrac{g'(x)}{g(x)}$  if $g(x) \ne 0$

$D_x (\log_b x) = \dfrac{1}{\ln b} \cdot \dfrac{1}{x}$

$D_x [\log_b g(x)] = \dfrac{1}{\ln b} \cdot \dfrac{g'(x)}{g(x)}$  if $g(x) \ne 0$

**Unlimited Exponential Growth or Decay**

If a function $A(t)$ satisfies $dA/dt = kA$, then $A(t) = A_0 e^{kt}$, where $A_0 = A(0)$ and $k$ is the growth constant if $k > 0$ or the decay constant if $k < 0$.

Doubling Time for Unlimited Exponential Growth:   $T_2 = (\ln 2)/k$

Half-Life for Unlimited Exponential Decay:   $T_{1/2} = (-\ln 2)/k$

**Limited Exponential Growth or Decay**

If a function $A(t)$ satisfies $dA/dt = k(M - A)$, then $A(t) = M(1 - Ce^{-kt})$, where $C$ is a constant.

**Logistic Growth**

If a function $A(t)$ satisfies $dA/dt = kA(M - A)$, then $A(t) = M/(1 + Ce^{-kMt})$, where $C$ is a constant and $M$ is the carrying capacity.

**CHAPTER 5  SHORT ANSWER REVIEW QUESTIONS**

1. (True or False) $2^x \le 3^x$ for $x$ in the interval $(-\infty, 0)$. _____

2. To five-decimal-place accuracy, $5^{\sqrt{7}} = $ _____.

3. (True or False) If $b > 0$, $b \ne 1$, the graph of $y = b^x$ is concave up. _____

4. The number $e$ is defined as $\lim\limits_{x \to \infty}$ _____.

5. (True or False) If $b > 1$, the function $g(x) = \log_b x$ is one-to-one. _____

6. If $2000 is invested in an account that yields 12% interest compounded monthly, the formula for the amount in this account after $t$ years is $A(t) = $ _____.

7. The value of $\log_8 4$ is (a) $\frac{2}{3}$; (b) $\frac{3}{2}$; (c) $\frac{1}{2}$; (d) 2.

8. (True or False) The expression $\log_b x - \log_b y$ may be simplified to $\log_b(x - y)$. _____

9. If $x \neq 0$, then $\log_b 3 + \log_b x^2 - \log_b x$ is equal to (a) $\log_b(3 + x^2 - x)$; (b) $\log_b(3x^2 - x)$; (c) $\log_b 3x$; (d) none of the above.

10. The derivative with respect to $x$ of $f(x) = e^2$ is (a) $2e^2$; (b) $2e^1$; (c) 0; (d) none of the above.

11. If $f(x) = e^{\ln x}$ and $x \neq 0$, then $f'(x)$ is equal to (a) $(1/x)e^{\ln x}$; (b) $[D_x(\ln x)]e^{\ln x}$; (c) 1; (d) all of the above.

12. A function $N(t)$ satisfies the differential equation $dN/dt = 0.03N$ and also satisfies $N(0) = 200$. The formula for $N(t)$ is therefore $N(t) = $ _____.

13. The rate of change of a function $A(t)$ is proportional to $(10000 - A)$. This function exhibits (a) unlimited exponential growth; (b) unlimited exponential decay; (c) logistic growth; (d) none of the above.

14. If a radioactive substance has a half-life of 50 years then (a) it will double in 100 years; (b) it will double in 200 years; (c) the substance is undergoing radioactive decay; (d) none of the above.

15. A common feature shared by the graphs of all the growth functions considered in Section 5.4 is that: (a) they are all concave up; (b) they all have a horizontal asymptote; (c) they are all increasing on $(0, \infty)$; (d) they all have a $t$-intercept of zero.

## CHAPTER 5 REVIEW EXERCISES

*Exercises marked with an asterisk (\*) constitute a Chapter Test.*

*In Exercises 1–4, simplify the given expressions.*

1. $8^{2/3}$
2. $\left(\frac{5}{9}\right)^3 \left(\frac{9}{5}\right)^{-1}$
3. $(7^{1/5})(49^{2/5})$
4. $(81x^4 y^8)^{3/4}$

*In Exercises 5–8, graph the given functions.*

\*5. $f(x) = 2^{-3x}$
6. $g(x) = e^{-0.5x}$
7. $f(x) = -2e^{(1/4)x}$
8. $g(x) = 4(5^{0.2x})$

9. (Interest) If you invest $10,000 today at an annual rate of 7%, how much will you have in five years if interest is compounded (a) annually? (b) quarterly? (c) monthly? (d) continuously?

10. (Continuous interest) If $5000 is invested in an account that pays an annual interest rate of 8% compounded continuously, after how many years will the account contain $12,500?

*In each of Exercises 11–14, evaluate the given expression.*

11. $\log_9 81$
12. $\log_3 81$
\*13. $\log_8 2$
14. $\log_3 \sqrt[3]{3}$

In Exercises 15–20, use a calculator to evaluate each expression accurate to four decimal places.

*15. $\ln \sqrt{5}$

16. $\ln\left(\frac{12}{7}\right)$

17. $\log(2 + \sqrt{7})$

18. $\log(8 - \sqrt[3]{5})$

*19. $\log_5 19$

20. $\log_7 196$

In each of Exercises 21–24, graph the given function.

21. $y = \log_5 \sqrt{x}$

22. $y = \log(10x)$

*23. $y = \ln x^3$

24. $y = \frac{1}{2} \ln(1/x)$

In Exercises 25–30, solve each given equation for $x$.

*25. $5e^{0.25x} = 30$

26. $7^x = 10^{2x-1}$

27. $e^{7x-3} = 10^{x+4}$

28. $10 \ln(3x) = 46$

29. $\ln(x^2/3) = 20$

30. $\log x - \log(x - 1) = 1$

*31. (Interest) If $20,000 is invested in a savings account that pays 8.75% annual interest, how long will it take for this investment to double if interest is compounded (a) quarterly? (b) daily? (c) continuously?

## Business and Economics

32. (Profit) The variable cost to a manufacturer of producing one VCR is $75. If total monthly fixed costs are $15000 and if the monthly demand function is $x = 5000e^{-0.15p}$, where $p$ is the price per unit in dollars, what should the manufacturer charge per VCR to maximize monthly profit?

In Exercises 33–50, find the derivative of the given functions.

33. $f(x) = 12e^{-7x+2}$

34. $f(x) = 2x - e^{x^2}$

*35. $f(x) = 5e^{3x^2} - 7x^3$

36. $f(x) = \sqrt{x} - \frac{1}{2}e^{\sqrt{x}}$

37. $f(x) = xe^{\sqrt[3]{x}}$

38. $f(x) = xe^{-x^2} + (1/e^x)$

39. $f(x) = 1/(e^x - 1)$

40. $f(x) = x/(2^x + 1)$

*41. $f(x) = 2^x - e^{2x}$

42. $f(x) = 10^{x^2+2x+1}$

43. $f(x) = \ln x^2 - 2 \ln x$

44. $f(x) = (\ln x)/x^2$

45. $f(x) = \ln \sqrt{x^2 + 5x + 7}$

46. $f(x) = e^x \ln(e^{-x})$

*47. $f(x) = \ln \dfrac{\sqrt{x^2 + 15}}{\sqrt[3]{x^5 + 7x + 3}}$

48. $f(x) = \ln \dfrac{(x - 5)^4}{(x + 2)^2(x - 1)^6}$

*49. $f(x) = x^3 \log_5(3x + 7)$

50. $f(x) = \log \dfrac{\sqrt{x}}{x^2 + 25}$

In Exercises 51–54, sketch the graph of the given function. Be sure to find relative extrema, intervals of concavity, and points of inflection.

51. $f(x) = xe^{-5x}$

52. $f(x) = x^2 e^{-2x}$

53. $f(x) = x \ln x + x$

54. $f(x) = x^2 \ln x$

55. (Dissolving chemical) A chemical is dissolving in water according to the formula for unlimited exponential decay. If initially 500 g of the undissolved chemical are placed in the water and if 2 minutes later 100 g still remain undissolved, find the amount of the chemical that is still not dissolved after 10 minutes.

**56.** (Interest) How long will it take an investment of $5000 to grow to $25000 if it is invested at an annual interest rate of 9% compounded continuously?

***57.** (Carbon dating) A bone fragment discovered by an archaeologist is found to have 10% of the original amount of $C^{14}$ still present. How old is the bone fragment? Assume that the half-life of $C^{14}$ is 5730 years.

**58.** (Radioactive decay) Nobelium 257 has a half-life of about 23 seconds. What percentage of an initial amount of $A_0$ remains after one minute?

**59.** (Population) A certain city's population is growing exponentially at 3% per year. How many years will it take for the population of the city to triple?

**60.** (Spread of a rumor) When the football coach of the local university decided to run for governor, the story was immediately picked up by radio and television. Within six hours after the story broke, one-half of the residents of the city knew about the news. Use the model of limited exponential growth to estimate when 95% of the residents will have heard the story.

### Social Sciences

***61.** (Student learning) After five days of study, Jane Chang has learned 250 French vocabulary words from a list of 10,000 words. After studying for an additional day, she has learned another 100 words. Experience has shown that even after many weeks of study, the best students of French can learn only about 9500 of the 10,000 words. Estimate the number of words that Jane will have learned after 20 days of study. Use the model of limited exponential growth.

### Life Sciences

**62.** (Pollution) After being contaminated with a pollutant, a parcel of land in Florida will bear only 10% of the normal level of vegetation. After one year has passed, the parcel of land is bearing only 25% of the normal level. Using a logistic growth function, estimate the percentage of normal vegetation that the land will bear after five years.

### OPTIONAL
### GRAPHING CALCULATOR/ COMPUTER EXERCISES

*In Exercises 63–67, sketch the graph of the given functions.*

**63.** $f(x) = x^3 - \sqrt{x}e^x$

**64.** $f(x) = (x^2 + 5)/(1 + e^x)$

**65.** $f(x) = e^x - 50(\ln x)^2$

**66.** $f(x) = (x^2 \ln x)/e^x$

**67.** $f(x) = 5(\log x)^3 - 2x$

*In Exercises 68–70, approximate the real solutions x of the given equations in [0, 10] to two-decimal-place accuracy.*

**68.** $3x^3 - e^x = 0$

**69.** $(\ln x)^{20} - x^2 = 3$

**70.** $-\log x + 5 \ln x - e^{2x} = -2$

# 6 INTEGRATION

## CHAPTER CONTENTS

**6.1** Antiderivatives and Indefinite Integrals

**6.2** Integration by Substitution

**6.3** The Definite Integral as a Net Change

**6.4** Integration by Parts

**6.5** Integration Using Tables

**6.6** Chapter 6 Review

## 6.1 ANTIDERIVATIVES AND INDEFINITE INTEGRALS

**Antiderivatives**

In this section we will begin the study of what is called integral calculus. If, in differential calculus, we view the process of finding a derivative as a forward process, then much of integral calculus is concerned with the opposite or backward procedure. That is, given a function $f(x)$, the primary problem of integral calculus is to find a function $F(x)$ such that $F'(x) = f(x)$. Such a function $F(x)$ is called an *antiderivative or indefinite integral* of $f(x)$.

---

A function $F(x)$ such that

$$F'(x) = f(x)$$

for every $x$ in the domain of $f$ is an *antiderivative* or *indefinite integral* of $f(x)$. The procedure of finding $F(x)$ given $f(x)$ is called *antidifferentiation* or *indefinite integration*.

---

To see how the idea of an antiderivative might arise in an applied setting, suppose that a manufacturer knows that the marginal cost of producing $x$ thousand units of his product is $C'(x) = 0.024x^2 - 0.08x + 3$, where $C(x)$ is measured in thousands of dollars and that his fixed cost is $3000 per month. If the manufacturer wants to determine his monthly cost function $C(x)$, he would have to find an antiderivative $C(x)$ of $C'(x)$ such that $C(0) = 3$ (thousand). We will solve this problem in Example 7.

In general, when we know the formula for the rate of change of a quantity and we desire a formula for the quantity itself, we must find an antiderivative. For example, we might want to find the population function given the rate of growth of the population; the learning function given the rate of learning; the position of an object moving in a straight line given the rate of change of position (that is, the velocity); or the profit function given the rate of change of profit (that is, the marginal profit).

As an example of finding an antiderivative of a function, let $f(x) = 2x$. To find an antiderivative of $f(x)$, we must find a function $F(x)$ such that the derivative of $F(x)$ is equal to $2x$. If $F(x) = x^2$, then $F'(x) = f(x) = 2x$, so that

$$F(x) = x^2 \text{ is an antiderivative of } f(x) = 2x.$$

Observe, however, that since

$$D_x(x^2 + 5) = 2x$$
$$D_x(x^2 - \sqrt{3}) = 2x$$
$$D_x(x^2 + 70.95) = 2x$$

each of the functions

$$x^2 + 5, \quad x^2 - \sqrt{3}, \quad \text{and} \quad x^2 + 70.95$$

is also an antiderivative of $f(x) = 2x$. We conclude that *an antiderivative of a function is not unique.* Note that all of the antiderivatives of $2x$ that we have obtained thus far have the form

$$F(x) = x^2 + C$$

where $C$ is a real constant. (Do not confuse the constant $C$ with the cost function $C(x)$ discussed earlier in this section.) It is natural to ask whether other antiderivatives of $2x$ exist that are not of this form. The next result (we omit the proof) answers this question in the negative.

**The General Antiderivative or Indefinite Integral**

> If $F(x)$ is one antiderivative of $f(x)$, then any other antiderivative $G(x)$ of $f(x)$ has the form
>
> $$G(x) = F(x) + C$$
>
> where $C$ is a constant. The form $F(x) + C$ is called the *general antiderivative* of $f(x)$ or the *indefinite integral* of $f(x)$.

**EXAMPLE 1** Find the general antiderivative of $f(x) = x^3$.

**Solution** We must find a function $F(x)$ such that $F'(x) = x^3$. Since the derivative of $\frac{1}{4}x^4$ is $x^3$, we observe that one antiderivative of $x^3$ is $F(x) = \frac{1}{4}x^4$. The most general antiderivative of $x^3$ is $G(x) = \frac{1}{4}x^4 + C$. ∎

The special notation used to denote the general antiderivative (or indefinite integral) of $f(x)$ is

$$\int f(x)\, dx$$

In this symbolism, $\int$ is an elongated **S** and is called the *integral sign*. In the next chapter we will see that $\int$ actually stands for "sum." The function $f(x)$ is called the *integrand*, and $dx$ is the differential of $x$ that we studied in Chapter 3; it identifies the independent variable or the *variable of integration* as $x$. As examples, $\int 2x\, dx = x^2 + C$, $\int 2z\, dz = z^2 + C$, and $\int ax\, dx = \frac{1}{2}ax^2 + C$. Another way to think about the symbol $dx$ is to view indefinite integration as the reverse procedure to finding the differential of a function. Recall that if $y = f(x)$ then $dy = f'(x)\, dx$ is called the differential of $y$. For example, the differential of $y = x^3 + C$ is $dy = d(x^3 + C) = 3x^2\, dx$, and so $\int 3x^2\, dx = x^3 + C$. We illustrate this as follows:

$$x^3 + C \underset{\int}{\overset{d}{\rightleftarrows}} 3x^2\, dx$$

This diagram illustrates that $d(x^3 + C) = 3x^2\, dx$ and $\int 3x^2\, dx = x^3 + C$: the operations of finding a differential and finding an indefinite integral are inverse processes provided we add the arbitrary constant $C$ when we find the indefinite integral. In general, we denote this diagram as follows:

$$F(x) + C \underset{\int}{\overset{d}{\rightleftarrows}} f(x)\, dx$$

**Properties of Indefinite Integrals**

Suppose that $F'(x) = f(x)$ and $G'(x) = g(x)$. Then, since

$$d(F(x) \pm G(x)) = [F'(x) \pm G'(x)]\, dx = [f(x) \pm g(x)]\, dx$$

and since $d(F(x)) \pm d(G(x)) = F'(x)\, dx \pm G'(x)\, dx = f(x)\, dx \pm g(x)\, dx$, we can write the following rule.

**THE SUM OR DIFFERENCE RULE FOR INTEGRALS**

$$\int [f(x) \pm g(x)]\, dx = \int f(x)\, dx \pm \int g(x)\, dx \quad \text{Property (1)}$$

This rule states that *the integral of a sum (or difference) of two functions $f$ and $g$ is equal to the sum (or difference) of the integrals of the functions.*

Also, if $c$ is constant, $d(cF(x)) = c[d(F(x))] = cf(x)\, dx$, we have another rule.

**THE CONSTANT MULTIPLE RULE FOR INTEGRALS**

$$\int cf(x)\, dx = c \int f(x)\, dx \quad \text{Property (2)}$$

where $c$ is a constant.

This rule states that *the integral of a constant times a function is the constant times the integral of the function.*

EXAMPLE 2

Find (a) $\int (x^2 + x^3 - x)\, dx$; (b) $\int 124x^4\, dx$.

Solution

(a) First we comment that Property (1) can be generalized and applies to any finite number of sums or differences of functions. That is, if $F_1'(x) = f_1(x)$, $F_2'(x) = f_2(x), \ldots, F_n'(x) = f_n(x)$, then

$$\int [f_1'(x) \pm f_2'(x) \pm \cdots \pm f_n'(x)]\, dx = \int f_1'(x)\, dx \pm \int f_2'(x)\, dx \pm \cdots \pm \int f_n'(x)\, dx$$

We can therefore write

$$\int (x^2 + x^3 - x)\, dx = \int x^2\, dx + \int x^3\, dx - \int x\, dx$$
$$= \left(\tfrac{1}{3}x^3 + C_1\right) + \left(\tfrac{1}{4}x^4 + C_2\right) - \left(\tfrac{1}{2}x^2 + C_3\right)$$
$$= \tfrac{1}{4}x^4 + \tfrac{1}{3}x^3 - \tfrac{1}{2}x^2 + (C_1 + C_2 - C_3)$$
$$= \tfrac{1}{4}x^4 + \tfrac{1}{3}x^3 - \tfrac{1}{2}x^2 + C$$

where $C = C_1 + C_2 - C_3$.

(b) $\displaystyle\int 124x^4\,dx = 124\int x^4\,dx$  (Using Property (2))

$\qquad\qquad\qquad = 124\left(\tfrac{1}{5}x^5 + C_1\right)$

$\qquad\qquad\qquad = \tfrac{124}{5}x^5 + 124C_1$

$\qquad\qquad\qquad = \tfrac{124}{5}x^5 + C$

where $C = 124C_1$. ■

In the future, whenever we do an indefinite integration, we will simply add a single arbitrary constant $C$ to our final result, since all of the individual constants that might result from the integration can always be combined into a single constant $C$.

## Some Integration Formulas

In the examples of indefinite integration that we have done so far in this section, we have used the idea of "thinking backwards" from differentiation. That is, to find $\int x^2\,dx$, for example, we reasoned that we were looking for a function $F(x)$ such that $F'(x) = x^2$. By "thinking backwards," we concluded that $F(x) = \tfrac{1}{3}x^3 + C$, since $D_x\!\left[\tfrac{1}{3}x^3 + C\right] = x^2$, and so $\int x^2\,dx = \tfrac{1}{3}x^3 + C$. In order to make the task of finding antiderivatives or indefinite integrals easier, we will develop several formulas in this section that, in combination with Properties (1) and (2), will allow us to find the indefinite integrals of certain functions.

One key point to keep in mind when looking for integration formulas is that there is an integration formula corresponding to every differentiation formula. This is because differentiation and integration are inverse processes. To obtain the integration formula corresponding to a given differentiation formula, we integrate both sides of the differentiation formula and simplify. For example, in differential notation, the power rule of differentiation can be written

$$d(x^{n+1}) = (n+1)x^n\,dx$$

To obtain the corresponding integration formula, we write

$$\int d(x^{n+1}) = \int (n+1)x^n\,dx$$

$$x^{n+1} + C_1 = (n+1)\int x^n\,dx$$

By solving for $\int x^n\,dx$ and letting $C = C_1/(n+1)$, we obtain the power rule for integrals.

**POWER RULE FOR INTEGRALS**

$$\int x^n\,dx = \frac{1}{n+1}x^{n+1} + C \quad \text{if } n \neq -1 \qquad \text{Property (3)}$$

We require that $n \neq -1$ in formula (3) so the quantity $1/(n+1)$ won't lead to division by zero. We illustrate the power rule for integrals in the following diagram.

390  CHAPTER 6  INTEGRATION

$$\frac{1}{n+1}x^{n+1} + C \underset{\int}{\overset{d}{\rightleftarrows}} x^n \, dx$$

(if $n \neq -1$)

**EXAMPLE 3**

Find (a) $\int x^5 \, dx$; (b) $\int 5\sqrt{x} \, dx$; (c) $\int (4 - \sqrt[3]{x}) \, dx$.

**Solution**

(a) $\int x^5 \, dx = \frac{1}{5+1}x^{5+1} + C = \frac{1}{6}x^6 + C$

(b) $\int 5\sqrt{x} \, dx = 5 \int \sqrt{x} \, dx$ (Using Property (2))

$\qquad = 5 \int x^{1/2} \, dx$ (Converting the radical to a fractional exponent)

$\qquad = 5\left(\frac{1}{\frac{1}{2}+1}x^{1/2+1}\right) + C = 5\left(\frac{1}{\frac{3}{2}}x^{3/2}\right) + C = \frac{10}{3}x^{3/2} + C$

(c) $\int (4 - \sqrt[3]{x}) \, dx = \int 4 \, dx - \int \sqrt[3]{x} \, dx$ (Using Property (1))

$\qquad = \int 4x^0 \, dx - \int x^{1/3} \, dx$

$\qquad = 4\left(\frac{1}{0+1}x^{0+1}\right) - \frac{1}{\frac{1}{3}+1}x^{1/3+1} + C$

$\qquad = 4x - \frac{3}{4}x^{4/3} + C$  ∎

**EXAMPLE 4 (Participative)**

Let's find (a) $\int (3x^2 - 1/\sqrt{x} + 2) \, dx$; (b) $\int (2x^3 - 5x^4)/x^2 \, dx$

**Solution**

(a) $\int \left(3x^2 - \frac{1}{\sqrt{x}} + 2\right) dx = \int 3x^2 \, dx - \int x^{-1/2} \, dx + \underline{\qquad}$ (Using Property (1))

$\qquad = 3 \int x^2 \, dx - \underline{\qquad} + 2x + C$ (By Properties (2) and (3))

$\qquad = \underline{\qquad}$

(b) $\int \left(\frac{2x^3 - 5x^4}{x^2}\right) dx = \int \left(\frac{2x^3}{x^2} - \frac{5x^4}{x^2}\right) dx$

$\qquad = \int (\underline{\qquad}) \, dx$

$\qquad = \int 2x \, dx - \int 5x^2 \, dx$ (Using Property (1))

$$= 2\int x\, dx - \underline{\hspace{2cm}} \quad \text{(Using Property (2))}$$

$$= \underline{\hspace{3cm}} \quad \text{(Using Property (3))}$$

∎

The power rule for integrals is valid for all real numbers $n$ except $n = -1$. Let us now consider

$$\int x^{-1}\, dx = \int \frac{1}{x}\, dx$$

To compute this indefinite integral, we recall the differentiation formula $d(\ln x) = (1/x)\, dx$. Integrating both sides of this formula yields

$$\int \frac{1}{x}\, dx = \ln x + C \tag{4}$$

Note carefully, however, that since $\ln x$ is defined only for $x > 0$, formula (4) is valid only for $x > 0$. For the case when $x < 0$,

$$d(\ln |x|) = \frac{d(|x|)}{|x|} = \frac{d(-x)}{-x} \quad \text{(Since } |x| = -x \text{ if } x < 0\text{)}$$

$$= \frac{-1}{-x} = \frac{1}{x}$$

Therefore if $x < 0$,

$$\int \frac{1}{x}\, dx = \ln |x| + C \tag{5}$$

Now, since $|x| = x$ if $x > 0$, the results of formulas (4) and (5) can be combined to assert that for any $x \neq 0$:

$$\int \frac{1}{x}\, dx = \ln |x| + C \qquad \text{Property (6)}$$

To compute $\int e^x\, dx$, recall that $d(e^x) = e^x\, dx$, so that

$$\int d(e^x) = \int e^x\, dx$$

$$e^x + C = \int e^x\, dx$$

We conclude that

$$\int e^x\, dx = e^x + C \qquad \text{Property (7)}$$

**SOLUTIONS TO EXAMPLE 4**

(a) $\int 2\, dx,\ \dfrac{x^{-1/2+1}}{-\frac{1}{2}+1},\ x^3 - 2x^{1/2} + 2x + C$; (b) $2x - 5x^2,\ 5\int x^2\, dx,\ x^2 - \frac{5}{3}x^3 + C$

## EXAMPLE 5

Find $\int (6x^4 - 4e^x + 2/x)\, dx$.

**Solution**

$$\int \left(6x^4 - 4e^x + \frac{2}{x}\right) dx = \int 6x^4\, dx - \int 4e^x\, dx + \int \frac{2}{x}\, dx \quad \text{(By Property (1))}$$

$$= 6\int x^4\, dx - 4\int e^x\, dx + 2\int \frac{1}{x}\, dx \quad \text{(By Property (2))}$$

$$= \tfrac{6}{5}x^5 - 4e^x + 2\ln|x| + C \quad \text{(By Properties (3), (7), and (6))} \quad \blacksquare$$

### Geometric Interpretation of Antiderivatives

How are the antiderivatives $F(x) + C$ related geometrically to $f(x)$ and how are these antiderivatives related geometrically to each other? Let us now address this important question.

There are three important facts concerning the geometry:

1. The slope of the tangent line to any of the antiderivatives $F(x) + C$ at any point $x$ in the domain of $f$ is equal to $f(x)$.
2. The graphs of $F(x) + C$ for various values of $C$ form a "family of parallel curves," meaning that each of them can be obtained geometrically by a vertical translation of the graph of $F(x)$.
3. Given any point $(a, b)$, one and only one curve of the family $F(x) + C$ passes through this point if $a$ is in the domain of $f$.

Let's consider an example.

## EXAMPLE 6

Consider the indefinite integration $\int 2x\, dx = x^2 + C$.

(a) Sketch the graphs of the antiderivatives $x^2 + C$ for $C = -2$, $C = 0$, and $C = 2$.
(b) Draw the tangent lines to the three curves sketched in part (a) at the points where $x = -1$, $x = 0$, and $x = 1$.
(c) Find and sketch the curve in the antiderivative family $x^2 + C$ that passes through the point $(2, 1)$.

**Solution**

(a) We are required to sketch the graphs of $y = x^2 - 2$, $y = x^2$, and $y = x^2 + 2$. The sketches are drawn in Figure 6.1.
(b) The tangent lines to each of the three curves at $x = -1$, $x = 0$, and $x = 1$ are sketched in Figure 6.1. These lines have slopes $-2, 0,$ and $2$, respectively, for each of the three graphs of $y = x^2 - 2$, $y = x^2$, and $y = x^2 + 2$.
(c) Let $G(x) = x^2 + C$ represent the general antiderivative. Since we require the member of the family that passes through $(2, 1)$, we require that $G(2) = 1$. But

$$G(2) = 2^2 + C = 1$$

so $C = -3$. The required antiderivative is $y = x^2 - 3$ and its graph is also sketched in Figure 6.1.

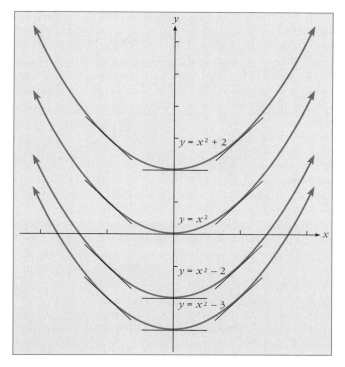

**FIGURE 6.1** Graphs of some members of the antiderivative family $G(x) = x^2 + C$.

## Applications

**EXAMPLE 7 (Participative)**

If a manufacturer has a monthly marginal cost function

$$C'(x) = 0.024x^2 - 0.08x + 3$$

where $C(x)$ is measured in thousands of dollars and $x$ is measured in thousands of units, and if fixed costs are $3000, let's find the monthly cost function for this manufacturer.

Solution

We are required to find an antiderivative $C(x)$ of $C'(x)$ such that $C(0) =$ _____. To begin, we note that

$$\begin{aligned} C(x) &= \int C'(x)\,dx \\ &= \int (0.024x^2 - 0.08x + 3)\,dx \\ &= \int 0.024x^2\,dx - \int \underline{\qquad} + \int 3\,dx \quad \text{(By Property (1))} \\ &= 0.024\int x^2\,dx - 0.08\int x\,dx + \underline{\qquad} + K \quad \text{(By Property (2); } K \text{ an arbitrary constant)} \\ &= \underline{\hspace{5cm}} \end{aligned}$$

To find $K$, we compute:

$$C(0) = 0.008(0)^3 - 0.04(0)^2 + 3(0) + K = 3$$

Solving for $K$, we get $K = \underline{\phantom{xxx}}$ so that $C(x) = \underline{\phantom{xxxxxxxxxxxxxxxxxxxxxxxxx}}$. ■

---

EXAMPLE 8

The rate of growth of the number of social workers in a certain city is given by $dN/dt = 5 + 4\sqrt[3]{t}$, where $t$ is time in years and $t = 0$ corresponds to 1990, when there were 150 social workers. Approximately how many social workers will the city have in the year 2000?

Solution

We begin by computing

$$N(t) = \int \frac{dN}{dt} dt = \int (5 + 4\sqrt[3]{t}) \, dt$$

$$= \int 5 \, dt + 4 \int t^{1/3} \, dt = 5t + \frac{4}{\frac{4}{3}} t^{4/3} + C$$

$$= 5t + 3t^{4/3} + C$$

Since $N(0) = 150$, we have

$$N(0) = 5(0) + 3(0)^{4/3} + C = 150$$

so that $C = 150$ and $N(t) = 5t + 3t^{4/3} + 150$. Since the year 2000 corresponds to $t = 10$, the approximate number of social workers in the year 2000 will be

$$N(10) = 5(10) + 3(10)^{4/3} + 150 \approx 265$$ ■

---

**SECTION 6.1**

**SHORT ANSWER EXERCISES**

1. (True or False) The function $F(x) = x^3$ is an antiderivative of $f(x) = x^2$. _____
2. (True or False) The function $f(x) = 1/x$ has a unique antiderivative given by $F(x) = \ln|x|$. _____
3. (True or False) An antiderivative $F(x)$ of $f(x)$ has the property that $F'(x) = f(x)$. Therefore we can always check that the result of an antiderivative is correct by differentiating $F(x)$. _____
4. Which of the following functions is an antiderivative of $f(x) = (2x + 1)^4$?
   (a) $\dfrac{(2x+1)^5}{5}$;   (b) $\dfrac{(2x+1)^5}{10}$;   (c) $\dfrac{(2x+1)^5}{4}$;   (d) $\dfrac{(2x+1)^4}{5}$
5. The result of $\int zx \, dz$ is _____.
6. The power rule for integrals is valid (a) for any number $n$; (b) only if $n$ is a positive integer; (c) only if $n \neq 1$; (d) none of the above.
7. (True or False) $\int \ln x \, dx = (1/x) + C$. _____
8. Given the antiderivative family $F(x) = \frac{1}{3}x^3 + C$, the slope of the tangent line to the graph of any member of this family at $x = 2$ is _____.
9. Given the general antiderivative $F(x) = e^x + C$, if we know that $F(0) = 2$, then $C = \underline{\phantom{xx}}$.

---

**SOLUTIONS TO EXAMPLE 7**   3, $0.08x \, dx$, $3x$, $0.008x^3 - 0.04x^2 + 3x + K$, 3, $0.008x^3 - 0.04x^2 + 3x + 3$

**10.** If an object moves along a straight line and has velocity $v(t) = 4t$, then $v(t) = ds/dt$, where $s(t)$ is the position of the object at time $t$, so that $s(t) = \int v(t)\, dt = \int 4t\, dt = $ _____ .

**11.** In Exercise 10, given that $s(0) = 0$, the constant $C$ is equal to _____ .

## SECTION 6.1
## EXERCISES

In Exercises 1–34, find the indefinite integral.

1. $\int 4\, dx$
2. $\int (-5)\, dx$
3. $\int x\, dx$
4. $\int x^2\, dx$
5. $\int 5x^3\, dx$
6. $\int 7x^4\, dx$
7. $\int (2x^2 - 7x)\, dx$
8. $\int (x^3 + 4x)\, dx$
9. $\int 2\sqrt{z}\, dz$
10. $\int 3\sqrt[3]{y}\, dy$
11. $\int (y^3 - 12y^2 - 4)\, dy$
12. $\int (z^4 + 5z^2 - 10)\, dz$
13. $\int (x^4 - 5x^3 + 3x^2 + 2x + 1)\, dx$
14. $\int (7x^3 - 10x^2 + 4x - 4)\, dx$
15. $\int \left(u + \frac{1}{u} - 2e^u\right) du$
16. $\int \left(u^2 - \frac{2}{u} + 4e^u\right) du$
17. $\int \left(\frac{3}{x^2} + \frac{2}{x^3} + 1\right) dx$
18. $\int \left(4 - \frac{2}{x^2} + \frac{5}{x^4}\right) dx$
19. $\int \left(2t^2 + \frac{4}{t^4}\right) dt$
20. $\int \left(3t + 2 - \frac{2}{t^3}\right) dt$
21. $\int \left(2e^x - \frac{1}{x}\right) dx$
22. $\int \left(\frac{6}{x} - 5e^x + 1\right) dx$
23. $\int (2\sqrt{x} - 5\sqrt[3]{x})\, dx$
24. $\int (\sqrt[3]{x} - \sqrt[4]{x})\, dx$
25. $\int \left(\frac{1}{\sqrt[4]{x}} - \frac{2}{\sqrt[3]{x}}\right) dx$
26. $\int \left(\frac{3}{\sqrt[4]{x}} + \frac{2}{\sqrt[3]{x^2}}\right) dx$
27. $\int \left(\frac{x^3 - 5x^2}{x}\right) dx$
28. $\int \left(\frac{2x^4 + 7x^2 + x}{x^2}\right) dx$
29. $\int \left(\frac{5x^3 - 7x^2 + 1}{2x}\right) dx$
30. $\int \left(\frac{2x - 5}{\sqrt{x}}\right) dx$
31. $\int (u^{4/5} - 7u^{2/3})\, du$
32. $\int (y^{4/3} + 3y^{1/5})\, dy$
33. $\int (2x + 1)^2\, dx$
34. $\int (5x - 2)^2\, dx$

In each of Exercises 35–39, find the function $F(x)$ that satisfies the given conditions.

**35.** $F(x) = \int (2x^3 - 3)\, dx$, for $F(0) = 2$

**36.** $F(x) = \int (7\sqrt{x} - 3)\, dx$, for $F(1) = 5$

**37.** $F(x) = \int \left(\frac{2}{x} - \frac{4}{x^3}\right) dx$, for $F(2) = -1$

**38.** $F(x) = \int \left(\frac{1}{\sqrt[3]{x}} - 6x + 1\right) dx$, for $F(-1) = 5$

39. $F(x) = \int (x^4 - \sqrt[5]{x} + 3)\, dx$, for $F(0) = -3$

40. (Business functions) A company has a monthly marginal cost function $C'(x) = 25$ and monthly marginal revenue function $R'(x) = 200$. If its fixed monthly costs are $600, find (a) the monthly cost function; (b) the monthly revenue function (*Hint:* $R(0) = 0$); (c) the monthly profit function; (d) the monthly profit from the sale of 18 units.

41. (Cost) A manufacturer has a monthly marginal cost function $C'(x) = 0.09x^2 - 0.2x + 2$, where $C(x)$ is measured in thousands of dollars and $x$ is measured in thousands of units. If monthly fixed costs are $2500, find the monthly cost function for this manufacturer.

42. (Profit) The marginal profit of a company is $P'(x) = -0.20x + 52$, where $x$ is the number of units sold per month. When no items are sold in a month, the company suffers a loss of $2500.
    (a) Find the monthly profit function $P(x)$.
    (b) Find the maximum monthly profit.

43. (Depreciation) The value $V$ (in dollars) of a certain piece of equipment decreases according to $dV/dt = -750$, where $t$ is time in years since it was purchased. If its purchase price was $11,250, find the function $V(t)$ that gives the value of the equipment $t$ years after its purchase.

44. (Real estate) The average price $P$ of new homes in a certain community has been increasing according to $P'(t) = 8/\sqrt{t}$, where $P(t)$ is measured in thousands of dollars, $t$ is time in years, and $t = 1$ corresponds to 1988. If the average price of a new home was $95,000 in 1988, estimate the average price in 1997.

45. (Revenue) The marginal revenue from the sale of $x$ units per month is $R'(x) = 150 - \tfrac{1}{4}x$.
    (a) Find the monthly revenue function $R(x)$. Assume that $R(0) = 0$.
    (b) Find the level of monthly sales that maximizes revenue.

46. (Demand) The monthly demand $x$ (in units) for a certain product satisfies $dx/dp = -0.04p$, where $p$ is the price per unit in dollars. When the price per unit is $20, monthly demand is 3992 units. What is monthly demand when the price per unit is $100?

**Business and Economics**

47. (Advertising and sales) The marketing department of a company predicts that $t$ days after the start of a new advertising campaign, daily sales will be increasing according to $S'(t) = 50\sqrt[3]{t}$ for $0 \le t \le 30$, where $S(t)$ is sales in units. What is the predicted number of units sold after eight days of advertising if daily sales before the campaign begins is 500 units.

48. (Velocity) An object moves along a straight line and its velocity is $v(t) = t^2 + t + 3$. If its position at time $t = 0$ is $s(0) = 2$, find its position at time $t = 5$.

49. (Falling object) A stone that is thrown upward from a height of six feet above the ground with an initial velocity of 64 ft/sec satisfies $v(t) = 64 - 32t$, where $t$ is time in seconds since release and $v(t)$ is the velocity at time $t$.
    (a) Find the position $s(t)$ of the stone at time $t$.
    (b) Find the maximum height reached by the stone.

50. (Employee training) After $t$ days of training, an average employee can assemble $N(t)$ units of a certain product per day, where $0 \le t \le 12$. If it is known that $N'(t) = \tfrac{1}{2}t + 1$, and if an average new employee can assemble ten units per day with no training at all, how many units can an average employee assemble after ten days of training?

Social Sciences

**51.** (Unemployment and self-esteem) After $t$ months of being unemployed, the self-esteem level $E$ of an individual changes according to $dE/dt = \frac{4}{3}(\frac{1}{3}t - 5)$, for $0 \leq t \leq 12$. In this situation $E$ is measured on a scale from 0 to 100 and just before the individual becomes unemployed, his self-esteem level is 70. Estimate the self-esteem level of the individual after six months of unemployment.

## 6.2 INTEGRATION BY SUBSTITUTION

### The Chain Rule for Integration

Integration by substitution is carried out essentially by reversing the chain rule for differentiation. Suppose that $F(x)$ is an antiderivative of $f(x)$ and that $g(x)$ is differentiable. Then, by the chain rule,

$$D_x F(g(x)) = F'(g(x))g'(x) = f(g(x))g'(x)$$

Integrating both sides of this equation gives us

**THE CHAIN RULE FOR INTEGRATION**

$$\int f(g(x))g'(x)\, dx = F(g(x)) + C \qquad (1)$$

Let's see how the chain rule for integration works in some examples.

**EXAMPLE 1** Find $\int (2x^2 + 1)^2 \cdot 4x\, dx$

Solution  Let $f(x) = x^2$ and $g(x) = 2x^2 + 1$. Then $f(g(x)) = (2x^2 + 1)^2$ and $g'(x) = 4x$. Furthermore, we can take $F(x) = \frac{1}{3}x^3$, since this is an antiderivative of $f(x) = x^2$. Then, applying the chain rule for integration, we obtain

$$\int \underbrace{(2x^2 + 1)^2}_{f(g(x))} \cdot \underbrace{4x\, dx}_{g'(x)\, dx} = \underbrace{\tfrac{1}{3}(2x^2 + 1)^3}_{F(g(x))} + C$$

We can check this result by observing that, by the chain rule for differentiation,

$$D_x\left[\tfrac{1}{3}(2x^2 + 1)^3 + C\right] = \tfrac{1}{3}(3)(2x^2 + 1)^2(4x) = (2x^2 + 1)^2 \cdot 4x \qquad \blacksquare$$

**EXAMPLE 2** Find $\int \sqrt{x^3 + 5} \cdot x^2\, dx$.

Solution  Let $f(x) = \sqrt{x}$ and $g(x) = x^3 + 5$. Then $f(g(x)) = \sqrt{x^3 + 5}$ and $g'(x) = 3x^2$. If we choose the arbitrary constant to be 0, then $F(x) = \int \sqrt{x}\, dx = \int x^{1/2}\, dx = \frac{2}{3}x^{3/2}$ is an antiderivative of $f(x) = \sqrt{x}$. We will add a single arbitrary constant $C$ to the completed indefinite integral.

A small problem arises in this example that did not occur in Example 1, namely, $g'(x) = 3x^2$, whereas in the integral, the factor $x^2$ appears without the 3. We'll have to make an adjustment in the integral by multiplying by $\frac{3}{3} = 1$:

$$\int \sqrt{x^3 + 5} \cdot x^2 \, dx = \int \sqrt{x^3 + 5} \cdot \left(\tfrac{3}{3}\right) x^2 \, dx$$

$$= \tfrac{1}{3} \int \underbrace{\sqrt{x^3 + 5}}_{f(g(x))} \cdot \underbrace{3x^2 \, dx}_{g'(x)dx} \quad \left(\text{Factoring the constant } \tfrac{1}{3} \text{ outside the integral}\right)$$

$$= \tfrac{1}{3}\left[\tfrac{2}{3}(x^3 + 5)^{3/2}\right] + C$$

$$= \tfrac{2}{9}(x^3 + 5)^{3/2} + C$$

Once again, this result can be checked by differentiation:

$$D_x\left[\tfrac{2}{9}(x^3 + 5)^{3/2} + C\right] = \tfrac{2}{9}\left(\tfrac{3}{2}\right)(x^3 + 5)^{1/2}(3x^2) = \sqrt{x^3 + 5} \cdot x^2 \qquad \blacksquare$$

### Integration by Substitution

The chain rule for integration can be written in a simpler form if we make the substitution $u = g(x)$. With this substitution, $du$ (the differential of $u$) is equal to $g'(x)dx$, so the chain rule for integration becomes

$$\int f(u) \, du = F(u) + C \qquad (\text{where } u = g(x))$$

This technique of making the substitution $u = g(x)$ and $du = g'(x)dx$ in an integral, then integrating the resulting integral in terms of $u$ and rewriting this result in terms of the original variable $x$ is called *integration by substitution*. Table 6.1 contains some important integration formulas written in terms of both the chain rule for integrals and integration by substitution.

| SOME IMPORTANT INTEGRATION FORMULAS ||
| --- | --- |
| Chain Rule for Integrals | Integration by Substitution |
| $\int [g(x)]^n g'(x) \, dx = \dfrac{[g(x)]^{n+1}}{n+1} + C$ | $\int u^n \, du = \dfrac{u^{n+1}}{n+1} + C$ |
| $\int e^{g(x)} g'(x) \, dx = e^{g(x)} + C$ | $\int e^u \, du = e^u + C$ |
| $\int \dfrac{g'(x)}{g(x)} \, dx = \ln|g(x)| + C$ | $\int \dfrac{du}{u} = \ln|u| + C$ |

TABLE 6.1

For each formula in this table we can convert from the form for the chain rule for integrals to the form for integration by substitution by making the substitutions $u = g(x)$ and $du = g'(x)dx$. Furthermore, each of these formulas can be proved using differentiation. For example, to prove the first formula in Table 6.1, we apply the chain rule for differentiation to obtain

$$D_x\left(\frac{[g(x)]^{n+1}}{n+1}\right) = \frac{1}{n+1} D_x([g(x)]^{n+1})$$
$$= \frac{1}{n+1} \cdot (n+1)[g(x)]^n g'(x)$$
$$= [g(x)]^n g'(x)$$

---

**EXAMPLE 3**

Find $\int xe^{x^2} dx$.

**Solution**

Inspecting the second formula in Table 6.1, we see that an appropriate substitution might be $u = x^2$, so that $du = 2xdx$. Since the expression $xdx$ appears in the integral, we solve this last equation for $xdx$ to obtain $xdx = \frac{1}{2}du$. Then, applying the technique of integration by substitution, we get:

$$\int xe^{x^2} dx = \int e^{x^2} \cdot x \, dx$$
$$= \int e^u \cdot \tfrac{1}{2} du \quad \left(\text{Note that } u = x^2, \text{ and } xdx = \tfrac{1}{2}du\right)$$
$$= \tfrac{1}{2}\int e^u \, du \quad \left(\text{Factoring the constant } \tfrac{1}{2} \text{ outside the integral}\right)$$
$$= \tfrac{1}{2}e^u + C \quad \text{(Using the second formula in Table 6.1)}$$
$$= \tfrac{1}{2}e^{x^2} + C \quad \left(\begin{array}{l}\text{Replacing } u \text{ by } x^2 \text{ to obtain the}\\ \text{result in the original variable } x\end{array}\right)$$

---

**EXAMPLE 4 (Participative)**

Let's rework Example 2 using the method of integration by substitution.

**Solution**

We want to integrate $\int \sqrt{x^3 + 5} \cdot x^2 \, dx = \int (x^3 + 5)^{1/2} \cdot x^2 \, dx$. The first formula in Table 6.1 suggests the substitution $u = $ _____, so that $du = $ _____. Since the expression $x^2 dx$ appears in the integral, we solve the last equation for $x^2 dx$ to get $x^2 dx = $ _____. Thus

$$\int (x^3 + 5)^{1/2} \cdot x^2 \, dx = \int u^{1/2} \cdot \tfrac{1}{3} du \quad \left(\begin{array}{l}\text{Note that } u = x^3 + 5\\ \text{and } x^2 dx = \tfrac{1}{3}du\end{array}\right)$$
$$= \tfrac{1}{3}\int \underline{\qquad} \quad \left(\begin{array}{l}\text{Factoring the constant } \tfrac{1}{3}\\ \text{outside the integral}\end{array}\right)$$
$$= \tfrac{1}{3} \cdot \frac{u^{3/2}}{\tfrac{3}{2}} + C \quad \left(\begin{array}{l}\text{Using the first formula}\\ \text{in Table 6.1}\end{array}\right)$$
$$= \tfrac{2}{9}u^{3/2} + C$$
$$= \underline{\qquad\qquad} \quad (\text{Replacing } u \text{ by } x^3 + 5)$$

---

**SOLUTIONS TO EXAMPLE 4**

$x^3 + 5$, $3x^2 \, dx$, $\tfrac{1}{3}du$, $u^{1/2} \, du$, $\tfrac{2}{9}(x^3 + 5)^{3/2} + C$

## EXAMPLE 5

Find $\int (x + 1)/(x^2 + 2x)\, dx$.

**Solution**

We first observe that $d(x^2 + 2x) = (2x + 2)dx = 2(x + 1)dx$. If we compare this with the third formula in Table 6.1, it suggests that we make the substitution $u = x^2 + 2x$ so that $du = 2(x + 1)dx$. Since the expression $(x + 1)dx$ occurs in the given integral, we solve the preceding equation for $(x + 1)dx$ to get $(x + 1)dx = \frac{1}{2}du$. Thus,

$$\int \frac{x + 1}{x^2 + 2x}\, dx = \int \frac{1}{x^2 + 2x} \cdot (x + 1)\, dx$$

$$= \int \frac{1}{u} \cdot \frac{1}{2} du \quad \left(\text{Note that } u = x^2 + 2x \text{ and } (x + 1)dx = \tfrac{1}{2}du\right)$$

$$= \frac{1}{2} \int \frac{du}{u} \quad \left(\text{Factoring the constant } \tfrac{1}{2} \text{ outside the integral}\right)$$

$$= \frac{1}{2} \ln |u| + C \quad \text{(Using the third formula in Table 6.1)}$$

$$= \frac{1}{2} \ln |x^2 + 2x| + C \quad (\text{Replacing } u \text{ by } x^2 + 2x) \quad \blacksquare$$

Some students have difficulty making an appropriate choice for $u$ in integration by substitution. Although there are no hard and fast rules, we should in general choose $u = g(x)$ to be some portion of the integral so that

(a) the original integral contains $du = g'(x)dx$ or a constant times this expression;

(b) the original integral can be converted into a new integral involving *only* $u$ and $du$ and *not* $x$ or $dx$;

(c) the new integral involving $u$ and $du$ is simpler than the original integral.

## EXAMPLE 6

For the indefinite integral $\int x\sqrt{3x^2 + 4}\, dx$, (a) attempt to use integration by substitution using $u = x^2$; (b) find the integral using a more appropriate substitution.

**Solution**

(a) Let $u = x^2$, so that $du = 2x\,dx$ and $x\,dx = \frac{1}{2}du$. Then

$$\int x\sqrt{3x^2 + 4}\, dx = \int \sqrt{3x^2 + 4} \cdot x\,dx = \int \sqrt{3u + 4} \cdot \tfrac{1}{2}du$$

$$= \tfrac{1}{2} \int \sqrt{3u + 4}\, du$$

In this attempt, since $du = 2x\,dx$ and $x\,dx = \frac{1}{2}du$, a constant times $du$ does appear in the original integral. Also, we have converted the original integral into a new integral involving only the variable $u$ and the differential $du$. Finally, although the new integral involving $u$ and $du$ is somewhat simpler than the original one involving $x$ and $dx$, it is not completely trivial. Although this problem can now be completed by a further substitution (see Exercise 81), we choose to use a better substitution in part (b) to finish the problem.

(b) Let $u = 3x^2 + 4$, so that $du = 6x\,dx$ and $x\,dx = \frac{1}{6}du$. Then we have

$$\int x\sqrt{3x^2 + 4}\, dx = \int \sqrt{3x^2 + 4} \cdot x\, dx$$
$$= \int \sqrt{u} \cdot \tfrac{1}{6} du = \tfrac{1}{6} \int u^{1/2}\, du$$
$$= \tfrac{1}{6} \cdot \frac{u^{3/2}}{3/2} + C = \tfrac{1}{9} u^{3/2} + C$$
$$= \tfrac{1}{9}(3x^2 + 4)^{3/2} + C$$

## EXAMPLE 7 (Participative)

Solution

Let's find $\int (\ln x)^3/x\, dx$.

If we let $u = \ln x$, then $du = $ _____ and the expression $1/x\, dx$ appears in the original integral. We write

$$\int \frac{(\ln x)^3}{x}\, dx = \int (\ln x)^3 \cdot \frac{1}{x}\, dx$$
$$= \int \underline{\hspace{2cm}} \qquad \text{(Making the substitution } u = \ln x,$$
$$\qquad\qquad\qquad\qquad\qquad du = 1/x\, dx)$$
$$= \frac{u^4}{4} + C$$
$$= \underline{\hspace{3cm}} \qquad \text{(Replacing } u \text{ by } \ln x)$$

Not every integral can be found by the technique of integration by substitution. As an example, consider $\int \sqrt{x^2 + 1}\, dx$. If we try the substitution $u = x^2 + 1$, then $du = 2x dx$ and $x dx = \tfrac{1}{2} du$. However, the expression $x dx$ does *not* appear in the original integral and both of the following attempts at a solution are *incorrect*:

1. $\displaystyle\int \sqrt{x^2 + 1}\, dx = \int \sqrt{u} \cdot \frac{du}{2x}$ (Integral involves both $u$ and $x$)

2. $\displaystyle\int \sqrt{x^2 + 1}\, dx = \int \sqrt{x^2 + 1} \cdot \frac{x}{x}\, dx$

$\displaystyle\qquad\qquad\qquad = \frac{1}{x} \int \sqrt{x^2 + 1} \cdot x dx$  (A variable cannot be factored through the integral sign)

In fact, the integral $\int \sqrt{x^2 + 1}\, dx$ cannot be found by the technique of integration by substitution. This integral requires what is called a *trigonometric substitution*—a technique beyond the scope of this book.

Sometimes a slight modification of the technique of integration by substitution is necessary to find an integral. In some problems we must solve the substitution equation $u = g(x)$ for $x$ in order to complete the integration successfully. The next example illustrates this procedure.

---

**SOLUTIONS TO EXAMPLE 7**   $\dfrac{1}{x}\, dx,\ u^3\, du,\ \tfrac{1}{4}(\ln x)^4 + C$

## EXAMPLE 8

Find $\int x/(2x + 1)\, dx$.

**Solution**

Let $u = 2x + 1$, so that $du = 2dx$ and $dx = \frac{1}{2}du$. In order to convert the original integral into a new integral involving only $u$ and $du$, we solve the equation $u = 2x + 1$ for $x$ to obtain $x = \frac{1}{2}u - \frac{1}{2}$. Thus

$$\int \frac{x}{2x+1}\, dx = \int \frac{\frac{1}{2}u - \frac{1}{2}}{u} \cdot \frac{1}{2}\, du$$

$$= \int \left(\frac{1}{2} - \frac{1}{2u}\right) \cdot \frac{1}{2}\, du$$

$$= \int \left(\frac{1}{4} - \frac{1}{4u}\right)\, du$$

$$= \int \frac{1}{4}\, du - \frac{1}{4}\int \frac{1}{u}\, du$$

$$= \frac{1}{4}u - \frac{1}{4}\ln|u| + C$$

$$= \frac{1}{4}(2x + 1) - \frac{1}{4}\ln|2x + 1| + C$$

$$= \frac{2x + 1 - \ln|2x + 1|}{4} + C \qquad \blacksquare$$

### Applications

## EXAMPLE 9

The marginal revenue from the sale of $x$ units of a certain product per month is given by

$$R'(x) = \frac{3x^2 + 3}{\sqrt{2x^3 + 6x}}$$

Find the revenue function $R(x)$, assuming that $R(0) = 0$.

**Solution**

We know that $R(x) = \int R'(x)\, dx$ and that $R(0) = 0$, meaning that no revenue is derived from the sale of zero units. Thus,

$$R(x) = \int \frac{3x^2 + 3}{\sqrt{2x^3 + 6x}}\, dx$$

To compute this integral, we let $u = 2x^3 + 6x$, so that $du = (6x^2 + 6)dx$. Since the integral contains the expression $(3x^2 + 3)dx$, we solve the equation $du = (6x^2 + 6)dx$ to obtain $(3x^2 + 3)dx = \frac{1}{2}du$. Making the appropriate substitutions and integrating, we get

$$R(x) = \int \frac{1}{\sqrt{2x^3 + 6x}} \cdot (3x^2 + 3)dx$$

$$= \int \frac{1}{\sqrt{u}} \cdot \frac{1}{2}du = \frac{1}{2}\int u^{-1/2}\, du$$

$$= \frac{1}{2}[2u^{1/2}] + C = u^{1/2} + C$$

$$= (2x^3 + 6x)^{1/2} + C$$
$$= \sqrt{2x^3 + 6x} + C$$

To find the constant $C$, we use the fact that $R(0) = 0$ to obtain
$$R(0) = \sqrt{2 \cdot 0^3 + 6 \cdot 0} + C = 0$$
so $C = 0$. Therefore the monthly revenue function is
$$R(x) = \sqrt{2x^3 + 6x}$$

Many other examples of applications are considered in the exercises. In all of these applications, the rate of change $Q'(x)$ of a quantity $Q(x)$ is known and can be integrated using the technique of integration by substitution to find the quantity $Q(x)$ plus an arbitrary constant $C$. Then the value of $Q(a)$ must be known for one number $a$ in the domain of $Q$ to find that constant $C$.

## SECTION 6.2
### SHORT ANSWER EXERCISES

1. (True or False) The technique of integration by substitution is just an application of the chain rule for integration. _____
2. To evaluate the integral $\int (x^3 + 7)^{20} \cdot 3x^2 \, dx$, you could use the substitution $u = $ _____.
3. To evaluate the integral $\int e^{x^4} \cdot 4x^3 \, dx$, you could use the substitution $u = $ _____.
4. To evaluate the integral $\int \dfrac{2x}{x^2 + 1} \, dx$, you could use the substitution $u = $ _____.
5. (True or False) $\int 2x\sqrt{x^2 + 1} \, dx = 2x \int \sqrt{x^2 + 1} \, dx$. _____
6. (True or False) $\int (5x + 1)^3 \, dx = \frac{1}{4}(5x + 1)^4 + C$. _____
7. The integral $\int \sqrt{x^2 + 1} \, dx$ (a) cannot be found; (b) can be found using integration by substitution; (c) requires a technique beyond the scope of this book; (d) none of the above.
8. If $Q'(x) = \dfrac{3x^2}{x^3 + 1}$, then $Q(x) = \int \dfrac{3x^2}{x^3 + 1} \, dx = \ln|x^3 + 1| + C$. What is the value of $C$ if $Q(0) = 10$? _____
9. The integral $\int \dfrac{2x}{4x + 1} \, dx$ (a) cannot be found; (b) can be found using integration by substitution; (c) requires a technique beyond the scope of this book; (d) none of the above.
10. In attempting to integrate by substitution, you obtain the expression $\int \sqrt{u} \cdot \dfrac{du}{3x}$. What is wrong with this expression? _____

## SECTION 6.2
### EXERCISES

In Exercises 1–10, find the indefinite integrals using the chain rule for integration.

1. $\int (3x + 1)^2 \cdot 3 \, dx$
2. $\int (1 - 5x)^3(-5 \, dx)$
3. $\int (2x^3 - 7)^5 \cdot 6x^2 \, dx$
4. $\int (x^4 - 5)^9 \cdot 4x^3 \, dx$
5. $\int (x^2 - 5)^4 \cdot x \, dx$
6. $\int (7x^5 - 20)^6 \cdot x^4 \, dx$
7. $\int 3x^2 e^{x^3} \, dx$
8. $\int e^{1-x} \, dx$

9. $\int xe^{1-x^2}\, dx$

10. $\int x^3 e^{5-x^4}\, dx$

*In Exercises 11–60, find the indefinite integrals using the technique of integration by substitution.*

11. $\int (7x-5)^4 \cdot 7\, dx$

12. $\int (5-4x)^5(-4\, dx)$

13. $\int (2x-3)^9\, dx$

14. $\int 2e^{4x}\, dx$

15. $\int x^2(2x^3+3)^7\, dx$

16. $\int 7x(x^2+9)^{12}\, dx$

17. $\int x^4\sqrt{x^5+16}\, dx$

18. $\int 2x^3\sqrt{3x^4+6}\, dx$

19. $\int xe^{4-x^2}\, dx$

20. $\int 3xe^{-x^2}\, dx$

21. $\int \dfrac{3e^{1/x}}{x^2}\, dx$

22. $\int \dfrac{x^2}{e^{x^3}}\, dx$

23. $\int 5x^4 e^{6-x^5}\, dx$

24. $\int e^x(2+3e^x)^5\, dx$

25. $\int e^{3x}(7+e^{3x})^4\, dx$

26. $\int (2+t)e^{4t+t^2}\, dt$

27. $\int x\sqrt{x^2+1}\, dx$

28. $\int x^2\sqrt{x^3+1}\, dx$

29. $\int \dfrac{2}{\sqrt{3x-5}}\, dx$

30. $\int \dfrac{5x}{\sqrt{x^2+16}}\, dx$

31. $\int \sqrt{x^2+4x}\,(x+2)\, dx$

32. $\int \sqrt{x^3+3x}\,(x^2+1)\, dx$

33. $\int 4x\sqrt[3]{1+2x^2}\, dx$

34. $\int x^2\sqrt[3]{x^3-10}\, dx$

35. $\int \dfrac{y}{(1+y^2)^4}\, dy$

36. $\int \dfrac{2z^2}{(z^3+12)^5}\, dz$

37. $\int \dfrac{2x+1}{(x^2+x+4)^2}\, dx$

38. $\int \dfrac{4x-7}{(2x^2-7x+3)^3}\, dx$

39. $\int \dfrac{2\, dx}{\sqrt{x}(1+\sqrt{x})^4}$

40. $\int \dfrac{5\, dx}{\sqrt{x}(1+3\sqrt{x})^5}$

41. $\int \dfrac{(\ln x)^4}{x}\, dx$

42. $\int \dfrac{2}{x\ln x}\, dx$

43. $\int \dfrac{1}{x(\ln x)^2}\, dx$

44. $\int (1+\ln x)^2\left(\dfrac{1}{x}\, dx\right)$

45. $\int \dfrac{\ln(3x)}{x}\, dx$

46. $\int \dfrac{\ln x^3}{2x}\, dx$

47. $\int \dfrac{dx}{x+1}$

48. $\int \dfrac{dx}{3x+2}$

49. $\int \dfrac{3x\, dx}{x^2+5}$

50. $\int \dfrac{x\, dx}{4-x^2}$

51. $\int \dfrac{1}{x\ln x}\, dx$

52. $\int \dfrac{e^x}{e^x+1}\, dx$

53. $\int \dfrac{2e^{2x}\ln(3+e^{2x})}{3+e^{2x}}\, dx$

54. $\int \dfrac{\ln(1+\sqrt{x})}{\sqrt{x}+x}\, dx$

55. $\int \dfrac{x}{x+1}\, dx$

56. $\int \dfrac{3x}{5-x}\, dx$

57. $\int \dfrac{2x}{6-5x}\,dx$

58. $\int \dfrac{7x}{7x+3}\,dx$

59. $\int \dfrac{x^2+1}{x+1}\,dx$

60. $\int \dfrac{1-2x^2}{2x+3}\,dx$

*In Exercises 61–68, try to use the method of integration by substitution to evaluate the given integrals. This method may or may not apply to these integrals!*

61. $\int \dfrac{\sqrt{\ln x + 1}}{x}\,dx$

62. $\int e^x \sqrt{3e^x + 5}\,dx$

63. $\int xe^{x^3}\,dx$

64. $\int \dfrac{dx}{\ln x}$

65. $\int x\sqrt{x^3+1}\,dx$

66. $\int \dfrac{\ln x}{x^2}\,dx$

67. $\int \dfrac{\ln(e^x)}{x^2}\,dx$

68. $\int \dfrac{x^3+1}{x+1}\,dx$

**Business and Economics**

69. (Cost) The marginal cost of producing $x$ units of a certain product per month is $C'(x) = 2x/\sqrt{x^2+2500}$. If fixed costs amount to \$2000 per month, find the monthly cost function $C(x)$.

70. (Revenue) The marginal revenue in hundreds of dollars from the sale of $x$ units of a certain product per month is $R'(x) = 4(x^3 + x)(3x^4 + 6x^2)^{-2/3}$.
   (a) Find the total monthly revenue function. Assume that $R(0) = 0$.
   (b) Find the revenue from the sale of 50 units.

71. (Oil spill) The rate (in thousands of gallons/hour) at which oil is spilling from a disabled tanker is given by

$$G'(t) = e^{0.5t}\sqrt{e^{0.5t}-1},\quad 0 \le t \le 10$$

Find the number of gallons that have spilled from the tanker during the first eight hours after the spill begins.

72. (Chemical production) The number of pounds of a certain chemical produced by a factory is predicted to grow according to the formula

$$\dfrac{dA}{dt} = \dfrac{40e^{25t}}{2e^{25t}+1}$$

lb/month. If current production is 20,000 lb/month what will production be in six months?

73. (Profit) The marginal profit from producing and selling $x$ units of a certain product per month is $P'(x) = 5x/\sqrt{x^2+100}$. If the profit from producing and selling 30 units is \$0, find the total monthly profit function.

74. (Revenue) If the marginal revenue from the sale of $x$ units of a certain product per month is $R'(x) = 1200/(2x+5)^3$, find the revenue as a function of the number of units sold per month. Assume that $R(0) = 0$.

**Life Sciences**

75. (Bacteria population) In $t$ hours the population of a certain culture of bacteria will be growing at the rate of $N'(t) = 2e^{0.09t}$ million bacteria per hour. If the culture contains 10 million bacteria at time $t = 0$, how many bacteria will the culture contain in five hours?

**76.** (Employee training)  A new computer programmer with $t$ days of training increases her daily output at the rate of $L'(t) = 2e^{-0.05t}$ lines of computer code per hour of training. If a certain new programmer can program 200 lines of code per day, how many lines can this programmer produce per day after ten days of training?

**77.** (Radioactive decay)  A certain radioactive substance decays according to $A'(t) = -e^{-0.02t}$ grams/year. If 50 g of the substance are present initially, how many grams will remain after 20 years?

**78.** (Real estate)  The value of homes in a certain subdivision is increasing at the rate of

$$V'(t) = 247.5\sqrt{165t + 400}$$

dollars per year for the next five years. If a home in this subdivision is presently worth $110,000, how much will it be worth in five years?

Social Sciences

**79.** (Voting)  During the period 1964–1988, the percentage $P$ of eligible voters who have turned out for gubernatorial and presidential elections has been changing (approximately) according to

$$\frac{dP}{dt} = \frac{-0.69113}{\sqrt[3]{(0.03399t + 1)^4}}, \quad 0 \le t \le 24$$

where $t = 0$ corresponds to 1964, when voter turnout was about 61%.
(a) What is the approximate percentage of voters who turned out in the 1986 gubernatorial elections?
(b) Assuming that the formula for $dP/dt$ remains approximately the same until 1996, predict the percentage of voters who will turn out in the 1996 presidential elections.

**80.** (Substance abuse)  During the period 1980–1990, the percentage $P$ of the adult population in a certain region who are alcohol or drug abusers has been changing according to

$$\frac{dP}{dt} = \frac{4.934}{3\sqrt[3]{2.467t + 27}}, \quad 0 \le t \le 15$$

where $t = 0$ corresponds to 1980, when the percentage was 9%.
(a) What percentage of the adult population consisted of alcohol or drug abusers in 1985?
(b) Assuming that the formula for $dP/dt$ remains approximately the same until 1995, predict the percentage of adults who will be alcohol or drug abusers in 1995.

**81.** Complete the solution begun in Example 6(a) by making the substitution $v = 3u + 4$ in the integral $\frac{1}{2} \int \sqrt{3u + 4} \, du$.

## 6.3  THE DEFINITE INTEGRAL AS A NET CHANGE

In all of the applied problems encountered in this chapter, we have been given the rate $F'(x)$ at which a quantity $F(x)$ changes. The solution of these problems required finding the quantity $F(x)$, and perhaps finding the amount of the

## 6.3 THE DEFINITE INTEGRAL AS A NET CHANGE

quantity $F(b)$ for some number $b$. A closely related problem is to find the **net change in the quantity $F(x)$ between $x = a$ and $x = b$: $F(b) - F(a)$**. This net change in $F(x)$ gives us the value of the definite integral of $F'(x) = f(x)$ from $a$ to $b$ and the difference $F(b) - F(a)$ is denoted by the symbol $F(x)]_a^b$.

**DEFINITION OF THE DEFINITE INTEGRAL**

Let $F(x)$ be an antiderivative of $f(x)$ for $a \leq x \leq b$. Then the definite integral of $f(x)$ from $a$ to $b$ is defined by

$$\int_a^b f(x)\, dx = F(x)]_a^b = F(b) - F(a)$$

This definite integral gives the net change in the antiderivative $F(x)$ from $a$ to $b$.

In the symbolism for the definite integral, $\int_a^b f(x)\, dx$, the numbers $a$ and $b$ are called the *lower and upper limits of integration*, respectively. The function $f(x)$ is the *integrand* and $dx$ is the *differential* of $x$.

In Chapter 7 we will see that there is an alternate definition of the definite integral in terms of the area under the graph of $f(x)$ from $x = a$ to $x = b$ when $f(x) \geq 0$. This alternate interpretation of the definite integral will yield important new applications. Right now, however, let's look at some applications that arise when we interpret the definite integral as the net change in $F(x)$ from $a$ to $b$.

### EXAMPLE 1

If $x$ thousand units are produced per month with monthly cost $C(x)$ measured in thousands of dollars, the marginal cost function of a certain manufacturer is $C'(x) = 0.03x^2 - 0.2x + 2$. Find the total change in cost if production is increased from 5000 units per month to 6000 units per month.

Solution

We are required to find the net change in $C(x)$ from $x = 5$ to $x = 6$, which is given by the definite integral

$$\int_5^6 C'(x)\, dx = C(6) - C(5) = C(x)]_5^6$$

To evaluate this definite integral, we need to find $C(x)$, the antiderivative of $C'(x)$. We proceed as follows:

$$C(x) = \int C'(x)\, dx = \int (0.03x^2 - 0.2x + 2)\, dx = 0.01x^3 - 0.1x^2 + 2x + K$$

where $K$ is a constant. Thus,

$$\int_5^6 C'(x)\, dx = \int_5^6 (0.03x^2 - 0.2x + 2)\, dx = 0.01x^3 - 0.1x^2 + 2x + K]_5^6$$

$$= [0.01(6)^3 - 0.1(6)^2 + 2(6) + K] - [0.01(5)^3 - 0.1(5)^2 + 2(5) + K]$$

$$= 2.16 - 3.6 + 12 + K - 1.25 + 2.5 - 10 - K = 1.81$$

Thus if monthly production is increased from 5000 units to 6000 units, monthly cost will increase by 1.81 thousand dollars = $1810.

You should observe that in Example 1, it was not necessary to determine the constant of integration $K$ since it cancelled out in the computation of $C(x)]_5^6$. This will be true in general for definite integrals, since if $F(x)$ is an antiderivative of $f(x)$, then

$$\int_a^b f(x)\,dx = [F(x) + K]_a^b = [F(b) + K] - [F(a) + K]$$
$$= F(b) + K - F(a) - K = F(b) - F(a)$$

Henceforth we won't bother to add the constant of integration to an antiderivative when we compute a *definite* integral. Alternatively, you can think of taking this constant to be zero in any problem involving a definite integral. Of course, we still must add a constant of integration to an *indefinite* integral:

$$\int f(x)\,dx = F(x) + C$$

## EXAMPLE 2

When a worker is assigned to a certain new job on an assembly line, the rate of change of the number of minutes required to produce the $x$th unit is

$$T'(x) = 10 - 0.03x^2, \qquad 0 \le x \le 18$$

Find the total time required to produce (a) the 5th unit; (b) the 15th unit.

Solution

(a) Since $T(x)$ is the number of minutes required to produce the first $x$ units, the number of minutes required to produce the fifth unit is $T(5) - T(4)$, or

$$\int_4^5 T'(x)\,dx = \int_4^5 (10 - 0.03x^2)\,dx$$
$$= 10x - 0.01x^3]_4^5 \qquad \text{(Let } C = 0 \text{ in the integration)}$$
$$= [10(5) - 0.01(5)^3] - [10(4) - 0.01(4)^3]$$
$$= 9.39 \text{ minutes}$$

(b) The time required to produce the 15th unit is $T(15) - T(14)$:

$$\int_{14}^{15} T'(x)\,dx = \int_{14}^{15} (10 - 0.03x^2)\,dx = 10x - 0.01x^3]_{14}^{15}$$
$$= [10(15) - 0.01(15)^3] - [10(14) - 0.01(14)^3]$$
$$= 3.69 \text{ minutes}$$

Observe that the on-the-job work experience has significantly improved the worker's performance between the 5th unit and the 15th unit. ■

## EXAMPLE 3

The number of homeless people in a city is increasing at the rate of $H'(t) = 800\sqrt[3]{t}$ people per year. If $t = 0$ corresponds to 1990, how many more homeless people will there be in 1998 than in 1990?

Solution

Note that $H(t)$ is the number of homeless people in the city $t$ years after 1990. So the net change in the number of homeless people between 1990 and 1998 is $H(8) - H(0)$, or

$$\int_0^8 H'(t)\,dt = \int_0^8 800\sqrt[3]{t}\,dt = \int_0^8 800 t^{1/3}\,dt$$
$$= 600 t^{4/3}\big]_0^8 = 600(8)^{4/3} - 600(0)^{4/3}$$
$$= 9600 \text{ people}$$

---

### EXAMPLE 4 (Participative)

**Solution**

Let's compute the definite integral $\int_{-1}^{2} (3x - x^2)\,dx$.

Let $F(x)$ be an antiderivative of $3x - x^2$. We are required to compute $F(2) -$ _____. Since an antiderivative of $3x - x^2$ is _____ (with $C = 0$), we have

$$\int_{-1}^{2} (3x - x^2)\,dx = \left(\tfrac{3}{2}x^2 - \tfrac{1}{3}x^3\right)\Big]_{-1}^{2}$$
$$= \left[\tfrac{3}{2}(2)^2 - \tfrac{1}{3}(2)^3\right] - [\underline{\phantom{XXXXXXXX}}]$$
$$= \underline{\phantom{XX}}$$

### Computing Definite Integrals Using Substitution

When computing a definite integral such as $\int_0^2 x(x^2 + 4)^3\,dx$ by substitution, we have a choice of methods available.

**Method 1** We can compute $F(x) = \int x(x^2 + 4)^3\,dx$ by using either the chain rule for integration or the method of substitution, *write our result in terms of the original variable $x$*, and then evaluate $F(x)$ at the given limits of integration:

$$\int_0^2 x(x^2 + 4)^3\,dx = F(2) - F(0)$$

**Method 2** We can compute the antiderivative $G(u) \int x(x^2 + 4)^3\,dx$ *in terms of $u$*, where $u = x^2 + 4$, and find the values of $u$ that correspond to the limits $x = 0$ and $x = 2$. These values, $u = 0^2 + 4 = 4$ and $u = 2^2 + 4 = 8$, are the new limits of integration:

$$\int_0^2 x(x^2 + 4)^3\,dx = G(8) - G(4)$$

### EXAMPLE 5

**Solution**

Compute $\int_0^2 x(x^2 + 4)^3\,dx$ using each of the methods just described.

**Method 1** We compute the *indefinite* integral $\int x(x^2 + 4)^3\,dx$ using the substitution $u = x^2 + 4$, so that $du = 2x\,dx$ and $x\,dx = \tfrac{1}{2}du$. Then

---

**SOLUTIONS TO EXAMPLE 4**  $F(-1), \tfrac{3}{2}x^2 - \tfrac{1}{3}x^3, \tfrac{3}{2}(-1)^2 - \tfrac{1}{3}(-1)^3, \tfrac{3}{2}$

$$\int x(x^2 + 4)^3 \, dx = \int (x^2 + 4)^3 \cdot x \, dx = \int u^3 \cdot \tfrac{1}{2} du$$

$$= \tfrac{1}{2} \int u^3 \, du = \tfrac{1}{8} u^4 \qquad \text{(Let } C = 0 \text{ since we will be computing a definite integral)}$$

$$= \tfrac{1}{8}(x^2 + 4)^4 \qquad \text{(Change back to the \textit{original} variable } x\text{)}$$

The given *definite* integral is now

$$\int_0^2 x(x^2 + 4)^3 \, dx = \tfrac{1}{8}(x^2 + 4)^4 \Big]_0^2$$

$$= \tfrac{1}{8}(2^2 + 4)^4 - \tfrac{1}{8}(0^2 + 4)^4$$

$$= 480$$

**Method 2** We compute the indefinite integral $\int x(x^2 + 4)^3 \, dx$ using the substitution $u = x^2 + 4$, so that $du = 2x\,dx$ and $x\,dx = \tfrac{1}{2} du$. Thus

$$\int x(x^2 + 4)^3 \, dx = \int (x^2 + 4)^3 \cdot x\,dx = \int u^3 \cdot \tfrac{1}{2} du$$

$$= \tfrac{1}{8} u^4 \qquad \text{(Let } C = 0; \text{ leave answer in terms of \textit{new} variable } u\text{)}$$

To find the limits that correspond to the variable $u$, we use the formula $u = x^2 + 4$ and note that when $x = 0$, we have $u = 0^2 + 4 = 4$, and when $x = 2$, we have $u = 2^2 + 4 = 8$. Then the given definite integral is

$$\int_0^2 x(x^2 + 4)^3 \, dx = \tfrac{1}{8} u^4 \Big]_4^8 \qquad \text{(The new limits 4 and 8 refer to the new variable } u\text{)}$$

$$= \tfrac{1}{8}(8)^4 - \tfrac{1}{8}(4)^4 = 480 \qquad \blacksquare$$

*Comment:* When computing this integral by Method 2, we can write the computation more compactly as follows:

$$\int_0^2 x(x^2 + 4)^3 \, dx = \int_4^8 u^3 \cdot \tfrac{1}{2} du = \tfrac{1}{8} u^4 \Big]_4^8 = \tfrac{1}{8}(8)^4 - \tfrac{1}{8}(4)^4 = 480$$

In general, when finding a definite integral $\int_a^b h(x) \, dx$ by substitution, we can do the problem in either of two ways:

**Method 1**
(a) Use either the chain rule for integration or the method of substitution to compute the indefinite integral $F(x) = \int h(x) \, dx$. Let $C = 0$ and write your answer in terms of the original variable $x$.
(b) Find the definite integral $\int_a^b h(x) \, dx = F(b) - F(a)$. Note that the original limits $a$ and $b$ are used in this computation, because the indefinite integral, $F(x)$, is written in terms of the original variable of integration, $x$.

**Method 2**
(a) Assuming that $\int h(x) \, dx = \int f(g(x)) g'(x) \, dx$, use the method of substitution with $u = g(x)$ and find the indefinite integral $F(u) = \int f(u) \, du$ in terms of the new variable $u$. Let $C = 0$.

**(b)** Find the definite integral

$$\int_a^b h(x)\,dx = \int_a^b f(g(x))g'(x)\,dx = \int_{g(a)}^{g(b)} f(u)\,du = F(u)\Big|_{g(a)}^{g(b)}$$

Note that the limits have been changed to $g(a)$ and $g(b)$ because the indefinite integral $F(u)$ is written in terms of the new variable $u$, so $u = g(a)$ when $x = a$, and $u = g(b)$ when $x = b$.

In this text we will usually do definite integrals by substitution using Method 2.

**EXAMPLE 6**

Compute $\int_1^e (\ln x / x)\,dx$.

**Solution**

We make the substitution $u = \ln x$, so that $du = (1/x)\,dx$. Using Method 2, we change the limits on $x$ to new limits on $u$. Thus when $x = 1$, we have $u = \ln 1 = 0$ and when $x = e$, we have $u = \ln e = 1$. Therefore

$$\int_1^e \frac{\ln x}{x}\,dx = \int_1^e \ln x \cdot \frac{1}{x}\,dx = \int_0^1 u\,du = \frac{u^2}{2}\Big|_0^1 = \frac{1^2}{2} - \frac{0^2}{2} = \frac{1}{2}$$

There are a number of important properties of the definite integral, which we state here.

**PROPERTIES OF THE DEFINITE INTEGRAL**

If $f(x)$ and $g(x)$ are continuous on the appropriate intervals, then

1. $\int_a^a f(x)\,dx = 0$

2. $\int_a^b f(x)\,dx = -\int_b^a f(x)\,dx$

3. $\int_a^b kf(x)\,dx = k\int_a^b f(x)\,dx$

4. $\int_a^b [f(x) \pm g(x)]\,dx = \int_a^b f(x)\,dx \pm \int_a^b g(x)\,dx$

5. $\int_a^c f(x)\,dx = \int_a^b f(x)\,dx + \int_b^c f(x)\,dx$

Observe that Property 2 extends the definition of the definite integral to include the case where the upper limit of integration is less than the lower limit of integration. In Property 5, the numbers $a, b,$ and $c$ do not have to be in increasing order. We will prove Properties 2 and 4 and leave the proofs of the remaining properties as exercises.

**Proof of Property 2**

Let $F(x)$ be an antiderivative of $f(x)$ on $[a, b]$. Then

$$\int_a^b f(x)\,dx = F(b) - F(a) = -[F(a) - F(b)] = -\int_b^a f(x)\,dx$$

**Proof of Property 4**

We will prove this property for the case when the integrand is $f(x) + g(x)$. Let $F(x)$ be an antiderivative of $f(x)$ and $G(x)$ an antiderivative of $g(x)$ on $[a, b]$.

Then, since $D_x[F(x) + G(x)] = f(x) + g(x)$, the sum $F(x) + G(x)$ is an antiderivative of $f(x) + g(x)$ on $[a, b]$. Thus

$$\int_a^b [f(x) + g(x)]\, dx = [F(x) + G(x)]_a^b = F(b) + G(b) - [F(a) + G(a)]$$
$$= [F(b) - F(a)] + [G(b) - G(a)]$$
$$= \int_a^b f(x)\, dx + \int_a^b g(x)\, dx$$

■

## EXAMPLE 7

Compute $\int_1^2 (1/x - 2e^x + 4x^2)\, dx$.

### Solution

We first use Property 4 extended to a sum or difference of three functions to write

$$\int_1^2 \left(\frac{1}{x} - 2e^x + 4x^2\right) dx = \int_1^2 \frac{1}{x}\, dx - \int_1^2 2e^x\, dx + \int_1^2 4x^2\, dx$$

Next we use Property 3 and then complete the solution using the formulas of Section 6.1.

$$\int_1^2 \left(\frac{1}{x} - 2e^x + 4x^2\right) dx = \int_1^2 \frac{1}{x}\, dx - 2\int_1^2 e^x\, dx + 4\int_1^2 x^2\, dx$$
$$= \ln |x|\Big|_1^2 - 2e^x\Big|_1^2 + \tfrac{4}{3}x^3\Big|_1^2$$
$$= (\ln 2 - \ln 1) - 2(e^2 - e^1) + \left[\tfrac{4}{3}(2)^3 - \tfrac{4}{3}(1)^3\right]$$
$$= \ln 2 - 2e^2 + 2e + \tfrac{28}{3} \approx 0.6849$$

■

## SECTION 6.3
### SHORT ANSWER EXERCISES

1. If $G(x)$ is an antiderivative of $g(x)$, then $\int_c^d g(x)\, dx =$ _____.
2. (True or False) If $f(x)$ is differentiable on $[0, 5]$, then $\int_0^5 f'(x)\, dx = f(5)$. _____
3. To compute $\int_0^1 e^x\, dx$, we have $\int_0^1 e^x\, dx = e^x]_0^1 =$ _____.
4. If $F'(x)$ is the rate of change of a quantity with respect to $x$, then the net change of $F(x)$ between $x = 2$ and $x = 5$ is given by: (a) $F(2) - F(5)$; (b) $\int F'(x)\, dx$; (c) $\int_2^5 F'(x)\, dx$; (d) $\int_2^5 F(x)\, dx$.
5. The definite integral $\int_a^b f(x)\, dx$ is typically a real number, whereas the indefinite integral $\int f(x)\, dx$ is typically a _____.
6. (True or False) If we use integration by substitution to compute a definite integral, we must always use the original limits to find the result. _____
7. In the integral $\int_2^3 2x^2 e^{x^3}\, dx$, we make the substitution $u = x^3$. If we use Method 2 of this section to find the result, the new limits on $u$ will be $u =$ _____ (lower limit) and $u =$ _____ (upper limit).
8. (True or False) If we use integration by substitution to compute a definite integral, it is correct both to change the limits and to retain the original variable in the antiderivative in computing the result. _____
9. (True or False) $\int_1^2 (2x^3 + xe^x)\, dx = 2\int_1^2 x^3\, dx + x\int_1^2 e^x\, dx$ _____
10. (True or False) If we make the substitution $u = \ln x$ to evaluate the integral $\int_1^2 \frac{\ln x}{x}\, dx$, then $\int_1^2 \frac{\ln x}{x}\, dx = \int_1^2 u\, du$. _____

## SECTION 6.3
## EXERCISES

*In Exercises 1–44, evaluate the definite integrals.*

1. $\int_{-1}^{2} 3 \, dx$
2. $\int_{-2}^{3} -4 \, dx$
3. $\int_{0}^{5} x \, dx$
4. $\int_{-1}^{4} 3x \, dx$
5. $\int_{-4}^{4} (1 - 2x) \, dx$
6. $\int_{-1}^{0} (2 - 4x) \, dx$
7. $\int_{1}^{3} (x^2 - 2x + 4) \, dx$
8. $\int_{-1}^{2} (3 - 5x - x^2) \, dx$
9. $\int_{-3}^{5} (2x^2 - x^3) \, dx$
10. $\int_{-3}^{-2} (2x^3 - 3x) \, dx$
11. $\int_{-1}^{1} (2y - 3) \, dy$
12. $\int_{-2}^{3} (z^2 - 2z + 3) \, dz$
13. $\int_{0}^{5} (t^2 - 3t) \, dt$
14. $\int_{1}^{4} (t^3 - 7t + 3) \, dt$
15. $\int_{-1}^{1} (12x^6 - x^4) \, dx$
16. $\int_{1}^{0} (7x - x^3 + x^5) \, dx$
17. $\int_{0}^{1} 3e^x \, dx$
18. $\int_{-1}^{2} e^{4x} \, dx$
19. $\int_{1}^{2} (-2e^{2x}) \, dx$
20. $\int_{1}^{3} (1/x) \, dx$
21. $\int_{2}^{4} \frac{6 \, dx}{x}$
22. $\int_{2}^{3} \left(\frac{1}{x} - \frac{1}{x^2}\right) dx$
23. $\int_{-2}^{-1} \left(\frac{2}{x^2} - \frac{1}{x^3}\right) dx$
24. $\int_{-4}^{-2} (3x^{-2} + x^{-4}) \, dx$
25. $\int_{0}^{4} 2\sqrt{y} \, dy$
26. $\int_{1}^{9} 4\sqrt[3]{y} \, dy$
27. $\int_{1}^{16} (u^{1/2} - u^{3/2}) \, du$
28. $\int_{8}^{81} (u^{2/3} + 2u^{1/3}) \, du$
29. $\int_{4}^{9} \sqrt{3x - 1} \, dx$
30. $\int_{2}^{5} \sqrt{5x + 2} \, dx$
31. $\int_{2}^{4} x\sqrt{x^2 - 1} \, dx$
32. $\int_{-1}^{2} 3x\sqrt{2x^2 + 5} \, dx$
33. $\int_{-1}^{3} 2x(x^2 - 4)^5 \, dx$
34. $\int_{-2}^{-1} x^2(x^3 - 5)^3 \, dx$
35. $\int_{0}^{1} \frac{2}{2x + 1} \, dx$
36. $\int_{1}^{3} \frac{2}{3x - 1} \, dx$
37. $\int_{e}^{e^2} \frac{(\ln x)^3}{x} \, dx$
38. $\int_{1}^{4} \frac{e^{\sqrt{x}}}{\sqrt{x}} \, dx$
39. $\int_{-1}^{1} (2t - 1)e^{(t^2 - t)} \, dt$
40. $\int_{e^2}^{e^3} \frac{2}{z \ln z} \, dz$
41. $\int_{-1}^{3} \frac{x \, dx}{3x^2 + 4}$
42. $\int_{2}^{3} \frac{x^2}{x^3 - 5} \, dx$

43. $\displaystyle\int_0^1 e^x \sqrt[3]{e^x + 5}\, dx$

44. $\displaystyle\int_0^1 \frac{\ln(x + 1)}{x + 1}\, dx$

45. **(Manufacturing cost)** A certain manufacturer has a marginal cost function $C'(x) = 5(2x - 9)^2$, where $C(x)$ is the total daily cost in dollars of manufacturing $x$ units. How much will monthly cost increase if the manufacturer raises output from 7 units/day to 12 units/day?

46. **(Repair and maintenance cost)** If $R(t)$ is the total (accumulated) cost of maintaining a school bus for $t$ years, then the rate of change of repair and maintenance costs (in dollars per year) for the school bus is given by $R'(t) = 3t^2 + 600$, where $t = 0$ corresponds to 1990, the year the bus was purchased. If the school district that purchased the bus plans to keep it for seven years, find the total maintenance and repair costs during that period.

47. **(Oil consumption)** A certain country will consume oil at the rate of $A'(t) = 1.4e^{0.07t}$ billions of gallons of oil per year for $0 \le t \le 5$. If $t = 0$ corresponds to 1991, how much oil will this country consume in 1996?

48. **(Production)** A manufacturer is expanding his facilities so that $t$ months from now, he expects to be producing units at the rate of $A'(t) = 2000e^{0.015t}$ units per month. What will the manufacturer's total increase in production be for the next 12 months?

**Life Sciences**

49. **(Spread of infection)** The rate at which people in a town are becoming infected with a certain disease is $I'(t) = 400t - 0.3t^3$ people per day, where $t = 0$ corresponds to the day that the disease begins to spread. How many people will be infected after one week?

**Social Sciences**

50. **(Memory)** After $t$ minutes of study, a certain history student can memorize $N(t)$ important historical dates. If the rate at which he learns these dates is approximately

$$N'(t) = \frac{1}{3}\sqrt[3]{t^2}, \qquad 0 \le t \le 15$$

how many dates does this student memorize (a) during the first five minutes of study? (b) during the interval $t = 5$ to $t = 10$? (c) during the interval $t = 10$ to $t = 15$?

51. **(Depreciation)** A piece of machinery is losing value at the rate of $V'(t) = -2000e^{-0.2t}$ dollars per year, where $t = 0$ corresponds to 1990, the year it was purchased. How much of its original value will this machine lose during the period from 1990 to 1998?

52. **(Gasoline demand)** The demand for gasoline in a certain community is increasing according to $A'(t) = 0.125e^{0.05t}$ millions of gallons of gasoline per year, where $t = 0$ corresponds to 1991. Find the total increase in consumption between 1991 and 1999.

53. **(Toxic waste)** Liquid toxic waste is leaking from a storage container at the rate of $G'(t) = 0.05t$ gal/hr. How many gallons will leak out during the first 24-hr period after the leakage begins?

**Business and Economics**

54. **(Profit)** The monthly marginal profit function of a certain firm is $P'(x) = (500 - x)/\sqrt{1000x - x^2}$ hundreds of dollars per unit for $0 < x < 1000$. Find the total increase in profit the firm will enjoy if it increases production and sales from 400 units to 500 units per month.

**55.** (Production costs) By investing in a new technology, a firm expects to save production costs at the rate of $A'(t) = 60000 - 4000t$ dollars per year, where $t = 0$ corresponds to 1992, the year in which the new technology is first used. Find the total savings on production costs from 1994 to 1998.

**56.** (Patient capacity) Due to an ongoing expansion program, the patient capacity $N$ of a mental health facility is increasing at the rate of $dN/dt = 825/\sqrt{165t + 400}$ patients per year for $0 \le t \le 6$. If $t = 0$ corresponds to the beginning of 1991, when the expansion began, find the total increase in the patient capacity of the clinic from 1991 to 1996.

**57.** (Traffic) Due to an increase in population, the average number $N$ of vehicles in a line of traffic at a certain highway exit at 6:00 P.M. on a working weekday has been increasing at the rate of $dN/dt = 181641/3\sqrt[3]{(181641t + 1000000)^2}$ vehicles per month for $0 \le t \le 36$. If $t = 0$ corresponds to the beginning of 1992, find the total increase in traffic at the exit from 1992 to 1994.

## 6.4 INTEGRATION BY PARTS

Imagine that the production department of your company is considering investing in new robotic technology. The *estimated* rate of savings in production costs due to this technology is $s'(t) = 6te^{-0.2t}$ thousands of dollars per year over the next five years. To compute the total savings over the next five years, we must compute $\int_0^5 s'(t)\, dt = \int_0^5 6te^{-0.2t}\, dt$. Although this integral cannot be computed by the techniques we have studied thus far, it is a routine problem using the technique of this section—integration by parts. The computation of this integral is carried out in Example 6.

The integration by parts formula is derived by reversing the product rule for differentiation. If $u(x)$ and $v(x)$ are differentiable functions of $x$, then by the product rule,

$$\frac{d}{dx}(uv) = u\frac{dv}{dx} + v\frac{du}{dx}$$

Writing this in differential form by multiplying the preceding equation by $dx$, the differential of $x$, we get

$$d(uv) = u\, dv + v\, du$$

Solving this equation for $u\, dv$ and then integrating, we obtain

$$u\, dv = d(uv) - v\, du$$

$$\int u\, dv = \int d(uv) - \int v\, du \qquad (1)$$

Now, $\int d(uv) = uv + C$. We will not bother to add the constant of integration $C$ at this point, however, since we will combine it with the constant obtained from the second integral on the right-hand side of equation (1). So, replacing $\int d(uv)$ by $uv$ in (1) gives us the *integration by parts formula*:

## CHAPTER 6 INTEGRATION

**INTEGRATION BY PARTS FORMULA**

$$\int u\, dv = uv - \int v\, du$$

It takes a bit of practice to be able to use this formula correctly. Let's look at two examples and then make some general observations.

---

**EXAMPLE 1** Find $\int x^2 \ln x\, dx$ for $x > 0$, using integration by parts.

**Solution** In the integration by parts formula, we must begin by making a choice for $u$ and $dv$ so that the product $u\, dv$ is the quantity $x^2 \ln x\, dx$. The factors $u$ and $dv$ constitute the "parts" into which we split the quantity $x^2 \ln x\, dx$. There are a number of different possibilities, three of which we list here:

$$\underbrace{x^2}_{u} \cdot \underbrace{\ln x\, dx}_{dv}, \qquad \underbrace{(\ln x)}_{u} \cdot \underbrace{x^2\, dx}_{dv}, \qquad \underbrace{x}_{u} \cdot \underbrace{x \ln x\, dx}_{dv}$$

Note that in the integration by parts formula, we *choose* the factors $u$ and $dv$ and we must *compute* the quantities $du$ and $v$. Thus if we choose, for example,

$$u = x^2 \quad \text{and} \quad dv = \ln x\, dx$$

then we must compute

$$du = 2x\, dx \quad \text{and} \quad v = \int dv = \int \ln x\, dx$$

Since we cannot find $\int \ln x\, dx$ easily, this choice of $u$ and $dv$ is not useful. Let's try the choice

$$u = \ln x \quad \text{and} \quad dv = x^2\, dx$$

We compute:

$$du = \frac{1}{x}\, dx \quad \text{and} \quad v = \int dv = \int x^2\, dx = \frac{1}{3}x^3$$

We need not add an arbitrary constant $C$ to $\frac{1}{3}x^3$ at this point, since it can be shown (see Exercise 45) that the choice $C = 0$ is always permissible when we compute $v$ from $dv$ in integration by parts. Now, the integration by parts formula gives us

$$\int u\, dv = uv - \int v\, du$$

$$\int \underbrace{(\ln x)}_{u} \cdot \underbrace{x^2\, dx}_{dv} = \underbrace{\ln x}_{u} \cdot \underbrace{\tfrac{1}{3}x^3}_{v} - \int \underbrace{\tfrac{1}{3}x^3}_{v} \cdot \underbrace{\tfrac{1}{x}\, dx}_{du}$$

$$= \tfrac{1}{3}x^3 \ln x - \tfrac{1}{3}\int x^2\, dx$$

Note that the integral on the right-hand side $\left(\text{that is, } \int v\, du = \frac{1}{3} \int x^2\, dx\right)$ is quite easy and we can now complete the problem as follows:

$$\int x^2 \ln x \, dx = \int (\ln x) \cdot x^2 \, dx = \frac{1}{3}x^3 \ln x - \frac{1}{3}\left(\frac{x^3}{3}\right) + C$$

$$= \frac{1}{3}x^3 \ln x - \frac{1}{9}x^3 + C \qquad \blacksquare$$

**EXAMPLE 2**  Use integration by parts to find $\int xe^x \, dx$.

**Solution**  The quantity $xe^x \, dx$ can be written as a product in several ways:

$$\underset{u\ \ \ dv}{x \cdot e^x \, dx}, \quad \underset{u\ \ \ dv}{e^x \cdot x \, dx}, \quad \underset{u\ \ \ dv}{xe^x \cdot dx}$$

Let's make the choice $u = x$ and $dv = e^x \, dx$. Then

$$du = 1 \cdot dx \quad \text{and} \quad v = \int e^x \, dx = e^x$$

The integration by parts formula gives:

$$\int u \, dv = uv - \int v \, du$$

$$\underset{u\ \ \ dv}{\int x \cdot e^x \, dx} = \underset{u\ \ \ v}{x \cdot e^x} - \underset{v\ \ \ du}{\int e^x \cdot dx} = xe^x - e^x + C$$

In this problem it would be inappropriate to let $u = e^x$ and $dv = x \, dx$, because $du = e^x \, dx$ and $v = \int x \, dx = \frac{1}{2}x^2$, and the integration by parts formula yields

$$\int u \, dv = uv - \int v \, du$$

$$\underset{u\ \ \ dv}{\int e^x \cdot x \, dx} = \underset{u\ \ \ v}{e^x \cdot \tfrac{1}{2}x^2} - \underset{v\ \ \ du}{\int \tfrac{1}{2}x^2 \cdot e^x \, dx}$$

Since the integral on the right-hand side is more difficult than the original integral, this choice of $u$ and $dv$ is a poor one. $\blacksquare$

The integration by parts formula is typically used on an integral of the form $\int f(x) \, dx$ where $f(x) \, dx$ can be expressed as a product $u \cdot dv$. In many cases the integral involves a power of $x$ multiplied by an exponential or logarithmic function, such as $\int x^3 e^{2x} \, dx$ or $\int x \ln(3x) \, dx$. However, integration by parts may also be applied to an integral involving a power of $x$ multiplied by a term that is linear in $x$ raised to a power, such as $\int x^2(2x + 1)^4 \, dx$, even though such integrals are perhaps more easily done by substitution. There are also other integrals involving trigonometric functions that are most easily handled by integration by parts, but we shall not consider integrals involving trigonometric functions here.

From our experience in Examples 1 and 2, we see that making an appropriate choice for $u$ and $dv$ in an integration by parts problem is critical. Since any choice for $u$ and $dv$ requires us to find $du$ and $v = \int dv$ and then to determine $\int v \, du$, we can state the following general guidelines for choosing $u$ and $dv$:

**GENERAL GUIDELINES FOR CHOOSING $u$ AND $dv$ IN INTEGRATION BY PARTS**

1. Choose $dv$ so that it can be easily integrated to determine $v$.
2. Choose $u$ so that $du$ is simpler than $u$.
3. Choose $u$ and $dv$ so that $\int v \, du$ is simpler than the original integral.

For example, in the computation of $\int x \ln(3x) \, dx$, an appropriate choice might be to let $u = \ln(3x)$ and $dv = x \, dx$. This choice is a good one, since

1. $v = \int dv = \int x \, dx = \frac{1}{2}x^2$ is readily found;

2. $du = \frac{3}{3x} dx = \frac{1}{x} dx$ is simpler than $\ln(3x)$;

3. $\int v \, du = \int \frac{1}{2}x^2 \cdot \frac{1}{x} dx = \frac{1}{2} \int x \, dx$ is simpler than the original integral.

To complete the computation of $\int x \ln(3x) \, dx$, we write

$$\int u \, dv = uv - \int v \, du$$

$$\int \underset{u}{\ln(3x)} \cdot \underset{dv}{x \, dx} = \underset{u}{[\ln(3x)]} \cdot \underset{v}{\tfrac{1}{2}x^2} - \int \underset{v}{\tfrac{1}{2}x^2} \cdot \underset{du}{\tfrac{1}{x} dx}$$

$$= \tfrac{1}{2}x^2 \ln(3x) - \tfrac{1}{2} \int x \, dx$$

$$= \tfrac{1}{2} x^2 \ln(3x) - \tfrac{1}{4} x^2 + C$$

Therefore, $\int x \ln(3x) \, dx = \tfrac{1}{2} x^2 \ln(3x) - \tfrac{1}{4} x^2 + C$.

When we apply the general guidelines for selecting $u$ and $dv$ to some specific types of integrals, we arrive at the following list of specific suggestions for choosing $u$ and $dv$.

**CHOOSING $u$ AND $dv$: SPECIFIC CASES OF INTEGRATION BY PARTS**

Let $m$ and $n$ be positive integers.

1. For an integral of the form $\int x^n (\ln x)^m \, dx$, let

$$u = (\ln x)^m \quad \text{and} \quad dv = x^n \, dx$$

2. For an integral of the form $\int x^n e^{ax} \, dx$, where $a$ is a constant, let

$$u = x^n \quad \text{and} \quad dv = e^{ax} \, dx$$

3. For an integral of the form $\int x^n (ax + b)^m \, dx$, let

$$u = x^n \quad \text{and} \quad dv = (ax + b)^m \, dx$$

## 6.4 INTEGRATION BY PARTS

*Comment:* If either *m* or *n* is not a positive integer, the suggested choice of *u* and *dv* in the preceding box may or may not be appropriate.

---

**EXAMPLE 3 (Participative)**

**Solution**

Let's integrate $\int x\sqrt{x + 4}\, dx$ in two ways: **(a)** by using integration by parts, and **(b)** by using substitution.

**(a)** Since $\int x\sqrt{x + 4}\, dx = \int x(x + 4)^{1/2}\, dx$ we follow the suggestions in the preceding box and let $u = x$ and $dv = $ _____. Then

$$du = dx \quad \text{and} \quad v = \int (x + 4)^{1/2}\, dx = \tfrac{2}{3}(x + 4)^{3/2}$$

The integration by parts formula then yields

$$\int u\, dv = uv - \int v\, du$$

$$\int x(x + 4)^{1/2}\, dx = \underline{\hspace{2cm}} - \int \tfrac{2}{3}(x + 4)^{3/2}\, dx$$

Finally, computing the integral on the right-hand side gives us

$$\int x\sqrt{x + 4}\, dx = \underline{\hspace{5cm}}$$

**(b)** To find $\int x(x + 4)^{1/2}\, dx$ by substitution, we let $u = x + 4$, so that $du = $ _____ and $x = $ _____. Then

$$\int x(x + 4)^{1/2}\, dx = \int (u - 4)u^{1/2}\, du = \int (u^{3/2} - 4u^{1/2})\, du$$

$$= \int u^{3/2}\, du - 4\int \underline{\hspace{1.5cm}}$$

$$= \tfrac{2}{5}u^{5/2} - \underline{\hspace{1.5cm}} + C$$

$$= \tfrac{2}{5}(x + 4)^{5/2} - \underline{\hspace{2cm}} + C$$

In Exercise 46 you are asked to show that the antiderivatives obtained in (a) and (b) above are equivalent by using a bit of algebra. ■

Sometimes the procedure of integration by parts must be repeated more than once to find a particular integral. This will be true, for example, in integrals of the form $\int x^n(\ln x)^m\, dx$ where $m \geq 2$ and in integrals of the form $\int x^n e^{ax}\, dx$ or $\int x^n(ax + b)^m\, dx$ where $n \geq 2$ and *m* and *n* are positive integers. Let's look at an example.

---

**EXAMPLE 4**

**Solution**

Use integration by parts to find $\int x(\ln x)^2\, dx$, where $x > 0$.

Since this integral has the form $\int x^n(\ln x)^m\, dx$, we let $u = (\ln x)^2$ and $dv = x\, dx$. Then

---

**SOLUTIONS TO EXAMPLE 3**

(a) $(x + 4)^{1/2}\, dx, \tfrac{2}{3}x(x + 4)^{3/2}, \tfrac{2}{3}x(x + 4)^{3/2} - \tfrac{4}{15}(x + 4)^{5/2} + C$

(b) $dx, u - 4, u^{1/2}\, du, \tfrac{8}{3}u^{3/2}, \tfrac{8}{3}(x + 4)^{3/2}$

$$du = \frac{2 \ln x}{x} dx \quad \text{and} \quad v = \int x\, dx = \frac{1}{2}x^2$$

The integration by parts formula now gives us

$$\int u\, dv = uv - \int v\, du$$

$$\int \underset{u}{(\ln x)^2} \cdot \underset{dv}{x\, dx} = \underset{u}{(\ln x)^2} \cdot \underset{v}{\tfrac{1}{2}x^2} - \int \underset{v}{\tfrac{1}{2}x^2} \cdot \underset{du}{\tfrac{2 \ln x}{x}\, dx}$$

$$= \tfrac{1}{2}x^2 (\ln x)^2 - \int x \ln x\, dx \qquad (2)$$

We will add a constant of integration to the final result later. To complete the problem, we compute $\int x \ln x\, dx$ using integration by parts a second time, with $u = \ln x$ and $dv = x\, dx$ so that

$$du = \frac{1}{x} dx \quad \text{and} \quad v = \int x\, dx = \tfrac{1}{2}x^2$$

The integration by parts formula gives us

$$\int \underset{u}{(\ln x)} \cdot \underset{dv}{x\, dx} = \underset{u}{\ln x} \cdot \underset{v}{\tfrac{1}{2}x^2} - \int \underset{v}{\tfrac{1}{2}x^2} \cdot \underset{du}{\tfrac{1}{x} dx}$$

$$= \tfrac{1}{2}x^2 \ln x - \tfrac{1}{2}\int x\, dx$$

$$= \tfrac{1}{2}x^2 \ln x - \tfrac{1}{4}x^2 \qquad (3)$$

Substituting (3) into (2) and adding a constant of integration, we get

$$\int x(\ln x)^2\, dx = \tfrac{1}{2}x^2(\ln x)^2 - \left[\tfrac{1}{2}x^2 \ln x - \tfrac{1}{4}x^2\right] + C$$

$$= \tfrac{1}{2}x^2(\ln x)^2 - \tfrac{1}{2}x^2 \ln x + \tfrac{1}{4}x^2 + C \qquad ∎$$

---

### EXAMPLE 5 (Participative)

Solution

Let us use integration by parts to find $\int \ln x\, dx$ for $x > 0$.

This is a somewhat unusual problem since integration by parts is typically used for products and this example does not contain a product of functions. However, if we let $u = \ln x$ and $dv = dx$, then $du = $ _____ and $v = \int dx = $ _____, so that the integration by parts formula yields

$$\int \underset{u}{\ln x} \cdot \underset{dv}{dx} = \underset{u}{\ln x} \cdot \underset{v}{\underline{\phantom{xxx}}} - \int \underset{v}{\underline{\phantom{xxx}}} \cdot \underset{du}{\underline{\phantom{xxx}}}$$

$$= x \ln x - \int dx$$

$$= \underline{\phantom{xxxxxxxx}} \qquad ∎$$

## EXAMPLE 6

Find $\int_0^5 6te^{-0.2t}\, dt$.

### Solution

Recall that this integral represents the total savings over five years enjoyed by a company with a new technology that produces an estimated savings rate of $s'(t) = 6te^{-0.2t}$ thousands of dollars per year in production costs. We write

$$\int_0^5 6te^{-0.2t}\, dt = 6\int_0^5 te^{-0.2t}\, dt$$

To compute the definite integral, we will first find $\int te^{-0.2t}\, dt$ by letting $u = t$ and $dv = e^{-0.2t}\, dt$, so that

$$du = dt \quad \text{and} \quad v = \int e^{-0.2t}\, dt = \frac{e^{-0.2t}}{-0.2} = -5e^{-0.2t}$$

The integration by parts formula gives us

$$\int u\, dv = uv - \int v\, du$$

$$\int t \cdot e^{-0.2t}\, dt = t \cdot (-5e^{-0.2t}) - \int (-5e^{-0.2t}) \cdot dt$$

$$= -5te^{-0.2t} + 5\int e^{-0.2t}\, dt$$

$$= -5te^{-0.2t} + 5\left[\frac{e^{-0.2t}}{-0.2}\right]$$

$$= -5te^{-0.2t} - 25e^{-0.2t}$$

Note that we have taken $C = 0$ here because we are going to compute a definite integral. Now,

$$\int_0^5 6te^{-0.2t}\, dt = 6(-5te^{-0.2t} - 25e^{-0.2t})\Big|_0^5$$

$$= 6[(-5 \cdot 5e^{-0.2(5)} - 25e^{-0.2(5)}) - (-5(0)e^{-0.2(0)} - 25e^{-0.2(0)})]$$

$$= 6(-25e^{-1} - 25e^{-1} + 25)$$

$$\approx 39.636$$

To answer our question, the company will save about $39,636 in production costs over the next five years. ∎

*Comment:* Integration by parts cannot be used to integrate every product of functions. For example, to integrate $\int xe^{x^2}\, dx$, we use the method of integration by substitution with $u = x^2$ and not integration by parts.

---

**SOLUTIONS TO EXAMPLE 5**

$\frac{1}{x}\, dx,\ x,\ x,\ x,\ \frac{1}{x}\, dx,\ x\ln x - x + C$

## SECTION 6.4
### SHORT ANSWER EXERCISES

1. Integration by parts is a reversal of the _____ rule of differentiation.
2. In the integration by parts formula $\int u\, dv = uv - \int v\, du$, we choose the "parts" $u$ and $dv$ and we must compute the quantities $du$ and _____.
3. In a certain integration by parts problem, you can choose $dv$ to be $x^3\, dx$ or $[\ln(x + 1)]^2\, dx$. According to the general guidelines for choosing $dv$ we should let $dv =$ _____.
4. (True or False) The method of integration by parts is always useful when products of functions must be integrated. _____
5. To use integration by parts on the integral $\int x^2(\ln x)^3\, dx$, we let $u =$ _____ and $dv =$ _____.
6. To use integration by parts on the integral $\int x^3 e^{5x}\, dx$, we let $u =$ _____ and $dv =$ _____.
7. To use integration by parts on the integral $\int x^2(3x - 4)^3\, dx$, we let $u =$ _____ and $dv =$ _____.
8. Given that $\int \ln x\, dx = x \ln x - x$, then $\int_1^e \ln x\, dx =$ _____.
9. (True or False) $\int xe^x\, dx = \frac{1}{2}x^2 e^x - \frac{1}{2}\int x^2 e^x\, dx$. Hint: See Example 2. _____
10. The integral $\int \dfrac{\ln x}{x}\, dx$ is most easily done using the method of _____ with $u =$ _____.

## SECTION 6.4
### EXERCISES

In Exercises 1–38, find the integrals. Some of these integrals may not require integration by parts.

1. $\int x \ln x\, dx$
2. $\int x^2 \ln x\, dx$
3. $\int x^3 \ln x\, dx$
4. $\int_0^2 xe^{-x}\, dx$
5. $\int_0^1 xe^{3x}\, dx$
6. $\int (x - 2)e^{-2x}\, dx$
7. $\int (3 - x)e^{-4x}\, dx$
8. $\int x^2 e^x\, dx$
9. $\int x^2 e^{-4x}\, dx$
10. $\int_1^e 2x^3(\ln x)^2\, dx$
11. $\int_e^{e^2} 7x^2(\ln x)\, dx$
12. $\int x(5x + 1)^6\, dx$
13. $\int 6x(6x - 5)^4\, dx$
14. $\int x\sqrt{x + 6}\, dx$
15. $\int 2x\sqrt{2x + 1}\, dx$
16. $\int x\sqrt[3]{7 - x}\, dx$
17. $\int 3x\sqrt[3]{5x + 3}\, dx$
18. $\int_6^8 x^2\sqrt{x - 5}\, dx$
19. $\int_{-1}^{0.5} 4x^2\sqrt{2x + 3}\, dx$
20. $\int (\ln x)^2\, dx$
21. $\int (\ln x)^3\, dx$
22. $\int x \ln 5x\, dx$
23. $\int_2^4 x^2 \ln x^3\, dx$
24. $\int \dfrac{\ln x}{x^2}\, dx$

25. $\int 2x^3 e^{x^2}\, dx$

26. $\int_{-2}^{3} (1-x)e^{-2x}\, dx$

27. $\int_0^2 \dfrac{3x}{\sqrt{x+1}}\, dx$

28. $\int \dfrac{5x}{\sqrt{3x-1}}\, dx$

29. $\int \dfrac{3-4x}{2e^{3x}}\, dx$

30. $\int_1^3 (2x-5)\ln x^3\, dx$

31. $\int \ln\sqrt{x}\, dx$

32. $\int \left(\dfrac{x}{e^{5x}} - x\ln x\right)\, dx$

33. $\int x^3 e^{x^4}\, dx$

34. $\int_e^{e^2} \dfrac{\ln x^4}{3x}\, dx$

35. $\int x^2 \sqrt[5]{x-3}\, dx$

36. $\int \sqrt{x}\, \ln\sqrt{5x}\, dx$

37. $\int_e^{e^3} \dfrac{1}{x(\ln x)^3}\, dx$

38. $\int x^3 e^{3x}\, dx$

## Business and Economics

39. (Profit) If the marginal profit for a manufacturer is $P'(x) = \ln(x+1)$ thousands of dollars per thousand units, find the monthly profit from the sale of 8000 units per month. Assume that the sale of no units per month results in a loss of $2000.

## Life Sciences

40. (Bacteria growth) During a six-hour period, bacteria in a certain culture are growing according to $N'(t) = 0.1te^t$ hundred thousand/hour. If the culture contains 100,000 bacteria initially, how many bacteria are present after five hours?

41. (Voting) The percentage $P(t)$ of voters in a certain community who are registered independents is projected to increase over the next eight years at a rate given by $P'(t) = 2te^{-0.10t}$. What percentage of the voters will be registered independents four years from now if currently, 10% of the voters are so registered?

42. (Employee training) After $t$ hours of training, a new employee can produce $4te^{-0.4t}$ units per hour. How many units does the employee produce during the first five hours of training if with no training she can produce no units?

43. (Ore production) It is estimated that a mine will produce ore at the rate of $2te^{-0.2t}$ thousand tons per week $t$ weeks from now. Find the total amount of ore produced during the first three months in which the mine operates.

## Social Sciences

44. (Unemployment) During the next year, the percentage of unemployed workers in a certain city is projected to increase according to $P'(t) = t^2 e^t$ percent per year. If the current unemployment level is 5%, what will the percentage of unemployed workers in this city be six months from now?

45. In applying the integration by parts formula, if $v$ is replaced by $v + C$, then

$$\int u\, dv = u(v+C) - \int (v+C)\, du$$

Simplify the expression on the right-hand side to obtain $uv - \int v\, du$. Conclude that adding a constant $C$ when determining $v$ from $dv$ will not affect the integration by parts results, so that choosing $C = 0$ in the integration $\int dv = v + C$ is always permissible.

46. By factoring out a common factor of $(x+4)^{3/2}$ from each expression and simplifying the results, show that the answers to parts (a) and (b) of Example 3 are equivalent.

## 6.5 INTEGRATION USING TABLES

### A Table of Integrals

An integral table is a list of integration formulas. Such a table can be useful when you must evaluate an integral that cannot be done by the techniques of integration presented in this text. As examples, each of the integrals

$$\int \frac{dx}{25 - x^2} \quad \text{and} \quad \int \sqrt{x^2 + 9}\, dx$$

requires a technique of integration beyond the scope of this book. However, integrals such as these are readily found in many mathematical handbooks containing integral tables. Such integral tables typically contain hundreds of formulas. In this section we present a short table of integrals containing 40 formulas and we illustrate its use. Some of the integrals in this table can be found using techniques we have discussed and others require more advanced methods. Of course, you can verify any of the formulas in the table by differentiating the right-hand side of the formula to obtain the integrand.

### Table of Integrals

- **Fundamental integrals**

1. $\int a\, du = au + C$

2. $\int [f(u) + g(u)]\, du = \int f(u)\, du + \int g(u)\, du$

3. $\int u^n\, du = \dfrac{u^{n+1}}{n+1} + C, \quad n \neq -1$

4. $\int \dfrac{du}{u} = \ln |u| + C$

- **Integrals containing $a + bu$, where $a \neq 0$ and $b \neq 0$**

5. $\int \dfrac{du}{a + bu} = \dfrac{1}{b} \ln |a + bu| + C$

6. $\int \dfrac{u\, du}{a + bu} = \dfrac{1}{b^2}(a + bu - a \ln |a + bu|) + C$

7. $\int \dfrac{u^2\, du}{a + bu} = \dfrac{1}{b^3}\left[\dfrac{1}{2}(a + bu)^2 - 2a(a + bu) + a^2 \ln |a + bu|\right] + C$

8. $\int \dfrac{u\, du}{(a + bu)^2} = \dfrac{1}{b^2}\left[\dfrac{a}{a + bu} + \ln |a + bu|\right] + C$

9. $\int \dfrac{du}{u(a + bu)} = \dfrac{1}{a} \ln \left|\dfrac{u}{a + bu}\right| + C$

10. $\int \dfrac{du}{u^2(a + bu)} = -\dfrac{1}{au} + \dfrac{b}{a^2} \ln \left|\dfrac{a + bu}{u}\right| + C$

- **Integrals containing $\sqrt{a + bu}$, where $a \neq 0$ and $b \neq 0$**

11. $\int \sqrt{a + bu}\, du = \dfrac{2(a + bu)^{3/2}}{3b} + C$

12. $\int u\sqrt{a + bu}\, du = \dfrac{2(3bu - 2a)}{15b^3}(a + bu)^{3/2} + C$

13. $\int u^2\sqrt{a + bu}\, du = \dfrac{2(15b^2u^2 - 12abu + 8a^2)}{105b^3}(a + bu)^{3/2} + C$

14. $\int \dfrac{u\, du}{\sqrt{a + bu}} = \dfrac{2(bu - 2a)}{3b^2}\sqrt{a + bu} + C$

15. $\int \dfrac{u^2\, du}{\sqrt{a + bu}} = \dfrac{2(3b^2u^2 - 4abu + 8a^2)}{15b^3}\sqrt{a + bu} + C$

- Integrals containing $a^2 \pm u^2$, where $a > 0$

16. $\int \dfrac{du}{a^2 - u^2} = \dfrac{1}{2a} \ln \left| \dfrac{u + a}{u - a} \right| + C$

17. $\int \dfrac{du}{u^2 - a^2} = \dfrac{1}{2a} \ln \left| \dfrac{u - a}{u + a} \right| + C$

- Integrals containing $a + bu$ and $c + du$, where $b \neq 0$, $d \neq 0$ and $ad - bc \neq 0$

18. $\int \dfrac{du}{(a + bu)(c + du)} = \dfrac{1}{ad - bc} \ln \left| \dfrac{c + du}{a + bu} \right| + C$

19. $\int \dfrac{u\, du}{(a + bu)(c + du)} = \dfrac{1}{ad - bc} \left( \dfrac{a}{b} \ln |a + bu| - \dfrac{c}{d} \ln |c + du| \right) + C$

20. $\int \dfrac{u^2\, du}{(a + bu)(c + du)}$
    $= \dfrac{u}{bd} - \dfrac{1}{ad - bc} \left( \dfrac{a^2}{b^2} \ln |a + bu| - \dfrac{c^2}{d^2} \ln |c + du| \right) + C$

- Integrals containing $\sqrt{u^2 \pm a^2}$, where $a > 0$

21. $\int \dfrac{du}{\sqrt{u^2 \pm a^2}} = \ln |u + \sqrt{u^2 \pm a^2}| + C$

22. $\int \sqrt{u^2 \pm a^2}\, du = \dfrac{u}{2}\sqrt{u^2 \pm a^2} \pm \dfrac{a^2}{2} \ln |u + \sqrt{u^2 \pm a^2}| + C$

23. $\int u^2\sqrt{u^2 \pm a^2}\, du =$
    $\dfrac{1}{8}\left(u(2u^2 \pm a^2)\sqrt{u^2 \pm a^2} - a^4 \ln |u + \sqrt{u^2 \pm a^2}|\right) + C$

24. $\int \dfrac{\sqrt{u^2 + a^2}}{u}\, du = \sqrt{u^2 + a^2} - a \ln \left| \dfrac{a + \sqrt{u^2 + a^2}}{u} \right| + C$

25. $\int \dfrac{\sqrt{u^2 \pm a^2}}{u^2}\, du = -\dfrac{\sqrt{u^2 \pm a^2}}{u} + \ln |u + \sqrt{u^2 \pm a^2}| + C$

26. $\int \dfrac{du}{u\sqrt{u^2 + a^2}} = -\dfrac{1}{a} \ln \left| \dfrac{a + \sqrt{u^2 + a^2}}{u} \right| + C$

27. $\int \dfrac{du}{u^2\sqrt{u^2 \pm a^2}} = -\dfrac{\sqrt{u^2 \pm a^2}}{\pm a^2 u} + C$

- Integrals containing $\sqrt{a^2 - u^2}$, where $a > 0$

28. $\int \dfrac{du}{u\sqrt{a^2 - u^2}} = -\dfrac{1}{a} \ln \left| \dfrac{a + \sqrt{a^2 - u^2}}{u} \right| + C$

29. $\displaystyle\int \frac{du}{u^2\sqrt{a^2 - u^2}} = -\frac{\sqrt{a^2 - u^2}}{a^2 u} + C$

30. $\displaystyle\int \frac{\sqrt{a^2 - u^2}}{u}\,du = \sqrt{a^2 - u^2} - a \ln\left|\frac{a + \sqrt{a^2 - u^2}}{u}\right| + C$

- Integrals containing $e^u$

31. $\displaystyle\int e^u\,du = e^u + C$

32. $\displaystyle\int u^n e^u\,du = u^n e^u - n \int u^{n-1} e^u\,du$

33. $\displaystyle\int \frac{e^u}{u^n}\,du = -\frac{e^u}{(n-1)u^{n-1}} + \frac{1}{n-1}\int \frac{e^u\,du}{u^{n-1}}$

34. $\displaystyle\int \frac{du}{a + be^u} = \frac{1}{a}(u - \ln|a + be^u|) + C, \quad a \neq 0$

- Integrals containing $\ln u$

35. $\displaystyle\int \ln u\,du = u \ln u - u + C$

36. $\displaystyle\int u^n \ln u\,du = \frac{u^{n+1}}{(n+1)} \ln u - \frac{u^{n+1}}{(n+1)^2} + C, \quad n \neq -1$

37. $\displaystyle\int \frac{du}{u \ln u} = \ln|\ln u| + C$

38. $\displaystyle\int \frac{du}{u(\ln u)^n} = \frac{-1}{(n-1)(\ln u)^{n-1}} + C, \quad n \neq 1$

39. $\displaystyle\int \frac{\ln u}{u}\,du = \tfrac{1}{2}(\ln u)^2 + C$

40. $\displaystyle\int (\ln u)^n\,du = u(\ln u)^n - n \int (\ln u)^{n-1}\,du$

### Using an Integral Table

To make a table of integrals easier to use, it is usually divided into sections, each of which contains certain types of integrals. For example, one of the sections in our table lists "fundamental integrals"; another lists "integrals containing $a + bu$"; and so forth. In larger tables these types of headings may save you the time of searching through hundreds of formulas before finding the one that applies to your problem.

In our table, notice that the variable of integration is $u$ and not $x$. This serves as a reminder that a substitution will usually be required in the given integral before it can be converted to the form of one of the integrals in the table. Sometimes the substitution will be the trivial substitution $u = x$ and sometimes a more complicated substitution will be required. Let's look at some examples.

---

**EXAMPLE 1**    Use the tables of integrals fo find $\int dx/(25 - x^2)$.

**Solution**    Since the integral contains the expression $25 - x^2 = 5^2 - x^2$, we look in the section of the table labeled "integrals containing $a^2 \pm u^2$, where $a > 0$." Formula 16 is

$$\int \frac{du}{a^2 - u^2} = \frac{1}{2a} \ln \left| \frac{u + a}{u - a} \right| + C$$

In our integral, we make the substitution $u = x$, so that $du = dx$. Also, since $a^2 = 5^2$, we take $a = 5$ to obtain

$$\int \frac{dx}{25 - x^2} = \int \frac{du}{5^2 - u^2}$$

$$= \frac{1}{2 \cdot 5} \ln \left| \frac{u + 5}{u - 5} \right| + C$$

$$= \frac{1}{10} \ln \left| \frac{x + 5}{x - 5} \right| + C \quad \text{(Since } u = x\text{)} \quad \blacksquare$$

---

EXAMPLE 2 (Participative)

Solution

Let's use the table of integrals to find $\int \sqrt{x^2 + 9} \, dx$.

Since this integral contains the expression $\sqrt{x^2 + 9} = \sqrt{x^2 + 3^2}$, we look in the section of the table labeled "integrals containing _____, where $a > 0$." The formula that seems applicable is number ____:

$$\int \sqrt{u^2 + a^2} \, du = \frac{u}{2} \sqrt{u^2 + a^2} + \frac{a^2}{2} \ln |u + \sqrt{u^2 + a^2}| + C$$

Notice that since our integrand, $\sqrt{x^2 + 9}$, has the $x^2$ and the 9 connected by a "+" sign rather than a "−" sign, we have chosen the "+" sign throughout the equation in Formula 22.

Now we make the substitution $u = x$, so that $du =$ _____, and we take $a =$ _____ to get

$$\int \sqrt{x^2 + 9} \, dx = \int \sqrt{u^2 + 3^2} \, du$$

$$= \frac{u}{2} \sqrt{u^2 + 3^2} + \underline{\hspace{2cm}} + C$$

$$= \underline{\hspace{5cm}}$$

(Replacing $u$ by $x$) $\blacksquare$

Sometimes the substitution required to convert a given integral into the form of one shown in a table of integrals is more complicated than simply $u = x$. The next example illustrates this point.

---

EXAMPLE 3

Find $\int dx/(x\sqrt{10 - 4x^2})$.

---

**SOLUTIONS TO EXAMPLE 2**

$\sqrt{u^2 \pm a^2}$, 22, $dx$, 3, $\frac{9}{2} \ln |u + \sqrt{u^2 + 9}|$, $\frac{x}{2}\sqrt{x^2 + 9} + \frac{9}{2} \ln |x + \sqrt{x^2 + 9}| + C$

**Solution** The integral contains the expression $\sqrt{10 - 4x^2}$ and will match Formula 28,

$$\int \frac{du}{u\sqrt{a^2 - u^2}} = -\frac{1}{a} \ln \left| \frac{a + \sqrt{a^2 - u^2}}{u} \right| + C$$

if we write $\sqrt{10 - 4x^2}$ as $\sqrt{(\sqrt{10})^2 - (2x)^2}$ and make the substitution $u = 2x$, so that $du = 2dx$. Letting $a = \sqrt{10}$, we get

$$\int \frac{dx}{x\sqrt{10 - 4x^2}} = \int \frac{1}{x\sqrt{(\sqrt{10})^2 - (2x)^2}} dx$$

$$= \int \frac{1}{\frac{u}{2}\sqrt{(\sqrt{10})^2 - u^2}} \cdot \frac{du}{2}$$

$$= \int \frac{1}{u\sqrt{(\sqrt{10})^2 - u^2}} du$$

$$= -\frac{1}{\sqrt{10}} \ln \left| \frac{\sqrt{10} + \sqrt{10 - u^2}}{u} \right| + C$$

$$= -\frac{1}{\sqrt{10}} \ln \left| \frac{\sqrt{10} + \sqrt{10 - 4x^2}}{2x} \right| + C \quad \text{(Since } u = 2x\text{)} \quad \blacksquare$$

The next example illustrates the computation of a definite integral using our table of integrals.

**EXAMPLE 4** Find $\int_0^2 x^5 \sqrt{1 + 9x^2} \, dx$.

**Solution** It appears that, since $\sqrt{1 + 9x^2} = \sqrt{(3x)^2 + 1}$, we should search in the table under "integrals containing $\sqrt{u^2 \pm a^2}$" and make the substitution $u = 3x$. Such a substitution leads to the indefinite integral

$$\int x^5 \sqrt{1 + 9x^2} \, dx = \int \left(\frac{u}{3}\right)^5 \sqrt{1 + u^2} \cdot \frac{du}{3} = \frac{1}{729} \int u^5 \sqrt{u^2 + 1} \, du$$

However, no integral of this last form can be found in our table!

Since the preceding approach was not successful, let's try viewing the quantity $\sqrt{1 + 9x^2}$ as $\sqrt{1 + 9u}$ by making the substitution $u = x^2$. Now we will have to search for an appropriate formula under "integrals containing $\sqrt{a + bu}$". Making the substitution $u = x^2$, so that $du = 2x \, dx$, we obtain

$$\int_0^2 x^5 \sqrt{1 + 9x^2} \, dx = \int_0^2 (x^2)^2 \sqrt{1 + 9x^2} \cdot x \, dx$$

$$= \int_0^4 u^2 \sqrt{1 + 9u} \cdot \tfrac{1}{2} du = \tfrac{1}{2} \int_0^4 u^2 \sqrt{1 + 9u} \, du$$

Observe that we have changed the limits of the definite integral to agree with the new variable, $u$. That is, since $u = x^2$, when $x = 0$, we have $u = 0^2 = 0$, and when $x = 2$, we have $u = 2^2 = 4$. It would now appear that Formula 13 applies:

$$\int u^2 \sqrt{a + bu} \, du = \frac{2(15b^2 u^2 - 12abu + 8a^2)}{105b^3} (a + bu)^{3/2} + C$$

Substituting $a = 1$ and $b = 9$, we obtain

$$\int_0^2 x^5\sqrt{1 + 9x^2}\, dx = \frac{1}{2}\left(\frac{2(15 \cdot 9^2 u^2 - 12 \cdot 1 \cdot 9u + 8 \cdot 1^2)}{105 \cdot 9^3}(1 + 9u)^{3/2}\right)\Big|_0^4$$

$$= \frac{1215u^2 - 108u + 8}{76545}(1 + 9u)^{3/2}\Big|_0^4$$

$$= \left[\frac{1215(4)^2 - 108(4) + 8}{76545}(1 + 36)^{3/2}\right] - \left[\frac{1215(0)^2 - 108(0) + 8}{76545}(1 + 0)^{3/2}\right]$$

$$= \frac{19016(37)^{3/2} - 8}{76545} \approx 55.9119 \qquad \blacksquare$$

### Reduction Formulas

Some of the integrals in our table express one integral in terms of another. See, for example, Formulas 32 and 40. In such cases it is anticipated that the integral on the right-hand side will be easier than the one on the left-hand side; this reduces our problem to a simpler one. In Formula 40, for example,

$$\int (\ln u)^n\, du = u(\ln u)^n - n\int (\ln u)^{n-1}\, du$$

the exponent on $\ln u$ has been reduced from $n$ to $n - 1$, resulting in an easier integral on the right-hand side. In many problems a *reduction formula* must be applied more than once to obtain the desired result.

**EXAMPLE 5**  Find $\int (\ln x)^3\, dx$.

**Solution**  In Formula 40, let $u = x$, so that $du = dx$. If we take $n = 3$, we obtain

$$\int (\ln x)^3\, dx = \int (\ln u)^3\, du = u(\ln u)^3 - 3\int (\ln u)^2\, du$$

Now we apply Formula 40 again to the integral on the right-hand side with $n = 2$:

$$\int (\ln x)^3\, dx = u(\ln u)^3 - 3\left[u(\ln u)^2 - 2\int (\ln u)^1\, du\right]$$

$$= u(\ln u)^3 - 3u(\ln u)^2 + 6\int \ln u\, du$$

The problem may now be completed by applying Formula 40 once again with $n = 1$ to the integral on the right-hand side. However, it is slightly more convenient simply to apply Formula 35 to obtain

$$\int (\ln x)^3\, dx = u(\ln u)^3 - 3u(\ln u)^2 + 6[u \ln u - u] + C$$

$$= x(\ln x)^3 - 3x(\ln x)^2 + 6x \ln x - 6x + C \qquad \blacksquare$$

### SECTION 6.5
### SHORT ANSWER EXERCISES

For each of the following integrals, identify the number of the formula in our table of integrals that you would use to evaluate it. Do not actually carry out the integration.

**1.** $\int x\sqrt{2 + 3x}\, dx$:  Formula _____

**2.** $\int 3x^4 e^x\, dx$:  Formula _____

3. $\int x^2\sqrt{x^2-10}$: Formula _____

4. $\int \dfrac{5}{x^2-20}\,dx$: Formula _____

5. $\int \dfrac{-2\,dx}{(1-x)(4+3x)}$: Formula _____

6. $\int \dfrac{x^2\,dx}{2+3x+x^2}$: Formula _____

7. $\int \dfrac{\sqrt{12-5x^2}}{x}\,dx$: Formula _____

8. $\int x^3 \ln(5x)\,dx$: Formula _____

9. $\int \dfrac{\sqrt{x^4+16}}{x^2}(2x\,dx)$: Formula _____

10. $\int \dfrac{e^{x^3}}{x^6}(3x^2\,dx)$: Formula _____

## SECTION 6.5
### EXERCISES

In Exercises 1–28, use the integral table given in this section to evaluate the indefinite integral.

1. $\int \dfrac{dx}{2+3x}$

2. $\int \dfrac{dx}{2x-5}$

3. $\int \dfrac{3x\,dx}{2-4x}$

4. $\int \dfrac{4x\,dx}{6-7x}$

5. $\int \dfrac{x\,dx}{\sqrt{3x-1}}$

6. $\int 4x\sqrt{7x+5}\,dx$

7. $\int \dfrac{x\,dx}{1-25x^2}$

8. $\int \dfrac{x\,dx}{2x^2+x-3}$

9. $\int \dfrac{dx}{(4-x)(5+x)}$

10. $\int \dfrac{\sqrt{x^2+9}}{3x}\,dx$

11. $\int \dfrac{\sqrt{16-x^2}}{2x}\,dx$

12. $\int \dfrac{\sqrt{x^2-25}}{5x^2}\,dx$

13. $\int \dfrac{dx}{2-3e^x}$

14. $\int \dfrac{du}{u(\ln u)^2}$

15. $\int \dfrac{x\,dx}{\sqrt{4x-9}}$

16. $\int \dfrac{\sqrt{36-9x^2}}{x}\,dx$

17. $\int 2x^2\sqrt{25x^2-9}\,dx$

18. $\int \dfrac{2\,dx}{81-49x^2}$

19. $\int \dfrac{7\,dx}{36x^2-25}$

20. $\int \dfrac{dx}{x\sqrt{100-25x^2}}$

21. $\int x^4\sqrt{3+4x^2}\cdot 2x\,dx$

22. $\int \dfrac{2x\,dx}{x^4\sqrt{x^4-4}}$

23. $\int \dfrac{x^3}{\sqrt{4-5x^2}}\,dx$

24. $\int \dfrac{x^3}{3+4x^2+x^4}\,dx$

25. $\int \dfrac{\sqrt{e^{2x}-16}}{e^x}\,dx$

26. $\int \dfrac{dx}{x(\ln x)^2\sqrt{9-(\ln x)^2}}$

27. $\int x^5 e^{x^2}\,dx$

28. $\int \sqrt{x}\,\ln\sqrt{x}\,dx$

In Exercises 29–40, evaluate the definite integral.

29. $\displaystyle\int_0^2 \dfrac{dx}{(5-x)(3-x)}$

30. $\displaystyle\int_1^3 \dfrac{3x\,dx}{\sqrt{4x-1}}$

31. $\int_{-1}^{0} \dfrac{4x^2\, dx}{4 - 4x - 3x^2}$

32. $\int_{-4}^{-1} \dfrac{3\, dx}{x^2\sqrt{20 - x^2}}$

33. $\int_{4}^{5} \dfrac{\sqrt{4x^2 - 39}}{3x^2}\, dx$

34. $\int_{-1}^{1} 6x^2\sqrt{10x^2 + 49}\, dx$

35. $\int_{-1}^{0} \dfrac{2x\, dx}{\sqrt{(x^2 + 1)^2 + 4}}$

36. $\int_{0}^{1} \dfrac{x\, dx}{10 - (x^2 + 2)^2}$

37. $\int_{1}^{2} \dfrac{e^x\, dx}{2 + e^x - 3e^{2x}}$

38. $\int_{2}^{3} \dfrac{\ln x\, dx}{x\sqrt{2 + 5\ln x}}$

39. $\int_{1}^{4} (4x + 2)[\ln(x^2 + x)]^3\, dx$

40. $\int_{0}^{1} (x^2 + 2x)^3(x + 1)e^{x^2 + 2x}\, dx$

41. (Maintenance costs) If $M(t)$ is the total (accumulated) cost of maintaining a piece of industrial equipment for $t$ years, then the maintenance costs of a piece of equipment purchased at the beginning of 1991 are projected to increase at a rate of $M'(t) = t^2/(1 + t)$ hundreds of dollars per year until the year 1999. If $t = 0$ corresponds to 1991, compute the total maintenance costs of the machine from 1991 to 1999.

42. (Capital expenditures) Over the next ten years, your company projects that the amount they will invest for capital expenditures will increase at a rate of $E'(t) = 1/\sqrt{t^2 + 1}$ hundreds of thousands of dollars per year. Find the total increase in capital expenditures your company will make over the next ten years.

**Business and Economics**

43. (Profit) The monthly profit of a certain company is increasing at a rate of $P'(t) = 1/(1 + 11t + 10t^2)$ hundreds of thousands of dollars per month. Find the total increase in profit of this company over the next year.

**Social Sciences**

44. (Public school enrollment) The number of students in the public schools of a certain community is projected to change at the rate of

$$N'(t) = -\dfrac{10\sqrt{144 - (t + 1)}}{144(t + 1)}$$

thousands of students per year for $0 \leq t \leq 5$. If $t = 0$ corresponds to 1992, find the change in the number of students from 1992 to 1994.

# 6.6 CHAPTER SIX REVIEW

**KEY TERMS**

antiderivative
indefinite integral
integral sign
integrand
variable of integration
constant of integration
integration by substitution

the chain rule for integration
definite integral
lower limit of integration
upper limit of integration
integration by parts
integration using tables
reduction formula

**KEY FORMULAS AND RESULTS**

**Antiderivative or Indefinite Integral**

If $F'(x) = f(x)$, then $F(x)$ is an antiderivative or indefinite integral of $f(x)$. The *general* antiderivative or indefinite integral is $F(x) + C$, where $C$ is a constant.

**The Sum or Difference Rule**

$$\int [f(x) \pm g(x)] \, dx = \int f(x) \, dx \pm \int g(x) \, dx$$

**The Constant Multiple Rule**

$$\int cf(x) \, dx = c \int f(x) \, dx$$

**The Power Rule**

$$\int x^n \, dx = \frac{1}{n+1} x^{n+1} + C \quad \text{if } n \neq -1$$

**The Integral of $e^x$**

$$\int e^x \, dx = e^x + C$$

**The Integral of $\frac{1}{x}$**

$$\int \frac{1}{x} \, dx = \ln |x| + C$$

**The Chain Rule for Integration**

$$\int f(g(x)) g'(x) \, dx = F(g(x)) + C$$

**Integration by Substitution**

If $u = g(x)$, then $\int f(g(x))g'(x) \, dx = \int f(u) \, du = F(u) + C = F(g(x)) + C$.

**Substitution Formulas**

If $u = g(x)$, then

1. $\int u^n \, du = \dfrac{u^{n+1}}{n+1} + C \quad \text{if } n \neq -1$

2. $\int e^u \, du = e^u + C$

3. $\int \dfrac{du}{u} = \ln |u| + C$

## The Definite Integral

$$\int_a^b f(x)\,dx = F(x)]_a^b = F(b) - F(a)$$

## Properties of the Definite Integral

1. $\int_a^a f(x)\,dx = 0$
2. $\int_a^b f(x)\,dx = -\int_b^a f(x)\,dx$
3. $\int_a^b kf(x)\,dx = k\int_a^b f(x)\,dx$
4. $\int_a^b [f(x) \pm g(x)]\,dx = \int_a^b f(x)\,dx \pm \int_a^b g(x)\,dx$
5. $\int_a^c f(x)\,dx = \int_a^b f(x)\,dx + \int_b^c f(x)\,dx$

## Integration by Parts

$$\int u\,dv = uv - \int v\,du$$

## CHAPTER 6 SHORT ANSWER REVIEW QUESTIONS

1. If $F(x) = x^3 + C$ and $F(2) = 3$, then $C = $ _____.
2. (True or False) If $G(x) = e^x + C$ and $G(0) = 0$, then $C = 0$. _____
3. (True or False) The function $F(x) = \frac{1}{3}e^{3x}$ is an antiderivative of $f(x) = e^{3x}$. _____
4. To evaluate the integral $\int 3x^2\sqrt{2x^3 - 5}\,dx$, you could use the substitution $u = $ _____.
5. (True or False) The integral $\int_0^1 3xe^{x^2}\,dx$ can be found by integration by substitution. _____
6. In the integral $\int_3^4 3x\,dx/(2x - 1)$, we make the substitution $u = 2x - 1$. The new lower limit of integration is $u = $ _____.
7. When you evaluate $\dfrac{2e^{2x}}{e^x + 1}\Big]_0^2$, the result accurate to three decimal places is
   (a) 13.017; (b) 12.017; (c) 0; (d) none of the above.
8. (True or False) $\int_0^3 2xe^x\,dx = 2\int_0^3 ze^z\,dz$ _____
9. $\int_1^e \dfrac{1}{x}\,dx = \ln x]_1^e = $ _____
10. The integral $\int_0^5 xe^x\,dx$ (a) is an indefinite integral; (b) can be done by integration by parts; (c) can be done by the substitution $u = x$; (d) is beyond the scope of this book.
11. In an integration by parts problem, $u = x^3$ and $dv = e^{5x}\,dx$. Then $du = $ _____ and $v = $ _____.

12. A quantity grows according to $Q(t) = 3t$. What is the change in the amount of the quantity present between $t = 1$ and $t = 2$? (a) $\frac{9}{2}$; (b) 3; (c) 1; (d) none of the above.

13. To evaluate $\int \frac{3\,dx}{x^2\sqrt{16 - x^2}}$, you could use Formula _____ in the table of integrals.

14. To evaluate $\int \frac{2x\,dx}{3 + 7e^{x^2+1}}$, you could use Formula _____ in the table of integrals.

15. To evaluate $\int \frac{e^{2x}\,dx}{3 + 4e^{2x} + e^{4x}}$, you could use Formula _____ in the table of integrals.

## CHAPTER 6 REVIEW EXERCISES

Exercises marked with an asterisk (*) constitute a Chapter Test.

In Exercises 1–28, find the indefinite integrals.

1. $\int (x^2 - 7x)\,dx$
2. $\int (3 - x^3 + 7x)\,dx$
*3. $\int (\sqrt{x} - 3\sqrt[3]{x})\,dx$
4. $\int \left(\frac{2}{\sqrt{x}} + 2\sqrt[3]{x}\right)dx$
5. $\int \left(\frac{3x^3 - 2x^2 + 3}{2x}\right)dx$
6. $\int \left(\frac{x - 3\sqrt{x} + 1}{\sqrt{x}}\right)dx$
7. $\int (x^{2/5} - \sqrt{x^3})\,dx$
8. $\int \left(\frac{3}{x} - \sqrt[4]{x^5}\right)dx$
*9. $\int xe^{5x}\,dx$
10. $\int 3x^2 e^{4x}\,dx$
*11. $\int 2x\sqrt{5 - 3x^2}\,dx$
12. $\int 5x^2\sqrt{3x^2 + 20}\,dx$
13. $\int x^2\sqrt{x^2 + 5}\,dx$
14. $\int 4x^2\sqrt{7 - x^3}\,dx$
*15. $\int x(\ln x)^3\,dx$
16. $\int 3x^2(\ln 2x)^2\,dx$
*17. $\int 7x\sqrt{1 - 5x}\,dx$
18. $\int x^2\sqrt{5 + 2x}\,dx$
19. $\int \frac{2\,dx}{5 - x^2}$
20. $\int \frac{dx}{5x^2 - 12x + 4}$
21. $\int \frac{(\ln \sqrt{x})^4}{3x}\,dx$
22. $\int x^3 e^{2x^4+5}\,dx$
*23. $\int \frac{7x\,dx}{4x^2 + 5}$
24. $\int \frac{e^{5x}\,dx}{2e^{5x} - 3}$
25. $\int e^{2x}\sqrt{e^{2x} - 5}\,dx$
26. $\int \frac{\sqrt{6 - 3x^2}}{5x}\,dx$
27. $\int \frac{5}{x(\ln x + 3)}\,dx$
28. $\int \frac{dx}{(2x + 1)[\ln(2x + 1)]^2}$

In Exercises 29–48, find the definite integral.

29. $\int_{-1}^{1} (x^3 - 2x + 1)\,dx$
30. $\int_{-2}^{1} (2x^2 - 7x + 4)\,dx$

*31. $\int_{0}^{2} 2x\sqrt{5 + x^2}\, dx$   32. $\int_{-1}^{0} \dfrac{dx}{x + 2}$

33. $\int_{0}^{1} 3xe^{2-3x^2}\, dx$   34. $\int_{2}^{3} \dfrac{dx}{2x(\ln x)^2}$

35. $\int_{-1}^{0} xe^{-3x}\, dx$   36. $\int_{-1}^{1} \dfrac{dx}{x\sqrt{12 - x^2}}$

37. $\int_{1}^{e} (3x + 1)\ln(2x)\, dx$   38. $\int_{0}^{1} \dfrac{3x}{2x - 5}\, dx$

*39. $\int_{-1}^{1} \dfrac{12\, dx}{\sqrt{4x^2 + 25}}$   40. $\int_{0}^{2} \dfrac{2 + x}{1 + x}\, dx$

41. $\int_{e}^{e^2} (2x + 1)^2 \ln x\, dx$   42. $\int_{0}^{1} (1 - x)^2 e^{4x}\, dx$

*43. $\int_{2}^{3} \dfrac{\sqrt{(\ln x) + 2}}{x}\, dx$   44. $\int_{-1}^{1} \dfrac{3e^{2x}}{\sqrt{2e^{2x} + 3}}\, dx$

45. $\int_{e}^{e^2} \dfrac{dx}{x(\ln x)(1 + \ln x)}$   46. $\int_{-1}^{0} \dfrac{e^{2x}\, dx}{3 + e^x}$

47. $\int_{2}^{e^2} \dfrac{dx}{x(\ln x)^2}$   48. $\int_{2}^{3} 3x\sqrt[3]{5 - 4x}\, dx$

## Business and Economics

*49. (Cost) The monthly marginal cost of a certain firm is $C'(x) = 3x\sqrt{2x + 5}$ dollars per unit produced. If monthly fixed costs are $1000, **(a)** find the monthly cost function $C(x)$, and **(b)** find the cost of producing 100 units.

50. (Revenue) A certain company has a monthly revenue function of $R'(x) = 10x\sqrt[3]{(3x + 5)^2}$ dollars per unit sold. Assuming that $R(0) = 0$, **(a)** find the monthly revenue function $R(x)$, and **(b)** find the revenue derived from the sale of 100 units.

51. (Oil embargo) A small country that produces no oil has a reserve of 10 billion barrels of oil and uses the oil at a rate of $A'(t) = 5000000 e^{0.1t}$ barrels/week. If the country is subjected to an oil embargo and can purchase no more oil, how long will its reserve last?

52. (Social workers) The number of social workers needed in a city is projected to increase at a rate of $N'(t) = 3\sqrt{t}$ workers/year, where $t = 0$ corresponds to 1991. How many more social workers will be needed in the year 2000 than in 1991?

53. (Monthly marginal profit) The monthly marginal profit of a firm is given by $P'(x) = 250/(0.5x + 0.1)$ dollars per unit produced and sold. If the production and sale of zero units produces a loss of $1000, find the monthly profit from the production and sale of 1000 units.

54. (Population) The population of a city is increasing at the rate of $P'(t) = 6t/\sqrt{6t^2 + 7}$ thousands of people per year, where $t = 0$ corresponds to 1992, when the population was 50,000. Find the population of the city in 1999.

*55. (Advertising budget) Over the next 12 months, the advertising budget for one of a company's products is projected to increase at the rate of

$$B'(t) = \dfrac{1}{10} t \ln(t + 1)$$

thousands of dollars per month. Find the total increase in the advertising budget for this product over the next year.

**56.** (Depreciation) The value of a piece of industrial machinery is decreasing according to the formula $V'(t) = -22500/(t + 1)^{3/2}$ dollars per year. If the machinery was purchased for $50,000, find the value five years after purchase.

**57.** (Velocity) The velocity of an object moving in a straight line is $v(t) = \sqrt{4t^2 + 9}/t$ ft/sec for $t > 0$. Find its position $s(t)$ for $t > 0$ if $s(2) = 0$.

# 7.1 THE DEFINITE INTEGRAL AS AN AREA

## The Area Theorem

There is an interesting connection between the definite integral and the area under the graph of a function $y = f(x)$. If $f(x) \geq 0$ for $a \leq x \leq b$, then

$$\int_a^b f(x)\,dx = \text{area between the graph of } f(x) \text{ and the x-axis from } a \text{ to } b$$

See Figure 7.1 for an illustration of this area. We will refer to this fact—that the definite integral $\int_a^b f(x)\,dx$ represents the area shown in Figure 7.1 if $f$ is nonnegative over $[a, b]$—as the *area theorem*.

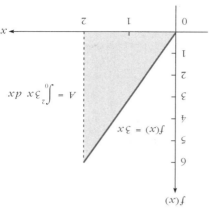

FIGURE 7.1 The definite integral is equal to the area under the graph of $f(x)$ from $a$ to $b$ if $f(x) \geq 0$ for $x$ between $a$ and $b$.

### THE AREA THEOREM

If $f$ is continuous and nonnegative on the interval $[a, b]$ and if $F$ is any antiderivative of $f$, then the area under the graph of $f$ from $a$ to $b$ is

$$\int_a^b f(x)\,dx = F(x)\Big|_a^b = F(b) - F(a)$$

Let's look at a simple example first and then investigate why this theorem is true.

**EXAMPLE 1**  Find the area under the graph of $f(x) = 3x$ from $x = 0$ to $x = 2$ using the area theorem.

**Solution**  The required area is sketched in Figure 7.2. In this case $f(x) = 3x$, $a = 0$, and $b = 2$, so that the area is given by

$$\int_0^2 3x\,dx = \left[\tfrac{3}{2}x^2\right]_0^2 = \tfrac{3}{2}(2)^2 - \tfrac{3}{2}(0)^2 = 6 \text{ square units}$$

Since the area in question is simply that of a triangle with base 2 and height 6, we can verify this result using the formula from geometry $A = \tfrac{1}{2}bb$ as follows:

$$A = \tfrac{1}{2}bb = \tfrac{1}{2}(2)(6) = 6 \text{ square units} \qquad \blacksquare$$

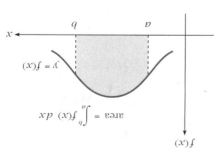

FIGURE 7.2 The area under $f(x) = 3x$ from $x = 0$ to $x = 2$.

### Proof of the Area Theorem

We will now outline a proof of the area theorem. Assume that $f(x) \geq 0$ for $x$ in $[a, b]$ and define $A(x)$ to be the area under the graph of $f$ from $a$ to $x$; see Figure 7.3. Observe that $A(x)$ is a function of $x$ and, for example, that $A(a)$ is the area under the graph of $f$ from $a$ to $a$, so $A(a) = 0$, and that $A(b)$ is the area under the graph of $f$ from $a$ to $b$. We will now show that $A(x)$ is an

# 7 ADDITIONAL INTEGRATION TOPICS

CHAPTER CONTENTS

7.1 The Definite Integral as an Area
7.2 The Definite Integral as the Limit of a Sum
7.3 Applications of the Definite Integral
7.4 Numerical Integration
7.5 Improper Integrals
7.6 Chapter 7 Review

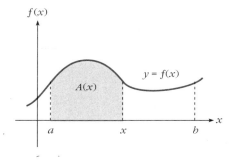

**FIGURE 7.3** The function $A(x)$ is the area from $a$ to $x$ under the graph of $f$.

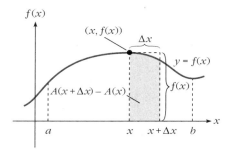

**FIGURE 7.4** The difference $A(x + \Delta x) - A(x)$ is the area under $f$ from $x$ to $x + \Delta x$ and can be approximated by the area $\Delta x \cdot f(x)$ of the rectangle shown.

antiderivative of $f(x)$ on $[a, b]$. For this, recall that by the definition of the derivative,

$$A'(x) = \lim_{\Delta x \to 0} \frac{A(x + \Delta x) - A(x)}{\Delta x}$$

The expression $A(x + \Delta x) - A(x)$ is the area under the curve from $x$ to $x + \Delta x$, as shown in Figure 7.4. This area can be approximated by the area of a rectangle with height $f(x)$ and base $\Delta x$ provided that $\Delta x$ is small. That is,

$$A(x + \Delta x) - A(x) \approx \Delta x \cdot f(x) \quad \text{if } \Delta x \text{ is small}$$

Dividing both sides by $\Delta x$, we get

$$\frac{A(x + \Delta x) - A(x)}{\Delta x} \approx f(x) \quad \text{if } \Delta x \text{ is small}$$

As we let $\Delta x$ approach zero, the approximation gets better and better, so that eventually,

$$\lim_{\Delta x \to 0} \frac{A(x + \Delta x) - A(x)}{\Delta x} = f(x)$$

But the left side is just $A'(x)$, and so

$$A'(x) = f(x)$$

Therefore $A(x)$ is an antiderivative of $f(x)$.

To complete the proof, let $F(x)$ be any antiderivative of $f(x)$. Then $F(x) = A(x) + C$, where $C$ is some constant. Therefore,

$$F(b) - F(a) = [A(b) + C] - [A(a) + C]$$
$$= A(b) - A(a) = A(b) \quad \text{(Since } A(a) = 0\text{)}$$
$$= \int_a^b f(x)\, dx$$

We have now shown that

$$\int_a^b f(x)\, dx = F(b) - F(a) = A(b)$$

Since $A(b)$ is the area under the graph of $f$ from $a$ to $b$, the proof of the theorem is complete. ∎

---

### EXAMPLE 2

Find the area of the region bounded by the graph of $f(x) = -x^2 + x + 6$ and the $x$-axis.

### Solution

We can factor $f(x)$ as

$$f(x) = -[x^2 - x - 6] = -(x + 2)(x - 3)$$

The $x$-intercepts of the graph are $x = -2$ and $x = 3$; the graph is sketched in Figure 7.5. Since the graph lies on or above the $x$-axis between $-2$ and $3$, we obtain the area of the region as the following definite integral:

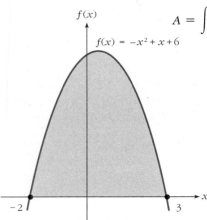

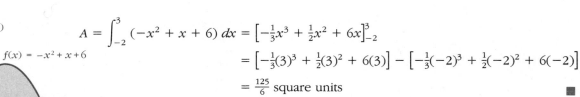

$$A = \int_{-2}^{3} (-x^2 + x + 6)\, dx = \left[-\tfrac{1}{3}x^3 + \tfrac{1}{2}x^2 + 6x\right]_{-2}^{3}$$
$$= \left[-\tfrac{1}{3}(3)^3 + \tfrac{1}{2}(3)^2 + 6(3)\right] - \left[-\tfrac{1}{3}(-2)^3 + \tfrac{1}{2}(-2)^2 + 6(-2)\right]$$
$$= \tfrac{125}{6} \text{ square units}$$

### Finding Areas When $f(x) \leq 0$

If the graph of $f$ lies below the $x$-axis between $a$ and $b$, as in Figure 7.6, then the definite integral $\int_a^b f(x)\, dx$ is a negative number. This number is the negative of the area between the graph and the $x$-axis from $a$ to $b$, so that the area is the absolute value of $\int_a^b f(x)\, dx$. In other words

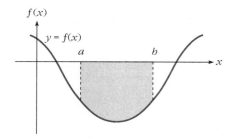

**FIGURE 7.5** The region bounded by $f(x) = -x^2 + x + 6$ and the $x$-axis.

**FIGURE 7.6** The function $f(x) \leq 0$ for $x$ between $a$ and $b$.

---

If $f(x) \leq 0$ for $x$ between $a$ and $b$, then the area between the graph of $f$ and the $x$-axis from $a$ to $b$ is given by

$$\left| \int_a^b f(x)\, dx \right| = |F(b) - F(a)|$$

---

### EXAMPLE 3

Find the area bounded by the graph of $f(x) = x^2 - 6x + 8$ and the $x$-axis.

### Solution

We write $f(x) = (x - 2)(x - 4)$, and so the $x$-intercepts are $x = 2$ and $x = 4$. The region, which is sketched in Figure 7.7, lies below the $x$-axis between $x = 2$ and $x = 4$. Thus

$$A = \left| \int_2^4 (x^2 - 6x + 8)\, dx \right| = \left| \left[\tfrac{1}{3}x^3 - 3x^2 + 8x\right]_2^4 \right|$$
$$= \left| \left[\tfrac{1}{3}(4)^3 - 3(4)^2 + 8(4)\right] - \left[\tfrac{1}{3}(2)^3 - 3(2)^2 + 8(2)\right] \right|$$
$$= \left| -\tfrac{4}{3} \right| = \tfrac{4}{3} \text{ square units}$$

In some situations we are required to find the area of a region bounded by the graph of $f$ and the $x$-axis between $a$ and $b$ such that $f$ lies above the $x$-axis on one portion of the interval $[a, b]$ and below the $x$-axis on another portion. Such a situation is illustrated in Figure 7.8, where the graph of $f(x)$ lies above

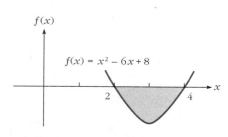

**FIGURE 7.7** The region bounded by $f(x) = x^2 - 6x + 8$ and the $x$-axis.

the $x$-axis for $x$ between $a$ and $c$ and below the $x$-axis for $x$ between $c$ and $b$. The shaded area in Figure 7.8 is given by

$$A = \int_a^c f(x)\, dx + \left| \int_c^b f(x)\, dx \right|$$

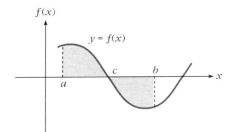

**FIGURE 7.8** The function $f(x)$ lies above the $x$-axis between $a$ and $c$ and below the $x$-axis between $c$ and $b$.

---

**EXAMPLE 4 (Participative)**

**Solution**

Let's find the area bounded by the graph of $y = x^3 - x$ and the $x$-axis between $x = 0$ and $x = 2$.

To determine the intervals over which $f(x)$ is positive and those over which $f(x)$ is negative, we first find the $x$-intercepts of $f(x)$ by solving the equation

$$x^3 - x = x(x+1)(x-1) = \underline{\phantom{aa}}$$
$$x = 0, \quad x = \underline{\phantom{aa}}, \quad x = \underline{\phantom{aa}}$$

Since $f(x)$ is continuous on $(-\infty, \infty)$, it can change sign only at the $x$-intercepts, namely, when $x = -1$, $x = 0$, or $x = 1$. This fact is a consequence of the intermediate value theorem. Since we are interested in the sign of $f$ over the interval $[0, 2]$, we choose as test values $x = \frac{1}{2}$ and $x = \frac{3}{2}$. We obtain

$$f\left(\tfrac{1}{2}\right) = \left(\tfrac{1}{2}\right)^3 - \tfrac{1}{2} = -\tfrac{3}{8} < 0$$

so $f(x) < 0$ on the interval $(0, 1)$. Also,

$$f\left(\tfrac{3}{2}\right) = \left(\tfrac{3}{2}\right)^3 - \underline{\phantom{aa}} = \underline{\phantom{aa}} > 0$$

so $f(x) > 0$ on the interval $(1, \infty)$. The sketch of $f(x) = x^3 - x$ drawn in Figure 7.9 bears out these conclusions. Now we calculate:

$$A = \left| \int_0^1 (x^3 - x)\, dx \right| + \underline{\phantom{aaaaaa}}$$

$$= \left| \left[ \left( \frac{x^4}{4} - \frac{x^2}{2} \right) \right]_0^1 \right| + \underline{\phantom{aaaaaa}}$$

$$= \left| \frac{1^4}{4} - \frac{1^2}{2} \right| + \underline{\phantom{aaaaaa}}$$

$$= \underline{\phantom{aaaaaa}}$$

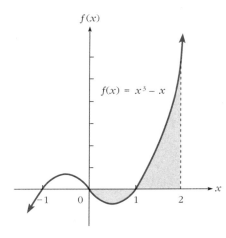

**FIGURE 7.9** The region bounded by the graph of $y = x^3 - x$ and the $x$-axis between 0 and 2.

---

**SOLUTIONS TO EXAMPLE 4**

$0, -1, 1, \frac{3}{2}, \frac{15}{8}, \int_1^2 (x^3 - x)\, dx, \left[ \left( \frac{x^4}{4} - \frac{x^2}{2} \right) \right]_1^2, \left( \frac{2^4}{4} - \frac{2^2}{2} \right) - \left( \frac{1^4}{4} - \frac{1^2}{2} \right), 2\frac{1}{2}$ square units

## The Area between Two Curves

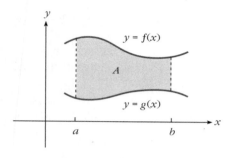

FIGURE 7.10  The area between the graphs of $f$ and $g$ on the interval $[a, b]$.

There are cases in applications where we must find the area between two curves. Figure 7.10 shows such a region. In this case $f(x) \geq g(x)$ on the interval $[a, b]$ and $f$ and $g$ are both continuous nonnegative functions. We will describe the region in Figure 7.10 as "the area between the graphs of $f$ and $g$ on the interval $[a, b]$."

As illustrated in Figure 7.11, this area can be obtained by finding the area under the upper curve and subtracting the area under the lower curve. That is,

[area between $f$ and $g$] = [area under $f$] − [area under $g$]

$$A = A_f - A_g$$
$$= \int_a^b f(x)\,dx - \int_a^b g(x)\,dx$$
$$= \int_a^b [f(x) - g(x)]\,dx$$

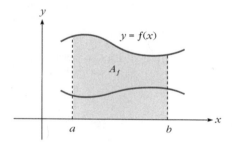 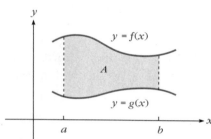 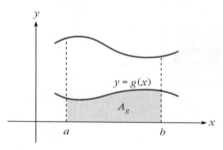

FIGURE 7.11  $A = A_f - A_g$, or $A = \int_a^b [f(x) - g(x)]\,dx$.

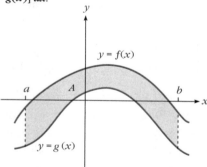

FIGURE 7.12  The formula $A = \int_a^b [f(x) - g(x)]\,dx$ remains valid even if $f(x)$ or $g(x)$ is not nonnegative.

It is left as an exercise to show that this formula is still valid if either $f$ or $g$ fails to be nonnegative. Such a situation is illustrated in Figure 7.12. We summarize:

**THE AREA BETWEEN TWO CURVES**

If $f(x)$ and $g(x)$ are continuous on the interval $[a, b]$, with $f(x) \geq g(x)$ for all $x$ in $[a, b]$, then the area between the graphs of $f$ and $g$ on $[a, b]$ is given by

$$A = \int_a^b [f(x) - g(x)]\,dx$$

## EXAMPLE 5

Find the area between the graphs of $f(x) = x^2 + 1$ and $g(x) = x$ on the interval $[0, 1]$.

### Solution

The graphs of $f(x)$ and $g(x)$ are sketched in Figure 7.13 and the required area $A$ is shaded. Since $f(x) \geq g(x)$ on the interval $[0, 1]$, we have

$$A = \int_a^b [f(x) - g(x)]\, dx = \int_0^1 [(x^2 + 1) - x]\, dx = \int_0^1 (x^2 - x + 1)\, dx$$
$$= \left[\left(\tfrac{1}{3}x^3 - \tfrac{1}{2}x^2 + x\right)\right]_0^1 = \left[\tfrac{1}{3}(1)^3 - \tfrac{1}{2}(1)^2 + 1\right] - \left[\tfrac{1}{3}(0)^3 - \tfrac{1}{2}(0)^2 + 0\right]$$
$$= \tfrac{5}{6} \text{ square units} \qquad \blacksquare$$

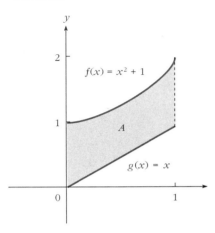

**FIGURE 7.13** The area between the graphs of $f(x) = x^2 + 1$ and $g(x) = x$ on the interval $[0, 1]$.

Sometimes we are required to find the area between two curves $f$ and $g$. Such a situation is illustrated in Figure 7.14. In this case it is important to find the $x$-coordinates of the points of intersection of $f$ and $g$, since these $x$-coordinates will determine the limits of integration.

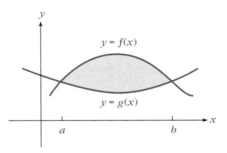

**FIGURE 7.14** The area between the graphs of $f$ and $g$.

## EXAMPLE 6

Find the area enclosed by the graphs of $f(x) = x^2 - 2x + 1$ and $g(x) = 2x + 1$.

### Solution

Since we are not given an interval over which to integrate, we begin by finding the $x$-coordinates of the points of intersection of the two curves. We do this by equating the two functions and solving for $x$:

$$x^2 - 2x + 1 = 2x + 1$$
$$x^2 - 4x = 0$$
$$x(x - 4) = 0$$
$$x = 0, \quad x = 4 \qquad (x\text{-coordinates of intersection points})$$

We use these values as an aid in producing the sketch drawn in Figure 7.15. Note that in this case the graph of $g(x) = 2x + 1$ lies above the graph of $f(x) = x^2 - 2x + 1$ over the interval $[0, 4]$. Remember that in finding the area between two curves, we find the area under the upper curve and subtract the area under the lower curve. Therefore

$$A = \int_0^4 [(2x + 1) - (x^2 - 2x + 1)] \, dx$$
$$= \int_0^4 (-x^2 + 4x) \, dx = \left(-\tfrac{1}{3}x^3 + 2x^2\right)\Big]_0^4$$
$$= \left[-\tfrac{1}{3}(4)^3 + 2(4)^2\right] - \left[-\tfrac{1}{3}(0)^3 + 2(0)^2\right]$$
$$= \frac{32}{3} \text{ square units}$$

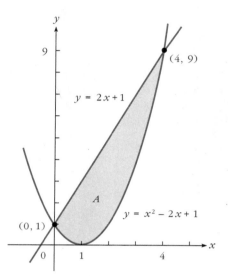

**FIGURE 7.15** The area between the graphs of $f(x) = x^2 - 2x + 1$ and $g(x) = 2x + 1$.

---

### EXAMPLE 7 (Participative)

Let's find the area bounded by the graphs of $f(x) = x^3$ and $g(x) = \sqrt[3]{x}$.

**Solution**

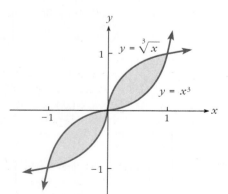

**FIGURE 7.16** The area enclosed by the graphs of $y = x^3$ and $y = \sqrt[3]{x}$.

To find the $x$-coordinates of the points of intersection of the two graphs, we equate the functions and solve for $x$:

$$x^3 = \sqrt[3]{x}$$
$$(x^3)^3 = (\sqrt[3]{x})^3 \quad \text{(Cubing both sides)}$$
$$x^9 = \underline{\phantom{xx}}$$
$$x^9 - x = 0$$
$$x(x^8 - 1) = 0$$

$x = 0$    or    $x^8 - 1 = 0$

$x = 0$    or    $x^8 = 1$

$x = 0,$    $x = \underline{\phantom{xx}},$    or    $x = \underline{\phantom{xx}}$

A sketch is drawn in Figure 7.16. Note that $x^3 \geq \sqrt[3]{x}$ for $x$ in $[-1, 0]$, whereas $\underline{\phantom{xx}} \geq \underline{\phantom{xx}}$ for $x$ in $[0, 1]$.

We find the area of the required region by evaluating two separate integrals.

$$A = \int_{-1}^{0} (x^3 - \sqrt[3]{x}) \, dx + \underline{\phantom{XXXXXXXXXXX}}$$

$$= \left[\left(\tfrac{1}{4}x^4 - \tfrac{3}{4}x^{4/3}\right)\right]_{-1}^{0} + \underline{\phantom{XXXXXXXXXXX}}$$

$$= \left[\tfrac{1}{4}(0)^4 - \tfrac{3}{4}(0)^{4/3}\right] - \left[\tfrac{1}{4}(-1)^4 - \tfrac{3}{4}(-1)^{4/3}\right] +$$

$$\underline{\phantom{XXXXXXXXXXXXXXXXXXXXXXXXXXXXXXXXXXXXX}}$$

$$= \tfrac{1}{2} + \underline{\phantom{XX}} = \underline{\phantom{XX}} \text{ square unit} \qquad \blacksquare$$

In Chapter 6 we learned that $\int_a^b f(x) \, dx$ could be interpreted as the net change in a quantity $F(x)$ between $x = a$ and $x = b$, where $f(x)$ is the rate of change of $F(x)$. In view of our new interpretation of $\int_a^b f(x) \, dx$ as an area when $f(x) \geq 0$ over the interval $[a, b]$, we may now make the following statement:

> If $f(x) \geq 0$ represents the rate at which a quantity $F(x)$ changes for $a \leq x \leq b$, then the net change in the quantity between $x = a$ and $x = b$ is given by the area under the graph of $f$ from $a$ to $b$.

Figure 7.17 gives a graphical illustration of this result. Let's look at an application.

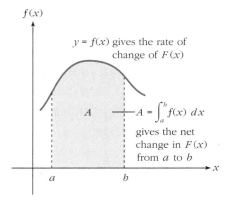

**FIGURE 7.17** The area under the rate-of-change function $f$ from $a$ to $b$ yields the net change in $F$ from $a$ to $b$.

EXAMPLE 8

Repair and maintenance costs for a certain new piece of industrial machinery placed into service at the beginning of 1991 are expected to change at a rate of $f(t) = 5t^2 + 8t + 1000$ dollars per year for the next ten years, where $t = 0$ corresponds to the beginning of 1991. Find the total amount of repair and maintenance costs for this machine from 1991 to 2001.

Solution

The quantity we want is $\int_0^{10} (5t^2 + 8t + 1000) \, dt$. This quantity is represented

---

**SOLUTIONS TO EXAMPLE 7**

$x, -1, 1, \sqrt[3]{x}, x^3, \int_0^1 (\sqrt[3]{x} - x^3) \, dx, \left[\tfrac{3}{4}x^{4/3} - \tfrac{1}{4}x^4\right]_0^1, \left[\tfrac{3}{4}(1)^{4/3} - \tfrac{1}{4}(1)^{4/3}\right] - \left[\tfrac{3}{4}(0)^{4/3} - \tfrac{1}{4}(0)^3\right], \tfrac{1}{2}, 1$

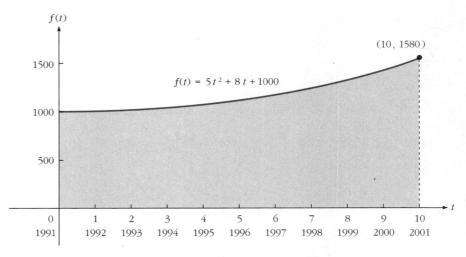

**FIGURE 7.18** The shaded area represents the total amount of repair and maintenance expense for a machine from 1991 to 2001.

by the area under the graph of $f(t) = 5t^2 + 8t + 1000$ from $t = 0$ to $t = 10$ and is sketched in Figure 7.18. We compute:

$$\int_0^{10} (5t^2 + 8t + 1000)\, dt = \left[\left(\tfrac{5}{3}t^3 + 4t^2 + 1000t\right)\right]_0^{10}$$
$$= \left[\tfrac{5}{3}(10)^3 + 4(10)^2 + 1000(10)\right] - \left[\tfrac{5}{3}(0)^3 + 4(0)^2 + 1000(0)\right]$$
$$\approx \$12{,}067$$

Thus the total amount of maintenance and repair costs for this machine from 1991 to 2001 will be approximately \$12,067. ■

## SECTION 7.1
### SHORT ANSWER EXERCISES

1. (True or False) The area under the graph of $y = x^2$ from $-1$ to $1$ is given by $\int_{-1}^{1} x^2\, dx$. _____

2. (True or False) The area under the graph of $y = x^3$ from $-1$ to $1$ is given by $\int_{-1}^{1} x^3\, dx$. _____

3. Given $f(x) = 5$, $a = 0$, and $b = 8$, the area function $A(x)$ defined in the proof of the area theorem is **(a)** $x^2$; **(b)** $5x^2$; **(c)** $5x$; **(d)** none of the above.

4. The area of the region bounded by the graph of $f(x) = 4 - x^2$ written as a definite integral is $A = $ _____.

5. (True or False) The definite integral $\int_a^b f(x)\, dx$ always gives the area bounded by the graph of $f(x)$, the $x$-axis, and the vertical lines $x = a$ and $x = b$ if $f(x)$ is nonnegative and continuous on $[a, b]$. _____

6. (True or False) If $f(x)$ is continuous on $[a, b]$, then the area bounded by the graph of $f(x)$, the $x$-axis, and the vertical lines $x = a$ and $x = b$ is $\left|\int_a^b f(x)\, dx\right|$. _____

7. If we wish to find the area between the graphs of $f(x) = x^2$ and $g(x) = x$ on $[0, 1]$, then the upper curve is _____.

8. Two continuous functions $f$ and $g$ intersect at $x = -2$ and $x = 3$ and nowhere else. If $f(0) = 3$ and $g(0) = 1$, the definite integral that gives the area between $f$ and $g$ is _____.
9. The graphs of $f(x) = x^2$ and $g(x) = x^3$ intersect when $x =$ ____ and $x =$ ____.
10. If a quantity $A$ changes at a rate of $A'(t)$, the net change in the quantity from $t = 1$ to $t = 3$ written as a definite integral is _____.

## SECTION 7.1 EXERCISES

In Exercises 1–18, find the area between the graph of $f$ and the x-axis from $a$ to $b$.

1. $f(x) = x$, $a = 0$, $b = 1$
2. $f(x) = 2x$, $a = 1$, $b = 3$
3. $f(x) = 2x - 1$, $a = 1$, $b = 4$
4. $f(x) = 10 - 3x$, $a = -2$, $b = 0$
5. $f(x) = 5x^2$, $a = -1$, $b = 2$
6. $f(x) = 16 - x^2$, $a = -2$, $b = 2$
7. $f(x) = 4x^2 + 3$, $a = 1$, $b = 2$
8. $f(x) = 2x^3$, $a = 0$, $b = 2$
9. $f(x) = 2/x$, $a = 1$, $b = 3$
10. $f(x) = 2e^x + 1$, $a = -1$, $b = 0$
11. $f(x) = 5x$, $a = -1$, $b = 0$
12. $f(x) = 2x^3$, $a = -2$, $b = 0$
13. $f(x) = 1 - \sqrt{x}$, $a = 0$, $b = 1$
14. $f(x) = e^x - 2$, $a = 0$, $b = \frac{1}{2}$
15. $f(x) = 4\sqrt[3]{x}$, $a = -8$, $b = 0$
16. $f(x) = x^2 - 4$, $a = -1$, $b = 1$
17. $f(x) = \ln x$, $a = 1$, $b = e$
18. $f(x) = xe^x$, $a = 0$, $b = 1$

In Exercises 19–30, find the area of the region bounded by the graph of $f$ and the x-axis.

19. $f(x) = 9 - x^2$
20. $f(x) = 50 - 2x^2$
21. $f(x) = -x^2 + 4x - 3$
22. $f(x) = -x^2 + 2x + 8$
23. $f(x) = \sqrt{x} - 1$
24. $f(x) = \sqrt[4]{x} - 2$
25. $f(x) = x^2 - 100$
26. $f(x) = x^2 + 5x + 6$
27. $f(x) = x^3 - x$
28. $f(x) = 3x^3 - 12x$
29. $f(x) = 2x^3 - 7x^2 - 4x$
30. $f(x) = 2x - x^2 - 3x^3$

In Exercises 31–38, find the area between the graphs of $f$ and $g$ on the given interval $I$.

31. $f(x) = x^2$, $g(x) = \sqrt{x}$, $I = \left[0, \frac{1}{2}\right]$
32. $f(x) = x^2 + 4$, $g(x) = 2x$, $I = [-1, 1]$
33. $f(x) = x^3 - 8$, $g(x) = 3x$, $I = [-1, 2]$
34. $f(x) = \ln x$, $g(x) = x^2$, $I = [2, e]$
35. $f(x) = e^x$, $g(x) = 1$, $I = [-1, 2]$
36. $f(x) = x^3$, $g(x) = \sqrt{x}$, $I = [0, 4]$
37. $f(x) = 1$, $g(x) = \ln x$, $I = \left[\frac{1}{2}, e\right]$
38. $f(x) = x^2$, $g(x) = x^3$, $I = [-1, 2]$

In Exercises 39–46, find the area enclosed by the graphs of $f$ and $g$.

39. $f(x) = x^2 - 10$, $g(x) = 3x$
40. $f(x) = 2x^2$, $g(x) = 5x + 12$
41. $f(x) = 4x^2 - 3$, $g(x) = x^2 + 7x + 3$
42. $f(x) = 4 + x - x^2$, $g(x) = 2 + 2x + 5x^2$
43. $f(x) = x^3$, $g(x) = 5x^2 - 6x$
44. $f(x) = x - 2x^3$, $g(x) = -x^2$
45. $f(x) = (x - 2)\ln x$, $g(x) = x - 2$
46. $f(x) = xe^x$, $g(x) = ex$

47. **(Cost)** The monthly marginal cost function for a certain firm is $C'(x) = 0.02x$ dollars per unit produced. If the fixed cost is $2000 per month, (a) find the monthly

cost function $C(x)$; **(b)** find the cost of producing 5000 units per month; **(c)** evaluate $\int_0^{5000} C'(x)\,dx$ (this integral represents the variable cost of producing 5000 units); **(d)** sketch the marginal cost function, and represent the cost found in part (c) as an area.

**48.** (Revenue) The monthly marginal revenue function for a certain firm is $R'(x) = 3\sqrt{x}$ dollars per unit sold. If $R(0) = 0$, **(a)** find the revenue generated from the sale of the first 1000 units; **(b)** sketch the graph of the marginal revenue function and represent the revenue found in part (a) as an area.

**49.** (Profit) A firm has monthly marginal cost function $C'(x) = 0.75$ dollars per unit produced and monthly marginal revenue function $R'(x) = 5 - 0.002x$ dollars per unit sold. If fixed costs are \$1500 per month and $R(0) = 0$, **(a)** find the monthly profit function $P(x)$; **(b)** find the profit derived from the manufacture and sale of 2500 units per month; **(c)** evaluate $\int_0^{2500} [R'(x) - C'(x)]\,dx$ (this integral represents the revenue derived from the sale of 2500 units less the variable cost of producing these units); **(d)** sketch the graphs of the marginal cost and marginal revenue functions and represent the quantity found in part (c) as an area.

**50.** (Depreciation) The value of an industrial machine is decreasing at a rate of $V'(t) = -200t$ dollars per year. If $t = 0$ corresponds to 1992, **(a)** find the loss in value of the machine during the years 1992 and 1997; **(b)** sketch the graph of $V'(t)$ for $0 \le t \le 10$, and represent the loss in value found in part (a) as an area.

### Social Sciences

**51.** (Population) The population of a city is increasing at the rate of $P'(t) = 4/\sqrt{t+1}$ thousands of people per year. If $t = 0$ corresponds to 1990, **(a)** find the increase in population between 1990 and 1998; **(b)** sketch the graph of $P'(t)$ and represent the increase in population found in part (a) as an area.

**52.** (Radioactive substance decay) A radioactive substance decays at a rate of $A'(t) = -2e^{-0.01t}$ g/yr.
**(a)** Find the amount of the substance that decays between $t = 5$ and $t = 10$.
**(b)** Sketch the graph of $A'(t)$ and represent the amount found in part (a) as an area.

**53.** (Motion) An object moves in a straight line, so that its velocity at time $t \ge 0$ is $v(t) = \ln(t + 1)$ ft/sec.
**(a)** Find the distance traveled by the object between $t = 1$ sec and $t = 5$ sec.
**(b)** Sketch the graph of $v(t)$ and represent the distance found in part (a) as an area.

**54.** (Self-perception) In a certain region of the country, an average person's self-perception $P$ of personal health, safety, and well-being decreases at the rate of

$$\frac{dP}{dt} = \frac{-0.0245}{\sqrt{(0.049t - 2.377)^3}}$$

units per year for people between the ages of $t = 50$ and $t = 90$, where $P$ is measured on a scale from 0 to 5.
**(a)** Find the loss in self-perception between the ages of 50 and 70.
**(b)** Sketch the graph of $dP/dt$ and represent the loss in self-perception found in part (a) as an area.

**55.** (Military budget) The yearly budget $B$ for a certain military weapons system is increasing at the rate of $B'(t) = 2\sqrt{t}$ millions of dollars per year, where $t = 0$ corresponds to 1990.
**(a)** Find the increase in the budget for this weapons system between 1991 and 1994.
**(b)** Sketch the graph of $B'(t)$ and represent the increase in the budget found in part (a) as an area.

## OPTIONAL
## GRAPHING CALCULATOR/ COMPUTER EXERCISES

In Exercises 56–59, find the area bounded by the graph of f and the x-axis. Estimate the x-intercepts of f to the nearest tenth of a unit.

56. $f(x) = 1 - x + x^2 + 2x^3 - x^4$
57. $f(x) = x^5 - 2x^4 + x^3 - 2x^2 + x + 1$
58. $f(x) = e^x - x - 2$
59. $f(x) = \ln(x + 1) - x + 1$

In Exercises 60–63, find the area enclosed by the graphs of f and g. Estimate the x-coordinates of the points of intersection of f and g to the nearest tenth of a unit.

60. $f(x) = x^3 + 3x$, $g(x) = x^4 - x^2 + 1$
61. $f(x) = \sqrt{x} - x + 2$, $g(x) = x^2$
62. $f(x) = 2 - x^2$, $g(x) = e^x$
63. $f(x) = 2x \ln(x^2 + 1)$, $g(x) = x^2 - 4$

## 7.2 THE DEFINITE INTEGRAL AS THE LIMIT OF A SUM

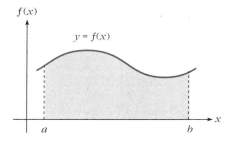

**FIGURE 7.19** The area under the graph of $f$ from $a$ to $b$.

### Approximating Areas Using Rectangles

In Chapter 6 we learned that the definite integral $\int_a^b f(x)\, dx$ can be interpreted as the net change in $F(x)$ between $a$ and $b$, where $F(x)$ is any antiderivative of $f(x)$. That is, $\int_a^b f(x)\, dx = F(b) - F(a)$. In this chapter we have seen that the definite integral $\int_a^b f(x)\, dx$ can also be interpreted as the area under the graph of $f$ from $a$ to $b$ if $f(x) \geq 0$ for $x$ in $[a, b]$. In this section we will consider another way of interpreting a definite integral—as the limit of what is called a Riemann sum. This new interpretation will be valuable for understanding many of the applications of the definite integral.

Let us begin by considering anew the problem of finding the area under the graph of a continuous nonnegative function from $a$ to $b$; see Figure 7.19. Our objective will be to approximate this area by a number of rectangles with bases on the x-axis between $a$ and $b$ and with heights that are the y-coordinates of certain points on the graph of $f(x)$, as shown in Figure 7.20.

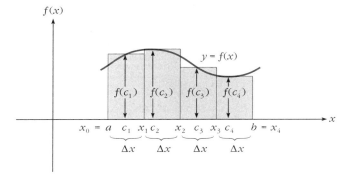

**FIGURE 7.20** The area under the graph of $f$ from $a$ to $b$ is approximated by the sum of the areas of the shaded rectangles.

In Figure 7.20, the interval $[a, b]$ has been divided into four equal subintervals: $[x_0, x_1], [x_1, x_2], [x_2, x_3]$, and $[x_3, x_4]$, where we have set $x_0 = a$ and $x_4 = b$. The length of each of these subintervals is designated by $\Delta x$, where $\Delta x = (b - a)/4$. In the four subintervals we have chosen arbitrary numbers $c_1, c_2, c_3$, and $c_4$. We have constructed four shaded rectangles with the four subintervals $[x_0, x_1], [x_1, x_2], [x_2, x_3]$, and $[x_3, x_4]$ for bases and with heights $f(c_1), f(c_2), f(c_3)$, and $f(c_4)$, respectively. Observe that the area under the graph of $f$ from $a$ to $b$ is approximated by the sum of the areas of the four shaded rectangles. Since the area of a rectangle is simply its base multiplied by its height, the sum of the areas of these four rectangles is

$$S_4 = \Delta x \cdot f(c_1) + \Delta x \cdot f(c_2) + \Delta x \cdot f(c_3) + \Delta x \cdot f(c_4)$$

where each term is the area of the 1st, 2nd, 3rd, and 4th rectangle respectively.

So if we let $A$ designate the exact area under the graph of $f$ from $a$ to $b$, then $A \approx S_4$.

**Riemann Sums**

The sum $S_4$ of the areas of the four rectangles is called a *Riemann sum* for $f$ on the interval $[a, b]$. The Riemann sum is named for the German mathematician Bernhard Riemann (1826–1866), who first devised it as a method of finding areas under curves.

Notice that our approximation of the exact area $A$ can be improved if we subdivide the interval $[a, b]$ into more than four subintervals. For example, Figure 7.21 shows the result of subdividing $[a, b]$ into eight equal subintervals and drawing the corresponding rectangle above each subinterval. We then obtain the Riemann sum

$$S_8 = \Delta x \cdot f(c_1) + \Delta x \cdot f(c_2) + \cdots + \Delta x \cdot f(c_7) + \Delta x \cdot f(c_8)$$

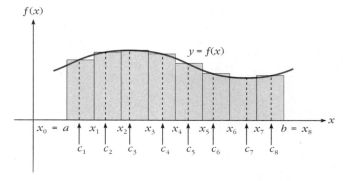

**FIGURE 7.21** A Riemann sum using $n = 8$ rectangles to approximate the area under the graph of $f$ from $a$ to $b$.

Of course, $\Delta x$ and $c_1, c_2, \ldots, c_8$ are different from those shown in Figure 7.20. The point is that $S_8$ approximates the true area $A$ more closely than $S_4$. Indeed, it is usually the case that the larger the number of rectangles we use to approximate the area under the graph of a function $f$ from $a$ to $b$, the more accurate the approximation becomes. It is not surprising, then, that as the number $n$ of subintervals increases without bound, $\Delta x$ approaches zero and the Riemann sum approaches the exact area $A$ under the graph of $f$ from $a$ to $b$, which is given by the definite integral $\int_a^b f(x)\, dx$ if $f(x) \geq 0$ on $[a, b]$. These results are summarized in the following box.

**THE DEFINITE INTEGRAL AS THE LIMIT OF A RIEMANN SUM**

Let $f(x)$ be continuous and nonnegative on $[a, b]$ and let $n$ be a positive integer. Partition the interval $[a, b]$ into $n$ subintervals of equal length: $[x_0, x_1], [x_1, x_2], \cdots, [x_{n-1}, x_n]$, where $x_0 = a$ and $x_n = b$ and where the length of each subinterval is $\Delta x = (b - a)/n$. Next, choose arbitrary numbers $c_1, c_2, \ldots, c_n$ in the intervals $[x_0, x_1], [x_1, x_2], \cdots, [x_{n-1}, x_n]$, respectively. Then the sum

$$S_n = \Delta x \cdot f(c_1) + \Delta x \cdot f(c_2) + \cdots + \Delta x \cdot f(c_n)$$
$$= [f(c_1) + f(c_2) + \cdots + f(c_n)]\Delta x$$

is called a *Riemann sum* for $f$ on $[a, b]$. Furthermore,

$$\int_a^b f(x)\, dx = \lim_{\Delta x \to 0} [f(c_1) + f(c_2) + \cdots + f(c_n)]\Delta x$$

Thus, the definite integral can be viewed as the limit of a Riemann sum. This result is valid even if $f(x)$ fails to be nonnegative on $[a, b]$.

Since we already know that $\int_a^b f(x)\, dx = F(b) - F(a)$, where $F(x)$ is an antiderivative of $f(x)$ on $[a, b]$, we can now state an important result, called the *fundamental theorem of calculus*, that shows the connection between the definite integral considered as the net change in the antiderivative $F(x)$ between $x = a$ and $x = b$ and the definite integral interpreted as the limit of a Riemann sum.

**THE FUNDAMENTAL THEOREM OF CALCULUS**

Let $f(x)$ be continuous on $[a, b]$ and let $\Delta x$ and $c_i$ be defined as before. If $F(x)$ is an antiderivative of $f(x)$ on $[a, b]$ then

$$\int_a^b f(x)\, dx = F(b) - F(a) = \lim_{\Delta x \to 0} [f(c_1) + f(c_2) + \cdots + f(c_n)]\Delta x$$

The interpretation of the definite integral as the limit of a Riemann sum is quite useful in applications. In many cases in such applications we wish to find some quantity $Q$ that can be approximated by a Riemann sum:

$$Q \approx [f(c_1) + f(c_2) + \cdots + f(c_n)]\Delta x$$

over an interval $[a, b]$. In this case the quantity $Q$ can be found as a limit:

$$Q = \lim_{\Delta x \to 0} [f(c_1) + f(c_2) + \cdots + f(c_n)]\Delta x = \int_a^b f(x)\, dx$$

Thus $Q$ can be found as the definite integral $\int_a^b f(x)\,dx$, so that if $F$ is an antiderivative of $f$ on $[a, b]$, then

$$Q = \int_a^b f(x)\,dx = F(b) - F(a)$$

We will consider several applications in the next section in which we apply the technique just described for finding the quantity $Q$. For now, let's consider just one typical application.

### The Average Value of a Function

There are many applied situations where it is required to find the average value of a function. We might be interested, for example, in finding the average value of money in an account over the next five years, the average population of a city over the next ten years, the average inventory on hand over the next six months, or the average monthly profit over the next year.

To begin, notice that the average of a simple set of *numbers* $y_1, y_2, \ldots, y_n$ is given by

$$\frac{y_1 + y_2 + \cdots + y_n}{n}$$

If $y = f(x)$, then we can *approximate* the average value of $f(x)$ on the interval $[a, b]$ by averaging the values $f(c_1), f(c_2), \ldots, f(c_n)$, where $c_1, c_2, \ldots, c_n$ are chosen in the $n$ subintervals $[x_0, x_1], [x_1, x_2], \ldots, [x_{n-1}, x_n]$, as in the definition of a Riemann sum. If $Q$ is the actual average value of $f(x)$ on $[a, b]$, then

$$Q \approx \frac{f(c_1) + f(c_2) + \cdots + f(c_n)}{n}$$

As we take $n$ to be larger and larger, this approximation should get better and better. We can write

$$Q \approx f(c_1) \cdot \frac{1}{n} + f(c_2) \cdot \frac{1}{n} + \cdots + f(c_n) \cdot \frac{1}{n}$$

$$\approx \frac{1}{b-a}\left[f(c_1) \cdot \frac{b-a}{n} + f(c_2) \cdot \frac{b-a}{n} + \cdots + f(c_n) \cdot \frac{b-a}{n}\right]$$

Recalling that $(b - a)/n = \Delta x$, we can write

$$Q \approx \frac{1}{b-a}[f(c_1)\Delta x + f(c_2)\Delta x + \cdots + f(c_n)\Delta x]$$

and so

$$(b - a)Q \approx [f(c_1) + f(c_2) + \cdots + f(c_n)]\Delta x$$

Notice that the expression on the right-hand side is a Riemann sum for $f$ on $[a, b]$, so that taking the limit as $\Delta x \to 0$ of both sides gives us

$$(b - a)Q = \int_a^b f(x)\,dx$$

Therefore

$$Q = \frac{1}{b-a}\int_a^b f(x)\,dx$$

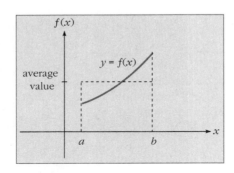

FIGURE 7.22 The rectangle with height equal to the average value of $f$ and with base $[a, b]$ has area equal to the area under $f$ from $a$ to $b$.

## THE AVERAGE VALUE OF A FUNCTION

If $f(x)$ is continuous on $[a, b]$, then the *average value* of $f$ over $[a, b]$ is

$$\text{average value} = \frac{1}{b-a} \int_a^b f(x)\, dx$$

Geometrically, if $f(x) \geq 0$ on $[a, b]$, the average value of $f(x)$ on $[a, b]$ is the height of the rectangle with base $[a, b]$ and area equal to the area under the graph of $f$ from $a$ to $b$. This can be seen by writing the average value formula in the form

$$(b - a)(\text{average value}) = \int_a^b f(x)\, dx$$

This is illustrated in Figure 7.22.

### EXAMPLE 1

Find the average value of $f(x) = x^2$ on the interval $[0, 2]$.

**Solution**

We have $a = 0$, $b = 2$, and $f(x) = x^2$, so that

$$\text{average value} = \frac{1}{2-0} \int_0^2 x^2\, dx = \frac{1}{2}\left(\frac{x^3}{3}\right)\Bigg|_0^2 = \frac{1}{2}\left(\frac{8}{3}\right) = \frac{4}{3}$$

For a geometric interpretation, see Figure 7.23, in which the rectangle with height $\frac{4}{3}$ and base $[0, 2]$ has the same area $\left(\frac{8}{3} \text{ square units}\right)$ as the area under the graph of $f(x) = x^2$ from 0 to 2.

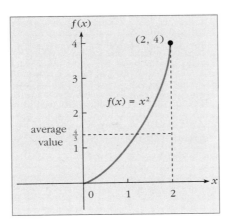

**FIGURE 7.23** The average value of $f(x) = x^2$ on $[0, 2]$ is $4/3$.

### EXAMPLE 2

The population of a certain city was 2.5 million at the beginning of 1990 and is growing exponentially at 2.8% per year. Find the average population between the beginning of 1990 and the beginning of 2001.

**Solution**

If $N(t)$ represents the population (in millions) of the city $t$ years after 1990, then $N(t) = 2.5e^{0.028t}$. We are required to find the average value of $N(t)$ from $t = 0$ (the year 1990) to $t = 11$ (the year 2001):

$$\text{average value} = \frac{1}{11-0}\int_0^{11} 2.5e^{0.028t}\,dt = \frac{1}{11}\left[\left(\frac{2.5}{0.028}e^{0.028t}\right)\right]_0^{11}$$

$$= \frac{1}{11}\left[\frac{2.5}{0.028}e^{0.028(11)} - \frac{2.5}{0.028}e^{0.028(0)}\right]$$

$$= \frac{1}{11}\left[\frac{2.5}{0.028}e^{0.308} - \frac{2.5}{0.028}\right] \approx 2.9$$

We conclude that the average population of this city between 1990 and 2001 will be about 2.9 million people.

## SECTION 7.2
### SHORT ANSWER EXERCISES

1. In forming a Riemann sum for a continuous nonnegative function $f(x)$ on $[a, b]$, we are approximating the area under the graph of $f$ by a sum of areas of certain _____.

2. For the interval $[1, 10]$, if the number of rectangles in a Riemann sum is 12, then $\Delta x = $ _____.

3. The exact value of $\int_0^2 x^2\,dx$ is $\frac{8}{3}$. The area represented by this integral can be approximated by a Riemann sum with $n = 2$, $c_1 = \frac{1}{2}$, and $c_2 = \frac{3}{2}$. The value of this sum is $S_2 = \Delta x \cdot f(c_1) + \Delta x \cdot f(c_2) = $ _____.

4. Would you expect a more accurate approximation of $\int_a^b f(x)\,dx$ by using a Riemann sum with $n = 10$ or with $n = 100$? _____

5. If we use a Riemann sum to approximate $\int_1^2 10\,dx$, then for any $n$, the approximation $S_n$ is (a) exact; (b) approximate; (c) not defined.

6. Suppose that $A \approx \left[c_1^3 + c_2^3 + \cdots + c_n^3\right]\Delta x$, where the approximation is valid over the interval $[0, 5]$ and $c_1, c_2, \ldots, c_n$ are chosen in subintervals of $[0, 5]$ of equal length. If $\Delta x = 5/n$ and the error in the approximation approaches zero as $\Delta x \to 0$, then $A = $ _____.

7. If $f(x)$ is continuous on $[2, 7]$ and if the average value of $f(x)$ on $[2, 7]$ is 5, then $\int_2^7 f(x)\,dx = $ _____.

8. The average value of a continuous nonnegative function $f(x)$ is equal to 3 over the interval $[0, 4]$. What is the area between the graph of $f(x)$ and the $x$-axis from $x = 0$ to $x = 4$? _____

9. (True or False) It is possible for the average value of a function over an interval to be zero. _____

10. (True or False) It is possible for the average value of a continuous function $f(x)$ over an interval $[a, b]$ to be larger than the absolute maximum value of $f(x)$ on $[a, b]$. _____

## SECTION 7.2
### EXERCISES

*In Exercises 1–8, find the average value of each function over its given interval.*

1. $f(x) = x^2 + 1$ over $[0, 2]$
2. $f(x) = x^3 + 2$ over $[1, 3]$
3. $f(x) = \sqrt{x}$ over $[0, 4]$
4. $f(x) = x^2 + 2x + 1$ over $[2, 4]$
5. $f(x) = 2x + 3$ over $[1, 4]$
6. $f(x) = \ln x$ over $[1, e]$
7. $f(x) = e^x$ over $[0, 1]$
8. $f(x) = x$ over $[-1, 1]$

9. (Velocity) The velocity of an object moving in a straight line is given by $v(t) = \sqrt[3]{t}$ ft/sec. Find the average velocity of this object between $t = 0$ and $t = 8$.

10. (Continuous interest) If $10,000 is invested in an account that yields 9% interest compounded continuously, find the average amount of money in the account over a period of five years.

11. (Cost) The monthly total cost function of a certain firm is given by $C(x) = 2000 + 0.1x + 0.01x^2$ for $0 \leq x \leq 5000$. Find the average value of this total cost function.

### Life Sciences

12. (Greenhouse temperature) The temperature in a greenhouse between 8:00 A.M. and 12:00 midnight is given by the function $f(t) = -0.15625t^2 + 2.5t + 75$ degrees Fahrenheit. If $t = 0$ corresponds to 8:00 A.M., what is the average temperature in this greenhouse between 8:00 A.M. and 12:00 midnight?

13. (Radioactive decay) The amount of a certain radioactive isotope present after $t$ years is given by $A(t) = 100e^{-0.0032t}$ grams. Find the average amount of the isotope present over the next 500 years if $t = 0$ corresponds to the present.

14. (Employee learning) After $t$ days on the job, an average new employee can assemble $N(t) = 3 + 10\sqrt{t}$ clock radios per day. Find the average number of clock radios produced by an employee during her first 10 days of employment.

15. (Learning) After $t$ days of training, a certain student in a word processing class can process $N(t) = 100 - 85e^{-0.025t}$ words per minute (wpm). Find the average number of wpm processed by a student during the first 30 days of training.

16. (Product reliability) A manufacturer of calculators estimates that after $t$ hours of use, the number of calculators still operating out of a batch of 10,000 will be $N(t) = 10000 - t \ln(t + 1)$. Find the average number of calculators still operating after 500 hours of use.

17. (Gasoline consumption) If $G(t)$ represents the yearly consumption of gasoline (in millions of gallons) by the people in a certain town and if $t = 0$ corresponds to the beginning of 1991, then $G(t) = 6.5\sqrt{t + 1}$. Find the average yearly consumption of gasoline by the residents of the town between the beginning of 1991 and the beginning of 1998.

### Business and Economics

18. (Inventory) At the beginning of 1991, the Maytex company has 20,000 tons of a certain chemical stored in its warehouse. If the chemical is used at a constant rate during the year, no orders for the chemical are placed during the year, and if 2000 tons remain at the beginning of 1992, find the average inventory for the year.

### Social Sciences

19. (Social satisfaction) The intellectual and social satisfaction of an individual $x$ years of age has been determined to be

$$I(x) = \frac{1}{\sqrt[3]{0.010995x - 0.63827}}, \quad 60 \leq x \leq 85$$

for individuals in a certain region of the country, where $I$ is measured on a scale of 0 to 5. Find the average level of intellectual and social satisfaction for people between the ages of 60 and 85.

## 7.3 APPLICATIONS OF THE DEFINITE INTEGRAL

### The Consumer's Willingness to Pay and Consumer's Surplus

Suppose that you go out to purchase a fan on a hot summer day. You pick one up at Bucko's, a local retail store, for $15. On your way home you notice an advertisement in Cheapo's store window displaying a similar fan for only $8. You stop in at Cheapo's and purchase two more fans.

Let's take a minute to analyze your actions from the point of view of economics. That first fan you purchased for $15 had a value of at least $15 to you—otherwise you would not have bought it. Perhaps you would have been willing to pay as much as $30 for that first fan. This $30 represents your *willingness to pay* for the first fan. Now, although it would have been convenient to have a second or even a third fan to put in different rooms of your home, you were not willing to pay as much as $15 to purchase a second fan—otherwise you would have bought it at Bucko's. You may have been willing to pay only $12 for the second fan and then only $9 for a third fan. That's why you bought the second and third fans at Cheapo's. Since you were willing to pay only $3 for a fourth fan, however, you did not purchase additional fans at Cheapo's.

Continuing the analysis, we have the following amounts:

The amount you were *willing to pay* for the 3 fans  = $30 + $12 + $9 = $51

The amount you *actually paid* for the 3 fans = $15 + $8 + $8 = $31

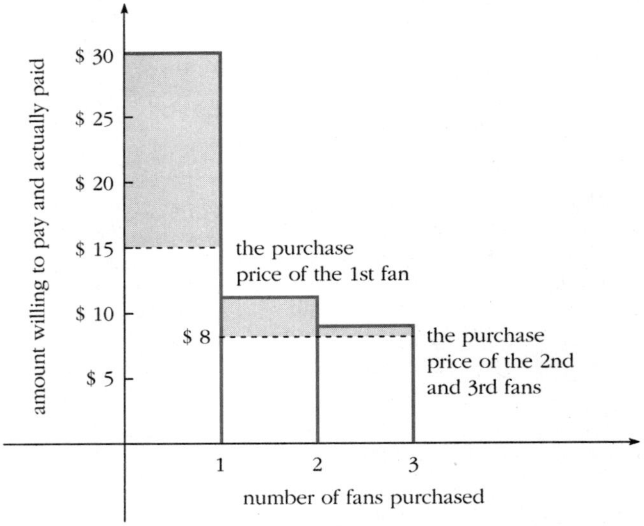

FIGURE 7.24 The shaded area represents your consumer surplus of $20 on the purchase of three fans.

The difference between these two amounts is called your *consumer surplus* on the transaction:

consumer surplus = $51 − $31 = $20

That is, you would have been willing to pay up to $20 more for the three fans than you actually paid. A graphical interpretation of this situation is presented in Figure 7.24, which shows that (a) the maximum amount you were willing to spend for the three fans can be represented by the total area of the three rectangles shown; (b) the amount you actually paid is the unshaded area of these three rectangles; (c) your consumer surplus is the shaded area of these three rectangles.

Now consider a demand function $p = D(x)$ that gives the total quantity $x$ of a commodity that consumers will purchase per unit time period if the price per unit is $p$ dollars. Typically, as price increases, demand decreases, so that price $p$ is a decreasing function of demand $x$. Figure 7.25 illustrates a typical consumer demand function.

In order to compute the total amount that consumers are willing to pay for $q_0$ units of this product, let's think about the product as being sold in small lots: see Figure 7.26. We divide the interval $[0, q_0]$ into $n$ equal subintervals of length $\Delta x = (q_0 - 0)/n = q_0/n$ and let $x_1, x_2, \ldots, x_n$ be the right-hand endpoint of each of these intervals. If only a small number of units, say $x_1 = \Delta x = q_0/n$, were available, consumers would be willing to pay a rather high price for them. In fact, if only $x_1$ units were available, consumers would be willing to pay at least $D(x_1)$ dollars per unit for these items. Thus, this first lot of $x_1 = \Delta x$ units could be sold for approximately $\Delta x \cdot D(x_1)$ dollars. Of course, if the price per unit is more than $D(x_1)$ dollars, no more than the first lot of $\Delta x$ units could be sold. If another lot of $\Delta x$ units becomes available, however, consumers are willing to pay at least $D(x_2)$ dollars per unit for these new items. For this second

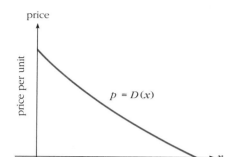

FIGURE 7.25 For ordinary goods, price is a decreasing function of demand.

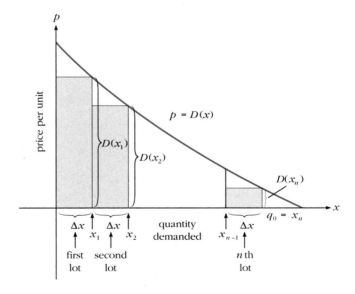

FIGURE 7.26 The sum of the shaded areas approximates the total willingness of consumers to pay for $q_0$ units of product.

lot, the total willingness to pay is approximated by $\Delta x \cdot D(x_2)$. Generalizing, if $n$ lots of size $\Delta x$ are available, consumers are willing to pay approximately

$$\Delta x \cdot D(x_1) + \Delta x \cdot D(x_2) + \cdots + \Delta x \cdot D(x_n)$$

dollars for these units. This sum, represented by the shaded area in Figure 7.26, is a Riemann sum for the function $p = D(x)$ on the interval $[0, q_0]$, where $c_1 = x_1, c_2 = x_2, \ldots, c_n = x_n$. Taking the limit as $\Delta x \to 0$ produces a definite integral $\int_0^{q_0} D(x)\,dx$ as a measure of the total amount consumers are willing to pay for $q_0$ units. This definite integral is simply the area under the graph of the demand curve $D(x)$ from $x = 0$ to $x = q_0$: see Figure 7.27. We summarize:

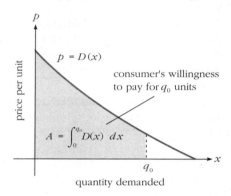

**FIGURE 7.27** The shaded area under the demand curve is the total amount that consumers are willing to pay for $q_0$ units.

**CONSUMER'S WILLINGNESS TO PAY**

The consumer's willingness to pay for a total of $q_0$ units of a product with demand function $p = D(x)$ is given by

$$\int_0^{q_0} D(x)\,dx$$

and is represented by the area under the graph of $D(x)$ from 0 to $q_0$.

---

**EXAMPLE 1**

In order to sell $x$ fans per week in the month of July, a retailer must charge a price of $p = 40 - 0.00001x^2$ dollars per fan. Find the total amount that consumers will be willing to pay for 1200 fans a week in July.

**Solution**

We compute:

$$\int_0^{1200} (40 - 0.00001x^2)\,dx = \left(40x - \frac{0.00001x^3}{3}\right)\Big]_0^{1200}$$

$$= \left[40(1200) - \frac{0.00001(1200)^3}{3}\right] - \left[40(0) - \frac{0.00001(0)^3}{3}\right]$$

$$= 48000 - 5760 = \$42{,}240$$

Thus consumers are willing to pay \$42,240 to purchase the 1200 fans. See Figure 7.28 for a graphical interpretation, in which the shaded area has a magnitude of 42,240.

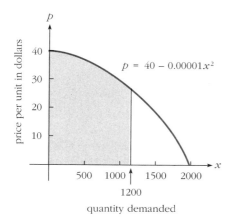

FIGURE 7.28 The shaded area under the demand curve represents the amount that consumers are willing to pay for 1200 fans, which is $42,240.

In Example 1, although consumers would be *willing* to pay $42,240 for those 1200 fans, in a free, purely competitive economy, the price paid by each consumer would be the *same* and would be determined by the market in accordance with the laws of supply and demand. That is, in a "purely competitive environment," there is no price discrimination in which one consumer pays more or less than another. All units of a commodity are offered to all consumers at the same price. Although *purely* competitive situations are fairly rare (as attested to by our initial example, where Bucko's charged more than Cheapo's for the same fan), economists still find the concept of pure competition useful in studying certain market structures and in approximating some "real-life" situations. If we assume that all consumers pay a price of, say, $15 each for the 1200 fans in Example 1, then their total actual expenditure is ($15)(1200) = $18,000. This actual expenditure is illustrated in Figure 7.29 as the area of the rectangle with base [0, 1200] along the $x$-axis and height 15 along the $p$-axis. The difference between what consumers are willing to pay and what they actually pay is the consumer's surplus:

consumer's surplus = $42,240 − $18,000 = $24,240

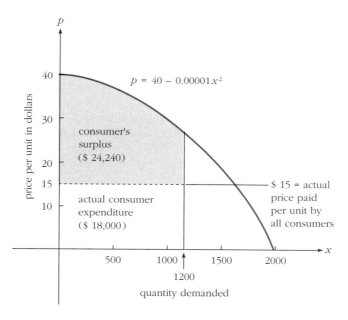

FIGURE 7.29 The consumer's surplus (shaded) and actual consumer expenditure when 1200 fans are purchased at $15 each.

The consumer's surplus is the shaded area in Figure 7.29 and represents the area between the demand function $p = D(x)$ and the horizontal line $p = 15$. This area can be given by the definite integral $\int_0^{1200} (40 - 0.00001x^2 - 15)\, dx$. These results are generalized in the following summary box. Refer to Figure 7.30 for a geometric interpretation.

## CONSUMER'S SURPLUS AND ACTUAL CONSUMER EXPENDITURE

If the demand function for a product is $p = D(x)$ and if $q_0$ units are purchased by consumers at a price per unit of $p_0$, then the *actual consumer expenditure* is

(price per unit)(number of units purchased) = $p_0 q_0$

and the *consumer's surplus* is

$$\text{consumer's surplus} = \int_0^{q_0} [D(x) - p_0]\, dx = \int_0^{q_0} D(x)\, dx - p_0 q_0$$

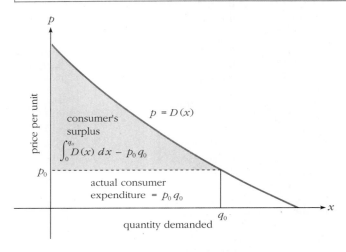

**FIGURE 7.30** A graphical illustration of consumer surplus and actual consumer expenditure when $q_0$ units are purchased at a price of $p_0$ each.

Observe that:

consumer's surplus = (consumer's willingness to pay) − (actual consumer expenditure)

---

### EXAMPLE 2 (Participative)

**Solution**

The monthly demand function for a certain product is $p = 100 - \sqrt{x}$. Let's find the actual consumer expenditure and the consumer's surplus if the price is $50 per unit.

Our first task is to find the number of units purchased per month if the price is $50 per unit. For this we solve the equation

$50 = 100 - \sqrt{x}$

$\sqrt{x} = \underline{\phantom{xxxx}}$

$x = \underline{\phantom{xxxx}}$

Thus $p_0 = \$50$ and $q_0 = \underline{\phantom{xxxx}}$, so the actual consumer expenditure on 2500 units is $p_0 q_0 = \underline{\phantom{xxxx}}$. The consumer's surplus is given by the expression

$$\int_0^{2500} (100 - \sqrt{x})\, dx - \underline{\phantom{xxxx}}$$

Computing the definite integral, we get

$$\left[100x - \tfrac{2}{3}x^{3/2}\right]_0^{2500} - 125000$$

Complete the computation on your own. The result is that the consumer's surplus is _____.

## The Producer's Surplus

Just as certain consumers benefit by not having to pay as much for certain products as they are willing to pay, some producers benefit by virtue of the fact that they are willing to sell a certain number of units of product for less than the market price they actually receive. Corresponding to the consumer's willingness to pay, the actual consumer expenditure, and the consumer's surplus on the demand side of the economic picture, we encounter the producer's willingness to accept, the actual producer revenue, and the producer's surplus on the supply side. The key results are summarized in the following box and illustrated in Figure 7.31, where the amount the producer is willing to accept is represented by the area under the supply curve $p = S(x)$ between $x = 0$ and $x = q_0$, actual producer's revenue is represented by the area of the rectangle with base $[0, q_0]$ along the x-axis and height $p_0$ along the p-axis, and the producer's surplus is the area between the horizontal line $p = p_0$ and the supply curve $p = S(x)$ on the interval $[0, q_0]$.

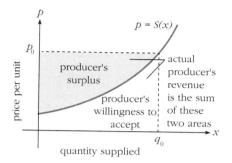

**FIGURE 7.31** A graphical illustration of producer's surplus, amount producers are willing to accept, and actual producer's revenue.

**THE PRODUCER'S SURPLUS**

Let $p = S(x)$ be a supply function, where $p$ is the price that must be offered to producers so that they will supply $x$ units of product per unit time period. If the market price is $p_0$ per unit and producers provide $q_0$ units at this price, then

$$\text{actual producer's revenue} = p_0 q_0$$

$$\text{producer's willingness to accept} = \int_0^{q_0} S(x)\, dx$$

$$\text{producer's surplus} = p_0 q_0 - \int_0^{q_0} S(x)\, dx$$

Observe that

producer's surplus = (actual producer's revenue) − (producer's willingness to accept)

---

**EXAMPLE 3**

Given the monthly supply function $S(x) = 50 + 2x$, find the actual producer's revenue, the amount producers are willing to accept, and the producer's surplus if the price per unit is $200.

**Solution**

If the unit price is $p_0 = \$200$, then $q_0$, the number of units offered by producers per month, is found by solving the equation

---

**SOLUTIONS TO EXAMPLE 2**    50, 2500, 2500, $125,000, 125000, $41,666.67

$$200 = 50 + 2x$$
$$2x = 150$$
$$x = 75$$

Therefore

actual producer's revenue $= (200)(75) = \$15{,}000$

The amount producers are willing to accept for 75 units is

$$\text{producer's willingness to accept} = \int_0^{75} (50 + 2x)\, dx = (50x + x^2)\Big|_0^{75}$$
$$= [50(75) + (75)^2] - [50(0) + 0^2] = \$9{,}375$$

The producer's surplus is then the difference:

producer's surplus $= \$15{,}000 - \$9{,}375 = \$5{,}625$

See Figure 7.32 for a graphical interpretation.

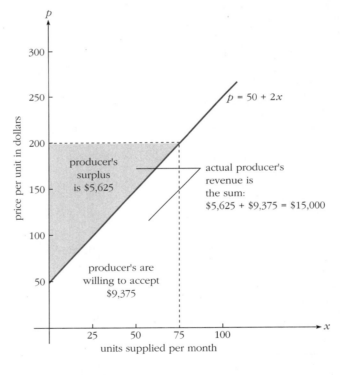

FIGURE 7.32  A graphical illustration of the results of Example 3.

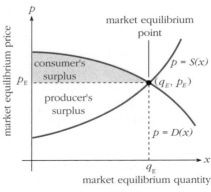

FIGURE 7.33  The supply and demand curves, market equilibrium, consumer's surplus (shaded), and producer's surplus.

If we know both the demand function $p = D(x)$ and the supply function $p = S(x)$ for a certain product, then the point at which these two curves intersect, denoted by $(q_E, p_E)$, is called the *market equilibrium point*, where the quantity $q_E$ is the *market equilibrium quantity* and the price $p_E$ is the *market equilibrium price*; see Figure 7.33. In a purely competitive environment, a product will have a tendency to be sold at its equilibrium price $p_E$, since then there will be neither a shortage of the product (if the market price is less than $p_E$) or a surplus of the product (if the market price is greater than $p_E$). The consumer's surplus is shown as the shaded area between the supply and demand curves in Figure 7.33 and the producer's surplus is the unshaded area between these curves.

## EXAMPLE 4

Given the supply function $S(x) = 1.5x$ and the demand function $D(x) = 100 - 0.025x^2$, find the consumer's surplus and the producer's surplus.

### Solution

First we must find the market equilibrium point $(q_E, p_E)$. Since the market equilibrium point occurs at the intersection of the supply and demand curves, we set $S(x)$ equal to $D(x)$ and solve the resulting equation:

$$1.5x = 100 - 0.025x^2$$
$$0.025x^2 + 1.5x - 100 = 0$$
$$x^2 + 60x - 4000 = 0 \quad \text{(Multiplying by 40)}$$
$$(x - 40)(x + 100) = 0$$
$$x = 40, \quad x = -100 \text{ (reject)}$$

We conclude that $q_E = 40$. To find $p_E$, we can substitute this value for $x$ into either $S(x)$ or $D(x)$. We choose to use $S(x)$, since it is simpler:

$$S(40) = (1.5)(40) = 60 = p_E$$

To continue, we can now find the consumer's surplus and the producer's surplus as follows:

$$\text{consumer's surplus} = \int_0^{q_E} D(x)\, dx - p_E q_E$$

$$= \int_0^{40} (100 - 0.025x^2)\, dx - (60)(40)$$

$$= \left(100x - \frac{0.025}{3}x^3\right)\Big|_0^{40} - 2400$$

$$= \left[100(40) - \frac{0.025}{3}(40)^3\right] - \left[100(0) - \frac{0.025}{3}(0)^3\right] - 2400$$

$$= \$1066.67$$

$$\text{producer's surplus} = p_E q_E - \int_0^{q_E} S(x)\, dx$$

$$= (60)(40) - \int_0^{40} 1.5x\, dx = 2400 - [0.75x^2]_0^{40}$$

$$= 2400 - [0.75(40)^2 - 0.75(0)^2] = \$1200$$

A sketch is drawn in Figure 7.34. ■

### Continuous Income Streams

Suppose you inherit a small coin-operated laundry, which is expected to earn money at the rate of $25,000 per year for the next four years. How much money

464  CHAPTER 7  ADDITIONAL INTEGRATION TOPICS

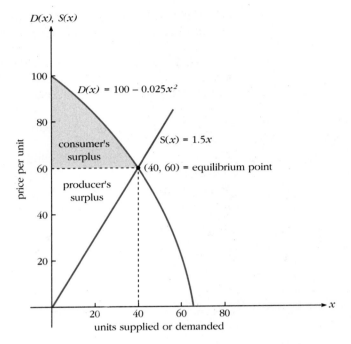

FIGURE 7.34  The consumer's surplus is the shaded area between $S(x)$ and $D(x)$. The producer's surplus is the unshaded area between $S(x)$ and $D(x)$.

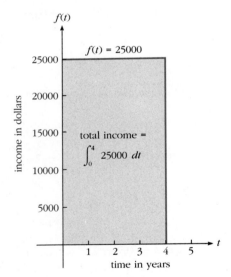

FIGURE 7.35  The total income for four years is the shaded area under the graph of $f(t) = 25000$.

will the laundry make over the four-year period? The answer to this question is

total amount of income = (income per year)(number of years)
= ($25,000)(4) = $100,000

Observe that since $25,000 per year is the (constant) rate of change of income, we can interpret this $100,000 as the area under the graph of $f(t) = 25000$ from $t = 0$ to $t = 4$, as pictured in Figure 7.35. Then the total income received over the four-year period is the definite integral

$$\text{total amount of income} = \int_0^4 25000\, dt = 25000t\Big]_0^4$$
$$= 25000(4) - 25000(0) = \$100,000$$

Of course, the $100,000 is not received in a single "lump sum" at the end of four years. In fact, you will probably want to pick up the money earned by the washers and dryers daily or even more often! Thus you will be privy to a "stream" of income. In situations like this, it is convenient to assume that this income stream is continuous or *steady*. In essence, we are assuming that the income "flows" in at a rate given by the continuous function $f(t)$. Note that $f(t)$ need *not* be a constant function, as we will see in the next example.

EXAMPLE 5

Based upon records that you recently found, you revise your estimate of the earning power of the laundry. You now expect to earn money at the rate of

$f(t) = 25000e^{0.05t}$ dollars per year for the next four years. Find your revised total earnings from the laundry for the next four years.

**Solution**

Note that the rate at which money flows in is no longer a constant. By the method of Section 6.3, since $f(t)$ is the rate of change of the total amount of money $A(t)$ earned in $t$ years, we compute the total amount earned in four years as follows:

$$A(4) - A(0) = \int_0^4 f(t)\,dt = \int_0^4 25000e^{0.05t}\,dt = 500000e^{0.05t}\Big]_0^4$$
$$= 500000e^{0.05(4)} - 500000e^{0.05(0)} = \$110{,}701.38$$

Of course, the definite integral $\int_0^4 25000e^{0.05t}\,dt$ may be interpreted as the area under the graph of $f(t) = 25000e^{0.05t}$ from $t = 0$ to $t = 4$, as shown in Figure 7.36. ∎

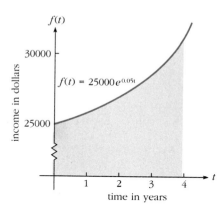

FIGURE 7.36 The area under the graph of $f(t)$ represents the total amount earned by the laundry in four years.

The results of Example 5 generalize easily.

### TOTAL INCOME GENERATED BY A CONTINUOUS INCOME STREAM

If $y = f(t)$ is a continuous function that gives the rate of flow of income from a business venture, then the total income derived from this venture between $t = a$ and $t = b$ is given by the definite integral

$$\int_a^b f(t)\,dt$$

and can be viewed as the area under the graph of the rate-of-flow function $f(t)$ between $a$ and $b$, as illustrated in Figure 7.37.

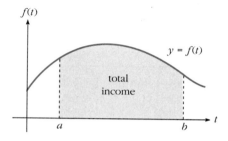

FIGURE 7.37 The area under the graph of $f(t)$ between $a$ and $b$ is the total income derived from the venture between times $a$ and $b$.

### Future Value of an Income Stream

In most cases, when a stream of income flows into a company, the money is invested rather than just stacked in a neat little pile! The problem that we now address is this: If the money from an income stream is continuously invested as it is received at an annual rate of $100r$ percent compounded continuously, what will be the total future value of the income stream at the end of $N$ years? To solve this problem, we begin by recalling that if $P$ dollars are invested at an annual rate of $100r$ percent compounded continuously, then the amount accumulated after $t$ years is $A = Pe^{rt}$.

Now suppose we divide the time interval $[0, N]$ into $n$ equal subintervals of length $\Delta t = (N - 0)/n = N/n$. We call these intervals $[t_0, t_1], [t_1, t_2], \ldots,$ $[t_{n-1}, t_n]$, where $t_0 = 0$ and $t_n = N$, and we let $c_1, c_2, \ldots, c_n$ be arbitrary numbers in each of the intervals $[t_0, t_1], [t_1, t_2], \ldots, [t_{n-1}, t_n]$, respectively. For the moment,

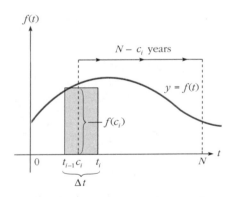

**FIGURE 7.38** The rectangle with base $[t_{i-1}, t_i]$ and height $f(c_i)$ has area $f(c_i) \cdot \Delta t$ and approximates the area under the graph of $f$ from $t_{i-1}$ to $t_i$.

let us concentrate on the single subinterval $[t_{i-1}, t_i]$. As shown in Figure 7.38, the amount of money earned by the venture in this time interval is approximately

$$\begin{pmatrix} \text{amount earned between} \\ t = t_{i-1} \text{ and } t = t_i \end{pmatrix} \approx \begin{pmatrix} \text{dollars per year} \\ \text{earned at time } c_i \end{pmatrix} \cdot \begin{pmatrix} \text{number of years} \\ \text{between } t_{i-1} \text{ and } t_i \end{pmatrix}$$

$$\approx f(c_i) \cdot \Delta t$$

This amount $f(c_i) \cdot \Delta t$ is invested immediately at $100r\%$ and will remain invested for approximately $N - c_i$ years (see Figure 7.38). Using the formula $A = Pe^{rt}$ with $P = f(c_i)\Delta t$ and $t = N - c_i$, the future value of this money invested during the time interval $[t_{i-1}, t_i]$ will be

$$\begin{pmatrix} \text{future value of money} \\ \text{invested during } [t_{i-1}, t_i] \end{pmatrix} \approx [f(c_i)\Delta t]e^{r(N-c_i)} = f(c_i)e^{r(N-c_i)}\Delta t$$

To obtain the approximate *total* future value of the income at the end of $N$ years, we sum up the future values of money invested during each of the $n$ subintervals $[t_{i-1}, t_i]$, $i = 1, 2, \ldots, n$. That is,

$$\begin{pmatrix} \text{future value of} \\ \text{income stream} \end{pmatrix} \approx f(c_1)e^{r(N-c_1)}\Delta t + f(c_2)e^{r(N-c_2)}\Delta t + \cdots + f(c_n)e^{r(N-c_n)}\Delta t$$

$$= \left[ f(c_1)e^{r(N-c_1)} + f(c_2)e^{r(N-c_2)} + \cdots + f(c_n)e^{r(N-c_n)} \right]\Delta t$$

The sum on the right-hand side is a Riemann sum for the function $f(t)e^{r(N-t)}$ on the interval $[0, N]$. Since the approximation improves as $\Delta t \to 0$, we obtain the *exact* future value of the income stream with the definite integral

$$\int_0^N f(t)e^{r(N-t)}\, dt$$

We summarize:

**FUTURE VALUE OF AN INCOME STREAM**

If income that flows in at the continuous rate of $f(t)$ dollars per year is invested at an annual rate of $100r\%$ compounded continuously, the total future value of this income stream after $N$ years is

$$\text{future value} = \int_0^N f(t)e^{r(N-t)}\, dt \tag{1}$$

**EXAMPLE 6**

Find the total value of the income stream from your laundry business over a period of four years if it earns money at the rate of $f(t) = 25000e^{0.05t}$ dollars per year and if this income is invested as it is earned in an account paying an 8% annual interest rate compounded continuously.

Solution

We must find the future value after $N = 4$ years of this income stream. Using the formula in the box with $f(t) = 25000e^{0.05t}$ and $r = 0.08$, we obtain

$$\text{future value} = \int_0^4 25000e^{0.05t} \cdot e^{0.08(4-t)}\, dt$$

$$= 25000 \int_0^4 e^{0.05t} \cdot e^{0.32} \cdot e^{-0.08t}\, dt$$

$$= 25000 e^{0.32} \int_0^4 e^{-0.03t}\, dt = 25000 e^{0.32} \left(\frac{e^{-0.03t}}{-0.03}\right)\bigg|_0^4$$

$$= 25000 e^{0.32} \left[\frac{e^{-0.03(4)}}{-0.03} - \left(\frac{e^{-0.03(0)}}{-0.03}\right)\right] \approx 25000 e^{0.32}[-29.5640 + 33.3333]$$

$$\approx 25000(1.3771)(3.7693) = \$129{,}767.58$$

You should notice that since, by Example 5, the income stream earns $110,701.38 if it is not invested and by what we have just found, it earns $129,767.58 if invested at 8% compounded continuously, the difference

$$\$129{,}767.58 - \$110{,}701.38 = \$19{,}066.20$$

represents the interest earned on the income stream. ∎

### Present Value of an Income Stream

In judging the merits of certain business ventures that yield income streams, it is important to know the *present value* of the venture. That is, what is the amount of money $P$ we must invest at the prevailing interest rate ($100r\%$ annual rate compounded continuously for some $r$) in order to have a future amount $F$ from the venture in $N$ years? This question is easily answered. Since a present amount $P$ invested at an annual rate of $100r\%$ compounded continuously for $N$ years will be worth $Pe^{rN}$ dollars in $N$ years, we must solve the equation $Pe^{rN} = F$ for $P$. Dividing both sides by $e^{rN}$ and substituting formula (1) for the future value $F$, we obtain

$$P = e^{-rN} \int_0^N f(t) e^{r(N-t)}\, dt = e^{-rN} \cdot e^{rN} \int_0^N f(t) e^{-rt}\, dt = \int_0^N f(t) e^{-rt}\, dt$$

**PRESENT VALUE OF AN INCOME STREAM**

If income from a venture flows in at the continuous rate of $f(t)$ dollars per year, the present value of this income stream at $100r\%$ compounded continuously for $N$ years is

$$\text{present value} = \int_0^N f(t) e^{-rt}\, dt \qquad (2)$$

---

EXAMPLE 7 (Participative)

Your wily uncle Will offers you $80,000 to be paid immediately to lease your laundry business for the next four years. Assuming that $f(t) = 25000 e^{0.05t}$ dollars per year is the rate of flow of income from the laundry, let's determine whether you should accept his offer if the prevailing interest rate is 8% compounded continuously. We will assume that the income from the business is net profit, neglecting income tax considerations.

**Solution**

If the present value of our laundry business for the next four years is (more than, less than) $80,000, we should accept Uncle Will's offer. To compute the present value, we use formula (2) with $N =$ _____ , $f(t) = 25000e^{0.05t}$, and $r =$ _____ to obtain

$$\text{present value} = \int_0^4 25000e^{0.05t} \cdot e^{-0.08t}\, dt = 25000 \int_0^4 e^{-0.03t}\, dt$$

$$= 25000[\underline{\hspace{2cm}}]_0^4 = 25000\left[\frac{e^{-0.03(4)}}{-0.03} - \left(\frac{e^{-0.03(0)}}{-0.03}\right)\right]$$

$$\approx \$\underline{\hspace{3cm}}$$

Should you accept Uncle Will's offer? _____

## SECTION 7.3
### SHORT ANSWER EXERCISES

1. If you would be willing to pay as much as $50 for two tickets to a concert but purchase the tickets for $30, your consumer's surplus on the transaction is _____.

2. The price of a certain calculator is $15. If you are willing to pay $40 for the first calculator, $18 for a second, and $5 for a third, how many calculators will you purchase? _____ What is your consumer's surplus on this transaction? _____

3. (True or False) Consumer's surplus is the difference between what you are willing to pay and what you actually pay for a product or service. _____

4. (True or False) Any two people will have the same consumer's surplus on a given transaction. _____

5. A producer is willing to sell his supply of 400 lb of oranges at $1.25/lb. If the market price of oranges is $1.50/lb, what is his producer's surplus if he sells all the oranges at the market price? _____

6. A producer receives $5000 in revenue on the sale of 200 units of his product. If his producer's surplus on this transaction is $1500, how much would he have been willing to accept for those 200 units? _____

7. (True or False) The sum of the consumer's surplus and the producer's surplus is a maximum when the price per unit is the market equilibrium price. _____

8. If a vending machine business has a rate of flow function $f(t) = 40000$ for $0 \le t \le 3$, how much will the business earn in three years? _____

9. You are offered $50,000 right now for a business venture that has a future value in four years of $67,000. Should you accept this offer? (a) yes; (b) no; (c) not enough information given to decide.

10. The prevailing interest rate is 8.5% compounded continuously. If you have $20,000 to invest, should you invest in a business venture with a future value in five years of $32,000 or simply bank your money? _____

**SOLUTIONS TO EXAMPLE 7**   less than, 4, 0.08, $\dfrac{e^{-0.03t}}{-0.03}$, 94,232.97, No

## SECTION 7.3
## EXERCISES

In Exercises 1–10, find (a) the consumer's willingness to pay; (b) the actual consumer expenditure; and (c) the consumer's surplus for each of the given demand functions at the given demand level $x$ or price level $p$. Assume that demand is measured in units and price in dollars.

1. $D(x) = 40 - 0.2x$ at $x = 80$
2. $D(x) = 500 - \frac{1}{20}x$ at $x = 6000$
3. $D(x) = 50 - \frac{1}{10}x^2$ at $p = 10$
4. $D(x) = 1000 - \frac{1}{4}x^2$ at $p = 375$
5. $D(x) = 0.002x^2 - 0.4x + 60$ at $x = 50$
6. $D(x) = 3000 - x - 0.1x^2$ at $x = 100$
7. $D(x) = 800/(x + 5)$ at $p = 2$
8. $D(x) = 1000/(3x + 8)$ at $p = 1$
9. $D(x) = 50/\sqrt{x + 1}$ at $x = 24$
10. $D(x) = 90/\sqrt{x + 4}$ at $x = 25$

In Exercises 11–20, find (a) the amount the producer is willing to accept; (b) the actual producer's revenue; and (c) the producer's surplus for each of the given supply functions at the given supply level $x$ or price level $p$. Assume that supply is measured in units and price in dollars.

11. $S(x) = 100 + 4x$ at $x = 30$
12. $S(x) = 20 + 0.05x$ at $x = 100$
13. $S(x) = 0.02x^2 + 4$ at $p = 6$
14. $S(x) = 0.1x^2 + 1$ at $p = 11$
15. $S(x) = 2 + \frac{1}{10}\sqrt{x}$ at $p = 12$
16. $S(x) = 4 + 0.01\sqrt{x^3}$ at $x = 100$
17. $S(x) = x\sqrt{x^2 + 1}$ at $x = 36$
18. $S(x) = e^{0.1x}$ at $p = e^3$
19. $S(x) = 0.1x \ln(x + 1)$ at $x = 10$
20. $S(x) = 0.2[\ln(x + 1)]^2$ at $x = 5$

In Exercises 21–24, find the consumer's surplus and the producer's surplus if the price is at market equilibrium for the given demand and supply functions. Assume that $x$ is measured in units and price in dollars.

21. $D(x) = 10 - 0.025x$
    $S(x) = 2 + 0.025x$
22. $D(x) = 10 - 0.095x$
    $S(x) = 0.00005x^2$
23. $D(x) = 600/(x + 10)$
    $S(x) = \frac{1}{5}x$
24. $D(x) = 10e^{-0.1x}$
    $S(x) = 2e^{0.4x}$

### Business and Economics

25. **(Income stream)** A company has a stream of income from one of its products of $250,000 per year. If this income flows in continuously and is immediately invested at 9 percent per year compounded continuously, find the total earnings from this income stream over the next four years.

26. **(Income stream)** Your rich uncle Ralph transfers money continuously into your bank account at the rate of $10,000 per year. If your account gives 8% annual interest compounded continuously and if you make no withdrawals, find the amount of money in the account after five years.

27. **(Income stream)** Your generous aunt Gerty wants to pay for your first year's tuition in the M.B.A. program at a prestigious university. If the tuition is expected to be $15,000, how much money must your aunt transfer into your bank account continuously and at a constant rate over the next three years if your account gives 7.5% annual interest compounded continuously?

28. **(Income stream)** Find the present value of a business venture that will yield income continuously at the rate of $12,000 per year for the next five years if the prevailing annual interest rate is 7% compounded continuously.

29. **(Income stream)** An income stream is expected to generate money over the next seven years at a rate of $25 + 4t$ thousand dollars per year. If the prevailing interest

470   CHAPTER 7   ADDITIONAL INTEGRATION TOPICS

rate is 9% annually compounded continuously, find **(a)** the present value of the income stream; and **(b)** the future value of the income stream.

**30.** (Income stream)   The income generated by an investment is expected to flow in at a continuous rate of $5000\sqrt{t+9}$ dollars per year for the next seven years. Find the total income derived from this investment during the last year it is held, that is, between the sixth and seventh years.

**31.** (Income stream)   An investment is expected to generate revenue at a continuous rate of $20000e^{0.08t}$ dollars per year for the next four years. Find the total revenue generated by this investment over the four-year period.

**32.** (Business decision)   Your cousin Kim has inherited a vending machine business that is expected to yield a continuous stream of income for the next four years at the rate of $30000e^{0.03t}$ dollars per year. She offers to trade this business for your laundry business, which is expected to earn money at the rate of $27000e^{0.05t}$ dollars per year. The trade would last for four years. If the prevailing interest rate is 9% annually compounded continuously, should you accept her offer?

---

**OPTIONAL**

**GRAPHING CALCULATOR/ COMPUTER EXERCISES**

In Exercises 33–36, find the consumer's surplus and the producer's surplus if the price is at market equilibrium for the given demand and supply functions. Estimate the x-coordinates of the point of intersection of $D(x)$ and $S(x)$ to the nearest tenth using a graphing calculator.

**33.** $D(x) = 400 - 5x$
$S(x) = 0.1x \ln(x+1)$

**34.** $D(x) = 500/(x+2)$
$S(x) = 0.01x^2$

**35.** $D(x) = 5/\sqrt{x+1}$
$S(x) = e^{0.2x}$

**36.** $D(x) = 100/\sqrt{x+1}$
$S(x) = 0.01x\sqrt{x^2+4}$

---

## 7.4   NUMERICAL INTEGRATION

The definite integral $\int_a^b f(x)\,dx$ is computed as follows:

$$\int_a^b f(x)\,dx = F(b) - F(a)$$

where $F(x)$ is an antiderivative of $f(x)$ on $[a, b]$. In practice, it is sometimes difficult or impossible to find an elementary antiderivative $F(x)$, so this method of evaluating the definite integral cannot be used. There are, however, various methods of *approximating* the definite integral $\int_a^b f(x)\,dx$ that do not require finding an antiderivative $F(x)$ of $f(x)$. These techniques of numerical integration include approximation by Riemann sums, the trapezoidal rule, and Simpson's rule.

### Riemann Sums

As we have already learned in Section 7.2, the definite integral $\int_a^b f(x)\,dx$ can be approximated by a sum of areas of rectangles if $f(x)$ is continuous and nonnegative on $[a, b]$. Divide the interval $[a, b]$ into $n$ equal subintervals

$[x_0, x_1], [x_1, x_2], \ldots, [x_{n-1}, x_n]$, where $a = x_0$ and $b = x_n$. The length of each of these subintervals is $\Delta x = (b - a)/n$. In the $n$ subintervals choose arbitrary numbers $c_1, c_2, \ldots, c_n$. Then we call the sum

$$R_n = f(c_1)\Delta x + f(c_2)\Delta x + \cdots + f(c_n)\Delta x$$

a *Riemann sum* for $f(x)$ on $[a, b]$ and it is an approximation to $\int_a^b f(x)\, dx$. Figure 7.39 illustrates the case where $n = 8$ and $f(x) \geq 0$ on $[a, b]$. We may write

$$\int_a^b f(x)\, dx \approx \Delta x[f(c_1) + f(c_2) + \cdots + f(c_n)] = \left(\frac{b - a}{n}\right)[f(c_1) + f(c_2) + \cdots + f(c_n)]$$

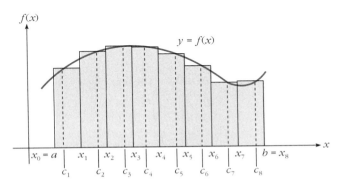

**FIGURE 7.39** A Riemann sum using $n = 8$ rectangles to approximate the area under the graph of $f$ from $a$ to $b$.

In summary, we have the following.

**APPROXIMATION BY A GENERAL RIEMANN SUM**

Let $f(x)$ be continuous on $[a, b]$. Divide the interval $[a, b]$ into $n$ subintervals of equal length: $[x_0, x_1], [x_1, x_2], \ldots, [x_{n-1}, x_n]$, where $x_0 = a$ and $x_n = b$ and the length of each subinterval is $\Delta x = (b - a)/n$. Next, choose arbitrary numbers $c_1, c_2, \ldots, c_n$ in the subintervals $[x_0, x_1], [x_1, x_2], \ldots, [x_{n-1}, x_n]$, respectively. Then

$$\int_a^b f(x)\, dx \approx \left(\frac{b - a}{n}\right)[f(c_1) + f(c_2) + \cdots + f(c_n)]$$

Note that we do *not* require that $f(x) \geq 0$ on $[a, b]$ in the preceding box. There are some common choices for $c_1, c_2, \ldots, c_n$:

(a) Let $c_i$ be the left-hand endpoint of the subinterval $[x_{i-1}, x_i]$; that is, $c_i = x_{i-1}$. We obtain a *left-hand Riemann sum*.
(b) Let $c_i$ be the right-hand endpoint of the subinterval $[x_{i-1}, x_i]$; that is, $c_i = x_i$. We obtain a *right-hand Riemann sum*.
(c) Let $c_i$ be the midpoint of the subinterval $[x_{i-1}, x_i]$; that is, $c_i = (x_{i-1} + x_i)/2$. We obtain a *midpoint Riemann sum*.

## EXAMPLE 1

Use $n = 4$ and find the left-hand Riemann sum, the right-hand Riemann sum, and the midpoint Riemann sum to approximate $\int_1^2 (1/x)\, dx$.

### Solution

We have $\Delta x = (2 - 1)/4 = \frac{1}{4}$, so that each of the four subintervals of $[1, 2]$ will have a length of $\frac{1}{4}$. Thus

$$x_0 = a = 1$$
$$x_1 = x_0 + \Delta x = 1 + \frac{1}{4} = \frac{5}{4}$$
$$x_2 = x_0 + 2\Delta x = 1 + \frac{2}{4} = \frac{3}{2}$$
$$x_3 = x_0 + 3\Delta x = 1 + \frac{3}{4} = \frac{7}{4}$$
$$x_4 = b = x_0 + 4\Delta x = 1 + 1 = 2$$

For the left-hand Riemann sum, we choose $c_1 = 1$, $c_2 = \frac{5}{4}$, $c_3 = \frac{3}{2}$, and $c_4 = \frac{7}{4}$ to obtain

$$\int_1^2 \frac{1}{x}\, dx \approx \frac{2-1}{4}\left(\frac{1}{1} + \frac{1}{\frac{5}{4}} + \frac{1}{\frac{3}{2}} + \frac{1}{\frac{7}{4}}\right)$$

$$= \frac{1}{4}\left(1 + \frac{4}{5} + \frac{2}{3} + \frac{4}{7}\right) \approx 0.759524$$

For the right-hand Riemann sum, we choose $c_1 = \frac{5}{4}$, $c_2 = \frac{3}{2}$, $c_3 = \frac{7}{4}$, and $c_4 = 2$ to obtain

$$\int_1^2 \frac{1}{x}\, dx \approx \frac{2-1}{4}\left(\frac{1}{\frac{5}{4}} + \frac{1}{\frac{3}{2}} + \frac{1}{\frac{7}{4}} + \frac{1}{2}\right)$$

$$= \frac{1}{4}\left(\frac{4}{5} + \frac{2}{3} + \frac{4}{7} + \frac{1}{2}\right) \approx 0.634524$$

For the midpoint Riemann sum, we choose $c_1 = \frac{1}{2}(1 + 5/4) = \frac{9}{8}$, $c_2 = \frac{1}{2}(5/4 + 3/2) = \frac{11}{8}$, $c_3 = \frac{1}{2}(3/2 + 7/4) = \frac{13}{8}$, and $c_4 = \frac{1}{2}(7/4 + 2) = \frac{15}{8}$ to obtain

$$\int_1^2 \frac{1}{x}\, dx \approx \frac{2-1}{4}\left(\frac{1}{\frac{9}{8}} + \frac{1}{\frac{11}{8}} + \frac{1}{\frac{13}{8}} + \frac{1}{\frac{15}{8}}\right)$$

$$= \frac{1}{4}\left(\frac{8}{9} + \frac{8}{11} + \frac{8}{13} + \frac{8}{15}\right) \approx 0.691220$$

Each of these three Riemann sums is illustrated in Figure 7.40.
Note that in this case, the value of the definite integral is easily found:

$$\int_1^2 \frac{1}{x}\, dx = \ln|x|\Big|_1^2 = \ln 2 - \ln 1 = \ln 2 \approx 0.693147 \quad \text{(to six decimal places)}$$

Upon comparison of this result with our three approximations, we see that the midpoint Riemann sum yields the best approximation in this case. Of course, all of the approximations will be improved if we choose a larger value for $n$. In Exercise 31 you will be asked to redo this example using $n = 8$. ∎

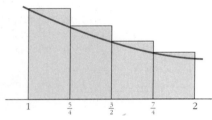

(a) Left-hand Riemann sum

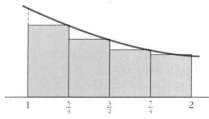

(b) Right-hand Riemann sum

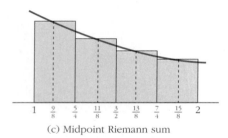

(c) Midpoint Riemann sum

**FIGURE 7.40** Three Riemann sums with $n = 4$ for $f(x) = 1/x$.

## 7.4 NUMERICAL INTEGRATION

In general, the method of approximating a definite integral by a Riemann sum requires values of $n$ that are too large for convenient computation by hand to achieve a reasonable degree of accuracy. Although the use of a computer or programmable calculator would take away the drudgery of these computations, more efficient methods exist for approximating definite integrals. Such methods are the trapezoidal rule and Simpson's rule.

### The Trapezoidal Rule

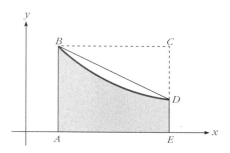

**FIGURE 7.41** The shaded area under the curve is better approximated by the trapezoid $ABDE$ than by the rectangle $ABCE$.

The trapezoidal rule is a numerical integration technique that uses a sum of areas of trapezoids to approximate $\int_a^b f(x)\,dx$. As illustrated in Figure 7.41, approximating the area under a portion of the graph of $f(x)$ by an appropriate trapezoid can often be more accurate than approximating the area by a rectangle. Recall that the area of a trapezoid with parallel bases $b$ and $B$ and altitude $h$ is equal to $A = \frac{1}{2}(b + B)h$. In Figure 7.42, the trapezoid is drawn so that its height $h$ runs horizontally.

To approximate $\int_a^b f(x)\,dx$, we divide $[a, b]$ into $n$ equal subintervals $[x_0, x_1]$, $[x_1, x_2], \ldots, [x_{n-1}, x_n]$, where $a = x_0$ and $b = x_n$ and the length of each subinterval is $\Delta x = (b - a)/n$. If we let $y_0 = f(x_0), y_1 = f(x_1), \ldots, y_n = f(x_n)$ then we connect successive points $(x_0, y_0)$ and $(x_1, y_1)$, $(x_1, y_1)$ and $(x_2, y_2)$, and so forth, with line segments as in Figure 7.43, where we have taken $f(x) \geq 0$ on $[a, b]$. The sum of the areas of the trapezoids shown in Figure 7.43 give an approximation of the area under the graph of $f$ from $a$ to $b$. Thus

$$\int_a^b f(x)\,dx \approx \text{sum of the areas of the } n \text{ trapezoids}$$

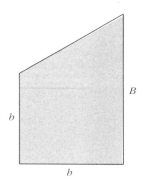

**FIGURE 7.42** The area of the trapezoid with parallel bases of length $b$ and $B$ and altitude of length $h$ is $A = \frac{1}{2}(b + B)h$.

The first trapezoid in Figure 7.43 has altitude $h = x_1 - x_0 = \Delta x$, so that its area is $\frac{1}{2}(y_0 + y_1)\Delta x$. The second trapezoid has altitude $x_2 - x_1$ and area $\frac{1}{2}(y_1 + y_2)\Delta x$. The last trapezoid has altitude $x_n - x_{n-1}$ and area $\frac{1}{2}(y_{n-1} + y_n)\Delta x$. If we designate the sum of the areas of the $n$ trapezoids by $T_n$ then

$$T_n = \left(\frac{y_0 + y_1}{2}\right)\Delta x + \left(\frac{y_1 + y_2}{2}\right)\Delta x + \left(\frac{y_2 + y_3}{2}\right)\Delta x + \cdots + \left(\frac{y_{n-1} + y_n}{2}\right)\Delta x$$

$$= \frac{\Delta x}{2}(y_0 + y_1 + y_1 + y_2 + y_2 + y_3 + \cdots + y_{n-1} + y_n)$$

$$= \left(\frac{b - a}{2n}\right)(y_0 + 2y_1 + 2y_2 + \cdots + 2y_{n-1} + y_n)$$

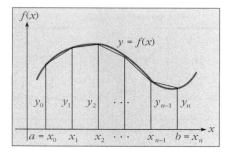

**FIGURE 7.43** The sum of the areas of the trapezoids approximates the area under the graph of $f$ from $a$ to $b$.

**THE TRAPEZOIDAL RULE**

Let $f(x)$ be continuous on $[a, b]$. Divide the interval $[a, b]$ into $n$ equal subintervals: $[x_0, x_1], [x_1, x_2], \ldots, [x_{n-1}, x_n]$, where the length of each subinterval is $\Delta x = (b - a)/n$. Let $y_0 = f(x_0), y_1 = f(x_1), \ldots, y_n = f(x_n)$. Then

$$\int_a^b f(x)\,dx \approx \left(\frac{b - a}{2n}\right)(y_0 + 2y_1 + 2y_2 + \cdots + 2y_{n-1} + y_n)$$

Note that the trapezoidal rule does *not* require that $f(x) \geq 0$ on $[a, b]$.

**EXAMPLE 2** Use the trapezoidal rule with $n = 5$ to approximate $\int_1^2 \ln x \, dx$.

**Solution** First note that $f(x) = (b - a)/n = (2 - 1)/5 = \frac{1}{5}$, so that $x_0 = a = 1$; $x_1 = x_0 + \Delta x = 1 + \frac{1}{5} = \frac{6}{5}$; $x_2 = x_0 + 2\Delta x = 1 + \frac{2}{5} = \frac{7}{5}$; $x_3 = x_0 + 3\Delta x = 1 + \frac{3}{5} = \frac{8}{5}$; $x_4 = x_0 + 4\Delta x = 1 + \frac{4}{5} = \frac{9}{5}$; $x_5 = b = x_0 + 5\Delta x = 1 + \frac{5}{5} = 2$. This scheme divides the interval $[1, 2]$ into five equal subintervals, as illustrated in Figure 7.44. Then we have

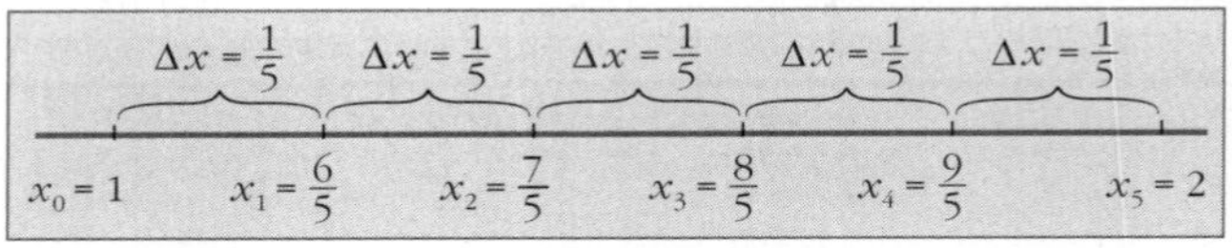

**FIGURE 7.44** The interval $[1, 2]$ has been divided into five subintervals, each having length $\Delta x = \frac{1}{5}$.

$$y_0 = f(x_0) = f(1) = \ln 1$$
$$y_1 = f(x_1) = f\left(\tfrac{6}{5}\right) = \ln \tfrac{6}{5}$$
$$y_2 = f(x_2) = f\left(\tfrac{7}{5}\right) = \ln \tfrac{7}{5}$$
$$y_3 = f(x_3) = f\left(\tfrac{8}{5}\right) = \ln \tfrac{8}{5}$$
$$y_4 = f(x_4) = f\left(\tfrac{9}{5}\right) = \ln \tfrac{9}{5}$$
$$y_5 = f(x_5) = f(2) = \ln 2$$

Using the trapezoidal rule, we obtain

$$\int_1^2 \ln x \, dx \approx \frac{2-1}{2 \cdot 5}\left(\ln 1 + 2 \ln \tfrac{6}{5} + 2 \ln \tfrac{7}{5} + 2 \ln \tfrac{8}{5} + 2 \ln \tfrac{9}{5} + \ln 2\right) \approx 0.384632$$

∎

Observe that in Example 2, we can find the definite integral by using integration by parts:

$$\int_1^2 \ln x \, dx = (x \ln x - x)\big]_1^2 = (2 \ln 2 - 2) - (\ln 1 - 1) \approx 0.386294$$

We see that our approximation in Example 2 was accurate (when rounded) to only one decimal place: $T_5 = 0.4$.

In cases when we cannot obtain the exact value of the definite integral $\int_a^b f(x) \, dx$, it is useful to have a formula that gives us the error in our estimate. The error is the difference between the exact value and the approximate value. The next result, which is stated without proof, is the appropriate formula for the trapezoidal rule.

## ERROR IN USING THE TRAPEZOIDAL RULE

If $T_n$ is the trapezoidal rule approximation of $\int_a^b f(x)\,dx$, then the error in this approximation, designated by $E(T_n)$, is $E(T_n) = \int_a^b f(x)\,dx - T_n$. Further, $E_n$ satisfies the inequality

$$|E(T_n)| \leq \frac{(b-a)^3 M}{12n^2} \qquad (3)$$

where $M$ is a number such that $|f''(x)| \leq M$ for $a \leq x \leq b$.

Formula (3) can be used to give us an upper bound for the error in an approximation we have already made or to tell us the value of $n$ that will yield a desired level of accuracy. The next example illustrates both of these uses.

### EXAMPLE 3

(a) If we use the trapezoidal rule with $n = 5$ to approximate $\int_1^2 \ln x\,dx$, find an upper bound for the error in the estimate.
(b) What value of $n$ guarantees an error of no more than 0.0005?

### Solution

(a) We are required to find an upper bound for $E(T_5)$. Using formula (3), we get

$$|E(T_5)| \leq \frac{(2-1)^3 M}{12(5)^2} \leq \frac{M}{300}$$

where $M$ is the maximum of $|f''(x)|$ on the interval $[1, 2]$. Since $f(x) = \ln x$, we have $f'(x) = 1/x$ and $f''(x) = -1/x^2$. Thus $|f''(x)| = 1/x^2$ and the maximum of $1/x^2$ on $[1, 2]$ is $M = 1/1^2 = 1$. (Recall how the graph looks.) Therefore

$$|E(T_5)| \leq \frac{1}{300} = 0.00\overline{3}$$

It is instructive to note that this number is an *upper bound* for the error. From the result of Example 2 and the comment following it, the actual error in this case is

$$0.386294 - 0.384632 = 0.001662 \quad \text{(to 6 decimal places)}$$

(b) We want to find $n$ such that $|E(T_n)| \leq 0.0005$, that is,

$$\frac{(b-a)^3 M}{12n^2} \leq 0.0005$$

Substituting the appropriate values for $a$, $b$, and $M$, we get

$$\frac{(2-1)^3}{12n^2} \leq 0.0005$$

$$0.006n^2 \geq 1$$

$$n^2 \geq 166.\overline{6}$$

$$n \geq 12.9 \text{ (approximately)}$$

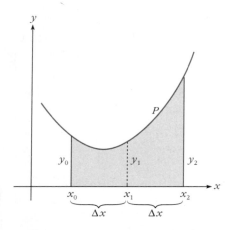

FIGURE 7.45 The shaded area under the parabola is $A = \dfrac{\Delta x}{2}(y_0 + 4y_1 + y_2)$.

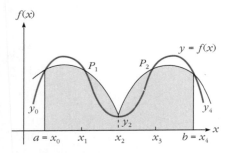

FIGURE 7.46 The shaded area under the two parabolas $P_1$ and $P_2$ approximates the area under the graph of $f$ from $a$ to $b$.

Since $n$ must be a whole number, the choice $n = 13$ guarantees an error of at most 0.0005. ∎

### Simpson's Rule

Simpson's rule uses a sum of areas under certain parabolas to approximate $\int_a^b f(x)\,dx$. It is frequently more accurate than the trapezoidal rule for the same value of $n$. Although we will not prove Simpson's rule in detail, we will outline the major steps in the proof.

Let $P$ be the parabola that passes through three points $(x_0, y_0)$, $(x_1, y_1)$, and $(x_2, y_2)$. Then the first step in deriving Simpson's rule is to verify that the area between the graph of $P$ and the $x$-axis from $x_0$ to $x_2$ is equal to

$$A = \frac{\Delta x}{3}[y_0 + 4y_1 + y_2]$$

where $\Delta x = x_1 - x_0 = x_2 - x_1$; see Figure 7.45. You are asked to verify this fact in Exercise 47.

Next, consider the situation where the area under the graph of $f(x)$ from $a$ to $b$ is approximated by the areas under the two parabolas, $P_1$ and $P_2$, as illustrated in Figure 7.46. We obtain

$$\int_a^b f(x)\,dx \approx \underbrace{\frac{\Delta x}{3}(y_0 + 4y_1 + y_2)}_{\text{area under } P_1} + \underbrace{\frac{\Delta x}{3}(y_2 + 4y_3 + y_4)}_{\text{area under } P_4}$$

$$= \frac{\Delta x}{3}(y_0 + 4y_1 + 2y_2 + 4y_3 + y_4)$$

Note the pattern of the coefficients of the $y_i$ terms carefully. This two-parabola case generalizes easily to the general case of Simpson's rule, where $n$ must be an even number (why?) and there will be $n/2$ parabolas.

**SIMPSON'S RULE**

Let $f(x)$ be continuous on $[a, b]$ and let $n$ be an even positive integer. Divide the interval $[a, b]$ into $n$ subintervals of equal lengths: $[x_0, x_1], [x_1, x_2], \ldots, [x_{n-1}, x_n]$, where the length of each subinterval is $\Delta x = (b - a)/n$. Let $y_0 = f(x_0), y_1 = f(x_1), \ldots, y_n = f(x_n)$. Then

$$\int_a^b f(x)\,dx \approx \left(\frac{b-a}{3n}\right)[y_0 + 4y_1 + 2y_2 + 4y_3 + \cdots + 2y_{n-2} + 4y_{n-1} + y_n]$$

The formula for finding the error in Simpson's rule approximations is given next. This error formula is derived in more advanced courses.

**ERROR IN USING SIMPSON'S RULE**

If $S_n$ is the Simpson's rule approximation of $\int_a^b f(x)\,dx$, then the error in this approximation, designated by $E(S_n)$, is $E(S_n) = \int_a^b f(x)\,dx - S_n$. Further, $S_n$ satisfies the inequality

$$|E(S_n)| \le \frac{(b-a)^5 M}{180 n^4}$$

where $M$ is a number such that $|f^{(4)}(x)| \le M$ for $a \le x \le b$.

---

**EXAMPLE 4 (Participative)**

Solution

Let's use Simpson's rule with $n = 4$ to approximate $\int_0^1 e^{-x^2}\,dx$.

Note that $\Delta x = (b-a)/n = $ _____, so that $x_0 = $ _____, $x_1 = $ _____, $x_2 = $ _____, $x_3 = $ _____, and $x_4 = $ _____. Therefore, $y_0 = f(0) = e^{-0^2} = 1$; $y_1 = f(\tfrac{1}{4}) = e^{-(1/4)^2} = e^{-1/16}$; $y_2 = $ _____; $y_3 = $ _____, and $y_4 = $ _____.
The formula for Simpson's rule for $n = 4$ is

$$\int_a^b f(x)\,dx \approx \left(\frac{b-a}{3n}\right)(y_0 + 4y_1 + 2y_2 + 4y_3 + y_4)$$

Substituting the appropriate quantities, we get

$$\int_0^1 e^{-x^2}\,dx \approx \left(\frac{1-0}{3 \cdot 4}\right)(1 + 4e^{-1/16} + 2e^{-1/4} + \underline{\hspace{2cm}})$$

Using a calculator to approximate the quantity on the right-hand side to seven decimal places, we get

$$\int_0^1 e^{-x^2}\,dx \approx \underline{\hspace{3cm}}$$

Unlike the other examples in this section, we cannot find an elementary antiderivative of $e^{-x^2}$, so we cannot compare the approximate value of $\int_0^1 e^{-x^2}\,dx$ with the exact value. The next example investigates the error in the Simpson's rule approximation carried out in Example 4. ∎

---

**EXAMPLE 5**

(a) Estimate the error in approximating $\int_0^1 e^{-x^2}\,dx$ by Simpson's rule with $n = 4$.
(b) What value of $n$ will guarantee that the error is no more than $0.00001$?

Solution

(a) Using formula (3) for the error in Simpson's rule, we have

$$|E_4(S)| \le \frac{(1-0)^5 M}{180(4)^4} = \frac{M}{46080}$$

---

**SOLUTIONS TO EXAMPLE 4**   $\tfrac{1}{4}$, $0$, $\tfrac{1}{4}$, $\tfrac{1}{2}$, $\tfrac{3}{4}$, $1$, $e^{-1/4}$, $e^{-9/16}$, $e^{-1}$, $4e^{-9/16} + e^{-1}$, $0.7468554$

where $M$ is the absolute maximum of $|f^{(4)}(x)|$ on $[a, b]$ and $f(x) = e^{-x^2}$. You can verify that

$$f'(x) = -2xe^{-x^2}$$
$$f''(x) = (4x^2 - 2)e^{-x^2}$$
$$f^{(3)}(x) = (12x - 8x^3)e^{-x^2}$$
$$f^{(4)}(x) = (12 - 48x^2 + 16x^4)e^{-x^2}$$

To find the maximum of $|f^{(4)}(x)|$ on $[a, b]$, we let $g(x) = f^{(4)}(x)$ and compute

$$g'(x) = f^{(5)}(x) = -8x(15 - 20x^2 + 4x^4)e^{-x^2}$$

To find the critical numbers of $g(x)$, we set $g'(x)$ equal to zero and obtain

$$-8x = 0 \quad \text{or} \quad 15 - 20x^2 + 4x^4 = 0 \quad \text{or} \quad e^{-x^2} = 0$$

The first of these equations has solution $x = 0$. The second equation can be expressed $4(x^2)^2 - 20x^2 + 15 = 0$ and the quadratic formula yields

$$x^2 = \frac{20 \pm \sqrt{(-20)^2 - 4(4)(15)}}{2(4)} = \frac{5}{2} \pm \frac{1}{2}\sqrt{10}$$

and so $x = \pm\sqrt{\frac{5}{2} \pm \frac{1}{2}\sqrt{10}}$. Of these four values of $x$, only $x = \sqrt{\frac{5}{2} - \frac{1}{2}\sqrt{10}} \approx 0.95857$ lies in the interval $[0, 1]$. Finally, since $e^z > 0$ for all $z$, the third equation has no solution. We conclude that the critical numbers of $g(x) = f^{(4)}(x)$ in $[0, 1]$ are $x = 0$ and $x \approx 0.95857$ so the absolute maximum of $|f^{(4)}(x)|$ is one of the three numbers

$$|f^{(4)}(0)| = 12 \quad \text{or} \quad |f^{(4)}(0.95857)| \approx 7.42 \quad \text{or} \quad |f^{(4)}(1)| = \frac{20}{e} \approx 7.36$$

We conclude that $M = 12$ and so

$$|E_4(S)| \leq \frac{12}{46080} \approx 0.00026$$

(b) We must find $n$ so that $|E_n(S)| \leq 0.00001$. Using the value of $M$ found in part (a), this will be true if

$$\frac{12}{180n^4} \leq 0.00001$$
$$0.0018n^4 \geq 12$$
$$n^4 \geq 6666.\overline{6}$$
$$n \geq \sqrt[4]{6666.\overline{6}} \approx 9.04$$

Since $n$ is required to be an even integer, the choice $n = 10$ will guarantee an error of no more than 0.00001. ∎

### An Application

Numerical integration can be used to approximate $\int_a^b f(x)\,dx$ when an explicit formula for $f(x)$ is unknown, but a number of values $f(x_i)$ are known for $a \leq x_i \leq b$. This technique is illustrated in the next example.

## EXAMPLE 6

A company estimates that its daily marginal cost at various production levels between 40 and 50 units is

| PRODUCTION LEVEL (units) | 40 | 42 | 44 | 46 | 48 | 50 |
|---|---|---|---|---|---|---|
| MARGINAL COST (dollars) | 130 | 128 | 126 | 129 | 134 | 140 |

Use the trapezoidal rule to approximate the company's cost of increasing production from 40 units per day to 50 units per day.

### Solution

If $x$ represents the production level in units and $C'(x)$ the marginal cost in dollars per unit, then the cost of increasing production from 40 to 50 units per day is given by

$$C(50) - C(40) = \int_{40}^{50} C'(x)\, dx$$

We approximate the definite integral by using the trapezoidal rule with $a = 40$, $b = 50$, and $n = 5$. Then we have $x_0 = 40$, $x_1 = 42$, $x_2 = 44$, $x_3 = 46$, $x_4 = 48$, and $x_5 = 50$. Further, $y_0 = C'(x_0) = C'(40) = 130$; $y_1 = C'(x_1) = C'(42) = 128$; $y_2 = C'(x_2) = C'(44) = 126$; $y_3 = C'(x_3) = C'(46) = 129$; $y_4 = C'(x_4) = C'(48) = 134$; and $y_5 = C'(x_5) = C'(50) = 140$. We obtain

$$\int_{40}^{50} C'(x)\, dx \approx \left(\frac{50 - 40}{2 \cdot 5}\right)[130 + 2(128) + 2(126) + 2(129) + 2(134) + 140]$$
$$= \$1304$$

See Figure 7.47 for a graphical interpretation.

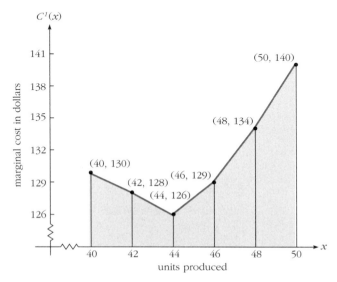

**FIGURE 7.47** The shaded area represents the approximate cost of increasing production from 40 units to 50 units per day.

## SECTION 7.4
**SHORT ANSWER EXERCISES**

1. In approximating $\int_0^5 \sqrt{x}e^{\sqrt{x}}\,dx$ by a midpoint Riemann sum with $n = 10$, we would use $\Delta x =$ _____ and $c_1 =$ _____.
2. (True or False) The numerical integration techniques of this section apply only to functions that are continuous and nonnegative on $[a, b]$. _____
3. (True or False) A midpoint Riemann sum always gives a better approximation to $\int_a^b f(x)\,dx$ than either a left-hand or a right-hand Riemann sum. _____
4. (True or False) The trapezoidal rule requires that the interval $[a, b]$ be divided into subintervals of equal length. _____
5. If $|f''(x)| \leq 24$ for $x$ in $[3, 6]$, the error in approximating $\int_3^6 f(x)\,dx$ using the trapezoidal rule with $n = 6$ is no more than _____.
6. If $|f^{(4)}(x)| \leq 4$ for $x$ in $[1, 3]$, the error in approximating $\int_1^3 f(x)\,dx$ using Simpson's rule with $n = 10$ is no more than _____.
7. If we approximate $\int_a^b f(x)\,dx$ using a numerical integration technique with $n = 10$, we would expect the best approximation to be given by (a) a left-hand Riemann sum; (b) a midpoint Riemann sum; (c) the trapezoidal rule; (d) Simpson's rule.
8. (True or False) The techniques of this section can be used to find indefinite integrals. _____
9. In approximating $\int_1^5 f(x)\,dx$ using the trapezoidal rule, we find that $T_2 = T_4$. A possible formula for $f(x)$ is (a) $f(x) = x$; (b) $f(x) = 3x + 1$; (c) $f(x) = 10$; (d) any of the above.
10. In approximating definite integrals, Riemann sums use rectangles, whereas Simpson's rule uses _____.

## SECTION 7.4
**EXERCISES**

*In Exercises 1–10, find* (a) *the left-hand Riemann sum;* (b) *the right-hand Riemann sum; and* (c) *the midpoint Riemann sum for the given value of $n$ to approximate the definite integral.*

1. $\int_0^2 (4x + 1)\,dx$ for $n = 4$
2. $\int_1^4 (3x - 1)\,dx$ for $n = 6$
3. $\int_1^2 x^2\,dx$ for $n = 6$
4. $\int_1^2 x^2\,dx$ for $n = 10$
5. $\int_0^3 \sqrt{x^3 + 1}\,dx$ for $n = 6$
6. $\int_1^2 \sqrt{x^4 + 1}\,dx$ for $n = 4$
7. $\int_2^3 \dfrac{1}{\ln x}\,dx$ for $n = 4$
8. $\int_0^1 e^{x^2}\,dx$ for $n = 4$
9. $\int_0^2 \sqrt{x}e^{\sqrt{x}}\,dx$ for $n = 8$
10. $\int_1^2 (\ln x)e^x\,dx$ for $n = 4$

**Exercises 11–20:** *for each of Exercises 1–10, use the trapezoidal rule to approximate the integral.*

**Exercises 21–30:** *for each of Exercises 1–10, use Simpson's rule to approximate the integral.*

31. Find the left-hand, right-hand, and midpoint Riemann sums which approximate $\int_1^2 (1/x)\,dx$ using $n = 8$. Compare these results with those of Example 1. What is the exact value of $\int_1^2 (1/x)\,dx$?

**32.** (a) Use the trapezoidal rule with $n = 6$ to approximate $\int_1^4 (1/x)\, dx$.
(b) Find an upper bound for the error in the approximation you obtained in part (a).
(c) What value of $n$ would guarantee an error of no more than 0.001 in the approximation of this integral by the trapezoidal rule?

**33.** Redo Exercise 32 using Simpson's rule.

**34.** (a) Use Simpson's rule with $n = 4$ to approximate $\int_2^3 \ln(x + 2)\, dx$.
(b) Find an upper bound for the error in the approximation you obtained in part (a).
(c) What value of $n$ guarantees an error of no more than $2 \times 10^{-10}$ in the approximation of this integral by Simpson's rule?

**35.** Redo Exercise 34 using the trapezoidal rule.

**36.** (a) Use the trapezoidal rule with $n = 4$ to approximate $\int_1^2 (1/x^2)\, dx$.
(b) Find an upper bound for the error in the approximation you obtained in part (a).
(c) What value of $n$ guarantees an error of no more than 0.00005 in the approximation of this integral by the trapezoidal rule?

**37.** Redo Exercise 36 using Simpson's rule.

**38.** (Monthly marginal revenue) A company has monthly marginal revenue function $R'(x) = \sqrt{x^2 + 1}\, \ln(2x + 3)$. If $x$ represents the number of units sold (in thousands) and $R(x)$ is the revenue in thousands of dollars, use the trapezoidal rule with $n = 6$ to approximate the revenue from the sale of 3000 units. Assume that $R(0) = 0$.

**39.** (Consumer's surplus) Given the demand function $D(x) = \sqrt[3]{4000 - 10x^2}$, use Simpson's rule with $n = 8$ to estimate the consumer's surplus when $x = 4$.

**40.** (Producer's surplus) Given the supply function $S(x) = \sqrt{x^3 + 8}$, use Simpson's rule with $n = 8$ to estimate the producer's surplus when $x = 2$.

**41.** (Learning) If a person learns vocabulary words at the rate of

$$N'(t) = 10\sqrt[3]{30 + \frac{1}{t + 1}}, \quad 0 \le t \le 3$$

words per hour, where $t$ represents the number of hours of study, use the trapezoidal rule with $n = 6$ to estimate the total number of words learned during the first three hours of study.

**42.** (Welfare) The number of people on welfare in a certain region is projected to increase at the rate of $N'(t) = \ln(t^3 + 2)$ thousand per year between 1991 (when $t = 0$) and 1998. Estimate the increase in the number of people on welfare between 1991 and 1995 using Simpson's rule with $n = 8$.

Social Sciences

**43.** (Child abuse) The number of cases of child abuse in a certain city is expected to increase at the rate of $N'(t) = \sqrt[3]{0.3t^2 + 1}$ thousand per year between 1993 (when $t = 0$) and 1998. Estimate the increase in the number of cases of child abuse between 1993 and 1997 using the trapezoidal rule with $n = 8$.

**44.** (Cost) A company estimates that its daily marginal cost at various levels of production between 10 and 16 units is

| PRODUCTION LEVEL (units) | 10 | 11 | 12 | 13 | 14 | 15 | 16 |
|---|---|---|---|---|---|---|---|
| MARGINAL COST (dollars) | 5 | 7 | 10 | 13 | 16 | 21 | 28 |

Use Simpson's rule to approximate the company's cost of increasing production from 10 to 16 units per day.

**Business and Economics**

**45.** (Profit) A company estimates that its daily marginal profit at various levels of production and sales between 20 and 32 units is given by the following schedule

| PRODUCTION/SALES LEVEL (units) | 20 | 22 | 24 | 26 | 28 | 30 | 32 |
|---|---|---|---|---|---|---|---|
| MARGINAL PROFIT (dollars) | 3.40 | 4.40 | 5.20 | 6.30 | 5.80 | 5.20 | 4.10 |

Use the trapezoidal rule with $n = 6$ to approximate the company's increase in profit when it raises its production and sales level from 20 units per day to 32 units per day.

**Life Sciences**

**46.** (Oil slick) The distance across an oil slick is measured every 100 ft (see Figure 7.48). If the distances are found to be those illustrated in the figure, approximate the total surface area of this oil slick using the trapezoidal rule.

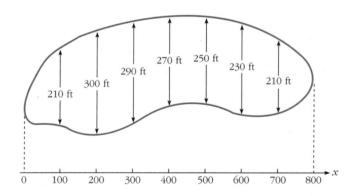

**FIGURE 7.48** Distances across an oil slick observed at intervals of 100 feet.

*Hint:* Let $D(x)$ be the distance across the oil slick $x$ feet from the left-hand edge, and approximate $\int_0^{800} D(x)\, dx$.

**47.** Let $P$ be a parabola passing through the three points $(x_0, y_0)$, $(x_1, y_1)$, and $(x_2, y_2)$, where $x_0 = -\Delta x$, $x_1 = 0$, and $x_2 = \Delta x$ for $\Delta x > 0$. Suppose that the equation of $P$ is $y = ax^2 + bx + c$.
(a) Show that
$$y_0 = a(\Delta x)^2 - b\Delta x + c$$
$$y_1 = c \quad \text{and} \quad y_2 = a(\Delta x)^2 + b\Delta x + c$$
(b) Show that
$$\int_{-\Delta x}^{\Delta x} (ax^2 + bx + c)\, dx = \frac{\Delta x}{3}[y_0 + 4y_1 + y_2]$$

# 7.5 IMPROPER INTEGRALS

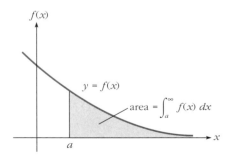

**FIGURE 7.49** The shaded area under the graph of $f$ to the right of $x = a$ is
$\int_a^\infty f(x)\, dx$.

## Improper Integrals: Various Types

The definite integrals we have considered so far have all been of the form $\int_a^b f(x)\, dx$, where $f(x)$ is continuous on the finite closed interval $[a, b]$. In this section, we consider integrals over intervals that are infinite in length, such as $[a, \infty)$, $(-\infty, b]$, and $(-\infty, \infty)$. Such integrals are called *improper* integrals and are designated by the symbolisms $\int_a^\infty f(x)\, dx$, $\int_{-\infty}^b f(x)\, dx$, and $\int_{-\infty}^\infty f(x)\, dx$, respectively.

Improper integrals arise in applications whenever we want to find the present value of a revenue stream that continues indefinitely. For example, we use an improper integral to compute the present value of a perpetual bond that pays the owner $1000 per year indefinitely (see Example 6).

In order to be consistent with our area interpretation of the definite integral, the improper integral $\int_a^\infty f(x)\, dx$ should represent the area under the graph of $f$ to the right of $x = a$ whenever $f(x) \geq 0$ on the interval $[a, \infty)$: see Figure 7.49. To see that such an "infinitely long" area can be finite and to get a better idea of how to define $\int_a^\infty f(x)\, dx$, let's look at an example.

## EXAMPLE 1

(a) Find the integral $\int_1^b (1/x^2)\, dx$ for $b > 1$ and interpret it as an area.
(b) Find $\lim_{b \to \infty} \int_1^b (1/x^2)\, dx$ and interpret it as an area.

### Solution

(a) $\int_1^b \frac{1}{x^2}\, dx = \int_1^b x^{-2}\, dx = -x^{-1}\Big|_1^b = -\frac{1}{b} + 1$

In Table 7.1, the values of $\int_1^b (1/x^2)\, dx = 1 - (1/b)$ are tabulated for various values of $b$. In Figure 7.50 we see that $\int_1^b (1/x^2)\, dx$ represents the area under the graph of $y = 1/x^2$ between $x = 1$ and $x = b$.

| $b$ | $\int_1^b \frac{1}{x^2}\, dx$ |
|---|---|
| 100 | 0.99 |
| 1000 | 0.999 |
| 100000 | 0.99999 |
| 1000000 | 0.999999 |
| $\to \infty$ | $\to 1$ |

**TABLE 7.1**

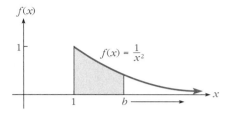

**FIGURE 7.50** The integral $\int_1^b (1/x^2)\, dx = 1 - (1/b)$ is the shaded area. As the arrow indicates, we are considering the situation where $b$ increases.

(b) By part (a), we have $\int_1^b (1/x^2)\, dx = 1 - (1/b)$. Thus

$$\lim_{b \to \infty} \int_1^b \frac{1}{x^2}\, dx = \lim_{b \to \infty} \left[1 - \frac{1}{b}\right] = 1$$

The quantity $\lim_{b \to \infty} \int_1^b (1/x^2)\, dx = 1$ represents the area under the graph of $f(x) = 1/x^2$ to the right of $x = 1$, as illustrated in Figure 7.51.

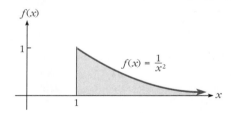

**FIGURE 7.51** The quantity
$\lim_{b \to \infty} \int_1^b (1/x^2)\, dx = 1$ represents the area under the graph of $f(x) = 1/x^2$ to the right of $x = 1$.

Motivated by the results of Example 1, we define $\int_1^\infty (1/x^2)\, dx$ to be $\lim_{b \to \infty} \int_1^b (1/x^2)\, dx$. In fact, we can generalize:

**IMPROPER INTEGRAL: UPPER LIMIT ∞**

If $f$ is continuous on $[a, \infty)$ and if the indicated limit exists, then

$$\int_a^\infty f(x)\, dx = \lim_{b \to \infty} \int_a^b f(x)\, dx \qquad (4)$$

If the limit in formula (4) exists, we say that the improper integral *converges*. If the limit does not exist, we say that the improper integral *diverges*.

---

**EXAMPLE 2 (Participative)**

Solution

Let's compute $\int_{-1}^\infty e^{-x}\, dx$.

By definition we have

$$\int_{-1}^\infty e^{-x}\, dx = \lim_{b \to \infty} \int_{-1}^b e^{-x}\, dx$$

Now, since $\int e^{-x}\, dx =$ _____, we continue as follows:

$$\int_{-1}^\infty e^{-x}\, dx = \lim_{b \to \infty} [-e^{-x}]_{-1}^b$$

$$= \lim_{b \to \infty} [-e^{-b} - (\underline{\qquad})]$$

$$= \lim_{b \to \infty} \left(e - \frac{1}{e^b}\right) = e - \underline{\qquad}$$

$$= e - \underline{\qquad} = \underline{\qquad}$$

Thus the improper integral $\int_{-1}^{\infty} e^{-x}\, dx$ converges to $e \approx 2.71828$. In terms of area, the area under the graph of $f(x) = e^{-x}$ to the right of $x = -1$ is equal to _____ square units; see Figure 7.52.

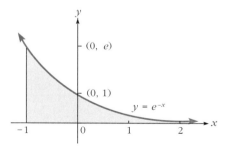

**FIGURE 7.52** The shaded region has an area of $\int_{-1}^{\infty} e^{-x}\, dx = e$ square units.

## EXAMPLE 3

Show that the improper integral $\int_{1}^{\infty} (1/x)\, dx$ diverges.

**Solution**

$$\int_{1}^{\infty} \frac{1}{x}\, dx = \lim_{b \to \infty} \int_{1}^{b} \frac{1}{x}\, dx = \lim_{b \to \infty} \ln |x| \Big|_{1}^{b}$$

$$= \lim_{b \to \infty} (\ln b - \ln 1) = \lim_{b \to \infty} \ln b$$

Since $\ln b$ increases without bound as $b \to \infty$, we conclude that

$$\int_{1}^{\infty} \frac{1}{x}\, dx = \lim_{b \to \infty} \ln b = \infty$$

and so $\int_{1}^{\infty} (1/x)\, dx$ diverges. Graphically, this means that the area under the graph of $f(x) = 1/x$ to the right of $x = 1$ is infinite; see Figure 7.53.

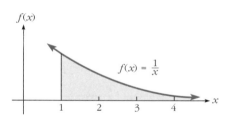

**FIGURE 7.53** The area under the graph of $f(x) = 1/x$ as given by $\int_{1}^{\infty} (1/x)\, dx$ is infinite.

Sometimes an improper integral involves an infinite *lower* limit of integration. This type of improper integral is defined as follows.

**IMPROPER INTEGRAL: LOWER LIMIT $-\infty$**

If $f$ is continuous on $(-\infty, b]$ and if the indicated limit exists, then

$$\int_{-\infty}^{b} f(x)\, dx = \lim_{a \to -\infty} \int_{a}^{b} f(x)\, dx$$

## EXAMPLE 4

Evaluate the improper integral $\int_{-\infty}^{0} e^{2x+1}\, dx$.

**Solution**

$$\int_{-\infty}^{0} e^{2x+1}\, dx = \lim_{a \to -\infty} \int_{a}^{0} e^{2x+1}\, dx$$

**SOLUTIONS TO EXAMPLE 2**  $-e^{-x}, -e^{-(-1)}, \lim_{b \to \infty} (1/e^b), 0, e, e$

## 486 CHAPTER 7 ADDITIONAL INTEGRATION TOPICS

You should observe that, by the substitution $u = 2x + 1$ and $du = 2\,dx$, we have $\int e^{2x+1}\,dx = \int e^u(\tfrac{1}{2}\,du) = \tfrac{1}{2}\int e^u\,du = \tfrac{1}{2}e^u = \tfrac{1}{2}e^{2x+1}$ (with $C = 0$). Thus

$$\int_{-\infty}^{0} e^{2x+1}\,dx = \lim_{a\to-\infty}\left[\tfrac{1}{2}e^{2x+1}\right]_a^0 = \lim_{a\to-\infty}\left(\tfrac{1}{2}e - \tfrac{1}{2}e^{2a+1}\right)$$

$$= \lim_{a\to-\infty} \tfrac{1}{2}e - \lim_{a\to-\infty} \tfrac{1}{2}e^{2a+1}$$

To compute the first of these limits, observe that since $\tfrac{1}{2}e$ is a constant, $\lim_{a\to-\infty}\tfrac{1}{2}e = \tfrac{1}{2}e$. To find the second limit, note that $2a + 1 \to -\infty$ as $a \to -\infty$, so that $e^{2a+1} \to 0$ and $\lim_{a\to-\infty}\tfrac{1}{2}e^{2a+1} = 0$. We conclude that

$$\int_{-\infty}^{0} e^{2x+1}\,dx = \tfrac{1}{2}e \qquad\blacksquare$$

Next we define the improper integral that involves infinite upper and lower limits of integration.

**IMPROPER INTEGRAL:**
**UPPER LIMIT $\infty$, LOWER LIMIT $-\infty$**

If $f$ is continuous on $(-\infty, \infty)$ and $c$ is any real number, then

$$\int_{-\infty}^{\infty} f(x)\,dx = \int_{-\infty}^{c} f(x)\,dx + \int_{c}^{\infty} f(x)\,dx \qquad (5)$$

provided each of the improper integrals on the right-hand side converges. If either of these two integrals diverges, then $\int_{-\infty}^{\infty} f(x)\,dx$ diverges.

---

**EXAMPLE 5**

Evaluate $\int_{-\infty}^{\infty} xe^{-x^2}\,dx$ if it converges.

**Solution**

Using formula (5) with $c = 0$ gives us

$$\int_{-\infty}^{\infty} xe^{-x^2}\,dx = \int_{-\infty}^{0} xe^{-x^2}\,dx + \int_{0}^{\infty} xe^{-x^2}\,dx$$

$$= \lim_{a\to-\infty}\int_{a}^{0} xe^{-x^2}\,dx + \lim_{b\to\infty}\int_{0}^{b} xe^{-x^2}\,dx$$

Now note that by the substitution $u = -x^2$ and $du = -2x\,dx$, we have $\int xe^{-x^2}\,dx = \int e^u(-\tfrac{1}{2}\,du) = -\tfrac{1}{2}\int e^u\,du = -\tfrac{1}{2}e^u = -\tfrac{1}{2}e^{-x^2}$, with $C = 0$. Thus

$$\int_{-\infty}^{\infty} xe^{-x^2}\,dx = \lim_{a\to-\infty}\left[-\tfrac{1}{2}e^{-x^2}\right]_a^0 + \lim_{b\to\infty}\left[-\tfrac{1}{2}e^{-x^2}\right]_0^b$$

$$= \lim_{a\to-\infty}\left[-\tfrac{1}{2} + \tfrac{1}{2}e^{-a^2}\right] + \lim_{b\to\infty}\left[-\tfrac{1}{2}e^{-b^2} + \tfrac{1}{2}\right]$$

$$= -\tfrac{1}{2} + \tfrac{1}{2} = 0$$

In terms of area, since $xe^{-x^2} < 0$ if $x < 0$, the quantity $\int_{-\infty}^{0} xe^{-x^2}\,dx = -\tfrac{1}{2}$ is the *negative* of the area between the curve $y = xe^{-x^2}$ and the $x$-axis to the left

## 7.5 IMPROPER INTEGRALS

of zero. Also, since $xe^{-x^2} > 0$ if $x > 0$, the quantity $\int_0^\infty xe^{-x^2}\,dx = \frac{1}{2}$ is the area between the curve $y = xe^{-x^2}$ and the $x$-axis to the right of zero. Thus

$$\int_{-\infty}^\infty xe^{-x^2}\,dx = \int_{-\infty}^0 xe^{-x^2}\,dx + \int_0^\infty xe^{-x^2}\,dx = -\frac{1}{2} + \frac{1}{2} = 0$$

See Figure 7.54.

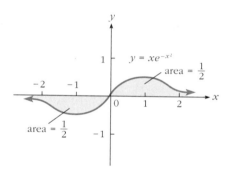

**FIGURE 7.54** The area of the shaded region is one square unit, although $\int_{-\infty}^\infty xe^{-x^2}\,dx = -\frac{1}{2} + \frac{1}{2} = 0$.

### An Application: Capitalized Value

In certain situations income from a business venture may form a stream of income that flows forever. Such is the case, for example, with a large stand of timber that is continuously harvested and replanted, or with perpetual bonds issued by a corporation. In Section 7.3 we learned that the present value of an income stream that generates $f(t)$ dollars per year at $100r\%$ compounded continuously for $N$ years is

$$\text{present value} = \int_0^N f(t)e^{-rt}\,dt$$

If the income stream flows forever, then to find the present value of the income stream, we merely replace $N$ by $\infty$ in the preceding formula to obtain what is called the *capitalized value* of the income stream.

**THE CAPITALIZED VALUE: PRESENT VALUE OF A PERPETUAL INCOME STREAM**

If income from a venture flows in at the continuous rate of $f(t)$ dollars per year forever, then the capitalized value of this income stream is the present value at $100r\%$ compounded continuously and is given by the improper integral

$$\text{capitalized value} = \int_0^\infty f(t)e^{-rt}\,dt$$

The capitalized value of an income stream is the amount of money that, if invested today at $100r\%$ compounded continuously, will generate the same amount of money as the income stream and hence may be considered as the present cash value of the income stream.

**EXAMPLE 6**

Your company is planning to issue perpetual bonds that will each pay the owner $1000 per year forever. Find the price at which the bonds should be offered if we assume that the annual rate of continuous compound interest remains fixed at 8%.

Solution

The annual payments from each bond yield a continuous income stream that generates $f(t) = 1000$ dollars per year and continues forever. Even though the income is received annually in a lump sum and not continuously throughout the year, it is standard practice to treat such a sequence of periodic payments as a continuous income stream. We are required to find the capitalized value (or the present value) of the stream:

$$\text{capitalized value} = \int_0^\infty 1000e^{-0.08t}\,dt = \lim_{b\to\infty} \int_0^b 1000e^{-0.08t}$$

$$= \lim_{b\to\infty}\left[\frac{1000e^{-0.08t}}{-0.08}\right]_0^b = \lim_{b\to\infty}[-12500e^{-0.08t}]_0^b$$

$$= \lim_{b\to\infty}\left[\frac{-12500}{e^{0.08b}} + 12500\right] = 12{,}500$$

Thus the bonds should be offered at a price of $12,500 each. ■

## SECTION 7.5
### SHORT ANSWER EXERCISES

1. (True or False) The types of improper integrals considered in this section have an infinite interval of integration. _____

2. The improper integral $\int_2^\infty (1/x^4)\,dx$ may be interpreted as the area under the graph of $y = 1/x^4$ to the right of $x = $ _____.

3. In evaluating a certain improper integral, the limit $\lim_{b\to\infty}(e^{-b} + 2)$ arises. The value of this limit is _____.

4. In evaluating a certain improper integral the limit $\lim_{a\to-\infty}[\ln |a| + 3]$ arises. The value of this limit is _____.

5. If $f(x)$ is continuous on $(-\infty, \infty)$ and if $\int_{-\infty}^2 f(x)\,dx = 4$ and $\int_2^\infty f(x)\,dx = 5$, then $\int_{-\infty}^\infty f(x)\,dx = $ _____.

6. If $f(x)$ is continuous on $(-\infty, \infty)$ and if $\int_{-\infty}^0 f(x)\,dx = -2$ and $\int_0^\infty f(x)\,dx$ diverges, then $\int_{-\infty}^\infty f(x)\,dx$ _____.

7. By interpreting $\int_0^\infty x\,dx$ as the area under the graph of $f(x) = x$ to the right of $x = 0$, we easily conclude that this improper integral _____.

8. Since $\int_{-\infty}^\infty e^x\,dx = \int_{-\infty}^0 e^x\,dx + \int_0^\infty e^x\,dx$, and since $\int_0^\infty e^x\,dx$ represents the area under the graph of $f(x) = e^x$ to the right of $x = 0$, we conclude that the improper integral $\int_{-\infty}^\infty e^x\,dx$ _____.

9. If the capitalized value of a certain business venture that yields a continuous stream of income indefinitely is $25,000 and the annual continuously compounded interest rate is 9% annually, what is the value of this business venture today? _____

10. The higher the fixed annual continuous compound interest rate we assume, the _____ (higher, lower) the capitalized value will be for a fixed income flow function $f(t)$.

## SECTION 7.5
### EXERCISES

In Exercises 1–26, evaluate each improper integral, if it converges.

1. $\int_2^\infty (1/x^3)\,dx$

2. $\int_1^\infty (1/x^4)\,dx$

3. $\int_1^\infty (1/\sqrt{x})\,dx$

4. $\int_2^\infty (1/\sqrt[3]{x})\,dx$

5. $\int_0^\infty 2e^{0.5x}\,dx$

6. $\int_0^\infty e^{0.05x}\,dx$

7. $\int_{-\infty}^{-1} (2/x^3)\,dx$

8. $\int_{-\infty}^{-2} (1/\sqrt[3]{x})\,dx$

9. $\int_2^\infty \frac{1}{(x-1)^2}\,dx$

10. $\int_5^\infty \frac{3}{(x-4)^3}\,dx$

11. $\int_{-1}^{\infty} (1/\sqrt[3]{x+2})\, dx$

12. $\int_{-2}^{\infty} (1/\sqrt{x+3})\, dx$

13. $\int_{-\infty}^{1} (1/\sqrt{2-x})\, dx$

14. $\int_{-\infty}^{2} 5/(x-3)^2\, dx$

15. $\int_{-\infty}^{0} 2e^{0.4x}\, dx$

16. $\int_{-\infty}^{0} e^{0.002x}\, dx$

17. $\int_{-\infty}^{-1} x^{-5/3}\, dx$

18. $\int_{-\infty}^{-1} x^{-1/5}\, dx$

19. $\int_{1}^{\infty} \dfrac{4x+1}{2x^2+x}\, dx$

20. $\int_{2}^{\infty} \dfrac{1}{(3x-1)^4}\, dx$

21. $\int_{0}^{\infty} x^3 e^{-x^4}\, dx$

22. $\int_{-\infty}^{1} \dfrac{3x}{2x^2+1}\, dx$

23. $\int_{-\infty}^{\infty} x^3 e^{-x^4}\, dx$

24. $\int_{-\infty}^{\infty} x^4 e^{-x^5}\, dx$

25. $\int_{-\infty}^{\infty} \dfrac{6x}{(x^2+2)^3}\, dx$

26. $\int_{-\infty}^{\infty} \dfrac{2x}{\sqrt{x^2+4}}\, dx$

In Exercises 27–34, the facts that if $k > 0$ and $n > 0$ imply $\lim\limits_{x \to \infty} x^n e^{-kx} = 0$ and $\lim\limits_{x \to \infty} x^{-n}(\ln x)^n = 0$; use this to evaluate each improper integral, if it converges.

27. $\int_{1}^{\infty} xe^{-x}\, dx$

28. $\int_{1}^{\infty} x^2 e^{-x}\, dx$

29. $\int_{0}^{\infty} x^2 e^{-2x}\, dx$

30. $\int_{-\infty}^{1} xe^x\, dx$

31. $\int_{1}^{\infty} \dfrac{\ln x}{x^2}\, dx$

32. $\int_{1}^{\infty} \dfrac{\ln x}{x^3}\, dx$

33. $\int_{e}^{\infty} x \ln x\, dx$

34. $\int_{2}^{\infty} \dfrac{(\ln x)^2}{x^3}\, dx$

In each of Exercises 35–38, find the area of the region bounded by the graph of $f(x)$ and the x-axis on the given interval.

35. $f(x) = \dfrac{1}{(x+1)^2}$ on $[0, \infty)$

36. $f(x) = \dfrac{x}{(x^2+1)^3}$ on $(-\infty, \infty)$

37. $f(x) = xe^x$ on $(-\infty, 0]$

38. $f(x) = 1/\sqrt{x^3}$ on $[1, \infty)$

39. (Capitalized value) Find the capitalized value of an asset that generates $15,000 per year indefinitely if we assume that the annual interest rate remains fixed at (a) 6% compounded continuously; (b) 8% compounded continuously.

40. (Capitalized value) A company has agreed to lease a parcel of land from you indefinitely for $20,000 a year. Find the capitalized value of the land, assuming an 8% annual interest rate compounded continuously.

41. (Capitalized value) Find the capitalized value of an income stream that produces income at the rate of $f(t) = 2500e^{0.04t}$ dollars per year forever. Assume an annual interest rate of 6% compounded continuously.

**Business and Economics**

42. (Business decision) It is estimated that $t$ years from now, your company's profit will be $250000e^{-0.15t}$ dollars per year. Assume that the company will remain in business indefinitely and that the interest rate is 9% annually compounded contin-

uously. If someone offers you $750,000 for the company today, should you accept this offer? Base your decision on the capitalized value of your company's profits and assume your company has no other value.

43. (Present value) Find the present value of a perpetual income stream that will produce $f(t) = 3000t$ dollars of income per year, assuming a 10% annual interest rate compounded continuously. *Hint:* $\lim_{x \to \infty} x^n e^{-kx} = 0$ if $k$ and $n$ are positive numbers.

44. (Ore production) A new mine is expected to produce ore at the rate of $25/(t + 1)^2$ millions of tons per year after $t$ years of operation. What is the total amount of ore that will be produced by the mine?

45. (Toxic waste) A new industrial plant will produce toxic waste as a by-product of its manufacturing process. Due to improving techniques of manufacturing, the rate at which this waste will be produced in the future is expected to be decreasing. After $t$ years, the plant will be producing toxic waste at the rate of $500000/(t + 1)^3$ gallons per year. Find the total amount of toxic waste the plant will produce.

**Life Sciences**

46. (Treatment of disease) If a new treatment method for a certain fatal disease is developed, then the rate at which deaths due to the disease occur will decrease. In fact, after $t$ years, the number of deaths per year due to the disease would be $8000000/(t + 2)^3$. Find the total number of deaths that would still occur in the future if the treatment method were developed today.

## 7.6 CHAPTER SEVEN REVIEW

**KEY TERMS**

area theorem
fundamental theorem of calculus
Riemann sum
average value of a function
consumer's willingness to pay
actual consumer expenditure
consumer's surplus
producer's willingness to accept
actual producer's revenue
producer's surplus
market equilibrium point

continuous income stream
future value of an income stream
present value of an income stream
left-hand Riemann sum
right-hand Riemann sum
midpoint Riemann sum
trapezoidal rule
Simpson's rule
improper integrals
capitalized value

**KEY FORMULAS AND RESULTS**

**The Area Theorem**

If $f$ is continuous and nonnegative on $[a, b]$ and if $F'(x) = f(x)$, then the area under the graph of $f$ from $a$ to $b$ is

$$\int_a^b f(x)\, dx = F(b) - F(a)$$

### Area between Two Curves

If $f(x) \geq g(x)$ for $x$ in $[a, b]$, then the area between the graphs of $f$ and $g$ on the interval $[a, b]$ is

$$\int_a^b [f(x) - g(x)] \, dx$$

### The Fundamental Theorem of Calculus

$$\int_a^b f(x) \, dx = \lim_{\Delta x \to 0} [f(c_1) + f(c_2) + \cdots + f(c_n)]\Delta x = F(b) - F(a)$$

where $\Delta x = (b-a)/n$, $x_i = a + i\Delta x$, $c_i$ is in $[x_{i-1}, x_i]$, and $F(x)$ is an antiderivative of $f(x)$ on $[a, b]$.

### The Average Value of a Function

$$\text{average value of } f \text{ over } [a, b] = \frac{1}{b-a} \int_a^b f(x) \, dx$$

### Supply and Demand

If $p = D(x)$ is a demand function and if $p_0 = D(q_0)$, then

(a) consumer's willingness to pay for $q_0$ units $= \int_0^{q_0} D(x) \, dx$

(b) actual consumer expenditure $= p_0 q_0$

(c) consumer's surplus $= \int_0^{q_0} D(x) \, dx - p_0 q_0$

If $p = S(x)$ is a supply function and if $p_0 = S(q_0)$, then

(a) producer's willingness to accept for $q_0$ units $= \int_0^{q_0} S(x) \, dx$

(b) actual producer's revenue $= p_0 q_0$

(c) producer's surplus $= p_0 q_0 - \int_0^{q_0} S(x) \, dx$

### Income Generated by a Continuous Stream

$$\text{total income} = \int_a^b f(t) \, dt$$

where $f(t)$ is the rate of flow.

### Future Value of an Income Stream after N Years

$$\text{future value} = \int_0^N f(t) e^{r(N-t)} \, dt$$

where $100r\%$ is the continuously compounded annual interest rate.

**Present Value of an Income Stream for N Years**

$$\text{present value} = \int_0^N f(t)e^{-rt}\, dt$$

**General Riemann Sum**

$$\int_a^b f(x)\, dx \approx \left(\frac{b-a}{n}\right)[f(c_1) + f(c_2) + \cdots + f(c_n)]$$

where $c_i$ is any number in $[x_{i-1}, x_i]$.

*Left-Hand Riemann Sum*    Choose $c_i = x_{i-1}$.
*Right-Hand Riemann Sum*    Choose $c_i = x_i$.
*Midpoint Riemann Sum*    Choose $c_i = \dfrac{x_{i-1} + x_i}{2}$

**Trapezoidal Rule**

$$\int_a^b f(x)\, dx \approx \left(\frac{b-a}{2n}\right)(y_0 + 2y_1 + 2y_2 + \cdots + 2y_{n-1} + y_n), \quad \text{where } y_i = f(x_i)$$

**Simpson's Rule (n even)**

$$\int_a^b f(x)\, dx \approx \left(\frac{b-a}{3n}\right)(y_0 + 4y_1 + 2y_2 + 4y_3 + \cdots + 2y_{n-2} + 4y_{n-1} + y_n)$$

**Error in the Trapezoidal Rule**

$$|E_n(T)| \leq \frac{(b-a)^3 M}{12n^2}, \quad \text{where } M = \text{maximum of } |f''(x)| \text{ on } [a, b]$$

**Error in Simpson's Rule**

$$|E_n(S)| \leq \frac{(b-a)^5 M}{180n^4}, \quad \text{where } M = \text{maximum of } |f^{(4)}(x)| \text{ on } [a, b]$$

**Improper Integrals**

**1.** If $f$ is continuous on $[a, \infty)$, then

$$\int_a^\infty f(x)\, dx = \lim_{b \to \infty} \int_a^b f(x)\, dx$$

**2.** If $f$ is continuous on $(-\infty, b]$, then

$$\int_{-\infty}^b f(x)\, dx = \lim_{a \to -\infty} \int_a^b f(x)\, dx$$

**3.** If $f$ is continuous on $(-\infty, \infty)$ and $c$ is any real number, then

$$\int_{-\infty}^\infty f(x)\, dx = \int_{-\infty}^c f(x)\, dx + \int_c^\infty f(x)\, dx$$

## Capitalized Value: Present Value of a Perpetual Income Stream

$$\text{capitalized value} = \int_0^\infty f(t)e^{-rt}\,dt$$

## CHAPTER 7 SHORT ANSWER REVIEW QUESTIONS

1. (True or False) The area under the graph of $f(x) = x^3$ from $-1$ to 1 is given by $\left|\int_{-1}^{1} x^3\,dx\right|$. _____

2. The graphs of $f(x) = x$ and $g(x) = \sqrt{x}$ intersect when $x = 0$ and $x = 1$. The area between these graphs, written as a definite integral, is $\int_0^1$ _____.

3. The area bounded by the graph of $f(x) = \ln x$, the x-axis, and the vertical lines $x = \frac{1}{2}$ and $x = 1$, written as a definite integral, is _____.

4. (True or False) The average value of $f(x) = 2x$ on $[0, 1]$ is 250. _____

5. (True or False) If $f(x)$ is continuous and nonnegative on $[a, b]$, then a Riemann sum approximation of $\int_a^b f(x)\,dx$ can never give the exact value of the integral. _____

6. If $\Delta x = 0.25$, $a = 2$, and $b = 4$, then $n =$ _____.

7. If consumers are willing to pay $20,000 to purchase 5000 units of a certain product and if the product is actually sold for $3 per unit, the consumer's surplus is _____.

8. (True or False) The producer's surplus on the sale of $q_0$ units at a price per unit of $p_0$ is the area under the supply curve $S(x)$ from 0 to $q_0$. _____

9. A certain business venture will generate a continuous stream of income amounting to $5000 per year for the next seven years. If a positive annual interest rate is assumed, the present value of this income stream is (a) more than $35,000; (b) less than $35,000; (c) equal to $35,000.

10. (True or False) For a certain function that is continuous and nonnegative on $[0, 3]$, we have $\int_0^3 f(x)\,dx = 10$, and a left-hand Riemann sum with $n = 6$ yields 9.5. We can conclude that $f(x)$ is decreasing on $[0, 3]$. _____

11. (True or False) The trapezoidal rule will always give an exact value for $\int_a^b f(x)\,dx$ if $f(x) = mx + b$, where $m$ and $b$ are constants. _____

12. In approximating $\int_1^2 f(x)\,dx$ using Simpson's rule with $n = 4$, if $|f^{(4)}(x)| \leq 8$ then an upper bound for the error is _____.

13. (True or False) $\int_{-\infty}^{\infty} dx/(x^2 + 1) = \int_{-\infty}^{75} dx/(x^2 + 1) + \int_{75}^{\infty} dx/(x^2 + 1)$. _____

14. Suppose that $f(x)$ is continuous on $(-\infty, \infty)$ and $\int_0^\infty f(x)\,dx = 2$. What can you say about $\int_{-1}^\infty f(x)\,dx$? (a) It is greater than 2. (b) It is less than 2. (c) It converges. (d) None of the above.

15. Let $f(x)$ and $g(x)$ be continuous and nonnegative on $(-\infty, \infty)$, with $g(x) \leq f(x)$ for all $x$ in $(-\infty, \infty)$. If $\int_{-\infty}^{-1} f(x)\,dx = 2$, what can you say about $\int_{-\infty}^{-1} g(x)\,dx$? (a) It is no larger than 2. (b) It may diverge. (c) It is negative. (d) None of the above.

# CHAPTER 7 REVIEW EXERCISES

*Exercises marked with an asterisk (\*) constitute a Chapter Test.*

In each of Exercises 1–4, find the area bounded by the graph of $f$ and the x-axis.

**\*1.** $f(x) = -x^2 + 2x + 3$
**2.** $f(x) = x^2 - 3x - 10$
**3.** $f(x) = x^3 - 3x^2 + 2x$
**4.** $f(x) = (x^2 - 1)(x^2 - 4)$

In Exercises 5–8, find the areas enclosed by the graphs of $f$ and $g$.

**\*5.** $f(x) = x^2 + 45;\ g(x) = -14x$
**6.** $f(x) = \sqrt{x};\ g(x) = \sqrt[3]{x}$
**7.** $f(x) = x^2 + 10x;\ g(x) = x^3 - 2x$
**8.** $f(x) = \sqrt{x};\ g(x) = 4x$

In Exercises 9–12, find the average values of the functions over the given interval.

**9.** $f(x) = 10$ on $[0, 2]$
**10.** $f(x) = \sqrt[3]{x}$ on $[0, 8]$
**\*11.** $f(x) = x^3 - 1$ on $[-1, 0]$
**12.** $f(x) = |x + 1|$ on $[0, 5]$

In each of Exercises 13–16, find the consumer's surplus at the given demand level $x$ or price level $p$. Assume that $x$ is measured in units and price in dollars.

**\*13.** $D(x) = 600 - 0.1x$ at $x = 1000$
**14.** $D(x) = \dfrac{5000}{x + 10}$ at $p = 10$
**15.** $D(x) = \dfrac{60}{\sqrt{x + 4}}$ at $p = 5$
**16.** $D(x) = \dfrac{1000x}{x^2 + 4}$ at $x = 25$

In each of Exercises 17–20, find the consumer's surplus and the producer's surplus if the price is at market equilibrium. Assume that $x$ is measured in units and price in dollars.

**17.** $D(x) = 5 - 0.01x$
$S(x) = 1 + 0.03x$
**18.** $D(x) = \sqrt{x} + 5$
$S(X) = -\sqrt{x} + 13$
**\*19.** $D(x) = 6 - 0.11x$
$S(x) = 0.001x^2$
**20.** $D(x) = 400/(x + 30)$
$S(x) = \tfrac{1}{10}x$

In Exercises 21–24, find (a) the left-hand Riemann sums; (b) the right-hand Riemann sums; and (c) the midpoint Riemann sums for the given value of $n$ to approximate the integrals.

**\*21.** $\int_{-1}^{2} \sqrt{\ln(x + 2)}\, dx$ for $n = 4$
**22.** $\int_{0}^{2} \sqrt[3]{x^2 + 1}\, dx$ for $n = 6$
**23.** $\int_{-1}^{1} e^{-x^2}\, dx$ for $n = 6$
**24.** $\int_{-1}^{3} xe^{-x^3}\, dx$ for $n = 8$

**Exercises 25–28:** *for each of Exercises 21–24, use the trapezoidal rule to approximate the integrals. Exercise 25 is a Chapter Test problem.*

**Exercises 29–32:** *for each of Exercises 21–24, use Simpson's rule to approximate the integrals. Exercise 29 is a Chapter Test problem.*

**\*33.** (a) Use the trapezoidal rule with $n = 4$ to approximate $\int_0^1 (1/\sqrt{2x + 3})\, dx$.
(b) Find an upper bound for the error in the approximation you obtained in part (a).
(c) What value of $n$ guarantees an error of no more than 0.0005 in the approximation of this integral by the trapezoidal rule?

**34.** Repeat the steps of Exercise 33 using Simpson's rule rather than the trapezoidal rule.

*In Exercises 35–40, evaluate each improper integral, if it converges.*

***35.** $\int_{4}^{\infty} e^{-4x}\, dx$

**36.** $\int_{-1}^{\infty} \frac{3}{(x+2)^3}\, dx$

***37.** $\int_{-\infty}^{0} xe^{x}\, dx$

**38.** $\int_{2}^{\infty} \frac{1}{x(\ln x)^5}\, dx$

**39.** $\int_{1}^{\infty} \frac{(e^x + 1)}{(e^x + x)^2}\, dx$

**40.** $\int_{-\infty}^{\infty} \frac{2x}{(x^2 + 1)^2}\, dx$

***41.** (Asset appreciation) A piece of land is increasing in value at the rate of $V'(t) = 3000e^{0.06t}$ dollars per year. If $t = 0$ corresponds to the beginning of 1991, find the amount by which the land appreciates in value between the beginning of 1991 and the beginning of 1999.

**42.** (Consumption of electricity) The amount of electricity used by a certain town is projected to increase at the rate of $A'(t) = 15\sqrt{t}$ millions of kilowatt hours per year. If $t = 0$ corresponds to the beginning of 1992, find the total projected increase in consumption of electricity from the beginning of 1992 to the beginning of 1996.

**43.** (Retail sales) Sales for 1990 at a certain store of a retail chain were 5 million dollars. Sales at this particular store are decreasing at a rate of $0.15e^{0.15t}$ millions of dollars per year. What is the first year that sales in this store will fall below 4 million dollars?

**44.** (Production requirements) Due to expansion of production, the amount of a certain chemical needed by a plant is expected to increase, so that $t$ years after 1992, the plant will require $A(t) = 9 + \sqrt{2t + 1}$ thousand pounds of this chemical yearly. Find the average amount of the chemical required between the beginning of 1992 and the beginning of 2000.

**45.** (Acceleration) The acceleration of an object moving in a straight line is given by $a(t) = 1/\sqrt{t + 1}$ ft/sec². Find the average acceleration of this object during the first minute of motion.

**46.** (Income stream) The proceeds from a silver mine form a continuous income stream amounting to $750,000/year. What is the total value of this income stream over five years if it is invested at 10%/year compounded continuously?

***47.** (Income stream) The income derived from a business venture is expected to flow in at a continuous rate of $20000 \ln(t + 1)$ dollars per year for the next five years. Find the total income derived from this venture over the five-year period.

**48.** (Investment decision) If the prevailing rate of interest is 10% annually compounded continuously, would you earn more money over a five-year period by investing $25,000 into a bank account or by investing the $25,000 into a business venture that will yield a stream of income amounting to $4000e^{0.08t}$ dollars per year for the next five years?

Business and Economics

**49.** (Bankruptcy) The number of small businesses that declare bankruptcy in a certain community is increasing at the rate of $N'(t) = 50 \ln(t^2 + 2)$ businesses per year. If $t = 0$ corresponds to the beginning of 1992, use Simpson's rule with $n = 6$ to estimate the increase in the number of bankruptcies from the beginning of 1992 to the beginning of 1995.

**50.** (Warehouse leasing) If your company can lease a warehouse forever for $25,000 a year or purchase it for $200,000 today, what should you do if the prevailing interest rate is 7% annually compounded continuously?

51. (Child disappearance) The number of cases of child disappearance in a certain city is expected to increase at the rate of $N'(t) = \sqrt[5]{t^2 + 2}$ hundred per year between 1993 (when $t = 0$) and 1999. Estimate the increase in the number of cases of child disappearance between 1993 and 1997 using the trapezoidal rule with $n = 8$.

52. (Religiosity of elderly) A sociologist claims that for people in a certain region of the country, between the ages of $t = 70$ and $t = 80$, religiosity $R$ increases with age according to

$$\frac{dR}{dt} = \frac{0.0093t}{\sqrt[3]{(0.01395t^2 - 46.3965)^2}}$$

where $R$ is measured on a scale from 0 to 4. Find the increase in religiosity from $t = 70$ to $t = 75$.

OPTIONAL

**GRAPHING CALCULATOR/ COMPUTER EXERCISES**

53. Find the area bounded by the graph of $f(x) = -x^3 + 3x^2 + 2x - 5$. Estimate the $x$-intercepts to the nearest tenth of a unit.

54. Find the area enclosed by the graphs of $f(x) = \ln(x + 3)$ and $g(x) = x - 0.1$. Estimate the $x$-coordinates of the points of intersection of $f$ and $g$ to the nearest hundredth of a unit.

55. Find the consumer's surplus and the producer's surplus if price is at market equilibrium, $D(x) = 2 - 0.001x^2$, and $S(x) = 0.02\sqrt{x} + 0.5$. Estimate the $x$-coordinate of the point of intersection of $D(x)$ and $S(x)$ to the nearest tenth of a unit.

# 8 FUNCTIONS OF MORE THAN ONE VARIABLE

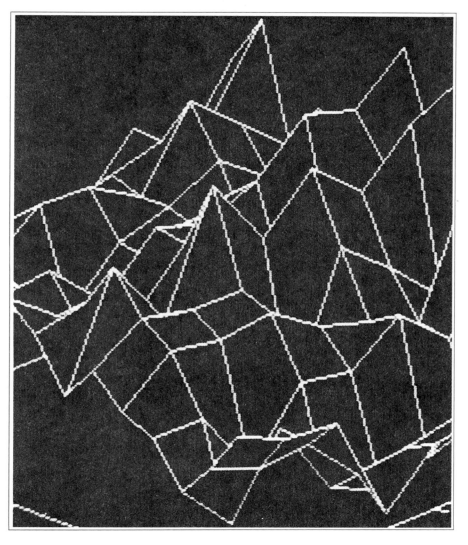

## CHAPTER CONTENTS

8.1 Introduction and Examples
8.2 Partial Derivatives
8.3 Maxima and Minima
8.4 Constrained Optimization: Lagrange Multipliers
8.5 Regression: The Method of Least Squares
8.6 Double Integrals
8.7 Chapter 8 Review

## 8.1 INTRODUCTION AND EXAMPLES

### Functions of More Than One Variable

Most of the functions considered thus far in this text have been functions of a single independent variable of the general form $y = f(x)$. In this general form, $x$ is the independent variable and $y$ is the dependent variable. In this chapter we consider the notion of a function of more than one independent variable. For example, the general form of a function of two independent variables ($x$ and $y$) with dependent variable $z$ is $z = f(x, y)$.

As an example of how functions of more than one variable might arise in an applied setting, suppose your company produces two products. If monthly fixed costs of production are $7000, variable costs of producing the first product are $10 per unit, and variable costs of producing the second product are $15 per unit, then the total monthly cost of producing $x$ units of the first product and $y$ units of the second product can be expressed by the total cost function

$$C(x, y) = 7000 + 10x + 15y$$

To continue with this example, if in a given month your company produces 500 units of the first product and 750 units of the second product, then total monthly production costs are expressed as $C(500, 750)$ and are computed to be

$$C(500, 750) = 7000 + 10(500) + 15(750) = \$23{,}250$$

**EXAMPLE 1 (Participative)**

**Solution**

For the cost function $C(x, y) = 7000 + 10x + 15y$, find and interpret
(a) $C(1000, 1200)$; (b) $C(0, 2000)$.

(a) $C(1000, 1200) = 7000 + 10(1000) + 15(\underline{\qquad}) = \$\underline{\qquad}$.
The monthly cost of producing 1000 units of the first product and 1200 units of the second product is $\underline{\qquad}$.

(b) $C(0, 2000) = 7000 + \underline{\qquad\qquad\qquad} = \underline{\qquad}$. The monthly cost of producing none of the first product and $\underline{\qquad}$ units of the second product is $\underline{\qquad}$. ■

Observe that in a general function of two variables $f(x, y) = z$, two input numbers $x$ and $y$ are required to produce a single unique output number $z$. Thus, unless otherwise specified, the *domain* of a function of two variables $f(x, y) = z$ is the set of ordered pairs of real numbers $(x, y)$ such that $z = f(x, y)$ is also a real number. The set of all values of $z$ form the *range* of the function of two variables $f(x, y) = z$.

Functions of more than two independent variables are easily defined. For example, if $f$ is a function of three independent variables $x$, $y$, and $z$, then we may write $w = f(x, y, z)$. The dependent variable here is $w$. In this text we will concentrate mainly on functions of two variables.

---

**SOLUTIONS TO EXAMPLE 1**    1200, 35,000, $35,000, 10(0) + 15(2000), 37,000, 2000, $37,000

## EXAMPLE 2

Let $z = f(x, y) = (x^2 + y^2)/(x - y)$ and $w = g(x, y, z) = \sqrt{x + y}/(xyz)$.

(a) Find $f(1, 2)$ and $g(-1, 2, 3)$.
(b) Find the domain of $f(x, y)$.
(c) Find the domain of $g(x, y, z)$.

**Solution**

(a) $f(1, 2) = \dfrac{1^2 + 2^2}{1 - 2} = -5;\quad g(-1, 2, 3) = \dfrac{\sqrt{-1 + 2}}{(-1)(2)(3)} = -\dfrac{1}{6}$

(b) The domain of $f(x, y) = (x^2 + y^2)/(x - y)$ consists of all ordered pairs $(x, y)$ of real numbers for which $x - y \neq 0$ or $x \neq y$, since if $x = y$ the denominator of the function will equal zero and division by zero is undefined.

(c) The domain of $g(x, y, z) = \sqrt{x + y}/(xyz)$ consists of all ordered triples $(x, y, z)$ of real numbers for which $x + y \geq 0$ and none of $x$, $y$, or $z$ is equal to zero. We require $x + y \geq 0$ because in computing $\sqrt{x + y}$, the quantity $x + y$ must be nonnegative. We require that none of $x$, $y$, or $z$ be equal to zero to avoid division by zero. ∎

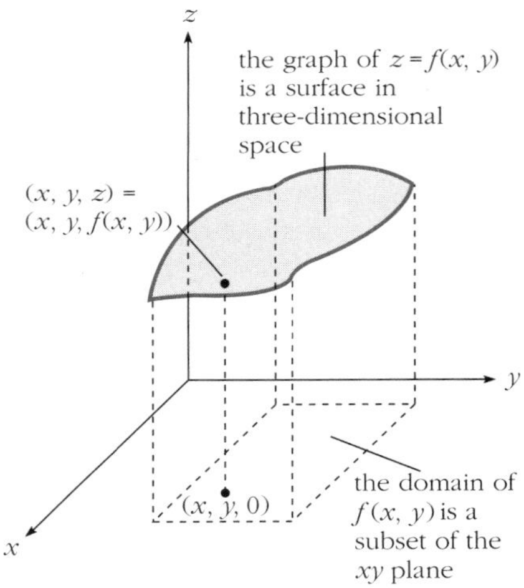

### The Graph of a Function of Two Variables

The graph of a function $z = f(x, y)$ of two variables is a *surface* in three-dimensional space. One such surface is illustrated in Figure 8.1. Observe that to obtain a three-dimensional coordinate system, we have added a third coordinate axis ($z$) to the usual $xy$-coordinate system. The $z$-axis is perpendicular to both the $x$- and $y$-axes and passes through the origin of the $xy$-coordinate system. We can view the $xyz$-coordinate system pictured in Figure 8.1 as the corner of a rectangular room. The $xy$-plane may be thought of as the floor of the room and the $yz$- and $xz$-planes as the walls in the room's corner.

**FIGURE 8.1** The graph of a function $z = f(x, y)$. A typical point $(x, y, z) = (x, y, f(x, y))$ is illustrated.

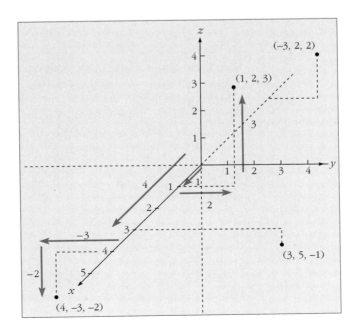

**FIGURE 8.2** Several points plotted in a three-dimensional coordinate system.

In Figure 8.2 we have located several points in a three-dimensional coordinate system. Note that to locate a point such as (1, 2, 3) in this system, we move one unit forward along the $x$-axis and then two units to the right parallel to the $y$-axis, and finally three units up, parallel to the $z$-axis. As another example, to locate $(4, -3, -2)$, we move four units forward on the $x$-axis, three units to the left parallel to the $y$-axis, and finally two units down, parallel to the $z$-axis. In a two-dimensional $xy$-coordinate system, the $x$- and $y$-axes divide the plane into four quadrants. In a three-dimensional coordinate system the $x$, $y$, and $z$-axes divide space into eight *octants*. The only named octant is the *first octant*, in which $x > 0$, $y > 0$, and $z > 0$.

### The Graph of a Plane

Recall that the graph of $ax + by + c = 0$ in the $xy$-plane is a straight line, if $a$ and $b$ are not both zero. In three-dimensional space, the graph of $ax + by + cz + d = 0$ is a *plane* if not all of $a$, $b$, and $c$ are equal to zero. Just as there is one and only one line that passes through two distinct points, there is one and only one plane which passes through three points that don't lie along the same straight line. So when graphing a plane in three-dimensional space, it is sufficient to find three points on the plane. In most cases it is easiest to find the three intercepts of the plane with the coordinate axes. The method of finding the intercepts is summarized in the following box.

**FINDING THE INTERCEPTS OF**
$ax + by + cz + d = 0$

| | |
|---|---|
| $x$-intercept: | set $y$ and $z$ equal to zero and solve for $x$ |
| $y$-intercept: | set $x$ and $z$ equal to zero and solve for $y$ |
| $z$-intercept: | set $x$ and $y$ equal to zero and solve for $z$ |

The rules in the box for finding intercepts can also be used to find the intercepts of a general surface $z = f(x, y)$. Let's look at a specific example of graphing a plane.

**EXAMPLE 3**

Sketch the graph of the plane with equation $2x + 3y + z = 6$.

**Solution**

To sketch this plane, we must find at least three points that lie on the plane. We will find the three intercepts of the plane and then find a fourth point on the plane as a check on our work. The intercepts are found by using the given procedure as follows:

| $x$-intercept | $y$-intercept | $z$-intercept |
|---|---|---|
| $2x + 3(0) + 0 = 6$ | $2(0) + 3y + 0 = 6$ | $2(0) + 3(0) + z = 6$ |
| $2x = 6$ | $3y = 6$ | $z = 6$ |
| $x = 3$ | $y = 2$ | |

We conclude that three points on the plane are (3, 0, 0), (0, 2, 0), and (0, 0, 6). To find a fourth point on the plane, we can assign arbitrary values to two of the variables in the equation and then solve for the value of the remaining variable. For example, if we assign the values $\frac{1}{2}$ and 1 to $x$ and $y$, respectively, then

$$2\left(\tfrac{1}{2}\right) + 3(1) + z = 6$$
$$z = 2$$

Thus a fourth point on this plane is $\left(\tfrac{1}{2}, 1, 2\right)$. The graph is sketched in Figure 8.3. This graph shows only the portion of the plane which lies in the first octant. ∎

### The Distance Formula in Three Dimensions

Recall that the distance between two points $(x_1, y_1)$ and $(x_2, y_2)$ is given by the formula

$$d = \sqrt{(x_2 - x_1)^2 + (y_2 - y_1)^2}$$

Although we will not prove it in this text, the distance formula generalizes easily to three dimensions by using the Pythagorean theorem.

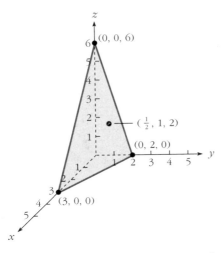

**FIGURE 8.3** The graph of $2x + 3y + z = 6$.

**THE DISTANCE FORMULA IN THREE DIMENSIONS**

The distance between two points $(x_1, y_1, z_1)$ and $(x_2, y_2, z_2)$ in three-dimensional space is given by

$$d = \sqrt{(x_2 - x_1)^2 + (y_2 - y_1)^2 + (z_2 - z_1)^2}$$

---

**EXAMPLE 4** Find the distance between $(1, -2, 4)$ and $(-3, 1, 2)$.

**Solution** Let $(x_1, y_1, z_1) = (1, -2, 4)$ and $(x_2, y_2, z_2) = (-3, 1, 2)$. Then
$$d = \sqrt{(-3 - 1)^2 + [1 - (-2)]^2 + (2 - 4)^2}$$
$$= \sqrt{16 + 9 + 4} = \sqrt{29}$$ ∎

### The Equation of a Sphere

The distance formula can be used to derive the equation of a sphere in three dimensions. Let $P(x, y, z)$ be an arbitrary point on the sphere that has center $(a, b, c)$ and radius $r$. Then the distance from the center of the sphere to $P$ is equal to $r$, that is,

$$\sqrt{(x - a)^2 + (y - b)^2 + (z - c)^2} = r$$

Squaring both sides of this equation gives us

**EQUATION OF A SPHERE WITH CENTER $(a, b, c)$ AND RADIUS $r$**

$$(x - a)^2 + (y - b)^2 + (z - c)^2 = r^2$$

---

**EXAMPLE 5** Find an equation of the sphere with center $(2, -3, 4)$ and radius 3; graph it.

**Solution** The center $(a, b, c)$ is $(2, -3, 4)$ and the radius $r$ is equal to 3. The equation is therefore

$$(x-2)^2 + [y-(-3)]^2 + (z-4)^2 = 3^2$$
$$(x-2)^2 + (y+3)^2 + (z-4)^2 = 9$$

This sphere is sketched in Figure 8.4.

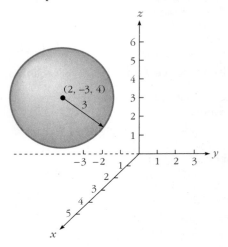

**FIGURE 8.4** The graph of a sphere with center $(2, -3, 4)$ and radius 3.

### Graphs of Other Surfaces—Traces

Sketching the graphs of surfaces other than planes and spheres in three-dimensional space can be a difficult task. In graphing a function $z = f(x, y)$, it is usually not feasible to plot the surface point by point because too many points would be required to obtain a reasonable sketch. Rather, we use what are called *traces* of the surface to obtain a skeletal structure of it. We will not dwell on sketching surfaces in three dimensions, but we'll illustrate the method of sketching using traces with one example.

> A *trace* of the graph of $z = f(x, y)$ is the graph of $f$ when one variable is held constant.

For example, if $z = 16 - x^2 - y^2$, then the trace in the plane $x = 0$ is the parabola $z = 16 - y^2$. Thus, when the surface with equation $z = 16 - x^2 - y^2$ is intersected with the plane with equation $x = 0$ (that is, the $yz$-plane), the resulting curve is a parabola. Let's continue this example.

**EXAMPLE 6** Sketch the graph of the surface $z = 16 - x^2 - y^2$ by finding traces in the planes $x = 0$, $y = 0$, $z = 16$, and $z = 0$.

**Solution** As we have already seen, the trace in the plane $x = 0$ is the parabola $z = 16 - y^2$ drawn in the $yz$-plane; see Figure 8.5. (Note that the scale on the $z$-axis is different from that on the $x$- and $y$-axes.) To obtain the trace in the plane $y = 0$, we have

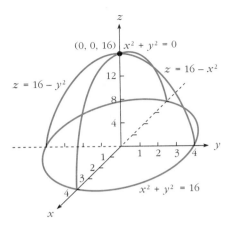

FIGURE 8.5 Some traces of the graph of $z = 16 - x^2 - y^2$.

$$z = 16 - x^2 - 0^2 = 16 - x^2$$

Once again this equation graphs as a parabola, this time in the $xz$-plane; see Figure 8.5.

If $z = 16$, we obtain

$$16 - x^2 - y^2 = 16$$
$$x^2 + y^2 = 0$$

Since the only values of $x$ and $y$ that satisfy $x^2 + y^2 = 0$ are $x = 0$ and $y = 0$, the trace of this surface in the plane $z = 16$ is the single point $(0, 0, 16)$; see Figure 8.5.

Finally, if $z = 0$, we obtain the trace in the $xy$-plane that is

$$16 - x^2 - y^2 = 0$$
$$x^2 + y^2 = 16$$

This equation represents a circle of radius 4 drawn in the $xy$-plane and centered at $(0, 0, 0)$; see Figure 8.5.

The resulting surface in Figure 8.5 is called a "circular paraboloid" because its traces are typically circles or parabolas. This skeletal structure may now be "fleshed in" to produce a final version of the surface, as shown in Figure 8.6.

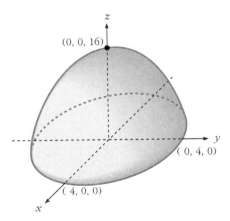

FIGURE 8.6 The final sketch of $z = 16 - x^2 - y^2$.

## SECTION 8.1
### SHORT ANSWER EXERCISES

1. In the function $V = lw^2$, the independent variables are ____ and ____ and the dependent variable is ____.
2. The domain of $f(x, y) = (x^2 + y^2)/(x - 2)$ does *not* include the ordered pair (____, y) for any value of y.
3. (True or False) The graph of $z = f(x, y)$ is a two-dimensional surface in three-dimensional space. ____
4. (True or False) The point $(3, -1, 2)$ lies in the first octant. ____
5. (True or False) Given that the points $(0, 0, 0)$, $(1, 2, 3)$, and $(-4, 7, 8)$ do not lie along the same striaght line, we can say that there is one and only one plane that passes through these points. ____
6. (True or False) The plane $-4x + 5y = 10$ has no z-intercept and thus is parallel to the z-axis. ____
7. Which of the following planes is parallel to the xy-plane in three-dimensional space? (a) $x = 5$; (b) $y = -2$; (c) $z = 4$; (d) none of these.
8. Which of the following points lie on the sphere with equation $(x - 2)^2 + (y + 3)^2 + (z - 4)^2 = 42$? (a) $(1, -1, 2)$; (b) $(0, 0, 0)$; (c) $(1, 1, -1)$; (d) none of these.
9. The trace of the graph of $x^2 + 4y^2 + z^2 = 16$ in the plane $y = 0$ is the circle with equation ____.
10. (True or False) In general, a trace of the graph of $z = f(x, y)$ is a curve, although a trace may also be a straight line or a point, and it is possible that certain traces may contain no points. ____

## SECTION 8.1
### EXERCISES

*In Exercises 1–10, find the indicated values of the given functions.*

1. $f(x, y) = 2x - 5y$
   (a) $f(1, 2)$ (b) $f(-1, 6)$ (c) $f(0, -5)$
2. $f(x, y) = x^2 - 2y$
   (a) $f(-1, 0)$ (b) $f(0, -2)$ (c) $f(3, 4)$
3. $f(x, y) = 2x^2 + 4y^2 - 3xy$
   (a) $f(1, 3)$ (b) $f(-1, 2)$ (c) $f(-2, 0)$
4. $f(x, y) = x^2 - 2y^2 - 2xy$
   (a) $f(-3, -3)$ (b) $f(1, 5)$ (c) $f(0, 1)$
5. $g(x, y) = \ln xy + \sqrt{2xy}$
   (a) $g(1, 1)$ (b) $g(-1, -2)$ (c) $g(e, e)$
6. $g(x, y) = e^x + 4e^{2y} - \sqrt[3]{x}$
   (a) $g(0, 1)$ (b) $g(1, \frac{1}{2})$ (c) $g(-8, -4)$
7. $f(x, y) = \dfrac{x^2 - y^2}{2x - 3y}$
   (a) $f(1, 0)$ (b) $f(-2, 3)$ (c) $f(a, 0)$, where $a \neq 0$
8. $f(x, y) = (xy - 3x^2)/\sqrt{x^2 + y^2}$
   (a) $f(-1, 2)$ (b) $f(1, 3)$ (c) $f(0, b)$, where $b > 0$
9. $f(x, y, z) = x^2 + y^2 + z^2 - 2xz + 10$
   (a) $f(1, 2, -3)$ (b) $f(0, 0, 0)$ (c) $f(-2, 4, 1)$

**10.** $f(x, y, z) = \dfrac{xy + xz}{x^2 + z}$

(a) $f(2, 0, -1)$  (b) $f(-1, 1, 1)$  (c) $f(0, 0, 1)$

*In each of Exercises 11–18, find the domain of the given function.*

**11.** $f(x, y) = 7x - 5y + 10$  
**12.** $f(x, y) = 2x - 5y + 3$  
**13.** $f(x, y) = \sqrt{x} + y^2$  
**14.** $f(x, y) = (2/y) + (3/x) - 10$  
**15.** $g(x, y) = 2xy/(x + 2y)$  
**16.** $g(x, y) = x\sqrt{y}/(x - 3)$  
**17.** $f(x, y, z) = \dfrac{2x + y}{x - z}$  
**18.** $f(x, y, z) = \dfrac{\sqrt{x} + y}{z}$

*In each of Exercises 19–26, sketch the graph of the given plane.*

**19.** $2x + 4y + z = 8$  
**20.** $x - y + 4z = 4$  
**21.** $x + y - 2z = 2$  
**22.** $3x - 2y - z = 6$  
**23.** $x + y - 2 = 0$  
**24.** $2x + z = 4$  
**25.** $2z - 6 = 0$  
**26.** $4y + 8 = 0$

*In each of Exercises 27–32, find the distance between the given points.*

**27.** $(0, 1, 2), (4, 2, 1)$  
**28.** $(-1, 2, 4), (0, 1, 4)$  
**29.** $(-2, -1, 1), (-3, 4, -4)$  
**30.** $(6, 4, -5), (-1, -3, 2)$  
**31.** $(-1, -2, -3), (a, b, c)$  
**32.** $(0, 4, -5), (a, b, c)$

*In Exercises 33–36, find an equation of each sphere with the given center and radius.*

**33.** center $(0, 0, 0)$, radius 4  
**34.** center $(0, 0, 0)$, radius 5  
**35.** center $(-1, 2, 4)$, radius 3  
**36.** center $(3, -2, -4)$, radius 10

*In Exercises 37–40, sketch the graph of each sphere with given equation.*

**37.** $x^2 + y^2 + z^2 = 9$  
**38.** $(x - 1)^2 + (y - 2)^2 + (z - 3)^2 = 16$  
**39.** $(x - 2)^2 + y^2 + (z + 3)^2 = 4$  
**40.** $(x + 2)^2 + (y - 3)^2 + (z + 4)^2 = 25$

*In Exercises 41–44, find the traces of each surface in the indicated planes and then sketch its graph in three dimensions.*

**41.** $f(x, y) = 25 - x^2 - y^2$  
$x = 0; y = 0; z = 0; z = 25$

**42.** $f(x, y) = 9 - x^2 - y^2$  
$x = 0, y = 0; z = 0; z = 9$

**43.** $f(x, y) = 2x^2 + y^2 + 1$  
$x = 0; y = 0; z = 1; z = 5$

**44.** $f(x, y) = x^2 + y^2$  
$x = 0; y = 0; z = 0; z = 4$

**45.** (Cost) A manufacturer produces two types of calculators: a standard model and a deluxe model. If variable costs are $8 for producing a standard calculator and $14 for a deluxe calculator and if fixed monthly costs are $5000, find (a) the general monthly cost function if the manufacturer produces $x$ standard and $y$ deluxe models per month; (b) the monthly cost of producing 4000 standard and 1500 deluxe models.

**46.** (Profit) The manufacturer in Exercise 45 sells the standard model calculator for $15 and the deluxe model for $24. Find (a) the monthly revenue function if the manufacturer sells $x$ standard and $y$ deluxe models per month; (b) the manufacturer's monthly profit function; (c) the monthly profit from the manufacture and sale of 5000 standard and 3000 deluxe models.

47. **(Demand)** If the unit price of product $A$ is $p$ dollars and the unit price of product $B$ is $q$ dollars, then the weekly demand for product $A$ is $A(p, q) = 400 - 6p + 5q$.
    (a) Find the weekly demand for product $A$ if $A$ is \$1 per unit and $B$ is \$2 per unit.
    (b) Suppose the price of $A$ is fixed at \$1.50 per unit. As the price of $B$ increases, does demand for $A$ increase or decrease?

48. **(Packaging)** A manufacturer makes cylindrical aluminum cans with volume given by $V(r, h) = \pi r^2 h$, where $r$ is the radius of the can and $h$ is its height. If $r$ and $h$ are measured in inches, find and interpret (a) $V(2, 6)$; (b) $V(3, 8)$.

49. **(Production function)** The output $z$ (in units) of a certain factory depends on the amount of labor, $x$ (in man-hours), and capital, $y$ (in thousands of dollars), according to the equation $z = 100x^{1/3}y^{2/3}$.
    (a) Compute the output of the factory if capital investment is \$64,000 and the amount of labor used is 2744 man-hours.
    (b) If capital investment is doubled to \$128,000 and the amount of labor is doubled to 5488 man-hours, what output will be produced?

50. **(Continuous interest)** If \$10,000 is invested in an account that pays $100r$ percent interest compounded continuously, then the amount in this account after $t$ years is $A(r, t) = 10000e^{rt}$. Find and interpret $A(0.06, 5)$, $A(0.075, 10)$, and $A(0.10, 15)$.

**Business and Economics**

51. A *Cobb–Douglas production function* has the form $z = f(x, y) = Cx^a y^{1-a}$, where $C$ and $a$ are constants, $C > 0$, $0 < a < 1$, and $x$ and $y$ are the amounts of labor and capital (respectively) needed to produce $z$ units of output. A production process that satisfies a Cobb–Douglas production function will have constant returns to scale in the sense that (for example) doubling both labor and capital will double output, tripling both labor and capital will triple output, and so forth.
    (a) Show that $f(2x, 2y) = 2f(x, y)$ and interpret this result.
    (b) Show that $f(3x, 3y) = 3f(x, y)$ and interpret this result.
    (c) If $n$ is any positive number, show that $f(nx, ny) = nf(x, y)$ and interpret this result.

## 8.2 PARTIAL DERIVATIVES

### First-Order Partial Derivatives

In Chapter 2 we learned that if $y = f(x)$ then $dy/dx$ represents the instantaneous rate of change of $y$ with respect to $x$, and so $dy/dx$ tells us the rate at which $f$ changes when $x$ changes. In dealing with a function of two variables $z = f(x, y)$, we will often be interested in the rate at which $f$ changes when the independent variables $x$ and $y$ change. In this situation we will concentrate on what are called the *partial derivatives* of $z = f(x, y)$.

The partial derivative of $f$ with respect to $x$ (denoted by $f_x(x, y)$ or $\partial f/\partial x$) represents the rate of change of $f$ with respect to $x$ *when $y$ is held fixed*. Similarly, the partial derivative of $f$ with respect to $y$ (denoted by $f_y(x, y)$ or $\partial f/\partial y$) represents the rate of change of $f$ with respect to $y$ *when $x$ is held fixed*. The formal definition of partial derivatives appears here.

**PARTIAL DERIVATIVES**

> If $z = f(x, y)$, then the partial derivative of $f$ with respect to $x$ is
> 
> $$f_x(x, y) = \frac{\partial f}{\partial x} = \lim_{\Delta x \to 0} \frac{f(x + \Delta x, y) - f(x, y)}{\Delta x}$$
> 
> provided the limit exists. The partial derivative of $f$ with respect to $y$ is
> 
> $$f_y(x, y) = \frac{\partial f}{\partial y} = \lim_{\Delta y \to 0} \frac{f(x, y + \Delta y) - f(x, y)}{\Delta y}$$
> 
> provided the limit exists.

We will not usually compute the partial derivatives using these definitions. Rather, observe that from the definition of $f_x(x, y)$, the partial derivative of $f$ with respect to $x$ can be found by assuming that $y$ is a constant and differentiating the function $f(x, y)$ with respect to $x$. Similarly, $f_y(x, y)$ can be found by assuming that $x$ is a constant and differentiating the function $f(x, y)$ with respect to $y$. Thus we can use the differentiation formulas presented in Chapter 3 for functions of one variable (the constant rule, the sum or difference rule, the product rule, the chain rule, etc.) to find partial derivatives of functions of more than one variable.

**EXAMPLE 1** If $f(x, y) = x^2 + 3y$, find $f_x(x, y)$ and $f_y(x, y)$.

**Solution** To find $f_x(x, y)$, we find the derivative of $f(x, y)$ with respect to $x$, treating $y$ as a constant. For this purpose we may view the function $f(x, y)$ as

$$f(x, y) = x^2 + \text{constant}$$

and so

$$f_x(x, y) = 2x + 0 = 2x$$

since the derivative of the "constant" ($3y$) with respect to $x$ is equal to zero.

To find $f_y(x, y)$, we find the derivative of $f(x, y)$ with respect to $y$, treating $x$ as a constant. Here we view the function $f(x, y)$ as $f(x, y) = \text{constant} + 3y$, so that $f_y(x, y) = 0 + 3 = 3$ because the derivative of the "constant" ($x^2$) with respect to $y$ is equal to zero. ∎

**EXAMPLE 2** For $z = x^2y - 2x + 5y^3 - 10$, find $\frac{\partial z}{\partial x}$ and $\frac{\partial z}{\partial y}$.

**Solution** To find $\partial z/\partial x$, we differentiate with respect to $x$ and treat $y$ as a constant. We get

$$\frac{\partial z}{\partial x} = 2xy - 2$$

To find $\partial z/\partial y$, we differentiate with respect to $y$ and treat $x$ as a constant. The result is

$$\frac{\partial z}{\partial y} = x^2 + 15y^2$$

The following notations are commonly employed to indicate the evaluation of partial derivatives at a point $(a, b)$ if $z = f(x, y)$:

$$f_x(a, b) = \left.\frac{\partial f}{\partial x}\right|_{(a,b)} = \left.\frac{\partial z}{\partial x}\right|_{(a,b)}$$

$$f_y(a, b) = \left.\frac{\partial f}{\partial y}\right|_{(a,b)} = \left.\frac{\partial z}{\partial y}\right|_{(a,b)}$$

### EXAMPLE 3

If $f(x, y) = x^2/y - x^2 y^3$, find (a) $f_y(x, y)$ and (b) $f_y(-1, 2)$.

**Solution**

(a) Let us write the given function as $f(x, y) = x^2 y^{-1} - x^2 y^3$. To find $f_y(x, y)$, we differentiate $f(x, y)$ with respect to $y$, assuming that $x$ is constant. We get

$$f_y(x, y) = -x^2 y^{-2} - 3x^2 y^2 = -\frac{x^2}{y^2} - 3x^2 y^2$$

(b) We compute $f_y(-1, 2)$ by replacing $x$ by $-1$ and $y$ by 2 in the expression for $f_y(x, y)$ we found in part (a). We obtain

$$f_y(-1, 2) = -\frac{(-1)^2}{2^2} - 3(-1)^2(2)^2 = -\frac{49}{4}$$

### EXAMPLE 4 (Participative)

If $z = xe^{xy} + \ln(x^2 + y^4)$, let's find (a) $\dfrac{\partial z}{\partial x}$; (b) $\left.\dfrac{\partial z}{\partial x}\right|_{(0,e)}$

**Solution**

(a) To find $\partial z/\partial x$, we differentiate $z$ with respect to $x$, assuming that ___ is constant. Recalling that

$$D_x[e^{f(x)}] = f'(x)e^{f(x)} \quad \text{and} \quad D_x[\ln f(x)] = \text{\_\_\_\_\_}$$

and using the product rule on the first term of $z$, we get

$$\frac{\partial z}{\partial x} = x \cdot \text{\_\_\_\_\_} + 1 \cdot e^{xy} + \frac{2x}{x^2 + y^4}$$

$$= xye^{xy} + e^{xy} + \frac{2x}{x^2 + y^4}$$

(b) We compute $\left.\dfrac{\partial z}{\partial x}\right|_{(0,e)}$ by replacing $x$ by 0 and $y$ by ___ in the expression for $\dfrac{\partial z}{\partial x}$ we found in part (a). We obtain

$$\left.\frac{\partial z}{\partial x}\right|_{(0,e)} = (0)(e)e^{(0)(e)} + e^{(0)(e)} + \text{\_\_\_\_\_} = \text{\_\_\_\_\_}$$

## A Geometric Interpretation

If $y = f(x)$ is a function of one variable, recall that in Chapter 2 we interpreted the derivative $f'(a)$ as the slope of the tangent line to the graph of $f$ at the point $(a, f(a))$. The partial derivatives $f_x(a, b)$ and $f_y(a, b)$ also have a geometric interpretation as the slopes of tangent lines to certain curves at the point $(a, b, f(a, b))$. This geometric interpretation is illustrated in Figures 8.7 and 8.8. As shown in Figure 8.7, $f_x(a, b)$ is the slope of the tangent line to the curve formed by the intersection between the surface $z = f(x, y)$ and the plane $y = b$. This partial derivative gives the rate of change of $f(x, y)$ in the $x$-direction if $y$ is held constant at $y = b$. In Figure 8.8, observe that $f_y(a, b)$ is the slope of the tangent line to the curve formed by the intersection of the surface $z = f(x, y)$ and the plane $x = a$. Thus $f_y(a, b)$ gives the rate of change of $f(x, y)$ in the $y$-direction when $x$ is held constant at $x = a$.

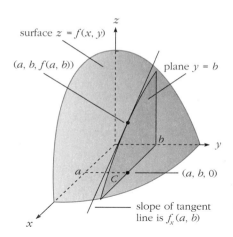

**FIGURE 8.7** The quantity $f_x(a, b)$ is the slope of the tangent line to the curve $C$ of intersection between the surface $z = f(x, y)$ and the plane $y = b$.

## Applications

Applications of the partial derivatives we will consider usually make use of the following facts.

---

If $z = f(x, y)$ then if $\Delta x$ and $\Delta y$ are small

(i) $f_x(a, b) \cdot \Delta x$ represents the approximate change in $f$ if $x$ changes from $a$ to $a + \Delta x$ and $y$ stays fixed at $b$;

(ii) $f_y(a, b) \cdot \Delta y$ represents the approximate change in $f$ if $y$ changes from $b$ to $b + \Delta y$ and $x$ stays fixed at $a$.

---

In particular, if we choose $\Delta x = 1$ in (i) and $\Delta y = 1$ in (ii), then if $z = f(x, y)$,

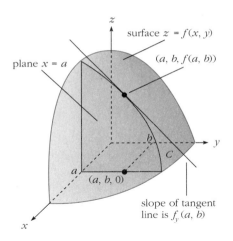

**FIGURE 8.8** The quantity $f_y(a, b)$ is the slope of the tangent line to the curve $C$ of intersection between the surface $z = f(x, y)$ and the plane $x = a$.

---

(iii) $f_x(a, b)$ represents the approximate change in $f$ if $x$ changes from $a$ to $a + 1$ and $y$ stays fixed at $b$;

(iv) $f_y(a, b)$ represents the approximate change in $f$ if $y$ changes from $b$ to $b + 1$ and $x$ stays fixed at $a$.

---

**EXAMPLE 5**   A company's monthly cost in dollars of producing $x$ units of product $A$ and $y$ units of product $B$ is given by $C(x, y) = 10,000 + 4x + 0.001x^2 + 5y + 0.0002y^2$. Find and interpret $C_x(100, 150)$ and $C_y(100, 150)$.

---

**SOLUTIONS TO EXAMPLE 4**   (a) $y, f'(x)/f(x), ye^{xy}$;   (b) $e, \dfrac{2(0)}{0^2 + e^4}, 1$

Solution

We compute that $C_x(x, y) = 4 + 0.002x$ and $C_y(x, y) = 5 + 0.0004y$. Thus

$$C_x(100, 150) = 4 + 0.002(100) = 4.20$$

We interpret this to mean that if we increase production of product A (when the production level is 100 units of A and 150 units of B) and hold production level of product B constant, then total cost will increase by about \$4.20 per additional unit of product A. It is instructive to compare this result with the exact cost of producing the 101st unit of product A while holding production of product B constant at 150 units. This exact cost is

$$\begin{aligned}C(101, 150) - C(100, 150) &= [10000 + 4(101) + 0.001(101)^2 + 5(150) + 0.0002(150)^2] \\ &\quad - [10000 + 4(100) + 0.001(100)^2 + 5(150) + 0.0002(150)^2] \\ &= 4.201\end{aligned}$$

Thus the approximation provided by $C_x(100, 150)$ is quite good and much more easily computed. We now find that

$$C_y(100, 150) = 5 + 0.0004(150) = 5.06$$

If we increase production of product B (when the production level is 100 units of A and 150 units of B) and hold the production level of product A constant, then total cost will increase by about \$5.06 per additional unit of product B. Let's compare this result to the exact cost of producing the 151st unit of product B while holding the production of product A constant at 100 units. This exact cost is

$$\begin{aligned}C(100, 151) - C(100, 150) &= [10000 + 4(100) + 0.001(100)^2 + 5(151) + 0.0002(151)^2] \\ &\quad - [10000 + 4(100) + 0.001(100)^2 + 5(150) + 0.0002(150)^2] \\ &= 5.0602\end{aligned}$$

Once again the partial derivative $C_y(100, 150)$ provides a good approximation to this exact value. ∎

In economics, some products can be substituted for one another because they satisfy similar consumer needs; these are called *substitute* or *competitive products*. Examples of substitute products are apples and pears. If the price of apples rises while the price of pears remains fixed, consumers will tend to purchase more of the substitute product, pears. In general, two products A and B are substitute or competitive products if a decrease in the quantity of A demanded can result in an increase in the quantity of B demanded.

Products that tend to be used together are called *complementary products*. Examples of complementary products are golf clubs and golf bags. If the price of golf clubs rises while the price of golf bags remains fixed, consumers will tend to purchase fewer golf clubs and hence fewer of the complementary golf bags. In general, two products A and B are complementary products if a decrease in the quantity of A demanded can result in a decrease in the quantity of B demanded.

We can determine whether two products are substitutes, complements, or neither by examining the signs of the partial derivatives of their demand functions. In particular, let $p$ denote the price per unit of product A and $s$ the price per unit of product B. Then if the demand function for product A is $x = f(p, s)$ and the demand function for product B is $y = g(p, s)$, and if A and B are substitutes, an increase in the price $p$ of $A$ (while the price $s$ of $B$ remains fixed) will result in an increase in demand $y$ for $B$, so that $\partial y/\partial p > 0$. Similarly, for substitute products, an increase in the price $s$ of $B$ (while the price $p$ of $A$ remains fixed) will result in an increase in demand $x$ for $A$, so that $\partial x/\partial s > 0$. Now, if $A$ and $B$ are complementary products, then an increase in the price $p$ of $A$ (while the price $s$ of $B$ remains fixed) will result in a decrease in quantity $y$ demanded of $B$, so that $\partial y/\partial p < 0$. Similarly, for complementary products, an increase in the price $s$ of $B$ (while the price $p$ of $A$ remains fixed) will result in a decrease in the quantity $x$ demanded for $A$, so that $\partial x/\partial s < 0$. We summarize these results.

**SUBSTITUTE AND COMPLEMENTARY PRODUCTS**

> $A$ and $B$ are substitutes if $\dfrac{\partial x}{\partial s} > 0$ and $\dfrac{\partial y}{\partial p} > 0$.
>
> $A$ and $B$ are complements if $\dfrac{\partial x}{\partial s} < 0$ and $\dfrac{\partial y}{\partial p} < 0$.
>
> $A$ and $B$ are neither substitutes nor complements if $\dfrac{\partial x}{\partial s}$ and $\dfrac{\partial y}{\partial p}$ differ in sign.

**EXAMPLE 6**

If $p$ is the price per pound for apples and $s$ is the price per pound for pears, then the weekly demand $x$ for apples at a certain grocery store is $x = 1000 - 150p + 200s$, and the weekly demand $y$ for pears at the store is $y = 750 + 100p - 120s$. Find and interpret $\partial x/\partial s$ and $\partial y/\partial p$.

**Solution**

We have $\partial x/\partial s = 200$ and $\partial y/\partial p = 100$. Since both $\partial x/\partial s$ and $\partial y/\partial p$ are positive, apples and pears are substitutes for one another. Further, since $\partial x/\partial s = 200$, for every \$1 increase in the price of pears, weekly demand for apples increases by about 200 pounds. Also, since $\partial y/\partial p = 100$, for every \$1 increase in the price of apples, weekly demand for pears increases by about 100 pounds. ∎

If the daily output $z$ (in units) of a company is related to both the number $x$ of labor-hours used per day and the number $y$ of dollars of capital used per day by $z = f(x, y)$, then $f_x(x, y)$ is called the *marginal productivity of labor* and $f_y(x, y)$ is called the *marginal productivity of capital*. The marginal productivity of labor $f_x(x, y)$ measures the approximate increase in production due to an increase of one labor-hour per day if capital remains fixed at $y$ dollars per day. The marginal productivity of capital $f_y(x, y)$ measures the approximate increase in production due to an increase of one dollar per day in capital if labor remains fixed at $x$ labor-hours per day.

**EXAMPLE 7**

The number $z$ of units produced daily by a certain factory is a function of the number $x$ of labor-hours used daily and the number $y$ of dollars of capital used daily; this function satisfies

$$z = f(x, y) = 0.2x^{4/5}y^{1/2}$$

Find and interpret $f_x(3125, 5625)$ and $f_y(3125, 5625)$.

Solution

We compute that $f_x(x, y) = 0.16x^{-1/5}y^{1/2}$, and so

$$f_x(3125, 5625) = 0.16(3125)^{-1/5}(5625)^{1/2} = 0.16(\tfrac{1}{5})(75) = 2.4$$

Thus the marginal productivity of labor when $x$ is 3125 labor-hours and $y$ is $5625 is 2.4 units of output per labor-hour. So if capital remains fixed at $5625 per day, increasing labor by one labor-hour per day will increase daily output by about 2.4 units.

Next we find that $f_y(x, y) = 0.1x^{4/5}y^{-1/2}$, so that

$$f_y(3125, 5625) = 0.1(3125)^{4/5}(5625)^{-1/2} = 0.1(625)(\tfrac{1}{75}) \approx 0.83$$

Therefore the marginal productivity of capital when $x$ is 3125 labor-hours and $y$ is $5625 is approximately 0.83 units of output per dollar of capital. So if labor remains constant at 3125 labor-hours, increasing capital by $1/day will increase daily output by about 0.83 units. ■

### Higher-Order Partial Derivatives

If $z = f(x, y)$, then $\partial z/\partial x = f_x(x, y)$ and $\partial z/\partial y = f_y(x, y)$ are sometimes called first-order partial derivatives because each of $\partial z/\partial x$ and $\partial z/\partial y$ involve only one differentiation with respect to an independent variable. Just as it is possible to compute higher-order derivatives of a function of one variable $y = f(x)$, we can compute higher-order partial derivatives of a function of more than one independent variable, such as $z = f(x, y)$. In the next section we will apply second-order partial derivatives to the problem of finding maxima and minima of functions of two variables. For now let's content ourselves with learning to find higher-order partial derivatives and mastering the associated notations.

**SECOND-ORDER PARTIAL DERIVATIVES**

If $z = f(x, y)$ is a function of two variables with first-order partial derivatives $\dfrac{\partial z}{\partial x} = f_x(x, y)$ and $\dfrac{\partial z}{\partial y} = f_y(x, y)$, then there are four second-order partial derivatives, given by

$$\frac{\partial^2 z}{\partial x^2} = \frac{\partial}{\partial x}\left(\frac{\partial z}{\partial x}\right) = f_{xx}(x, y)$$

$$\frac{\partial^2 z}{\partial y \partial x} = \frac{\partial}{\partial y}\left(\frac{\partial z}{\partial x}\right) = f_{xy}(x, y)$$

$$\frac{\partial^2 z}{\partial x \partial y} = \frac{\partial}{\partial x}\left(\frac{\partial z}{\partial y}\right) = f_{yx}(x, y)$$

$$\frac{\partial^2 z}{\partial y^2} = \frac{\partial}{\partial y}\left(\frac{\partial z}{\partial y}\right) = f_{yy}(x, y)$$

The first- and second-order partial derivatives of $z = f(x, y)$ are summarized in the following diagram.

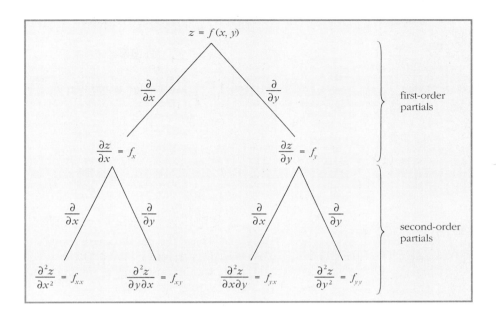

For example, the notation $\dfrac{\partial^2 z}{\partial y\,\partial x}$ means that we are to compute $\dfrac{\partial}{\partial y}\left(\dfrac{\partial z}{\partial x}\right)$. That is, first we find the partial derivative of $z$ with respect to $x$ and then we find the partial derivative of this result with respect to $y$. Observe that the order of the variables $x$ and $y$ in the notations $\dfrac{\partial^2 z}{\partial y\,\partial x}$ and $f_{xy}$ are different, even though both refer to the *same* second-order partial derivative. In each case, we differentiate with respect to $x$ first and then with respect to $y$, because the $x$ is closest to the $f$.

EXAMPLE 8 (Participative)

For $z = f(x, y) = 2x^3 + 3x^2 y - 4y^2$, let's find (a) $\dfrac{\partial^2 z}{\partial y\,\partial x}$ and $\dfrac{\partial^2 z}{\partial x\,\partial y}$; (b) $f_{yy}(x, y)$; (c) $f_{xx}(-1, 2)$.

Solution

(a) First, we compute

$$\dfrac{\partial z}{\partial x} = 6x^2 + 6xy \quad \text{and} \quad \dfrac{\partial z}{\partial y} = \underline{\qquad}$$

Then

$$\dfrac{\partial^2 z}{\partial y\,\partial x} = \dfrac{\partial}{\partial y}\left(\dfrac{\partial z}{\partial x}\right) = \dfrac{\partial}{\partial y}(6x^2 + 6xy) = \underline{\qquad}$$

and

$$\dfrac{\partial^2 z}{\partial x\,\partial y} = \dfrac{\partial}{\partial x}\left(\dfrac{\partial z}{\partial y}\right) = \dfrac{\partial}{\partial x}(\underline{\qquad}) = \underline{\qquad}$$

(b) From part (a) we know that $f_y(x, y) = \partial z/\partial y =$ _____, so that

$$f_{yy}(x, y) = \frac{\partial}{\partial y}\left(\frac{\partial z}{\partial y}\right) = \frac{\partial}{\partial y}(3x^2 - 8y) = \underline{\quad}$$

(c) From part (a) we know that $f_x(x, y) = \partial z/\partial x = 6x^2 + 6xy$. Thus

$$f_{xx}(x, y) = \frac{\partial}{\partial x}\left(\frac{\partial z}{\partial x}\right) = \frac{\partial}{\partial x}(\underline{\qquad}) = \underline{\qquad}$$

We conclude that

$$f_{xx}(-1, 2) = 12(-1) + 6(\underline{\quad}) = \underline{\quad}$$

■

Observe that in Example 8, the mixed partials $\dfrac{\partial^2 z}{\partial y \partial x} = f_{xy}$ and $\dfrac{\partial^2 z}{\partial x \partial y} = f_{yx}$ were equal. There is a result called Young's theorem that says that the mixed partial derivatives $f_{xy}$ and $f_{yx}$ are equal at any point $(x_0, y_0)$ provided that both $f_{xy}$ and $f_{yx}$ are continuous at that point. In this text, the functions that we use will always satisfy $f_{xy}(x, y) = f_{yx}(x, y)$.

## SECTION 8.2
**SHORT ANSWER EXERCISES**

1. (True or False) If $f(x, y) = 10x^2 + 7x + 5$, then $f_y(x, y) = 0$. _____
2. (True or False) $\dfrac{\partial}{\partial x}(xy) = x + y$. _____
3. (True or False) For a certain function $z = f(x, y)$, it is known that $f_x(x, y) = 0$. We can conclude that $f(x, y)$ is a constant function. _____
4. Given that $f(1, 2) = 4$ and $f_x(1, 2) = 2$, an approximate value of $f(1.5, 2)$ is _____.
5. A plane and a surface in three dimensions can intersect in (a) a sphere; (b) a curve; (c) a circular paraboloid; (d) none of the above.
6. If $f_x(x, y) = 2x$ and $f_y(x, y) = 2y$, then a possible formula for $f(x, y)$ is
   (a) $f(x, y) = x^2 + 2y$; (b) $f(x, y) = y^2 + 2x$; (c) $f(x, y) = x^2 + y^2 - 10$; (d) $f(x, y) = x^2y + y^2x$.
7. A company's monthly cost in dollars of producing $x$ units of product $A$ and $y$ units of product $B$ is $C(x, y)$, where $C_x(600, 800) = 6.50$ and $C_y(600, 800) = 7.20$. If the company can sell a unit of product $A$ for $\$8.50$ and a unit of $B$ for $\$9.00$ at its current production levels of 600 units of $A$ and 800 units of $B$, is it more profitable to expand production of product $A$ or of product $B$? _____
8. If $x$ pounds of beef are demanded per day by consumers when the price of beef is $p$ dollars per pound and $y$ pounds of pork are demanded per day when the price of pork is $s$ dollars per pound, then $\partial x/\partial s$ is (positive, negative) and $\partial y/\partial p$ is (positive, negative), since beef and pork are (substitute, complementary) products.
9. If $f(x, y) = x + y^5$ and $g(x, y) = y + x^5$, then $f_{xx} + g_{yy} = $ _____.

**SOLUTIONS TO EXAMPLE 8** (a) $3x^2 - 8y$, $6x$, $3x^2 - 8y$, $6x$; (b) $3x^2 - 8y$, $-8$; (c) $6x^2 + 6xy$, $12x + 6y$, $2$, $0$

**10.** Given a production function $z = f(x, y)$, where $z$ is the output produced by using $x$ labor-hours and $y$ dollars of capital, if $f_x(100, 200) > f_y(100, 200)$, in order to expand production by a small amount, should we employ more labor or more capital? _____

## SECTION 8.2
## EXERCISES

*In Exercises 1–16, find $f_x(x, y)$ and $f_y(x, y)$.*

**1.** $f(x, y) = x^2 + y^2 - 3$
**2.** $f(x, y) = 3x^3 - 2y^4 + 5$
**3.** $f(x, y) = 4xy + 3x^2y^2 - x$
**4.** $f(x, y) = 6x^3 - 7xy^2 + y^2$
**5.** $f(x, y) = (x/y) + y^4$
**6.** $f(x, y) = (2y/x) - 7xy$
**7.** $f(x, y) = \sqrt{x^2 + y^2}$
**8.** $f(x, y) = \sqrt[3]{2x^2 + 7y^4 + 3}$
**9.** $f(x, y) = 2x\sqrt{y} - \sqrt{x}$
**10.** $f(x, y) = 3\sqrt{xy} + 5\sqrt{y}$
**11.** $f(x, y) = \dfrac{x - 2y}{2x + 3y}$
**12.** $f(x, y) = \dfrac{x^2 + 15xy}{2y^2 - 3}$
**13.** $f(x, y) = e^{2x-y}$
**14.** $f(x, y) = xe^{2y+3x}$
**15.** $f(x, y) = ye^x + \ln(x + 2y)$
**16.** $f(x, y) = e^y \ln x + \ln(x^2 + y^2)$

**17.** Find $f_x(1, 2)$ and $f_y(-3, 4)$ if $f(x, y) = 12x^2y - y^3$.

**18.** Find $f_x(-2, 3)$ and $f_y(-1, -2)$ if $f(x, y) = xy^3 - 7xy + 10$.

**19.** Find $\left.\dfrac{\partial z}{\partial x}\right|_{(1,3)}$ and $\left.\dfrac{\partial z}{\partial y}\right|_{(-2,4)}$ if $z = xy\sqrt{x^2 + 3y^2}$.

**20.** Find $\left.\dfrac{\partial z}{\partial x}\right|_{(-1,2)}$ and $\left.\dfrac{\partial z}{\partial y}\right|_{(3,4)}$ if $z = 3x + y - x^2 + 2y^2 + 6$.

**21.** Find $f_x(-4, 3)$ if $f(x, y) = \dfrac{x^2}{y} + \dfrac{y^2}{x}$.
**22.** Find $f_y(6, -2)$ if $f(x, y) = \dfrac{x + y + 1}{2x + 3y}$.

**23.** Find $\left.\dfrac{\partial z}{\partial y}\right|_{(0,1)}$ if $z = e^{xy} - \ln(x + y)$.
**24.** Find $\left.\dfrac{\partial z}{\partial x}\right|_{(1,1)}$ if $z = x^2e^y - \ln\left(\dfrac{x}{y}\right)$.

*In Exercises 25–34, find all the second-order partial derivatives of the given function.*

**25.** $f(x, y) = 2x - 3y^2 + xy$
**26.** $f(x, y) = 3y^3 - 7x^2 + x^2y$
**27.** $f(x, y) = x^3y - y^3x$
**28.** $f(x, y) = 2x^2y^2 + 3xy^2$
**29.** $f(x, y) = x/y$
**30.** $f(x, y) = 2x/(y + 1)$
**31.** $f(x, y) = \sqrt{2x + 1}/(5y)$
**32.** $f(x, y) = \sqrt{4x + 2y}/(3x^2)$
**33.** $f(x, y) = xe^y + x \ln x$
**34.** $f(x, y) = ye^{x^2} - \ln(x^2 + y^4)$

**35.** If $f(x, y) = x^2 - y^3 + 3xy^2$, find **(a)** $f_{xx}(-3, 1)$; **(b)** $f_{xy}(-1, 2)$; **(c)** $f_{yy}(3, 0)$.

**36.** If $f(x, y) = (x/y^2) - 7x^3y$, find **(a)** $f_{xx}(2, -1)$; **(b)** $f_{yx}(0, 1)$; **(c)** $f_{yy}(-1, 3)$.

**37.** If $z = x^2\sqrt{y} + \dfrac{\ln x}{y}$, find **(a)** $\left.\dfrac{\partial^2 z}{\partial x^2}\right|_{(1,1)}$; **(b)** $\left.\dfrac{\partial^2 z}{\partial x \partial y}\right|_{(e,4)}$; **(c)** $\left.\dfrac{\partial^2 z}{\partial y^2}\right|_{(1,9)}$.

**38.** If $z = e^{x^2-y^2} + \ln(x + 2y)$, find **(a)** $\left.\dfrac{\partial^2 z}{\partial x^2}\right|_{(1,0)}$; **(b)** $\left.\dfrac{\partial^2 z}{\partial y \partial x}\right|_{(3,-1)}$; **(c)** $\left.\dfrac{\partial^2 z}{\partial y^2}\right|_{(0,\frac{1}{2})}$.

**39.** (Cost) A manufacturer's cost in dollars of producing $x$ units of product $A$ and $y$ units of product $B$ is $C(x, y) = 8000 + 20x + 25y$. Find and interpret **(a)** $C_x(200, 150)$ and **(b)** $C_y(200, 150)$. Compare each of these values with the exact cost of producing the 201st and the 151st units.

**40.** (Cost to produce microcomputer) The cost in dollars of producing a certain microcomputer depends on the cost of labor and materials. If the cost of a labor-hour is $x$ dollars and the cost of material is $y$ dollars, the cost $C$ of making a microcomputer is $C(x, y) = 5x + 7y^2 - 2xy + 12$. Find and interpret (a) $\left.\frac{\partial C}{\partial x}\right|_{(5,10)}$ and (b) $\left.\frac{\partial C}{\partial y}\right|_{(5,10)}$.

**41.** (Complementary products) If $x$ sets of golf clubs and $y$ golf bags are manufactured and sold per month, the cost and revenue functions of a manufacturer are $C(x, y) = 6000 + 65x + 12y$ and $R(x, y) = 110x + 42y + 0.001xy$, respectively.
(a) Find the profit function $P(x, y)$. (b) Find and interpret $C_x(300, 250)$.
(c) Find and interpret $R_x(300, 250)$. (d) Find and interpret $P_x(300, 250)$.

**42.** (Production) The production function of a certain company is given by $f(x, y) = x^2 + \frac{1}{2}x + 50y^2 + 20y + \frac{1}{2}xy$, where $x$ is labor-hours and $y$ is thousands of dollars of capital. Find and interpret (a) the marginal productivity of labor when $x = 100$ and $y = 200$; and (b) the marginal productivity of capital when $x = 100$ and $y = 200$.

**43.** (Production) Given the Cobb–Douglas production function (see Exercise 51 of Section 8.1) $f(x, y) = 25x^{4/5}y^{1/5}$, (a) find the marginal productivity of labor when the labor level is $x = 32$ and the capital level is $y = 243$; (b) find the marginal productivity of capital when the capital level is $y = 243$ and the labor level is $x = 32$.

**44.** (Demand) The monthly demand for a certain type of athletic shoe is $Q(x, y)$, where $x$ is the list price of a pair of shoes and $y$ is the amount of money spent on advertising. What is the likely sign of $Q_x(x, y)$? Explain. What is the likely sign of $Q_y(x, y)$? Explain.

## Business and Economics

**45.** (Demand) The monthly demand $D$ for a certain type of compact disc player is a function of its list price $x$, the amount $y$ of advertising of the product, and the prices $p$ and $s$ of its two major competitors, where

$$D(x, y, p, s) = 0.5\sqrt{y} - 0.2x - 0.001x^2 + 0.5\sqrt{p} + 0.6\sqrt{s} + 4000$$

If $x$, $y$, $p$, and $s$ are all measured in dollars, find and interpret
(a) $D_x(150, 2500, 144, 169)$; (b) $D_y(150, 2500, 144, 169)$;
(c) $D_p(150, 2500, 144, 169)$; (d) $D_s(150, 2500, 144, 169)$.

**46.** (Retail sales) Monthly dollar sales of a product in a company's product line is a function of monthly outlays for television advertising, $x$, for newspaper advertising, $y$, and the list price $p$ of the product, according to

$$S(x, y, p) = 10000 + 2x + y + 0.0001x^2 + 0.0002y^2 + 0.004xy - 4p - 0.01p^2$$

where $x$, $y$, and $p$ are all measured in dollars. Currently the price of this product is $50 and $5000 is being spent on television advertising and $2000 on newspaper advertising. If the price next month is to remain fixed at $50, will sales increase more if the company increases television advertising or if it increases newspaper advertising? Use partial derivatives to support and explain your answer.

**47.** (Demand) If $p$ is the price per container of brand $A$ of a certain floor wax and $s$ is the price per container of brand $B$, then the weekly demand $x$ for brand $A$ in a department store is $x = 800 - 30p + 20s$, and the weekly demand $y$ for brand $B$ in that store is $y = 1200 + 15p - 25s$.
(a) Find and interpret $\frac{\partial x}{\partial s}$ and $\frac{\partial y}{\partial p}$.
(b) Are these two products complements, substitutes, or neither?

**48. (Demand)** Monthly demand for product A is $x = 2000 - \frac{1}{2}\sqrt{p} - \frac{1}{5}\sqrt{s}$, where $p$ is the price per unit of $A$ and $s$ is the price per unit of another product, $B$. If monthly demand for $B$ is $y = 5000 - \frac{1}{4}\sqrt{p} - \frac{1}{2}\sqrt{s}$, are products $A$ and $B$ complements, substitutes, or neither? Explain.

## 8.3 MAXIMA AND MINIMA

In Chapter 4 we learned how to find maxima and minima of functions of one variable. We will address the theory of finding extrema of functions of two variables in this section. We will also consider some applications of this theory at the end of the section.

We begin by assuming that $z = f(x, y)$ is a function of two variables and all second-order partial derivatives $f_{xx}, f_{yy}, f_{xy}$, and $f_{yx}$ are defined in an open circular domain of the $xy$-plane. This assumption guarantees that the surface $z = f(x, y)$ is "smooth," i.e., has no sharp points, breaks, or edges; see Figure 8.9.

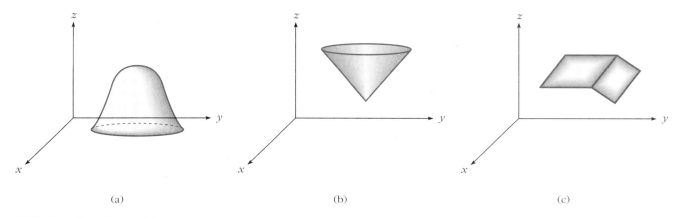

**FIGURE 8.9** The surface in (a) represents a smooth surface. The surface in (b) has a sharp point, and the surface in (c) has an edge.

A surface with a relative maximum point is illustrated in Figure 8.10 and a surface with a relative minimum point is illustrated in Figure 8.11.

Informally, if you were standing at the relative maximum point $(a, b, f(a, b))$ on the surface of Figure 8.10, you would be at the "top of the mountain," and if you were to take a small step *in any direction*, your altitude with respect to the $xy$-plane would decrease. Similarly, if you were standing at the relative minimum point $(a, b, f(a, b))$ on the surface of Figure 8.11, you would be in the "bottom of the valley," and if you were to take a small step *in any direction*, your altitude with respect to the $xy$-plane would increase.

The graph in Figure 8.10 has the property that if $(x, y, 0)$ is *any* point in the open circular domain you see in the $xy$-plane of the illustration, then $f(x, y) \leq f(a, b)$. That is, the point $(a, b, f(a, b))$ is the highest point on the surface in a small neighborhood of $(a, b, f(a, b))$. Similarly, the graph in Figure 8.11 has the property that if $(x, y, 0)$ is *any* point in the open circular domain you see in the $xy$-plane of the illustration, then $f(a, b) \leq f(x, y)$. Thus the point

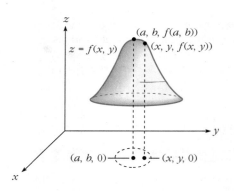

**FIGURE 8.10** The graph of the surface $z = f(x, y)$ has a relative maximum point at $(a, b, f(a, b))$.

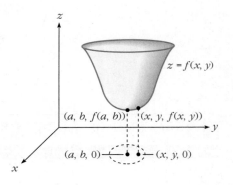

**FIGURE 8.11** The graph of the surface $z = f(x, y)$ has a relative minimum point at $(a, b, f(a, b))$.

$(a, b, f(a, b))$ is the lowest point on the surface in a small neighborhood of $(a, b, f(a, b))$. We use these ideas in the formal definition of relative extrema.

**RELATIVE EXTREMA OF A FUNCTION OF TWO VARIABLES**

(i) A point $(a, b, f(a, b))$ is a *relative maximum point of the surface representing* $f(x, y)$ if there is an open circular domain in the *xy*-plane with center $(a, b, 0)$ such that for any point $(x, y, 0)$ inside this domain,

$$f(x, y) \leq f(a, b)$$

(ii) A point $(a, b, f(a, b))$ is a *relative minimum point of the surface representing* $f(x, y)$ if there is an open circular domain in the *xy*-plane with center $(a, b, 0)$ such that for any point $(x, y, 0)$ inside this domain,

$$f(a, b) \leq f(x, y)$$

By inspecting Figure 8.12, you can see that a necessary condition for a relative maximum of $z = f(x, y)$ to occur at $(a, b)$ is that the tangent lines to the surface in both the *x*- and *y*-directions must be horizontal. That is, we must have $f_x(a, b) = 0$ and $f_y(a, b) = 0$. A similar result is true if a relative minimum occurs at $(a, b)$. If a point $(a, b)$ in the domain of $f$ has the property $f_x(a, b) = 0$ and $f_y(a, b) = 0$, we call it a *critical point* of $z = f(x, y)$. We can therefore state the following result.*

If $(a, b, f(a, b))$ is a relative extreme point of $z = f(x, y)$, then $(a, b)$ is a critical point of $f(x, y)$.

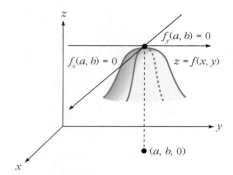

**FIGURE 8.12** If $(a, b, f(a, b))$ is a relative maximum point, then $f_x(a, b) = 0$ and $f_y(a, b) = 0$.

*In this text we will not discuss critical points for which one or both of the partial derivatives $f_x(a, b)$ and $f_y(a, b)$ do not exist, since in that case at least one of the second partials does not exist and the resulting surface is not smooth.

## EXAMPLE 1

Find the critical points of $f(x, y) = x^2 - 4x + y^2 - 6y + 17$.

### Solution

We compute that $f_x = 2x - 4$ and $f_y = 2y - 6$. Equating $f_x$ and $f_y$ to zero yields $2x - 4 = 0$ and $2y - 6 = 0$, so that $x = 2$ and $y = 3$. We conclude that $(2, 3)$ is the only critical point of $f(x, y)$. The graph of $f(x, y)$, which is sketched in Figure 8.13, reveals that a relative minimum of $f(x, y)$ occurs at the critical point $(2, 3)$. The relative minimum value of $f(x, y)$ at $(2, 3)$ is

$$f(2, 3) = 2^2 - 4(2) + 3^2 - 6(3) + 17 = 4$$

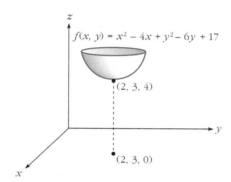

**FIGURE 8.13** The graph of $f(x, y) = x^2 - 4x + y^2 - 6y + 17$ near the point $(2, 3, 4)$ shows that this point is a relative minimum point.

## EXAMPLE 2

Find the critical points of $f(x, y) = y^2 - x^2$.

### Solution

First, we compute that $f_x = -2x$ and $f_y = 2y$. Next, we set each of these partial derivatives equal to zero to obtain $-2x = 0$ and $2y = 0$, so that $x = 0$ and $y = 0$.

Thus, $(0, 0)$ is the critical point of $f(x, y) = y^2 - x^2$. In this case the (saddle-shaped) graph of $f(x, y)$ which is sketched in Figure 8.14 indicates that no relative extremum occurs at the critical point $(0, 0)$. This is true because if you were standing at $(0, 0, 0)$ on the surface and took a small step parallel to the $y$-axis, you would step up along the parabola $z = y^2$, increasing your altitude with respect to the $xy$-plane. On the other hand, if you took a small step parallel to the $x$-axis, you would step down along the parabola $z = -x^2$, decreasing your altitude with respect to the $xy$-plane. Indeed, the point $(0, 0, 0)$ is a relative minimum for the parabola $z = y^2$ and a relative maximum for the parabola $z = -x^2$, but it is not a relative extremum for the surface $z = y^2 - x^2$. The point $(0, 0, 0)$ is called a *saddle point* of $f(x, y) = y^2 - x^2$.

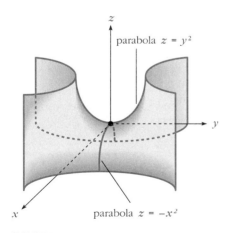

**FIGURE 8.14** The graph of $z = y^2 - x^2$ has no relative extremum at the critical point $(0, 0)$. The point $(0, 0, 0)$ is called a *saddle point* of the surface.

Just as in the case of a function of one variable, Example 2 illustrates that although the critical points give all possible candidates of points where relative extrema of $f(x, y)$ can occur, not every candidate critical point need result in a relative extremum.

The next result (which is analogous to the second derivative test for a function of one variable) is stated without proof and can be used to classify critical points as relative maxima, relative minima, or saddle points.

**SECOND PARTIALS TEST FOR RELATIVE EXTREMA**

> Suppose that all of the second partials of $f(x, y)$ are continuous in an open circular domain in the $xy$-plane with center $(a, b, 0)$, and suppose further that $f_x(a, b) = f_y(a, b) = 0$. Let
>
> $$D(x, y) = f_{xx}(x, y) \cdot f_{yy}(x, y) - [f_{xy}(x, y)]^2$$
>
> 1. If $D(a, b) > 0$ and $f_{xx}(a, b) > 0$, then $f(a, b)$ is a relative minimum value.
> 2. If $D(a, b) > 0$ and $f_{xx}(a, b) < 0$, then $f(a, b)$ is a relative maximum value.
> 3. If $D(a, b) < 0$, then $(a, b, f(a, b))$ is a saddle point.
> 4. If $D(a, b) = 0$, the test fails and no conclusion can be made about the nature of the graph at $(a, b, f(a, b))$.

The function $D(x, y)$ of the second partials test is called the *discriminant* of $f(x, y)$ because it *discriminates*: it allows us to determine the nature of a critical point $(a, b)$. Because of symmetry, we may use $f_{yy}$ in Cases 1 and 2 instead of $f_{xx}$ to reach the same conclusions.

---

### EXAMPLE 3

Use the second partials test to verify that the critical point $(2, 3)$ of $f(x, y) = x^2 - 4x + y^2 - 6y + 17$ found in Example 1 corresponds to a relative minimum.

**Solution**

Recall that $f_x = 2x - 4$ and $f_y = 2y - 6$. Computing second partials, we get

$$f_{xx} = 2, \quad f_{yy} = 2, \quad \text{and} \quad f_{xy} = 0.$$

Thus the discriminant function $D(x, y)$ is

$$D(x, y) = f_{xx} \cdot f_{yy} - [f_{xy}]^2 = (2)(2) - 0^2 = 4$$

So, evaluating $D(x, y)$ at the critical point $(2, 3)$, we get $D(2, 3) = 4 > 0$. Also, since $f_{xx}(x, y) = 2$, observe that $f_{xx}(2, 3) = 2 > 0$. Therefore Case 1 of the second partials test applies and $(2, 3)$ corresponds to a relative minimum. ∎

---

### EXAMPLE 4

Find any relative extrema or saddle points of $f(x, y) = xy - x^2 - y^2 - x + 11y + 2$.

**Solution**

We compute that $f_x = y - 2x - 1$ and $f_y = x - 2y + 11$. To find the critical point(s), we set these partial derivatives equal to zero and solve the resulting system of two equations in the two unknowns $x$ and $y$:

$$y - 2x - 1 = 0$$
$$x - 2y + 11 = 0$$

Solving the first equation for $y$, we get $y = 2x + 1$. Substituting this into the second equation yields

$$x - 2(2x + 1) + 11 = 0$$
$$x - 4x - 2 + 11 = 0$$
$$-3x = -9$$
$$x = 3$$

Now, substituting $x = 3$ into the relationship $y = 2x + 1$, we get $y = 2(3) + 1 = 7$. The only critical point of $f(x, y)$ is, therefore, $(3, 7)$. To determine the nature of the graph of $f$ at the critical point, we use the second partials test. We compute

$$f_{xx} = -2, \quad f_{yy} = -2, \quad \text{and} \quad f_{xy} = 1$$

The discriminant $D(x, y)$ is

$$D(x, y) = f_{xx} \cdot f_{yy} - [f_{xy}]^2 = (-2)(-2) - 1^2 = 3$$

Thus $D(3, 7) = 3 > 0$. Also, since $f_{xx}(3, 7) = -2 < 0$, Case 2 of the second partials test indicates that the critical point $(3, 7)$ corresponds to a relative maximum. ■

### EXAMPLE 5 (Participative)

Let's find any relative extrema or saddle points of

$$f(x, y) = x^3 - 3xy - y^3.$$

**Solution**

To begin, we compute

$$f_x = 3x^2 - 3y$$

$$f_y = \underline{\hspace{2cm}}$$

To find the critical points, we equate each of these partials to zero to obtain the system of equations

$$3x^2 - 3y = 0$$
$$-3x - 3y^2 = 0$$

Solving the first of these equations for $y$, we get $y = \underline{\hspace{1cm}}$. Substituting this expression for $y$ into the second equation gives us

$$-3x - 3(\underline{\hspace{1cm}})^2 = 0$$
$$-3x - 3x^4 = 0$$

We proceed to solve this equation for $x$:

$$-3x(\underline{\hspace{2cm}}) = 0 \quad \text{(Factoring)}$$
$$-3x = 0 \quad \text{or} \quad 1 + x^3 = 0 \quad \text{(Setting each factor equal to 0)}$$
$$x = \underline{\hspace{1cm}} \quad \text{or} \quad x = \underline{\hspace{1cm}}$$

These values represent the $x$-coordinates of the critical points of $f(x, y)$. To find the corresponding $y$-coordinates, we use the relationship $y = x^2$. In fact, if $x = 0$ then $y = 0^2 = 0$, and if $x = -1$ then $y = (-1)^2 = 1$. The two critical points of $f(x, y)$ are therefore $(0, 0)$ and $\underline{\hspace{2cm}}$.

To determine the nature of the graph of $f(x, y)$ at the critical points, we use the second partials test. We compute

$$f_{xx} = 6x, \quad f_{yy} = -6y, \quad \text{and} \quad f_{xy} = \underline{\hspace{1cm}}$$

The discriminant function is

$$D(x, y) = (6x)(\underline{\hspace{1cm}}) - (-3)^2 = -36xy - 9$$

For the critical point (0, 0) we find that $D(0, 0) = $ _____ $< 0$. Case _____ of the second partials test can now be applied to conclude that a saddle point of $f(x, y)$ occurs at (0, 0). Since $f(0, 0) = 0^3 - 3(0)(0) - 0^3 = 0$, the saddle point on this surface is (0, 0, 0). For the critical point (−1, 1), we compute $D(-1, 1) = $ _____ $> 0$. Also, since $f_{xx}(-1, 1) = $ _____ $< 0$, Case 2 of the second partials test indicates that the critical point (−1, 1) corresponds to a relative _____. The relative maximum value of $f(x, y)$ is

$$f(-1, 1) = (-1)^3 - 3(-1)(1) - (1)^3 = \underline{\quad}$$

## Applications

We will now look at two examples of applications of finding relative extrema of a function of two variables $f(x, y)$. In most applications the points of interest are generally the *absolute* extrema of $f(x, y)$, rather than the relative extrema. Since we will not discuss methods for finding absolute extrema of functions of two variables, we will assume for purposes of the following examples and exercises that the absolute extrema of $f(x, y)$ can be found among the relative extrema.

**EXAMPLE 6**

The marketing department of a company has determined that if the company spends $x$ thousand dollars on television advertising per month and $y$ thousand dollars on newspaper and magazine advertising per month, then monthly sales in hundreds of units is

$$S(x, y) = -0.4x^2 + 80x - 0.01y^3 + 75y$$

Find the advertising mix that maximizes monthly unit sales and find the maximum monthly unit sales.

**Solution**

We compute that $S_x = -0.8x + 80$ and $S_y = -0.03y^2 + 75$. To find the critical point(s) of $S(x, y)$, we must solve the system of equations

$$-0.8x + 80 = 0$$
$$-0.03y^2 + 75 = 0$$

From the first equation we get $x = 100$ and from the second we get $y = 50$, so that the only critical point of $S(x, y)$ is (100, 50). To verify that a relative maximum of $S(x, y)$ occurs at this point we find

$$S_{xx} = -0.8, \quad S_{yy} = -0.06y, \quad \text{and} \quad S_{xy} = 0$$

The discriminant function is therefore

$$D(x, y) = (-0.8)(-0.06y) - (0)^2 = 0.048y$$

Note that $D(100, 50) = 0.048(50) = 2.4 > 0$ and $S_{xx}(100, 50) = -0.8 < 0$, so

---

**SOLUTIONS TO EXAMPLE 5**     $-3x - 3y^2, \dot{x}^2, x^2, 1 + x^3, 0, -1, (-1, 1), -3, -6y, -9, 3, 27, -6$, maximum, 1

that by Case 2 of the second partials test, the critical point (100, 50) corresponds to a relative maximum. We conclude that to maximize monthly unit sales, the company should spend $100,000 on television advertising and $50,000 on newspaper and magazine advertising. The maximum monthly unit sales will then be

$$S(100, 50) = -0.4(100)^2 + 80(100) - 0.01(50)^3 + 75(50)$$
$$= 6500 \text{ (hundred units)} = 650{,}000 \text{ units} \qquad \blacksquare$$

## EXAMPLE 7

A manufacturer makes and sells two products, $A$ and $B$, that have the following yearly demand equations:

$$x = 250 - 3p - s \qquad \text{(product } A\text{)}$$
$$y = 200 - p - 4s \qquad \text{(product } B\text{)}$$

where $p$ is the price per unit (in dollars) of product $A$, $s$ is the price per unit (in dollars) of product $B$, and $x$ and $y$ are the yearly demands (in thousands of units) for $A$ and $B$, respectively. If it costs the manufacturer $20 per unit to manufacture product $A$ and $15 per unit to manufacture product $B$, determine the price that should be charged for $A$ and $B$ to maximize yearly profit.

### Solution

Let $R$ denote the yearly revenue (in thousands of dollars) from the manufacture and sale of $x$ thousand units of product $A$ at $p$ dollars per unit and $y$ thousand units of product $B$ at $s$ dollars per unit. Then since

revenue = (price per unit of $A$)(number of units of $A$ sold)
+ (price per unit of $B$)(number of units of $B$ sold)

we have

$$R = p(250 - 3p - s) + s(200 - p - 4s)$$

Simplifying, we obtain

$$R = 250p - 3p^2 - ps + 200s - ps - 4s^2$$
$$= 250p + 200s - 3p^2 - 4s^2 - 2ps$$

The total cost (in thousands of dollars) of producing $x$ thousand units of $A$ and $y$ thousand units of $B$ is

$$C = 20x + 15y = 20(250 - 3p - s) + 15(200 - p - 4s)$$
$$= 5000 - 60p - 20s + 3000 - 15p - 60s = 8000 - 75p - 80s$$

Using the basic relationship

profit = revenue − cost

we get (using the letter $M$ to represent profit in thousands of dollars)

$$M = (250p + 200s - 3p^2 - 4s^2 - 2ps) - (8000 - 75p - 80s)$$
$$= 325p + 280s - 3p^2 - 4s^2 - 2ps - 8000$$

To find the maximum of $M$, we compute

$$M_p = 325 - 6p - 2s \qquad \text{and} \qquad M_s = 280 - 8s - 2p$$

Equating both partial derivatives to zero, we obtain the system of equations

$$325 - 6p - 2s = 0$$
$$280 - 8s - 2p = 0$$

Solving the first of these equations for $s$, we get $s = 162.5 - 3p$. Substituting this into the second equation gives us

$$280 - 8(162.5 - 3p) - 2p = 0$$
$$280 - 1300 + 24p - 2p = 0$$
$$22p = 1020$$
$$p = \tfrac{510}{11} \approx 46.36$$

To find $s$ we substitute the value we just found for $p$ into $s = 162.5 - 3p$ to get

$$s = 162.5 - 3\left(\tfrac{510}{11}\right) \approx 23.41$$

The only critical point $(p, s)$ of the function $M$ is therefore (approximately) (46.36, 23.41). To verify that a relative maximum of $M$ occurs at this critical point, note that

$$M_{pp} = -6, \qquad M_{ss} = -8, \qquad \text{and} \qquad M_{ps} = -2$$

so that

$$D(p, s) = (-6)(-8) - (-2)^2 = 44$$

We conclude that $D(46.36, 23.41) = 44 > 0$. You should also observe that $M_{pp}(46.36, 23.41) = -6 < 0$, so that a relative maximum of $M$ occurs at (46.36, 23.41) by Case 2 of the second partials test. We conclude that to maximize yearly profit, the manufacturer should charge approximately $46.36 per unit for product $A$ and $23.41 per unit for product $B$. ∎

## SECTION 8.3
### SHORT ANSWER EXERCISES

1. (True or False) Every surface represented by a function of two variables $z = f(x, y)$ has at least one relative extremum. _____

2. If (1, 2) is a critical point of $f(x, y)$ and $f_{xx}(1, 2) = 4$ and $D(1, 2) = 7$, where $D(x, y) = f_{xx}f_{yy} - [f_{xy}]^2$, then (1, 2) corresponds to a _____ of $f(x, y)$.

3. The point $(-1, 3, 7)$ is a relative maximum point on the graph of $z = f(x, y)$. One critical point of $z = f(x, y)$ is _____.

4. (True or False) If $f_x = 2x - 2$ and $f_y = 7$, then (1, 0) is a critical point of $f(x, y)$. _____

5. If $(-4, 2)$ is a critical point of $f(x, y)$ and $D(-4, 2) = -3$, where $D(x, y) = f_{xx}f_{yy} - [f_{xy}]^2$, then $(-4, 2)$ corresponds to a _____ of $f(x, y)$.

6. (True or False) If the point $(2, -3)$ corresponds to a relative maximum of $f(x, y)$, then $f(0, 0) \leq f(2, -3)$. _____

7. For the function $f(x, y) = x^2 + y^2$, observe that $f(x, y) > 0$ for all $x \neq 0$ and $y \neq 0$. A relative minimum point on the graph of $f(x, y)$ is therefore _____.

8. The fixed cost of a certain company is $10,000 per month. If the company produces two products ($A$ and $B$) and if the variable cost per unit of $A$ is $30 and the

variable cost per unit of B is $50, then the total monthly cost of producing $x$ units of A and $y$ units of B is _____.

9. (True or False) Two functions $f(x, y)$ and $g(x, y)$ that have the same relative extrema must be identical. _____

10. Given that $f_x = 3x + 6$ and $f_y = 4y - 12$, we can say that the point _____ is a critical point of $f(x, y)$ and it corresponds to a relative _____ of $f(x, y)$.

## SECTION 8.3 EXERCISES

In Exercises 1–18, find any relative extrema or saddle points of the function.

1. $f(x, y) = x^2 + y^2 + 5$
2. $f(x, y) = x^2 - 2x + y^2 + 1$
3. $f(x, y) = 10 - 3x^2 + 3x - y^2 - 2y$
4. $f(x, y) = 2xy + x^2$
5. $f(x, y) = y^2 + 2x^2 + 2x - 3y + xy - 1$
6. $f(x, y) = x^2 - 2xy - y^2 + 4x + 3$
7. $f(x, y) = x^3 + y^2 - 12x$
8. $f(x, y) = y^3 + 3x^2 - 75y + 4$
9. $f(x, y) = y^2 + 2x^3 - 3xy + 7$
10. $f(x, y) = x^2 + y^4 - 5xy - 3$
11. $f(x, y) = x^2y - 2x - xy + 3$
12. $f(x, y) = y^2 + x^3 + 2xy + 4x - 3y$
13. $f(x, y) = xy + x^3 - y^2 - 2$
14. $f(x, y) = 2y^3 - x^2 + 3xy + 4$
15. $f(x, y) = xy - 3x + 2y - y^2 + 1$
16. $f(x, y) = 4x + 2y + 2xy - 3x^2 + 2$
17. $f(x, y) = e^{2xy} + 3$
18. $f(x, y) = e^{2x^2+4x+y^2-2y}$

### Business and Economics

19. (Profit) A manufacturer has determined that his monthly profit (in thousands of dollars) from the sale of $x$ thousand units of product A and $y$ thousand units of product B is

$$P(x, y) = 60x + 80y - \tfrac{1}{2}x^2 - y^2 - xy$$

Find how many of each product should be sold by this manufacturer to maximize monthly profit.

20. (Profit) The owner of a sporting goods store sells two brands of tennis balls. Both brand A and brand B cost the owner $0.25 per tennis ball. The balls are each packaged and sold in cartons of three. If brand A is priced at $p$ dollars a carton and brand B at $s$ dollars a carton, the owner will sell $100 - 50p + 20s$ cartons of brand A and $90 + 20p - 40s$ cartons of brand B during a typical week in the summer. What should the owner charge per carton for each brand to maximize weekly profit during the summer?

21. (Profit) A chain of hardware stores sells two brands of circular power saws—Cutall and Sawzall. Both brands cost the hardware store $20 per saw. If Cutall saws are priced at $p$ dollars each and Sawzall saws are priced at $s$ dollars each, the chain will sell $10000 - 250p + 75s$ Cutall saws and $8000 + 90p - 270s$ Sawzall saws per week. What should the store charge per saw for each brand to maximize weekly profit?

22. (Revenue) A company sells a product both in the United States and in Japan. The monthly demand functions for the product are

$p = 200 - 0.01x$ (U.S.)

$s = 150 - 0.02y$ (Japan)

where $x$ and $y$ are the quantities demanded in the U.S. and Japan, respectively, when the prices (in dollars) are $p$ and $s$. What price should the company charge in each market to maximize revenue?

**23. (Production)** In a production process, if a company uses $x$ thousand dollars worth of labor and $y$ thousand dollars worth of capital, its weekly production (in units) is given by

$$P(x, y) = 54x + 90y - x^2 - 2y^2 - xy$$

Find the mix of labor and capital that will maximize weekly production.

**24. (Advertising and sales)** If a company spends $x$ hundred dollars on television advertising and $y$ hundred dollars on coupon promotion of one of its products, its annual sales in units is given by

$$S(x, y) = 100x + 80y - 0.02x^2 - 0.03y^2 - 0.01xy$$

Determine the optimal amounts to be spent on each marketing activity to maximize annual unit sales.

**25. (Cost)** A company has a product that is produced at two different factories. The cost of producing $x$ units daily at factory $A$ is $C(x) = 0.001x^3 + 10x + 300$, while the cost of producing $y$ units daily at factory $B$ is $C(y) = 0.002y^3 + 8y + 250$. If the company can sell this product for \$90 a unit, determine the number of units the company should produce each day at each location to maximize daily profit.

**26. (Packaging)** A rectangular metal box with an open top is to be constructed and must have a volume of 4 ft$^3$. Find the dimensions of the box that will use the least amount of metal.

**27. (Packaging)** A rectangular box with an open top is to be constructed with a divider in the middle. If the material for the divider costs 0.02¢/in.$^2$ and the material for the rest of the box costs 0.04¢/in.$^2$, find the minimum cost of materials for constructing such a box with a volume of 48 in$^3$.

**28. (Architecture)** A rectangular warehouse with a volume of 1,200,000 ft$^3$ is to be constructed. The cost of the material for the roof will be twice as much as the material for the four sides, and the cost of material for the floor will be half as much as the material for the sides. Find the dimensions of the warehouse that will minimize the cost of materials.

## Life Sciences

**29. (Optimum location)** A new hospital is to be constructed serving three communities, $A$, $B$, and $C$, with locations in an $xy$-coordinate plane at $A(0, 0)$, $B(10, 15)$, $C(-20, 5)$. (In Figure 8.15 the distances are measured in miles.) The population of these communities (in thousands) are 20, 25, and 12 for $A$, $B$, and $C$, respectively. The hospital will be built at a location that minimizes the sum of the products of the number of people in a given community with the square of the distance between that community and the hospital. Find the location where the hospital will be built.

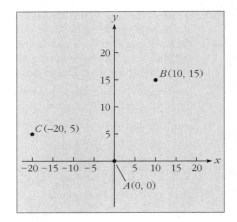

**FIGURE 8.15** The relative location of the three communities, $A$, $B$, and $C$. Distances are measured in miles.

## 8.4 CONSTRAINED OPTIMIZATION: LAGRANGE MULTIPLIERS

In Section 8.3 we considered the problem of finding the relative extrema of a function of two variables $z = f(x, y)$ without any restrictions (or *constraints*) on our choice of $x$ and $y$. For example, the unconstrained minimum of $f(x, y) = x^2 + y^2$ is $f(0, 0) = 0$; see Figure 8.16. In this section we consider the problem of finding the relative maximum and minimum of a function $z = f(x, y)$ when $x$ and $y$ are required to satisfy an additional condition (called a constraint) of the form $g(x, y) = 0$. For example, we might want to find the minimum (optimum) of $f(x, y) = x^2 + y^2$ subject to the constraint $x + y - 4 = 0$; see Figure 8.17. In this case we want to find the lowest point on the surface $z = f(x, y)$ that lies above the constraint line $x + y - 4 = 0$ in the $xy$-plane. In other words, we want to find the minimum $z$-value on the curve $C$ of intersection of the surface $f(x, y) = x^2 + y^2$ and the plane $x + y - 4 = 0$.

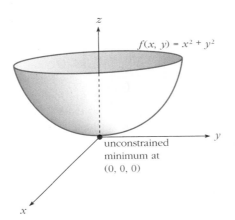

**FIGURE 8.16** The function $f(x, y) = x^2 + y^2$ has an unconstrained minimum value of $f(0, 0) = 0$.

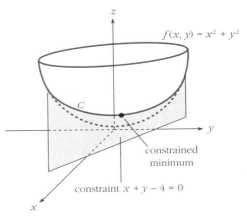

**FIGURE 8.17** The minimum of $f(x, y)$ subject to the constraint $x + y - 4 = 0$ is the minimum value of $z$ on the curve $C$ of intersection of the surface $f(x, y) = x^2 + y^2$ and the plane $x + y - 4 = 0$.

The general problem of constrained optimization for a function of two variables and a single constraint is stated as follows.

**CONSTRAINED OPTIMIZATION: FUNCTION OF TWO VARIABLES AND ONE CONSTRAINT**

| Maximize (or minimize): | $z = f(x, y)$ |
|---|---|
| Subject to: | $g(x, y) = 0$ |

The function $z = f(x, y)$ is often called the *objective function* and the equation $g(x, y) = 0$ the *constraint*.

One way to solve a constrained optimization problem of this type is to solve the constraint for either $x$ or $y$ and substitute this expression into the function $z = f(x, y)$ to obtain a function of a single independent variable. We then use the methods of Chapter 4 to determine the relative extrema of this function.

**EXAMPLE 1** Use the method we have just described to find the relative minimum of $f(x, y) = x^2 + y^2$ subject to the constraint $x + y - 4 = 0$.

**Solution** Solving the constraint for $y$, we obtain $y = 4 - x$. Substituting into $f(x, y)$ yields the function of one variable

$$H(x) = x^2 + (4 - x)^2 = 2x^2 - 8x + 16$$

To determine the relative extrema of this function we compute the derivative $H'(x) = 4x - 8$. Equating this derivative to zero, we get $4x - 8 = 0$ or $x = 2$ as the only critical number. Since $H''(x) = 4$, we see that $H''(2) = 4 > 0$, so that a relative minimum of $H(x)$ occurs at $x = 2$. Because $y = 4 - x$, when $x = 2$ we obtain $y = 4 - 2 = 2$. Thus a constrained minimum of $f(x, y) = x^2 + y^2$ occurs at the point $(2, 2)$ and the constrained minimum value is $f(2, 2) = 2^2 + 2^2 = 8$. ∎

Unfortunately, the method of solution used in Example 1 may not always work, because the constraint $g(x, y) = 0$ may be difficult or impossible to solve for either $x$ or $y$. Furthermore, even if $g(x, y) = 0$ can be solved for one of the variables, the function of one variable obtained when this expression is substituted into $z = f(x, y)$ can be rather complicated. There is another method, due to the French mathematician Joseph Louis Lagrange (1736–1813), that avoids these difficulties. It has the additional advantage of generalizing easily to functions of more than two variables that are subject to more than one constraint.

The method of Lagrange makes use of a *new* function of *three* variables:

$$F(x, y, \lambda) = f(x, y) + \lambda g(x, y)$$

where $\lambda$ is the Greek letter lambda and is called a *Lagrange multiplier* in this context. The next result (which is stated without proof) is the basis for the method of Lagrange multipliers.

**THE METHOD OF LAGRANGE MULTIPLIERS: FUNCTION OF TWO VARIABLES AND ONE CONSTRAINT**

> If the first partial derivatives of $f(x, y)$ and $g(x, y)$ exist, then the relative extrema of $f(x, y)$ subject to the constraint $g(x, y) = 0$ must occur at points $(a, b)$ such that $(a, b, c)$ is a solution of the following system of three equations in the three unknowns $x$, $y$, and $\lambda$:
>
> $$F_x(x, y, \lambda) = 0$$
> $$F_y(x, y, \lambda) = 0$$
> $$F_\lambda(x, y, \lambda) = 0$$

In the next example we again solve the problem of Exercise 1, this time using Lagrange multipliers.

**EXAMPLE 2** Use the method of Lagrange multipliers to find the relative minimum of $f(x, y) = x^2 + y^2$ subject to the constraint $x + y - 4 = 0$.

**Solution**

We form the function

$$F(x, y, \lambda) = f(x, y) + \lambda g(x, y)$$

using $f(x, y) = x^2 + y^2$ and $g(x, y) = x + y - 4$. We then obtain

$$F(x, y, \lambda) = x^2 + y^2 + \lambda(x + y - 4)$$

Computing the three first partials of $F$ and equating each to zero, we obtain the system of equations

$$F_x(x, y, \lambda) = 2x + \lambda = 0$$
$$F_y(x, y, \lambda) = 2y + \lambda = 0$$
$$F_\lambda(x, y, \lambda) = x + y - 4 = 0$$

Solving the first two of these equations for $\lambda$ we get $\lambda = -2x$ and $\lambda = -2y$. Therefore $-2x = -2y$, so that $x = y$.

Substituting $x$ for $y$ in the third equation of the system gives us $x + x - 4 = 0$ or $x = 2$. Since $x = y$, we conclude that $y = 2$. Also, since $\lambda = -2x$, we get $\lambda = -2(2) = -4$. The resulting point $(x, y, \lambda) = (2, 2, -4)$ is a solution of the simultaneous system of equations and is a critical point of $F(x, y, \lambda)$, since it is a point where all three first partials of $F(x, y, \lambda)$ are equal to zero. By the preceding boxed result, we know that if $f(x, y) = x^2 + y^2$ subject to $x + y - 4 = 0$ has a relative extremum, this extremum must occur at $(2, 2)$. Note that $f(2, 2) = 2^2 + 2^2 = 8$. By computing some functional values at points near $(2, 2)$ that satisfy the constraint we get, for example,

$$f(2.1, 1.9) = (2.1)^2 + (1.9)^2 = 8.02 > f(2, 2) = 8$$
$$f(2.05, 1.95) = (2.05)^2 + (1.95)^2 = 8.005 > f(2, 2) = 8$$
$$f(1.99, 2.01) = (1.99)^2 + (2.01)^2 = 8.0002 > f(2, 2) = 8$$

Our conclusion is that $f(2, 2) = 8$ is the constrained relative minimum value.

If you worked through Example 2 carefully, you probably already noticed that once we have found a critical point $(a, b, c)$ of $F(x, y, \lambda)$, we don't have a formal test for determining whether a constrained relative maximum, minimum, or neither occurs at $(a, b)$. Although such a test (which involves first and second partials of $F$) exists, it is quite complicated and will not be stated here. Rather, in all of the examples and exercises in this text, you may assume that the required constrained extrema occur among the points $(a, b)$ found by the method of Lagrange multipliers. In certain instances you may want to compute some functional values at points near $(a, b)$ that satisfy the constraint in order to determine whether $(a, b)$ is a constrained relative maximum or minimum.

In addition to the three-dimensional graphical interpretation of the result of Example 2 provided by Figure 8.17, an interesting two-dimensional interpretation can be made. For a given function of two variables $z = f(x, y)$, any curve $f(x, y) = C$, where $C$ is a constant, is called a *level curve* of $f(x, y)$. Level curves are sketched in the $xy$-plane. In Figure 8.18, several level curves of $f(x, y) = x^2 + y^2$ have been sketched along with the (constraint) line $x + y - 4 = 0$. As you can see from the figure, the level curves $x^2 + y^2 = C$ of $f(x, y)$ are all circles centered at the origin and with radius $\sqrt{C}$. As $C$ decreases, the radius of the circle decreases. The smallest value of $C$ for which the level curve $x^2 + y^2 = C$ intersects the constraint line is $C = 8$; this represents the minimum of $x^2 + y^2$ subject to the constraint $x + y - 4 = 0$.

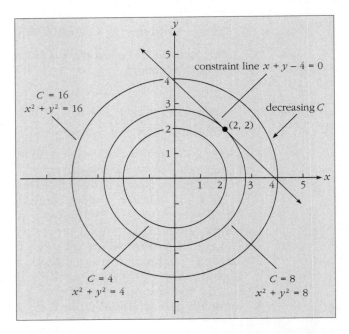

FIGURE 8.18 The level curve $x^2 + y^2 = C$ with the smallest value of $C$ that intersects the constraint line $x + y - 4 = 0$ has $C = 8$, so the constrained minimum value of $f(x, y) = x^2 + y^2$ is 8.

Before beginning the next example, let's summarize the steps in the method of Lagrange multipliers for a function of two variables and a single constraint.

**STEPS IN THE METHOD OF LAGRANGE MULTIPLIERS**

To solve the problem of optimizing $f(x, y)$ subject to the constraint $g(x, y) = 0$, follow these steps.

Step 1 Form the function $F(x, y, \lambda) = f(x, y) + \lambda g(x, y)$.

Step 2 Find $F_x$, $F_y$, and $F_\lambda$.

Step 3 Form a system of three equations in three unknowns by setting each of $F_x$, $F_y$, and $F_\lambda$ equal to zero.

Step 4 Solve the system formed in Step 3.

Step 5 The constrained optimum values of $f(x, y)$ must occur at a point $(a, b)$ such that $(a, b, c)$ is a solution of the system found in Step 4.

Step 4 in this procedure can usually be accomplished by solving each of the equations $F_x = 0$ and $F_y = 0$ for $\lambda$, equating these expressions, and solving for either $x$ or $y$ in terms of the other, and then substituting the result in $F_\lambda = 0$. This yields an equation in a single variable ($x$ or $y$) that is usually easily solved.

## 8.4 CONSTRAINED OPTIMIZATION: LAGRANGE MULTIPLIERS

**EXAMPLE 3 (Participative)**

**Solution**

Let's find the relative extrema of $f(x, y) = 4xy$ subject to the constraint $x^2 + y^2 = 1$.

We begin by writing the constraint in the form $g(x, y) = 0$. In this form the constraint is _____, so that $g(x, y) = x^2 + y^2 - 1$. Next we form the function of three variables

$$F(x, y, \lambda) = f(x, y) + \lambda g(x, y) = 4xy + \lambda(\underline{\hspace{2cm}})$$

Now we compute all of the first partials of $F$ and equate each of them to zero:

$$F_x = 4y + 2x\lambda = 0$$
$$F_y = \underline{\hspace{2cm}} = 0$$
$$F_\lambda = \underline{\hspace{2cm}} = 0$$

which is a simultaneous system of three equations in the three unknowns $x$, $y$, and $\lambda$. The solutions of this system are the _____ points of $F$. To solve the system, let's solve the first two equations for $\lambda$; we obtain $\lambda = -2y/x$ and $\lambda = \underline{\hspace{1cm}}$.

Observe that in solving for $\lambda$ we divided by $x$ and $y$. So for the preceding equations to be valid, we must assume that $x \neq 0$ and $y \neq 0$. However, by the first equation of the system, if $x = 0$ then $y = 0$ and then by the third equation, $0^2 + 0^2 - 1 = 0$, which is a contradiction. We conclude that a solution $(x, y, z)$ of the system cannot have $x = 0$. A similar argument shows that for any solution $(x, y, z)$ of the system, $y \neq 0$. So our assumption that $x \neq 0$ and $y \neq 0$ does not hinder us from finding any valid critical points.

From the two equations for $\lambda$ we get $-2y/x = -2x/y$ or $x^2 = \underline{\hspace{1cm}}$. Substituting $x^2$ for $y^2$ into the third equation of the system gives us

$$x^2 + x^2 - 1 = 0$$
$$2x^2 = 1$$
$$x = \pm\sqrt{\frac{1}{2}} = \pm\frac{\sqrt{2}}{2}$$

Since $y^2 = x^2$, when $x = +\sqrt{2}/2$ we get $y^2 = \underline{\hspace{1cm}}$ so $y = \underline{\hspace{1cm}}$. When $x = -\sqrt{2}/2$ we get $y^2 = \underline{\hspace{1cm}}$, so that $y = \underline{\hspace{1cm}}$. The $x$ and $y$ coordinates of the four critical points of $F$ are therefore $\left(\frac{\sqrt{2}}{2}, \frac{\sqrt{2}}{2}\right)$,

$\left(\frac{\sqrt{2}}{2}, \underline{\hspace{1cm}}\right)$, $\left(-\frac{\sqrt{2}}{2}, \frac{\sqrt{2}}{2}\right)$ and $\left(-\frac{\sqrt{2}}{2}, -\frac{\sqrt{2}}{2}\right)$. It is not necessary to find the corresponding values of $\lambda$ to complete the problem. Next we compute

$$f\left(\frac{\sqrt{2}}{2}, \frac{\sqrt{2}}{2}\right) = 4\left(\frac{\sqrt{2}}{2}\right)\left(\frac{\sqrt{2}}{2}\right) = 2$$

$$f\left(\frac{\sqrt{2}}{2}, -\frac{\sqrt{2}}{2}\right) = \underline{\hspace{1cm}}$$

$$f\left(-\frac{\sqrt{2}}{2}, \frac{\sqrt{2}}{2}\right) = \underline{\hspace{1cm}}$$

$$f\left(-\frac{\sqrt{2}}{2}, -\frac{\sqrt{2}}{2}\right) = \underline{\hspace{1cm}}$$

We conclude that the maximum value of $f(x, y) = 4xy$ subject to the constraint $x^2 + y^2 = 1$ is \_\_\_\_ and this maximum occurs at the points $\left(\frac{\sqrt{2}}{2}, \frac{\sqrt{2}}{2}\right)$ and _____. The minimum value is \_\_\_\_\_ and occurs at the points $\left(\frac{\sqrt{2}}{2}, -\frac{\sqrt{2}}{2}\right)$ and _____. See Figure 8.19 for a (two-dimensional) graphical illustration of these results. In Figure 8.19 the constraint is the circle $x^2 + y^2 = 1$ centered at the origin and with a radius of 1, while the level curves $4xy = C$ are hyperbolas if $C \neq 0$. In this diagram the effect of increasing $C$ to a value larger than $C = 2$ is to produce a level

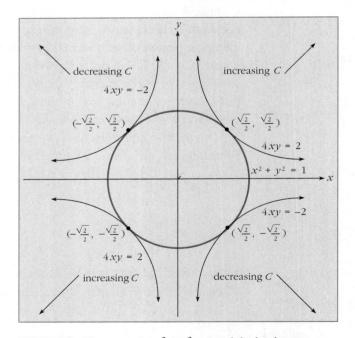

**FIGURE 8.19** The constraint $x^2 + y^2 = 1$ and the level curves $xy = C$ for $C = 2$ and $C = -2$. The effect of increasing $C$ from 2 or decreasing $C$ from $-2$ on the level curves is to move the level curves in the direction of the arrows.

curve (a hyperbola) that does not intersect the constraint circle, so the maximum value of $4xy$ subject to the constraint $x^2 + y^2 = 1$ is 2 and this occurs at the points $\left(\frac{\sqrt{2}}{2}, \frac{\sqrt{2}}{2}\right)$ and $\left(-\frac{\sqrt{2}}{2}, -\frac{\sqrt{2}}{2}\right)$. Similarly, the effect of decreasing $C$ to a value less than $-2$ is to produce a level curve that does not intersect the constraint curve. Thus the minimum value of $f(x, y) = 4xy$ subject to the constraint $x^2 + y^2 = 1$ is $-2$ and this occurs at the points $\left(-\frac{\sqrt{2}}{2}, \frac{\sqrt{2}}{2}\right)$ and $\left(\frac{\sqrt{2}}{2}, -\frac{\sqrt{2}}{2}\right)$. ∎

The next example illustrates the use of Lagrange multipliers in an application to a business problem.

EXAMPLE 4

If $x$ units of labor and $y$ units of capital are used per week in the production of a certain product, the number of units produced is $f(x, y)$, where $f(x, y) = 600x^{2/3}y^{1/3}$. If the cost of labor is $80 per unit and the cost of capital is $100 per unit, and if $24,000 is budgeted for next week's production, find the mix of labor and capital that will maximize production next week.

Solution

We are required to maximize production subject to a budgetary constraint. The problem here is to maximize $f(x, y) = 600x^{2/3}y^{1/3}$ subject to $80x + 100y = 24000$. We begin by forming the function

$$F(x, y, \lambda) = 600x^{2/3}y^{1/3} + \lambda(80x + 100y - 24000)$$

Next we equate the first partial derivatives of $F$ to zero, yielding the following system of equations:

$$F_x = 400x^{-1/3}y^{1/3} + 80\lambda = 0$$
$$F_y = 200x^{2/3}y^{-2/3} + 100\lambda = 0$$
$$F_\lambda = 80x + 100y - 24000 = 0$$

From the first two equations we get $\lambda = -5x^{-1/3}y^{1/3}$ and $\lambda = -2x^{2/3}y^{-2/3}$, and so $5x^{-1/3}y^{1/3} = 2x^{2/3}y^{-2/3}$. By multiplying this equation by $\frac{1}{5}x^{1/3}y^{2/3}$, we obtain $y = \frac{2}{5}x$. Substituting this into the third equation of the system gives us

$$80x + 100\left(\tfrac{2}{5}x\right) - 24000 = 0$$

from which we obtain $x = 200$. Substituting into $y = \frac{2}{5}x$ gives us $y = \frac{2}{5}(200) = 80$. Therefore the mix of labor and capital that will maximize production next week subject to the budget of $24,000 is 200 units of labor and 80 units of capital. The maximum production will be

$$f(200, 80) = 600(200)^{2/3}(80)^{1/3} \approx 88{,}417 \text{ units}$$

SOLUTIONS TO EXAMPLE 3

$x^2 + y^2 - 1 = 0, x^2 + y^2 - 1, 4x + 2y\lambda, x^2 + y^2 - 1$, critical, $-2x/y, y^2, \frac{1}{2}, \pm\sqrt{2}/2, \frac{1}{2}, \pm\sqrt{2}/2, -\sqrt{2}/2, -2, -2, 2, 2, \left(-\frac{\sqrt{2}}{2}, -\frac{\sqrt{2}}{2}\right), -2, \left(-\frac{\sqrt{2}}{2}, \frac{\sqrt{2}}{2}\right)$

Although it is not necessary to determine $\lambda$, since we have already solved the problem, it turns out that $\lambda$ has an important practical interpretation that we will explain. Substituting $x = 200$ and $y = 80$ into the expression $\lambda = -5x^{-1/3}y^{1/3}$ we get

$$\lambda = -5(200)^{-1/3}(80)^{1/3} \approx -3.68$$ ∎

As we indicated at the end of Example 4, the number $\lambda$ can be given an important interpretation in many applied problems.

> In general, $|\lambda|$ can be interpreted as the approximate change from the optimum value that results from an increase of one constraint unit.

Applying this interpretation to the result we obtained in Example 4, we can say that if one more dollar were available in next week's production budget, the maximum production of 88,417 could be increased by about $|-3.68| = 3.68$ units. In the context of this type of problem, $|\lambda|$ is sometimes called the *marginal productivity of money*.

As indicated earlier in this section, the method of Lagrange multipliers can be generalized to cover the case of optimizing a function of more than two independent variables subject to one or more constraints. We will now consider the use of Lagrange multipliers to optimize a function of three independent variables subject to a single constraint. This problem is formally stated in the following box.

**CONSTRAINED OPTIMIZATION: FUNCTION OF THREE VARIABLES AND ONE CONSTRAINT**

> Maximize (or minimize)　　$w = f(x, y, z)$
> Subject to:　　$g(x, y, z) = 0$

To solve this problem, we introduce a new function of four variables:

$$F(x, y, z, \lambda) = f(x, y, z) + \lambda g(x, y, z)$$

The next result provides the basis for the solution of this optimization problem.

**THE METHOD OF LAGRANGE MULTIPLIERS: FUNCTION OF THREE VARIABLES AND ONE CONSTRAINT**

> If the first partial derivatives of $f(x, y, z)$ and $g(x, y, z)$ exist, then the relative extrema of $f(x, y, z)$ subject to the constraint $g(x, y, z) = 0$ must occur at points $(a, b, c)$ such that $(a, b, c, d)$ is a solution of the following system of four equations in the four unknowns $x$, $y$, $z$, and $\lambda$:
>
> $$F_x(x, y, z, \lambda) = 0$$
> $$F_y(x, y, z, \lambda) = 0$$
> $$F_z(x, y, z, \lambda) = 0$$
> $$F_\lambda(x, y, z, \lambda) = 0$$

## EXAMPLE 5

A rectangular box without a top is to be constructed from 588 in$^2$ of material. Find the maximum volume of such a box.

### Solution

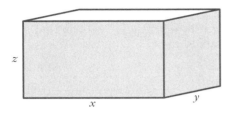

**FIGURE 8.20** A rectangular box without a top.

Let the dimensions of the box be $x$, $y$, and $z$, as shown in Figure 8.20. Then the volume is $f(x, y, z) = xyz$. The amount of material used in constructing the box is

$$\underset{\underset{\text{bottom}}{\uparrow}}{xy} + \underset{\underset{\substack{\text{front} \\ \text{and back}}}{\uparrow}}{2xz} + \underset{\underset{\text{sides}}{\uparrow}}{2yz} = \underset{\underset{\substack{\text{total material} \\ \text{available}}}{\uparrow}}{588}$$

The problem is thus to maximize $f(x, y, z) = xyz$ subject to the material constraint $xy + 2xz + 2yz - 588 = 0$. We form the function

$$F(x, y, z, \lambda) = xyz + \lambda(xy + 2xz + 2yz - 588)$$

Computing partial derivatives and equating them to zero, we obtain the system of equations

$$F_x = yz + \lambda(y + 2z) = 0$$
$$F_y = xz + \lambda(x + 2z) = 0$$
$$F_z = xy + \lambda(2x + 2y) = 0$$
$$F_\lambda = xy + 2xz + 2yz - 588 = 0$$

Solving each of the first three equations for $\lambda$ gives us

$$\lambda = -\frac{yz}{y + 2z}, \quad \lambda = -\frac{xz}{x + 2z}, \quad \lambda = -\frac{xy}{2x + 2y}$$

Equating the first and second of these expressions for $\lambda$ gives us

$$\frac{yz}{y + 2z} = \frac{xz}{x + 2z}$$
$$\frac{y}{y + 2z} = \frac{x}{x + 2z}$$
$$xy + 2yz = xy + 2xz$$
$$2yz = 2xz$$
$$y = x$$

Equating the second and third of the expressions for $\lambda$ gives us

$$\frac{xz}{x + 2z} = \frac{xy}{2x + 2y}$$
$$\frac{z}{x + 2z} = \frac{y}{2x + 2y}$$
$$2xz + 2yz = xy + 2yz$$
$$2xz = xy$$
$$z = \tfrac{1}{2}y$$

Substituting $x = y$ and $z = \frac{1}{2}y$ into the fourth equation of the original system gives us

$$y^2 + 2y\left(\tfrac{1}{2}y\right) + 2y\left(\tfrac{1}{2}y\right) - 588 = 0$$
$$3y^2 = 588$$
$$y = 14 \quad \text{(We reject the negative solution)}$$

Therefore, since $x = y$ and $z = \frac{1}{2}y$, when $y = 14$ we obtain $x = 14$ and $z = 7$. So the maximum volume of the resulting box is

$$f(14, 14, 7) = (14)(14)(7) = 1372 \text{ in}^3$$

Observe that from the equation $\lambda = -yz/(y + 2z)$, we obtain (with $x = 14$, $y = 14$, and $z = 7$)

$$\lambda = -\frac{(14)(7)}{14 + 2(7)} = -3.5$$

Therefore, if the amount of material available for the construction of the box increases by 1 in$^2$, the maximum volume will increase by approximately $|-3.5| = 3.5$ in$^3$. ∎

## SECTION 8.4
### SHORT ANSWER EXERCISES

1. (True or False) The minimum of $f(x, y) = x^2 + y^2$ subject to the constraint $x = y$ is 0. _____

2. The surface represented by $f(x, y)$ and the plane $x - 2y - 6 = 0$ intersect in a curve $C$ that is highest at the point $(2, -2, 1)$. The maximum value of $f(x, y)$ subject to the constraint $x - 2y = 6$ is _____.

3. If we apply the method of Lagrange multipliers to the problem of minimizing $f(x, y) = x^2 + y^2 + 4$ subject to the constraint $y = 3x + 5$, we would begin by writing $F(x, y, \lambda) =$ _____.

4. Given that $(-1, 2, 3)$ is a critical point of $F(x, y, \lambda)$ and that $f(-1, 2) = 3$, $f(-0.9, 1.1) = 2.9$ and $f(-1.1, 2.1) = 2.93$, where the points $(-0.9, 1.1)$ and $(-1.1, 2.1)$ satisfy the constraint $g(x, y) = 0$, we conclude that 3 is the relative _____ value of $f(x, y)$ subject to $g(x, y) = 0$.

5. Given the data of Exercise 4 and that $f(x + h, y + k) > f(x, y)$ for all positive values $x, y, h$, and $k$, what is the (approximate) relative maximum value of $f(x, y)$ subject to the constraint $g(x, y) = 1$? _____

6. The level curves of $f(x, y) = 2x + 3y$ are all (a) circles; (b) ellipses; (c) hyperbolas; (d) straight lines.

7. (True or False) In solving the problem of finding extrema of $f(x, y)$ subject to $g(x, y) = 0$, every critical point of $F(x, y, \lambda)$ corresponds to either a constrained relative maximum or a constrained relative minimum. _____

8. In maximizing a production function $f(x, y)$ subject to a budgetary constraint $g(x, y) = 0$, the value of $\lambda$ at the critical point $F(x, y, \lambda)$ that yields the maximum is $\lambda = -10$. If production is measured in units and the budget in dollars, and if one more dollar is allotted to production, about _____ more units can be produced.

9. Solve the following problem by inspection; do *not* use Lagrange multipliers:

   Maximize:    $f(x, y) = \sqrt{9 - x^2 - y^2}$

   Subject to:    $x - y = 0$

10. (True or False) In the problem of finding the extrema of $f(x, y)$ subject to the constraint $g(x, y) = 0$, it is always true that $F_\lambda = g(x, y)$. _____

## SECTION 8.4
## EXERCISES

*In Exercises 1–19, use the method of Lagrange multipliers to solve each problem by finding the relative extrema of a function subject to a constraint.*

1. Minimize $f(x, y) = 2x^2 + y^2 + 5$ subject to the constraint $x + 2y = 4$.
2. Maximize $f(x, y) = 2xy$ subject to the constraint $2x + y = 6$.
3. Find the extrema of $f(x, y) = 2xy$ subject to the constraint $x^2 + y^2 = 64$.
4. Find the extrema of $f(x, y) = x^2 - 6x + y^2 - 2y - 2$ subject to the constraint $x + y = 1$.
5. Find the extrema of $f(x, y) = y^2 + x$ subject to the constraint $x^2 + y^2 = 25$.
6. Maximize $f(x, y) = 36 - x^2 - y^2$ subject to the constraint $x + 2y = 4$.
7. Maximize $f(x, y) = x^2 + 2xy - 4y$ subject to the constraint $x + y = 20$.
8. Find the extrema of $f(x, y) = y^2 - x^2$ subject to the constraint $x^2 + y^2 = 49$.
9. Find the minimum of $f(x, y) = x^2 + y^2$ subject to the constraint $xy = 9$.
10. Find the maximum of $f(x, y) = 2xy^2$ subject to the constraint $2x - 3y = 6$.
11. Find the extrema of $f(x, y) = x^2y^2$ subject to the constraint $x^2 + y^2 = 8$.
12. Find the extrema of $f(x, y) = 2xy$ subject to the constraint $x^2 + 4y^2 = 32$.
13. If the sum of two numbers must be 12, maximize the product of the numbers.
14. If the difference of two numbers must be 18, minimize the product of the numbers.
15. Find the minimum of $f(x, y, z) = x^2 + y^2 + z^2$ subject to the constraint $x - 2y - 3z = 12$.
16. Find the maximum of $f(x, y, z) = xyz$ subject to the constraint $x + 2y + z = 4$.
17. Find the maximum of $f(x, y, z) = 2x - x^2 + 4y - y^2 + 6z - z^2$ subject to the constraint $x + y + z = 12$.
18. Find the extrema of $f(x, y, z) = x + y + z$ subject to the constraint $x^2 + y^2 + z^2 = 27$.
19. Find three numbers with sum 60 and product a maximum.
20. (Minimum cost) A company produces two products, A and B. The total monthly cost of producing $x$ units of product A and $y$ units of product B is $C(x, y) = x^2 + 2y^2 + 900$. If next month's production of A and B must satisfy the condition $2x + y = 400$, find the number of units of A and B that the company should make next month to minimize cost.
21. (Minimum cost) A company produces the same product at two different factories. The total weekly cost of producing $x$ units at Factory 1 and $y$ units at Factory 2 is given by $C(x, y) = \frac{1}{2}x^2 - xy + \frac{1}{4}y^2 + 300$. The combined production of the two factories must be 700 units in order to fill a certain order. How many units should be produced by each factory in order to minimize the cost of production?
22. (Sales) Monthly sales of one of a company's products depend upon the level of advertising and direct incentives (coupons, rebates, etc.). If the company spends $x$

thousand dollars on advertising and $y$ thousand dollars on direct incentives, its monthly sales in thousands of dollars will be

$$S(x, y) = 0.01x^2 + 2x + 0.05y^2 + y + 50$$

If next month's budget is $130,000, how should this money be allocated to maximize sales?

**23.** (Production) If $x$ units of labor and $y$ units of capital are used per week in the production of a certain product, the number of units produced is $f(x, y)$, where $f(x, y) = 800x^{1/2}y^{3/4}$. If the cost of labor is $50 per unit and the cost of capital is $100 per unit, and if $40,000 is budgeted for next week's production, find the optimal mix of labor and capital for the coming week. What is the marginal productivity of money for this production process?

**24.** (Maximum profit) A company that makes lawn mowers manufactures both a standard model and a deluxe model. If $x$ standard models and $y$ deluxe models are manufactured in a day, the daily profit from the sale of these mowers is

$$P(x, y) = 50x + 75y - 0.25x^2 - 0.75y^2$$

A union contract requires that the total daily output of standard and deluxe models must be 150 mowers. How many of each type of mower should this company make to optimize daily profit?

## Business and Economics

**25.** (Utility function) A consumer who purchases two goods, $A$ and $B$, will derive a certain amount of satisfaction, or utility, from $x$ units of $A$ and $y$ units of $B$. In economics, this utility is measured by a *utility function*. Of course, utility functions are personal, since they vary according to a person's tastes. One consumer with $10 to spend on meat or vegetables might spend $8 on meat and $2 on vegetables, whereas another consumer might spend $3 on meat and $7 on vegetables. Consumers tend to make the choice that will maximize total utility. Also, consumers are constrained by the amount of money they have to spend on the goods $A$ and $B$. In our meat and vegetables example, the total amount to be spent on these goods was $10, so if meat cost $2 a pound and vegetables cost $1 a pound, the budgetary constraint in this example is $2x + y = 10$, where $x$ is the number of pounds of meat and $y$ is the number of pounds of vegetables. Suppose the utility function of a certain consumer is $U(x, y) = 30x^{2/3}y^{1/3}$. How many pounds each of meat and vegetables will the consumer purchase to maximize utility?

**26.** (Packaging) A closed rectangular box is to be constructed using $192 worth of material. If the material for the top and bottom costs $4/ ft² and the material for the four sides costs $2/ft², find the dimensions of the box of largest possible volume and find that maximum volume.

**27.** (Postal regulations) In order for a certain postal service to deliver a rectangular box, the sum of its length and girth (distance around) must be no more than 120 in. Find the dimensions of the box of maximum volume that this postal service will deliver.

**28.** (Equimarginal rule) If $x$ units of labor and $y$ units of capital are used in the manufacture of $f(x, y)$ units of a certain product, the quantities $f_x(x, y)$ and $f_y(x, y)$ are the marginal productivity of labor and the marginal productivity of capital, respectively. Suppose that labor costs $l$ dollars per unit and capital costs $c$ dollars per unit, and suppose the budget for labor and capital is $B$ dollars, so that $lx + cy = B$. If the maximum production subject to this budgetary constraint occurs at $(a, b)$, prove the *equimarginal rule*:

$$\frac{f_x(a, b)}{f_y(a, b)} = \frac{l}{c}$$

OPTIONAL

**GRAPHING CALCULATOR/ COMPUTER EXERCISES**

*In the following exercises, verify graphically that the solutions of the indicated exercises are correct by:* (a) *graphing the constraint equation;* (b) *graphing the level curves* $f(x, y) = k$ *for several values of the constant* $k$; (c) *observing that the extrema occur for the values of* $k$ *you found earlier by the method of Lagrange multipliers.*

**29.** Exercise 3  **30.** Exercise 2

**31.** Exercise 9  **32.** Exercise 6

## 8.5 REGRESSION: THE METHOD OF LEAST SQUARES

### The Least-Squares Method

In order to apply the methods of calculus you have learned so far to practical problems, it is usually necessary to have a functional relationship available between two or more of the variables under consideration. Up to this point, these functional relationships have usually been given in the statement of the problem. How are such formulas derived in the real world?

Let's confine ourselves to the relationship between two variables $x$ and $y$ and assume that the relation is given by a function $y = f(x)$. To determine the (unknown) functional relationship between $x$ and $y$ in practice, data is collected and the *data points* $(x_i, y_i)$ are plotted in what is called a *scatter diagram*. Then the researcher attempts to fit the "best" straight line or curve to the data points, that is, the one that (in a sense that will be explained later in this section) deviates from the data points as little as possible. The formula for this straight line or curve gives the desired relationship. As an example, consider the data points $(x_i, y_i)$ in Table 8.1, which are observations generated by the marketing department of a company concerning the monthly level of advertising $(x)$ for a product and the corresponding monthly sales $(y)$.

The four data points $(x_1, y_1) = (25, 100)$, $(x_2, y_2) = (30, 120)$, $(x_3, y_3) = (35, 150)$, and $(x_4, y_4) = (40, 160)$ are plotted in a scatter diagram in Figure 8.21. Note that although no straight line will pass exactly through all four data points, we could probably sketch in a straight line that fits these data points pretty well. One such line is sketched in Figure 8.22. What we would like, of course, is a mathematical procedure by which we can find the line that *best* fits these data points.

What will the criterion be for the best-fitting line? We want this line to be as close as possible to all the data points, but how can we measure this closeness? One possible way is to consider the vertical *deviations* $d_i$ of the data points $(x_i, y_i)$ from the line $y = mx + b$, where

$$d_i = y_i - (mx_i + b)$$

(See Figure 8.22.) You might initially think that a measure of overall closeness of the data points to the line is obtained simply by summing the four deviations $d_1$, $d_2$, $d_3$, and $d_4$. However, such a measure of closeness is not a good one, since $d_i$ will be positive if the data point is above the line and negative if the data point is below the line, and these positive and negative deviations can

| ADVERTISING IN THOUSANDS OF DOLLARS ($x$) | SALES IN THOUSANDS OF DOLLARS ($y$) |
|---|---|
| 25 | 100 |
| 30 | 120 |
| 35 | 150 |
| 40 | 160 |

**TABLE 8.1**

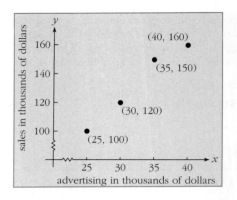
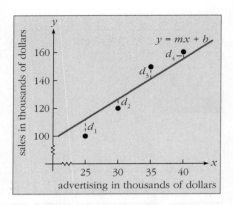

**FIGURE 8.21** A scatter diagram of four data points.

**FIGURE 8.22** A straight-line approximation of the four data points in Figure 8.21 and the four vertical deviations $d_i$ of the data points from the line.

cancel one another and yield a small sum, even when the line fits the data rather poorly. Statisticians generally prefer to measure overall closeness by the sum of the *squares* of the deviations. Squaring each deviation makes the resulting numbers positive. Thus let

$$S = d_1^2 + d_2^2 + d_3^2 + d_4^2$$
$$= (y_1 - mx_1 - b)^2 + (y_2 - mx_2 - b)^2 + (y_3 - mx_3 - b)^2$$
$$+ (y_4 - mx_4 - b)^2$$

Substituting the values of $x_i$ and $y_i$ for our four data points in the preceding expression gives us

$$S = (100 - 25m - b)^2 + (120 - 30m - b)^2 + (150 - 35m - b)^2$$
$$+ (160 - 40m - b)^2$$

Observe that $S$ is a function of the two variables $m$ and $b$, that is, $S = f(m, b)$. The objective is now to choose $m$ and $b$ so that $S = f(m, b)$ is a minimum, hence we minimize the sum of the squared deviations as represented by $S$. We may use the technique of Section 8.3 to find this minimum. We compute:

$$f_m = 2(100 - 25m - b)(-25) + 2(120 - 30m - b)(-30)$$
$$+ 2(150 - 35m - b)(-35) + 2(160 - 40m - b)(-40)$$
$$= -35500 + 8700m + 260b$$
$$f_b = 2(100 - 25m - b)(-1) + 2(120 - 30m - b)(-1)$$
$$+ 2(150 - 35m - b)(-1) + 2(160 - 40m - b)(-1)$$
$$= -1060 + 260m + 8b$$

Equating $f_m$ and $f_b$ to zero to find the critical points, we obtain the system of equations

$$8700m + 260b = 35500$$
$$260m + 8b = 1060$$

One way to solve this system is to solve the second equation for $b$ to obtain

$$b = \frac{1060 - 260m}{8} = \frac{265 - 65m}{2}$$

We then substitute this expression for $b$ into the first equation to get

$$8700m + 260\left(\frac{265 - 65m}{2}\right) = 35500$$

$$8700m + 34450 - 8450m = 35500$$

$$250m = 1050$$

$$m = \frac{1050}{250} = \frac{21}{5}$$

To find $b$ we can substitute this value for $m$ into the second equation of the system to get

$$260\left(\tfrac{21}{5}\right) + 8b = 1060$$

$$8b = -32$$

$$b = -4$$

Therefore $(m, b) = \left(\tfrac{21}{5}, -4\right)$ is a critical point of $f(m, b)$. To verify that $S = f(m, b)$ attains a relative minimum value at $\left(\tfrac{21}{5}, -4\right)$, we apply the second partials test. We compute

$$f_{mm} = 8700, \qquad f_{mb} = 260, \qquad f_{bb} = 8$$

The discriminant function is

$$D(m, b) = f_{mm}f_{bb} - [f_{mb}]^2 = (8700)(8) - (260)^2 = 2000 > 0$$

Since $f_{mm} = 8700 > 0$, this verifies that $f(m, b)$ has a relative minimum at $\left(\tfrac{21}{5}, -4\right)$. The equation of the required line is therefore $y = \tfrac{21}{5}x - 4$ and this line is sketched in Figure 8.23. The technique we used to find this line is called *method of least squares* and the line is sometimes called the *line of regression* or the *least-squares line* for the given set of data points.

The equation $y = \tfrac{21}{5}x - 4$ can be used, for example, to predict sales when monthly advertising is $38,000. The predicted sales level is

$$y = \frac{21}{5}(38) - 4 = 155.6 \text{ thousand dollars.}$$

We would *not* expect the equation to give us a very accurate estimate for sales if monthly advertising is $100,000, however, since our regression line is based on advertising data in the range $25 \leq x \leq 40$. Hence we can expect good predictions of corresponding sales levels only for advertising levels in or near this range. In general, attempting to predict values of $y$ from a fitted line (or curve) $y = f(x)$ given a value of $x$ that is not inside or nearly inside the range of $x$-values in the data set is called *extrapolation* and is risky business. In the present example, most students of marketing are aware that increasing the level of advertising beyond a certain point does not pay off, since such an increase will not generate enough sales to cover the additional advertising expense. This

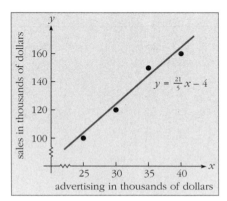

**FIGURE 8.23** The least-squares line $y = \tfrac{21}{5}x - 4$, which best fits the four data points.

phenomenon is *not* taken into account by our linear model $y = \frac{21}{5}x - 4$, however, since it says that every increase of $1,000 in advertising will produce an increase of $\frac{21}{5} = \$4,200$ in sales! The point is that although our linear model may be quite accurate within the range of $25 \le x \le 40$, it may be terribly inaccurate for advertising levels $x$ outside of this range.

To consider the general case of finding the equation of the line of regression corresponding to the $n$ data points $(x_1, y_1), (x_2, y_2), \ldots, (x_n, y_n)$, it is convenient to introduce the following symbolism to represent a sum of terms:

$$x_1 + x_2 + \cdots + x_n = \sum_{i=1}^{n} x_i$$

The letter $\Sigma$ is the upper-case Greek letter sigma, and the letter $i$ is called the *index of summation*.

If $y = mx + b$ is the line of regression for the $n$ data points $(x_1, y_1), (x_2, y_2), \ldots, (x_n, y_n)$, then $m$ and $b$ must be chosen to minimize

$$S = f(m, b) = \sum_{i=1}^{n} d_i^2 = \sum_{i=1}^{n} (y_i - mx_i - b)^2$$

That is, we must choose $m$ and $b$ to minimize the sum of the squares of the deviations of the data points from the line of regression. Using techniques similar to those illustrated in the example with four data points, it can be shown that the equations $f_m = 0$ and $f_b = 0$ simplify to yield the following system of two equations in the two unknowns $m$ and $b$:

$$\left(\sum_{i=1}^{n} x_i\right)m + nb = \sum_{i=1}^{n} y_i$$

$$\left(\sum_{i=1}^{n} x_i^2\right)m + \left(\sum_{i=1}^{n} x_i\right)b = \sum_{i=1}^{n} x_i y_i$$

These equations are sometimes called the *normal equations*. It can be shown further that the solution of this system produces a minimum of $f(m, b)$. This solution is presented in the following box.

**LEAST-SQUARES LINE FORMULAS**

> Given the data set $(x_1, y_1), (x_2, y_2), \ldots, (x_n, y_n)$, the equation of the least-squares line for this data set is $y = mx + b$, where
>
> $$m = \frac{n \sum_{i=1}^{n} x_i y_i - \left(\sum_{i=1}^{n} x_i\right)\left(\sum_{i=1}^{n} y_i\right)}{n \sum_{i=1}^{n} x_i^2 - \left(\sum_{i=1}^{n} x_i\right)^2}$$
>
> $$b = \frac{1}{n}\left(\sum_{i=1}^{n} y_i - m \sum_{i=1}^{n} x_i\right)$$

Observe that these formulas require that we first compute $m$, since the value of $m$ is needed in the formula for $b$.

## EXAMPLE 1

Use the given formulas to find the least-squares line for the data in Table 8.2.

### Solution

It is convenient to use a table to summarize the computation of the sums necessary for use in the formulas for $m$ and $b$. The necessary sums are illustrated in Table 8.3.

| $x$ | $y$ |
|---|---|
| 1 | 5 |
| 2 | 8 |
| 4 | 16 |
| 6 | 18 |
| 9 | 24 |
| 10 | 25 |

**TABLE 8.2**

| $x_i$ | $y_i$ | $x_i y_i$ | $x_i^2$ |
|---|---|---|---|
| 1 | 5 | 5 | 1 |
| 2 | 8 | 16 | 4 |
| 4 | 16 | 64 | 16 |
| 6 | 18 | 108 | 36 |
| 9 | 24 | 216 | 81 |
| 10 | 25 | 250 | 100 |
| Sum: 32 | 96 | 659 | 238 |

**TABLE 8.3**

Therefore $\sum_{i=1}^{6} x_i = 32$, $\sum_{i=1}^{6} y_i = 96$, $\sum_{i=1}^{6} x_i y_i = 659$, and $\sum_{i=1}^{6} x_i^2 = 238$. Using the least-squares line formulas, we now compute:

$$m = \frac{n \sum_{i=1}^{n} x_i y_i - \left(\sum_{i=1}^{n} x_i\right)\left(\sum_{i=1}^{n} y_i\right)}{n \sum_{i=1}^{n} x_i^2 - \left(\sum_{i=1}^{n} x_i\right)^2}$$

$$= \frac{6(659) - (32)(96)}{6(238) - (32)^2} = \frac{882}{404} \approx 2.183$$

$$b = \frac{1}{n}\left(\sum_{i=1}^{n} y_i - m \sum_{i=1}^{n} x_i\right)$$

$$= \frac{1}{6}\left[96 - \frac{882}{404}(32)\right] = \frac{1}{6}\left[\frac{10560}{404}\right] = \frac{10560}{2424} \approx 4.356$$

So the equation of the least-squares line for the given data is (approximately) $y = 2.183x + 4.356$. The data points and this least-squares line are sketched in Figure 8.24. ∎

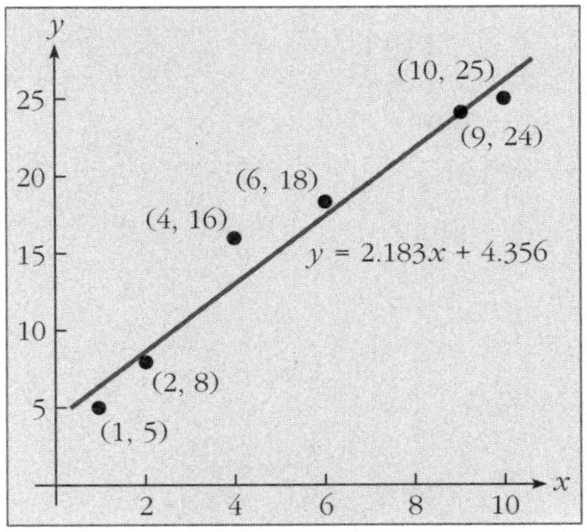

### Applications

As indicated earlier in this section, the method of least squares can be used to derive a formula for the relationship between two variables $x$ and $y$. In the remainder of this chapter and in the exercises, we will assume that this relationship is linear unless otherwise specified. Thus the formula relating $x$ and $y$ has the form $y = f(x) = mx + b$ and we may use the least-squares line formulas to find $m$ and $b$.

**FIGURE 8.24** The data points and regression line of Example 1.

**EXAMPLE 2** The placement-test scores and final course grade scores for ten students in a precalculus class are given in Table 8.4.

(a) Find the least-squares line $y = mx + b$.
(b) Use the equation found in part (a) to predict the final course grade of a student who scored 75 on the placement test.
(c) If a passing grade in the course is 60, what is the predicted minimum placement-test score for a student who hopes to pass the course?

Solution

(a) Let the least-squares line be given by $y = mx + b$. The necessary sums for the least-squares line formulas are computed in Table 8.5.

| PLACEMENT TEST (x) | FINAL COURSE GRADE (y) |
|---|---|
| 62 | 72 |
| 48 | 54 |
| 71 | 80 |
| 35 | 30 |
| 82 | 75 |
| 85 | 90 |
| 26 | 40 |
| 95 | 98 |
| 54 | 63 |
| 38 | 50 |

**TABLE 8.4**

| $x_i$ | $y_i$ | $x_i y_i$ | $x_i^2$ |
|---|---|---|---|
| 62 | 72 | 4464 | 3844 |
| 48 | 54 | 2592 | 2304 |
| 71 | 80 | 5680 | 5041 |
| 35 | 30 | 1050 | 1225 |
| 82 | 75 | 6150 | 6724 |
| 85 | 90 | 7650 | 7225 |
| 26 | 40 | 1040 | 676 |
| 95 | 98 | 9310 | 9025 |
| 54 | 63 | 3402 | 2916 |
| 38 | 50 | 1900 | 1444 |
| Sum: 596 | 652 | 43238 | 40424 |

**TABLE 8.5**

Therefore $\sum_{i=1}^{10} x_i = 596$, $\sum_{i=1}^{10} y_i = 652$, $\sum_{i=1}^{10} x_i y_i = 43238$, and $\sum_{i=1}^{10} x_i^2 = 40424$. Using the least-squares line formulas for $m$ and $b$, we get

$$m = \frac{10(43238) - (596)(652)}{10(40424) - (596)^2} = \frac{43788}{49024} \approx 0.8932$$

$$b = \frac{1}{10}\left[652 - \frac{43788}{49024}(596)\right] = \frac{1}{10}\left[\frac{5866000}{49024}\right] = \frac{586600}{49024} \approx 11.966$$

The least-squares line is therefore approximately

$$y = 0.8932x + 11.966$$

(b) If $x = 75$, the predicted final course grade is

$$y = 0.8932(75) + 11.966 = 78.956 \approx 79 \quad \text{(rounding to the nearest integer)}$$

(c) In this case we are given $y = 60$ and we're asked for the corresponding value of $x$. We obtain

$$60 = 0.8932x + 11.966$$
$$0.8932x = 48.034$$
$$x = 53.777 \approx 54$$

So a student who hopes to pass the precalculus course should score at least 54 on the placement test. The data points and the least-squares line are sketched in Figure 8.25 for reference.

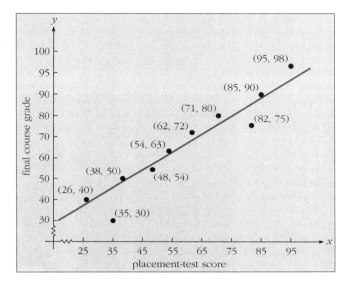

**FIGURE 8.25** The data points and regression line of Example 2.

---

**EXAMPLE 3 (Participative)**

Solution

In a psychology experiment, a person is given a subset of the words in the set {of, for, off, to, two, too, on} to memorize. The person is then presented with one of these words and asked to decide as quickly as possible whether the word is one of the words in her subset. The time required for the subject to respond depends on the number of words in her subset. The data in Table 8.6 were recorded for a particular subject.* Let's find the line of regression corresponding to this data.

We begin by forming a table to help compute the sums for the least-squares line formulas. Fill in the sums at the bottom of Table 8.7.

---

*The results of a similar experiment were reported by S. Sternberg (1966) in the article, "High-Speed Scanning in Human Memory" in *Science*, 153, pp. 652–664.

| NUMBER OF WORDS IN THE SUBSET (x) | TIME REQUIRED TO RESPOND IN SECONDS (y) |
|---|---|
| 1 | 0.500 |
| 2 | 0.521 |
| 3 | 0.564 |
| 4 | 0.598 |
| 5 | 0.630 |
| 6 | 0.640 |

**TABLE 8.6**

| $x_i$ | $y_i$ | $x_i y_i$ | $x_i^2$ |
|---|---|---|---|
| 1 | 0.500 | 0.500 | 1 |
| 2 | 0.521 | 1.042 | 4 |
| 3 | 0.564 | 1.692 | 9 |
| 4 | 0.598 | 2.392 | 16 |
| 5 | 0.630 | 3.150 | 25 |
| 6 | 0.640 | 3.840 | 36 |
| Sum: ___ | ___ | ___ | ___ |

**TABLE 8.7**

Thus $\sum_{i=1}^{6} x_i = 21$, $\sum_{i=1}^{6} y_i = 3.453$, $\sum_{i=1}^{6} x_i y_i = 12.616$, and $\sum_{i=1}^{6} x_i^2 = 91$. Using the least-squares line formulas for $m$ and $b$, we obtain

$$m = \frac{6(12.616) - (\underline{\quad})(\underline{\quad})}{6(91) - (21)^2} = \frac{3.183}{105}$$

$$\approx \underline{\quad} \quad \text{(to four decimal places)}$$

$$b = \frac{1}{6}\left[3.453 - \left(\frac{3.183}{105}\right)(\underline{\quad})\right]$$

$$= \underline{\quad}$$

The equation of the least-squares line is therefore approximately
$y = \underline{\qquad\qquad\qquad}$.  ■

## SECTION 8.5
### SHORT ANSWER EXERCISES

1. (True or False) For a set of data points $(x_1, y_1), (x_2, y_2), \ldots, (x_n, y_n)$, statisticians generally use the sum of the deviations $d_i = y_i - (mx_i + b)$ as a measure of the overall closeness of the data set to the line $y = mx + b$. _____

2. The sum of the squared deviations of three data points from the line $y = mx + b$ is given by

   $S = f(m, b) = (10 - 4m - b)^2 + (15 - 7m - b)^2 + (20 - 10m - b)^2$

   What are the data points? _____

3. For the function $f(m, b)$ of Exercise 2, find and simplify $f_m$.
   _____

4. For the function $f(m, b)$ of Exercise 2, find and simplify $f_b$.
   _____

5. Given the least-squares line $y = 0.5730x + 2.7150$ that is based on data within the range $0 \leq x \leq 10$, predict the value of $y$ when $x$ is 3.5. _____

---

**SOLUTIONS TO EXAMPLE 3**     21, 3.453, 12.616, 91, 21, 3.453, 0.0303, 21, 0.4694, 0.0303x + 0.4694

6. (True or False) Given a set of data points $(x_1, y_1), (x_2, y_2), \ldots, (x_n, y_n)$, where $n \geq 3$, it is possible that the line of regression passes through all $n$ data points. _____

7. If the normal equations for a set of data points are

   $16m + 3b = 28$
   $90m + 16b = 158$

   for $n = 3$, then $\sum_{i=1}^{3} x_i =$ _____, $\sum_{i=1}^{3} y_i =$ _____, $\sum_{i=1}^{3} x_i^2 =$ _____, and $\sum_{i=1}^{3} x_i y_i =$ _____.

8. Let $x$ represent the temperature in degrees Fahrenheit and $y$ the shelf-life of a certain brand of milk. If a set of observations $(x_1, y_1), (x_2, y_2), \ldots, (x_n, y_n)$ is made and the least-squares line $y = mx + b$ is computed, then the expected sign of $m$ is _____.

9. If $x$ represents the size of a car engine in liters and $y$ represents the horsepower of the car, a set of observations $(x_1, y_1), (x_2, y_2), \ldots, (x_n, y_n)$ is made and the least-squares line $y = mx + b$ is computed. The expected sign of $m$ is _____.

10. (True or False) For any set of data points $(x_1, y_1), (x_2, y_2), \ldots, (x_n, y_n)$, the least-squares line $y = mx + b$ is always the function $y = f(x)$ that best describes the relationship between $x$ and $y$. _____

## SECTION 8.5 EXERCISES

In each of Exercises 1–8, find the line of regression for the given data points by finding $S = f(m, b)$, solving the system of equations $f_m = 0$ and $f_b = 0$, and verifying by the second partials test that the critical point corresponds to a relative minimum of $f(m, b)$.

1. | x | 1  | 5  | 12 |
   |---|----|----|----|
   | y | 12 | 20 | 25 |

2. | x | 0 | 4 | 10 |
   |---|---|---|----|
   | y | 1 | 7 | 11 |

3. | x | 3  | 9  | 15 |
   |---|----|----|----|
   | y | 12 | 14 | 17 |

4. | x | 22 | 28  | 35  |
   |---|----|-----|-----|
   | y | 60 | 100 | 125 |

5. | x | −3 | 2 | 5 |
   |---|----|---|---|
   | y | −1 | 4 | 9 |

6. | x | −6 | −3 | 2  |
   |---|----|----|----|
   | y | 4  | 8  | 15 |

7. | x | −2 | 1  | 3 | 5 |
   |---|----|----|---|---|
   | y | 14 | 10 | 6 | 0 |

8. | x | −2 | 2 | 4 | 6  |
   |---|----|---|---|----|
   | y | 3  | 1 | 0 | −1 |

In each of Exercises 9–14, find the least-squares line for the given set of data points by using the least-squares line formulas.

9. | x | 3 | 6  | 8  | 10 |
   |---|---|----|----|----|
   | y | 4 | 12 | 16 | 22 |

10. | x | 14 | 16 | 10 | 11 | 18 |
    |---|----|----|----|----|----|
    | y | 30 | 32 | 5  | 10 | 44 |

**11.**

| x | 6 | 10 | 15 | 22 | 27 | 29 | 31 |
|---|---|----|----|----|----|----|----|
| y | 110 | 107 | 105 | 99 | 95 | 94 | 91 |

**12.**

| x | 12 | 15 | 20 | 23 | 27 | 30 | 34 | 39 |
|---|----|----|----|----|----|----|----|----|
| y | 52 | 49 | 46 | 41 | 35 | 31 | 29 | 26 |

**13.**

| x | 0.3 | 0.7 | 1.1 | 1.6 | 1.9 | 2.1 | 2.6 | 2.8 | 3.0 | 3.2 |
|---|-----|-----|-----|-----|-----|-----|-----|-----|-----|-----|
| y | 6.7 | 7.4 | 8.1 | 8.9 | 9.6 | 10.2 | 10.4 | 11.4 | 12.0 | 12.6 |

**14.**

| x | 0.4 | 0.5 | 0.6 | 0.7 | 0.8 | 0.9 | 1.0 | 1.1 | 1.2 | 1.3 |
|---|-----|-----|-----|-----|-----|-----|-----|-----|-----|-----|
| y | −6.3 | −6.1 | −5.8 | −5.4 | −5.3 | −5.2 | −5.0 | −4.7 | −4.3 | −4.2 |

*In Exercises 15–20, find the predicted values of y corresponding to the given values of x, using the equations for the least-squares lines you computed in Exercises 9–14. Which of these predicted y-values do you consider accurate?*

**15.** Use the data set of Exercise 9 with (a) $x = 5$; (b) $x = 7$; (c) $x = 20$.

**16.** Use the data set of Exercise 10 with (a) $x = 12$; (b) $x = 19$; (c) $x = 1$.

**17.** Use the data set of Exercise 11 with (a) $x = 150$; (b) $x = 20$; (c) $x = 25$.

**18.** Use the data set of Exercise 12 with (a) $x = 0$; (b) $x = 35$; (c) $x = 50$.

**19.** Use the data set of Exercise 13 with (a) $x = 0.2$; (b) $x = 1.5$; (c) $x = 5.3$.

**20.** Use the data set of Exercise 14 with (a) $x = 0.55$; (b) $x = 1.25$; (c) $x = 1.95$.

**21.** (Demand) In order to determine the demand for a new product, a company tested the product in five test markets of approximately equal populations and obtained the following data.

| PRICE IN DOLLARS (x) | DEMAND IN THOUSANDS OF UNITS (y) |
|---|---|
| 5.00 | 40 |
| 4.50 | 60 |
| 4.75 | 52 |
| 5.25 | 34 |
| 5.50 | 28 |

(a) Find the least-squares line $y = mx + b$ for this data. The result is the (approximate) demand function for the product.

(b) Estimate demand in a market of the same population if the price is set at $5.10.

**22.** (Cost) Production records from a certain factory yield the following data, which show the cost in dollars of producing $x$ units of a certain product per day.

| DAILY PRODUCTION IN UNITS (x) | COST IN DOLLARS (y) |
|---|---|
| 600 | 3400 |
| 650 | 3650 |
| 700 | 3875 |
| 750 | 4200 |
| 800 | 4550 |

(a) Use the least-squares line formulas to obtain the regression equation $y = mx + b$.

(b) Use the equation you found in part (a) to predict the cost of producing 825 units per day.

## Business and Economics

| YEAR | COST OF MATERIALS (thousands of dollars) |
|---|---|
| 1987 | 770 |
| 1988 | 790 |
| 1989 | 800 |
| 1990 | 812 |

23. (Cost of materials) During the years 1987–1990, a factory has produced the same number of units of a product. Due to fluctuations in the prices of materials, the cost of materials has varied from year to year, as shown in the table.

(a) Let the year 1987 be designated as year 1, 1988 as year 2, and so forth. Find the line of regression $y = mx + b$, where $x$ is the year and $y$ is the cost of materials.

(b) Use the least-squares line you found in part (a) to predict the cost of materials in 1991 (which is year 5).

24. (Advertising and sales) A company that spends $x$ thousand dollars a month for advertising a certain product has $y$ thousand dollars in sales. The observed values of sales for several levels of advertising are listed in the following table.

| ADVERTISING IN THOUSANDS OF DOLLARS (x) | SALES IN THOUSANDS OF DOLLARS (y) | ADVERTISING IN THOUSANDS OF DOLLARS (x) | SALES IN THOUSANDS OF DOLLARS (y) |
|---|---|---|---|
| 70 | 360 | 90 | 430 |
| 65 | 320 | 95 | 442 |
| 80 | 390 | 100 | 455 |

(a) Find the least-squares line for this data set.

(b) Use the least-squares line to predict monthly sales if monthly advertising is $105,000.

(c) Would it be appropriate to use the least-squares line to predict monthly sales for a monthly advertising level of $175,000?

25. (Quality control) In a quality control study, 12 assembly line workers are observed for various lengths of time immediately following a break, and the number of units of a certain product they assemble in the time interval is recorded. These data are summarized as follows:

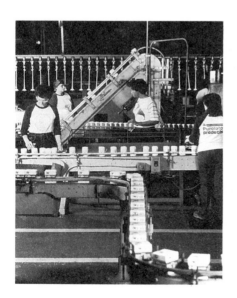

| TIME IN MINUTES (x) | NUMBER OF UNITS ASSEMBLED (y) | TIME IN MINUTES (x) | NUMBER OF UNITS ASSEMBLED (y) |
|---|---|---|---|
| 15 | 20 | 45 | 90 |
| 15 | 24 | 60 | 105 |
| 30 | 54 | 60 | 110 |
| 30 | 60 | 60 | 115 |
| 45 | 80 | 75 | 130 |
| 45 | 82 | 75 | 140 |

(a) Find the least-squares line for these data.
(b) Use the least-squares line to estimate the number of units assembled by a worker in the first 80 minutes following a break.

Social Sciences

| TEMPERATURE IN DEGREES FAHRENHEIT (x) | NUMBER OF HOMELESS PEOPLE (y) |
|---|---|
| 45 | 6 |
| 35 | 10 |
| 25 | 15 |
| 15 | 23 |
| 5 | 30 |

26. (Homeless people) The number of homeless people who arrive at a certain shelter on a given evening is a function of the outdoor temperature. Some observations made by the director of the shelter are summarized here.

(a) Compute the least-squares line for these data.
(b) Use the least-squares line to estimate the number of homeless people who will arrive at the shelter if the temperature is 0°F.
(c) The capacity of the shelter is 40 homeless people. According to the least-squares line, what outdoor temperature will result in the shelter being filled to capacity?

27. (Retention and graduation) In a study on retention and graduation of students at a large school of technology, data was collected concerning ten students' final grades in a freshman mathematics course and the students' overall grade point average at graduation or termination (suspension, withdrawal, etc.). These data are summarized in the table.

| GRADE IN FRESHMAN MATHEMATICS (x) | FINAL G.P.A. (y) | GRADE IN FRESHMAN MATHEMATICS (x) | FINAL G.P.A. (y) |
|---|---|---|---|
| 75 | 2.1 | 32 | 0.5 |
| 40 | 1.1 | 60 | 1.3 |
| 92 | 3.3 | 81 | 2.8 |
| 79 | 2.4 | 65 | 1.8 |
| 85 | 2.9 | 54 | 1.0 |

(a) Compute the least-squares line for these data.
(b) Predict the final G.P.A. of a student whose grade in freshman mathematics is 90.
(c) If the school requires at least a 2.0 final G.P.A. for a student to graduate, what is the lowest grade in freshman mathematics that predicts that a student is likely to have a high enough G.P.A. to graduate?

| YEAR (x) | CONSUMER PRICE INDEX (y) |
|---|---|
| 1977 | 181.5 |
| 1978 | 195.4 |
| 1979 | 217.4 |
| 1980 | 246.8 |
| 1981 | 272.4 |

**28.** (Consumer price index) The consumer price index for a number of years between 1977 and 1981 is given in the accompanying table.
   (a) Find the least-squares line for these data. *Hint:* In computing the least-squares line, it is simpler to let 1977 represent $x = 1$, to let 1978 represent $x = 2$, and so forth.
   (b) Use the least-squares line to estimate the consumer price index for 1982. The actual consumer price index for 1982 was 289.1.

**29.** The least-squares line is not the only curve that can be fit to a set of data points. For example, we might try to fit a parabola $y = ax^2 + bx + c$ or an exponential curve $y = ae^{bx}$ to a data set. If more than one curve is fit to the same set of data, we usually choose as the final model the curve for which the sum of the squared deviation terms is least, since this sum represents the total error in using the curve to approximate the data. For the data (1, 3), (2, 8), (3, 15), (4, 26), (5, 44), **(a)** find the least-squares line $y = mx + b$ by finding the minimum value of

$$f(m, b) = (3 - m - b)^2 + (8 - 2m - b)^2 + (15 - 3m - b)^2 + (26 - 4m - b)^2$$
$$+ (44 - 5m - b)^2$$

   **(b)** Find the *least-squares parabola* $y = ax^2 + bx + c$ by finding the minimum value of

$$g(a, b, c) = (3 - a - b - c)^2 + (8 - 4a - 2b - c)^2 + (15 - 9a - 3b - c)^2$$
$$+ (26 - 16a - 4b - c)^2 + (44 - 25a - 5b - c)^2$$

   *Hint:* Set each of $g_a$, $g_b$, and $g_c$ equal to zero, solve the resulting system, and assume that this critical point corresponds to the minimum.
   **(c)** Which of these two curves fits the data more closely? Explain.

OPTIONAL

GRAPHING CALCULATOR/
COMPUTER EXERCISE

**30.** In practice, data sets from which regression lines are computed may contain hundreds or even thousands of data points. These types of problems are best handled using a computer and the appropriate statistical software. Many pocket calculators also have statistical capabilities that allow you to compute $m$ and $b$ in the regression equation $y = mx + b$ for a limited number of data points. Some graphing calculators will even plot the scatter diagram and the least-squares line. If you have access to a calculator or a computer with statistical capabilities, find the least-squares line for the following data set: (7, 9), (6, 4), (5, 2), (6, 5), (7, 8), (5, 6), (9, 20), (9, 24), (10, 30), (10, 31), (10, 34), (10, 26), (11, 40), (11, 45), (11, 34), (11, 39), (12, 47), (12, 53), (12, 54), (13, 49), (13, 58), (13, 66), (14, 70), (14, 75), (14, 69), (15, 80), (15, 75), (15, 83), (16, 84), (16, 88).

# 8.6 DOUBLE INTEGRALS

## Partial Integration

In Section 8.2, we learned how to extend the concept of finding the derivative of a function to functions of two variables. For example, we found the partial derivative of a function $f(x, y)$ with respect to $x$ by treating $y$ as a constant and finding the "ordinary" derivative with respect to $x$. There is a similar way of extending the concept of finding an antiderivative (or indefinite integral) of a function to functions of two or more variables. The resulting procedure is referred to as *partial integration* or *partial antidifferentiation*.

As an example of partial integration with respect to $x$, let $f(x, y) = 6xy^2 + 4xy$ and consider

$$\int (6xy^2 + 4xy)\, dx$$

In this integral, the symbol "$dx$" indicates that we are to integrate $6xy^2 + 4xy$ with respect to $x$ and to treat $y$ as a constant. The result is

$$\int (6xy^2 + 4xy)\, dx = \frac{6x^2}{2} \cdot y^2 + \frac{4x^2}{2} \cdot y + C(y)$$
$$= 3x^2y^2 + 2x^2y + C(y)$$

Note that we added an arbitrary function of $y$, namely $C(y)$, to the result, rather than just a simple arbitrary constant $C$. The reason we do this is that since $\frac{\partial}{\partial x}[C(y)] = 0$, we have

$$\frac{\partial}{\partial x}[3x^2y^2 + 2x^2y + C(y)] = 6xy^2 + 4xy$$

It can be shown that $F(x, y) = 3x^2y^2 + 2x^2y + C(y)$ is the most general function having the property $\frac{\partial}{\partial x} F(x, y) = f(x, y) = 6xy^2 + 4xy$. We say that $F(x, y)$ is the *general partial antiderivative of $f(x, y)$ with respect to $x$.*

We can also compute the general partial antiderivative of $f(x, y) = 6xy^2 + 4xy$ with respect to $y$. This is given by

$$\int (6xy^2 + 4xy)\, dy = 6x \cdot \frac{y^3}{3} + 4x \cdot \frac{y^2}{2} + C(x)$$
$$= 2xy^3 + 2xy^2 + C(x)$$

where $C(x)$ is an arbitrary function of $x$.

The fundamental theorem of calculus can be used to evaluate an expression like $\int_0^1 (6xy^2 + 4xy)\, dx$ as follows:

$$\int_0^1 (6xy^2 + 4xy)\, dx = [3x^2y^2 + 2x^2y + C(y)]_{x=0}^{x=1}$$
$$= [3(1)^2y^2 + 2(1)^2y + C(y)] - [3(0)^2y^2 + 2(0)^2y + C(y)]$$
$$= 3y^2 + 2y$$

Notice that after computing the antiderivative with respect to $x$, we substitute the upper limit, 1, for $x$ in the antiderivative and subtract from this the result of substituting the lower limit, 0, for $x$ in the antiderivative. In the process, $C(y)$, the arbitrary function of $y$, cancels out. Since this is always the case in computing such integrals, we typically choose $C(y) = 0$ when evaluating an integral of the form $\int_a^b f(x, y)\, dx$, and we choose $C(x) = 0$ when evaluating an integral of the form $\int_c^d f(x, y)\, dy$.

## EXAMPLE 1

Find $\int_2^3 (6xy^2 + 4xy)\, dy$.

**Solution**

$$\int_2^3 (6xy^2 + 4xy)\, dy = [2xy^3 + 2xy^2]_{y=2}^{y=3} \quad (\text{Choose } C(x) = 0)$$
$$= [2x(3)^3 + 2x(3)^2] - [2x(2)^3 + 2x(2)^2]$$
$$= (54x + 18x) - (16x + 8x) = 48x \quad \blacksquare$$

### Iterated Integrals

Since $\int_0^1 (6xy^2 + 4xy)\, dx = 3y^2 + 2y$ is a function of $y$ alone, this function may be integrated with respect to $y$. For example, consider

$$\int_2^3 \left[ \int_0^1 (6xy^2 + 4xy)\, dx \right] dy$$

An integral like this is called an *iterated integral* because we must integrate twice to obtain the final result. To compute this integral, we work from the inside out, as follows:

$$\int_2^3 \left[ \int_0^1 (6xy^2 + 4xy)\, dx \right] dy = \int_2^3 \{[3x^2y^2 + 2x^2y]_{x=0}^{x=1}\}\, dy$$
$$= \int_2^3 (3y^2 + 2y)\, dy$$
$$= [y^3 + y^2]_2^3 = (3^3 + 3^2) - (2^3 + 2^2) = 24$$

## EXAMPLE 2

Compute $\int_0^1 [\int_2^3 (6xy^2 + 4xy)\, dy]\, dx$.

**Solution**

$$\int_0^1 \left[ \int_2^3 (6xy^2 + 4xy)\, dy \right] dx = \int_0^1 \{[2xy^3 + 2xy^2]_{y=2}^{y=3}\}\, dx$$
$$= \int_0^1 48x\, dx \quad (\text{See Example 1})$$
$$= [24x^2]_0^1 = 24(1)^2 - 24(0)^2 = 24 \quad \blacksquare$$

### Double Integrals over Rectangular Regions

You should observe that the preceding results show that

$$\int_0^1 \left[ \int_2^3 (6xy^2 + 4xy)\, dy \right] dx = \int_2^3 \left[ \int_0^1 (6xy^2 + 4xy)\, dx \right] dy = 24$$

In fact, for all of the functions we will consider in this text, it will be true that if $a$, $b$, $c$, and $d$ are constants, then

$$\int_a^b \left[ \int_c^d f(x, y)\, dy \right] dx = \int_c^d \left[ \int_a^b f(x, y)\, dx \right] dy$$

We will usually write an iterated integral without the brackets. That is, we will write $\int_a^b \int_c^d f(x, y)\, dy\, dx$ instead of $\int_a^b [\int_c^d f(x, y)\, dy]\, dx$ and $\int_c^d \int_a^b f(x, y)\, dx\, dy$ instead of $\int_c^d [\int_a^b f(x, y)\, dx]\, dy$.

By the statement in the box, the result is the same whether the order of integration is "*dydx*" (with respect to $y$ first and then with respect to $x$) or "*dxdy*" (with respect to $x$ first and then with respect to $y$). Thus we may define the *double integral* over a rectangular region $R$ as follows.

**DOUBLE INTEGRAL OVER A RECTANGULAR REGION**

Let $R$ be the rectangular region formed by the set of points $(x, y)$ such that $a \leq x \leq b$ and $c \leq y \leq d$ (see Figure 8.26). Then the *double integral* of $f(x, y)$ over $R$ is

$$\iint_R f(x, y)\, dA = \int_a^b \int_c^d f(x, y)\, dy\, dx = \int_c^d \int_a^b f(x, y)\, dx\, dy$$

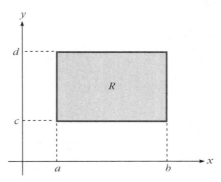

**FIGURE 8.26** The rectangular region $R$ defined by $a \leq x \leq b$ and $c \leq y \leq d$.

In the symbolism $\iint_R f(x, y)\, dA$, the letter $R$ denotes the region of integration, $f(x, y)$ is the integrand, and $dA$ is a "differential of area," indicating that the double integral is computed over a two-dimensional region.

**EXAMPLE 3**

Find $\iint_R \left(3 - \tfrac{1}{2}x - \tfrac{1}{4}y\right) dA$, where $R$ is the rectangular region formed by the set of points $(x, y)$ that satisfy $0 \leq x \leq 2$ and $0 \leq y \leq 4$.

**Solution**

The rectangular region $R$ is sketched in Figure 8.27. Integrating in the order "*dydx*," we have

$$\iint_R \left(3 - \tfrac{1}{2}x - \tfrac{1}{4}y\right) dA = \int_0^2 \int_0^4 \left(3 - \tfrac{1}{2}x - \tfrac{1}{4}y\right) dy\, dx$$

$$= \int_0^2 \left\{ \left[3y - \tfrac{1}{2}xy - \tfrac{1}{8}y^2\right]_{y=0}^{y=4} \right\} dx$$

$$= \int_0^2 \left[3(4) - \tfrac{1}{2}x(4) - \tfrac{1}{8}(4)^2\right] dx$$

$$= \int_0^2 (10 - 2x)\, dx = [10x - x^2]_0^2$$

$$= 10(2) - 2^2 = 16$$

8.6 DOUBLE INTEGRALS   555

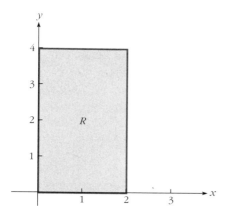

FIGURE 8.27 The rectangular region $R$ defined by $0 \leq x \leq 2$ and $0 \leq y \leq 4$.

We also could have integrated in the order "$dxdy$" to obtain the same result. To illustrate this and as a check on our work we compute

$$\iint_R \left(3 - \tfrac{1}{2}x - \tfrac{1}{4}y\right) dA = \int_0^4 \int_0^2 \left(3 - \tfrac{1}{2}x - \tfrac{1}{4}y\right) dxdy$$

$$= \int_0^4 \left\{\left[3x - \tfrac{1}{4}x^2 - \tfrac{1}{4}yx\right]_{x=0}^{x=2}\right\} dy$$

$$= \int_0^4 \left[3(2) - \tfrac{1}{4}(2)^2 - \tfrac{1}{4}y(2)\right] dy$$

$$= \int_0^4 \left(5 - \tfrac{1}{2}y\right) dy = \left[5y - \tfrac{1}{4}y^2\right]_0^4$$

$$= 5(4) - \tfrac{1}{4}(4)^2 = 16 \qquad \blacksquare$$

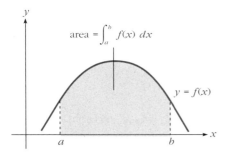

FIGURE 8.28 The definite integral $\int_a^b f(x)\, dx$ gives the area under the graph of $f(x)$ between $a$ and $b$ if $f(x) \geq 0$ on $[a, b]$.

### The Double Integral over a Rectangular Region as a Volume

Recall that if a function $f(x)$ is continuous and nonnegative on an interval $[a, b]$, then the (single) integral $\int_a^b f(x)\, dx$ gives the *area* under the graph of $f(x)$ from $a$ to $b$; see Figure 8.28. It can be shown that if $f(x, y)$ is continuous and nonnegative for all $(x, y)$ in the rectangle $R$ defined by $a \leq x \leq b$ and $c \leq y \leq d$, then the double integral $\iint_R f(x, y)\, dA$ represents the *volume* of the solid with top the surface $f(x, y)$ and with base the rectangular region $R$; see Figure 8.29.

As an example of finding the volume of a solid using double integration, recall that in Example 3 we found that if $R$ is the rectangular region formed by the set of points $(x, y)$ satisfying $0 \leq x \leq 2$ and $0 \leq y \leq 4$, then

$$\iint_R \left(3 - \frac{1}{2}x - \frac{1}{4}y\right) dA = 16$$

Since $f(x, y) = 3 - \tfrac{1}{2}x - \tfrac{1}{4}y \geq 0$ for all $(x, y)$ in $R$, the value of this double integral represents the volume of the solid that has a top consisting of the plane $f(x, y) = 3 - \tfrac{1}{2}x - \tfrac{1}{4}y$ and a base consisting of the rectangular region $R$. This solid is sketched in Figure 8.30 and has a volume of 16 cubic units.

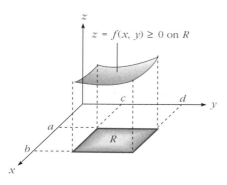

FIGURE 8.29 The volume of the solid with base the rectangular region $R$ and with top the surface $z = f(x, y)$ is given by $\iint_R f(x, y)\, dA$.

FIGURE 8.30 The solid that has a base consisting of the rectangular region $R$ defined by $0 \leq x \leq 2$ and $0 \leq y \leq 4$ and a top consisting of the plane $f(x, y) = 3 - \tfrac{1}{2}x - \tfrac{1}{4}y$ has a volume of 16 cubic units.

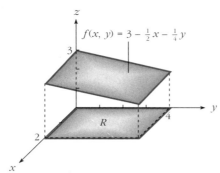

### Iterated Integrals with Variable Limits of Integration

The "inner limits" on an iterated integral may be functions of the "outer variable" of integration. For a function $f(x, y)$, we define two types of iterated integrals:

$$\int_a^b \int_{g_1(x)}^{g_2(x)} f(x, y) \, dy\,dx \qquad \text{(Type I)}$$

and

$$\int_c^d \int_{h_1(y)}^{h_2(y)} f(x, y) \, dx\,dy \qquad \text{(Type II)}$$

Note that in a Type I iterated integral, the outer variable of integration is $x$, the inner limits ($g_1(x)$ and $g_2(x)$) are functions of $x$, and the outer limits are constants ($a$ and $b$). In a Type II iterated integral, the outer variable of integration is $y$, the inner limits ($h_1(y)$ and $h_2(y)$) are functions of $y$, and the outer limits are constants ($c$ and $d$). The result of evaluating either a Type I or a Type II integral is a number.

Let's look at an example of each type of iterated integral.

---

**EXAMPLE 4**

Evaluate the following iterated integrals: (a) $\int_0^1 \int_{x^2}^{\sqrt{x}} 2xy \, dy\,dx$; (b) $\int_0^1 \int_{2y}^2 e^{x+y} \, dx\,dy$.

**Solution**

(a) This is a Type I integral with order of integration "$dy\,dx$" and so it is treated as follows:

$$\int_0^1 \int_{x^2}^{\sqrt{x}} 2x \, dy\,dx = \int_0^1 \left[ \int_{x^2}^{\sqrt{x}} 2xy \, dy \right] dx$$

$$= \int_0^1 \{[xy^2]_{y=x^2}^{y=\sqrt{x}}\} \, dx = \int_0^1 [x(\sqrt{x})^2 - x(x^2)^2] \, dx$$

$$= \int_0^1 (x^2 - x^5) \, dx = \left(\tfrac{1}{3}x^3 - \tfrac{1}{6}x^6\right)_0^1 = \tfrac{1}{3}(1)^3 - \tfrac{1}{6}(1)^6 = \tfrac{1}{6}$$

(b) This iterated integral is a Type II integral with order of integration "$dx\,dy$," and so

$$\int_0^1 \int_{2y}^2 e^{x+y} \, dx\,dy = \int_0^1 \left[\int_{2y}^2 e^{x+y} \, dx\right] dy$$

$$= \int_0^1 \{[e^{x+y}]_{x=2y}^{x=2}\} \, dy = \int_0^1 (e^{y+2} - e^{3y}) \, dy$$

$$= \left[e^{y+2} - \tfrac{1}{3}e^{3y}\right]_0^1 = \left(e^{1+2} - \tfrac{1}{3}e^{3(1)}\right) - \left(e^{0+2} - \tfrac{1}{3}e^{3(0)}\right)$$

$$= \tfrac{2}{3}e^3 - e^2 + \tfrac{1}{3} \qquad \blacksquare$$

### The Double Integral over Nonrectangular Regions

The concept of the double integral can be generalized to allow us to find the double integral of a function $f(x, y)$ over certain types of nonrectangular regions. Such regions fall into two categories: Type I and Type II; see Figure 8.31. The definitions of these regions are as follows.

## 8.6 DOUBLE INTEGRALS

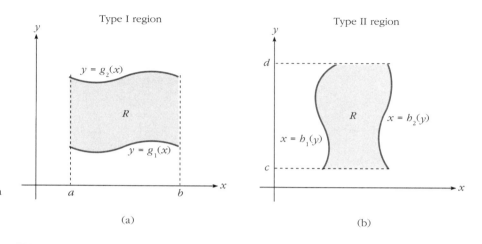

FIGURE 8.31 A Type I region is sketched in part (a) and a Type II region is sketched in part (b).

**TYPE I AND TYPE II REGIONS IN TWO DIMENSIONS**

A set of points $(x, y)$ is called a *Type I region* if $x$ and $y$ satisfy $a \leq x \leq b$ and $g_1(x) \leq y \leq g_2(x)$ for some functions $g_1(x)$ and $g_2(x)$ that are continuous on $[a, b]$.

A set of points $(x, y)$ is called a *Type II region* if $x$ and $y$ satisfy $c \leq y \leq d$ and $h_1(y) \leq x \leq h_2(y)$ for some functions $h_1(y)$ and $h_2(y)$ that are continuous on $[c, d]$.

We can now define the double integral over these types of regions in terms of Type I and Type II iterated integrals as follows:

**DOUBLE INTEGRALS OVER TYPE I AND TYPE II REGIONS**

Let $R$ be the Type I region of Figure 8.31(a) and suppose $f(x, y)$ is continuous at all points $(x, y)$ in $R$. Then the double integral of $f(x, y)$ over $R$ is

$$\iint_R f(x, y) \, dA = \int_a^b \int_{g_1(x)}^{g_2(x)} f(x, y) \, dy dx$$

Let $R$ be the Type II region of Figure 8.31(b) and suppose $f(x, y)$ is continuous at all points $(x, y)$ in $R$. Then the double integral of $f(x, y)$ over $R$ is

$$\iint_R f(x, y) \, dA = \int_c^d \int_{h_1(y)}^{h_2(y)} f(x, y) \, dx dy$$

---

**EXAMPLE 5**

Evaluate $\iint_R (2x + y) \, dA$ where $R$ is the finite region enclosed by the graphs of $y = x^2$ and $y = x$. Treat $R$ as a Type I region.

**Solution**

The region $R$ is sketched in Figure 8.32. The points of intersection of the curves $y = x^2$ and $y = x$ are found by solving the equation $x^2 = x$ for $x$ to obtain $x = 0$ and $x = 1$. Thus when $x = 0$ we get $y = 0$, and when $x = 1$ we obtain $y = 1$, so that the points of intersection are $(0, 0)$ and $(1, 1)$. Treating this region

## 558  CHAPTER 8  FUNCTIONS OF MORE THAN ONE VARIABLE

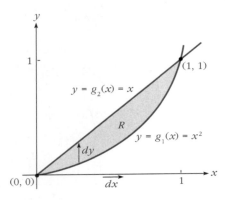

FIGURE 8.32  The region of Example 5 viewed as a Type I region. In this case we integrate in the order *dydx*: first vertically from $y = g_1(x) = x^2$ to $y = g_2(x) = x$ and then horizontally from $x = 0$ to $x = 1$.

as a Type I region, we see that the lower curve is $y = g_1(x) = x^2$ and the upper curve is $y = g_2(x) = x$. Thus we integrate in the vertical direction first (as indicated by the vertical arrow in the figure) from the lower curve, $g_1(x) = x^2$, to the upper curve, $g_2(x) = x$. Then we integrate in the horizontal direction (as indicated by the horizontal arrow on the $x$-axis) from $x = 0$ to $x = 1$. The result is

$$\iint_R (2x + y)\, dA = \int_0^1 \int_{x^2}^{x} (2x + y)\, dy\, dx$$

We now proceed to evaluate the iterated integral on the right-hand side as follows:

$$\iint_R (2x + y)\, dA = \int_0^1 \left[ \int_{x^2}^{x} (2x + y)\, dy \right] dx = \int_0^1 \left\{ \left[ 2xy + \tfrac{1}{2}y^2 \right]_{y=x^2}^{y=x} \right\} dx$$

$$= \int_0^1 \left[ \left(2x \cdot x + \tfrac{1}{2}x^2\right) - \left(2x \cdot x^2 + \tfrac{1}{2}(x^2)^2\right) \right] dx$$

$$= \int_0^1 \left( \tfrac{5}{2}x^2 - 2x^3 - \tfrac{1}{2}x^4 \right) dx = \left[ \tfrac{5}{6}x^3 - \tfrac{1}{2}x^4 - \tfrac{1}{10}x^5 \right]_0^1$$

$$= \tfrac{5}{6}(1)^3 - \tfrac{1}{2}(1)^4 - \tfrac{1}{10}(1)^5 = \tfrac{7}{30} \qquad ■$$

---

EXAMPLE 6 (Participative)

Solution

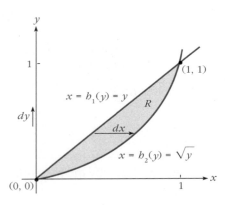

FIGURE 8.33  The region of Example 6 viewed as a Type II region. We integrate in the order *dxdy*: first horizontally from $x = h_1(y) = y$ to $x = h_2(y) = \sqrt{y}$ and then vertically from $y = 0$ to $y = 1$.

Let's evaluate $\iint_R (2x + y)\, dA$, where $R$ is the finite region enclosed by the graphs of $y = x^2$ and $y = x$ and let's treat it as a Type II region.

The region $R$ is sketched in Figure 8.33. Observe that this region $R$ and the integrand $f(x, y) = 2x + y$ are the same as those of Example 5. The difference here is that we will now view $R$ as a Type II region rather than a Type I region. In order to do this, we must express $y = x^2$ and $y = x$ as functions of $x$ by solving each of them for $x$ in terms of $y$. Thus, if $y = x^2$ then $x = h_2(y) = \sqrt{y}$. We choose the positive square root to define our function because $R$ clearly lies in quadrant I, where $y$ is _____. Also if $y = x$, then $x = h_1(y) = y$. So the left curve is $x = h_1(y) = y$ and the right curve is $x = h_2(y) = \sqrt{y}$. We integrate in the _____ direction first (as indicated by the horizontal arrow in this figure) from the left curve $h_1(y) = y$ to the right curve $h_2(y) = \sqrt{y}$. Then we integrate in the _____ direction (as indicated by the vertical arrow on the $y$-axis) from $y = 0$ to $y = $ _____. We obtain

$$\iint_R (2x + y)\, dA = \underline{\hspace{4cm}}$$

To evaluate the iterated integral on the right-hand side, we proceed as follows:

$$\iint_R (2x + y)\, dA = \int_0^1 \left[ \int_y^{\sqrt{y}} (2x + y)\, dx \right] dy$$

$$= \int_0^1 \left\{ \left[ \underline{\phantom{xxxxxx}} \right]_{x=y}^{x=\sqrt{y}} \right\} dy \qquad \text{(Integrating with respect to } x\text{)}$$

$$= \int_0^1 \{[(\sqrt{y})^2 + \sqrt{y}(y)] - (\underline{\phantom{xxxx}})\}\, dy \qquad \text{(Evaluating at the limits } x = \sqrt{y} \text{ and } x = y\text{)}$$

$$= \int_0^1 (\underline{\phantom{xxxxxxxx}})\, dy \qquad \text{(Simplifying)}$$

$$= \left[ \tfrac{1}{2}y^2 + \tfrac{2}{5}y^{5/2} - \tfrac{2}{3}y^3 \right]_0^1 \qquad \text{(Integrating with respect to } y\text{)}$$

$$= \tfrac{1}{2}(1)^2 + \tfrac{2}{5}(1)^{5/2} - \tfrac{2}{3}(1)^3 \qquad \text{(Evaluating at the limits } y = 1 \text{ and } y = 0\text{)}$$

$$= \underline{\phantom{xxxx}}. \qquad \blacksquare$$

You should notice that the evaluation of $\iint_R (2x + y)\, dA$ in Examples 5 and 6 produced the same result. In general, it is true that when a region can be viewed as both a Type I region and a Type II region, the resulting iterated integrals produce the same result. However, in certain instances it may be difficult or impossible to evaluate a double integral when $R$ is viewed as one type of region and relatively easy to evaluate it when $R$ is viewed as the other type. The next example illustrates this situation.

**EXAMPLE 7**  Find $\iint_R e^{y^2}\, dA$, where $R$ is the region enclosed by the graphs of $y = 4$, $y = 2x$, and $x = 0$.

**Solution**  The given region $R$ is sketched in Figure 8.34. If we consider $R$ to be a Type I region, then we obtain

$$\iint_R e^{y^2}\, dA = \int_0^2 \int_{2x}^4 e^{y^2}\, dy\, dx$$

However, in this case we must compute $\int e^{y^2}\, dy$, which does not integrate into any finite combination of elementary functions. Let us therefore view $R$ as a Type II region and use the integration order $dx\, dy$:

---

**SOLUTIONS TO EXAMPLE 6**

positive, horizontal, vertical, 1, $\int_0^1 \int_y^{\sqrt{y}} (2x + y)\, dx\, dy$, $x^2 + yx$, $y^2 + y^2$, $y + y^{3/2} - 2y^2$, $\tfrac{7}{30}$

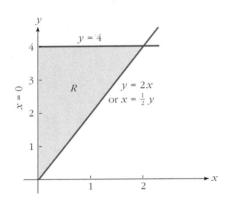

**FIGURE 8.34** The region of Example 7. The computation of the integral $\iint_R e^{y^2} dA$ is easier if we view $R$ as a Type II region.

$$\iint_R e^{y^2} dA = \int_0^4 \int_0^{y/2} e^{y^2} dx dy = \int_0^4 \{[xe^{y^2}]_{x=0}^{x=y/2}\} dy$$

$$= \int_0^4 \tfrac{1}{2} y e^{y^2} dy = \tfrac{1}{2} \int_0^4 e^{y^2} \cdot y \, dy$$

Now we can make the substitution $u = y^2$, so that $du = 2y\,dy$ and $y\,dy = \tfrac{1}{2} du$. Then when $y = 0$, we have $u = 0$, and when $y = 4$, we have $u = 16$. Thus

$$\tfrac{1}{2} \int_0^4 e^{y^2} \cdot y \, dy = \tfrac{1}{2} \int_0^{16} e^u \cdot \tfrac{1}{2} du = \tfrac{1}{4}[e^u]_0^{16} = \tfrac{1}{4}(e^{16} - 1) \quad \blacksquare$$

### The Double Integral over a Nonrectangular Region as a Volume

We have seen earlier in this section that if $f(x,y)$ is continuous and nonnegative on a rectangular region $R$, then $\iint_R f(x,y) \, dA$ represents the volume of the solid with base $R$ and top $f(x,y)$. We can extend this idea by generalizing $R$ to a Type I or Type II region.

**THE DOUBLE INTEGRAL AS A VOLUME**

> Let $R$ be a Type I or a Type II region in the $xy$-plane and suppose that $f(x,y)$ is continuous and nonnegative for all $(x,y)$ in $R$. Then $\iint_R f(x,y) \, dA$ is the volume of the solid that has a top consisting of the surface $f(x,y)$ and a base consisting of the region $R$.

Figure 8.35 illustrates the boxed result for a Type I region. The sketch for a Type II region is analogous. Let's look at an example of finding a volume using double integration.

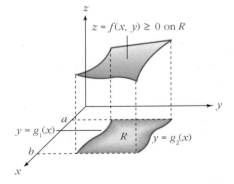

**FIGURE 8.35** The solid that has a base consisting of the Type I region of all points $(x, y)$ satisfying $a \leq x \leq b$ and $g_1(x) \leq y \leq g_2(x)$ and a top consisting of the surface $z = f(x, y)$ has volume $\iint_R f(x, y) \, dA$.

**EXAMPLE 8**  Find the volume in the first octant bounded by the three coordinate planes and the plane $z = 6 - 3x - 2y$.

**Solution**  The solid is sketched in Figure 8.36 and is a four-sided solid with faces that are all triangles. Such a solid is called a tetrahedron. In this case the intersection of the plane $z = 6 - 3x - 2y$ and the $xy$-plane is the straight line $0 = 6 - 3x - 2y$ or $3x + 2y = 6$; it forms one boundary of the triangular

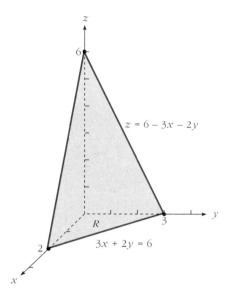
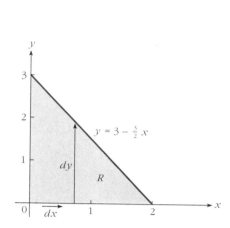

**FIGURE 8.36** The solid in the first octant bounded by the three coordinate planes and the plane $z = 6 - 3x - 2y$ is a tetrahedron.

**FIGURE 8.37** The region $R$ of Example 8 drawn in two dimensions and viewed as a Type I region.

region $R$ in the $xy$-plane. This region is sketched in two dimensions in Figure 8.37; we will view it as a Type I region. The top of the tetrahedron is the surface $z = f(x, y) = 6 - 3x - 2y$, and the volume is

$$\iint_R (6 - 3x - 2y)\, dA = \int_0^2 \int_0^{3-3x/2} (6 - 3x - 2y)\, dy\, dx$$

$$= \int_0^2 \{[6y - 3xy - y^2]_{y=0}^{y=3-3x/2}\}\, dx$$

$$= \int_0^2 \left[6\left(3 - \tfrac{3}{2}x\right) - 3x\left(3 - \tfrac{3}{2}x\right) - \left(3 - \tfrac{3}{2}x\right)^2\right] dx$$

$$= \int_0^2 \left(9 - 9x + \tfrac{9}{4}x^2\right) dx = \left[9x - \tfrac{9}{2}x^2 + \tfrac{3}{4}x^3\right]_0^2$$

$$= 9(2) - \tfrac{9}{2}(2)^2 + \tfrac{3}{4}(2)^3 = 6$$

Therefore the volume of the solid is 6 cubic units. As an exercise, find the volume of the tetrahedron by viewing the triangular region $R$ in the $xy$-plane as a Type II region. ∎

## EXAMPLE 9 (Participative)

Let's find the volume of the solid with top equal to the surface $z = 2x\sqrt{y}$ and base the triangular region in the $xy$-plane with vertices are $(0, 0)$, $(2, 1)$, and $(4, 0)$.

**Solution**

In the form of a double integral, the required volume is _____.
The triangular region $R$ is sketched in Figure 8.38. First let's find the equation

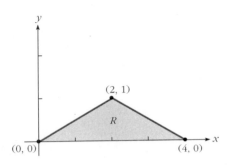

**FIGURE 8.38** The triangular region $R$ with vertices $(0, 0)$, $(2, 1)$, and $(4, 0)$.

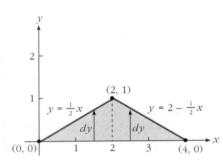

**FIGURE 8.39** The region of Example 9 viewed as a Type I region: evaluating a double integral requires the use of two iterated integrals.

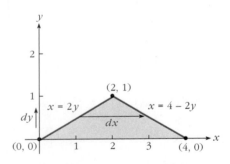

**FIGURE 8.40** The region of Example 9 viewed as a Type II region: evaluating a double integral requires only one iterated integral.

of the line segment joining $(0, 0)$ and $(2, 1)$. For this we can use the two-point form of the equation of a straight line passing through $(x_1, y_1)$ and $(x_2, y_2)$:

$$y - y_1 = \frac{y_2 - y_1}{x_2 - x_1}(x - x_1)$$

If we let $(x_1, y_1) = (0, 0)$ and $(x_2, y_2) = (2, 1)$, the equation of the line segment is _____ if $0 \leq x \leq 2$. To find the equation of the line segment joining $(2, 1)$ and $(4, 0)$, we use the two-point formula again with $(x_1, y_1) = (2, 1)$ and $(x_2, y_2) = (4, 0)$. The result is $y =$ _____ if $2 \leq x \leq 4$.

As indicated in Figure 8.39, if we view the region $R$ as a Type I region, we must evaluate *two* iterated integrals in order to find the required volume. That is,

$$\text{volume} = \iint_R 2x\sqrt{y}\, dA = \int_0^2 \int_0^{x/2} 2x\sqrt{y}\, dy\,dx + \underline{\hspace{2cm}}$$

Rather than find the volume in this way, let's see what happens if we view $R$ as a Type II region; see Figure 8.40. Notice that in order to view $R$ as a Type II region, we have solved $y = \frac{1}{2}x$ for $x$ to obtain $x =$ _____ and we have solved $y = 2 - \frac{1}{2}x$ for $x$ to obtain $x =$ _____. We find that

$$\text{volume} = \iint_R 2x\sqrt{y}\, dA = \underline{\hspace{2cm}}$$

$$= \int_0^1 \{[\underline{\hspace{1cm}}]_{x=2y}^{x=4-2y}\}\, dy$$

$$= \int_0^1 [(4 - 2y)^2\sqrt{y} - (2y)^2\sqrt{y}]\, dy$$

$$= \int_0^1 (\underline{\hspace{2cm}})\, dy \quad \text{(Simplifying the integrand)}$$

$$= \left[\frac{32}{3}y^{3/2} - \frac{32}{5}y^{5/2}\right]_0^1 = \underline{\hspace{1cm}}$$

The volume of the solid is _____ cubic inches. As an exercise, verify that you obtain the same result by viewing $R$ as a Type I region and evaluating the two iterated integrals. ∎

### Average Value of a Function of Two Variables

Recall that in Section 7.2, we defined the average value of a function $y = f(x)$ over the interval $[a, b]$ to be

$$\text{average value} = \frac{1}{b - a}\int_a^b f(x)\, dx$$

---

**SOLUTIONS TO EXAMPLE 9**   $\iint_R 2x\sqrt{y}\, dA$, $y = \frac{1}{2}x$, $2 - \frac{1}{2}x$, $\int_2^4 \int_0^{2-x/2} 2x\sqrt{y}\, dy\,dx$, $2y$, $4 - 2y$, $\int_0^1 \int_{2y}^{4-2y} 2x\sqrt{y}\, dx\,dy$, $x^2\sqrt{y}$, $16y^{1/2} - 16y^{3/2}$, $\frac{64}{15}$, $\frac{64}{15}$

provided that $f$ is continuous on $[a, b]$. This concept generalizes easily to a function of two variables $z = f(x, y)$ defined over a region $R$ in the $xy$-plane.

**THE AVERAGE VALUE OF A FUNCTION OF TWO VARIABLES**

Let $z = f(x, y)$ be continuous for all points $(x, y)$ in a region $R$. Then if $R$ is either a Type I region or a Type II region, the average value of $f(x, y)$ over $R$ is defined to be

$$\text{average value} = \frac{1}{\text{area of } R} \iint_R f(x, y) \, dA$$

The area of the region $R$ in this definition can be found by using a single integral. For example, referring to the regions in Figure 8.31,

$$\text{area of a Type I region} = \int_a^b [g_2(x) - g_1(x)] \, dx$$

and

$$\text{area of a Type II region} = \int_c^d [h_2(y) - h_1(y)] \, dy$$

## EXAMPLE 10

The weekly cost (in dollars) to a certain company of producing $x$ units of product $A$ and $y$ units of product $B$ is given by

$$C(x, y) = 0.3x^2 + 0.21y^2 + 0.06xy + 400$$

If the number of units of product $A$ produced per week varies between 50 and 70 units and if the number of units of $B$ produced per week varies between 100 and twice the weekly production of $A$, estimate the average weekly cost of producing the two products.

### Solution

The region of integration for the required double integral is sketched in Figure 8.41. Observe that this region is defined by the inequalities $50 \leq x \leq 70$ and $100 \leq y \leq 2x$; we will view it as a Type I region. The area of this region can be computed by evaluating the single integral $\int_{50}^{70} (2x - 100) \, dx$. The result is that the area of $R$ is 400 square units. Observe that since $R$ is a triangular region, a more elementary way to compute the area is by using the formula from geometry:

$$\text{area of } R = \tfrac{1}{2}(\text{base})(\text{height})$$

We obtain again

$$\text{area of } R = \tfrac{1}{2}(20)(40) = 400 \text{ square units}$$

Thus the average weekly cost is given by the expression $\frac{1}{400} \iint_R C(x, y) \, dA$. Substituting $C(x, y) = 0.3x^2 + 0.21y^2 + 0.06xy + 400$ and expressing the double integral as an iterated integral, the average weekly cost is

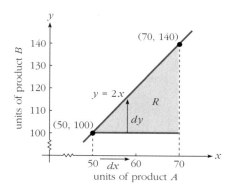

**FIGURE 8.41** The region of integration for Example 10 viewed as a Type I region.

$$\frac{1}{400}\int_{50}^{70}\int_{100}^{2x}(0.3x^2 + 0.21y^2 + 0.06xy + 400)\,dydx$$

$$= \frac{1}{400}\int_{50}^{70}\{[0.3x^2y + 0.07y^3 + 0.03xy^2 + 400y]_{y=100}^{y=2x}\}\,dx$$

$$= \frac{1}{400}\int_{50}^{70}\{[0.3x^2(2x) + 0.07(2x)^3 + 0.03x(2x)^2 + 400(2x)]$$
$$- [0.3x^2(100) + 0.07(100)^3 + 0.03x(100)^2 + 400(100)]\}\,dx$$

$$= \frac{1}{400}\int_{50}^{70}(0.6x^3 + 0.56x^3 + 0.12x^3 + 800x - 30x^2 - 70000 - 300x - 40000)\,dx$$

$$= \frac{1}{400}\int_{50}^{70}(1.28x^3 - 30x^2 + 500x - 110000)\,dx$$

$$= \frac{1}{400}[0.32x^4 - 10x^3 + 250x^2 - 110000x]_{50}^{70}$$

$$= \frac{1}{400}\{[0.32(70)^4 - 10(70)^3 + 250(70)^2 - 110000(70)]$$
$$- [0.32(50)^4 - 10(50)^3 + 250(50)^2 - 110000(50)]\}$$

$$= \frac{1903200}{400} = 4758$$

So the average weekly cost of producing the two products is $4758 per week. ∎

## SECTION 8.6
### SHORT ANSWER EXERCISES

1. (True or False) $\int (1/x)\,dy = (y/x) + C(x)$, where $C(x)$ is an arbitrary function of $x$. _____

2. (True or False) The result of the integration $\int_x^{x^2}(xy^3 + y^2)\,dx$ is some function of $y$. _____

3. (True or False) The integral $\int_{-1}^{1}\int_{-1}^{1}(x - y)\,dA$ represents the volume of the solid that has a base consisting of the rectangular region defined by $-1 \leq x \leq 1$ and $-1 \leq y \leq 1$ and a top consisting of the plane $z = x - y$. _____

4. (True or False) $\int_{-7}^{12}\int_{-5}^{10}\sqrt{x^2 + y^2}\,dxdy = \int_{-5}^{10}\int_{-7}^{12}\sqrt{x^2 + y^2}\,dydx$. _____

5. If we view the region of integration $R$ as a Type II region, the order of integration in $\iint_R f(x, y)\,dA$ is _____.

6. When viewed as a Type I region, the region $R$ can be described by the inequalities $0 \leq x \leq 3$ and $\frac{2}{3}x \leq y \leq 2$. Find the inequalities that describe $R$ as a Type II region. _____

7. (True or False) $\int_0^1\int_x^{x^2} x^2y\,dydx = \int_0^1\int_y^{y^2} x^2y\,dxdy$. _____

8. (True or False) The area of the Type I region sketched in Figure 8.31(a) can be expressed $\iint_R 1\,dA = \int_a^b\int_{g_1(x)}^{g_2(x)} 1\,dydx$. _____

9. Express the area of the Type II region sketched in Figure 8.31(b) as a double integral and as an iterated integral. _____

10. If $f(x, y) \geq 0$ for all $(x, y)$ in region $R$, then the average value of $f(x, y)$ over $R$ is the volume of the solid with base $R$ and top the surface $f(x, y)$ divided by _____.

## SECTION 8.6
## EXERCISES

*In Exercises 1–6, find the partial integrals. Check your answers by using partial differentiation.*

1. $\int (xy + y)\, dx$
2. $\int (x^2/y)\, dx$
3. $\int y\sqrt{y^2 + x}\, dy$
4. $\int (2x/\sqrt{x^2 + 3y^2})\, dx$
5. $\int ye^{2xy}\, dx$
6. $\int ye^{2xy}\, dy$

*In Exercises 7–14, evaluate the integrals.*

7. $\int_0^2 (4x^2y + xy^3)\, dy$
8. $\int_{-1}^1 (6xy + y - 5)\, dx$
9. $\int_{-2}^3 \dfrac{x}{x^2 + y^2}\, dx$
10. $\int_1^x \dfrac{6x^2y + 2y}{3x^2y^2 + y^2}\, dy$
11. $\int_x^{x+1} (2xy - 3x^2y^2)\, dy$
12. $\int_1^4 e^{2x+3y+1}\, dx$
13. $\int_{\sqrt{y}}^{\sqrt{y+2}} (\sqrt{y} + 2x\sqrt{y+2})\, dx$
14. $\int_{y^2+1}^{4y^2+4} \dfrac{y+2}{y^2+1}\, dx$

*In Exercises 15–30, evaluate the iterated integrals.*

15. $\int_0^1 \int_0^2 (xy + x)\, dx\, dy$
16. $\int_{-1}^1 \int_1^2 (2x + y)\, dy\, dx$
17. $\int_{-1}^1 \int_0^1 (3y^2 - 2x)\, dy\, dx$
18. $\int_0^1 \int_0^1 2x\sqrt{x^2 + y}\, dx\, dy$
19. $\int_1^2 \int_{-1}^1 [(x/y) + 1]\, dx\, dy$
20. $\int_0^1 \int_1^2 e^{2x+3y}\, dy\, dx$
21. $\int_0^1 \int_x^1 3\, dy\, dx$
22. $\int_0^1 \int_{2y}^2 2y\, dx\, dy$
23. $\int_1^4 \int_{\sqrt{y}}^2 xy\, dx\, dy$
24. $\int_0^2 \int_{x^2}^{2x} (x^2y - 1)\, dy\, dx$
25. $\int_0^1 \int_{\sqrt{x^2+1}}^1 (xy/\sqrt{x^2+1})\, dy\, dx$
26. $\int_0^{16} \int_{y/4}^{\sqrt{y}} (x^2 - y^2)\, dx\, dy$
27. $\int_1^e \int_y^1 (1/y)\, dx\, dy$
28. $\int_0^1 \int_{-2x}^{2x} e^{2x+y}\, dy\, dx$
29. $\int_1^2 \int_{-x}^x e^{x-y+1}\, dy\, dx$
30. $\int_0^1 \int_{\sqrt{y}}^{\sqrt[3]{y}} (3x^2 + y^2)\, dx\, dy$

*In Exercises 31–36, find the volume of the solid with top the surface $f(x, y)$ and with base the rectangular region R described by the given inequalities.*

31. $f(x, y) = 10 - x - y;\ 0 \le x \le 3,\ 0 \le y \le 4$
32. $f(x, y) = 15 - 2x - 3y;\ 1 \le x \le 2,\ 2 \le y \le 3$
33. $f(x, y) = 25 - x^2 - y^2;\ 1 \le x \le 2,\ 2 \le y \le 4$
34. $f(x, y) = x^2y^2 + 1;\ -1 \le x \le 2,\ -2 \le y \le 1$
35. $f(x, y) = 10 - x - 2y - xy;\ -1 \le x \le 1,\ -1 \le y \le 1$

**36.** $f(x, y) = x^2y + \sqrt{xy}$; $1 \le x \le 2$, $2 \le y \le 4$

**37.** Evaluate $\iint_R (x - 3y) \, dA$, where $R$ is the finite region enclosed by the graphs of $y = x^2$ and $y = x^3$. View $R$ as a Type I region.

**38.** Evaluate $\iint_R (xy - 4x + 5) \, dA$, where $R$ is the finite region enclosed by the graphs of $y = x^2$ and $y = 2x$. View $R$ as a Type I region.

**39.** Redo Exercise 37, treating $R$ as a Type II region.

**40.** Redo Exercise 38, treating $R$ as a Type II region.

**41.** Evaluate $\iint_R (x/y) \, dA$, where $R$ is the triangular region with vertices $(-1, 1)$, $(0, 3)$, and $(1, 1)$.

**42.** Evaluate $\iint_R e^{x+y} \, dA$, where $R$ is the triangular region with vertices $(2, 1)$, $(4, 1)$, and $(3, 3)$.

**43.** Evaluate $\iint_R xe^{y^2} \, dA$, where $R$ is the region defined by $0 \le x \le 1$ and $x^2 \le y \le 1$.

**44.** Evaluate $\iint_R (x^3y + x\sqrt{y}) \, dA$, where $R$ is the region defined by $0 \le y \le 4$ and $y \le x \le 4y$.

**45.** Evaluate $\iint_R 2y \, dA$, where $R$ is the region bounded by $y = \sqrt{x + 3}$, $y = \sqrt{x}$, $x = 1$, and $x = 4$.

**46.** Evaluate $\iint_R e^{4x} \, dA$, where $R$ is the region bounded by $y = x$, $y = x - 2$, $x = 0$, and $x = 1$.

**47.** Find the volume of the solid that has the surface $f(x, y) = 20 - x - y^2$ for a top and the triangular region in the $xy$-plane with vertices $(0, 0)$, $(2, 0)$, and $(0, 1)$ for a base.

**48.** Find the volume of the solid that has the surface $f(x, y) = x^2\sqrt{y}$ for a top and the triangular region in the $xy$-plane with vertices $(-2, 0)$, $(0, 4)$, and $(2, 0)$ for a base.

**49.** Find the volume of the solid that has the surface $f(x, y) = x^2 + y^2$ for a top and the finite region in the $xy$-plane formed by the intersection of the curves $y = x^2 - 2$ and $y = -x$ for a base.

**50.** Find the volume of the solid that has the surface $f(x, y) = 2\sqrt{xy}$ for a top and the finite region in the $xy$-plane formed by the intersection of the curves $y = \frac{1}{2}x$ and $y = \sqrt{x}$ for a base.

**51.** Find the volume in the first octant bounded by the three coordinate planes and the plane $z = 10 - x - 2y$.

**52.** Find the volume in the first octant bounded by the three coordinate planes and the plane $\frac{1}{2}x + \frac{1}{3}y + z = 1$.

**53.** (Volume) A storage shed with a rectangular base 4 ft × 8 ft has a sloping roof. If we think of the base as lying in the $xy$-plane with vertices $(0, 0)$, $(4, 0)$, $(0, 8)$, and $(4, 8)$, then the roof lies along the plane $z = 8 - \frac{3}{8}y$. Find the volume of the shed.

**54.** (Average cost) The weekly cost in dollars to a certain company of producing $x$ units of product $A$ and $y$ units of product $B$ is given by $C(x, y) = 0.6x^2 + 0.32y^2 + 0.012xy + 300$. If the number of units of $B$ produced per week varies between 40 and 70 and the number of units $A$ varies between 60 and one and one-half times the production of $B$, estimate the weekly average weekly cost of producing the two products.

**55. (Average production)** Given the weekly production function $z = 200x^{2/3}y^{1/3}$, where $x$ is the number of units of labor, $y$ is the number of units of capital, and $z$ is the weekly production in units, find the average weekly production if $x$ varies between 27 and 64 and $y$ varies between 8 and 27.

**56. (Average profit)** The weekly profit in dollars of a certain company from the sale of $x$ units of product $A$ and $y$ units of product $B$ is

$$P(x, y) = 70x + 80y - xy - \tfrac{1}{2}x^2 - \tfrac{1}{4}y^2$$

If weekly sales of product $A$ vary between 30 and 60 and weekly sales of product $B$ vary between 10 and 40, estimate the average weekly profit.

Business and Economics

**57. (Average revenue)** A manufacturer makes and sells two products, $A$ and $B$, with yearly demand equations

$$x = 300 - 3p - s \quad \text{(product A)}$$
$$y = 250 - p - 5s \quad \text{(product B)}$$

where $p$ is the price per unit (in dollars) of product $A$, $s$ is the price per unit (in dollars) of product $B$, and $x$ and $y$ are the yearly demands (in thousands of units) for $A$ and $B$, respectively. If the price of $A$ varies between 40 and 50 dollars and the price of $B$ varies between 15 and 30 dollars, find the average yearly revenue.

**58. (Manufacturing)** A manufacturing process requires that a small hole be drilled in a triangular piece of sheet metal at the centroid, which is the point at which the piece of sheet metal would balance if placed on a pointed object, such as a nail. If the vertices of the triangle are $(0, 0)$, $(5, 10)$, and $(8, 0)$ and if the coordinates $(\bar{x}, \bar{y})$ of the centroid are defined by

$$\bar{x} = \frac{\iint\limits_R x \, dA}{\iint\limits_R dA}, \quad \bar{y} = \frac{\iint\limits_R y \, dA}{\iint\limits_R dA}$$

where $R$ is the (triangular) region, find $(\bar{x}, \bar{y})$.

Life Sciences

**59. (Concentration of deer)** An ancient tribe once hunted a rectangular region surrounding their campground for deer. The vertices of the hunting ground (considered as points in the $xy$-plane) were $(-10, 5)$, $(-10, -5)$, $(10, 5)$, and $(10, -5)$, where distances are given in miles and the tribe's campground was located at $(0, 0)$. If the concentration of deer per square mile at a point $d$ miles from $(0, 0)$ was given by $\tfrac{1}{15}d^2$, find the average concentration of deer in the hunting ground.

OPTIONAL

GRAPHING CALCULATOR/
COMPUTER EXERCISES

*In Exercises 60–63, estimate $\iint\limits_R (3x + 2y) \, dA$, where $R$ is the finite region bounded by the graphs of $f(x)$ and $g(x)$. Estimate the x- and y-coordinates of the points of intersection of the graphs of $f$ and $g$ to two decimal places.*

**60.** $f(x) = x^3; \quad g(x) = x^4 - 4$ 

**61.** $f(x) = x^2; \quad g(x) = \sqrt{x + 2}$

**62.** $f(x) = x^2 + 4x - 2; \quad g(x) = e^x$ (in quadrant I)

**63.** $f(x) = \sqrt{x}; \quad g(x) = x^3 - x + 1$

## 8.7 CHAPTER EIGHT REVIEW

**KEY TERMS**

function of more than one variable
surface
first octant
plane
distance formula in three dimensions
sphere
trace
partial derivative
substitute products
complementary products
marginal productivity of labor
marginal productivity of capital
second-order partial derivatives
relative maximum point
relative minimum point
saddle point
objective function
constraint

Lagrange multiplier
level curve
constrained optimization
data points
scatter diagram
least-squares line
line of regression
extrapolation
normal equations
partial integration
general partial antiderivative
iterated integral
double integral
volume of a solid
Type I region
Type II region
average value

**KEY FORMULAS AND RESULTS**

### The Equation of a Plane

The graph of $ax + by + cz + d = 0$ is a plane in three-dimensional space. It can be graphed (in most cases) by finding the $x$-intercept (set $y$ and $z$ equal to zero and solve for $x$), the $y$-intercept (set $x$ and $z$ equal to zero and solve for $y$), and the $z$-intercept (set $x$ and $y$ equal to zero and solve for $z$).

### The Distance Formula in Three Dimensions

The distance between two points $(x_1, y_1, z_1)$ and $(x_2, y_2, z_2)$ is

$$d = \sqrt{(x_2 - x_1)^2 + (y_2 - y_1)^2 + (z_2 - z_1)^2}$$

### The Equation of a Sphere

The graph of $(x - a)^2 + (y - b)^2 + (z - c)^2 = r^2$ is a sphere in three-dimensional space with center $(a, b, c)$ and radius $r$.

### Partial Derivatives: The Limit Definition

Provided the limits exist,

$$\frac{\partial f}{\partial x} = \lim_{\Delta x \to 0} \frac{f(x + \Delta x, y) - f(x, y)}{\Delta x}$$

and

$$\frac{\partial f}{\partial y} = \lim_{\Delta y \to 0} \frac{f(x, y + \Delta y) - f(x, y)}{\Delta y}$$

### Substitute Products and Complementary Products

If $p$ is the price per unit of product $A$ and $s$ is the price per unit of product $B$ and if the demand functions for $A$ and $B$ are

$x = f(p, s)$     (demand function for $A$)
$y = g(p, s)$     (demand function for $B$)

then

i. $A$ and $B$ are substitutes if $\frac{\partial x}{\partial s} > 0$ and $\frac{\partial y}{\partial p} > 0$;

ii. $A$ and $B$ are complements if $\frac{\partial x}{\partial s} < 0$ and $\frac{\partial y}{\partial p} < 0$;

iii. $A$ and $B$ are neither substitutes nor complements if $\frac{\partial x}{\partial s}$ and $\frac{\partial y}{\partial p}$ differ in sign.

### Second-Order Partial Derivatives: Notation

If $z = f(x, y)$, then the four second-order partial derivatives of $f$ are denoted by

$$\frac{\partial^2 z}{\partial x^2} = \frac{\partial}{\partial x}\left(\frac{\partial z}{\partial x}\right) = f_{xx}(x, y)$$

$$\frac{\partial^2 z}{\partial y \partial x} = \frac{\partial}{\partial y}\left(\frac{\partial z}{\partial x}\right) = f_{xy}(x, y)$$

$$\frac{\partial^2 z}{\partial x \partial y} = \frac{\partial}{\partial x}\left(\frac{\partial z}{\partial y}\right) = f_{yx}(x, y)$$

$$\frac{\partial^2 z}{\partial y^2} = \frac{\partial}{\partial y}\left(\frac{\partial z}{\partial y}\right) = f_{yy}(x, y)$$

### Second Partials Test for Relative Extrema

If $f_x(a, b) = f_y(a, b) = 0$ and $D = f_{xx}f_{yy} - (f_{xy})^2$, then

i. if $D(a, b) > 0$ then $f(a, b)$ is a relative minimum value if $f_{xx}(a, b) > 0$ and $f(a, b)$ is a relative maximum value if $f_{xx}(a, b) < 0$;
ii. if $D(a, b) < 0$ then $(a, b, f(a, b))$ is a saddle point;
iii. if $D(a, b) = 0$, the test gives no conclusion.

### The Method of Lagrange Multipliers

The relative extrema of $f(x, y)$ subject to the constraint $g(x, y) = 0$ occur at points $(a, b)$ for which $(a, b, c)$ is a solution of the system of equations

$$F_x(x, y, \lambda) = 0$$
$$F_y(x, y, \lambda) = 0$$
$$F_\lambda(x, y, \lambda) = 0$$

where $F(x, y, \lambda) = f(x, y) + \lambda g(x, y)$.

### Least-Squares Line Formulas

The equation of the least-squares line for the data set $(x_1, y_1), (x_2, y_2), \ldots, (x_n, y_n)$ is $y = mx + b$, where

$$m = \frac{n \sum_{i=1}^{n} x_i y_i - \left(\sum_{i=1}^{n} x_i\right)\left(\sum_{i=1}^{n} y_i\right)}{n \sum_{i=1}^{n} x_i^2 - \left(\sum_{i=1}^{n} x_i\right)^2}$$

and

$$b = \frac{1}{n}\left[\sum_{i=1}^{n} y_i - m \sum_{i=1}^{n} x_i\right]$$

### Type I Region

A Type I region is defined by two inequalities $a \leq x \leq b$ and $g_1(x) \leq y \leq g_2(x)$.

### Type II Region

A Type II region is defined by two inequalities $c \leq y \leq d$ and $h_1(y) \leq x \leq h_2(y)$.

### Double Integral

If $R$ is a Type I region, then

$$\iint_R f(x, y) \, dA = \int_a^b \int_{g_1(x)}^{g_2(x)} f(x, y) \, dy\, dx$$

If $R$ is a Type II region, then

$$\iint_R f(x, y) \, dA = \int_c^d \int_{h_1(x)}^{h_2(x)} f(x, y) \, dx\, dy$$

### The Double Integral as a Volume

If $R$ is a Type I region or a Type II region and if $f(x, y) \geq 0$ for all $(x, y)$ in $R$, where $f(x, y)$ is continuous on $R$, then $\iint_R f(x, y) \, dA$ is the volume of the solid that has a top consisting of the surface $f(x, y)$ and a base consisting of the region $R$.

### The Average Value

If $R$ is a Type I region or a Type II region, then if $f(x, y)$ is continuous on $R$, the average value of $f(x, y)$ over $R$ is

$$\text{average value} = \frac{1}{\text{area of } R} \iint_R f(x, y) \, dA.$$

**CHAPTER 8 SHORT ANSWER REVIEW QUESTIONS**

1. If $f(x, y) = xe^x + \ln(xy)$, then $f(1, e) = $ _____.
2. (True or False) The distance of the point $(1, 1, 1)$ from the origin is 1 unit. _____
3. A sphere and a plane can intersect in (a) a circle; (b) a point; (c) no points; (d) all of the above.
4. If $R(x, y)$ is the revenue from the sale of $x$ units of product $A$ and $y$ units of product $B$ and if $R_x(50, 60) = 5$ and $R(50, 60) = 720$, then an approximate value of $R(51, 60)$ is _____.
5. If $f(x, y) = \ln(xy)$, then $f_{xx}(x, y) = $ _____.
6. If $f_x = 3x - 6$ and $f_y = 2y + 4$, then a critical point of $f$ is _____ and this point corresponds to a relative _____.
7. (True or False) Because the graph $f(x, y) = ax + by + c$ is a plane and since a plane has no relative extrema, we conclude that $f(x, y)$ has no critical points. _____
8. For a certain function $f(x, y)$, it is true that $f(-1, 3) \geq f(x, y)$ for all points $(x, y)$ lying within a circle of radius 0.01 about $(-1, 3)$. We conclude that $f(-1, 3)$ is (a) a relative maximum value; (b) a relative minimum value; (c) a saddle point; (d) none of these.
9. (True or False) Given that the absolute maximum of $f(x, y)$ subject to the constraint $g(x, y) = 0$ is 10 and that $g(-3, 4) = 0$, we can conclude that $f(-3, 4) \leq 10$. _____
10. If the maximum of $f(x, y) = x^2 + 2y^2$ subject to the constraint $g(x, y) = 0$ is $f(2, 5) = 54$, and if $(2, 5, 7)$ is a critical point of $F(x, y, \lambda)$, what is the approximate maximum value of $f(x, y)$ subject to the constraint $g(x, y) = 1$? _____
11. (True or False) The least-squares line for a given set of data points always passes through at least two of the points. _____
12. Given the data set $(1, 3)$, $(2, 5)$, $(3, 6.6)$, the least-squares line has the equation $y = 1.8x + 1.267$. If the point $(4, 7)$ is added to this set, the slope of the line will (a) increase; (b) decrease; (c) stay the same.
13. (True or False) The integral $\int_{-1}^{1}\int_{-1}^{1} x^2 y^2 \, dA$ represents the volume of the solid with rectangular base defined by $-1 \leq x \leq 1$ and $-1 \leq y \leq 1$ and top by the surface $f(x, y) = x^2 y^2$. _____
14. The average value of the constant function $f(x, y) = 10$ over any Type I or Type II region is _____.

15. The region sketched in the accompanying figure is (a) a Type I region but not a Type II region; (b) a Type II region but not a Type I region; (c) both a Type I region and a Type II region; (d) neither type of region.

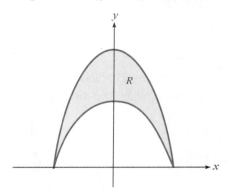

FIGURE 8.42

## CHAPTER 8 REVIEW EXERCISES

Exercises marked with an asterisk (*) constitute a Chapter Test.

In each of Exercises 1–4, find the domain of the given function.

*1. $f(x, y) = \sqrt{25 - x^2 - y^2}$      2. $f(x, y) = \sqrt{x - y}$

3. $f(x, y) = \dfrac{2}{x} + \dfrac{3}{y - 1}$      4. $f(x, y) = \dfrac{x^2 + y^2}{x + y + 1}$

In Exercises 5–8, sketch the graphs of the given planes in three dimensions.

*5. $x + 2y + 3z = 6$      6. $2x - y + z = 4$

7. $5x - 10 = 0$      8. $x + y = 1$

*9. One point on a sphere with center $(-1, 2, -4)$ is $(1, 3, 2)$. Find an equation of the sphere.

10. One point on a sphere with center $(2, -1, 3)$ is $(0, 1, -2)$. Find an equation of the sphere.

*11. (Packaging) The surface area of a cylindrical container is given by $S(r, h) = 2\pi r^2 + 2\pi r h$. Container $A$ has a radius of 2 in. and a height of 8 in. and container $B$ has a radius of 2.5 in. and a height of 5 in. Which container has the larger surface area? If material for the containers costs 2¢/in², find the cost of materials for each container.

In each of Exercises 12–15, find $\partial z/\partial x$ and $\partial z/\partial y$.

12. $z = (2x/y) - 3x^2 y$      13. $z = x\sqrt{x + y} - y\sqrt{2x}$

14. $z = xe^{2xy}$      *15. $z = \dfrac{2x - y}{x^2 + y^2}$

16. Find $f_x(-1, 1)$ and $f_y(-2, 2)$ if $f(x, y) = (7x/y) - x \ln y$.

17. Find $f_x(0, 1)$ and $f_y(0, 4)$ if $f(x, y) = \sqrt{2x + y} + xe^x$.

18. Find $\left.\dfrac{\partial z}{\partial x}\right|_{(1, -1)}$ and $\left.\dfrac{\partial z}{\partial y}\right|_{(-2, -1)}$ if $z = xy - x^3 y^2 + x^2 - y^2 - 1$.

19. Find $\left.\dfrac{\partial z}{\partial x}\right|_{(-1, 2)}$ and $\left.\dfrac{\partial z}{\partial y}\right|_{(-1, 2)}$ if $z = xe^{2x+y}$.

In Exercises 20–23, find all the second-order partial derivatives of each function.

**20.** $f(x, y) = 4y - 5x^2 - x^2y^4$

**21.** $f(x, y) = 12 - 20x^3 + (x/y) - y^2$

**22.** $f(x, y) = xe^{y^2+2y} - ye^{x^2+2x}$

**\*23.** $f(x, y) = \ln(2x + 3y) + \sqrt{2x + 3y}$

**24.** If $f(x, y) = 3x^3 - x^2y^2 + y^3$, find (a) $f_{xx}(0, 1)$; (b) $f_{xy}(-2, -1)$; (c) $f_{yy}(2, -3)$.

**25.** If $f(x, y) = (x/y) - (y/x)$, find (a) $f_{xx}(-1, -1)$; (b) $f_{xy}(2, 3)$; (c) $f_{yy}(-2, 1)$.

**26.** (Revenue) The revenue to a company from the manufacture and sale of $x$ computers and $y$ printers is given by $R(x, y) = 300x + 125y + 0.02xy$.
 (a) Find and interpret $R_x(400, 300)$.
 (b) Find and interpret $R_y(400, 300)$.

Business and Economics

**27.** (Profit) The monthly profit to a company from the manufacture and sale of $x$ units of product A and $y$ units of product B is given by $P(x, y) = 10x + 15y - 0.002xy - 2000$, where $P(x, y)$ is measured in dollars.
 (a) Find and interpret $P_x(x, y)$.
 (b) Find and interpret $P_y(x, y)$.

**28.** (Demand for two products) If the price per unit of product A is $p$ and the price per unit of product B is $s$, then weekly demand for A is $x = 1000 - 4p + s$, and weekly demand for B is $y = 1400 + 2p - 5s$.
 (a) Find and interpret $\partial x/\partial s$ and $\partial y/\partial p$.
 (b) Are these two products complements, substitutes, or neither?

In Exercises 29–34, find any relative extrema or saddle points of each function.

**29.** $f(x, y) = xy + y^2 - x^2$

**30.** $f(x, y) = x^2y + y^2x - y$

**31.** $f(x, y) = 3x^2 + y^2 + x + xy - y + 1$

**32.** $f(x, y) = 4x - x^2 + 5y - 2y^2 + 6$

**33.** $f(x, y) = e^{x^2-y^2}$

**34.** $f(x, y) = (\ln x)(\ln y)$

**\*35.** (Maximum profit) The monthly profit in hundreds of dollars from the sale of $x$ hundred units of product A and $y$ hundred units of product B is

$$P(x, y) = 80x - \tfrac{1}{135}x^3 + 125y - \tfrac{1}{240}y^3 - 500$$

Find the number of units of each product that should be sold to maximize profit; find the maximum profit.

**36.** (Demand in two countries) A company sells the same product in two different countries, A and B. The monthly demand functions are

$p = 400 - 0.01x$ (country A)

$s = 250 - 0.15y$ (country B)

where $x$ and $y$ are the quantities demanded in country A and country B, respectively, when the prices (in dollars) are $p$ and $s$. What price should the company charge in each country to maximize total revenue?

**37.** (Packaging) Find the dimensions of a closed rectangular box having a volume of 729 in³ and requiring the least amount of material to construct.

In Exercises 38–43 use the method of Lagrange multipliers to solve the problem by finding the relative extrema of the function subject to the constraint.

**38.** Minimize $f(x, y) = x^2 + 2y^2 - 1$ subject to the constraint $2x + 3y = 6$.

**39.** Find the extrema of $f(x, y) = 4xy$ subject to the constraint $x^2 + y^2 = 25$.

**40.** Find the extrema of $f(x, y) = 2x + y + 3$ subject to the constraint $x^2 + y^2 = 1$.

41. Find the extrema of $f(x, y) = 10x - x^2 + 20y - y^2 + 8$ subject to the constraint $2x + y = 4$.

42. Find the extrema of $f(x, y, z) = xy + 2xz + 2yz$ subject to the constraint $xyz = 108$.

*43. Find the extrema of $f(x, y, z) = 2x + 3y + z$ subject to the constraint $x^2 + y^2 + z^2 = 9$.

44. (Production) If $x$ units of labor and $y$ units of capital are used per week in the production of a certain product, the number of units produced is $f(x, y)$, where $f(x, y) = 1000x^{1/5}y^{4/5}$. If the cost of labor is $20 per unit and the cost of capital is $80 per unit, and if $36,000 is budgeted for next week's production, find the optimal mix of labor and capital for the coming week's production. What is the marginal productivity of money for this production process?

*In Exercises 45–48, find the least-squares line for the given set of data points by using the least-squares line formulas.*

45.

| x | 0 | 5 | 12 | 15 | 21 |
|---|---|---|---|---|---|
| y | 40 | 38 | 34 | 32 | 30 |

46.

| x | 4 | 1 | 7 | 12 | 9 |
|---|---|---|---|---|---|
| y | 13 | 10 | 18 | 23 | 16 |

47.

| x | 6.4 | 6.5 | 6.6 | 6.7 | 6.8 | 6.9 | 7.0 | 7.1 | 7.2 |
|---|---|---|---|---|---|---|---|---|---|
| y | 0.5 | 12.5 | 20.5 | 27.5 | 40.0 | 37.5 | 50.5 | 56.5 | 66.5 |

48.

| x | 60 | 70 | 80 | 90 | 100 | 110 | 120 | 130 | 140 |
|---|---|---|---|---|---|---|---|---|---|
| y | 250 | 243 | 239 | 232 | 230 | 224 | 218 | 212 | 206 |

*49. (Advertising and sales) In order to investigate the relationship between weekly sales and advertising for a product, a company collected the data on the left.
   (a) Find the least-squares line $y = mx + b$ for these data.
   (b) Estimate the weekly sales if the advertising level is set at $15,000.

| ADVERTISING LEVEL IN THOUSANDS OF DOLLARS (x) | WEEKLY SALES IN THOUSANDS OF DOLLARS (y) |
|---|---|
| 5 | 38 |
| 8 | 48 |
| 10 | 53 |
| 12 | 59 |
| 16 | 66 |

Business and Economics

50. (Price-to-earnings ratios) The following data give the approximate Dow Jones price-to-earnings ratios for stocks during a certain period in 1988.

| MONTH | May | June | July | August |
|---|---|---|---|---|
| P.E. RATIO (y) | 14.0 | 13.2 | 12.4 | 12.0 |

   (a) Letting $x = 1$ represent May, $x = 2$ represent June, and so forth, find the least-squares line $y = mx + b$ for these data.
   (b) Estimate the price-to-earnings ratio for September 1988. The actual value for September 1988 was 11.8.
   (c) Estimate the price-to-earnings ratio for September 1989. The actual value for September 1989 was about 11.9. Note that the poor estimate here was a result of unwarranted extrapolation.

*In Exercises 51–54, evaluate the iterated integrals.*

**51.** $\int_0^1 \int_{-1}^1 (x^2y - y^2x) \, dy \, dx$

**52.** $\int_{-1}^0 \int_{-2}^1 (x^3y - 3xy^2) \, dx \, dy$

**\*53.** $\int_{-1}^1 \int_{-y^2}^{y^2} (x^2 + y^2) \, dx \, dy$

**54.** $\int_0^1 \int_{2x}^{5x} e^{x+y} \, dy \, dx$

**55.** Evaluate $\iint_R (x^2y - 4xy) \, dA$, where $R$ is the finite region enclosed by the graphs of $y = x^3$ and $y = \sqrt[3]{x}$ in the first quadrant.

**56.** Evaluate $\iint_R (y^2/x) \, dA$, where $R$ is the triangular region with vertices $(-1, 1)$, $(0, 2)$, and $(1, 1)$.

**57.** Find the volume of the solid that has the surface $f(x, y) = 10xy + y^2$ for a top and the rectangular region described by the inequalities $0 \le x \le 3$ and $1 \le y \le 2$ for a base.

**58.** Find the volume of the solid that has the surface $f(x, y) = 12x^2y + 2$ for a top and the trapezoidal region in the $xy$-plane with vertices $(0, 0)$, $(0, 2)$, $(1, 3)$, and $(2, 0)$ for a base. The evaluation of this volume requires two double integrals.

**\*59.** Find the volume of the solid that has the surface $f(x, y) = 4 - x^2 - y$ for a top and the region in the first quadrant bounded by the $x$-axis, the $y$-axis, and the line $x + y = 1$ for a base.

**60.** Find the volume in the first octant bounded by the three coordinate planes and the plane $z = 20 - 4x - 5y$.

**61.** Find the average value of $f(x, y) = x + y$ over the rectangular region described by the inequalities $0 \le x \le 2$ and $0 \le y \le 4$.

**62.** (Average profit) The weekly profit in dollars of a certain company from the sale of $x$ units of product $A$ and $y$ units of related product, $B$, is given by

$$P(x, y) = 100x + 40y - 0.4xy - 0.4x^2 - 0.05y^2$$

If weekly sales of product $A$ vary between 40 and 60 units and weekly sales of product $B$ vary between 60 and one and one-half times the sales of product $A$, estimate the average weekly profit.

# 9 DIFFERENTIAL EQUATIONS

## CHAPTER CONTENTS

- 9.1 Introduction to Differential Equations
- 9.2 Separation of Variables
- 9.3 First-Order Linear Differential Equations
- 9.4 Applications of Differential Equations
- 9.5 Chapter 9 Review

# 9.1 INTRODUCTION TO DIFFERENTIAL EQUATIONS

## Solutions of Differential Equations

A *differential equation* is an equation that involves one or more derivatives of an unknown function. We have already seen several applied examples of differential equations in Section 5.4, where we studied differential equations of the forms

$$(1)\ \frac{dA}{dt} = kA, \qquad (2)\ \frac{dA}{dt} = k(M - A), \qquad \text{and} \qquad (3)\ \frac{dA}{dt} = kA(M - A)$$

where $k$ and $M$ were known constants. In each of these equations, the unknown function is $A(t)$, and the corresponding solutions are

$$(1)\ A(t) = Ce^{kt}, \qquad (2)\ A(t) = M(1 - Ce^{-kt}), \qquad (3)\ A(t) = \frac{M}{1 + Ce^{-kMt}}$$

where $C$ is an arbitrary constant that we can determine if we are given an initial condition $A(0) = A_0$.

In this chapter we simply denote the unknown function in a differential equation by $y$ and the independent variable by $x$, so that $y = f(x)$. The order of the highest derivative appearing in the differential equation is called the *order* of the differential equation. For example, the differential equation

$$\frac{dy}{dx} = 2y \tag{1}$$

has order one, since the highest derivative in the equation is the first derivative, whereas the differential equation

$$\frac{d^2y}{dx^2} + 3\frac{dy}{dx} - 4y = 0$$

has order two, since the highest derivative in this equation is the second derivative.

A *solution* of a differential equation is a function $y = f(x)$ that (along with its derivatives), when substituted into the equation, produces a statement that is true for all $x$. For example, the function $y = 5e^{2x}$ is a solution of the differential equation (1), since taking the derivative yields $dy/dx = 10e^{2x}$ and substituting this into (1) gives us

$$\underset{\underset{\frac{dy}{dx}}{\uparrow}}{10e^{2x}} \stackrel{?}{=} \underset{\underset{y}{\uparrow}}{2(5e^{2x})}$$

$$10e^{2x} \stackrel{\checkmark}{=} 10e^{2x}$$

which is true for all $x$.

## EXAMPLE 1

Show that $y = 7e^{2x}$ is a solution of the differential equation

$$\frac{dy}{dx} = 2y$$

but that $y = 4e^{3x}$ is not a solution of the equation.

**Solution**

If $y = 7e^{2x}$, then $dy/dx = 14e^{2x}$. Substituting this into the differential equation gives us

$$14e^{2x} \stackrel{?}{=} 2(7e^{2x})$$
$$14e^{2x} \stackrel{\checkmark}{=} 14e^{2x}$$

which is true for all $x$. This shows that $y = 7e^{2x}$ is a solution. If $y = 4e^{3x}$, then $dy/dx = 12e^{3x}$. Substituting this into the differential equation gives us

$$12e^{3x} \stackrel{?}{=} 2(4e^{3x}) = 8e^{3x}$$

which is certainly not true for all $x$. Thus $y = 4e^{3x}$ is not a solution. ■

We see that a differential equation can have more than one solution since, for example, the differential equation $dy/dx = 2y$ has the solutions $y = 10e^{2x}$ and $y = 7e^{2x}$. In fact, the *general solution* of this differential equation is $y = Ce^{2x}$, where $C$ is an arbitrary constant, because any solution of $dy/dx = 2y$ has the form $y = Ce^{2x}$. The solutions $y = 10e^{2x}$ and $7e^{2x}$ are called *particular solutions* because they are formed by choosing a particular value for $C$ in the general solution.

The solutions $y = Ce^{2x}$ of the differential equation $dy/dx = 2y$ form a one-parameter *family* of solutions, of which $y = 10e^{2x}$ and $y = 7e^{2x}$ are members and the parameter is $C$. In the theory of differential equations, a parameter is an arbitrary constant that appears in the general solution of the equation. If each of the parameters in the general solution is replaced by a specific real number, we obtain a particular solution of the differential equation. In most (but not all) cases, a first-order differential equation has a general solution containing a single arbitrary constant (parameter); a second-order differential equation's general solution contains two arbitrary constants (parameters); and so forth.

## EXAMPLE 2 (Participative)

Let's show that $y = C_1 e^x + C_2 e^{-4x}$ is a solution of the second-order differential equation $\dfrac{d^2y}{dx^2} + 3 \dfrac{dy}{dx} - 4y = 0$.

**Solution**

We compute:

$$\frac{dy}{dx} = C_1 e^x - 4C_2 e^{-4x}$$

$$\frac{d^2y}{dx^2} = \underline{\qquad\qquad}$$

Substituting this into the differential equation gives us

$$(C_1 e^x + 16 C_2 e^{-4x}) + 3(\underline{\qquad \underset{\uparrow}{\phantom{xxx}} \qquad}) - 4(\underline{\qquad \underset{\uparrow}{\phantom{xxx}} \qquad}) \stackrel{?}{=} 0$$

$$\phantom{xxxxxxxxxxxxxxxx} \frac{d^2 y}{dx^2} \phantom{xxxxxxxx} \frac{dy}{dx} \phantom{xxxxxxx} y$$

$$C_1 e^x + 16 C_2 e^{-4x} + 3 C_1 e^x - 12 C_2 e^{-4x} - 4 C_1 e^x - 4 C_2 e^{-4x} \stackrel{?}{=} 0$$

Combining terms on the left-hand side, we get

$$\underline{\phantom{xxxx}} \stackrel{?}{=} 0$$

which shows that $y$ *is* a solution of the given differential equation. In fact, it can be shown that $y$ is the *general* solution of the differential equation, but that uses methods beyond the scope of this text. That is, *every* solution of this differential equation can be written in the form $y = C_1 e^x + C_2 e^{-4x}$ for some particular choice of the constants $C_1$ and $C_2$. ∎

For the remainder of this chapter, we will concentrate on first-order differential equations, and we will assume that we have found a general solution to the equation when we have a solution involving a single arbitrary constant. The next example shows how a particular solution can be chosen from among the family of solutions given by the general solution. We will use the notation $y(x_0)$ to denote $f(x_0)$.

---

**EXAMPLE 3**

Show that $y = Cx^2$ is the general solution of the differential equation $2y = x\, dy/dx$. Find the particular solution that satisfies the condition $y(2) = 4$.

**Solution**

Differentiating $y = Cx^2$, we obtain $dy/dx = 2Cx$. Substituting this into the differential equation gives us

$$2(Cx^2) \stackrel{?}{=} x(2Cx)$$
$$2Cx^2 \stackrel{?}{=} 2Cx^2$$

Thus $y = Cx^2$ is a solution to the given first-order differential equation; it contains a single arbitrary constant and so is the general solution of the differential equation.

To find the particular solution that satisfies the condition $y(2) = 4$, we substitute 2 for $x$ and 4 for $y$ in the general solution $y = Cx^2$ to obtain

$$4 = C(2)^2$$
$$4C = 4$$
$$C = 1$$

Thus the particular solution that satisfies $y(2) = 4$ is $y = x^2$. ∎

Some of the particular solutions corresponding to the general solution $y = Cx^2$ of the differential equation of Example 3 are sketched in Figure 9.1. Each

---

**SOLUTIONS TO EXAMPLE 2**  $C_1 e^x + 16 C_2 e^{-4x},\ C_1 e^x - 4 C_2 e^{-4x},\ C_1 e^x + C_2 e^{-4x},\ 0$

of these curves is a parabola with the exception of the case $C = 0$. Note that the particular solution determined by the condition $y(2) = 4$ passes through the point (2, 4) and has $C = 1$. Also observe that the curve corresponding to $C = 0$ is the $x$-axis.

The condition $y(2) = 4$, which was specified in Example 3, is called an *initial condition*. The reason for the name "initial condition" is that if the independent variable is time, $t$, then the condition $y(0) = y_0$ gives the state of the system described by the differential equation initially; that is, at time $t = 0$. A first-order differential equation together with an initial condition is called an *initial-value problem*. Typically, an initial condition $y(x_0) = y_0$ allows us to find the constant $C$ in the general solution of a first-order equation by specifying that the solution curve pass through the point $(x_0, y_0)$. Thus the solution of an initial-value problem is usually a particular solution. There are exceptions to this, however, as you will see in the next example.

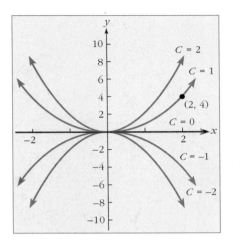

**FIGURE 9.1** Some particular solution curves $y = Cx^2$ for $C = -2, -1, 0, 1,$ and 2.

---

EXAMPLE 4

For the differential equation $2y = x\, dy/dx$, which has the general solution $y = Cx^2$, find (if possible) the particular solution that satisfies the initial condition **(a)** $y(0) = 1$; **(b)** $y(0) = 0$.

Solution

**(a)** To find the particular solution that satisfies $y(0) = 1$, we substitute 0 for $x$ and 1 for $y$ in the general solution $y = Cx^2$ to obtain $1 = C(0)^2$. Since this equation is never true for any choice of $C$, there is no particular solution satisfying $y(0) = 1$. In Figure 9.1, notice that none of the solution curves passes through the point (0, 1).

**(b)** To find the particular solution that satisfies $y(0) = 0$, we substitute 0 for $x$ and 0 for $y$ in the general solution $y = Cx^2$ to obtain $0 = C(0)^2$. This equation is true for any constant $C$. We conclude that all solutions $y = Cx^2$ satisfy $y(0) = 0$, so that no unique particular solution is determined. In Figure 9.1, notice that all of the solution curves pass through the point (0, 0). ∎

### Solving the Differential Equation $\frac{dy}{dx} = g(x)$

Thus far we have not actually solved any differential equations but we have merely verified that certain given functions are solutions. Although no single method of solution works for all differential equations, one particularly easy type of equation to solve is one that can be put into the form $dy/dx = g(x)$, because the solution is simply $y = \int g(x)\, dx$.

SOLUTION OF $\frac{dy}{dx} = g(x)$

> The general solution of the differential equation
> $$\frac{dy}{dx} = g(x) \quad \text{is} \quad y = \int g(x)\, dx$$

## EXAMPLE 5

Solve the initial-value problem

$$\frac{dy}{dx} = 3x^2 + 2x - 1, \quad y(1) = 3$$

**Solution**

Since the given differential equation has the form $dy/dx = g(x)$, where $g(x) = 3x^2 + 2x - 1$, the general solution is

$$y = \int (3x^2 + 2x - 1)\, dx = x^3 + x^2 - x + C$$

To determine the particular solution that satisfies the initial condition $y(1) = 3$, we substitute 1 for $x$ and 3 for $y$ in the general solution $y = x^3 + x^2 - x + C$ to obtain

$$3 = 1^3 + 1^2 - 1 + C$$
$$C = 2$$

The solution of the initial-value problem is therefore

$$y = x^3 + x^2 - x + 2 \quad \blacksquare$$

### Applications

Differential equations arise in many applied problems in business and the sciences. The next two examples illustrate typical elementary applications for which we can solve the differential equation that arises. Thus, the differential equation in question will have the form $dy/dx = g(x)$. In Section 9.4 we'll consider further applications after we have learned to solve some other types of differential equations.

## EXAMPLE 6

The monthly marginal cost of a certain company is proportional to the square root of the number of units produced per month. If monthly fixed costs are $1500 and if it costs $2500 to produce 100 units per month, find the company's total cost function.

**Solution**

Let $x$ be the number of units produced per month and $C(x)$ the total cost of producing $x$ units. We are given that $dC/dx = k\sqrt{x}$, where $k$ is a constant of proportionality. This differential equation is easily solved by integration:

$$C(x) = \int k\sqrt{x}\, dx = \tfrac{2}{3}kx^{3/2} + K$$

where $K$ is the constant of integration. Since fixed costs are $1500, we have $C(0) = 1500$, so that

$$1500 = \tfrac{2}{3}k(0)^{3/2} + K = K$$

We conclude that $C(x) = \tfrac{2}{3}kx^{3/2} + 1500$. We also know that $C(100) = 2500$, so that

$$2500 = \tfrac{2}{3}k(100)^{3/2} + 1500$$
$$2500 = \tfrac{2}{3}k(1000) + 1500$$
$$1000 = \tfrac{2}{3}k(1000)$$
$$k = \tfrac{3}{2}$$

The company's monthly total cost function is therefore $C(x) = x^{3/2} + 1500$. ∎

### EXAMPLE 7 (Participative)

In a certain psychological test, the rate of change of the number of tasks done by a subject is inversely proportional to the cube root of the time allotted, for times between 0 and 20 minutes. If an average subject can perform 24 tasks in the first 8 minutes, let's find the function that gives the number of tasks that can be done for any time $t$, $0 \le t \le 20$.

**Solution**

Let $N(t)$ be the number of tasks done in the first $t$ minutes, for $0 \le t \le 20$. We are told that $dN/dt = \underline{\qquad}$, where $k$ is a constant of proportionality. Integrating both sides of this differential equation, we get

$$N(t) = \tfrac{3}{2}kt^{2/3} + C \qquad (2)$$

where $C$ is the constant of integration. Since no tasks are done in 0 minutes, $N(0) = \underline{\qquad}$, and so substituting this into (2) gives us

$$0 = \tfrac{3}{2}k(0)^{2/3} + C$$
$$C = \underline{\qquad}$$

Thus $N(t) = \tfrac{3}{2}kt^{2/3}$. We are given that $N(8) = \underline{\qquad}$, so that

$$24 = \tfrac{3}{2}k(8)^{2/3} = \underline{\qquad} k$$
$$k = \underline{\qquad}$$

We conclude that $N(t) = \underline{\qquad}$ for $0 \le t \le 20$. ∎

---

### SECTION 9.1
### SHORT ANSWER EXERCISES

1. The order of the differential equation $dy/dx = xy^2$ is $\underline{\qquad}$.
2. The order of the differential equation $dA/dt = 0.03t$ is $\underline{\qquad}$.
3. The order of the differential equation $y^3 - 7\dfrac{d^2y}{dx^2} = xy^4 \dfrac{dy}{dx}$ is $\underline{\qquad}$.
4. (True or False) The constant function $y = 0$ is a solution of the differential equation
$$3\dfrac{d^2y}{dx^2} - 7\dfrac{dy}{dx} + 10y = \dfrac{d^3y}{dx^3}$$
$\underline{\qquad}$
5. (True or False) The constant function $y = 0$ is a solution of every differential equation. $\underline{\qquad}$

---

**SOLUTIONS TO EXAMPLE 7**  $k/\sqrt[3]{t}$, 0, 0, 24, 6, 4, $6t^{2/3}$

6. The general solution of a first-order differential equation is $y = Cx^3 + 2$. If it is known that $y(1) = 0$, then $C = $ \_\_\_\_.

7. A first-order differential equation and an initial condition $y(x_0) = y_0$ is called an _____.

8. (True or False) An initial-value problem can always be solved to find a unique particular solution. _____

9. If marginal profit is proportional to the cube root of the number of units produced, then $dP/dx = $ _____.

10. If $s(t)$ is the position of an object moving in a straight line and if its velocity is inversely proportional to time for $t > 0$, then $ds/dt = $ _____.

## SECTION 9.1
## EXERCISES

*In each of Exercises 1–8, verify that the given function is a solution of the differential equation.*

1. $\dfrac{dy}{dx} = 7x$;  $y = \tfrac{7}{2}x^2 + 5$

2. $\dfrac{dy}{dx} = \dfrac{1}{x}$;  $y = \ln|x| + 10$

3. $\dfrac{dy}{dx} = 20e^{2x}$;  $y = 10e^{2x} - 2$

4. $\dfrac{dy}{dx} = 3x^2 + 2$;  $y = x^3 + 2x - 4$

5. $\dfrac{dy}{dx} = 4y$;  $y = 7e^{4x}$

6. $x\dfrac{dy}{dx} - 15 = 3y$;  $y = x^3 - 5$

7. $x\dfrac{dy}{dx} - x = y$;  $y = x \ln x$

8. $\dfrac{dy}{dx} - y = e^x$;  $y = xe^x$

*In each of Exercises 9–16, verify that the given function is a solution of the differential equation for any choice of the constants.*

9. $\dfrac{dy}{dx} = -3y$;  $y = Ce^{-3x}$

10. $\dfrac{dy}{dx} = 5(100 - y)$;  $y = 100(1 - Ce^{-5x})$

11. $\dfrac{dy}{dx} = y(1 - y)$;  $y = \dfrac{1}{1 + Ce^{-x}}$

12. $x\dfrac{dy}{dx} - 4y = 12$;  $y = Cx^4 - 3$

13. $\dfrac{d^2y}{dx^2} + \dfrac{dy}{dx} - 6y = 0$;  $y = C_1e^{2x} + C_2e^{-3x}$

14. $\dfrac{d^2y}{dx^2} - 8\dfrac{dy}{dx} + 16y = 0$;  $y = C_1e^{4x} + C_2xe^{4x}$

15. $x\dfrac{dy}{dx} - 2xy = y$;  $y = Cxe^{2x}$

16. $2y = (x \ln x)\dfrac{dy}{dx}$;  $y = C(\ln x)^2$

*In Exercises 17–24, use the initial conditions to determine a particular solution for each given general solution when possible.*

17. $y = Ce^{-3x}$;  $y(0) = 2$

18. $y = 100(1 - Ce^{-5x})$;  $y(0) = 3$

19. $y = Cx^4 + 5$;  $y(0) = 1$

20. $y = C\sqrt{x} - 4$;  $y(0) = -4$

21. $y = C_1e^x + C_2e^{-x}$;  $y(0) = 1$, $y'(0) = 2$

22. $y = C_1e^{2x} + C_2xe^{2x}$;  $y(0) = 0$; $y'(0) = 4$

23. $y = Cxe^x - e$;  $y(1) = e$

24. $y = Cx - \ln x$;  $y(1) = -2$

*In Exercises 25–36, find the general solution of each differential equation.*

25. $\dfrac{dy}{dx} = x$

26. $\dfrac{dy}{dx} = 3x^2 - x + 1$

27. $\dfrac{dy}{dx} = e^{-x} + 1$

28. $\dfrac{dy}{dx} = \ln x - 5, \; x > 0$

29. $\dfrac{dy}{dx} = \sqrt{x} - 2$

30. $\dfrac{dy}{dx} = \sqrt{2x + 1}$

31. $\dfrac{dy}{dx} = xe^x$

32. $\dfrac{dy}{dx} - 2x = 1$

33. $2\dfrac{dy}{dx} - e^{3x} = x$

34. $\dfrac{d^2y}{dx^2} = 12x$ (*Hint:* Integrate twice.)

35. $\dfrac{d^2y}{dx^2} = \sqrt{x}$ (*Hint:* Integrate twice.)

36. $\dfrac{d^2y}{dx^2} - e^{2x} = 3$

*In Exercises 37–46, solve the given initial-value problems.*

37. $\dfrac{dy}{dx} = 6x$ for $y(0) = 1$

38. $\dfrac{dy}{dx} = 1 - 4x$ for $y(0) = -2$

39. $\dfrac{dy}{dx} = \dfrac{4}{x} - 2$ for $y(1) = 3$

40. $\dfrac{dy}{dx} = e^{3x} - x + 2$ for $y(0) = 0$

41. $3\dfrac{dy}{dx} = \sqrt[3]{x}$ for $y(-1) = -1$

42. $6\dfrac{dy}{dx} - 12x^2 = 0$ for $y(-2) = 0$

43. $2\dfrac{dy}{dx} - 1 = 2x$ for $y(1) = -2$

44. $x^2\dfrac{dy}{dx} - x = 2$ for $y(1) = 4$

45. $e^x\dfrac{dy}{dx} - x = 2$ for $y(0) = 2$

46. $\dfrac{dy}{dx} = x \ln x$ for $y(1) = 3$

**Life Sciences**

47. (Bacteria growth) Bacteria in a certain culture are growing at the rate of $2000e^{4t}$ bacteria per hour. If $y(t)$ is the number of bacteria in the culture after $t$ hours and if 5000 bacteria are present initially, (a) write the initial-value problem describing the situation; and (b) find an expression for $y(t)$, the number of bacteria present at time $t$.

48. (Revenue) The monthly marginal revenue of a certain firm is proportional to the cube root of the number of units sold per month. Assuming that no sales produces no revenue and that the sale of 125 units yields a revenue of $600, find the firm's revenue function.

**Business and Economics**

49. (Bank loans) The monthly rate at which a bank generates new loans is inversely proportional to the square of the average monthly interest rate it charges. If the bank charges 10% for loans, it makes 1000 loans, and if it charges 15%, it makes 500 loans. Find an expression for the number $N$ of loans this bank makes per month as a function of the interest rate $r\%$ that it charges.

50. (Concert tickets) The demand for tickets to a concert satisfies the differential equation $dx/dp = -1.25p^{3/2}$, where $x$ gives the number of tickets sold when the price per ticket is $p$ dollars. At a price of $16 per ticket, 1000 tickets can be sold. How many tickets can be sold at $9 per ticket?

Social Sciences

**51.** (Political awareness) The number $N$ of people in a town who are aware of a political candidate $t$ days after the beginning of an advertising campaign satisfies the differential equation $dN/dt = 6t^2 + 2t + 10$. If 100 people were aware of the candidate before the advertising campaign began, how many people will be aware of the candidate after 10 days of advertising?

**52.** (Advertising and sales) The rate of sales $S$ of a company in units per month is proportional to the square root of its monthly advertising costs $A$ in dollars. If the company can sell 500 units per month with no advertising and 2000 units per month by spending $10,000 on advertising, what is the monthly sales level generated by a monthly advertising expense of $22,500?

## 9.2 SEPARATION OF VARIABLES

The technique of separation of variables can be used to solve differential equations that can be put into the form

**SEPARABLE DIFFERENTIAL EQUATION**

$$g(y) \frac{dy}{dx} = f(x) \qquad \text{(Derivative form)}$$

An alternative form of this equation can be written by considering $dy/dx$ to be the quotient of the two differentials $dy$ and $dx$ and multiplying both sides of the equation by $dx$. We then obtain

**SEPARABLE DIFFERENTIAL EQUATION**

$$g(y) \, dy = f(x) \, dx \qquad \text{(Differential form)}$$

In this second form it is evident that the variables have been separated. That is, the left-hand side is written entirely in terms of the dependent variable $y$ and the right-hand side entirely in terms of the independent variable $x$. We can now solve the equation by integrating both sides to obtain

$$\int g(y) \, dy = \int f(x) \, dx$$

Thus if $G(y)$ is an antiderivative of $g(y)$ and $F(x)$ is an antiderivative of $f(x)$, the solution can be written $G(y) = F(x) + C$

**SEPARATION OF VARIABLES**

If a differential equation can be written in the form $g(y) \, dy = f(x) \, dx$ and if $G'(y) = g(y)$ and $F'(x) = f(x)$, then the general solution of the differential equation is

$$G(y) = F(x) + C$$

## 9.2 SEPARATION OF VARIABLES

In general, the solution we obtain by using the technique of separation of variables is an *implicit solution*, since the technique usually does not yield an explicit formula for $y$ in terms of $x$. This is in contrast to the *explicit solutions* we obtained in the previous section, where we solved equations of the form $dy/dx = g(x)$ to obtain a solution $y = G(x) + C$. In general, unless it is quite easy to solve the implicit solution $G(y) = F(x) + C$ for $y$, we will leave the solution in this implicit form. Let's look at some examples.

---

**EXAMPLE 1**

Find the general solution of $3y^2 \dfrac{dy}{dx} - x = 0$.

**Solution**

We begin the solution by separating the variables:

$$3y^2 \frac{dy}{dx} - x = 0$$

$$3y^2 \frac{dy}{dx} = x$$

$$3y^2 \, dy = x \, dx \quad \text{(Multiplying both sides by } dx\text{)}$$

Since the variables have now been separated, we can complete the solution by integrating both sides:

$$\int 3y^2 \, dy = \int x \, dx$$

$$y^3 + C_1 = \tfrac{1}{2}x^2 + C_2$$

$$y^3 = \tfrac{1}{2}x^2 + (C_2 - C_1)$$

$$y^3 = \tfrac{1}{2}x^2 + C \quad \text{(where } C = C_2 - C_1\text{)}$$

Notice that the two constants $C_1$ and $C_2$ from the integrations can always be combined into a single constant $C$. In the future we will not show these separate constants $C_1$ and $C_2$.

Our solution in implicit form is $y^3 = \tfrac{1}{2}x^2 + C$. In this case we can easily solve this equation explicitly for $y$ in terms of $x$ to obtain the explicit solution

$$y = \sqrt[3]{\tfrac{1}{2}x^2 + C}$$

∎

---

**EXAMPLE 2**

Find the general solution of $xy \dfrac{dy}{dx} - 2 = 0$ for $x \neq 0$.

**Solution**

Our method of solution will be to separate the variables and then integrate both sides of the equation, as follows:

$$xy \frac{dy}{dx} - 2 = 0$$

$$xy \frac{dy}{dx} = 2$$

$$y\,dy = \frac{2}{x}\,dx \qquad \left(\text{Multiplying both sides by } \frac{dx}{x}\right)$$

$$\int y\,dy = \int \frac{2}{x}\,dx \qquad \text{(Integrating both sides)}$$

$$\tfrac{1}{2}y^2 = 2\ln|x| + C$$

$$y^2 = 4\ln|x| + 2C$$

$$y^2 = 4\ln|x| + K \qquad \text{(Setting } K = 2C\text{)}$$

This is our solution in implicit form. ∎

---

**EXAMPLE 3**   Find the general solution of $\dfrac{dy}{dx} = 3x^2 y^2$ for $y \neq 0$.

**Solution**   Beginning with $dy/dx = 3x^2 y^2$, we can separate variables by multiplying both sides of the equation by $dx/y^2$ to obtain (assuming $y \neq 0$)

$$\frac{dy}{y^2} = 3x^2\,dx$$

Next, we integrate both sides:

$$\int \frac{dy}{y^2} = \int 3x^2\,dx$$

$$-\frac{1}{y} = x^3 + C$$

This is our solution in implicit form, and it is easily solved for $y$ to get the explicit solution

$$y = \frac{-1}{x^3 + C}$$
∎

The explicit solution $y = -1/(x^3 + C)$ we obtained in Example 3 does *not* give all solutions of the differential equation. In the solution of this problem we assumed that $y$ was not equal to zero, since we divided by $y^2$ in the process of solving the equation. We must now investigate whether $y = 0$ is a solution of the original differential equation $dy/dx = 3x^2 y^2$. If $y = 0$, then $dy/dx = 0$, and substituting this into the equation yields

$$0 \overset{?}{=} 3x^2(0)^2$$

$$0 \overset{\checkmark}{=} 0$$

Therefore, $y = 0$ is indeed a solution of this differential equation, one that cannot be obtained from the expression $y = -1/(x^3 + C)$ for any choice of constant $C$. A solution of this kind is called a *singular solution* of the differential equation. When we use the technique of separation of variables and in the process divide by some function of $y$, say, $h(y)$, we must assume that $h(y) \neq 0$. Thus any values of $y$ for which $h(y) = 0$ must be checked later in the original differential equation, since these values may lead to singular solutions.

## EXAMPLE 4 (Participative)

**Solution**

Let us solve the initial-value problem $y\frac{dy}{dx} - x^2 = 5$ for $y(0) = 2$.

In this case we will find a particular solution of the given differential equation. We begin by separating the variables to obtain the general solution.

$$y\frac{dy}{dx} - x^2 = 5$$

$$y\frac{dy}{dx} = \underline{\qquad} \qquad \text{(Adding } x^2 \text{ to both sides)}$$

$$y\,dy = \underline{\qquad} \qquad \text{(Multiplying both sides by } dx\text{)}$$

$$\int y\,dy = \underline{\qquad} \qquad \text{(Integrating both sides)}$$

$$\tfrac{1}{2}y^2 = \underline{\qquad}$$

The last line represents the general solution in implicit form. To determine the particular solution corresponding to the initial condition $y(0) = 2$, we substitute $x = 0$ and $y = 2$ in this general solution to obtain

$$\tfrac{1}{2}(2)^2 = \underline{\qquad}$$

so that $C = \underline{\qquad}$ and the particular solution is $\tfrac{1}{2}y^2 = \underline{\qquad}$. ■

## EXAMPLE 5

**Solution**

Find the general solution of $\dfrac{dy}{dx} = \dfrac{2xy}{x^2 + 1}$.

We separate the variables and then integrate both sides, as follows:

$$\frac{dy}{dx} = \frac{2xy}{x^2 + 1}$$

$$\frac{dy}{y} = \frac{2x}{x^2 + 1}\,dx \qquad \left(\text{Multiplying both sides by }\frac{dx}{y},\text{ assuming } y \neq 0\right)$$

$$\int \frac{dy}{y} = \int \frac{2x}{x^2 + 1}\,dx \qquad \text{(Integrating both sides)}$$

$$\ln|y| = \ln|x^2 + 1| + C \qquad \left(\text{Recalling that } \int \frac{du}{u} = \ln|u| + C\right)$$

The last line represents our general solution if $y \neq 0$. To solve for $y$, we first note that since $x^2 + 1 \geq 0$ for all $x$, we can write $|x^2 + 1| = x^2 + 1$. Then, applying the exponential function to both sides of the general solution, we get

---

**SOLUTIONS TO EXAMPLE 4**  $x^2 + 5$, $(x^2 + 5)\,dx$, $\int (x^2 + 5)\,dx$, $\tfrac{1}{3}x^3 + 5x + C$, $\tfrac{1}{3}(0)^3 + 5(0) + C$, $2$, $\tfrac{1}{3}x^3 + 5x + 2$

$$e^{\ln |y|} = e^{\ln(x^2+1)+C}$$
$$|y| = e^C \cdot e^{\ln(x^2+1)} \quad \text{(Recalling that } e^{\ln z} = z\text{)}$$
$$|y| = e^C(x^2 + 1)$$
$$y = \pm e^C(x^2 + 1)$$

Now, $e^C$ is a positive arbitrary constant, so the expression $\pm e^C$ represents an arbitrary constant that may be positive or negative. If we assume that the solution function $y$ is continuous, we can replace $\pm e^C$ by a single arbitrary constant $K$, where $K$ may be positive or negative. Thus, for $y \neq 0$, we obtain $y = K(x^2 + 1)$. Since we divided by $y$ in the process of finding this solution, we assumed that $y \neq 0$ and we must now check to see if $y = 0$ is a solution of the differential equation

$$\frac{dy}{dx} = \frac{2xy}{x^2 + 1}$$

If $y = 0$ then $dy/dx = 0$, so that substitution into this equation yields

$$0 \stackrel{?}{=} \frac{2x(0)}{x^2 + 1}$$
$$0 \stackrel{\checkmark}{=} 0$$

Thus $y = 0$ is a solution of the differential equation. Note, however, that in this case $y = 0$ is *not* a singular solution, since it can be obtained from the expression $y = K(x^2 + 1)$ by choosing $K = 0$. We conclude that the general solution of the differential equation is $y = K(x^2 + 1)$. ∎

### A Mathematical Application

We will consider applications of separable differential equations to problems in business and the sciences in Section 9.4. Right now, let's look at a mathematical problem that can be solved using the technique of separation of variables.

**EXAMPLE 6** Find an equation of the curve that passes through the point $(1, 1)$ and such that the slope at each point $(x, y)$ satisfies the differential equation $dy/dx = -y/x$.

**Solution** We are required to solve the initial-value problem

$$\frac{dy}{dx} = -\frac{y}{x}, \quad y(1) = 1 \tag{1}$$

We begin by separating the variables:

$$\frac{dy}{y} = -\frac{dx}{x} \quad \left(\text{Multiplying both sides of (1) by } \frac{dx}{y}\right)$$
$$\int \frac{dy}{y} = \int \left(-\frac{dx}{x}\right) \quad \text{(Integrating both sides)}$$
$$\ln |y| = -\ln |x| + C$$

Since in the particular solution we require $x = 1$, we may assume that $x > 0$, so that the preceding equation becomes $\ln |y| = -\ln x + C$. As in Example 5, in order to solve this equation for $y$, we apply the exponential function to both sides:

$$e^{\ln|y|} = e^{-\ln x + C}$$
$$|y| = e^C(e^{\ln x})^{-1} = e^C x^{-1}$$
$$y = \pm\frac{e^C}{x}$$

Assuming that the solution function $y$ is continuous for $x > 0$, we may replace $\pm e^C$ by a single constant $K$ to obtain $y = K/x$. Then, using the initial condition $y(1) = 1$, we obtain $1 = K/1 = K$. The required curve is therefore $y = 1/x$. This curve is sketched in Figure 9.2 for reference. You may verify that $y = 1/x$ actually satisfies (1) for all $x \neq 0$.

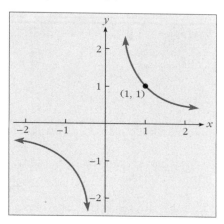

**FIGURE 9.2** The graph of $y = 1/x$.

## SECTION 9.2
**SHORT ANSWER EXERCISES**

*In Exercises 1–6, decide whether the differential equation is separable or not separable. Do not solve the equation.*

1. $\dfrac{dy}{dx} - 5x^2y = 0$ _____

2. $2y^2 - \dfrac{dy}{dx} = 4$ _____

3. $xy\dfrac{dy}{dx} - x + y = 1$ _____

4. $\dfrac{dA}{dt} = 10(1000 - A)$ _____

5. $x^2 \, dy = y^3 \, dx$ _____

6. $xy \, dy + x^2 y^3 \, dx = 1$ _____

7. An implicit solution of a certain separable equation is $y^3 - 1 = Cx^2$. An explicit solution is $y = $ _____.

8. Is $y = 1$ a solution of the separable equation $\dfrac{dy}{dx} = x(y-1)^2$? _____

9. An implicit solution of a certain separable equation is $\ln|y| = x + C$. Assuming that the solution function $y$ is continuous for all $x$, an explicit solution is _____.

10. (True or False) A singular solution of a first-order differential equation can be obtained by making a particular choice of the arbitrary constant $C$ in the general solution. _____

## SECTION 9.2
## EXERCISES

In each of Exercises 1–12, find the general solution for the given differential equation in explicit form. Be sure to determine any singular solutions.

1. $y^2 \dfrac{dy}{dx} = x^2$

2. $\dfrac{dy}{dx} = 2x^2$

3. $\dfrac{dy}{dx} = 3x^2 y^2$

4. $\dfrac{dy}{dx} - x^2 y^4 = 0$

5. $y \dfrac{dy}{dx} = 4x^3$

6. $2y^5 \dfrac{dy}{dx} = 5x$

7. $x^2 y^2 \dfrac{dy}{dx} = 3$

8. $\dfrac{dy}{dx} - 4x^3 y^2 = 0$

9. $y^2 \dfrac{dy}{dx} - e^x = 1$

10. $\dfrac{dy}{dx} = 3x^2(y - 2)$

11. $\dfrac{dy}{dx} - 2x(y + 2) = 0$

12. $\dfrac{1}{y} \dfrac{dy}{dx} - x^2 = e^{2x} + 1$

In Exercises 13–22, solve each initial-value problem by finding an explicit particular solution.

13. $y^2 \dfrac{dy}{dx} - x = 0$ for $y(0) = 2$

14. $y \dfrac{dy}{dx} = x + 2$ for $y(0) = 4$

15. $\dfrac{dy}{dx} - \dfrac{3x^2}{y^2} = 0$ for $y(1) = -3$

16. $\dfrac{dy}{dx} = \dfrac{x}{e^y}$ for $y(0) = 1$

17. $y^3 \dfrac{dy}{dx} - 2xe^{x^2} = 0$ for $y(0) = 4$

18. $\dfrac{dy}{dx} = (2x - 1)(y + 1)$ for $y(1) = 0$

19. $\dfrac{1}{y} \dfrac{dy}{dx} = 3x - \sqrt{x}$ for $y(1) = e$

20. $\dfrac{dy}{dx} = \dfrac{2y}{x}$ for $y(1) = 2e$

21. $\dfrac{dy}{dx} = xy + x + y + 1$ for $y(1) = 0$

22. $\dfrac{dy}{dx} = \dfrac{x^3 + x}{\sqrt{y}}$ for $y(0) = 1$

In Exercises 23–30, find the general solution for each given differential equation. You may leave your solution in implicit form.

23. $\dfrac{dy}{dx} = \dfrac{x^2}{y + 1}$

24. $\dfrac{dy}{dx} = \dfrac{e^x + 1}{e^y + 1}$

25. $x(y^2 + 1) \dfrac{dy}{dx} = 2 - x^2$

26. $y(x^2 + 1) \dfrac{dy}{dx} = x(y^2 + 4)$

27. $\dfrac{dy}{dx} = \dfrac{xe^x}{y^2 + 1}$

28. $y \ln y \dfrac{dy}{dx} = x \ln x$

29. $e^y \dfrac{dy}{dx} = \dfrac{x \ln x}{y}$

30. $\dfrac{x}{y + 2} = \dfrac{dy}{dx} (x^2 + 2)$

In Exercises 31–36, solve each initial-value problem by finding a particular solution of the differential equation. You may leave your particular solution in implicit form.

31. $\dfrac{dy}{dx} = \dfrac{2x}{y + 2}$ for $y(1) = 0$

32. $\dfrac{y}{x} \dfrac{dy}{dx} = \dfrac{x + 1}{3y + 1}$ for $y(0) = 2$

33. $x\dfrac{dy}{dx} = \dfrac{x+3}{3y^2+1}$ for $y(1) = 1$

34. $\dfrac{y}{\sqrt{x}}\dfrac{dy}{dx} = \dfrac{x-1}{\sqrt{y^2+1}}$ for $y(1) = 0$

35. $\dfrac{e^y+1}{e^x}\dfrac{dy}{dx} = x$ for $y(0) = 0$

36. $(e^{0.1y}+y)\dfrac{dy}{dx} = x^3$ for $y(0) = 0$

*In each of Exercises 37–40, find an equation of the curve that has a slope at each point $(x, y)$ that satisfies the given differential equation and that passes through the given point.*

37. $\dfrac{dy}{dx} = -\dfrac{x}{y}$; $(3, 4)$

38. $\dfrac{dy}{dx} = -\dfrac{9x}{4y}$; $(0, 3)$

39. $\dfrac{dy}{dx} = \dfrac{x}{9y}$; $(1, 0)$

40. $\dfrac{dy}{dx} = \dfrac{2}{y}$; $(1, -2)$

## 9.3 FIRST-ORDER LINEAR DIFFERENTIAL EQUATIONS

A *first-order linear differential equation* is a differential equation that can be put into the form

$$\dfrac{dy}{dx} + P(x)y = Q(x) \qquad (1)$$

where $P(x)$ and $Q(x)$ are functions of $x$. Such an equation is first-order because the highest derivative in the equation is the first derivative. It is linear because it involves only $y$ (and not $y^2$, or $\sqrt{y}$, and so forth). In general, a first-order linear equation cannot be solved by the method of separation of variables. In this section we will concentrate on solving this type of a differential equation. In Section 9.4 we will consider some applications of first-order linear equations.

As an example of a first-order linear differential equation, consider

$$\dfrac{dy}{dx} + y = e^{-x} \qquad (2)$$

In this equation $P(x) = 1$ and $Q(x) = e^{-x}$. To solve it, we begin by multiplying both sides by $e^x$; the reason for this will be explained later. We obtain

$$e^x\dfrac{dy}{dx} + e^x y = e^x \cdot e^{-x}$$

$$e^x\dfrac{dy}{dx} + e^x y = 1 \qquad (3)$$

By the product rule, the left-hand side of equation (3) is the derivative of $e^x y$, namely, $\dfrac{d}{dx}(e^x y)$. Thus we can rewrite equation (3)

$$\dfrac{d}{dx}(e^x y) = 1$$

Integrating both sides gives us

$$\int \frac{d}{dx}(e^x y)\,dx = \int 1\,dx$$

$$e^x y = x + C$$

Solving for $y$, we obtain $y = xe^{-x} + Ce^{-x}$, which is the general solution of the original differential equation.

The factor $e^x$ is called an *integrating factor* of the differential equation (2), because multiplying by this factor produces an expression on the left-hand side that is trivial to integrate, and this integration leads to the general solution. How did we know that multiplying both sides of (2) by the factor $e^x$ would lead to a solution?

To answer this question, let's return to (1), the general form of a first-order linear differential equation,

$$\frac{dy}{dx} + P(x)y = Q(x)$$

and suppose $I(x)$ is an integrating factor of this equation. Then, multiplying both sides by $I(x)$ yields

$$I(x)\frac{dy}{dx} + I(x)P(x)y = I(x)Q(x) \tag{4}$$

Comparing with our specific example, which used $I(x) = e^x$, the left-hand side of this equation should be $\frac{d}{dx}[I(x)y] = I(x)\frac{dy}{dx} + \frac{dI}{dx}y$. Comparing the left-hand side of (4) and this last expression, we see that we must have

$$\frac{dI}{dx} = I(x)P(x)$$

Thus, to find $I(x)$, we must solve this (separable) differential equation. We have

$$\frac{dI}{I} = P(x)\,dx \qquad \left(\text{Multiplying both sides by } \frac{dx}{I}\right)$$

$$\int \frac{dI}{I} = \int P(x)\,dx \qquad \text{(Integrating both sides)}$$

$$\ln|I| = \int P(x)\,dx + C$$

$$|I| = e^{\int P(x)\,dx + C} \qquad \text{(Writing in exponential form)}$$

$$|I| = e^C \cdot e^{\int P(x)\,dx}$$

$$I = \pm e^C \cdot e^{\int P(x)\,dx}$$

If we let $K = \pm e^C$, then $I = Ke^{\int P(x)\,dx}$. By choosing $K = 1$, the integrating factor becomes $I(x) = e^{\int P(x)\,dx}$. In our specific example, equation (2), we had $P(x) = 1$, and so $I(x) = e^{\int 1\,dx} = e^x$. Notice that in the integration $\int P(x)\,dx$, we chose the constant of integration to be zero.

Summarizing, the steps in solving a first-order linear differential equation $\frac{dy}{dx} + P(x)y = Q(x)$ are as follows.

## SOLVING THE EQUATION
$\frac{dy}{dx} + P(x)y = Q(x)$

**Step 1** Put the equation into the form $\frac{dy}{dx} + P(x)y = Q(x)$ and identify $P(x)$ and $Q(x)$.

**Step 2** Find the integrating factor $I(x) = e^{\int P(x)\, dx}$. In the integration $\int P(x)\, dx$, choose the constant of integration to be zero.

**Step 3** Multiply both sides of the differential equation by $I(x)$. The equation can then be put into the form

$$\frac{d}{dx}[I(x)y] = I(x)Q(x)$$

**Step 4** Integrate both sides of the equation to obtain

$$I(x)y = \int I(x)Q(x)\, dx + C$$

Observe that an arbitrary constant $C$ has been added to the right-hand side.

**Step 5** To find an explicit solution, divide both sides of the equation by $I(x)$ to obtain

$$y = \frac{1}{I(x)} \int I(x)Q(x)\, dx + \frac{C}{I(x)}$$

Let's look at some examples of solving first-order linear differential equations.

---

### EXAMPLE 1

Find the general solution of $\frac{dy}{dx} + 2xy = x$.

**Solution**

The equation $dy/dx + 2xy = x$ is in the form $dy/dx + P(x)y = Q(x)$, with $P(x) = 2x$ and $Q(x) = x$. The integrating factor is

$$I(x) = e^{\int P(x)\, dx} = e^{\int 2x\, dx} = e^{x^2}$$

Multiplying both sides of the equation by the integrating factor gives us

$$e^{x^2} \frac{dy}{dx} + 2xe^{x^2} y = xe^{x^2}$$

$$\frac{d}{dx}(e^{x^2} y) = xe^{x^2}$$

Integrating both sides, we obtain

$$\int \frac{d}{dx}(e^{x^2} y)\, dx = \int xe^{x^2}\, dx$$

$$e^{x^2} y = \frac{1}{2} \int e^{x^2}(2x\, dx) = \frac{1}{2} e^{x^2} + C$$

Solving for $y$ gives us $y = \frac{1}{2} + Ce^{-x^2}$, which is the general solution. ∎

## EXAMPLE 2 (Participative)

Let's find the general solution of: $x \, (dy/dx) + 2y - 3x = 0$.

**Solution**

We begin by putting the differential equation in the standard form $dy/dx + P(x)y = Q(x)$:

$$x \frac{dy}{dx} + 2y - 3x = 0$$

$$x \frac{dy}{dx} + 2y = \underline{\phantom{xxx}} \quad \text{(Adding } 3x \text{ to both sides)}$$

$$\frac{dy}{dx} + \frac{2}{x} \cdot y = \underline{\phantom{xxx}} \quad \text{(Dividing both sides by } x, \text{ assuming } x \neq 0\text{)}$$

We observe that $P(x) = \underline{\phantom{xxx}}$ and $Q(x) = \underline{\phantom{xxx}}$. The integrating factor is

$$I(x) = e^{\int P(x) \, dx} = e^{\int (2/x) \, dx} = e^{2 \ln |x|} = \left(e^{\ln |x|}\right)^2 = |x|^2 = x^2$$

Multiplying both sides of the standard form of the differential equation by $x^2$, we now obtain

$$x^2 \frac{dy}{dx} + 2xy = \underline{\phantom{xxx}}$$

$$\frac{d}{dx}(\underline{\phantom{xxx}}) = 3x^2$$

Integrating both sides gives us

$$\int \frac{d}{dx}[x^2 y] \, dx = \int 3x^2 \, dx$$

$$x^2 y = \underline{\phantom{xxx}}$$

Solving this last equation for $y$ by dividing both sides by $x^2$ yields the general solution in explicit form:

$$y = \underline{\phantom{xxx}} \qquad \blacksquare$$

In Example 2, we had $P(x) = 2/x$, so that $\int (2/x) \, dx = 2 \ln |x|$ and the integrating factor was $e^{2 \ln |x|} = |x|^2 = x^2$. In this case the absolute value sign in the integrating factor was easily removed. However, if $P(x)$ had been $1/x$, then $\int (1/x) \, dx = \ln |x|$ and the integrating factor would have been $e^{\ln |x|} = |x|$. To avoid the difficulties of working with an integrating factor involving absolute values, we will assume in a case like this that $x > 0$, so that the integrating factor becomes $|x| = x$. Indeed, in all the problems of this section, whenever an expression of the form $e^{\ln |f(x)|}$ appears in the solution of a first-order linear equation, we will assume that the values of $x$ are restricted (if necessary), so that $f(x) > 0$ and $|f(x)| = f(x)$.

---

**SOLUTIONS TO EXAMPLE 2**    $3x, \, 3, \, 2/x, \, 3, \, 3x^2, \, x^2 y, \, x^3 + C, \, x + (C/x^2)$

**EXAMPLE 3** Solve the initial value problem: $x^2 (dy/dx) + 3xy = e^{2x}$ for $y(1) = 0$.

**Solution** We can put the given differential equation into standard form by dividing both sides by $x^2$ to obtain

$$\frac{dy}{dx} + \frac{3}{x} \cdot y = \frac{e^{2x}}{x^2}$$

Note that $P(x) = 3/x$ and $Q(x) = e^{2x}/x^2$. The integrating factor is therefore

$$I(x) = e^{\int P(x)\, dx} = e^{\int (3/x)\, dx} = e^{3 \ln |x|}$$

Assuming that $x > 0$, we write

$$I(x) = e^{3 \ln x} = (e^{\ln x})^3 = x^3$$

Next we multiply both sides of the standard form of the differential equation by $x^3$ to get

$$x^3 \frac{dy}{dx} + 3x^2 y = xe^{2x}$$

$$\frac{d}{dx}[x^3 y] = xe^{2x}$$

Integrating both sides, we obtain

$$\int \frac{d}{dx}[x^3 y]\, dx = \int xe^{2x}\, dx$$

$$x^3 y = \int xe^{2x}\, dx$$

The right-hand side requires an integration by parts. We choose $u = x$ and $dv = e^{2x}$, so that $du = dx$ and $v = \frac{1}{2}e^{2x}$, and therefore

$$x^3 y = x\left(\tfrac{1}{2}e^{2x}\right) - \int \left(\tfrac{1}{2}e^{2x}\right) dx$$
$$\quad\quad\uparrow\ \uparrow\quad\quad\quad\uparrow\ \uparrow$$
$$\quad\quad u\ \ v\quad\quad\quad\ v\ \ du$$

$$= \tfrac{1}{2}xe^{2x} - \tfrac{1}{4}e^{2x} + C$$

To obtain an explicit solution, we divide both sides by $x^3$ to get the general solution:

$$y = \frac{1}{2x^2}e^{2x} - \frac{1}{4x^3}e^{2x} + \frac{C}{x^3}$$

Next, to find the particular solution corresponding to the initial condition $y(1) = 0$, we substitute $x = 1$ and $y = 0$ in the general solution:

$$0 = \frac{1}{2(1)^2}e^{2(1)} - \frac{1}{4(1)^3}e^{2(1)} + \frac{C}{1^3}$$

$$= \tfrac{1}{2}e^2 - \tfrac{1}{4}e^2 + C$$

$$C = -\tfrac{1}{4}e^2$$

The required particular solution is therefore

$$y = \frac{1}{2x^2}e^{2x} - \frac{1}{4x^3}e^{2x} - \frac{e^2}{4x^3}$$

## SECTION 9.3
### SHORT ANSWER EXERCISES

1. An integrating factor for the differential equation $\dfrac{dy}{dx} - y = x$ is _____.

2. An integrating factor for the differential equation $x^2 \dfrac{dy}{dx} - y = 2$ is _____.

3. (True or False) The differential equation $\dfrac{dy}{dx} = \dfrac{2+y}{x^2}$ is a first-order linear equation. _____

4. (True or False) The differential equation $\dfrac{dy}{dx} = \dfrac{2+y}{x^2}$ is a separable equation. _____

5. (True or False) A first-order linear differential equation may be separable. _____

6. In a first-order linear differential equation, if $Q(x) = 0$ and the integrating factor is $I(x) = x$, the general solution is $y = $ _____.

7. In a first-order linear differential equation, if $P(x) = 0$ and $Q(x) = 2x$, the general solution is $y = $ _____.

8. In a first-order linear differential equation, $P(x) = 2/x^2$ and $Q(x) = x$. Is this sufficient information to find the general solution? _____

9. The differential equation $x^2 \dfrac{dy}{dx} + xy^2 = x^3$ (is, is not) first-order linear.

10. The differential equation $x \dfrac{dy}{dx} + x^2 = x^2 y$ (is, is not) first-order linear.

## SECTION 9.3
### EXERCISES

In Exercises 1–18, find the general solution of each differential equation in explicit form.

1. $\dfrac{dy}{dx} + 2y = e^{-2x}$

2. $\dfrac{dy}{dx} - 3y = e^{3x}$

3. $\dfrac{dy}{dx} + 4xy = 3x$

4. $\dfrac{dy}{dx} - xy = 2x$

5. $2x \dfrac{dy}{dx} + x^2 y = 2x^2$

6. $x \dfrac{dy}{dx} + 3y - x^2 = 0$

7. $x \dfrac{dy}{dx} + y = \dfrac{1}{x^2}$

8. $\dfrac{dy}{dx} - y - x = 0$

9. $\dfrac{dy}{dx} - 2y = 4$

10. $x \dfrac{dy}{dx} = x^3 - 2x - 3y$

11. $x^2 \dfrac{dy}{dx} + y = 1$

12. $\sqrt{x} \dfrac{dy}{dx} + y - 2 = 0$

13. $\dfrac{dy}{dx} - 2y = e^{5x}$

14. $x^2 \dfrac{dy}{dx} + y = e^{-1/x}$

15. $\dfrac{dy}{dx} + (2x+1)y = 6x + 3$

16. $\dfrac{dy}{dx} + \dfrac{y}{x} = e^{-x}$

17. $\dfrac{dy}{dx} + \dfrac{y}{2x+1} = 2$

18. $\dfrac{dy}{dx} + \dfrac{y}{x+1} = \sqrt[3]{x}$

*In Exercises 19–26, solve each initial-value problem.*

19. $6\dfrac{dy}{dx} - 12y = e^{2x}$ for $y(0) = 1$

20. $\dfrac{dy}{dx} + 2xy + x = 0$ for $y(0) = 2$

21. $\dfrac{dy}{dx} - e^x = -y$ for $y(0) = 0$

22. $x\dfrac{dy}{dx} + 2y = 3x\sqrt{x^3 + 1}$ for $y(-1) = 3$

23. $x^2\dfrac{dy}{dx} - 4xy = x^6$ for $y(-1) = -1$

24. $x\dfrac{dy}{dx} + y = x \ln x$ for $y(1) = 0$

25. $(x^2 + 4)\dfrac{dy}{dx} + xy = 0$ for $y(0) = 1$

26. $\sqrt[3]{x}\dfrac{dy}{dx} + \tfrac{5}{3}xy = \dfrac{\sqrt[3]{x}}{e\sqrt[3]{x^5}}$ for $y(1) = 1$

*In Exercises 27–30, use* (a) *the method of solution by separation of variables and* (b) *the method of this section to find the general solution of each differential equation.*

27. $\dfrac{dy}{dx} = \dfrac{2 + y}{x^2}$

28. $\dfrac{dy}{dx} = x(1 + y)$

29. $\dfrac{dy}{dx} = \dfrac{4 - y}{x}$

30. $\dfrac{dy}{dx} = \dfrac{y + 3}{x + 2}$

## 9.4 APPLICATIONS OF DIFFERENTIAL EQUATIONS

### Separation of Variables

In Section 5.4 we studied the solutions of several types of differential equations involving growth and decay:

| Differential Equation | General Solution | Type of Growth/Decay |
|---|---|---|
| $\dfrac{dA}{dt} = kA$ | $A(t) = Ce^{kt}$ | Unlimited |
| $\dfrac{dA}{dt} = k(M - A)$ | $A(t) = M(1 - Ce^{-kt})$ | Limited |
| $\dfrac{dA}{dt} = kA(M - A)$ | $A(t) = \dfrac{M}{1 + Ce^{-kMt}}$ | Logistic |

The general solution of each of these three types of equations can be found by the technique of separation of variables. We will now show how these solutions can be obtained.

Recall that if a quantity grows or decays so that the rate of change of the amount $A(t)$ of the quantity present at time $t$ is proportional to the amount present at time $t$, we write

$$\dfrac{dA}{dt} = kA$$

**EXAMPLE 1** A certain colony of bacteria is growing so that the rate of change of the number of bacteria present at any time is proportional to the size of the colony at that time. (a) If initially $A_0$ bacteria are present, find a formula for the number present at time $t$. (b) If 100 bacteria are present initially and 400 are present after two hours, find the formula for the number present at time $t$.

**Solution** (a) Let $A(t)$ be the number of bacteria present at time $t$. We must solve the initial-value problem

$$\frac{dA}{dt} = kA, \quad k > 0, \quad A(0) = A_0$$

To find the general solution of $dA/dt = kA$, we begin by separating the variables by multiplying both sides by $dt/A$.

$$\frac{dA}{A} = k\, dt$$

$$\int \frac{dA}{A} = \int k\, dt \qquad \text{(Integrating both sides)}$$

$$\ln A = kt + C_1 \qquad \begin{array}{l}\text{(Observe that } A > 0, \\ C_1 \text{ is an arbitrary constant)}\end{array}$$

$$A = e^{kt+C_1} = e^{C_1}e^{kt}$$

$$= Ce^{kt} \qquad \text{(Where } C = e^{C_1}\text{)}$$

Observe that we have just shown that the general solution of $dA/dt = kA$ is $A = Ce^{kt}$ when $A > 0$. To find the particular solution that satisfies $A(0) = A_0$, we substitute $t = 0$ and $A = A_0$ into the general solution to obtain $A_0 = Ce^{k(0)}$, so that $C = A_0$. The particular solution is therefore

$$A(t) = A_0 e^{kt} \qquad (1)$$

(b) We are given that $A(0) = 100$ and $A(2) = 400$. Thus $A_0 = 100$, so that our solution (1) from part (a) becomes $A(t) = 100e^{kt}$. Since $A(2) = 400$, we have

$$400 = A(2) = 100e^{k(2)}$$

$$e^{2k} = 4$$

$$\ln e^{2k} = \ln 4$$

$$2k = \ln 4$$

$$k = \frac{\ln 4}{2} = \frac{\ln 2^2}{2} = \frac{2 \ln 2}{2} = \ln 2 \approx 0.6931$$

Thus the formula for the number of bacteria present at time $t$ is $A(t) = 100e^{(\ln 2)t}$. The graph of $A(t)$ is sketched in Figure 9.3 for reference. Note that $100e^{(\ln 2)t}$ can also be written as $100(2)^t$. ■

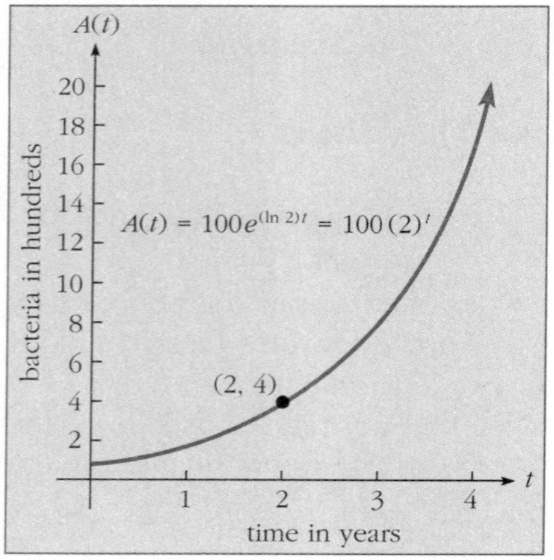

If a quantity grows or decays so that the rate of change of the amount present at time $t$ is proportional to the difference between the maximum possible amount, $M$, and the current amount, $A$, then we write

$$\frac{dA}{dt} = k(M - A)$$

**FIGURE 9.3** The graph of $A(t) = A_0 e^{kt}$ for $A_0 = 100$ and $k = \ln 2 > 0$.

## 9.4 APPLICATIONS OF DIFFERENTIAL EQUATIONS

**EXAMPLE 2**

The yearly dollar sales (in millions) of a company are growing at a rate that is proportional to the difference between current sales and maximum attainable yearly sales $M$. **(a)** Find a formula for yearly sales during year $t$. **(b)** If annual sales initially are two million dollars and annual sales after one year are \$4,000,000, find a formula for annual sales after $t$ years if $M = \$15{,}000{,}000$.

**Solution**

**(a)** If $A(t)$ is annual dollar sales in year $t$, then

$$\frac{dA}{dt} = k(M - A) \qquad (2)$$

We begin by multiplying both sides of equation (2) by $\dfrac{dt}{M - A}$ to separate variables:

$$\frac{dA}{M - A} = k\,dt$$

$$\int \frac{dA}{M - A} = \int k\,dt$$

$$-\ln |M - A| = kt + C_1$$

$$-\ln(M - A) = kt + C_1 \qquad (M > A \text{ so } |M - A| = M - A)$$

$$\ln(M - A) = -kt - C_1$$

$$M - A = e^{-kt - C_1} \qquad \text{(Converting to exponential form)}$$

$$A = M - e^{-C_1} e^{-kt}$$

$$= M - Ke^{-kt} \qquad \text{(Where } K = e^{-C_1} \text{ is a constant)}$$

$$A(t) = M(1 - Ce^{-kt}) \qquad \text{(Where } C = K/M \text{ is a constant)}$$

We have just shown that the general solution of the differential equation $dA/dt = k(M - A)$ is $A = M(1 - Ce^{-kt})$.

**(b)** We are given that $A(0) = 2$, $A(1) = 4$, and $M = 15$. Substituting $M = 15$ into our solution from part (a), we get $A(t) = 15(1 - Ce^{-kt})$. Using the initial condition $A(0) = 2$, we have

$$2 = A(0) = 15(1 - Ce^{-k(0)}) = 15(1 - C)$$

$$15C = 13$$

$$C = \frac{13}{15}$$

Thus our solution becomes $A(t) = 15\left(1 - \frac{13}{15}e^{-kt}\right)$. Using the condition $A(1) = 4$, we get

$$4 = A(1) = 15\left(1 - \tfrac{13}{15}e^{-k(1)}\right) = 15 - 13e^{-k}$$

$$13e^{-k} = 11$$

$$e^{-k} = \frac{11}{13}$$

$$-k = \ln\left(\frac{11}{13}\right)$$

$$k = -\ln\left(\frac{11}{13}\right) \approx 0.1671$$

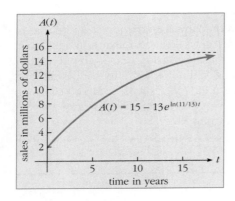

FIGURE 9.4 The graph of $A(t) = M(1 - Ce^{-kt})$ for $M = 15$, $C = \frac{13}{15}$, and $k = -\ln\left(\frac{11}{13}\right)$.

The formula for annual sales after $t$ years is therefore

$$A(t) = 15\left(1 - \tfrac{13}{15}e^{\ln(11/13)t}\right)$$
$$= 15 - 13e^{\ln(11/13)t}$$

The graph of $A(t)$ is sketched in Figure 9.4 for reference. ∎

If a quantity grows so that the rate of change of the amount $A(t)$ present at time $t$ is proportional to the product of $A$ and the difference between a maximum amount, $M$, and $A$, then we write

$$\frac{dA}{dt} = kA(M - A)$$

## EXAMPLE 3

The rate at which a rumor spreads throughout a population of $M$ people is proportional to the number of people who have heard the rumor times the number who have not heard it. (a) Find a formula for the number of people who have heard the rumor by time $t$. (b) If initially 20 people have heard the rumor and after one day 50 people have heard it, find a formula for the number of people who have heard the rumor after $t$ days if at most 400 people can hear the rumor.

Solution

(a) If $A(t)$ is the number of people who have heard the rumor after $t$ days, then

$$\frac{dA}{dt} = kA(M - A)$$

To find the general solution of this separable equation, we multiply both sides of the equation by $dt/A(M - A)$ and integrate to get

$$\frac{dA}{A(M - A)} = k\, dt$$

$$\int \frac{dA}{A(M - A)} = \int k\, dt \qquad (3)$$

We now make use of the following algebraic identity:

$$\frac{1}{A(M - A)} = \frac{1}{M}\left[\frac{(M - A) + A}{A(M - A)}\right] = \frac{1}{M}\left[\frac{1}{A} + \frac{1}{M - A}\right]$$

Using this identity on the left-hand side of (3), we obtain

$$\frac{1}{M}\int \frac{dA}{A} + \frac{1}{M}\int \frac{dA}{M - A} = \frac{1}{M}\ln|A| - \frac{1}{M}\ln|M - A|$$
$$= \frac{1}{M}[\ln A - \ln(M - A)] \qquad (A > 0 \text{ and } M - A > 0)$$
$$= kt + C_1$$

$$\ln A - \ln(M - A) = Mkt + C_1 M$$

$$\ln\left(\frac{A}{M-A}\right) = Mkt + C_2 \qquad \text{(Where } C_2 = C_1 M \text{ is a constant)}$$

$$\frac{A}{M-A} = e^{Mkt + C_2} \qquad \text{(Changing to exponential form)}$$

$$A = (M - A)\left[e^{C_2} \cdot e^{Mkt}\right]$$

$$= (M - A)\left[C_3 e^{Mkt}\right] \qquad \text{(Where } C_3 = e^{C_2}\text{)}$$

$$A + A C_3 e^{Mkt} = M C_3 e^{Mkt}$$

$$(1 + C_3 e^{Mkt})A = M C_3 e^{Mkt}$$

$$A = \frac{M C_3 e^{Mkt}}{1 + C_3 e^{Mkt}}$$

Dividing the numerator and denominator of the fraction on the right-hand side by $C_3 e^{Mkt}$, we get

$$A = \frac{M}{\dfrac{1}{C_3 e^{Mkt}} + 1} = \frac{M}{1 + \dfrac{1}{C_3} e^{-kMt}}$$

If we let $C = \dfrac{1}{C_3}$, then the formula becomes

$$A(t) = \frac{M}{1 + Ce^{-kMt}} \tag{4}$$

We have just shown that the general solution of the differential equation $dA/dt = kA(M - A)$, where $A > 0$ and $M - A > 0$, is given by equation (4).

(b) We are given that $M = 400$, so that equation (4) becomes $A(t) = 400/(1 + Ce^{-400kt})$. Using the initial condition $A(0) = 20$, we obtain

$$20 = A(0) = \frac{400}{1 + Ce^{-400k(0)}} = \frac{400}{1 + C}$$

$$20 + 20C = 400$$

$$C = 19$$

Our solution becomes $A(t) = 400/(1 + 19e^{-400kt})$. Using the condition $A(1) = 50$, we get

$$50 = A(1) = \frac{400}{1 + 19e^{-400k(1)}}$$

$$50 + 950e^{-400k} = 400$$

$$e^{-400k} = \frac{7}{19}$$

$$-400k = \ln\left(\frac{7}{19}\right)$$

$$k = -\frac{\ln\left(\frac{7}{19}\right)}{400}$$

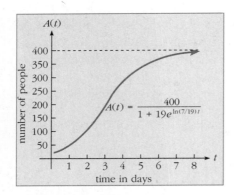

**FIGURE 9.5** The graph of $A(t) = M/(1 + Ce^{-kMt})$ for $M = 400$, $C = 19$, and $k = -\frac{1}{400}\ln\left(\frac{7}{19}\right)$.

The formula for the number of people who have heard the rumor after $t$ days is

$$A(t) = \frac{400}{1 + 19e^{\ln(7/19)t}}$$

The graph of $A(t)$ is sketched in Figure 9.5 for reference. ∎

## First-Order Linear Equations

We will now consider some applied problems that lead to first-order linear differential equations. Many of these applications can be conceptualized as finding the amount $A(t)$ of a substance in a compartment at time $t$, given the rate at which the substance flows into the compartment, the rate at which the substance flows out of the compartment, and the initial amount of the substance in the compartment (see Figure 9.6). The rate of change of the amount of the substance in the compartment at time $t$ must be the difference between the rate at which it flows in and the rate at which it flows out. That is,

$$\frac{dA}{dt} = \text{rate in} - \text{rate out}$$

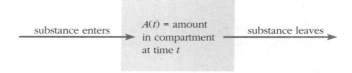

**FIGURE 9.6** A substance enters a compartment at one rate and simultaneously leaves the compartment at another rate.

If, for example, the substance enters the compartment at a rate proportional to the amount $A$ present and leaves at a constant rate $b$, then the governing differential equation is

$$\frac{dA}{dt} = kA - b$$

where $k$ is a constant of proportionality. If we write this equation as

$$\frac{dA}{dt} - kA = -b$$

we recognize it as a first-order linear equation. The next example illustrates such a situation.

---

**EXAMPLE 4 (Participative)**

Suppose $12,000 is deposited in a bank account that gives 5% annual interest compounded continuously. If money is withdrawn at frequent intervals at a constant rate of $1000 per year, let's find **(a)** the amount in the account at any time $t$ before it is depleted; **(b)** the time when the account will be depleted; and **(c)** the total amount of all the withdrawals.

## 9.4 APPLICATIONS OF DIFFERENTIAL EQUATIONS

**Solution**

(a) Let $A(t)$ be the amount in the account at any time $t$ before depletion. Then $A(0) = $ _____. Since the account earns interest at 5% compounded continuously, money is continuously entering the account at a rate equal to 5% of the money in the account. That is,

rate in $= 0.05A$ dollars per year

On the other hand, the rate at which money leaves the account is constant and is simply

rate out $=$ _____ dollars per year

(See Figure 9.7.) Using the rule $dA/dt = $ rate in $-$ rate out, we obtain

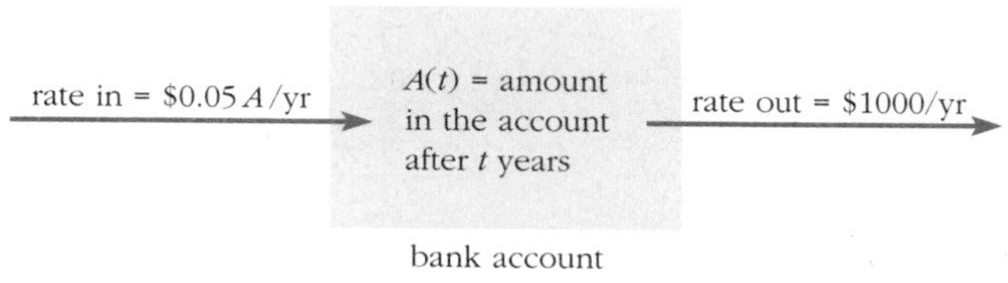

rate in = \$0.05 A/yr  →  $A(t)$ = amount in the account after $t$ years  →  rate out = \$1000/yr

bank account

**FIGURE 9.7** Money enters the account at the rate of $0.05A$ dollars per year and is simultaneously withdrawn at the rate of \$1000/year.

$$\frac{dA}{dt} = \underline{\hspace{2cm}}$$

which can be rewritten

$$\frac{dA}{dt} - 0.05A = -1000 \tag{5}$$

This differential equation is first-order linear, with $P(t) = $ _____ and $Q(t) = -1000$. The integrating factor is

$$I(t) = e^{\int P(t)\,dt} = e^{\int -0.05\,dt} = \underline{\hspace{2cm}}$$

Multiplying both sides of the differential equation (5) by $I(t)$, we get

$$e^{-0.05t}\frac{dA}{dt} - 0.05e^{-0.05t}A = -1000e^{-0.05t}$$

$$\frac{d}{dt}[\underline{\hspace{2cm}}] = -1000e^{-0.05t}$$

Integrating both sides, we have

$$e^{-0.05t}A = \int(-1000e^{-0.05t})\,dt = \underline{\hspace{3cm}}$$

$$A = 20000 + Ce^{0.05t}$$

To find $C$, we apply the initial condition $A(0) = 12000$ to obtain

$$12000 = A(0) = 20000 + Ce^{0.05(0)}$$

$$C = \underline{\hspace{2cm}}$$

Thus the amount in the account at any time $t$ before it is depleted is

$$A(t) = \underline{\hspace{3cm}}$$

(b) The account will be depleted at the time $T$ when $A(T) = 0$. Thus we must solve the equation

$$20000 - 8000e^{0.05T} = 0$$

for $T$. We have

$$8000e^{0.05T} = 20000$$

$$e^{0.05T} = \underline{\phantom{xxxx}}$$

$$0.05T = \underline{\phantom{xxxxxx}}$$

$$T = \underline{\phantom{xxxxxxxxxxxx}} \qquad \text{(To 3 decimal places)}$$

So the account will be depleted after 18.326 years, or approximately 18 years, 119 days.

(c) Since $1000 is withdrawn per year for 18.326 years before the account is depleted, the total amount of all the withdrawals is

$$(\underline{\phantom{xxxx}} \text{ years})(\$1000 \text{ per year}) = \$\underline{\phantom{xxxx}}$$

## EXAMPLE 5

Initially, a large tank that holds 1000 gal contains 200 gal of water in which 50 lb of salt are dissolved. A brine solution that contains 1 lb of salt per gallon is pumped into the tank at a rate of 3 gal/min and is well mixed with the water in the tank; see Figure 9.8. The mixture is simultaneously pumped out of the tank at the rate of 2 gal/min. Find (a) the total amount of salt in the tank at any time; and (b) the amount of salt in the tank when it is just at the point of overflow.

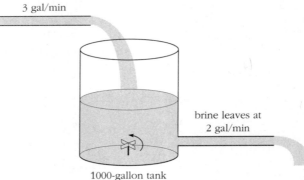

**FIGURE 9.8** The tank initially contains 200 gal of water in which 50 lb of salt are dissolved.

### Solution

(a) Let $A(t)$ be the number of pounds of salt in the tank at time $t$. Then the rate of change of $A(t)$ is

$$\frac{dA}{dt} = \text{rate in} - \text{rate out}$$

---

**SOLUTIONS TO EXAMPLE 4**

(a) $12000, 1000, $0.05A - 1000$, $-0.05$, $e^{-0.05t}$, $e^{-0.05t}A$, $20000e^{-0.05t} + C$, $-8000$, $20000 - 8000e^{0.05t}$; (b) 2.5, $\ln(2.5)$, $20 \ln(2.5) \approx 18.326$; (c) 18.326, 18326

Since brine containing 1 lb of salt per gallon is pumped into the tank at 3 gal/minute,

rate in = (1 lb/gal) · (3 gal/min) = 3 lb/min

The well-mixed solution is leaving the tank at the slower rate of 2 gal/min. Thus the amount of brine in the tank is increasing at the rate of

3 gal/min − 2 gal/min = 1 gal/min

Thus after $t$ minutes the tank will contain $200 + t$ gallons of brine. These $200 + t$ gallons of brine will contain $A = A(t)$ lb of salt, so that the concentration of salt in the brine solution leaving the tank is

$$\frac{A \text{ lb of salt}}{(200 + t) \text{ gal of solution}} = \frac{A}{200 + t} \text{ lb/gal}$$

Since brine is leaving the tank at 2 gal/min,

$$\text{rate out} = (2 \text{ gal/min})\left(\frac{A}{200 + t} \text{ lb/gal}\right) = \frac{2A}{200 + t} \text{ lb/min}$$

We obtain the initial-value problem

$$\frac{dA}{dt} = 3 - \frac{2A}{200 + t} \quad \text{for} \quad A(0) = 50$$

We can write the differential equation in the standard form of a first-order linear equation:

$$\frac{dA}{dt} + \left(\frac{2}{200 + t}\right)A = 3 \tag{6}$$

so that $P(t) = 2/(200 + t)$ and $Q(t) = 3$. The integrating factor is

$$I(t) = e^{\int P(t)\, dt} = e^{\int 2/(200+t)\, dt} = e^{2 \ln(200+t)} = (200 + t)^2$$

Multiplying both sides of equation (6) by $I(t)$ gives us

$$(200 + t)^2 \frac{dA}{dt} + 2(200 + t)A = 3(200 + t)^2$$

$$\frac{d}{dt}[(200 + t)^2 A] = 3(200 + t)^2$$

Integrating both sides gives us

$$(200 + t)^2 A = \int 3(200 + t)^2 \, dt = (200 + t)^3 + C$$

Solving for $A$, we now get the general solution

$$A(t) = (200 + t) + \frac{C}{(200 + t)^2}$$

To find $C$, we use the initial condition $A(0) = 50$ to get

$$50 = A(0) = (200 + 0) + \frac{C}{(200 + 0)^2}$$

$$\frac{C}{40000} = -150$$

$$C = -6000000$$

The total amount of salt in the tank after $t$ minutes is therefore

$$A(t) = 200 + t - \frac{6000000}{(200 + t)^2}$$

(b) From part (a), recall that after $t$ minutes the tank will contain $200 + t$ gallons of brine. Since the capacity of the tank is 1000 gallons, it will be at the point of overflow when

$$200 + t = 1000 \quad \text{or} \quad t = 800 \text{ minutes}$$

After 800 minutes the total amount of salt in the tank will be

$$A(800) = 200 + 800 - \frac{6000000}{(200 + 800)^2} = 994 \text{ pounds}$$

## SECTION 9.4
### SHORT ANSWER EXERCISES

1. If the amount $A$ of a certain quantity present at time $t$ satisfies the differential equation $dA/dt = 0.03A$, then the quantity is undergoing _____ growth.
2. A quantity grows so that its rate of change at time $t$ is proportional to the difference between 1000 and the current amount $A$. The differential equation satisfied by this quantity is _____.
3. If a quantity exhibits logistic growth and satisfies the differential equation $dA/dt = kA(500 - A)$, then the quantity is growing most rapidly when $A =$ _____.
4. (True or False) A quantity that undergoes limited exponential growth is always increasing at a decreasing rate. _____
5. (True or False) A quantity that undergoes unlimited exponential growth is always increasing at an increasing rate. _____
6. (True or False) A quantity that undergoes logistic exponential growth is always increasing at a decreasing rate. _____
7. A substance enters a compartment at the rate $0.04A(t)$ lb/min, where $A$ is the amount of the substance in the compartment at time $t$. Simultaneously, the substance leaves the compartment at the rate of 50 lb/min. Thus $dA/dt =$ _____.
8. If a substance enters a compartment at the rate of $0.05A(t)$ gal/hr and leaves the compartment at the rate of $0.04A(t)$ gal/hr, then $dA/dt =$ _____.
9. At time $t$, a large tank contains $100 + 2t$ gal of a solution made up of $A(t)$ pounds of a certain chemical. If solution leaves the tank at a rate of 5 gal/min, the rate at which the chemical is leaving the tank is _____ lb/min.
10. Water flows into a 500-gal tank at 2 gal/min and out of the tank at 1.5 gal/min. If the tank contains 200 gal at time $t = 0$, the tank will be at the point of overflow after _____ hours.

## SECTION 9.4
### EXERCISES

1. (Growth of a fungus) A fungus grows at a rate that is proportional to the amount present. If initially 2 g of this fungus are present and 5 hr later, 3 g are present, (a) how much of the fungus is present after 20 hours? (b) How long will it take before the amount present is 15 g?
2. (Continuous compounding) If you put $10,000 into a bank account that pays 9.8% compounded continuously, how much money will be in the account after eight years?

3. **(Doubling time)** Find the time required for a sum of money to double if it is invested at an annual rate of 10% per year compounded continuously.

4. **(Boll weevil extermination)** Due to an extermination program, the number of boll weevils in a certain southern region of the United States is expected to decline by 15%/yr over the next 20 years. How long will it take for the boll weevil population to be reduced to 1% of its current level?

5. **(Radioactive decay)** The half-life of a certain radioactive isotope is 128 days. If 100 g of this substance is present initially, how much will be present in one year if the substance decays at a rate proportional to the amount present?

6. **(Population)** The population of a certain town increases at a rate proportional to the number of residents at any time. If the population of this town was 20,000 in 1980 and 28,000 in 1990, what will the population be in 2000?

7. **(Chemical reaction)** In a certain chemical reaction, the rate at which one chemical is converted into another is proportional to the amount of the first chemical present at any time. At the end of one hour, 80 g of the first chemical remain. At the end of 4 hr, 25 g remain.
   (a) What amount of the first chemical was present initially?
   (b) How many grams of the first chemical will be present at the end of eight hours?
   (c) When will only 1 g of the first chemical remain?

8. **(Newton's law of cooling)** Newton's law of cooling states that the rate at which the temperature of a cooling object changes is proportional to the difference between the temperature of the object and the (constant) temperature of the medium surrounding the object. If an object cools from 200°F to 150°F in 5 min in air at a temperature of 75°F, what will the temperature of the object be after 20 min?

9. **(Newton's law of cooling)** See Exercise 8 for the statement of Newton's law of cooling. A roast turkey is taken from an oven when its temperature is 325°F, and it is placed to cool in a room that has a temperature of 68°F. After 5 min, the temperature of the turkey is 300°F.
   (a) What will the temperature be after 30 min?
   (b) When will the temperature be 200°F?

10. **(Company grapevine)** The rate at which the 20,000 employees of a certain company hear about a change in management is proportional to the number of people who have not yet heard about the change. If initially 5 people have heard the news and if 200 employees have heard the news after one day, when will 90% of the employees have heard the news?

**Business and Economics**

11. **(Product awareness)** The rate at which the target market of 10 million people become aware of a new product advertised by your company is proportional to the number of people who are not aware of the product. At the start of the advertising campaign, no one in the target market is aware of the product, and after one week, 500,000 people are aware of the product.
    (a) How many people will be aware of the product after three months?
    (b) When will 95% of the target market be aware of the product?

12. **(Psychology)** In a psychology experiment, a person who has memorized 100 facts is asked to recall them. If the rate at which the person recalls these facts is proportional to the number not yet recalled, and if the person recalls ten facts after four minutes, after how many minutes will the person recall 50 facts?

13. **(Employee learning)** The rate at which 1000 employees in a factory learn to use a new machine is proportional to the number of employees who have not yet learned to use the machine. If initially only one employee can operate the machine and five

days later, 50 employees can operate the machine, when will half of the employees know how to operate the machine?

14. (New product sales) The dollar sales of a product introduced at the beginning of 1990 are increasing at a rate proportional to the product of current sales times the difference between current sales and 5 million dollars. If in 1990, dollar sales were $500,000 and in 1991, dollars sales were 1.5 million dollars, in what year will sales exceed 4 million dollars?

**Social Sciences**

15. (Psychology) The Weber–Fechner law in psychology states that a person's perception $p$ of a stimulus and the strength $s$ of the stimulus are related by the differential equation $dp/ds = k/s$, where $k$ is a constant. Find the general solution of this equation.

16. (Spread of a rumor) The rate at which a rumor spreads among 1000 people is proportional to the product of the number of people who have heard the rumor and those who have not. If initially only one person has heard the rumor and if, after two days, 400 people have heard the rumor, find a formula that gives the number of people who have heard it after $t$ days.

**Life Sciences**

17. (Spread of a disease) The rate at which a flu epidemic spreads is proportional to the product of the number of people who have contracted the flu and those who have not. In a school of 1200 students, initially five students have the flu. After two days, 75 students have caught it. Find a formula that gives the number of students who have contracted the flu after $t$ days.

18. (Sales growth) The rate of growth in sales of a new product is proportional to the product of current sales times the difference between current sales and maximum sales of 10 million dollars. If sales in 1990 were 1 million dollars and sales in 1991 were 2.5 million dollars, when will sales exceed 8 million dollars?

19. (Bank account withdrawals) Suppose that $15,000 is invested in a bank account that earns 9.5% annual interest compounded continuously. If money is withdrawn at frequent intervals at a constant rate of $1500 per year, find (a) the amount in the account at any time $t$ before it is depleted; (b) the time when the account will be depleted; and (c) the total amount of all the withdrawals.

20. (Saving) If $25,000 is initially invested in a bank account that earns 8% annual interest compounded continuously, and if money is deposited into the account continuously at the rate of $10,000 per year, after how many years will the account contain 1 million dollars?

21. (Saving) If an initial investment of $20,000 is made in an account that pays 10% annual interest compounded continuously, how much money must be added continuously to the account each year to have $250,000 at the end of 20 years?

22. (Home mortgage) You are considering the purchase of a home for which you will have to take out a mortgage of $80,000 at an interest rate of 9% per year. What will your monthly payment be if you wish to pay off (or *amortize*) the loan in (a) 15 years? (b) 30 years?

23. (Business loan) To expand your small business, you are investigating the possibility of taking out a loan of $25,000 at an interest rate of 12% per year. If the loan must be repaid in 10 years with a series of equal monthly payments, what is the monthly payment?

24. (Mixing problem) A 500-gal tank initially contains 100 gallons of fresh water. Starting at time $t = 0$, a brine solution containing 4 lb of salt/gal is pumped into the

tank at the rate of 2 gal/min. The well-stirred mixture leaves the tank at the same rate.
  (a) Find the amount of salt in the tank at any time $t > 0$.
  (b) Find the amount of salt in the tank after 30 min.
  (c) If $A(t)$ is the amount of salt in the tank at time $t > 0$, find and interpret $\lim_{t \to \infty} A(t)$.

25. (Mixing problem) Solve Exercise 24 (a) and (b), assuming that the brine solution leaves the tank at the rate of 1 gal/min.

26. (Mixing problem) A 500-liter tank initially contains 200 liters (l) of a solution composed of 2 g of a certain chemical per liter. Fluid containing 10 g of this chemical per liter flows into the tank at the rate of 4 l/min, and the well-stirred mixture leaves the tank at the rate of 5 l/min. When the tank contains 50 l of fluid, find the amount of the chemical in the tank.

27. (Pollution concentration) A 500-liter tank initially contains 100 l of a solution made up of 0.5 kg of a certain pollutant per liter. If fluid containing 2 kg of the pollutant per liter flows into the tank at the rate of 50 l/hr and the well-stirred mixture flows out of the tank at the rate of 25 l/hr, find (a) the amount of the pollutant in the tank after 10 hr; (b) the concentration (in kg/l) of the pollutant in the tank after 10 hr.

28. (Air quality) The air in a room with volume 5000 ft$^3$ contains 500 ft$^3$ of carbon dioxide ($CO_2$). If outside air, which contains 4% $CO_2$, enters the room at the rate of 400 ft$^3$/min, and well-mixed air leaves the room at the same rate, (a) how many cubic feet of $CO_2$ does the room contain after 10 min? (b) When will the air in the room contain only 5% $CO_2$?

29. (Gompertz growth) The Gompertz growth equation is a model of limited exponential growth in which the amount $A(t)$ of a substance present at time $t$ satisfies the differential equation

$$\frac{dA}{dt} = A(a - b \ln A)$$

where $a$ and $b$ are constants. Use the technique of separation of variables to show that the general solution of this equation is

$$A(t) = e^{a/b} e^{-Ce^{-bt}}$$

where $C$ is an arbitrary constant.

## 9.5 CHAPTER NINE REVIEW

**KEY TERMS**

differential equation
general solution
initial condition
particular solution
initial-value problem
order

separation of variables
explicit solution
implicit solution
singular solution
first-order linear equation
integrating factor

**KEY FORMULAS AND RESULTS**

### The Solution of $dy/dx = g(x)$

The general solution is $y = \int g(x)\, dx$.

### Separable Differential Equation

A separable differential equation is one that can be put into one of the forms

$$g(y)\frac{dy}{dx} = f(x) \quad \text{(Derivative form)}$$

or

$$g(y)\, dy = f(x)\, dx \quad \text{(Differential form)}$$

### The Technique of Separation of Variables

If $g(y)\, dy = f(x)\, dx$ and $G'(y) = g(y)$ and $F'(x) = f(x)$, then the general solution of the differential equation is

$$G(y) = F(x) + C$$

### First-Order Linear Differential Equation

A first-order linear differential equation is one that can be put into the form

$$\frac{dy}{dx} + P(x)y = Q(x)$$

### The Solution of $dy/dx + P(x)y = Q(x)$

Let $I(x) = e^{\int P(x)\, dx}$. Then the general solution of $\frac{dy}{dx} + P(x)y = Q(x)$ is

$$y = \frac{1}{I(x)} \int I(x) Q(x)\, dx + \frac{C}{I(x)}$$

The General Solution of $\frac{dA}{dt} = kA$ is $A(t) = Ce^{kt}$

The General Solution of $\frac{dA}{dt} = k(M - A)$ is $A(t) = M(1 - Ce^{-kt})$

The General Solution of $\frac{dA}{dt} = kA(M - A)$ is $A(t) = \dfrac{M}{1 + Ce^{-kMt}}$

### Compartmental Analysis

If $A(t)$ is the amount of a substance in a compartment at time $t$, then

$$\frac{dA}{dt} = \text{rate in} - \text{rate out}$$

**CHAPTER 9 SHORT ANSWER REVIEW QUESTIONS**

1. (True or False) The general solution of the differential equation $dy/dx = 3x^2$ is $y = x^3$. _____

2. The general solution of a first-order differential equation is $y = (C/x^2) + 3$. Given the initial condition $y(1) = 1$, we find that $C =$ _____.

3. If marginal revenue is proportional to the square root of the number of units sold, then $dR/dx =$ _____.

4. (True or False) The differential equation $x \dfrac{dy}{dx} = \dfrac{1}{y^2}$ is separable. _____

5. (True or False) The differential equation $y^3 \dfrac{dy}{dx} = 2x + 3$ is separable. _____

6. The general solution of a certain separable differential equation is $y^2 + 1 = Cx^2$. Given the initial condition $y(2) = 3$, we find that $C =$ _____.

7. An integrating factor for the differential equation $x \dfrac{dy}{dx} - x^2 y = 5$ is _____.

8. The differential equation $(x^3 + 1) \dfrac{dy}{dx} - e^x = x^4 y$ (is, is not) first-order linear.

9. (True or False) In the first-order linear equation $\dfrac{dy}{dx} + P(x)y = Q(x)$, if $P(x) = 1$ and $Q(x) = 1$, the equation is separable. _____

10. (True or False) In the solution $A(t) = M(1 - Ce^{-kt})$ of the differential equation $dA/dt = k(M - A)$, the constant $C$ is equal to $A(0)$. _____

11. In the solution $A(t) = Ce^{kt}$ of the differential equation $dA/dt = kA$, if $A(0) = 100$, then $C =$ _____.

12. If you put $10,000 into an account that pays 8% annual interest compounded continuously and you withdraw money continuously at the rate of $500 per year, if $A(t)$ is the amount of money in the account after $t$ years, then $dA/dt =$ _____.

## CHAPTER 9 REVIEW EXERCISES

Exercises marked with an asterisk (*) constitute a Chapter Test.

In Exercises 1–6, determine whether each given function is a solution of the differential equation.

1. $\dfrac{dy}{dx} - 2xy = 0; \quad y = e^{x^2}$

2. $\dfrac{dy}{dx} \ln x - \dfrac{y}{x} = 0; \quad y = \ln x$

3. $\dfrac{dy}{dx} - \dfrac{y}{x} = \dfrac{2}{x}; \quad y = Cx^3 + 2$

4. $\dfrac{d^2y}{dx^2} + 4y + 4 = 0; \quad y = C_1 e^{2x} + C_2 x e^{2x}$

*5. $\dfrac{dy}{dx} - e^x y = x^2; \quad y = e^{2x} + C$

6. $y \dfrac{dy}{dx} = \dfrac{1}{4}; \quad y = C\sqrt{x}$

In Exercises 7–16, find the general solution of each differential equation.

7. $x \dfrac{dy}{dx} = 1$

8. $\dfrac{dy}{dx} = \dfrac{x}{y}$

*9. $\dfrac{dy}{dx} = \dfrac{e^x}{2y}$

10. $\dfrac{dy}{dx} + y = 2$

11. $(x + 2) \dfrac{dy}{dx} - 2 = 0$

12. $\dfrac{dy}{dx} - \dfrac{2y^3}{x^2} = 0$

*13. $\dfrac{dy}{dx} = x + 2y$  

14. $\dfrac{dy}{dx} = \dfrac{2+y}{4+x}$

*15. $x^2 y \dfrac{dy}{dx} = x^3 + 1$

16. $x\dfrac{dy}{dx} + y = x \ln x$

In Exercises 17–26, solve the initial-value problems.

17. $\dfrac{dy}{dx} = \dfrac{2}{x}$ for $y(1) = 2$

18. $x\dfrac{dy}{dx} = x^2(1 - x^3)$ for $y(1) = 0$

*19. $(x - 1)y^2 - x^2(y^2 + 2)\dfrac{dy}{dx} = 0$ for $y(2) = 1$

20. $\dfrac{dy}{dx} + 2y = x^2 e^{-2x}$ for $y(0) = 1$

21. $\dfrac{dy}{dx} = \dfrac{2y - 2xy}{2xy - x}$ for $y(1) = 1$

22. $\dfrac{dy}{dx} + 2xy = 0$ for $y(0) = -1$

*23. $x\dfrac{dy}{dx} + 2y - x^4 = 0$ for $y(-1) = 1$

24. $\dfrac{dy}{dx} = \dfrac{e^{x-y}}{1 + e^x}$ for $y(0) = 0$

*25. $(x + 1)\left(\dfrac{dy}{dx} + y\right) = 2e^{-x}$ for $y(0) = -1$

26. $\dfrac{dy}{dx} = e^{2x+y+2}$ for $y(0) = 0$

*27. (Stock price) The price per share of the stock of a newly formed company is $40 initially and $50 after one month. Suppose the price $p$ is projected to increase monthly at a rate proportional to the product of $p$ times the difference between $p$ and the maximum of $200.
 (a) Find a formula that gives the price of this stock at any time $t$.
 (b) What will the price per share be after one year?
 (c) How long will it take for the price per share to be more than $100?

28. (Break-even analysis) The monthly marginal profit of a company is proportional to the square root of the number of units produced and sold per month. (Assume all units produced are sold.) If monthly sales of no units produces a loss of $5000 and monthly sales of 25 units produces a loss of $2000, find the company's break-even point.

29. (Crime rate) The rate at which crimes are committed each evening in a certain urban area is inversely proportional to the square root of the number of patrolmen on duty in the area. If only one patrolman is assigned to the area, 100 crimes are committed. If 25 patrolmen are assigned to the area, only 10 crimes are committed. If 16 patrolmen are assigned to this area on a particular evening, approximately how many crimes will be committed?

30. (Mathematical application) Find the equation of the curve that passes through the point (2, 1) and such that the slope at each point $(x, y)$ satisfies the differential equation

$$\dfrac{dy}{dx} = \dfrac{-y}{x - 1}$$

*31. (Bank account withdrawals) Your rich uncle deposits $25,000 in an account in your name at 10% annual interest compounded continuously. You withdraw money from the account continuously at the rate of $4000/year. After how many years will the account be empty?

Business and Economics

32. (Lottery winnings) You have just won $1,000,000 in a lottery. Your winnings will be paid to you continuously at the rate of $50,000/year for the next 20 years. How much money must the lottery officials place in an account at 12% interest compounded continuously in order for the account to be depleted by your withdrawals in 20 years?

*33. (Loan amortization) What equal monthly payments are required to repay a loan of $30,000 over a period of ten years if the annual interest rate is 9% compounded continuously?

34. (Newton's law of cooling) The rate at which an object cools is proportional to the difference between its temperature and the temperature of the surrounding medium. If a cup of tea has a temperature of 190°F initially and 150°F four minutes later, find its temperature after 15 minutes if the temperature in the room is 70°F.

*35. (Mixing problem) A tank contains 100 gal of brine, each gallon of which contains 2 lb of salt. Fresh water is added to the tank at the rate of 5 gal/min, and the well-mixed solution is drained off at the rate of 4 gal/min. If the capacity of the tank is 500 gal, is it possible by this procedure to obtain a brine solution in the tank containing less than $\frac{1}{4}$ lb of salt per gallon?

# 10 APPLICATIONS OF CALCULUS TO PROBABILITY AND STATISTICS

**CHAPTER CONTENTS**

10.1 Discrete Random Variables

10.2 Continuous Random Variables and the Uniform Distribution

10.3 Expected Value, Variance, and Median of Continuous Random Variables

10.4 The Normal Distribution

10.5 Chapter 10 Review

## 10.1 DISCRETE RANDOM VARIABLES

Many problems in business and the sciences are not completely *deterministic* in that they involve elements of uncertainty or randomness that are not taken into account by the methods you have studied thus far in this text. The concept of *probability* is essential in modeling situations where uncertainty or randomness must be taken into account.

As an example of a problem involving uncertainty, suppose a business has a choice of two investment options, $A$ and $B$. The cost of investing in each option, the total revenue from the investments, and the probabilities of these revenues are summarized in Table 10.1. If we assume that the original cost of each investment is not recoverable, which investment option should this company choose given that the decision is to be based on the largest expected profit? This problem is solved in Example 6.

| OPTION A: COST = $1 MILLION | | | | OPTION B: COST = $1.5 MILLION | | | |
|---|---|---|---|---|---|---|---|
| Revenue (in millions) | $0.75 | $1.5 | $2.1 | Revenue (in millions) | $1.2 | $2.7 | $3.0 |
| Probability | 0.10 | 0.60 | 0.30 | Probability | 0.35 | 0.50 | 0.15 |

**TABLE 10.1**

### Some Preliminaries

In probability theory, a *random experiment* is any process that generates data that is recorded and that may not yield the same results when done repeatedly under identical conditions. A simple example of a random experiment is tossing a fair coin twice in succession and observing whether the coin lands heads up or tails up on each toss. (The word "fair" here implies that a head and a tail have the same probability of occurring on any single toss. This probability is $\frac{1}{2}$ or 0.50.)

The set of all possible outcomes of a random experiment is called the *sample space* and is designated by the symbol $S$. Each element of the sample space is called a *sample point* or a *simple event*. In general, an *event* is any subset of the sample space. For example, when a fair coin is tossed twice, the sample space is

$$S = \{HH, HT, TH, TT\}$$

where, for example, HT means that the result of the first toss is a head and the result of the second toss is a tail. Notice that this sample space lists all possible outcomes of this random experiment, discounting the possibility of the coin landing on its edge. By not including that in our sample space, we are tacitly assigning a probability of zero to this event. Thus the sample space contains the following four sample points or simple events:

HH:   heads on both tosses

HT:   heads on the first toss and tails on the second

TH: tails on the first toss and heads on the second
TT: tails on both tosses

It is possible to define other events associated with this random experiment. As examples, consider events $E$, $F$, and $G$ defined as follows:

$E = \{\text{HH, HT, TH}\}$: at least one head is tossed
$F = \{\text{HH, TT}\}$: two heads or two tails are tossed
$G = \{\text{HH, HT, TH}\}$: at most one tail is tossed

Observe that $E = G$, since the event "at least one head is tossed" is the same as the event "at most one tail is tossed."

Because the coin in our random experiment is fair, the probability of each simple event in the sample space $S = \{\text{HH, HT, TH, TT}\}$ is the same. Using the symbol $P$ to designate "the probability of," we conclude that

$$P\{\text{HH}\} = P\{\text{HT}\} = P\{\text{TH}\} = P\{\text{TT}\} = \tfrac{1}{4}$$

> In general, if $S$ is a sample space consisting of $n$ equally likely sample points $e_1, e_2, \ldots, e_n$, then
> $$P\{e_1\} = P\{e_2\} = \cdots = P\{e_n\} = \frac{1}{n}$$

In our example, to find the probability of the event $E$, "at least one head is tossed," we observe that $E = \{\text{HH, HT, TH}\}$, so that $P(E) = \tfrac{3}{4}$. There are two ways that you can think about obtaining $P(E) = \tfrac{3}{4}$ in this case. First, if we denote the number of elements in a set $A$ by $n(A)$, then

$$P(E) = \frac{3}{4} = \frac{n(E)}{n(S)}$$

It turns out that this method of computing a probability is appropriate in the case where the simple events of $S$ are all equally likely.

### Probability of an Event: Equally Likely Case

> Let $S = \{e_1, e_2, \ldots, e_k\}$, where each $e_i$ is a simple event. Then if all of the simple events $e_i$ are equally likely and $E$ is any event,
> $$P(E) = \frac{n(E)}{n(S)}$$

Another way that we can obtain $P(E) = \tfrac{3}{4}$ in our example is

$$P(E) = P\{\text{HH, HT, TH}\} = P\{\text{HH}\} + P\{\text{HT}\} + P\{\text{TH}\} = \tfrac{1}{4} + \tfrac{1}{4} + \tfrac{1}{4} = \tfrac{3}{4}$$

Notice that in this case, to obtain the probability of the event $E$, we have simply added up the probabilities of all the simple events that comprise $E$. This is a more general way to compute a probability and can be employed to find the probability of any event, whether or not the simple events in $S$ are equally likely.

**Probability of an Event: General Case**

Let $E = \{f_1, f_2, \ldots, f_k\}$, where each $f_i$ is a simple event. Then
$$P(E) = P\{f_1\} + P\{f_2\} + \cdots + P\{f_k\}.$$

---

**EXAMPLE 1**

A box contains two red and four black balls. A ball is picked at random from the box and the color is observed.

(a) Find the sample space $S$ for this experiment.
(b) Assign reasonable probabilities to each simple event.
(c) Find $P(S)$.

**Solution**

(a) Since we select either a red ball (R) or a black ball (B), we have $S = \{R, B\}$.
(b) Since the box contains two red balls out of a total of six balls, $P(R) = \frac{2}{6} = \frac{1}{3}$. Also the box contains four black balls out of a total of six balls, so $P(B) = \frac{4}{6} = \frac{2}{3}$.
(c) $P(S) = P\{R, B\} = P\{R\} + P\{B\} = \frac{1}{3} + \frac{2}{3} = 1$  ∎

In Example 1 we saw that $P(S)$ was equal to 1, where $S$ was the sample space of the experiment. This is just a a specific instance of the general case:

If $S$ is any finite sample space, then $P(S) = 1$.

A way to think about this boxed result is that since the sample space contains all possible outcomes, it is certain that something in the sample space will occur.
Of course, if $E$ is any event, then $E$ is a subset of $S$, so that

$$0 \leq P(E) \leq 1$$

This boxed result is just the statement that the probability of any event must be a number between 0 and 1.

**Random Variables**

As we have seen, the outcomes of a random experiment need not be numbers. For example, the outcomes in our coin tossing experiment are given by
$$S = \{HH, HT, TH, TT\}$$

A *random variable* $X$ associates a number with each simple event of the sample space. For example, if we let $X$ be the number of tails that occur when a fair coin is tossed twice, then this assignment of a number to each simple event is summarized in the following table. Formally, we say that a random variable in this case defines a function with a domain equal to the sample space $S = \{HH, HT, TH, TT\}$ and range equal to the finite set of numbers $\{0, 1, 2\}$.

| Simple Event | Value of X |
| --- | --- |
| HH | 0 |
| HT | 1 |
| TH | 1 |
| TT | 2 |

We write $P(X = 0)$ for the probability that $X$ is equal to zero, $P(X = 1)$ for the probability that $X$ is equal to 1, and $P(X = 2)$ for the probability that $X$ is equal to 2. Then

$$P(X = 0) = P\{HH\} = \tfrac{1}{4}$$
$$P(X = 1) = P\{HT, TH\} = P\{HT\} + P\{TH\} = \tfrac{1}{4} + \tfrac{1}{4} = \tfrac{1}{2}$$
$$P(X = 2) = P\{TT\} = \tfrac{1}{4}.$$

We use the notation $P(X = x)$ when $X$ represents the random variable and $x$ one of the particular values (0, 1, or 2 in this case) of the random variable and summarize our results in Table 10.2. This table gives what is called the *probability distribution* for the random variable $X$, since it indicates how the probabilities are distributed among the possible values of $X$. The data in Table 10.2 is graphed in Figure 10.1. This graph is referred to as a *probability histogram*. Notice that the possible outcomes $x$ serve as the midpoints of the bases of the rectangles and that the heights of the rectangles are the corresponding probabilities. Furthermore, the area of the rectangle centered at $x = 0$ is $\tfrac{1}{4}$, which is equal to $P(X = 0)$. Similar statements are true for the areas of the rectangles centered at $x = 1$ and $x = 2$.

The random variable $X$ that corresponds to the histogram in Figure 10.1 is called a *discrete* random variable because its value must be one of the integers 0, 1, or 2. In general, a *discrete random variable* $X$ is one with values which are integers. A discrete random variable usually counts something: the number of tails in two tosses of a coin, the number of aces in a hand of five cards dealt from a standard deck of 52 cards, or the number of defective units that occur daily in a certain manufacturing process. Let's consider another example of a discrete random variable.

| OUTCOME ($x$) | PROBABILITY $P(X = x)$ |
| --- | --- |
| 0 | $\tfrac{1}{4}$ |
| 1 | $\tfrac{1}{2}$ |
| 2 | $\tfrac{1}{4}$ |

TABLE 10.2

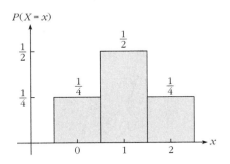

FIGURE 10.1 The probability histogram of the random variable $X$ that represents the number of tails when a fair coin is tossed twice.

EXAMPLE 2

A box contains three balls numbered 1, 2, and 3. A ball is drawn at random from the box and the number on the ball is observed. This ball is then replaced into the box and a second ball is drawn at random. If $X$ is the sum of the numbers on the two balls, find the probability distribution of $X$ and sketch the probability histogram.

Solution

The sample space of this random experiment consists of all possible ordered pairs of the form $(a, b)$, where $a$ represents the number observed on the first draw and $b$ the number observed on the second draw. Thus,

$$S = \{(1, 1), (1, 2), (1, 3), (2, 1), (2, 2), (2, 3), (3, 1), (3, 2), (3, 3)\}$$

Each of the nine sample points in the sample space is equally likely, so the probability of each simple event is $\frac{1}{9}$. Since $X$ represents the sum of the numbers on the two balls, the possible values of $X$ are 2, 3, 4, 5, and 6. Further,

$$P(X = 2) = P\{(1, 1)\} = \tfrac{1}{9}$$

$$P(X = 3) = P\{(1, 2), (2, 1)\} = P\{(1, 2)\} + P\{(2, 1)\} = \tfrac{1}{9} + \tfrac{1}{9} = \tfrac{2}{9}$$

$$P(X = 4) = P\{(1, 3), (2, 2), (3, 1)\} = P\{(1, 3)\} + P\{(2, 2)\} + P\{(3, 1)\}$$
$$= \tfrac{1}{9} + \tfrac{1}{9} + \tfrac{1}{9} = \tfrac{1}{3}$$

$$P(X = 5) = P\{(2, 3), (3, 2)\} = P\{(2, 3)\} + P\{(3, 2)\} = \tfrac{1}{9} + \tfrac{1}{9} = \tfrac{2}{9}$$

$$P(X = 6) = P\{(3, 3)\} = \tfrac{1}{9}$$

Summarizing, we get the probability distribution of $X$ in Table 10.3. The probability histogram of $X$ is sketched in Figure 10.2.

| OUTCOME ($x$) | PROBABILITY $P(X = x)$ |
|---|---|
| 2 | $\tfrac{1}{9}$ |
| 3 | $\tfrac{2}{9}$ |
| 4 | $\tfrac{1}{3}$ |
| 5 | $\tfrac{2}{9}$ |
| 6 | $\tfrac{1}{9}$ |

TABLE 10.3

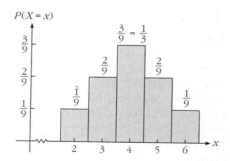

FIGURE 10.2 The probability histogram of the random variable $X$ of Example 2.

## EXAMPLE 3 (Participative)

Let's find the probability distribution of the random variable $X$ of Example 2 if we assume that the first ball is *not* replaced into the box before the second is drawn.

Solution

Since the first ball is not replaced before drawing the second one, the sample space of this random experiment is

$$S = \{(1, 2), (1, \underline{\hphantom{xx}}), (2, 1), (2, \underline{\hphantom{xx}}), (3, 1), (3, \underline{\hphantom{xx}})\}$$

Each of the six sample points in this sample space is equally likely, so the probability of each simple event is $\underline{\hphantom{xxxx}}$. If $X$ represents the sum of the numbers on the two balls, the possible values of $X$ are $\underline{\hphantom{xxxxxx}}$. To find the probability distribution of $X$, we compute

$$P(X = 3) = P\{(1, 2), (2, 1)\} = P\{(1, 2)\} + P\{(2, 1)\} = \tfrac{1}{6} + \tfrac{1}{6} = \tfrac{1}{3}$$

$$P(X = 4) = P\{(1, 3), (3, 1)\} = P\{(1, 3)\} + P\{(3, 1)\} = \tfrac{1}{6} + \tfrac{1}{6} = \tfrac{1}{3}$$

$$P(X = 5) = \underline{\hphantom{xxxxxxxxxxxxxxxxxxxx}}$$

Complete this example on your own by tabulating the probability distribution of $X$ and checking your results with those given at the bottom of the page. ■

### Expected Value or Mean of a Random Variable

Suppose you roll a fair six-sided die a large number of times and record the number of dots showing on the upper face after each roll. If you compute the average (or arithmetic mean) of these numbers, what value can you expect to get for this average? The answer to this question turns out to be 3.5, and the number 3.5 is called the *expected value* or the *mean* of the random variable $X$, if $X$ represents the number of dots showing on any single roll. Let's investigate why the appropriate number is 3.5.

First of all, the possible outcomes of $X$ are 1, 2, 3, 4, 5, and 6, and each of these outcomes is equally likely, since the die is fair. The probability distribution of $X$ is summarized in Table 10.4.

| OUTCOME $x$ | 1 | 2 | 3 | 4 | 5 | 6 |
|---|---|---|---|---|---|---|
| PROBABILITY $P(X = x)$ | $\frac{1}{6}$ | $\frac{1}{6}$ | $\frac{1}{6}$ | $\frac{1}{6}$ | $\frac{1}{6}$ | $\frac{1}{6}$ |

**TABLE 10.4**

Consider an experiment where you roll the die, say, 600 times in succession. Since the probability of the die showing one dot on any particular roll is $\frac{1}{6}$, we anticipate that approximately $\frac{1}{6}(600) = 100$ of the rolls will result in $X = 1$. Similarly, we expect that about $\frac{1}{6}(600) = 100$ rolls will result in $X = 2$, since $P(X = 2) = \frac{1}{6}$. A similar argument can be applied for $X = 3$, $X = 4$, $X = 5$, and $X = 6$, and we anticipate that about 100 rolls will result for each. This situation is summarized in Table 10.5. If we now find the average of these 600 occurrences, we have

$$\text{average} \approx \frac{1 \cdot 100 + 2 \cdot 100 + 3 \cdot 100 + 4 \cdot 100 + 5 \cdot 100 + 6 \cdot 100}{600}$$

$$= 1\left(\frac{100}{600}\right) + 2\left(\frac{100}{600}\right) + 3\left(\frac{100}{600}\right) + 4\left(\frac{100}{600}\right) + 5\left(\frac{100}{600}\right) + 6\left(\frac{100}{600}\right)$$

$$= 1 \cdot \frac{1}{6} + 2 \cdot \frac{1}{6} + 3 \cdot \frac{1}{6} + 4 \cdot \frac{1}{6} + 5 \cdot \frac{1}{6} + 6 \cdot \frac{1}{6}$$

$$= 3.5$$

---

**SOLUTIONS TO EXAMPLE 3**

3, 3, 2, $\frac{1}{6}$, 3 or 4 or 5, $P\{(2, 3), (3, 2)\} = P\{(2, 3)\} + P\{(3, 2)\} = \frac{1}{6} + \frac{1}{6} = \frac{1}{3}$

| OUTCOME $(x)$ | 3 | 4 | 5 |
|---|---|---|---|
| PROBABILITY $P(X = x)$ | $\frac{1}{3}$ | $\frac{1}{3}$ | $\frac{1}{3}$ |

| OUTCOME x | 1 | 2 | 3 | 4 | 5 | 6 |
|---|---|---|---|---|---|---|
| APPROXIMATE NUMBER OF OCCURRENCES OUT OF 600 | 100 | 100 | 100 | 100 | 100 | 100 |

**TABLE 10.5**

This average value is the expected value of $X$, and we write $\mu = E(X) = 3.5$. The letter $\mu$ is the Greek letter "mu" and it stands for "mean." The capital $E$ stands for expected value. Notice that $E(X) = 3.5$ can be obtained by multiplying each possible outcome $x$ of $X$ by its corresponding probability and then adding the results. This is true in general for a discrete random variable $X$.

**EXPECTED VALUE OF A DISCRETE RANDOM VARIABLE**

If $X$ is a discrete random variable with probability distribution given by Table 10.6, then the *expected value* or *mean* of $X$ is given by

$$E(X) = x_1 p_1 + x_2 p_2 + \cdots + x_n p_n$$

where the $x_i$ and $p_i$ are defined by Table 10.6.

| OUTCOME x | $x_1$ | $x_2$ | $x_3$ | $\cdots$ | $x_n$ |
|---|---|---|---|---|---|
| PROBABILITY $P(X = x)$ | $p_1$ | $p_2$ | $p_3$ | $\cdots$ | $p_n$ |

**TABLE 10.6**

It is important to observe that $E(X)$ need not be one of the possible outcomes of the random variable $X$. The expected value of $X$ is what you would expect as an *average* value of $X$ if the experiment were repeated a large number of times. Thus, just as the average of a set of numbers need not be one of the numbers in the set, the expected value of a random variable $X$ need not be one of the possible outcomes of $X$.

**EXAMPLE 4**

A committee of three students is chosen at random from a group of three freshmen and two juniors. Find the expected number of freshmen on this committee.

**Solution**

Let the three freshmen be denoted by $F_1$, $F_2$, and $F_3$ and the two juniors by $J_1$ and $J_2$. Then one possible committee is $F_1 F_2 J_1$. Notice that the order in which these letters are written is immaterial here, since the committee $F_2 J_1 F_1$ consists of the same three people and so is the same committee. The sample space of this random experiment is

$$S = \{F_1 F_2 F_3,\ F_1 F_2 J_1,\ F_1 F_2 J_2,\ F_1 F_3 J_1,\ F_1 F_3 J_2,\ F_2 F_3 J_1,\ F_2 F_3 J_2,\ F_1 J_1 J_2,\ F_2 J_1 J_2,\ F_3 J_1 J_2\}$$

and consists of ten equally likely sample points. If $X$ represents the number of freshmen on the committee, then the possible values of $X$ are 1, 2, and 3, and

| OUTCOME $x$ | PROBABILITY $P(X = x)$ |
|---|---|
| 1 | $\frac{3}{10}$ |
| 2 | $\frac{6}{10}$ |
| 3 | $\frac{1}{10}$ |

TABLE 10.7

$P(X = 1) = P\{F_1J_1J_2, F_2J_1J_2, F_3J_1J_2\} = \frac{3}{10}$

$P(X = 2) = P\{F_1F_2J_1, F_1F_2J_2, F_1F_3J_1, F_1F_3J_2, F_2F_3J_1, F_2F_3J_2\} = \frac{6}{10}$

$P(X = 3) = P\{F_1F_2F_3\} = \frac{1}{10}$

The probability distribution of $X$ is displayed in Table 10.7. Therefore

$$\mu = E(X) = 1 \cdot \frac{3}{10} + 2 \cdot \frac{6}{10} + 3 \cdot \frac{1}{10} = 1.8$$

So if we select a committee of three students at random from three freshmen and two juniors and repeat the procedure a large number of times, the committees so chosen will contain an average of 1.8 freshmen. ∎

### EXAMPLE 5

A standard deck of 52 cards contains four aces. A gambler bets $1 and draws a single card from the deck. If the card is an ace, she wins $10 and the $1 bet is returned to her. If the card is not an ace, she loses the $1 bet. Find the gambler's expected winnings.

### Solution

Let $W$ be the winnings on any play of the game. The possible values of $W$ are 10 and $-1$. The negative sign indicates a loss of $1. Since the deck contains four aces and 52 cards, the probability of drawing an ace is $\frac{4}{52} = \frac{1}{13}$. Also, since 48 of the 52 cards in the deck are not aces, the probability of not drawing an ace is $\frac{48}{52} = \frac{12}{13}$. We conclude that

$P(W = 10) = \frac{1}{13}$

$P(W = -1) = \frac{12}{13}$

The probability distribution of the random variable $W$ is displayed in Table 10.8. Thus the expected winnings are

$$E(W) = 10 \cdot \frac{1}{13} + (-1) \cdot \frac{12}{13} = -\frac{2}{13} \approx -0.154$$

| OUTCOME $w$ | PROBABILITY $P(W = w)$ |
|---|---|
| 10 | $\frac{1}{13}$ |
| $-1$ | $\frac{12}{13}$ |

TABLE 10.8

So this gambler can expect to lose an average of about 15 cents on any play of the game. For example, if she plays the game 100 times, she can expect to lose about $(100)(0.154) = \$15.40$. ∎

In general, a game is *fair* if $E(W) = 0$, where $W$ is the player's expected winnings. The game described in Example 5 is not fair because the "house" has an advantage. In a fair game, a player's average long-run winnings will be equal to zero, since neither the player nor the house has any advantage.

The concept of expected value can be used as an aid in making certain decisions for which the outcomes and their probabilities are known. The next example illustrates this use of expected value.

### EXAMPLE 6

Refer to the investment decision considered at the beginning of this section and summarized by Table 10.1. Basing your decision on the largest expected profit, which investment option should the company choose?

**Solution**

Let $M$ be the company's profit: then $M$ is a random variable. Table 10.1 can be quickly converted into a table with outcomes that are values of $M$ by using the basic relationship

$$\text{profit} = \text{revenue} - \text{cost}$$

This is done in Table 10.9, where profit is measured in millions of dollars.

Option A

| OUTCOME M | −$0.25 | $0.50 | $1.10 |
|---|---|---|---|
| PROBABILITY $P(M = m)$ | 0.10 | 0.60 | 0.30 |

Option B

| OUTCOME M | −$0.30 | $1.20 | $1.50 |
|---|---|---|---|
| PROBABILITY $P(M = m)$ | 0.35 | 0.50 | 0.15 |

TABLE 10.9

Computing expected values, we obtain

Option A:  $E(M) = (-0.25)(0.10) + (0.50)(0.60) + (1.10)(0.30) = \$0.605$
Option B:  $E(M) = (-0.30)(0.35) + (1.20)(0.50) + (1.50)(0.15) = \$0.72$

Thus the company should choose investment option $B$ with an expected profit of $720,000, since the expected profit for option $A$ is only $605,000.

The expected value or mean of a random variable $X$ can be thought of as a single number that represents the "center" of the values of $X$, just as the average of a set of numbers is a single number that represents the center of the entire set. For this reason it is sometimes referred to as a measure of *central tendency*. Geometrically, if we think of hanging weights of magnitude $p_i$ at location $x_i$ on a uniform metal rod, then this system balances on a wedge placed under the rod at the mean $\mu$. This is illustrated in Figure 10.3.

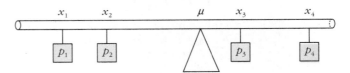

**FIGURE 10.3** The system of weights will balance on a wedge placed at the mean $\mu$.

### Variance and Standard Deviation of a Random Variable

Just as the mean is a measure of the center of the probability distribution of a random variable, another measure called the *variance* is a measure of the "dispersion" or "spread" of the probability distribution away from the mean. Given a random variable $X$ with probability distribution appearing in Table 10.10 and with mean is $\mu$, we employ the following steps to measure the dispersion about the mean:

Step 1  Compute $(x_i - \mu)$, the deviation of each outcome $x_i$ from the mean $\mu$.

Step 2  Square each of the deviations from the mean that were found in Step 1 to produce a set of positive squared deviations from the mean: $(x_i - \mu)^2$.

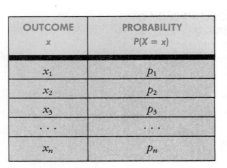

TABLE 10.10

Step 3  Multiply each squared deviation from the mean $(x_i - \mu)^2$ by the probability $p_i$ of outcome $x_i$ to obtain $(x_i - \mu)^2 p_i$.

Step 4  Compute the sum of these products:

$$(x_1 - \mu)^2 p_1 + (x_2 - \mu)^2 p_2 + \cdots + (x_n - \mu)^2 p_n$$

The result of Step 4 is the variance of $X$, and is denoted by $V(X)$ or $\sigma^2$, where $\sigma$ is the Greek letter "sigma." Notice that in Step 2 we squared the deviations from the mean. The result is that if $X$ is measured in units, then the variance of $X$ is measured in square units. To produce a measure with the same units as $X$, we employ the square root of the variance, which is called the *standard deviation* of $X$. These results are summarized in the following box.

**VARIANCE AND STANDARD DEVIATION**

If $X$ is a discrete random variable with mean $\mu$ and probability distribution in Table 10.10, then the *variance of $X$* is

$$V(X) = \sigma^2 = (x_1 - \mu)^2 p_1 + (x_2 - \mu)^2 p_2 + \cdots + (x_n - \mu)^2 p_n$$

The *standard deviation of $X$* is the square root of the variance and is denoted by $\sigma$.

**EXAMPLE 7**

Compute the variance and the standard deviation of each of the random variables $X$, $Y$, and $Z$ with probability distributions appearing in Table 10.11.

| x | 0 | 1 | 2 | 3 | 4 | 5 | 6 |
|---|---|---|---|---|---|---|---|
| P(X = x) | 0.05 | 0.05 | 0.1 | 0.6 | 0.1 | 0.05 | 0.05 |

(a)

| y | 0 | 1 | 2 | 3 | 4 | 5 | 6 |
|---|---|---|---|---|---|---|---|
| P(Y = y) | $\frac{1}{7}$ | $\frac{1}{7}$ | $\frac{1}{7}$ | $\frac{1}{7}$ | $\frac{1}{7}$ | $\frac{1}{7}$ | $\frac{1}{7}$ |

(b)

| z | 0 | 1 | 2 | 3 | 4 | 5 | 6 |
|---|---|---|---|---|---|---|---|
| P(Z = z) | 0.21 | 0.15 | 0.1 | 0.08 | 0.1 | 0.15 | 0.21 |

(c)

**TABLE 10.11**

Solution

You should verify that the mean of each of the random variables $X$, $Y$, and $Z$ is equal to 3. As an example computation, the mean of $X$ is

$$E(X) = (0)(0.05) + 1(0.05) + 2(0.1) + 3(0.6) + 4(0.1) + 5(0.05) + 6(0.05)$$
$$= 3$$

Similarly, it can be shown that $E(Y) = E(Z) = 3$. To compute the variances we proceed as follows:

$$V(X) = (0-3)^2(0.05) + (1-3)^2(0.05) + (2-3)^2(0.1) + (3-3)^2(0.6)$$
$$+ (4-3)^2(0.1) + (5-3)^2(0.05) + (6-3)^2(0.05)$$
$$= 1.5$$

Similar computations show that $V(Y) = 4$ and $V(Z) = 5.18$. Thus the standard deviations are

$$\sigma_X = \sqrt{V(X)} = \sqrt{1.5} \approx 1.2247$$
$$\sigma_Y = \sqrt{V(Y)} = \sqrt{4} = 2.0000$$
$$\sigma_Z = \sqrt{V(Z)} = \sqrt{5.18} \approx 2.2760$$

The probability histograms of the random variables $X$, $Y$, and $Z$ are sketched in Figure 10.4 for reference. Notice that $\sigma_X < \sigma_Y < \sigma_Z$ and that the probability histogram of $X$ shows the least dispersion about the mean, whereas the probability distribution of $Z$ shows the most dispersion about the mean.

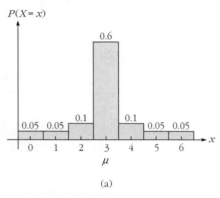
(a)

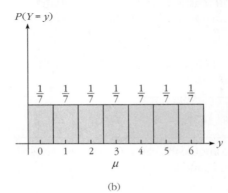
(b)

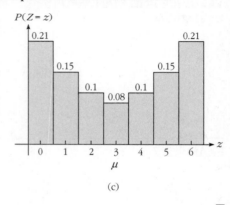
(c)

**FIGURE 10.4** The probability histograms of the random variables $X$, $Y$, and $Z$. ■

## SECTION 10.1
**SHORT ANSWER EXERCISES**

1. How many sample points are contained in the sample space $S = \{HD, DH, HH, DD\}$? ____

2. (True or False) The set $\{HD, CC\}$ is an event corresponding to the sample space $S = \{HD, DH, HH, DD\}$. ____

3. (True or False) All simple events in a sample space are equally likely. ____

4. The probability of any event must be a number between ____ and ____, inclusive.

5. The random variable $X$ has possible outcomes 1, 2, and 3. Given that $P(X = 1) = 0.1$ and $P(X = 2) = 0.6$, then $P(X = 3) =$ ____.

6. Which of the following random variables $X$ are discrete? (a) $X$ is the number of clubs in a bridge hand; (b) $X$ is the number of phone calls you receive during a day; (c) $X$ is the barometric pressure on a given day; (d) all of the above.

7. The expected winnings $W$ in a certain game are $E(W) = \$2$. If you play the game 50 times, you can expect to win about ____ dollars.

8. The random variable $X$ has three equally likely outcomes, $-3$, 0, and 3. What is the mean of $X$? ____

9. (True or False) The expected value of $X$ is the outcome of $X$ that we would expect to occur on any single repetition of the experiment. _____
10. (True or False) If the outcomes of a random variable are all equally likely, then the variance of $X$ is equal to zero. _____

## SECTION 10.1
## EXERCISES

*In Exercises 1–7, the experiment consists of tossing a fair coin three times.*

1. Find the sample space $S$.
2. Identify the following events as subsets of $S$:
   (a) tails on exactly two tosses
   (b) heads on more than one toss
   (c) tails on the first toss
   (d) heads on the first and last tosses
   (e) no heads
3. Find the probability of each simple event in $S$.
4. Find the probability of each event listed in Exercise 2.
5. Let $X$ be the number of heads that occur in the three tosses. Find the probability distribution of $X$ and sketch the probability histogram of $X$.
6. If $X$ is the number of heads, find $E(X)$.
7. If $X$ is the number of heads, find $V(X)$ and $\sigma_X$.

*In Exercises 8–14, the experiment consists of rolling two fair dice, a red die and a green die.*

8. Find the sample space $S$.
9. Identify the following events as subsets of $S$:
   (a) The red die shows a 3 and the green die shows a 4.
   (b) The red die shows a number less than 5 and the green die shows a 2.
   (c) Both dice show an even number.
   (d) The red die shows a number greater than 3 and the green die shows a number less than 5.
   (e) One die shows an even number and the other shows an odd number.
10. Find the probability of each sample point in $S$.
11. Find the probability of each event listed in Exercise 9.
12. Let $X$ be the sum of the number of dots that show on the faces of the two dice. Find the probability distribution of $X$ and sketch the probability histogram of $X$.
13. If $X$ is defined as in Exercise 12, find the mean of $X$.
14. If $X$ is defined as in Exercise 12, find the standard deviation of $X$.

*In Exercises 15–21, the experiment consists of drawing two balls from a box containing four balls numbered 5, 10, 15, and 20. The first ball is replaced into the box before the second ball is drawn.*

15. Find the sample space $S$.
16. Identify the following events as subsets of $S$:
    (a) The first ball is numbered 10 and the second ball is numbered less than 15.
    (b) The first ball is numbered at least 10 and the second ball is numbered with an odd number.
    (c) Neither ball is numbered more than 10.
    (d) Both balls are numbered with an even number.
    (e) The number on one ball is less than the number on the other ball.
17. Find the probability of each sample point in $S$.

18. Find the probability of each event listed in Exercise 16.

19. Let $X$ be the sum of the numbers on the two balls. Find the probability distribution of $X$ and sketch the probability histogram of $X$.

20. If $X$ is the sum of the numbers on the two balls, find $E(X)$.

21. If $X$ is the sum of the numbers on the two balls, find $V(X)$.

22–28. Redo Exercises 15–21 assuming that the first ball is *not* replaced into the box before the second is drawn.

*In Exercises 29–35, the experiment consists of randomly choosing two televisions from a group of eight in which six are good and two are defective.*

29. Find the sample space $S$.

30. Identify the following events as subsets of $S$:
    (a) at least one television is defective
    (b) both televisions are good
    (c) one television is good and one is defective
    (d) both televisions are defective
    (e) at least one television is good

31. Find the probability of each sample point in $S$.

32. Find the probability of each event listed in Exercise 30.

33. Let $X$ be the number of defective television sets chosen. Find the probability distribution of $X$ and graph the probability histogram of $X$.

34. Find the mean number of defective television sets chosen.

35. If $X$ is the number of defective television sets chosen, find the standard deviation of $X$.

*In Exercises 36–40, a gambler bets $1 on each play of the game described. If the gambler wins, the $1 bet is returned to her and in addition she wins the amount described in the exercise. If the gamber loses, she loses the $1 bet. In each exercise let W be the expected winnings. (a) Find $E(W)$. (b) Determine if the game is fair.*

36. The game consists of rolling a pair of dice twice. If the sum of the number of dots showing on the faces of the dice is 7 or 11, the gambler wins $5; otherwise she loses.

37. The game consists of tossing a fair coin four times. If three or more heads are tossed, the gambler wins $10; otherwise she loses.

38. The game consists of tossing a fair coin four times. The gambler wins $2 for each head that is tossed. If no heads are tossed, she loses.

39. The game consists of the gambler choosing a four-digit number from 0000 to 9999. Four balls are then drawn from a box containing 10 balls numbered 0 through 9, where each ball is replaced before the next is drawn. If the number chosen by the gambler matches the four-digit number generated by the four balls, she wins $800; otherwise she loses.

40. The game consists of drawing a single card from a standard deck of 52 cards. The deck contains four aces and 12 face cards consisting of four jacks, four queens, and four kings. If the gambler draws an ace, she wins $10. If the gambler draws a face card she wins $5; otherwise she loses.

Business and Economics

41. (Investment decision)   A business has a choice of two investment options, $A$ and $B$. Option $A$ will cost $5 million and option $B$ will cost $7 million. Option $A$ will

produce revenues of $6 million, $8 million, and $12 million with probabilities 0.1, 0.6, and 0.3, respectively. Option B will produce revenues of $4 million, $10 million, and $15 million with probabilities 0.25, 0.50, and 0.25, respectively. If the investment decision is based on largest expected profit, which investment option will the business choose?

42. (Investment decision)  A company has $100,000 to invest for a period of one year. There are three investment options:
   Option A: Put the money in a one-year certificate of deposit yielding 10% annual interest compounded continuously;
   Option B: Invest in a stock at a price of $50 per share and which has a probability of 0.4 of rising to $75 per share after one year and a probability of 0.6 of falling to $45 per share after one year;
   Option C: Invest in a piece of land that may increase dramatically to $500,000 in value if a major shopping mall locates there. The probability of this happening is 0.05. If the shopping mall does not locate there the property will be worth only $75,000 after one year.
   Basing your decision on the largest expected profit, which option should the company choose?

43. (Insurance)  The annual insurance premium for fire insurance on a piece of property you own is $500. If the property is worth $120,000 and the probability of a 25% loss is 0.003 and the probability of a 100% loss is 0.002, compute your expected annual gain or loss if you decide to insure this property. Ignore all other partial losses.

44. (Insurance)  In writing an insurance policy for a client, an insurance underwriter must make a net profit of 10% of the premium for the company. In insuring a $500,000 building, the underwriter estimates that the probability of a 50% loss is 0.004 and the probability of a 100% loss is 0.002. If administrative costs are 20% of the premium, how much should the annual premium be?

## 10.2 CONTINUOUS RANDOM VARIABLES AND THE UNIFORM DISTRIBUTION

### Continuous Random Variables

A *continuous random variable* $X$ is one with a value that can be any number in an interval of real numbers. The interval of numbers that represent the outcomes of a continuous random variable $X$ may be finite (such as $[0, 30]$) or infinite (such as $(2, \infty)$). An interval such as $[0, 30]$ is considered to be a *continuum* of real numbers rather than a set of discrete points.

Continuous random variables typically measure rather than count. In so measuring, we assume that we can do so to any degree of accuracy. As an example, the height in inches of a person chosen at random from a certain town is a continuous random variable $X$. We could ask, for example, what the probability is that a person chosen at random from the town is between 65 inches and 72 inches tall, inclusive. This is written symbolically as $P(65 \leq X \leq 72)$.

An unusual property of a continuous random variable $X$ defined on an interval $[a, b]$ is that the probability that $X$ is exactly equal to any value in $[a, b]$ is zero.

That is, if $c$ is a number in $[a, b]$, then $P(X = c) = 0$. To get an intuitive idea of why this should be true, consider the random variable $X$ that measures the height of an individual chosen at random from a town, and let $c = 68$. The probability that the person selected has a height of *exactly* 68 inches and not 68.0001 inches or 67.99992 inches or any of the infinitely many other measures that are close to 68 inches is so extremely small that we assign a value of zero to this event. That is, $P(X = 68) = 0$.

### The Probability Density Function

Since for any number $c$ in $[a, b]$ we have $P(X = c) = 0$, we certainly cannot write down the probability distribution function of a continuous random variable in tabular form, as we did for discrete random variables in Section 10.1. We need another means of defining the probability of an event.

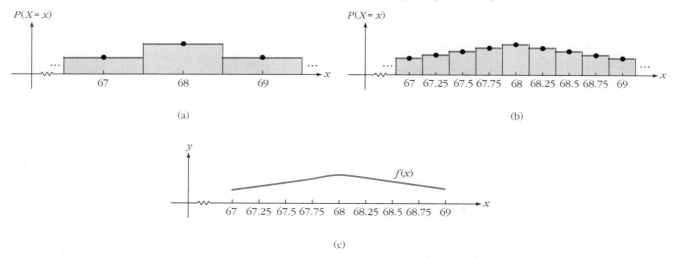

**FIGURE 10.5** The derivation of a probability density function $f(x)$ of a continuous random variable $X$.

Recall the interpretation of probability as the area of certain rectangles in the probability histogram of a random variable. If we round all heights to the nearest inch, then we are dealing with a *discrete* random variable $X$; a portion of its probability histogram is shown in Figure 10.5(a). In this diagram, probabilities are given by the areas of rectangles. If we round all heights to the nearest quarter of an inch, we still have a discrete random variable in which probabilities are given by areas of rectangles, as shown in Figure 10.5(b). If we round to smaller and smaller units—tenths of an inch, hundredths of an inch, and so forth—the discrete random variable approaches a continuous one and the probability histograms of rectangles approach the continuous curve illustrated in Figure 10.5(c). This curve can be visualized as the one that passes through the dots centered on top of the histograms as the rectangles get smaller and smaller. In Figure 10.5(c), the probability that the continuous random variable $X$ falls within any interval, say between 67.2 and 68.3, is given by the *area* under the curve between $x = 67.2$ and $68.3$. The continuous curve in Figure 10.5(c) is a portion

## 10.2 CONTINUOUS RANDOM VARIABLES AND THE UNIFORM DISTRIBUTION

of what is called the *probability density function* for the continuous random variable X. The general definition of a probability density function is as follows.

**PROBABILITY DENSITY FUNCTION**

A continuous function $f(x)$ is a *probability density function* for a continuous random variable $X$ on $[a, b]$ if

i. $f(x) \geq 0$ for all $x$ in $[a, b]$;

ii. $\int_a^b f(x)\, dx = 1$; and

iii. if $a \leq c \leq d \leq b$, then $P(c \leq X \leq d) = \int_c^d f(x)\, dx =$ the area under $f$ from $c$ to $d$.

Parts (ii) and (iii) of this definition are illustrated in Figure 10.6. The definition also applies to a continuous random variable $X$ defined on an unbounded interval such as $[0, \infty)$ or $(-\infty, \infty)$.

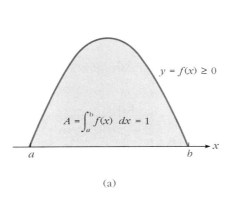
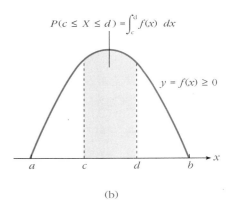

**FIGURE 10.6** In (a), the area under the entire density function must be 1. In (b), the probability that $X$ is between $c$ and $d$ is the area under the density function from $c$ to $d$.

### EXAMPLE 1

The time between the arrival of trains at a particular train station is 20 minutes. If $X$ is the amount of time that a person arriving at a random time must wait for a train, then the probability density function of $X$ is given by

$$f(x) = \begin{cases} \frac{1}{20} & 0 \leq x \leq 20 \\ 0 & \text{otherwise} \end{cases}$$

Find **(a)** $P(0 \leq X \leq 4)$; **(b)** $P(7 \leq X \leq 11)$; **(c)** $P(X \geq 10)$.

**Solution**

**(a)** $P(0 \leq X \leq 4) = \int_0^4 \frac{1}{20}\, dx = \frac{1}{20}x \Big]_0^4 = \frac{1}{20}(4) - \frac{1}{20}(0) = 0.20$

**(b)** $P(7 \leq X \leq 11) = \int_7^{11} \frac{1}{20}\, dx = \frac{1}{20}x \Big]_7^{11} = \frac{1}{20}(11) - \frac{1}{20}(7) = 0.20$

**(c)** $P(X \geq 10) = \int_{10}^{20} \frac{1}{20}\, dx = \frac{1}{20}x \Big]_{10}^{20} = \frac{1}{20}(20) - \frac{1}{20}(10) = 0.50$ ∎

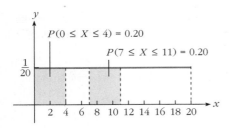

FIGURE 10.7 The uniform density function of Example 1.

The probabilities computed in parts (a) and (b) of Example 1 are illustrated in Figure 10.7. Observe that the probability that a person will have to wait for a train between 0 and 4 minutes is the same as the probability of having to wait between 7 and 11 minutes and is equal to 0.20. In fact, the probability of having to wait between $t$ and $t + 4$ minutes is equal to 0.20 for any time $t \leq 16$.

In Example 1, the probability that $X$ lies in an interval of length $l$ within $[0, 20]$ is $l/20$ and does not depend on the exact location of the interval. The density function of Example 1 is a *uniform probability density function*, since $X$ is equally likely to lie in any small subinterval of length $l$.

## THE UNIFORM DISTRIBUTION

If the continuous random variable $X$ defined on $[a, b]$ has the property that $X$ is equally likely to lie in any small subinterval of length $l$ within $[a, b]$, then $X$ is said to be *uniformly distributed* on $[a, b]$ and has probability density function

$$f(x) = \begin{cases} \dfrac{1}{b-a} & a \leq x \leq b \\ 0 & \text{otherwise} \end{cases}$$

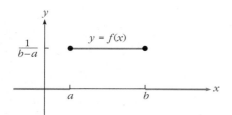

FIGURE 10.8 The uniform probability density function.

A sketch of the uniform probability density function is drawn in Figure 10.8.

---

EXAMPLE 2

For what value of $k$ is $f(x) = kx^3$ a probability density function on the interval $[1, 2]$?

Solution

We must find the number $k$ such that $\int_1^2 kx^3 \, dx = 1$. We proceed as follows:

$$1 = \int_1^2 kx^3 \, dx = \left.\frac{kx^4}{4}\right|_1^2 = \frac{k(2)^4}{4} - \frac{k(1)^4}{4}$$

$$1 = \frac{15}{4}k$$

$$k = \frac{4}{15}$$

Thus $f(x) = \frac{4}{15}x^3$ is a probability density function on the interval $[1, 2]$, because $f(x) \geq 0$ for $x$ in $[1, 2]$ and $\int_1^2 \frac{4}{15}x^3 \, dx = 1$. ∎

One consequence of the fact that $P(X = c) = 0$ for a continuous random variable $X$ is this:

If $X$ is a continuous random variable, then

$$P(a \leq X \leq b) = P(a < X \leq b) = P(a \leq X < b) = P(a < X < b)$$

This statement is true because the probability that the continuous random variable $X$ is exactly equal to either of the endpoints $a$ or $b$ of the interval is zero. For example,

$$P(a \leq X \leq b) = P(X = a) + P(a < X \leq b) = 0 + P(a < X \leq b)$$
$$= P(a < X \leq b)$$

The other equalities in the boxed result follow by similar arguments.

---

EXAMPLE 3 (Participative)

The number of gallons of coffee sold weekly by a chain of convenience stores varies randomly between 5,000 and 23,000 gallons. If $X$ represents the number of gallons sold weekly and $X$ is measured in thousands of gallons then $5 \leq X \leq 23$. Further, suppose it is known that the probability density function of $X$ is

$$f(x) = \begin{cases} \dfrac{8}{(x+1)^2} & 5 \leq x \leq 23 \\ 0 & \text{otherwise} \end{cases}$$

Let's find the probability that weekly sales of coffee will exceed 20,000 gallons.

Solution

We require the probability that $X > $ . Since $f(x) = 0$ if $x > 23$, this probability is given by

$$P(X > 20) = \int_{20}^{23} \frac{8}{(x+1)^2}\, dx$$

Making the substitution $u = x + 1$, we have $du = $ _____. Further, when $x = 20$, we have $u = 20 + 1 = 21$, and when $x = 23$, we have $u = $ _____. Thus the integral becomes

$$P(X > 20) = \int_{21}^{24} 8u^{-2}\, du = \left. \frac{-8}{u} \right]_{21}^{24} = \underline{\quad\quad}$$

The graph of the density function is sketched in Figure 10.9 for reference. ∎

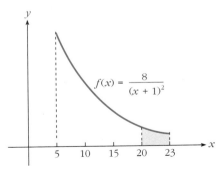

FIGURE 10.9 The graph of the density function of Example 3. The shaded area represents $P(X > 20)$.

## The Probability Distribution Function

If $f(x)$ is a probability density function of a continuous random variable $X$, then $P(c \leq X \leq d)$ is the area under the graph of $f$ between $c$ and $d$; see Figure 10.6(b). It follows that

$$P(c \leq X \leq d) = P(X \leq d) - P(X \leq c)$$

If we now define $F(x)$ to be

$$F(x) = P(X \leq x) = \int_{-\infty}^{x} f(t)\, dt$$

then

$$P(c \leq X \leq d) = F(d) - F(c)$$

---

**SOLUTIONS TO EXAMPLE 3**    20, $dx$, 24, $\frac{1}{21}$

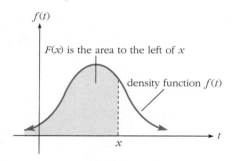

**FIGURE 10.10** The function $F(x)$ gives the accumulated area under the density function $f(t)$ to the left of $x$.

Since $P(c \le X \le d) = \int_c^d f(x)\,dx$, we conclude that $\int_c^d f(x)\,dx = F(d) - F(c)$, so that $F(x)$ is an antiderivative of $f(x)$. That is, $F'(x) = f(x)$. The function $F(x)$ is called the *probability distribution function* of $X$.

> If $f(x)$ is the probability density function for the continuous random variable $X$, then the *probability distribution function* of $X$ is defined by
> $$F(x) = P(X \le x) = \int_{-\infty}^{x} f(t)\,dt$$
> It follows that
> $$P(c \le X \le d) = F(d) - F(c)$$

As shown in Figure 10.10, the probability distribution function $F(x)$ represents the area accumulated under the density function $f(t)$ to the left of $x$. For this reason it is sometimes called the *cumulative probability distribution function* of $X$.

---

**EXAMPLE 4**

Let $X$ be a continuous random variable defined on $[0, 4]$ with probability density function $f(x) = \frac{1}{8}x$.

(a) Find the probability distribution function $F(x)$ of $X$.
(b) Use the distribution function to find $P(X \le 2)$.
(c) Use the distribution function to find $P(1 < X \le 3)$.

**Solution**

(a) For $0 \le x \le 4$ we have
$$F(x) = \int_{-\infty}^{x} f(t)\,dt = \int_0^x \tfrac{1}{8}t\,dt = \tfrac{1}{16}t^2 \Big]_0^x = \tfrac{1}{16}x^2$$

If $x < 0$ then $f(x) = 0$, so $F(x) = \int_{-\infty}^{0} 0\,dt = 0$. If $x > 4$ then
$$F(x) = \int_0^4 \tfrac{1}{8}t\,dt + \int_4^x 0\,dt = \tfrac{1}{16}t^2 \Big]_0^4 = 1 \quad \text{(Since } f(t) = 0 \text{ if } x > 4\text{)}$$

We conclude that
$$F(x) = \begin{cases} 0 & \text{if } x < 0 \\ \tfrac{1}{16}x^2 & \text{if } 0 \le x \le 4 \\ 1 & \text{if } x > 4 \end{cases}$$

The graph of $F(x)$ is sketched in Figure 10.11.

(b) $P(X \le 2) = F(2) = \tfrac{1}{16}(2)^2 = \tfrac{1}{4}$; see Figure 10.11.
(c) $P(1 < X \le 3) = P(1 \le X \le 3)$    (Since $X$ is continuous)
$$= F(3) - F(1) = \tfrac{1}{16}(3)^2 - \tfrac{1}{16}(1)^2 = \tfrac{1}{2}$$

Figure 10.12 shows the relationship between the probability density function $f(x)$ and the probability distribution function $F(x)$. ∎

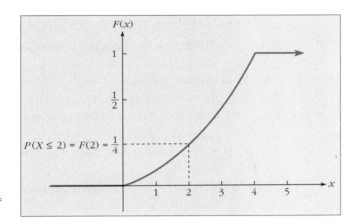

**FIGURE 10.11** The cumulative distribution function of Example 4 showing $P(X \leq 2) = F(2) = \frac{1}{4}$.

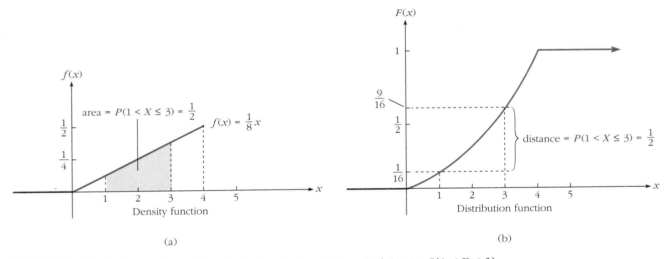

**FIGURE 10.12** The density function and the distribution function of Example 4 showing $P(1 < X \leq 3)$.

---

**EXAMPLE 5**  If a factory has a daily production such that $X$ is the fraction of units that have no defects, and if $X$ has the probability density function

$$f(x) = \begin{cases} 702x^{25}(1-x) & 0 \leq x \leq 1 \\ 0 & \text{otherwise} \end{cases}$$

find the probability that at least nine-tenths of the daily output from this factory is free of defects.

**Solution**  We must find $P(X \geq 0.90)$. Thus,

$$P(X \geq 0.90) = \int_{0.90}^{1} 702x^{25}(1-x)\,dx = 702 \int_{0.90}^{1} (x^{25} - x^{26})\,dx$$

$$= 702\left(\frac{x^{26}}{26} - \frac{x^{27}}{27}\right)\Big]_{0.90}^{1}$$

$$= 702\left[\left(\frac{1}{26} - \frac{1}{27}\right) - \left(\frac{(0.90)^{26}}{26} - \frac{(0.90)^{27}}{27}\right)\right] \approx 0.77$$

∎

## SECTION 10.2
### SHORT ANSWER EXERCISES

In Exercises 1–4, classify the random variable X as discrete or continuous.

1. X is the number of defective units found in a production run of 500 units. _____

2. X is the percentage of voters in the United States who favor a particular candidate for election as President. _____

3. X is the time required for a student to learn at least 50 out of 100 facts. _____

4. X is the number of students present for a certain professor's calculus class on a given day. _____

5. If X represents the amount of soda (in liters) that a machine puts into a two-liter bottle, then $P(X = 2) =$ _____.

6. If a continuous random variable X is uniformly distributed on [0, 10], then $P(X \leq 5) =$ _____.

7. (True or False) If X is a continuous random variable, then $P(X > a) = 1 - P(X \leq a)$. _____

8. (True or False) If X is a continuous random variable defined on [1, 5], then $P(2 \leq X \leq 3) = P(X < 3)$. _____

9. If $F(x)$ is the distribution function for a continuous random variable X and if $F(10) = 1$, then $F(11) =$ _____.

10. If $F(x)$ is the cumulative distribution function for a continuous random variable X and if $F(10) = 0.90$ and $F(6) = 0.40$, the $P(6 \leq X < 10) =$ _____.

## SECTION 10.2
### EXERCISES

In Exercises 1–6, verify that the given function f is a probability density function over the given interval. Sketch the graph of each density function.

1. $f(x) = \frac{1}{5}$ over [5, 10]
2. $f(x) = \frac{3}{64}x^2$ over [0, 4]
3. $f(x) = 1/(2\sqrt{x + 1})$ over [0, 3]
4. $f(x) = \frac{3}{14}\sqrt{x + 1}$ over [0, 3]
5. $f(x) = \frac{1}{4}e^{-x/4}$ over [0, ∞)
6. $f(x) = 30x^4(1 - x)$ over [0, 1]

7. Use the density function of Exercise 1 to find (a) $P(X \leq 6)$; (b) $P(6 < X \leq 10)$.
8. Use the density function of Exercise 2 to find (a) $P(X > 1)$; (b) $P(1 \leq X < 3)$.
9. Use the density function of Exercise 3 to find (a) $P(0 \leq X \leq 2)$; (b) $P(X \geq 2)$.
10. Use the density function of Exercise 4 to find (a) $P(1 < X < 2)$; (b) $P(X < 2)$.
11. Use the density function of Exercise 5 to find (a) $P(X \leq 4)$; (b) $P(2 < X \leq 10)$.
12. Use the density function of Exercise 6 to find (a) $P(X \leq 0.5)$; (b) $P(X > 0.75)$.
13. (a) Find the distribution function for the density function of Exercise 1.
    (b) Use the distribution function found in part (a) to determine
        (i) $P(X \leq 9)$; (ii) $P(X > 8)$.
14. (a) Find the distribution function for the density function of Exercise 2.
    (b) Use the distribution function found in part (a) to determine
        (i) $P(X > 2)$; (ii) $P(1 < X < 3)$.
15. (a) Find the distribution function for the density function of Exercise 3.
    (b) Use the distribution function found in part (a) to determine
        (i) $P(X > 1.5)$; (ii) $P(1 < X \leq 2)$.

**16.** (a) Find the distribution function for the density function of Exercise 4.
  (b) Use the distribution function found in part (a) to determine
     (i) $P(0.5 \leq X < 1.5)$; (ii) $P(X > 3)$.

**17.** (a) Find the distribution function for the density function of Exercise 5.
  (b) Use the distribution function found in part (a) to determine
     (i) $P(X \leq 10)$; (ii) $\int_5^\infty \frac{1}{4} e^{-x/4}\, dx$.

**18.** (a) Find the distribution function for the density function of Exercise 6.
  (b) Use the distribution function found in part (a) to determine
     (i) $P(X > 0.2)$; (ii) $\int_{0.4}^1 30x^4 (1 - x)\, dx$.

*In Exercises 19–26, find the value of k that makes the given function a probability density function on the given interval.*

**19.** $f(x) = kx$ on $[2, 10]$
**20.** $f(x) = k(2 - x)$ on $[0, 2]$
**21.** $f(x) = k(3 - x)^2$ on $[0, 3]$
**22.** $f(x) = k(25 - x^2)$ on $[-5, 5]$
**23.** $f(x) = kx^2(4 - x)$ on $[0, 4]$
**24.** $f(x) = k/(2 + x)^2$ on $[1, 6]$
**25.** $f(x) = k \ln x$ on $[1, e]$
**26.** $f(x) = ke^x$ on $[0, 1]$

*In Exercises 27–30, the given function F(x) is the distribution function for a continuous random variable X. In each case, find the density function of X.*

**27.** $F(x) = \begin{cases} 0 & x < 1 \\ \dfrac{x-1}{2} & 1 \leq x \leq 3 \\ 1 & x > 3 \end{cases}$

**28.** $F(x) = \begin{cases} 0 & x < 0 \\ \dfrac{x^2}{25} & 0 \leq x \leq 5 \\ 1 & x > 5 \end{cases}$

**29.** $F(x) = \begin{cases} 0 & x < 0 \\ 12x^{11} - 11x^{12} & 0 \leq x \leq 1 \\ 1 & x > 1 \end{cases}$

**30.** $F(x) = \begin{cases} 0 & x < 1 \\ \dfrac{x^4}{15} & 1 \leq x \leq 2 \\ 1 & x > 2 \end{cases}$

**31.** (Time to complete bank transactions) The time in minutes required to complete a transaction at a local bank is a continuous random variable $X$ with density function $f(x) = 2(x + 4)^{-3/2}$ defined on the interval $[0, 12]$. Find the probability that a transaction takes less than five minutes.

**32.** (Transportation) The time between arrival of buses at a particular bus stop is 15 minutes. Find the probability that a person arriving at a random time must wait more than 10 minutes for a bus. Assume that waiting time is a uniformly distributed continuous random variable.

**33.** (Fast-food service) The waiting time in minutes to be served in a fast-food restaurant is uniformly distributed between 0 and 10 minutes. Find the probability that a person arriving at a random time must wait less than two minutes for service.

**Social Sciences**

**34.** (Psychology) The time in seconds that it takes a subject to respond to a stimulus in a psychology experiment has the density function

$$f(x) = 6/(5x^2), \quad 1 \leq x \leq 6$$

Find the probability that the subject responds to the stimulus in two seconds or less.

## Business and Economics

**35.** (Demand)  The daily demand (in pounds) for a chemical used in a production process is a continuous random variable with density function

$$f(x) = 0.02 - 0.0002x, \qquad 0 \le x \le 100$$

  (a) Find the probability that daily demand will be less than 50 pounds.
  (b) If 90 pounds of the chemical are available on a certain day, what is the probability of running short of this chemical in the production process?

**36.** (Useful life of an asset)  The useful life (in years) of a certain machine used in a production process is a continuous random variable with density function $f(x) = \frac{1}{5}e^{-x/5}$ for $x \ge 0$. Find the probability that the useful life of a machine will be less than five years.

**37.** (Food sales)  The number of pounds of chicken sold daily by a chain of grocery stores varies randomly between 1,000 lb and 5,000 lb. If the continuous random variable $X$ (measured in thousands of pounds) represents daily sales of chicken, the probability density function of $X$ is of the form $f(x) = k/x^2$.
  (a) Find the appropriate value of $k$.
  (b) Find the probability that daily sales of chicken will exceed 4,500 lb.

**38.** (Employee training)  The time in hours that it takes a new employee to learn to operate a certain machine correctly on a production line is a continuous random variable $X$ with density function $f(x) = \frac{1}{4}e^{-x/4}$ for $x \ge 0$. Find the probability that an employee learns to operate the machine correctly in less than eight hours.

**39.** (Retention of teachers)  The fraction of newly hired teachers who are still teaching in a particular school district after four years is a continuous random variable with density function

$$f(x) = \begin{cases} 272x^{15}(1-x) & 0 \le x \le 1 \\ 0 & \text{otherwise} \end{cases}$$

What is the probability that the school district will retain at least $\frac{3}{4}$ of its new teachers for four years?

## OPTIONAL
### GRAPHING CALCULATOR/ COMPUTER EXERCISES

**40.** The *beta density function* has the form

$$f(x) = \begin{cases} \dfrac{(\alpha + \beta - 1)!}{(\alpha - 1)!(\beta - 1)!} x^{\alpha-1}(1-x)^{\beta-1} & 0 \le x \le 1 \\ 0 & \text{otherwise} \end{cases}$$

where $\alpha$ and $\beta$ are positive integers. Sketch the graphs of the beta density function on the same coordinate axes for (a) $\alpha = 3$ and $\beta = 2, 3, 4, 5, 10,$ and 12; (b) $\alpha = 11$ and $\beta = 2, 3, 4, 5,$ and 10.

**41.** The *Rayleigh density function* has the form

$$f(x) = \begin{cases} 2\alpha x e^{-\alpha x^2} & \text{for } x > 0 \\ 0 & \text{otherwise} \end{cases}$$

where $\alpha$ is a positive constant. Sketch the graphs of the Rayleigh density function on the same coordinate axes for (a) $\alpha = \frac{1}{2}, \alpha = 1, \alpha = 2,$ and $\alpha = 5$; (b) $\alpha = 10, \alpha = 20,$ and $\alpha = 50$.

## 10.3 EXPECTED VALUE, VARIANCE, AND MEDIAN OF CONTINUOUS RANDOM VARIABLES

### The Expected Value and Variance

In Section 10.1 we defined the expected value (or mean) and the variance of a discrete random variable $X$. Here are the analogous definitions for a continuous random variable $X$.

**EXPECTED VALUE AND VARIANCE OF A CONTINUOUS RANDOM VARIABLE**

Let $X$ be a continuous random variable with density function $f(x)$ defined on $[a, b]$. The *expected value* or *mean of* $X$ is defined by

$$E(X) = \mu = \int_a^b x f(x)\, dx \quad (1)$$

The *variance of* $X$ is defined by

$$V(X) = \sigma^2 = \int_a^b (x - \mu)^2 f(x)\, dx \quad (2)$$

The *standard deviation of* $X$ is defined by

$$\sigma = \sqrt{V(X)} \quad (3)$$

This definition also applies if the continuous random variable $X$ is defined on an unbounded interval, such as $[0, \infty)$ or $(-\infty, \infty)$.

### EXAMPLE 1

Find the mean, variance, and standard deviation of the random variable $X$ that has the uniform density function

$$f(x) = \begin{cases} \frac{1}{20} & 0 \leq x \leq 20 \\ 0 & \text{otherwise} \end{cases}$$

We considered this random variable in Example 1 of Section 10.2: it represents the amount of time that a person arriving at a random time must wait for a train if the time between arrivals of trains is 20 minutes.

**Solution**

$$E(X) = \mu = \int_0^{20} x\left(\tfrac{1}{20}\right) dx = \tfrac{1}{20} \int_0^{20} x\, dx$$

$$= \tfrac{1}{20}\left[\frac{x^2}{2}\right]_0^{20} = \tfrac{1}{20}\left(\frac{20^2}{2} - \frac{0^2}{2}\right) = 10$$

This result is not surprising, for if the time between arrival of trains is 20 minutes, we expect that a person arriving at a random time will have to wait an average of about 10 minutes.

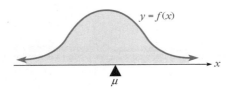

**FIGURE 10.13** The mean $\mu$ is the point on the $x$-axis where the probability density function balances on a wedge.

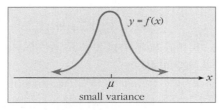

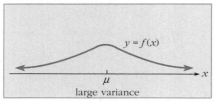

**FIGURE 10.14** The probability density function with the large variance is more dispersed about the mean $\mu$.

## CHEBYSHEV'S THEOREM

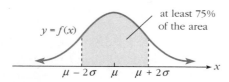

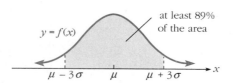

**FIGURE 10.15** At least 75% of the area lies within two standard deviations of $\mu$ and at least 89% within three standard deviations of $\mu$ if $f(x)$ is the density function of the continuous random variable $X$.

$$V(X) = \sigma^2 = \int_0^{20} (x-10)^2 \left(\tfrac{1}{20}\right) dx = \tfrac{1}{20} \int_0^{20} (x^2 - 20x + 100)\, dx$$

$$= \frac{1}{20}\left[\frac{x^3}{3} - 10x^2 + 100x\right]_0^{20}$$

$$= \frac{1}{20}\left[\left(\frac{20^3}{3} - 10(20)^2 + 100(20)\right) - \left(\frac{0^3}{3} - 10(0)^2 + 100(0)\right)\right]$$

$$= \frac{1}{20}\left[\frac{2000}{3}\right] = \frac{100}{3}$$

We conclude that the standard deviation of $X$ is $\sigma = \sqrt{V(X)} = 10/\sqrt{3} \approx 5.77$. ∎

### Interpretation of the Mean and Variance: Chebyshev's Theorem

Just as in the case of a discrete random variable, the expected value of a continuous random variable $X$ represents an average value of $X$ over the long run. Thus $E(X)$ is a measure of central tendency of the probability density function of $X$. On the other hand, the variance of a continuous random variable $X$ is a measure of the dispersion or spread of the density function about the mean.

Geometrically, the mean of a continuous random variable is the point on the $x$-axis where the probability density function of $X$ balances on a wedge, as illustrated in Figure 10.13.

In Figure 10.14, two probability density functions are illustrated with the same mean $\mu$ and different variances. Note how the distribution with the large variance is more spread out about the mean $\mu$.

As a more specific explanation of the geometric significance of the variance and standard deviation, the following theorem was proven by the nineteenth-century Russian mathematican P. L. Chebyshev.

> Let $\mu$ and $\sigma$ be the mean and the standard deviation, respectively, of a random variable $X$ and let $k$ be a positive constant. Then the probability that $X$ will take on a value within $k$ standard deviations of the mean is *at least* $1 - 1/k^2$. In symbols:
> 
> $$P(\mu - k\sigma \leq X \leq \mu + k\sigma) \geq 1 - \frac{1}{k^2}$$

Chebyshev's theorem holds for both discrete and continuous random variables and its proof may be found in any good statistics text. If we apply it to a continuous random variable $X$ with probability density function $f(x)$ and interpret probability in terms of area under the density function, then, using $k = 2$ and $k = 3$, for example, we can assert the following.

> (a) At least $\tfrac{3}{4} = 75\%$ of the area under the density function $f(x)$ lies within two standard deviations of the mean.
> (b) At least $\tfrac{8}{9} \approx 89\%$ of the area under the density function $f(x)$ lies within three standard deviations of the mean.

## 10.3 EXPECTED VALUE, VARIANCE, AND MEDIAN OF CONTINUOUS RANDOM VARIABLES

See Figure 10.15 for a graphical illustration of these facts.

You should note that Chebyshev's theorem provides a *lower bound* for the amount of area within $k$ standard deviations of the mean under the density function $f(x)$. For many density functions the amount of area within two or three standard deviations of the mean will be substantially more than 75% and 89%, respectively. Let's look at an example.

**EXAMPLE 2**

Let $X$ be a continuous random variable with density function

$$f(x) = \begin{cases} \frac{1}{72}x^2 & 0 \le x \le 6 \\ 0 & \text{otherwise} \end{cases}$$

(a) Find the mean, variance, and standard deviation of $X$.
(b) Find the area under the density function within two standard deviations of the mean and also within three standard deviations of the mean.

**Solution**

(a) $E(X) = \mu = \int_0^6 x\left(\frac{1}{72}x^2\right) dx = \frac{1}{72} \int_0^6 x^3 \, dx$

$= \frac{1}{72}\left[\frac{x^4}{4}\right]_0^6 = \frac{1}{72}\left[\frac{6^4}{4} - \frac{0^4}{4}\right] = 4.5$

$V(X) = \int_0^6 (x - 4.5)^2 \left(\frac{1}{72}x^2\right) dx = \frac{1}{72} \int_0^6 (x^2 - 9x + 20.25)x^2 \, dx$

$= \frac{1}{72} \int_0^6 (x^4 - 9x^3 + 20.25x^2) \, dx$

$= \frac{1}{72}\left[\frac{x^5}{5} - \frac{9x^4}{4} + \frac{20.25x^3}{3}\right]_0^6$

$= \frac{1}{72}\left[\left(\frac{6^5}{5} - \frac{9(6)^4}{4} + \frac{20.25(6)^3}{3}\right) - \left(\frac{0^5}{5} - \frac{9(0)^4}{4} + \frac{20.25(0)^3}{3}\right)\right]$

$= 1.35$

$\sigma = \sqrt{V(X)} \approx 1.16$

(b) The area under the density function within two standard deviations of the mean is

$P(\mu - 2\sigma \le X \le \mu + 2\sigma) \approx P(4.5 - 2(1.16) \le X \le 4.5 + 2(1.16))$

$= P(2.18 \le X \le 6.82)$

$= P(2.18 \le X \le 6)$ (Since $f(x) = 0$ if $x > 6$)

$= \int_{2.18}^6 \frac{1}{72}x^2 \, dx = \frac{1}{72} \int_{2.18}^6 x^2 \, dx$

$= \frac{1}{72}\left[\frac{x^3}{3}\right]_{2.18}^6 = \frac{1}{72}\left[\frac{6^3}{3} - \frac{(2.18)^3}{3}\right] = 0.952$

Thus in this case about 95% of the area under the density function lies within two standard deviations of the mean; see Figure 10.16(a). The area under the density function within three standard deviations of the mean is

$$P(\mu - 3\sigma \le X \le \mu + 3\sigma) \approx P(4.5 - 3(1.16) \le X \le 4.5 + 3(1.16))$$
$$= P(1.02 \le X \le 7.98) = P(1.02 \le X \le 6)$$
$$= \int_{1.02}^{6} \tfrac{1}{72} x^2 \, dx = \tfrac{1}{72} \int_{1.02}^{6} x^2 \, dx$$
$$= \frac{1}{72}\left[\frac{x^3}{3}\right]_{1.02}^{6} = \frac{1}{72}\left[\frac{6^3}{3} - \frac{(1.02)^3}{3}\right] = 0.995$$

So for this density function, about 99.5% of the area under the density function lies within three standard deviations of the mean; see Figure 10.16(b).

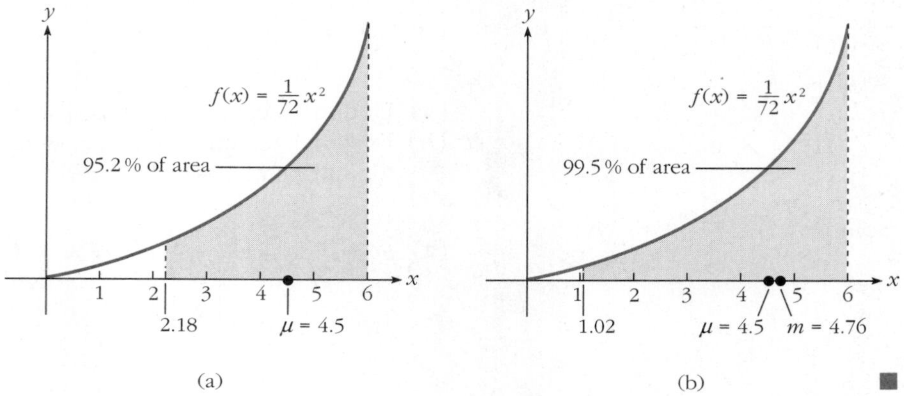

**FIGURE 10.16** For the density function of Example 2, about 95% of the area lies within two standard deviations of the mean, and about 99.5% of the area lies within three standard deviations. The number $m$ in part (b) is called the median and will be explained in Example 6.

## EXAMPLE 3

The weekly amount spent for a certain chemical used in a manufacturing process has a mean of $1200 and a variance of 900. Find an interval that contains the weekly chemical costs (a) at least 75% of the time; (b) at least 95% of the time.

Solution

(a) We know from Chebyshev's theorem that if $X$ is weekly chemical costs, then $X$ will lie in the interval $[\mu - 2\sigma, \mu + 2\sigma]$ at least $\tfrac{3}{4} = 75\%$ of the time. Using $\mu = \$1200$ and $\sigma = \sqrt{900} = \$30$, we obtain the interval $[1200 - 2(30), 1200 + 2(30)] = [\$1140, \$1260]$.

(b) Again using Chebyshev's theorem, we must first find the positive number $k$ such that $1 - (1/k^2) = 0.95$. Solving this equation for $k$, we get

$$\frac{1}{k^2} = 0.05$$
$$k^2 = 20$$
$$k = \sqrt{20} \approx 4.47$$

Thus at least 95% of the time, the interval $[1200 - 4.47(30), 1200 + 4.47(30)] = [\$1065.90, \$1334.10]$ contains the weekly expenditure for the chemical.

An alternate formula for the variance can be derived from formula (2), which was given at the beginning of this section. The derivation of this formula is omitted.

## 10.3 EXPECTED VALUE, VARIANCE, AND MEDIAN OF CONTINUOUS RANDOM VARIABLES

**AN ALTERNATE FORMULA FOR THE VARIANCE**

Let $X$ be a continuous random variable with density function $f(x)$ defined on $[a, b]$. Then
$$V(X) = \int_a^b x^2 f(x)\, dx - \mu^2 = E(X^2) - [E(X)]^2$$

---

EXAMPLE 4 (Participative)

During a 40-hour work week, the amount of time that an employee's microcomputer is in use is a continuous random variable $X$ with density function

$$f(x) = \begin{cases} \frac{1}{1728}x^2(12 - x) & 0 \le x \le 12 \\ 0 & \text{otherwise} \end{cases}$$

Let's find (a) the expected number of hours that the computer is in use; and (b) the variance of the random variable $X$.

Solution

(a) $\mu = E(X) = \displaystyle\int_a^b x f(x)\, dx$    (Formula (1))

$= \displaystyle\int_0^{12} x(\underline{\hspace{2cm}})\, dx$

$= \dfrac{1}{1728} \displaystyle\int_0^{12} x[x^2(12 - x)]\, dx$

$= \dfrac{1}{1728} \displaystyle\int_0^{12} (\underline{\hspace{2cm}})\, dx$   (Expanding the integrand)

$= \dfrac{1}{1728}[\underline{\hspace{2cm}}]_0^{12}$   (Integrating)

$= \dfrac{1}{1728}\left[\left(3(12)^4 - \dfrac{12^5}{5}\right) - \left(3(0)^4 - \dfrac{0^5}{5}\right)\right]$

$= \underline{\hspace{2cm}}$

(b) To find the variance of $X$, let's use the alternate formula for variance,

$V(X) = \displaystyle\int_a^b x^2 f(x)\, dx - \mu^2$. From part (a), we know that $\mu = \underline{\hspace{1cm}}$, so that

$V(X) = \displaystyle\int_0^{12} x^2\left[\dfrac{1}{1728}x^2(12 - x)\right] dx - [\underline{\hspace{0.5cm}}]^2$

$= \dfrac{1}{1728} \displaystyle\int_0^{12} (\underline{\hspace{2cm}})\, dx - 51.84$   (Expanding the integrand)

$= \dfrac{1}{1728}[\underline{\hspace{2cm}}]_0^{12} - 51.84$   (Integrating)

$= \dfrac{1}{1728}\left[\left(\dfrac{12(12)^5}{5} - \dfrac{12^6}{6}\right) - \left(\dfrac{12(0)^5}{5} - \dfrac{0^6}{6}\right)\right] - 51.84$

$= \underline{\hspace{2cm}} - 51.84$

$= \underline{\hspace{2cm}}$

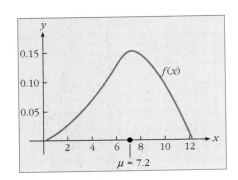

FIGURE 10.17 The density function of Example 4.

The graph of the density function for this example is displayed in Figure 10.17 for reference. ■

EXAMPLE 5

Find the expected value and the variance of the continuous random variable $X$ with density function

$$f(x) = \begin{cases} \dfrac{4}{x^5} & x \geq 1 \\ 0 & \text{otherwise} \end{cases}$$

Solution

The density function is defined on the unbounded interval $[1, \infty)$. For the expected value, we obtain

$$\begin{aligned} E(X) = \mu &= \int_1^{\infty} x \cdot \frac{4}{x^5}\, dx = \int_1^{\infty} \frac{4}{x^4}\, dx \\ &= \lim_{b\to\infty} \int_1^b 4x^{-4}\, dx = \lim_{b\to\infty}\left[-\frac{4}{3x^3}\right]_1^b \\ &= \lim_{b\to\infty}\left[-\frac{4}{3b^3} + \frac{4}{3}\right] = \frac{4}{3} \end{aligned}$$

To compute the variance of $X$ we will use the alternate formula. First, we find

$$\begin{aligned} E(X^2) &= \int_1^{\infty} x^2 \cdot \frac{4}{x^5}\, dx = \int_1^{\infty} \frac{4}{x^3}\, dx \\ &= \lim_{b\to\infty} \int_1^b 4x^{-3}\, dx = \lim_{b\to\infty}\left[-\frac{2}{x^2}\right]_1^b \\ &= \lim_{b\to\infty}\left[-\frac{2}{b^2} + 2\right] = 2 \end{aligned}$$

Using the formula $V(X) = E(X^2) - [E(X)]^2$, we conclude that

$$V(X) = 2 - \left(\frac{4}{3}\right)^2 = \frac{2}{9}$$ ■

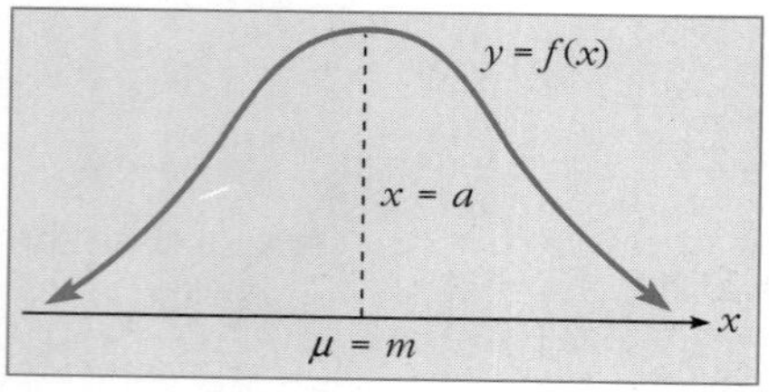

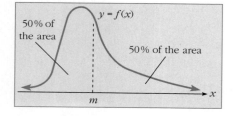

**FIGURE 10.18** The median $m$ divides the area under the density function into two equal portions.

## The Median of a Continuous Random Variable

The *median* of a random variable is another measure of central tendency. The median $m$ of a continuous random variable $X$ is the value of $X$ that divides the area under the graph of the probability density function into two equal areas, each of magnitude 0.5. Thus the median $m$ must satisfy

$$P(X \leq m) = \int_{-\infty}^{m} f(x)\, dx = 0.5$$

**FIGURE 10.19** The graph of the density function $f(x)$ is symmetric about the vertical line $x = a$. In this case the mean and median are both equal to $a$.

See Figure 10.18 for a graphical interpretation of the median. If the graph of the probability density function is symmetric about some vertical line $x = a$, then the mean and the median of $X$ are both equal to $a$; see Figure 10.19. In general, however, the mean and the median are *not* equal, as is illustrated in the next example.

**SOLUTIONS TO EXAMPLE 4**

**(a)** $\frac{1}{1728}x^2(12 - x)$, $12x^3 - x^4$, $3x^4 - \frac{1}{5}x^5$, 7.2; **(b)** 7.2, 7.2, $12x^4 - x^5$, $\frac{12}{5}x^5 - \frac{1}{6}x^6$, 57.60, 5.76

### EXAMPLE 6

Find the median of the continuous random variable with density function

$$f(x) = \begin{cases} \frac{1}{72}x^2 & 0 \le x \le 6 \\ 0 & \text{otherwise} \end{cases}$$

**Solution**

This is the random variable of Example 2, where we showed that $\mu = 4.5$. To find the median, we have

$$0.5 = \int_{-\infty}^{m} f(x)\, dx = \int_{0}^{m} \frac{1}{72}x^2\, dx$$

$$= \frac{1}{72}\left[\frac{x^3}{3}\right]_0^m = \frac{1}{72}\left[\frac{m^3}{3} - \frac{0^3}{3}\right] = \frac{m^3}{216}$$

$$m^3 = 108$$
$$m = \sqrt[3]{108} \approx 4.76$$

We conclude that the median is about 4.76 and that 50% of the area under the density function lies to the left of 4.76 and 50% to the right of 4.76. See Figure 10.16(b). ■

---

### SECTION 10.3
**SHORT ANSWER EXERCISES**

1. A continuous random variable $X$ is uniformly distributed over the interval $[0, 30]$. The mean of $X$ is equal to _____.
2. (True or False) The graph of a probability density function will always balance on a wedge placed on the $x$-axis at a point $b$ such that 50% of the area lies to the left of $b$ and 50% of the area to the right. _____
3. According to Chebyshev's theorem, what percentage of the area under the density function must lie within four standard deviations of the mean? _____
4. (True or False) The variance and standard deviation of a continuous random variable $X$ must both be nonnegative numbers. _____
5. (True or False) The mean of a continuous random variable $X$ must be a nonnegative number. _____
6. A continuous random variable $X$ has a mean $\mu = 3$ and $E(X^2) = 25$. The standard deviation of $X$ is _____.
7. The variance of a continuous random variable $X$ is 144 and $E(X^2) = 169$. The mean of $X$ is _____.
8. A continuous probability density function defined on $[0, 10]$ is decreasing on $[0, 10]$. The mean of $X$ is (a) greater than 5; (b) less than 5; (c) equal to 5; (d) none of the above.
9. A continuous probability density function defined on $[-4, 4]$ is increasing on $[-4, 4]$. The median of $X$ is (a) greater than zero; (b) less than zero; (c) equal to zero; (d) none of the above.
10. (True or False) A continuous random variable with mean $\mu = 0$ can have a density function $f(x)$ that is nonzero on $[0, 10]$ and zero otherwise. _____

## SECTION 10.3
## EXERCISES

In Exercises 1–14, find the mean, variance, and standard deviation of each of the continuous random variables with given density function. Use the alternate formula for variance in Exercises 7–14.

1. $f(x) = \begin{cases} \frac{1}{5} & 0 \le x \le 5 \\ 0 & \text{otherwise} \end{cases}$

2. $f(x) = \begin{cases} \frac{1}{4} & 5 \le x \le 9 \\ 0 & \text{otherwise} \end{cases}$

3. $f(x) = \begin{cases} \frac{1}{8}x & 0 \le x \le 4 \\ 0 & \text{otherwise} \end{cases}$

4. $f(x) = \begin{cases} \frac{1}{9}x^2 & 0 \le x \le 3 \\ 0 & \text{otherwise} \end{cases}$

5. $f(x) = \begin{cases} 4x^3 & 0 \le x \le 1 \\ 0 & \text{otherwise} \end{cases}$

6. $f(x) = \begin{cases} \frac{1}{14} + \frac{1}{7}x & 2 \le x \le 4 \\ 0 & \text{otherwise} \end{cases}$

7. $f(x) = \begin{cases} 3/x^4 & x \ge 1 \\ 0 & \text{otherwise} \end{cases}$

8. $f(x) = \begin{cases} e^{-x} & x \ge 0 \\ 0 & \text{otherwise} \end{cases}$

9. $f(x) = \begin{cases} 4e^{-4x} & x \ge 0 \\ 0 & \text{otherwise} \end{cases}$
   Hint: $\lim_{x \to \infty} x^n e^{-kx} = 0$ if $k$ and $n$ are positive

10. $f(x) = \begin{cases} \frac{5}{2}x^{-7/2} & x \ge 1 \\ 0 & \text{otherwise} \end{cases}$

11. $f(x) = \begin{cases} -\frac{3}{32}(4 - x^2) & 2 \le x \le 4 \\ 0 & \text{otherwise} \end{cases}$

12. $f(x) = \begin{cases} \frac{1}{36}x(6 - x) & 0 \le x \le 6 \\ 0 & \text{otherwise} \end{cases}$

13. $f(x) = \begin{cases} 1/(2\sqrt{x + 1}) & 0 \le x \le 3 \\ 0 & \text{otherwise} \end{cases}$

14. $f(x) = \begin{cases} 1/(x + 1) & 0 \le x \le e - 1 \\ 0 & \text{otherwise} \end{cases}$

In Exercises 15–28, find the median of the continuous random variables with given density functions.

15. $f(x) = \begin{cases} \frac{1}{18}x & 0 \le x \le 6 \\ 0 & \text{otherwise} \end{cases}$

16. $f(x) = \begin{cases} \frac{1}{32}x & 0 \le x \le 8 \\ 0 & \text{otherwise} \end{cases}$

17. $f(x) = \begin{cases} 3x^2 & 0 \le x \le 1 \\ 0 & \text{otherwise} \end{cases}$

18. $f(x) = \begin{cases} 4x^3 & 0 \le x \le 1 \\ 0 & \text{otherwise} \end{cases}$

19. $f(x) = \begin{cases} 1/(6\sqrt{x}) & 0 \le x \le 9 \\ 0 & \text{otherwise} \end{cases}$

20. $f(x) = \begin{cases} 1/(6\sqrt[3]{x}) & 0 \le x \le 8 \\ 0 & \text{otherwise} \end{cases}$

21. $f(x) = \begin{cases} 8/x^3 & x \ge 2 \\ 0 & \text{otherwise} \end{cases}$

22. $f(x) = \begin{cases} 24/x^4 & x \ge 2 \\ 0 & \text{otherwise} \end{cases}$

23. $f(x) = \begin{cases} \frac{4}{3} - \frac{2}{3}x & 0 \le x \le 1 \\ 0 & \text{otherwise} \end{cases}$

24. $f(x) = \begin{cases} \frac{5}{4} - \frac{1}{2}x & 0 \le x \le 1 \\ 0 & \text{otherwise} \end{cases}$

25. $f(x) = \begin{cases} 4e^{-4x} & x \ge 0 \\ 0 & \text{otherwise} \end{cases}$

26. $f(x) = \begin{cases} \frac{1}{2}e^{-(1/2)x} & x \ge 0 \\ 0 & \text{otherwise} \end{cases}$

27. $f(x) = \begin{cases} 2/(9\sqrt[3]{x + 1}) & 0 \le x \le 7 \\ 0 & \text{otherwise} \end{cases}$

28. $f(x) = \begin{cases} 1/(2\sqrt{x + 2}) & 0 \le x \le 5 \\ 0 & \text{otherwise} \end{cases}$

29. (Worker productivity) The time in minutes required for a worker to complete a certain task on an assembly line is a continuous random variable with probability density function

$$f(x) = \begin{cases} -\frac{2}{9}(x - 2)(x - 5) & 2 \le x \le 5 \\ 0 & \text{otherwise} \end{cases}$$

Find the expected time to complete this task.

30. (Sales) Monthly sales (in thousands of dollars) of one of your company's products is a continuous random variable with probability density function

$$f(x) = \begin{cases} \frac{1}{450}(80 - x) & 50 \le x \le 80 \\ 0 & \text{otherwise} \end{cases}$$

(a) Find the expected monthly sales of this product.
(b) Find the median monthly sales.
(c) Find the variance of $X$.

Business and Economics

**31.** (Retail sales returns) The amount of time (in days) that passes before a customer returns an item for a refund at a retail store is a random variable $X$ with probability density function

$$f(x) = \begin{cases} \frac{1}{450}(30 - x) & 0 \le x \le 30 \\ 0 & \text{otherwise} \end{cases}$$

(a) Find the expected time for a customer to return an item.
(b) Find the median time.
(c) Find the variance of $X$.

**32.** (Retail sales returns) The proportion of items sold that will not eventually be returned to the retail store for a refund is a random variable $X$ with density function

$$f(x) = \begin{cases} 56x^6(1 - x) & 0 \le x \le 1 \\ 0 & \text{otherwise} \end{cases}$$

(a) Find the expected proportion of items that will not be returned for a refund.
(b) Find the variance of $X$.

**33.** (Small business failures) The proportion of small businesses that fail during their first two years of operation in a certain region is a continuous random variable $X$ with density function

$$f(x) = \begin{cases} \frac{231}{100}\sqrt[10]{x}(1 - x) & 0 \le x \le 1 \\ 0 & \text{otherwise} \end{cases}$$

Find the expected proportion of small businesses in this region that fail during their first two years of operation.

**34.** (Profit) Monthly profit (in thousands of dollars) from a certain product in a company's product line is a continuous random variable $X$ with a mean of 50 and a variance of 16. Find an interval that contains the monthly profit (a) at least 75% of the time; (b) at least 90% of the time.

Social Sciences

**35.** (Psychology experiment) The amount of time it takes for a subject in a psychology experiment to complete a certain task is 15 minutes with a variance of four minutes. Find an interval that contains the time required to complete this task (a) at least 75% of the time; (b) at least 85% of the time.

**36.** (Useful life of an asset) The useful life (in years) of a certain small piece of machinery is a continuous random variable $X$ with density function

$$f(x) = \begin{cases} \frac{1}{8}x & 0 \le x \le 4 \\ 0 & \text{otherwise} \end{cases}$$

(a) Find the expected useful life of this piece of machinery.
(b) Find the median useful life.
(c) Find the standard deviation of $X$.

**37. (Product reliability)** The time (in years) until failure of a certain electrical component in a machine is a continuous random variable $X$ with density function

$$f(x) = \begin{cases} \frac{1}{2}e^{-x/2} & x \geq 0 \\ 0 & \text{otherwise} \end{cases}$$

Find the expected time until failure. *Hint:* $\lim_{x \to \infty} x^n e^{-kx} = 0$ if $k$ and $n$ are positive.

**38. (Product lifetime)** The lifetime (in thousands of miles) of a certain type of radial tire is a continuous random variable $X$ with density function

$$f(x) = \begin{cases} \frac{1}{40}e^{-x/40} & x \geq 0 \\ 0 & \text{otherwise} \end{cases}$$

Find the expected lifetime of this type of tire.

# 10.4 THE NORMAL DISTRIBUTION

In this section we will discuss an important continuous distribution that arises in numerous applications. This distribution is the normal distribution.

The normal distribution is undoubtedly the most important probability distribution in statistics. This distribution can be used to model quantities such as the height or weight of a population of people or animals, the length of a part produced by a machine in a production process, and the scores on many standardized tests, such as IQ tests, the SAT, or the GMAT.

The density function of a normally distributed random variable $X$ is given in the following box.

**THE NORMAL DENSITY FUNCTION**

$$f(x) = \frac{1}{\sigma\sqrt{2\pi}} e^{(-1/2)((x-\mu)/\sigma)^2}, \qquad -\infty < x < \infty \qquad (3)$$

The normal density function contains two parameters, $\mu$ and $\sigma^2$. The results in the following box can be proved using techniques beyond the scope of this text.

**MEAN AND VARIANCE OF A NORMAL RANDOM VARIABLE**

If $X$ is a continuous random variable with density function

$$f(x) = \frac{1}{\sigma\sqrt{2\pi}} e^{(-1/2)((x-\mu)/\sigma)^2}, \qquad -\infty < x < \infty$$

then $E(X) = \mu$ and $V(X) = \sigma^2$.

The graph of the normal density function has a characteristic shape that looks like the cross-section of a bell (see Figure 10.20). It is important to observe certain features of this graph, some of which you will be asked to verify in the exercises of this section.

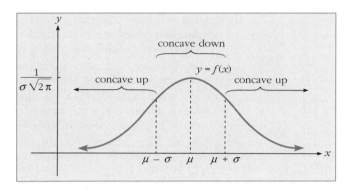

**FIGURE 10.20** The graph of the normal density function.

1. The graph is symmetric about the vertical line $x = \mu$ and hence the median of $X$ is equal to $\mu$.
2. The point $(\mu, 1/(\sigma\sqrt{2\pi}))$ is a relative and absolute maximum point on the graph.
3. The graph is concave up on the intervals $(-\infty, \mu - \sigma)$ and $(\mu + \sigma, \infty)$ and is concave down on the interval $(\mu - \sigma, \mu + \sigma)$.
4. The $x$-axis is a horizontal asymptote of the graph.

The mean $\mu$ determines the center of the distribution and the variance $\sigma^2$ determines its spread or dispersion about the mean $\mu$. In Figure 10.21, two normal density functions are sketched on the same coordinate axes. Notice that the curve with the smaller variance is less dispersed about the mean, producing a curve that is taller and narrower than the curve with the larger variance, which is shorter and flatter.

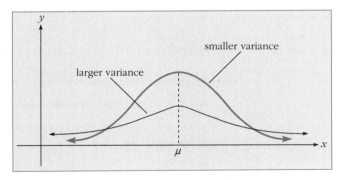

**FIGURE 10.21** A comparison of two normal density functions with the same mean $\mu$ and two different variances.

Unfortunately, the integral of the normal density function defined by equation (1) cannot be expressed as a finite combination of functions studied in this text. Therefore we cannot compute a probability such as

$$P(c \leq X \leq d) = \int_c^d \frac{1}{\sigma\sqrt{2\pi}} e^{(-1/2)((x-\mu)/\sigma)^2} \, dx \qquad (2)$$

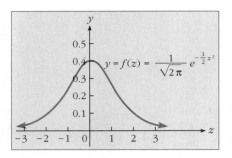

**FIGURE 10.22** The graph of the standard normal density function.

by finding an antiderivative of the integrand and using the fundamental theorem of calculus. We *could* use a numerical integration technique (such as Simpson's rule) to find a probability resembling (2). In most cases, however, we will refer to a table that has been constructed using numerical integration, such as the one on the back endsheets. This table refers to a normal density function with mean $\mu = 0$ and variance $\sigma^2 = 1$. A random variable that is normally distributed with a mean of zero and a variance of one is called a *standard normal* random variable and is usually designated by the letter $Z$. The table gives approximate values of the cumulative distribution function of $Z$,

$$F(z) = P(Z \leq z) = \int_{-\infty}^{z} \frac{1}{\sqrt{2\pi}} e^{-(1/2)t^2} \, dt$$

for selected values of $z$ from $-3.49$ to $3.49$. Thus the values of the body of this table give the area under the graph of the standard normal density function $f(z) = e^{-(1/2)z^2}/\sqrt{2\pi}$ to the left of $z$; see Figure 10.22.

### EXAMPLE 1

Use the table on the back endsheets to compute the following probabilities for a standard normal random variable $Z$: **(a)** $P(Z \leq 1.53)$; **(b)** $P(Z > 1.53)$; **(c)** $P(-1 \leq Z \leq 0.92)$.

### Solution

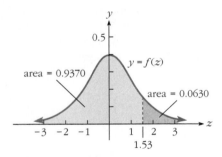

**FIGURE 10.23** An illustration of Example 1(a) and (b).

**(a)** To find $P(Z \leq 1.53)$, from the table find $z = 1.5$ in the $z$-column of the table and move over to the right in this row until you are under the column headed "0.03." The resulting number is 0.9370 and so

$$P(Z \leq 1.53) = 0.9370$$

**(b)** To find $P(Z > 1.53)$ we use the fact that $P(Z \leq 1.53) + P(Z > 1.53) = 1$. Therefore

$$P(Z > 1.53) = 1 - P(Z \leq 1.53)$$
$$= 1 - 0.9370 \quad \text{(Using part (a))}$$
$$= 0.0630$$

Parts (a) and (b) are illustrated in Figure 10.23.

**(c)** To compute $P(-1 \leq Z \leq 0.92)$, observe that

$$P(-1 \leq Z \leq 0.92) = P(Z \leq 0.92) - P(Z \leq -1)$$

See Figure 10.24. The two probabilities on the right-hand side can be found in the table; we have

$$P(-1 \leq Z \leq 0.92) = 0.8212 - 0.1587 = 0.6625 \quad \blacksquare$$

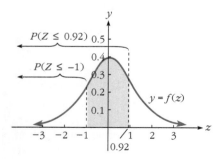

**FIGURE 10.24** The shaded area is $P(-1 \leq Z \leq 0.92)$ and is obtained from $P(Z \leq 0.92) - P(Z \leq -1)$.

Of course, the table allows us to compute probabilities only for a *standard* normal random variable $Z$, which has a mean of zero and a variance of one. Suppose now that $X$ is a normal random variable with mean $\mu$ and variance $\sigma^2$. Then

$$P(X \leq b) = \int_{-\infty}^{b} \frac{1}{\sigma\sqrt{2\pi}} e^{-(1/2)((x-\mu)/\sigma)^2} \, dx$$

## 10.4 THE NORMAL DISTRIBUTION

Making the substitution $z = (x - \mu)/\sigma$ in this integral, we obtain $dz = (1/\sigma) dx$ and the integral becomes

$$P(X \le b) = \int_{-\infty}^{(b-\mu)/\sigma} \frac{1}{\sqrt{2\pi}} e^{-(1/2)z^2} \, dz = P\left(Z \le \frac{b - \mu}{\sigma}\right)$$

It follows that for any $x$,

$$P(X \le x) = P\left(Z \le \frac{x - \mu}{\sigma}\right)$$

This result is illustrated in Figure 10.25 and is summarized in the following box.

---

If $X$ is a normal random variable with mean $\mu$ and variance $\sigma^2$, and if $Z$ is a standard normal random variable, then

$$P(X \le x) = P\left(Z \le \frac{x - \mu}{\sigma}\right)$$

---

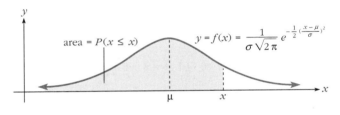

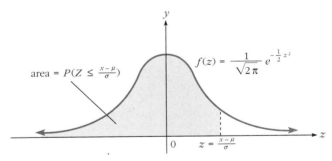

**FIGURE 10.25** The two shaded areas are equal, since $P(X \le x) = P\left(Z \le \frac{x - \mu}{\sigma}\right)$.

---

**EXAMPLE 2**  Let $X$ be a normally distributed random variable with mean 20 and variance 16. Find (a) $P(X \le 22)$; (b) $P(X > 17.4)$; (c) $P(15 \le X \le 21)$.

**Solution**  (a) We have $\mu = 20$ and $\sigma^2 = 16$, so $\sigma = 4$. Thus

$$P(X \le 22) = P\left(Z \le \frac{22 - 20}{4}\right) = P(Z \le 0.5)$$

$$= 0.6915 \quad \text{(Using the table)}$$

(b) $P(X > 17.4) = 1 - P(X \le 17.4) = 1 - P\left(Z \le \dfrac{17.4 - 20}{4}\right)$

$\qquad\qquad\qquad\ = 1 - P(Z \le -0.65) = 1 - 0.2578 = 0.7422$

(c) $P(15 \le X \le 21) = P(X \le 21) - P(X \le 15)$

$\qquad\qquad\qquad\ = P\left(Z \le \dfrac{21 - 20}{4}\right) - P\left(Z \le \dfrac{15 - 20}{4}\right)$

$\qquad\qquad\qquad\ = P(Z \le 0.25) - P(Z \le -1.25)$

$\qquad\qquad\qquad\ = 0.5987 - 0.1056 = 0.4931$

## EXAMPLE 3

A machine fills bottles of soda with a mean of 16 oz of soda with a variance of 0.04. Find the probability that the machine fills a bottle with less than 15.75 oz of soda. Assume that the amounts of soda the machine puts into the bottles are normally distributed.

**Solution**

Let $X$ be the amount of soda (in ounces) the machine puts into a bottle. Then $X$ is a normal random variable with $\mu = 16$ and $\sigma = \sqrt{0.04} = 0.2$. Thus the probability that the machine fills a bottle with less than 15.75 oz of soda is

$$P(X < 15.75) = P\left(Z \le \dfrac{15.75 - 16}{0.2}\right) = P(Z < -1.25) = 0.1056$$

See Figure 10.26 for a graphical illustration.

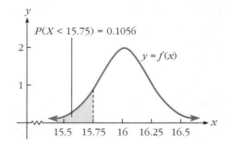

**FIGURE 10.26** The probability that the machine in Example 3 fills a bottle with less than 15.75 ounces is represented by the shaded area.

## EXAMPLE 4 (Participative)

Scores on a certain standardized test are normally distributed with a mean of 500 and a standard deviation of 100. Let's find (a) the probability that a randomly selected person taking the test will score between 475 and 645; and (b) a score that is greater than 90% of all scores.

**Solution**

(a) Let $X$ be the score that a randomly selected person receives on the test. Thus $X$ is normally distributed with a mean of 500 and a standard deviation of 100. Mathematically,

$$P(475 \le X \le 645) = P\left(\dfrac{475 - 500}{100} \le Z \le \underline{\qquad}\right)$$

$$= P(Z \le 1.45) - \underline{\qquad}$$

$$= 0.9265 - \underline{\qquad}$$

$$= \underline{\qquad}$$

(b) Let $x$ be a score that is greater than 90% of all scores. Then $P(X \leq x) =$ _____. Thus

$$P\left(Z \leq \frac{x - 500}{100}\right) = P(Z \leq z) = 0.90$$

where $z = (x - 500)/100$. From the table, the $z$-value of 1.28 satisfies $P(Z \leq 1.28) =$ _____, so we choose $z = 1.28$. Thus $1.28 = (x - 500)/100$. Solving this equation for $x$ yields $x =$ _____ as a score that is greater than 90% of all scores. ∎

## SECTION 10.4
### SHORT ANSWER EXERCISES

1. (True or False) For a normal random variable $X$ with mean $\mu$, $P(X \leq \mu) = 0.50$. _____

2. If $Z$ is a standard normal random variable, then $P(Z < -1.23) =$ _____.

3. If $X$ and $Y$ are both normal random variables with mean 100, and if the variance of $X$ is larger than the variance of $Y$, which is larger, $P(X \geq 105)$ or $P(Y \geq 105)$? _____

4. (True or False) If $X$ is a normal random variable with mean 10 and variance 16, then $P(X \leq 10) = P(X \geq 10)$. _____

5. (True or False) If $Z$ is a standard normal random variable and if $a \leq b$, then $P(a \leq Z \leq b) = P(Z \leq b) - P(Z \leq a)$. _____

6. The value $x = 50$ is two standard deviations to the right of the mean of a normally distributed random variable $X$. What is $P(X \leq 50)$? _____

## SECTION 10.4
### EXERCISES

In Exercises 1–10, assume that $Z$ is a standard normal variable. Use the table on the back endsheets to find the indicated probabilities.

1. $P(Z < 0.5)$
2. $P(Z < -1.42)$
3. $P(Z \leq -0.32)$
4. $P(Z \geq 3.10)$
5. $P(Z > -1.23)$
6. $P(Z > 0.79)$
7. $P(-1.2 \leq Z \leq 1.5)$
8. $P(-2.5 \leq Z \leq 0.41)$
9. $P(1.32 < Z \leq 2.45)$
10. $P(0.68 \leq Z < 1.64)$

In Exercises 11–18, assume that $X$ is a normal random variable with mean 40 and variance 81. Find the indicated probabilities.

11. $P(X < 40)$
12. $P(X \leq 49)$
13. $P(30 \leq X \leq 50)$
14. $P(X > 30)$
15. $P(31 < X < 49)$
16. $P(22 \leq X \leq 58)$
17. $P(X \geq 60)$
18. $P(25 \leq X < 42)$

In Exercises 19–26, assume that $X$ is a normal random variable with mean 10 and standard deviation 5. Find the indicated probabilities.

19. $P(X \geq 15)$
20. $P(X \leq 5)$
21. $P(X < 17.4)$
22. $P(X > 7.62)$

**SOLUTIONS TO EXAMPLE 4**

(a) $\frac{645 - 500}{100}$, $P(Z \leq -0.25)$, 0.4013, 0.5252; (b) 0.90, 0.8997, 628

23. $P(X > -2.3)$
24. $P(X < -3.71)$
25. $P(-0.7 \leq X < 10)$
26. $P(-5 \leq X \leq 20)$

27. **(Product lifetime)** A certain type of automobile battery has a lifetime that is normally distributed with a mean of four years and a standard deviation of 0.5 year. Find the probability that the lifetime of a battery will be **(a)** less than one year; **(b)** between three and four years; **(c)** more than four years.

28. **(Quality control)** A production process produces a machine part with an average length of 6.5 in. and a standard deviation of 0.05 in. Any part that differs in length from the mean by more than 0.1 in. will be rejected by a quality control inspector. Assuming that the lengths are normally distributed, what percentage of these parts will be rejected?

### Business and Economics

29. **(Profit)** The monthly profit (in thousands of dollars) of your company is normally distributed with a mean of 100 and a standard deviation of 15. Find the probability that next month's profit will be **(a)** less than $80,000; **(b)** between $90,000 and $110,000; **(c)** greater than $125,000.

### Social Sciences

30. **(Standardized testing)** The SAT mathematics scores of freshmen entering a large university are normally distributed with a mean of 550 and a standard deviation of 70. Find the probability that the mathematics SAT score of a randomly chosen freshman will be **(a)** less than 450; **(b)** between 500 and 600; **(c)** greater than 650.

31. **(Grade point averages)** The grade point averages of undergraduate students at a large university are normally distributed with a mean of 2.4 and a standard deviation of 0.9. If 20,000 undergraduates attend the university, approximately how many of these students will have a grade point average between 2.0 and 3.0?

32. **(Traffic tickets)** The speed limit on a certain highway is 55 mph. If the speed of cars on the highway is normally distributed with a mean of 60 mph and a standard deviation of 3 mph, and if the police decide to ticket the fastest 10% of the cars, what is the fastest speed a car can travel without getting a ticket?

33. **(Psychology experiment)** The time required for a subject to complete a certain sequence of tasks in a psychology experiment is normally distributed, with a mean of eight minutes and a standard deviation of 45 seconds. Find the probability that an individual will perform the sequence of tasks **(a)** in less than six minutes; **(b)** in less than seven minutes.

34. Referring to Exercise 33, find the time below which 95% of randomly chosen individuals will complete the sequence of tasks.

35. **(Employee heights)** The heights of a group of 10,000 employees of a certain company are normally distributed, with a mean of 5.6 feet and a standard deviation of 0.2 feet. How many of these employees would you expect to have heights of **(a)** less than 5.4 feet; **(b)** between 5.4 and 5.9 feet; **(c)** greater than 6.0 feet?

36. Verify that the point $(\mu, 1/(\sigma\sqrt{2\pi}))$ is a relative maximum point of the graph of the normal density function.

37. Verify that the graph of the normal density function is concave up on the intervals $(-\infty, \mu - \sigma)$ and $(\mu + \sigma, \infty)$ and is concave down on the interval $(\mu - \sigma, \mu + \sigma)$.

**OPTIONAL**

**GRAPHING CALCULATOR/ COMPUTER EXERCISES**

38. On the same coordinate axes, sketch the graphs of the density function of three normally distributed random variables, each with mean 50 and with standard deviations 5, 10, and 15.

39. On the same coordinate axes, sketch the graphs of the density function of three normally distributed random variables, each with mean 3 and with standard deviations 1, 2, and 3.

## 10.5 CHAPTER TEN REVIEW

**KEY TERMS**

random experiment
sample space
sample point
simple event
event
random variable
probability distribution
probability histogram
discrete random variable
expected value or mean
fair game

variance of a random variable
standard deviation of a random variable
continuous random variable
probability density function
uniform distribution
probability distribution function
median of a random variable
Chebyshev's theorem
normal distribution
standard normal random variable

**KEY FORMULAS AND RESULTS**

**Probability of an Event: Equally Likely Case**

If each simple event $e_i$ is equally likely, where $S = \{e_1, e_2, \ldots, e_k\}$ is the sample space and $E$ is any event, then

$$P(E) = \frac{n(E)}{n(S)}$$

**Probability of an Event: General Case**

If $E = \{f_1, f_2, \ldots, f_k\}$, where each $f_i$ is a simple event, then

$$P(E) = P\{f_1\} + P\{f_2\} + \cdots + P\{f_k\}$$

**Properties of Probability**

For any sample space $S$ and event $E$,

(a) $P(S) = 1$;
(b) $0 \leq P(E) \leq 1$.

### The Expected Value, Variance, and Standard Deviation of a Discrete Random Variable

If $X$ is a discrete random variable having $P(X = x_i) = p_i$ for $i = 1, 2, \ldots, n$, then

(a) $E(X) = \mu = x_1 p_1 + x_2 p_2 + \cdots + x_n p_n$ is the expected value or mean of $X$;
(b) $V(X) = \sigma^2 = (x_1 - \mu)^2 p_1 + (x_2 - \mu)^2 p_2 + \cdots + (x_n - \mu)^2 p_n$ is the variance of $X$;
(c) $\sqrt{V(X)} = \sigma$ is the standard deviation of $X$.

### Properties of a Probability Density Function

Let $f(x)$ be a continuous probability density function defined on $[a, b]$. Then

(a) $f(x) \geq 0$ for all $x$ in $[a, b]$;
(b) $\int_a^b f(x)\, dx = 1$;
(c) If $a \leq c \leq d \leq b$, then

$$P(c \leq X \leq d) = \int_c^d f(x)\, dx = \text{the area under } f \text{ from } a \text{ to } b$$

### The Uniform Density Function

$$f(x) = \begin{cases} \dfrac{1}{b-a} & a \leq x \leq b \\ 0 & \text{otherwise} \end{cases}$$

### The Probability of a Continuous Random Variable at a Point

If $X$ is a continuous random variable, then

(a) $P(X = k) = 0$;
(b) $P(a \leq X \leq b) = P(a < X \leq b) = P(a \leq X < b) = P(a < X < b)$.

### The Probability Distribution Function

If $X$ is a continuous random variable with density function $f(x)$, then the probability distribution function of $X$ is

$$F(x) = P(X \leq x) = \int_{-\infty}^{x} f(t)\, dt$$

For the distribution function $F(x)$,

$$P(c \leq X \leq d) = F(d) - F(c)$$

### Expected Value and Variance of a Continuous Random Variable

If $X$ is a continuous random variable with density function $f(x)$ defined on $[a, b]$, then

$$E(X) = \mu = \int_a^b xf(x)\,dx$$

$$V(X) = \sigma^2 = \int_a^b (x - \mu)^2 f(x)\,dx$$

$$\sigma = \sqrt{V(X)}$$

### Chebyshev's Theorem

The probability that a random variable $X$ will take on a value that lies within $k$ standard deviations of the mean is at least $1 - (1/k^2)$.

### An Alternate Formula for the Variance

If $X$ is a continuous random variable with density function $f(x)$ defined on $[a, b]$, then

$$V(X) = \int_a^b x^2 f(x)\,dx - \mu^2 = E(X^2) - [E(X)]^2$$

### The Median of a Continuous Random Variable

The median $m$ of a continuous random variable must satisfy $P(X \le m) = 0.5$.

### The Normal Distribution

Let $X$ be a random variable that is normally distributed.

Density function: $f(x) = \dfrac{1}{\sigma\sqrt{2\pi}} e^{-(1/2)((x-\mu)/\sigma)^2}, \quad -\infty < x < \infty$

Mean: $E(X) = \mu$

Variance: $V(X) = \sigma^2$

Median: $m = \mu$

### Standard Normal Distribution

Let $Z$ be a random variable that has a standard normal distribution.

Density function: $f(z) = \dfrac{1}{\sqrt{2\pi}} e^{-(1/2)z^2}, \quad -\infty < z < \infty$

Mean: $E(X) = 0$

Variance: $V(X) = 1$

Median: $m = 0$

Distribution function: See the table on the back endsheets

### Relationship between Normal and Standard Normal Random Variables

$$P(X \le x) = P\left(Z \le \frac{x - \mu}{\sigma}\right)$$

## CHAPTER 10 SHORT ANSWER REVIEW QUESTIONS

1. (True or False) If a sample space consists of two sample points, there are four different events corresponding to this sample space. _____
2. (True or False) If the means of two discrete random variables $X$ and $Y$ are equal, then their variances must also be equal. _____
3. A coin is biased so that the probability of getting a head when the coin is flipped once is 0.6. The probability of a tail is _____.
4. If you play a fair game a large number of times, how much would you expect to win overall? _____
5. A random variable $X$ defined on [0, 10] has a uniform distribution. What is $P(0 \le X \le 1)$? (a) $\frac{1}{5}$; (b) $\frac{1}{10}$; (c) $\frac{3}{10}$; (d) not enough information given to determine.
6. The distribution function of a continuous random variable defined on [0, 1] is $F(x) = x^2$. The probability density function of $X$ is $f(x) =$ _____.
7. If the distribution function of a continuous random variable $X$ defined on [0, 2] is $F(x) = \frac{3}{8}x^3$, then $P(X \le 1) =$ _____.
8. (True or False) The random variable $X$ that represents the diameter of a ball bearing produced by a certain machine is a discrete random variable. _____
9. If $F(x)$ is the distribution function for a continuous random variable $X$ and if $F(2) = 0.5$, then the median of $X$ is _____.
10. (True or False) The function $f(x) = \frac{1}{3}x^3$ is a probability density function on [−1, 1]. _____
11. If $X$ is a continuous random variable with mean 10 and if $E(X^2) = 164$, then the standard deviation of $X$ is _____.
12. If $X$ is a continuous random variable defined on [0, 20] with mean 12 and standard deviation 2, then according to Chebyshev's theorem, $P(8 \le X \le 16) \ge$ _____.
13. If $X$ is normally distributed with mean 50 and variance 36, then $P(X \ge 56) = P(Z \ge$ _____).
14. (True or False) If $X$ and $Y$ are normally distributed random variables with density functions $f(x)$ and $g(x)$, respectively, then the area under the graph of $f(x)$ within two standard deviations of the mean of $X$ is equal to the area under the graph of $g(x)$ within two standard deviations of the mean of $Y$. _____

## CHAPTER 10 REVIEW EXERCISES

*The exercises marked with an asterisk (\*) constitute a Chapter Test.*

*In Exercises 1–7, the experiment consists of choosing two electrical components at random and without replacement from a box containing six components of which five are good and one is defective.*

*1. Find the sample space $S$ of this experiment.
2. Identify the following events as subsets of $S$: (a) at least one component is defective; (b) both components are good; (c) neither component is good.
*3. Find the probability of each simple event in $S$.
4. Find the probability of each event listed in Exercise 2.

**\*5.** Let $X$ be the number of good components selected. Find the probability distribution of $X$ and sketch the probability histogram.

**6.** If $X$ is the number of good components selected, find $E(X)$.

**\*7.** If $X$ is the number of good components selected, find $V(X)$.

**8.** (Gambling) A game consists of tossing a pair of dice once. A gambler bets $1 on each toss of the dice, and if he wins the game the $1 bet is returned to him. The gambler picks a number from 1 to 6, and if both dice show the selected number he wins $10; if one of the two dice show the selected number he wins $5; otherwise he loses. Find the expected winnings of the gambler. Is this game fair?

**\*9.** (Investment decision) A business can invest in either stock $A$ or stock $B$. If it invests in stock $A$, it can expect a 20% gain with probability 0.7 and a 10% loss with probability 0.3. If it invests in stock B, it can expect a 15% gain with probability 0.6 and a 6% loss with probability 0.4. If the investment decision is based on largest expected percentage gain, which stock should the company purchase?

**10.** (Insurance premium) The annual premium for insurance on a $200,000 diamond necklace is $400. If the probability of loss or theft of the necklace is 0.0015, what is the expected annual gain or loss to the owner who decides to insure the jewelery?

*In Exercises 11–14, verify that each function f is a probability density function over the given interval.*

**11.** $f(x) = 3/(2x^2)$ on $[1, 3]$

**12.** $f(x) = \frac{1}{50}x$ on $[0, 10]$

**13.** $f(x) = \frac{3}{16}\sqrt{x}$ on $[0, 4]$

**14.** $f(x) = \ln x$ on $[1, e]$

**\*15.** Use the density function of Exercise 11 to find (a) $P(X \leq 2)$; (b) $P(1.5 \leq X \leq 2.5)$.

**16.** Use the density function of Exercise 12 to find (a) $P(X > 6)$; (b) $P(0 \leq X < 5)$.

**17.** Use the density function of Exercise 13 to find (a) $P(X \geq 3)$; (b) $P(X < 2)$.

**18.** Use the density function of Exercise 14 to find (a) $P(X > 2)$; (b) $P(1.5 < X < 2.1)$.

**\*19.** (a) Find the distribution function for the density function of Exercise 11.
  (b) Use the distribution function found in part (a) to determine (i) $P(X > 2)$; (ii) the median of $X$.
  (c) Find $E(X)$ and $V(X)$.

**20.** (a) Find the distribution function for the density function of Exercise 12.
  (b) Use the distribution function found in part (a) to determine (i) $P(1 < X < 6)$; (ii) the median of $X$.
  (c) Find $E(X)$ and $V(X)$.

**21.** (a) Find the distribution function for the density function of Exercise 13.
  (b) Use the distribution function found in part (a) to determine (i) $P(1 < X \leq 3)$; (ii) the median of $X$.
  (c) Find $E(X)$ and $V(X)$.

**22.** (a) Find the distribution function for the density function of Exercise 14.
  (b) Use the distribution function found in part (a) to determine (i) $P(X \geq 2)$; (ii) $P(X < 1.5)$.
  (c) Find $E(X)$ and $V(X)$.

**23.** (Traffic control) A traffic light is red for a period of 45 seconds. If you arrive at the light when it is red, what is the probability that you will have to wait less than 10 seconds before the light changes to green?

Life Sciences

**24. (Demand)** The weekly demand (in thousands of units) for a company's product can be considered to be a continuous random variable $X$ with density function

$$f(x) = \begin{cases} 1/(2\sqrt{x}) & 4 \leq x \leq 9 \\ 0 & \text{otherwise} \end{cases}$$

Find the probability that the demand during a randomly selected week will be **(a)** less than 6000 units; **(b)** between 4000 and 5000 units; **(c)** greater than 8000 units.

**\*25. (Small business success)** The proportion of newly opened restaurants that will still be open after two years is a continuous random variable $X$ with density function

$$f(x) = \begin{cases} 20x^3(1-x) & 0 \leq x \leq 1 \\ 0 & \text{otherwise} \end{cases}$$

Find the probability that at least 50% of all new restaurants will still be open after two years.

**26. (Sales)** The amount of monthly sales (in thousands of dollars) of a certain product in a company's product line is a continuous random variable $X$ with a mean of 75 and a variance of 64. Find an interval that contains the amount of monthly sales **(a)** at least 75% of the time; **(b)** at least 95% of the time.

*In Exercises 27–30, assume that $X$ is a normally distributed random variable with mean 100 and variance 81. Find the indicated probabilities.*

**\*27.** $P(X > 105)$      **28.** $P(X \leq 95)$

**\*29.** $P(94.5 < X \leq 102.4)$      **30.** $P(91.3 \leq X < 107.9)$

**\*31. (Pollution in a lake)** The amount of a certain pollutant (in parts per million) found in a large lake is normally distributed with a mean of 45 and a standard deviation of 12. If the level of the pollutant in the lake exceeds 65 parts per million, the lake is considered unsuitable for swimming. On a certain day you go to the lake to swim. What is the probability that the water will be unsuitable?

**32. (Quality control)** The number of fluid ounces of fruit punch dispensed into a 64-fluid-ounce container is normally distributed with a mean of 64.4 fluid ounces and a standard deviation of 0.3 fluid ounce.
**(a)** Find the probability that a container has less than 64 fluid ounces of fruit punch.
**(b)** Find the probability that a container has more than 65 fluid ounces of fruit punch.

**\*33. (Quality control)** A machine punches a circular hole in a piece of metal to produce a component part for a certain type of lawn furniture. The diameter of the holes is normally distributed with a mean of 1 cm and a standard deviation of 0.1 cm. If the diameter of the hole varies from the mean by more than 0.2 cm, the part is considered unacceptable. In a production run of 10,000 parts, about how many will be unacceptable?

# ANSWERS TO ODD-NUMBERED EXERCISES

## 1.1 EXERCISES

**Short Answer**

1. $x; y$  3. range  5. 7; 950
7. false  9. 1

**Exercises**

1. (a) 0 (b) 300 (c) 500 (d) $-200$
 (e) $100n$ (f) $100x + 100h$
3. (a) 1 (b) 1 (c) 11 (d) 11
 (e) $n^2 - 3n + 1$
 (f) $x^2 + 2xh + h^2 - 3x - 3h + 1$
5. (a) 1 (b) 7 (c) 91 (d) 7
 (e) $n^3 - 7n + 1$
 (f) $x^3 + 3x^2h + 3xh^2 + h^3 - 7x - 7h + 1$
7. (a) 5 (b) 5 (c) 5 (d) 5 (e) 5 (f) 5
9. (a) $-\frac{3}{2}$ (b) $-6$ (c) $-\frac{22}{3}$ (d) $\frac{1}{4}$
 (e) $\dfrac{-n^2 + 3}{n - 2}$ (f) $\dfrac{-x^2 - 2xh - h^2 + 3}{x + h - 2}$
11. domain: $\{x \mid x \text{ is a real number}\}$; range: $\{y \mid y \text{ is a real number}\}$
13. domain: $\{x \mid x \text{ is a real number}\}$; range: $\{75\}$
15. $\{x \mid x \neq 5\}$  17. $\{x \mid x \geq \frac{5}{3}\}$
19. $\{x \mid x > 2\}$  21. no  23. yes
25. $1100; $600: this last cost is a fixed cost for rent, insurance, electricity, etc., which must be paid even when 0 ounces are produced.
27. (a) $C(6) = $60$; $C(13) = $123.50$; $C(30) = $270$
 (b) If you buy less than 12, the cost is $10 per part; if you buy at least 12 but less than 24, the cost is $9.50 per part; if you buy at least 24, the cost is $9 per part.
29. $C(499) = $988$; $C(500) = $975$: it costs more to order 499 parts than 500 parts.
31. (a) $2000; $2000; $4000; $8000; $14,000
 (b) $22,000
33. $P(x) = 20000 + 2x, \quad 0 \leq x \leq 50000$

## 1.2 EXERCISES

**Short Answer**

1. 2, 1, 2  3. false  5. domain
7. (a) and (d)

**Exercises**

1. domain: $\{x \mid x \text{ is a real number}\}$; range: $\{y \mid y \text{ is a real number}\}$

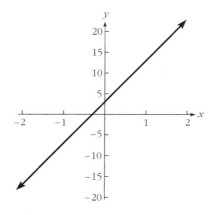

3. domain: $\{x \mid x \text{ is a real number}\}$; range: $\{y \mid y \geq 2\}$

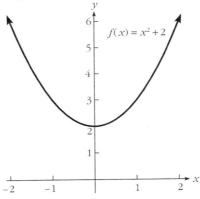

5. domain: $\{x \mid x \text{ is a real number}\}$; range: $\{y \mid y \geq -1\}$

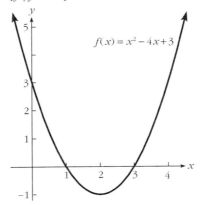

7. domain: $\{x \mid x \text{ is a real number}\}$; range: $\{y \mid y \text{ is a real number}\}$

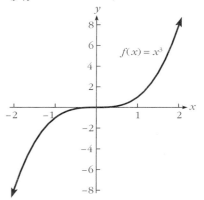

9. domain: $\{x \mid x \geq -3\}$; range: $\{y \mid y \geq 0\}$

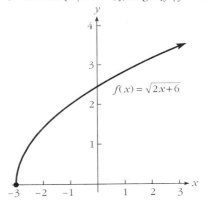

11. domain: $\left\{x \mid x \leq \frac{3}{5}\right\}$; range: $\{y \mid y \geq 0\}$

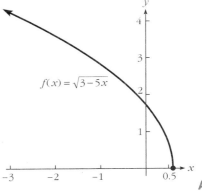

A1

**13.** domain: $\{x \mid x \neq 2\}$; range: $\{y \mid y \neq 0\}$

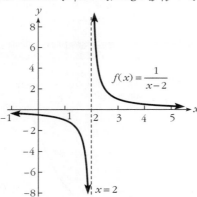

**15.** domain: $\{x \mid x \neq -2\}$; range: $\{y \mid y \neq 0\}$

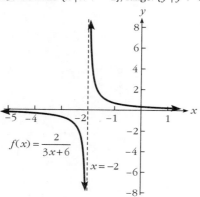

**17.** domain: $\{x \mid x \neq 0\}$; range: $\{y \mid y \leq -2\sqrt{2} \text{ or } y \geq 2\sqrt{2}\}$

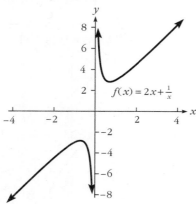

**19.**

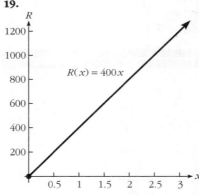

**21.**

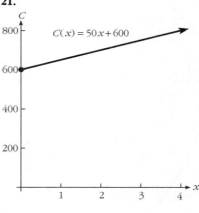

**23.**

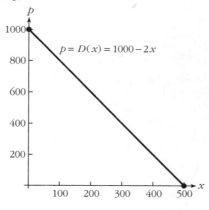

**25. (a)** $C(x) = \begin{cases} 10x & \text{if } 0 \leq x < 12 \\ 9.5x & \text{if } 12 \leq x < 24 \\ 9x & \text{if } x \geq 24 \end{cases}$

**(b)**

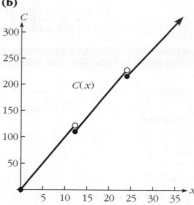

**27.**

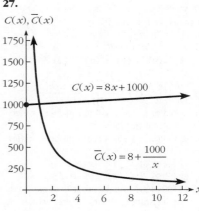

**29. (a)**

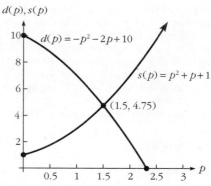

**(b)** (1.5, 4.75)

**31. (a)**

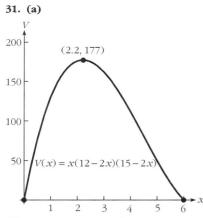

**(b)** Length of square to be cut is about 2.2 in.; approximate maximum volume is 177 in³.

**33.** domain: $\{x \mid x \text{ is a real number}\}$; range: $\{y \mid y \geq -46.86\}$

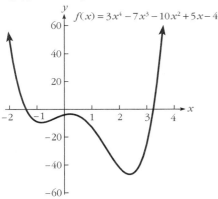

**35.** domain: $\{x \mid x \neq 3\}$; range: $\{y \mid y \leq -0.68 \text{ or } y > 1.73\}$

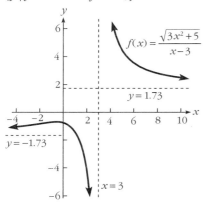

**37.** domain: $\{x \mid x \text{ is a real number}\}$; range: $\{y \mid -0.73 \leq y < 0.50\}$

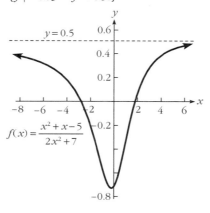

### 1.3 EXERCISES

**Short Answer**

**1.** $x^2 + 2x$    **3.** $x^4 - x^3$    **5.** 49
**7.** false    **9.** 144; 48
**11.** In general they are different.

**Exercises**

**1.** $f(x) + g(x) = 5x - 4$; $f(x) - g(x) = x - 6$; $f(x) \cdot g(x) = 6x^2 - 7x - 5$; $f(x)/g(x) = (3x - 5)/(2x + 1)$
**3.** $f(x) + g(x) = 0$; $f(x) - g(x) = 2\sqrt{x^2 + 1}$; $f(x) \cdot g(x) = -x^2 - 1$; $f(x)/g(x) = -1$
**5.** $f(x) + g(x) = (1 + 2x^4)/x$; $f(x) - g(x) = (1 - 2x^4)/x$; $f(x) \cdot g(x) = 2x^2$; $f(x)/g(x) = 1/(2x^4)$
**7.** $f(3) + g(3) = 11$; $f(3) - g(3) = -3$; $f(3) \cdot g(3) = 28$; $f(3)/g(3) = \frac{4}{7}$
**9.** $f(3) + g(3) = 0$; $f(3) - g(3) = 2\sqrt{10}$; $f(3) \cdot g(3) = -10$; $f(3)/g(3) = -1$
**11.** $f(3) + g(3) = \frac{163}{3}$; $f(3) - g(3) = -\frac{161}{3}$; $f(3) \cdot g(3) = 18$; $f(3)/g(3) = \frac{1}{162}$
**13.** $(f \circ g)(x) = 6x + 10$; domain: $\{x \mid x \text{ is a real number}\}$; $(g \circ f)(x) = 6x + 5$
**15.** $(f \circ g)(x) = \sqrt{2x - 5x^2}/x$; domain: $\{x \mid 0 < x \leq \frac{2}{5}\}$; $(g \circ f)(x) = \sqrt{2x - 5}/(2x - 5)$
**17.** $(f \circ g)(x) = 4x^2/(x^2 - 1)$; domain: $\{x \mid x \neq 1, -1\}$; $(g \circ f)(x) = 16x^2/(x - 1)^2$
**19.** $(f \circ g)(5) = 40$; $(f \circ g)(5) = 35$; $f^2(5) = 100$
**21.** $(f \circ g)(5)$ is undefined; $(g \circ f)(5) = \frac{1}{5}\sqrt{5}$; $f^2(5) = 5$
**23.** $(f \circ g)(0) = 0$; $(g \circ f)(0) = 0$; $f^2(0) = 0$
**25.** $(f \circ g)(\frac{1}{3}) = \sqrt{5}$    **27.** $f(g(2)) = \frac{16}{3}$

**29. (a)** $P(x) = 50x - 300$   **(b)** $1200
**31. (a)** $P(p) = -9p^2 + 3600p - 300000$
**(b)** $37,500   **(c)** 150 units
**33. (a)** $O = 10\sqrt{5x}$   **(b)** 50
**35. (a)** $S = 0.002V - 200000$
**(b)** $7,800,000
**37.** The profit maximizing price is about $200, maximum profit is about $60,000, and weekly demand is about 300 units.
**39.** $x = 3868$, $T = $3873$ (approximately)

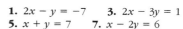

### 1.4 EXERCISES

**Short Answer**

**1.** 20; −600    **3.** −2    **5.** $-4x + 5$
**7.** $20x - 1000$    **9.** 10; 4    **11.** $500
**13.** $\frac{2}{3}$

**Exercises**

**1.** $2x - y = -7$    **3.** $2x - 3y = 1$
**5.** $x + y = 7$    **7.** $x - 2y = 6$
**9.** linear    **11.** linear    **13.** linear
**15.** nonlinear
**17.** slope: 3; y-intercept: 2

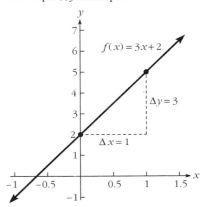

**19.** slope: $-2$; $y$-intercept: 1

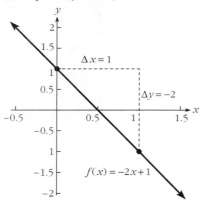

**21.** slope: $\frac{2}{3}$; $y$-intercept: $-1$

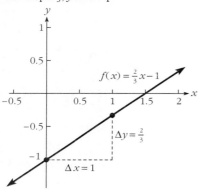

**23.** slope: 100; $y$-intercept: $-400$

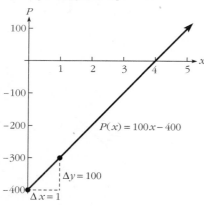

**25.** slope: 50; $y$-intercept: 200

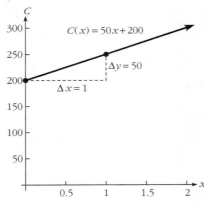

**27.** slope: $-1000$; $y$-intercept: 5000

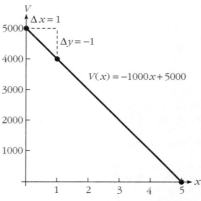

**29.** perpendicular   **31.** perpendicular
**33.** parallel   **35.** perpendicular
**37.** $3x - 5y = 19$   **39.** $3x + 7y = 2$
**41.** (a) $500 (b) 8 units is the break-even point (c) $3750
**43.** (a) $q(p) = -150p + 850$ (b) 400 units
**45.** (a) $V(t) = 75000 - 5000t$, $0 \le t \le 15$ (b) $40,000
(c)

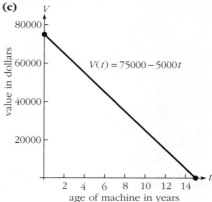

**47.** $(x, p) = (100, 500)$
**49.** (a) $F = 1.8C + 32$ (b) $212°$ Fahrenheit
**51.** (a) $g(p) = \frac{8}{15}p + \frac{106}{3}$ (b) $75\frac{1}{3}$ (c) 65
**53.** Exact point of intersection is $\left(\frac{260}{79}, -\frac{22}{79}\right)$.

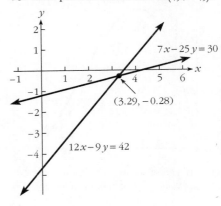

**55.** Exact point of intersection is $\left(\frac{99943}{72225}, -\frac{3272}{72225}\right)$.

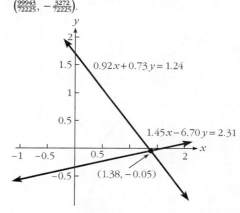

## 1.5 EXERCISES

### Short Answer

**1.** (b)   **3.** $-100$   **5.** $0.25p^2 + 500p$
**7.** 1   **9.** low   **11.** $x$   **13.** false
**15.** true   **17.** false

### Exercises

**1.** $x$-intercepts: 2 and $-1$; $y$-intercept: $-2$; vertex: $\left(\frac{1}{2}, -\frac{9}{4}\right)$; axis: $x = \frac{1}{2}$

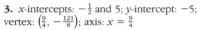

**3.** $x$-intercepts: $-\frac{1}{2}$ and 5; $y$-intercept: $-5$; vertex: $\left(\frac{9}{4}, -\frac{121}{8}\right)$; axis: $x = \frac{9}{4}$

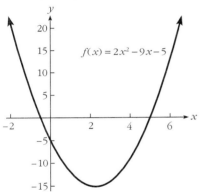

**5.** $x$-intercepts: $(7 \pm \sqrt{37})/2$; $y$-intercept: 3; vertex: $\left(\frac{7}{2}, -\frac{37}{4}\right)$; axis: $x = \frac{7}{2}$

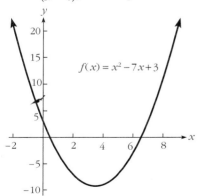

**7.** $x$-intercepts: none; $y$-intercept: 5; vertex: $\left(-\frac{3}{4}, \frac{31}{8}\right)$; axis: $x = -\frac{3}{4}$

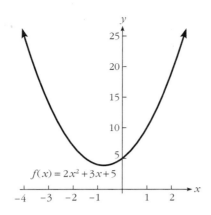

**9.** $x$-intercepts: $-2$ and 5; $y$-intercept: 10; vertex: $\left(\frac{3}{2}, \frac{49}{4}\right)$; axis: $x = \frac{3}{2}$

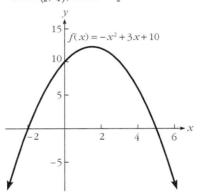

**11.** $x$-intercepts: $(-1 \pm \sqrt{73})/6$; $y$-intercept: 6; vertex: $\left(-\frac{1}{6}, \frac{73}{12}\right)$; axis: $x = -\frac{1}{6}$

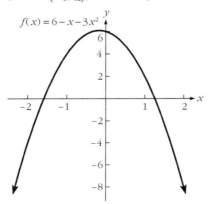

**13. (a)**

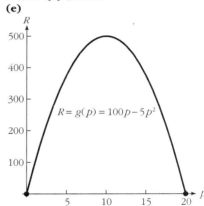

**(b)** $R = g(p) = 100p - 5p^2$ **(c)** Price to maximize revenue is $10; maximum revenue is $500. **(d)** 50 units
**(e)**

**15. (a)** Data is linear: equal increases in price of $25 are accompanied by equal decreases in demand of 400 units.
**(b)** $x = -16p + 2800$, $50 \leq p \leq 100$

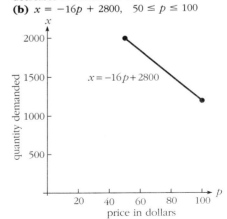

**(c)** $R = -16p^2 + 2800p$, $50 \le p \le 1000$
**(d)** $P = f(p) = -16p^2 + 3120p - 56800$, $50 \le p \le 100$ **(e)** optimal price is $97.50; maximum profit is $95,300; quantity demanded is 1240 units **(f)**

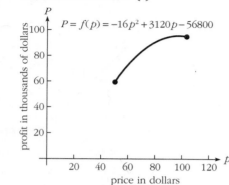

**17.** Although a formal analysis yields a price of $181.25 to optimize profit, this price would require us to produce 21 ounces per week, which is impossible. The profit-optimizing price is, therefore, $187.50 per ounce, since at this price we can sell our production capacity of 20 ounces per week and make a maximum profit of $2150 per week.
**19.** A production level of 2000 units per week produces a minimum cost of $4000.

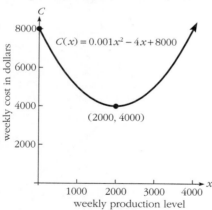

**21.** To obtain a maximum profit of $17,120, the company should spend $12,000 on advertising.
**23.** **(a)** $C(x) = 30x + 2000$
**(c)** $R(x) = 65x - 0.013x^2$
**(d)** $P(x) = -0.013x^2 + 35x - 2000$
**(e)** 1000 saws, which yield a profit of $20,000 **(f)** 1346 saws, which yield a profit of $21,557.69
**25.** 9 in. by 9 in.
**27.**

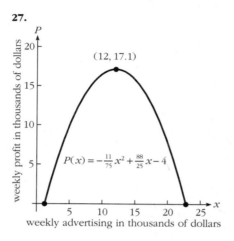

**29.**

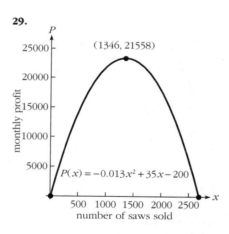

### 1.6 EXERCISES

**Short Answer**

**1.** $x; R$    **3.** $1625
**5.** $C(x) = 25x + 1200$    **7.** **(a)** 8 **(b)** 10
**(c)** $\frac{2}{3}$ **(d)** 11    **9.** $-\frac{1}{3}$    **11.** (a)
**13.** 20    **15.** (c)

**Exercises**

**1. (a)** $-5$ **(b)** $-20$ **(c)** $3a - 5$
**3. (a)** 0 **(b)** $\sqrt{11}$ **(c)** $\sqrt{1 - 2x - 2h}$

**5.** domain: $\{x \mid x \text{ is a real number}\}$; range: $\{y \mid y \le 3\}$
**7.** $\{x \mid 0 \le x < 3 \text{ or } x > 3\}$
**9.** $C(x) = 30x + 3000$

**11.**

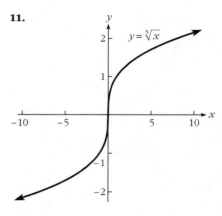

**13.**

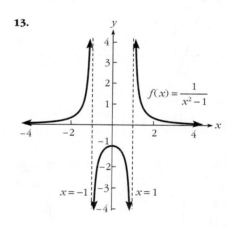

**15.**

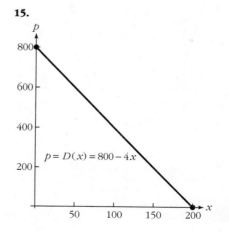

**17. (a)** $P(x) = 20x - 1000$ **(b)** $P(x) = \$1000$ **(c)** 50 units, since $P(50) = 0$
**19. (a)** 38 **(b)** $-4$ **(c)** 9 **(d)** $\sqrt{3}/25$
**21.** $(f \circ g)(x) = x - 3$; domain: $\{x \mid x \geq 5\}$; $(f \circ g)(6) = 3$
**23. (a)** $P(p) = -25p^2 + 10500p - 920000$
**(b)** $P(100) = -\$120,000$ (a loss!)
**(c)** Maximum profit of $182,500 is obtained if we charge $210 per unit, and 450 units will be demanded at this price.
**25. (a)** $y = 500$ **(b)** $x = 200$
**27.** $y = -3x$ **29.** $x + 5y = 2005$

**31.**

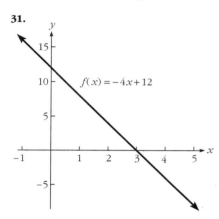

**33.** $y = 30x + 700$
**35. (a)** $x = -100p + 15000$
**(b)** 6500 units
**37.** $x$-intercepts: $\frac{1}{3}$ and 2; $y$-intercept: 2; vertex: $\left(\frac{7}{6}, -\frac{25}{12}\right)$

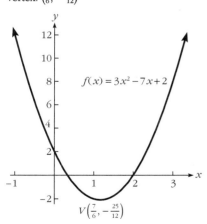

**39. (a)** $R(p) = 1000p - 20p^2$ **(b)** The price that maximizes revenue is $25 per unit; maximum revenue is $R(25) = \$12,500$.
**(c)** 500 units

**41.**

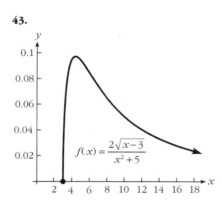

**43.**

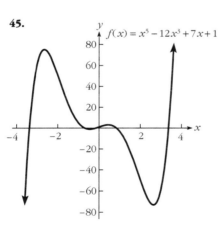

**45.**

## 2.1 EXERCISES

**Short Answer**

**1.** 5; 37 **3.** 10 **5.** (d) **7.** false
**9.** 7 **11.** 1

**Exercises**

**1.** 20 **3.** 1 **5.** 6 **7.** $-7$
**9.** does not exist **11.** $-4$ **13.** 0
**15.** 0 **17.** $-\frac{3}{2}$ **19.** $\frac{21}{8}$
**21.** $\lim_{x \to 2^-} f(x) = 8$; $\lim_{x \to 2^+} f(x) = 10$; $\lim_{x \to 2} f(x)$ does not exist

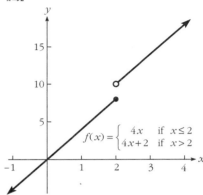

**23.** $\lim_{x \to 0^-} f(x) = 5$; $\lim_{x \to 0^+} f(x) = 5$; $\lim_{x \to 0} f(x) = 5$

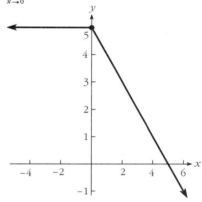

**25.** $\lim_{x \to 0^-} f(x) = 0$; $\lim_{x \to 0^+} f(x) = 0$; $\lim_{x \to 0} f(x) = 0$

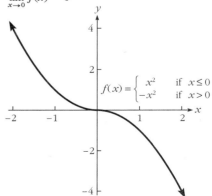

**27.** 7  **29.** 0  **31.** 19  **33.** −15
**35.** $\frac{16}{9}$  **37.** $\frac{54}{23}$  **39.** −14  **41.** $\sqrt[3]{30}$
**43.** 3  **45.** 1  **47.** $-\frac{1}{6}$  **49.** $\frac{3}{4}$
**51.** −1  **53.** $\frac{1}{2}$  **55.** 0  **57.** 35
**59.** $-\frac{22}{7}$  **61.** $\sqrt{7}$

## 2.2 EXERCISES

### Short Answer

**1.** (a)  **3.** true  **5.** (c)  **7.** −300
**9.** true  **11.** false  **13.** 0

### Exercises

**1.** 0  **3.** 3  **5.** $\frac{5}{2}$  **7.** 0  **9.** 2
**11.** 0  **13.** ∞  **15.** (b)  **17.** (a)
**19.** (c)  **21.** (a)  **23.** (b)  **25.** (b)
**27.** (b)
**29.** horizontal asymptote: $y = 0$; vertical asymptote: $x = 0$

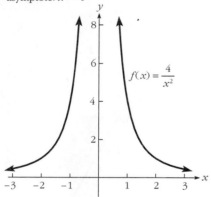

**31.** horizontal asymptote: $y = -\frac{7}{2}$; vertical asymptote: $x = 2$

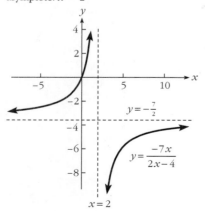

**33.** horizontal asymptote: $y = 0$; vertical asymptote: $x = 6$

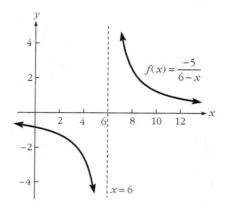

**35.** horizontal asymptote: $y = 0$; vertical asymptotes: $x = -6, x = 6$
**37.** horizontal asymptote: none; vertical asymptotes: $x = -5, x = 3$
**39.** horizontal asymptote: $y = -2$; vertical asymptote: $x = -1$
**41.** $A(x) = 50000\left(1 + \frac{0.10}{x}\right)^x$

| x | A(x) |
|---|---|
| 1 | 55000 |
| 2 | 55125 |
| 4 | 55190.64 |
| 12 | 55235.65 |
| 365 | 55257.79 |
| 1000 | 55258.27 |
| 100000 | 55258.54 |
| 1000000 | 55258.55 |
| 10000000 | 55258.55 |

$\lim_{x \to \infty} A(x) \approx 55258.55$

**43.** (a) $\overline{C}(x) = 50 + (600/x)$
(b) $\overline{C}(0.01) = \$60050$ is the average cost of producing 0.01 units, $\overline{C}(1) = \$650$ is the average cost of producing one unit, $\overline{C}(10) = \$110$ is the average cost of producing 10 units, and $\overline{C}(10000) = \$50.06$ is the average cost of producing 10000 units
(c) horizontal asymptote: $y = 50$; vertical asymptote: $x = 0$  (d) See graph.
(e) The average cost of producing very low levels of output (for example, 0.01 units) is very large, while the average cost of producing high levels of output (for example, 10000 units) is very close to $50.

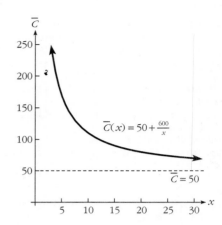

**45.** $\lim_{t \to \infty} N(t) = 500$ parts per day is the number of parts an average new employee will be able to assemble after many hours of training.
**47.** $\lim_{t \to \infty} N(t) = 25$; so after many days, about 25,000 people will have heard the rumor.
**49.** horizontal asymptote: $y = 2$; vertical asymptotes: $x = -7, x = 3$;
$\lim_{x \to -7^-} f(x) = \infty$; $\lim_{x \to -7^+} f(x) = -\infty$;
$\lim_{x \to 3^-} f(x) = -\infty$; $\lim_{x \to 3^+} f(x) = \infty$

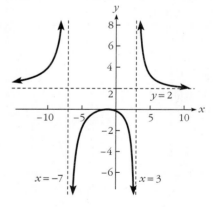

**51.** horizontal asymptote: none; vertical asymptotes: $x = -1, x = 1$;
$\lim_{x \to -1^-} f(x) = -\infty$; $\lim_{x \to -1^+} f(x) = \infty$;
$\lim_{x \to 1^-} f(x) = -\infty$; $\lim_{x \to 1^+} f(x) = \infty$

ANSWERS TO ODD-NUMBERED EXERCISES  A9

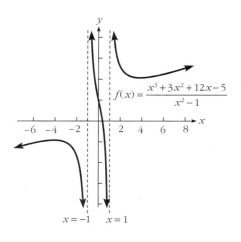
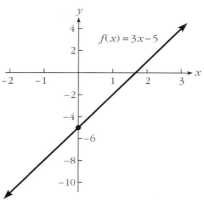
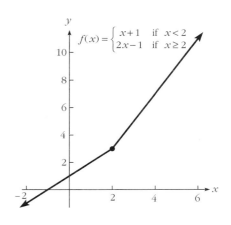

**3.** $x = -1$: yes; $x = 0$: no; $x = 1$: yes

**53.** horizontal asymptote: $y = -1$; vertical asymptotes: $x = -\frac{3}{2} - \frac{1}{2}\sqrt{5} = r_1$; $x = -\frac{3}{2} + \frac{1}{2}\sqrt{5} = r_2$; $\lim_{x \to r_1^-} g(x) = \infty$; $\lim_{x \to r_1^+} g(x) = -\infty$; $\lim_{x \to r_2^-} g(x) = -\infty$; $\lim_{x \to r_2^+} g(x) = \infty$

**9.** continuous at $x = 0$

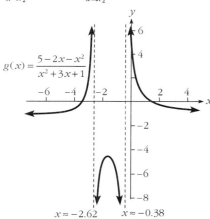
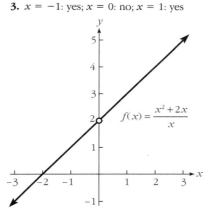
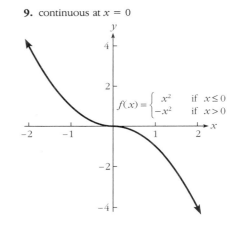

**5.** $x = -2$: yes; $x = 0$: yes; $x = 2$: no

**11.** continuous at $x = 2$

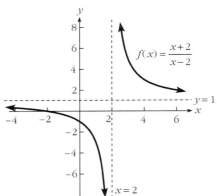
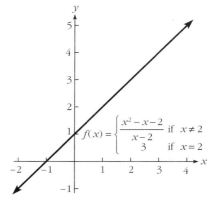

## 2.3 EXERCISES

**Short Answer**

**1.** no   **3.** −3   **5.** (b)
**7.** 250; yes   **9.** (d)

**Exercises**

**1.** $x = -2$: yes; $x = 0$: yes; $x = 2$: yes

**7.** continuous at $x = 2$

**13.** not continuous at $x = 7$

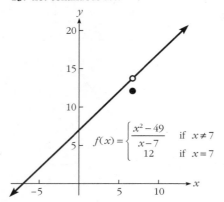

$$f(x) = \begin{cases} \dfrac{x^2 - 49}{x - 7} & \text{if } x \ne 7 \\ 12 & \text{if } x = 7 \end{cases}$$

**15.** $x = 0$: no; $x = 3$: no; $x = 4$: yes

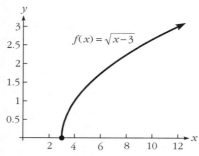

**17.** $x = -2$: yes; $x = -1$: no; $x = 0$: no
**19.** $x = -1$: yes; $x = 0$: yes; $x = 1$: yes
**21.** $(-\infty, \infty)$   **23.** $(-\infty, 0)$ and $(0, \infty)$
**25.** $(-\infty, 2)$ and $(2, \infty)$   **27.** $(-\infty, \infty)$
**29.** $(-\infty, \infty)$   **31.** $(-\infty, \infty)$
**33.** $(-\infty, 7)$ and $(7, \infty)$   **35.** $[3, \infty)$
**37.** $(-\infty, -1), (-1, 0),$ and $(0, \infty)$
**39.** $(-\infty, \infty)$

**41. (a)** $C(x) = \begin{cases} 5 & \text{if } 0 < x < 20 \\ 15 & \text{if } 20 \le x < 50 \\ 25 & \text{if } 50 \le x < 100 \end{cases}$

**(b)**

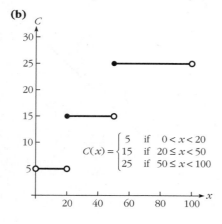

$$C(x) = \begin{cases} 5 & \text{if } 0 < x < 20 \\ 15 & \text{if } 20 \le x < 50 \\ 25 & \text{if } 50 \le x < 100 \end{cases}$$

**(c)** $C(x)$ is continuous at every real number $x$, $0 < x < 100$, except $x = 20$ and $x = 50$.
**(d)** $C(x)$ is continuous on the intervals $(0, 20), [20, 50),$ and $[50, 100)$.

**43. (a)** $C(x) = \begin{cases} 10x & \text{if } 0 \le x \le 100 \\ 9.5x + 50 & \text{if } x > 100 \end{cases}$

**(b)**

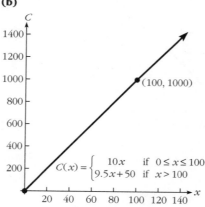

$$C(x) = \begin{cases} 10x & \text{if } 0 \le x \le 100 \\ 9.5x + 50 & \text{if } x > 100 \end{cases}$$

**(c)** $C(x)$ is continuous at every real number $x > 0$.
**(d)** $C(x)$ is continuous on $[0, \infty)$.

**45. (a)** $P(x) = \begin{cases} 5x & \text{if } 0 \le x \le 10000 \\ 4.5x + 5000 & \text{if } x > 10000 \end{cases}$

**(b)**

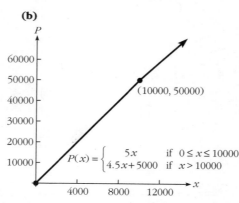

$$P(x) = \begin{cases} 5x & \text{if } 0 \le x \le 10000 \\ 4.5x + 5000 & \text{if } x > 10000 \end{cases}$$

**(c)** $P(x)$ is continuous at every real number $x > 0$ and on the interval $[0, \infty)$.

## 2.4 EXERCISES

### Short Answer

**1.** false   **3.** horizontal   **5.** 3   **7.** (c)
**9.** (b)

### Exercises

**1.** $m_{sec} = -2 + \Delta x$ if $\Delta x \ne 0$. If $\Delta x = 1$, then $m_{sec} = -1$; if $\Delta x = 0.1$, then $m_{sec} = -1.9$; if $\Delta x = 0.01$, then $m_{sec} = -1.99$. If $\Delta x = -1$, $m_{sec} = -3$; if $\Delta x = -0.1$, then $m_{sec} = -2.1$; if $\Delta x = -0.01$, then $m_{sec} = -2.01$.

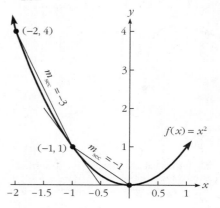

**3.** $m_{sec} = \Delta x$ if $\Delta x \ne 0$. If $\Delta x = 1$, then $m_{sec} = 1$; if $\Delta x = 0.1$, then $m_{sec} = 0.1$; if $\Delta x = 0.01$, then $m_{sec} = 0.01$. If $\Delta x = -1$, $m_{sec} = -1$; if $\Delta x = -0.1$, then $m_{sec} = -0.1$; if $\Delta x = -0.01$, then $m_{sec} = -0.01$.

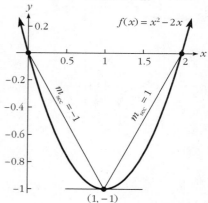

**5.** $m_{sec} = 3 + 3\Delta x + (\Delta x)^2$ if $\Delta x \ne 0$. If $\Delta x = 1$, then $m_{sec} = 7$; if $\Delta x = 0.1$, then $m_{sec} = 3.31$; if $\Delta x = 0.01$, then $m_{sec} = 3.0301$. If $\Delta x = -1$, $m_{sec} = 1$; if $\Delta x = -0.1$, then $m_{sec} = 2.71$; if $\Delta x = -0.01$, then $m_{sec} = 2.9701$.

# ANSWERS TO ODD-NUMBERED EXERCISES  A11

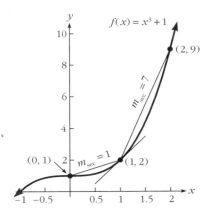

**11.** $m_{tan} = 4$

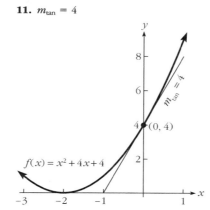

**(b)** The slope of tangent line is $-1$; see figure for sketch. The cost will decrease.
**(c)** The slope of tangent line is 3; see figure for sketch. The slope will increase. **(d)** The slope of tangent line is 0; see figure for sketch. The cost-minimizing production level is 1500 units. **(e)** The minimum average cost is 6.75 thousand dollars per thousand units or \$6.75 per unit.

**37. (a)**

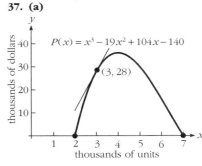

**7.** $m_{sec} = -1/(2(\Delta x + 2))$ if $\Delta x \neq 0$. If $\Delta x = 1$, then $m_{sec} = -\frac{1}{6}$; if $\Delta x = 0.1$, then $m_{sec} = -\frac{5}{21}$; if $\Delta x = 0.01$, then $m_{sec} = -\frac{50}{201}$. If $\Delta x = -1$, then $m_{sec} = -\frac{1}{2}$; if $\Delta x = -0.1$, then $m_{sec} = -\frac{5}{19}$; if $\Delta x = -0.01$, then $m_{sec} = -\frac{50}{199}$.

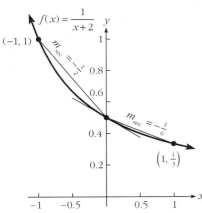

**13.** $m_{tan} = 4$
**15.** $m_{tan} = -\frac{1}{4}$

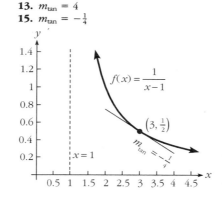

**17.** $m_{tan} = \frac{1}{4}$  **19.** $m_{tan} = 0$
**21.** $m_{tan} = 1$  **23.** $m_{tan}$ is undefined
**25.** $x + y = 3$  **27.** $x - 2y = -2$
**29.** $3x - y = -9$  **31.** does not exist
**33.** $2x + 2y = 1$
**35. (a)**

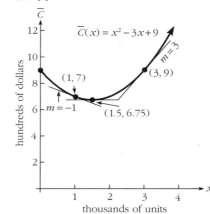

**(b)** The slope is 17; see figure for sketch; yes. **(c)** 4000 units **(d)** 0 **(e)** \$36,000

## 2.5 EXERCISES

**Short Answer**

**1.** true  **3.** $2 - x^2 - 2x\Delta x - (\Delta x)^2$
**5.** $6x$  **7.** true  **9.** (c)  **11.** (d)

**Exercises**

**1.** $f'(x) = 2x, f'(-2) = -4, f'(0) = 0, f'(2) = 4$

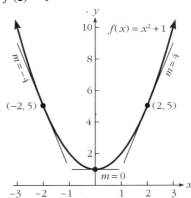

**9.** $m_{tan} = -2$

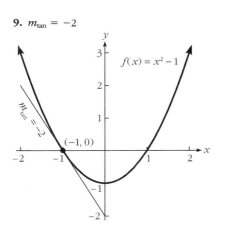

**A12** ANSWERS TO ODD-NUMBERED EXERCISES

**3.** $f'(x) = 4x + 1, f'(-1) = -3, f'(0) = 1,$
$f'(2) = 9$

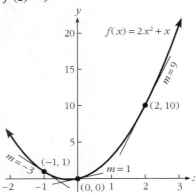

**5.** $f'(x) = 2x - 6, f'(0) = -6, f'(3) = 0,$
$f'(5) = 4$

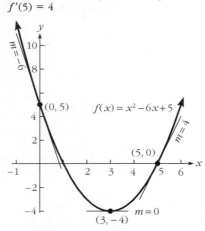

**7.** $f'(x) = 3x^2, f'(-1) = 3, f'(0) = 0,$
$f'(1) = 3$
**9.** $f'(x) = -3/x^2, f'(-1) = -3,$
$f'(3) = -\frac{1}{3}$
**11.** $f'(x) = 1/\sqrt{x}, f'(1) = 1, f'(4) = \frac{1}{2}$
**13.** $\dfrac{dy}{dx} = -1 - 6x$
**15.** $\dfrac{dy}{dx} = 3x^2 + 6x - 1$
**17.** $\dfrac{dy}{dx} = \dfrac{-3}{(x - 1)^2}$
**19.** $f(x)$ is not differentiable at $x = 2$.

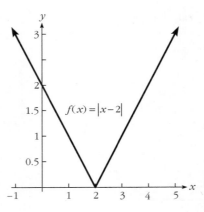

**21.** $f(x)$ is not differentiable at $x = 0$.

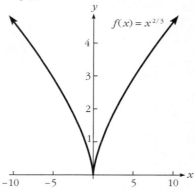

**23.** $f(x)$ is not differentiable at $x = 2$.

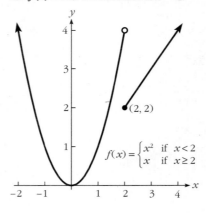

**25.** $f'(x) = -2 - 6x$
**27.** $f'(x) = -6/x^3$
**29.** $f'(x) = -3/(x + 4)^2$
**31. (a)** not continuous at $x = -1$ and $x = 2$ **(b)** not differentiable at $x = -1, x = 2,$
$x = 4,$ and $x = 5$

**33. (a)** 4844 students **(b)** $N'(t) = -25 - 18t$ **(c)** $N'(2) = -61, N'(3) = -79$

## 2.6 EXERCISES

### Short Answer

**1.** 3, average  **3.** true  **5.** (c)
**7.** (d)  **9.** true

### Exercises

**1.** 6

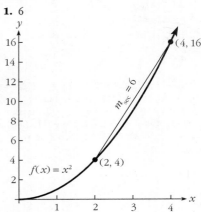

**3.** 3

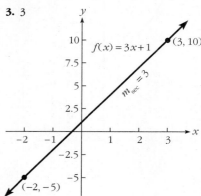

**5.** 0

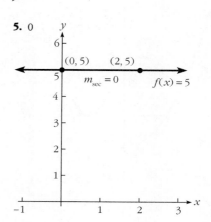

**7.** 4  **9.** −7
**11.** 2

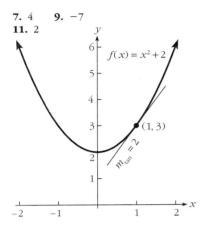

**13.** 9

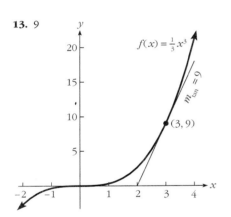

**15.** −3

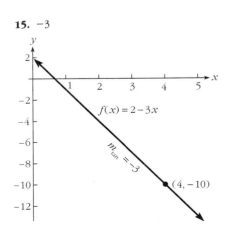

**17.** 1

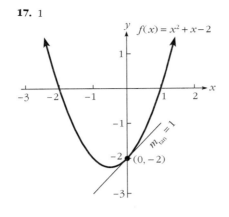

**19.** −6

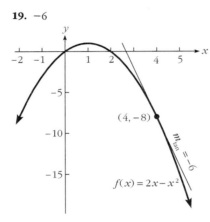

**21.** $-\frac{1}{4}$

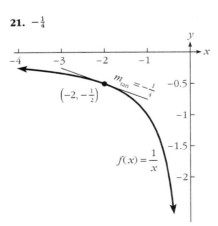

**23. (a)** 128 ft/sec  **(b)** 96 ft/sec
**(c)**

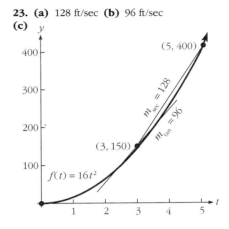

**25. (a)** $6/unit  **(b)** $4/unit
**(c)**

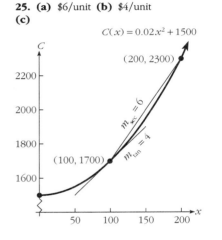

**27. (a)** 0.16%/month  **(b)** −0.1%/month
**29. (a)** 7 questions/hour of instruction
**(b)** 9 questions/hour of instruction
**(c)**

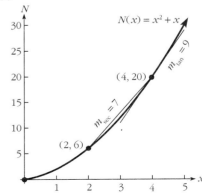

**31. (a)** −0.16 gal of ethanol per adult per year  **(b)** −0.2 gal of ethanol per adult per year

**33.**

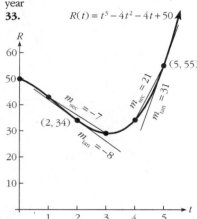

**43.** continuous at $x = -2$ and on the interval $(-\infty, \infty)$

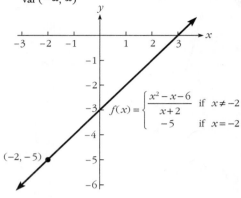

**49.** not continuous at $x = 0$; continuous on the intervals $(-\infty, 0)$, and $(0, \infty)$

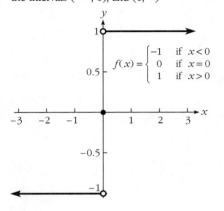

## 2.7 EXERCISES

### Short Answer

**1.** (c)   **3.** false   **5.** (c)   **7.** (c)
**9.** (a)   **11.** $\frac{1}{500}x - 2 + (1500/x)$
**13.** false   **15.** true   **17.** (b)
**19.** (d)   **21.** −2 units of $y$ per unit of $x$
**23.** (c)

### Exercises

**1.** 1   **3.** 3   **5.** −36   **7.** 1   **9.** 0
**11.** $\infty$   **13.** $\sqrt{21}$   **15.** $\frac{5}{4}$   **17.** $\frac{1}{12}$
**19.** 3   **21.** −1   **23.** 0   **25.** −$\infty$
**27.** −$\infty$   **29.** $\infty$   **31.** does not exist
**33.** $\infty$
**35.** horizontal asymptote: $y = 0$; vertical asymptotes: $x = -2, x = 2$
**37.** horizontal asymptote: $y = 1$; vertical asymptote: $x = -1$
**39.** horizontal asymptote: $y = 0$; vertical asymptote: $x = 6$
**41.** not continuous at $x = 1$; continuous on the intervals $(-\infty, 0)$, $(0, 1)$, and $(1, \infty)$

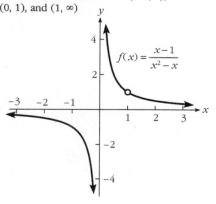

**45.** not continuous at $x = -1$; continuous on the intervals $(-\infty, -1)$, and $(-1, \infty)$

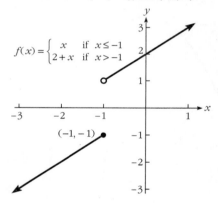

**47.** not continuous at $x = 2$; continuous on the interval $(-\infty, 2]$

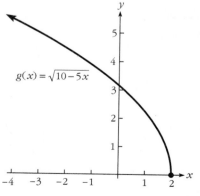

**51.** $m_{sec} = 4, m_{tan} = 2$
**53.** $m_{sec} = \frac{1}{3}, m_{tan} = \frac{1}{2}$
**55.** $dy/dx = 4$   **57.** $dy/dx = 6x + 1$
**59.** $dy/dx = 6x^2 - 2$
**61.** $dy/dx = 1/(2\sqrt{x+1})$
**63.** no   **65.** no
**67.** Average rate of change is 1, instantaneous rate of change is −3.
**69.** Average rate of change is 4.9, instantaneous rate of change is 5.
**71.** Average rate of change is −$\frac{400}{903}$, instantaneous rate of change is −$\frac{4}{9}$.
**73. (a)** horizontal asymptote: $A(Q) = 0$; vertical asymptote: $Q = 0$
**(b)**

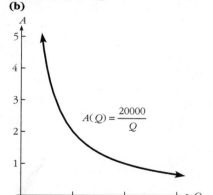

**75. (a)** $S(x) = \begin{cases} 10x & \text{if } x \leq 6 \\ 12.5x - 15 & \text{if } 6 < x \leq 24 \end{cases}$

**(b)**

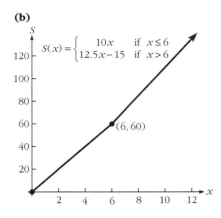

**(c)** $S(x)$ is continuous at $x = 6$.
**77. (a)** $\frac{25}{22}$ ft/sec  **(b)** $\frac{25}{121}$ ft/sec
**79.** The average rate between $t = 0$ and $t = 2$ is about 35.7 words per hour, and the average rate between $t = 8$ and $t = 10$ is about 14.4 words per hour.

## 3.1 EXERCISES

**Short Answer**

**1.** 0  **3.** $x^{-3}$  **5.** (b)  **7.** 35
**9.** (c)  **11.** $\dfrac{v(x)u'(x) - u(x)v'(x)}{[v(x)]^2}$

**Exercises**

**1.** 0  **3.** $6x^5$  **5.** $-4/x^5$  **7.** $9t^2$
**9.** $t$  **11.** 2  **13.** $1/(3\sqrt[3]{x^2})$
**15.** $-4/t^5$  **17.** $-1/\sqrt{x^3}$  **19.** $2x - 2$
**21.** $4x - 5$  **23.** $-3 - 12t$  **25.** $-2p$
**27.** $12x^3 - 12x^2 + 4x$
**29.** $2 - \frac{4}{3}\sqrt[3]{x} - (1/x^2)$
**31.** $-\dfrac{8}{x^3} - \dfrac{5}{2\sqrt{x}} + \dfrac{4}{3\sqrt[3]{x}}$  **33.** $4x + 7$
**35.** $32x^3 - 4x$
**37.** $-4x^3 - 3x^2 - 16x - 12$
**39.** $\dfrac{-2}{(x+1)^2}$  **41.** $\dfrac{4}{(x+2)^2}$
**43.** $\dfrac{2x - x^4}{(x^3 + 1)^2}$  **45.** $\dfrac{2x^2 + 10x + 7}{(x^2 + 3x + 4)^2}$
**47. (a)** 25.93 hundreds of dollars/hundred units or 25.93 dollars/unit  **(b)** yes
**49.** 0.09796 dollars/hundred units
**51.** Knowledge is decreasing at 50.25 percent per month.
**53.** The percentage of divorces is increasing at the rate of 3%/year.

**55.**

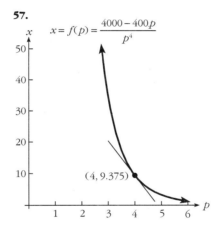

**57.**

## 3.2 EXERCISES

**Short Answer**

**1.** $x^3 + x$; $x^5$  **3.** (c)  **5.** false
**7.** true  **9.** $\dfrac{dy}{du} \cdot \dfrac{du}{dx}$

**Exercises**

**1.** $g(x) = 2x - 5$; $f(x) = x^6$
**3.** $g(x) = 5 - x$; $f(x) = x^{-4}$
**5.** $g(x) = x^2 + 1$; $f(x) = \sqrt{x}$
**7.** $g(x) = x^2 + 4x$; $f(x) = x^{3/5}$
**9.** $g(x) = x^2 + 10$; $f(x) = -5x^{-2/3}$
**11.** $12(2x - 5)^5$  **13.** $4/(5 - x)^5$
**15.** $x/\sqrt{x^2 + 1}$
**17.** $(6x + 12)/(5\sqrt[5]{(x^2 + 4x)^2})$
**19.** $20x/(3\sqrt[3]{(x^2 + 10)^5})$
**21.** $60x(2x^2 - 5)^{14}$
**23.** $(-12x - 2)/(3x^2 + x)^3$
**25.** $(-6x - 6)/(x^2 + 2x)^4$
**27.** $3x/\sqrt{3x^2 + 5}$
**29.** $-12x/\sqrt{(3x^2 + 6)^3}$
**31.** $(2x^2 + 8)/\sqrt[3]{(x^3 + 12x)^4}$

**33.** $-2x(2x - 5)^9(12x - 5)$
**35.** $x(7x + 10)/\sqrt[3]{(3x + 5)^2}$
**37.** $-21x(x^3 + 4)/\sqrt[3]{(x^3 + 6)^2}$
**39.** $2430/(5 - 2x)^6$
**41.** $1/(3\sqrt[3]{(x + 1)^2(2x + 3)^4})$
**43.** For Exercise 21, $F'(1) = 286{,}978{,}140$. For Exercise 22, $F'(1) = 140$.
**45.** For Exercise 38, $y'(0) = 0$. For Exercise 39, $y'(0) = \frac{486}{3125}$.
**47. (a)** $140 - 2p$  **(b)** $140 - 2p$  **(c)** 40 dollars of profit/dollar increase in price
**49. (a)** $500\left(1 + \dfrac{r}{400}\right)^{19}$

**(b)** $\left.\dfrac{dA}{dr}\right|_{r=6} = \$663.48$/percentage point;
$\left.\dfrac{dA}{dr}\right|_{r=8} = \$728.41$/percentage point

**51.** approximately 27.3 persons per day
**53.** approximately 6.85%/day
**55.**

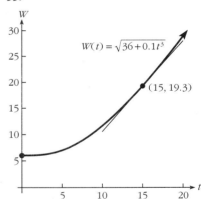

**57.**

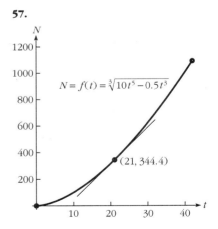

## 3.3 EXERCISES

**Short Answer**

**1.** 0; undefined  **3.** $6x$  **5.** (b)
**7.** false  **9.** 3

## Exercises

**1.** $f''(x) = 6x - 8$; $f''(2) = 4$
**3.** $f''(x) = -2/(9\sqrt[3]{x^5}) + 10$; $f''(2) = 10 - 1/(9\sqrt[3]{4})$
**5.** $f''(x) = 4/(2x - 1)^3$; $f''(2) = \frac{4}{27}$
**7.** $\frac{d^3y}{dx^3} = \frac{-12}{x^4}$; $\left.\frac{d^3y}{dx^3}\right|_{x=-1} = -12$
**9.** $\frac{d^3y}{dx^3} = \frac{60}{(x-1)^6}$; $\left.\frac{d^3y}{dx^3}\right|_{x=-1} = \frac{15}{16}$
**11.** $D_x^2 y = 12x$; $D_x^2 y|_{x=4} = 48$
**13.** $D_x^2 y = 24(x-5)^2$; $D_x^2 y|_{x=4} = 24$
**15.** $D_x^2 y = \frac{-216}{(3x+2)^5}$; $D_x^2 y|_{x=4} = \frac{-27}{67228}$
**17.** $360/x^6$   **19.** $315/(16\sqrt{x^9})$
**21.** $v(2) = 19$ ft/sec; $a(2) = 10$ ft/sec²
**23.** $v(2) = 64$ ft/sec; $a(2) = 32$ ft/sec²
**25.** $v(5) = 40$ ft/sec; $a(5) = -32$ ft/sec²
**(b)** 625 ft **(c)** after 12.5 seconds
**(d)**

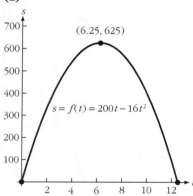

**27. (a)** $f'(1) = 0.1875$; thus at the end of one month, rates will be increasing at 0.1875% per month. $f'(3) \approx 0.3248$; thus at the end of three months, rates will be increasing at 0.3248% per month. $f'(5) \approx 0.4193$; thus at the end of five months, rates will be increasing at 0.4193% per month.
**(b)** $3/(32\sqrt{t})$ **(c)** Mortgage rates will increase at an increasing rate.
**29. (a)** $\left.\frac{dN}{dt}\right|_{t=1} = 4.5$; thus after one hour of instruction, an average student is improving his performance on the test at the rate of 4.5 questions per hour. $\left.\frac{dN}{dt}\right|_{t=16} \approx 1.42$; thus after 16 hours of instruction, an average student is improving his performance on the test at the rate of 1.42 questions per hour.
**(b)** $\frac{d^2N}{dt^2} = \frac{-5}{4\sqrt{t^3}} - \frac{2}{3\sqrt[3]{t^4}}$ **(c)** Learning is increasing at a decreasing rate.

**31.**

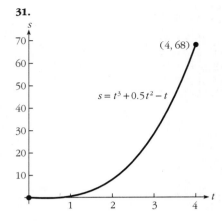

**33.**

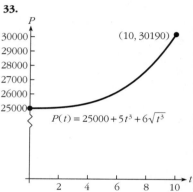

$\frac{d^2P}{dt^2}$ is positive.

## 3.4 EXERCISES

### Short Answer

**1.** false   **3.** false   **5.** $7x\frac{dy}{dx} + 7y$
**7.** $-\frac{9}{2}$   **9.** (b)

### Exercises

**1.** 2   **3.** $\frac{3}{4}$
**5.** $\frac{2x - 10xy}{5x^2 + 5}$ or $\frac{-6x}{5(x^2+1)^2}$
**7.** $\frac{2}{4y - 1}$ or $\pm\frac{2}{\sqrt{16x + 1}}$
**9.** $\frac{dy}{dx} = -4x$; $\left.\frac{dy}{dx}\right|_{(1,1)} = -4$
**11.** $\frac{dy}{dx} = \frac{1}{3y^2 + 1}$; $\left.\frac{dy}{dx}\right|_{(-1,0)} = 1$
**13.** $\frac{dy}{dx} = \frac{2 - 3x^2y^2 - 7y}{2x^3y + 7x}$; $\left.\frac{dy}{dx}\right|_{(2,-3/2)} = \frac{29}{20}$
**15.** $\frac{dy}{dx} = \frac{y}{x}$; $\left.\frac{dy}{dx}\right|_{(-1,3)} = -3$
**17.** $\frac{dy}{dx} = \frac{2\sqrt{y}}{\sqrt{x}(2\sqrt{y} - 1)}$; $\left.\frac{dy}{dx}\right|_{(1,4)} = \frac{4}{3}$
**19.** $4y - x = 1$   **21.** $156x - y = 280$
**23.** $3x + 4y = 25$ and $3x - 4y = 25$
**25.** $y = -4$ and $y = -x$
**27.** $\frac{23}{12}$ hundred pounds (or about 191.7 lb), $\frac{41}{144}$ hundred pounds (or about 28.5 pounds)
**(b)** Weekly sales of margarine are decreasing at the rate of about 88.9 lb/week.
**29.** $\frac{dp}{dx} = \frac{\sqrt{5000 - 2p^2}}{-p}$; $\left.\frac{dp}{dx}\right|_{x=35} \approx -9.9$. So when demand is 35 units per week, price is decreasing at about $9.90 per unit of additional demand beyond 35 units.
**31.** $\frac{dp}{dx} = \frac{(p^2 + 5)^2}{200(5 - p^2)}$; $\left.\frac{dp}{dx}\right|_{x=35} \approx -0.21$. So when demand is 35 units per week, price is decreasing at about $0.21 per unit of additional demand beyond 35 units.
**33. (a)** $y = \frac{10x^2}{25x + 4}$
**(b)** and **(c)**

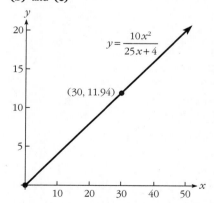

**35. (a)** $p = -5 + \frac{1}{2}\sqrt{24100 - 30x}$
**(b)** and **(c)**

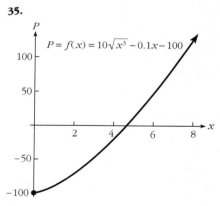

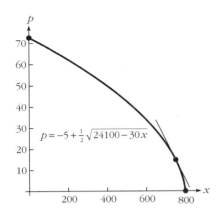

**37.** (a) $p = \dfrac{100 + \sqrt{10000 - 5x^2}}{x}$

For $x = 35$ we must use $p = \dfrac{100 + \sqrt{10000 - 5x^2}}{x}$, since $p \geq 4$.

(b) and (c)

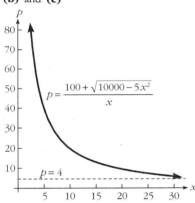

## 3.5 EXERCISES

**Short Answer**

**1.** 4 ft/min  **3.** true  **5.** (b)
**7.** (a)  **9.** $2\dfrac{dw}{dt}$

**Exercises**

**1.** $-8$ ft/min  **3.** $5.25/day
**5.** $\dfrac{dc}{dt} = 1000 \dfrac{\text{dollars}}{\text{month}}$; $\dfrac{dR}{dt} = 4000 \dfrac{\text{dollars}}{\text{month}}$; $\dfrac{dP}{dt} = 3000 \dfrac{\text{dollars}}{\text{month}}$
**7.** (a) $-16 \dfrac{\text{dollars}}{\text{week}}$  (b) $0 \dfrac{\text{dollars}}{\text{week}}$

(c) $16 \dfrac{\text{dollars}}{\text{week}}$
**9.** $40\pi$ ft$^2$/sec
**11.** approximately 1.51 ft/sec
**13.** It is decreasing at the rate of 100 mph
**15.** approximately 63,355 units per week
**17.** $p(t) = 250 - 2t$; $q(t) = 375 + t$

## 3.6 EXERCISES

**Short Answer**

**1.** 0.3  **3.** (d)  **5.** (a)  **7.** true
**9.** 25; $-1$

**Exercises**

**1.** $5\,dx$  **3.** $(6x + 1)\,dx$
**5.** $\dfrac{5x}{\sqrt{5x^2 + 21}}\,dx$
**7.** $\left(\dfrac{-2}{x^2} - \dfrac{1}{2\sqrt{x}}\right) dx$
**9.** $\Delta y = 19$; $dy = 12$

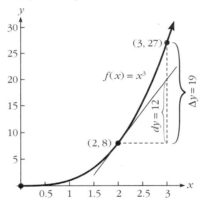

**11.** $\Delta y = -2$; $dy = -2$

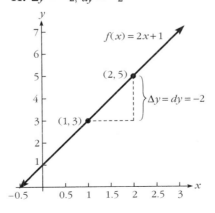

**13.** $6\tfrac{1}{12}$  **15.** $9\tfrac{9}{10}$  **17.** $2\tfrac{1}{32}$
**19.** $\pm 0.44$ in$^2$  **21.** $11.52\pi$ in$^3$
**23.** $1800
**25.** Demand will increase by approximately 100 units.
**27.** $0.83
**29.** (a) $dy|_{x=0,\,\Delta x=1} = 1$  (b) $dy|_{x=0,\,\Delta x=1} = \dfrac{1}{3}$

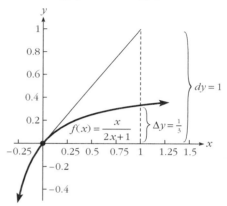

**31.**

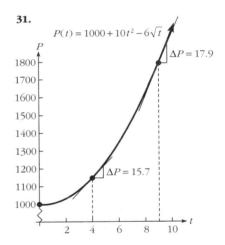

## 3.7 EXERCISES

**Short Answer**

**1.** 0  **3.** $2/(3x + 2)^2$  **5.** (d)
**7.** false  **9.** (c)  **11.** false  **13.** $-7$
**15.** (b)  **17.** (b)  **19.** true

**Exercises**

**1.** $6x + 2$
**3.** $72x^3 - 12x^2$  **5.** $3\sqrt{t} + 1/(3\sqrt[3]{t^4})$

7. $12x^3 - 5/(3\sqrt[3]{x^2})$   9. $12x + 19$
11. $32t^3 + 54t^2 - 54t - 47$
13. $2/(4x + 1)^2$   15. $-\dfrac{2t^2 + 10t + 5}{(5 - t - 3t^2)^2}$
17. $-12(2 - 3x)^3$   19. $\dfrac{-5(4t + 1)}{(2t^2 + t)^6}$
21. $\dfrac{7}{3\sqrt[3]{(7x + 10)^2}} + \dfrac{1}{3\sqrt[3]{x^2}}$
23. $\dfrac{4x^2 + 10x - 15}{3\sqrt[3]{(x^2 - 3x)^2}(4x + 5)^4}$
25. $\dfrac{-16x^3 - 14x^2}{\sqrt[5]{(x^3 + x^2)^4}}$
27. $\dfrac{d^2y}{dx^2} = -\dfrac{18}{25\sqrt[5]{x^8}}$; $\dfrac{d^2y}{dx^2}\bigg|_{x=2} = -\dfrac{9}{25\sqrt[5]{8}}$
29. $\dfrac{d^2y}{dx^2} = -\dfrac{400}{(2x - 5)^6}$; $\dfrac{d^2y}{dx^2}\bigg|_{x=2} = -400$
31. $\dfrac{d^2y}{dx^2} = \dfrac{98}{(5 - 7x)^3} + \dfrac{36}{25\sqrt[5]{(3x - 14)^9}}$;
$\dfrac{d^2y}{dx^2}\bigg|_{x=2} \approx -0.1685$
33. $\dfrac{6x}{8y^3 + 1}$   35. $\dfrac{3x^2 - 2y}{4y^3 + 2x}$
37. $\dfrac{48x^3y^2 - 2xy}{6y^3 - x^2}$   39. $\dfrac{-3y^2}{\sqrt{xy^3} + 2\sqrt{x}}$
41. $\dfrac{-y\dfrac{dy}{dt}}{x}$   43. $\dfrac{50x}{6\sqrt[3]{y^2}\sqrt{x} - 75y}\dfrac{dy}{dt}$
45. $\dfrac{4x\sqrt{x}}{6x^3y\sqrt{xy} + y\sqrt{y}}\dfrac{dy}{dt}$
47. $\Delta y \approx -0.04591$, $dy = -0.05$
49. $\Delta y \approx -0.9037$, $dy \approx -0.8536$
51. $\Delta y = -\dfrac{2786}{15}$, $dy = -\dfrac{133}{9}$
53. (a) $m_{\tan} = \dfrac{250}{(x + 5)^2}$
(b) $\dfrac{dy}{dx}\bigg|_{x=1} = \dfrac{125}{18}$; $\dfrac{dy}{dx}\bigg|_{x=10} = \dfrac{10}{9}$;
$\dfrac{dy}{dx}\bigg|_{x=25} = \dfrac{5}{18}$
(c) When the level of advertising is $1000, the rate of change of sales with respect to advertising is $\dfrac{125}{18}$ or about 6.944 thousands of dollars of sales per thousand dollars of advertising. Similar interpretations hold for the other two derivatives.
55. $\dfrac{dA}{dr} = 150\left(1 + \dfrac{r}{5200}\right)^{155}$;
$\dfrac{dA}{dr}\bigg|_{r=5} = 174.10$; $\dfrac{dA}{dr}\bigg|_{r=10} = 202.03$
57. (a) decreasing (b) increasing (c) 5000 units
59. 0.4 in³/min

61.

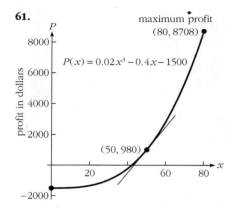

63.

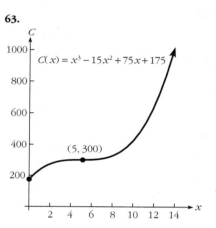

## 4.1 EXERCISES

**Short Answer**

1. increasing   3. (d)   5. true
7. false   9. intermediate value theorem

**Exercises**

1. increasing on $(-\infty, \infty)$
3. decreasing on $\left(-\infty, -\tfrac{1}{2}\right)$ and increasing on $\left(-\tfrac{1}{2}, \infty\right)$
5. increasing on $\left(-\infty, \tfrac{1}{6}\right)$ and decreasing on $\left(\tfrac{1}{6}, \infty\right)$
7. increasing on $\left(-\infty, -\tfrac{1}{3}\right)$ and on $(1, \infty)$, and decreasing on $\left(-\tfrac{1}{3}, 1\right)$
9. increasing on $\left(-\infty, -\tfrac{1}{2} - \tfrac{1}{6}\sqrt{15}\right)$ and on $\left(-\tfrac{1}{2} + \tfrac{1}{6}\sqrt{15}, \infty\right)$, and decreasing on $\left(-\tfrac{1}{2} - \tfrac{1}{6}\sqrt{15}, -\tfrac{1}{2} + \tfrac{1}{6}\sqrt{15}\right)$
11. decreasing on $(-\infty, 0)$ and increasing on $(0, \infty)$
13. increasing on $(-\infty, -\sqrt{2})$ and on $(0, \sqrt{2})$, and decreasing on $(-\sqrt{2}, 0)$ and on $(\sqrt{2}, \infty)$
15. decreasing on $(-\infty, 0)$ and on $\left(1, \tfrac{3}{2}\right)$, and increasing on $(0, 1)$ and on $\left(\tfrac{3}{2}, \infty\right)$
17. increasing on $(-\infty, 0)$ and decreasing on $(0, \infty)$
19. decreasing on $(-\infty, 0)$ and on $(0, \infty)$
21. decreasing on $(-\infty, 2)$ and on $(2, \infty)$
23. increasing on $(-\infty, 0)$ and on $(0, \infty)$
25. Efficiency increasing on $(0, 1.2)$ and decreasing on $(1.2, 3)$
27. decreasing on $(0, \sqrt[3]{500})$ and increasing on $(\sqrt[3]{500}, \infty)$
29. (a) $C(x) = 800 + 5x + \dfrac{x^2}{10000}$
(b) $\overline{C}(x) = \dfrac{800}{x} + 5 + \dfrac{x}{10000}$
(c) decreasing on $(0, 2000\sqrt{2})$ and increasing on $(2000\sqrt{2}, \infty)$
31. increasing on $(0, 14)$ and on $(28, 42)$, and decreasing on $(14, 28)$
33.

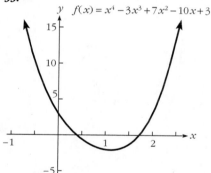

Decreasing on $(-\infty, 1.12)$ and increasing on $(1.12, \infty)$

35.

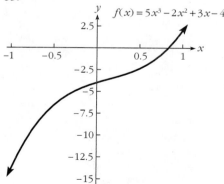

Increasing on $(-\infty, \infty)$

**37.**

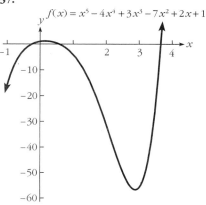

$f(x) = x^5 - 4x^4 + 3x^3 - 7x^2 + 2x + 1$

Increasing on $(-\infty, 0.15)$ and on $(2.90, \infty)$, and decreasing on $(0.15, 2.90)$

## 4.2 EXERCISES

### Short Answer

**1.** false  **3.** true  **5.** (c)  **7.** (d)
**9.** 2000

### Exercises

**1.** Relative minimum point is $(1, -1)$.

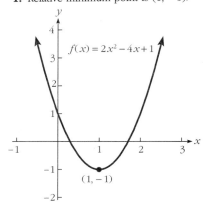

**3.** Relative maximum point is $\left(-\frac{3}{2}, \frac{17}{4}\right)$.

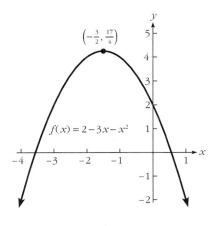

**5.** Relative maximum point is $(1, 20)$. Relative minimum point is $(6, -105)$.

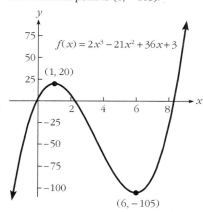

**7.** No relative extrema.

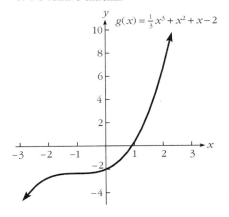

**9.** Relative minimum point is $(-5, -265)$. Relative maximum point is $(2, 78)$.

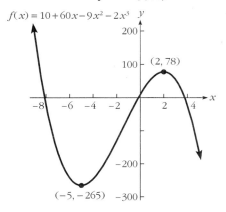

**11.** No relative extrema.

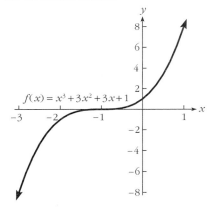

**13.** Relative maximum points are $\left(\pm\frac{1}{2}\sqrt{3}, \frac{9}{4}\right)$. Relative minimum point is $(0, 0)$.

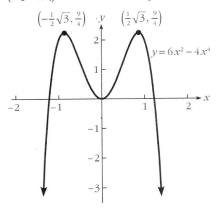

**15.** Relative minimum points are $\left(-\frac{2}{3}, -\frac{167}{27}\right)$ and $\left(\frac{1}{3}, -\frac{140}{27}\right)$. Relative maximum point is $(0, -5)$.

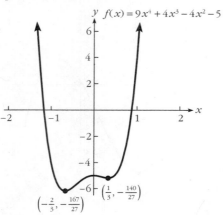

**17.** Relative maximum point is $(-2, 136)$. Relative minimum point is $(2, -120)$.

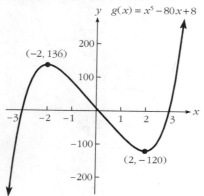

**19.** No relative extrema.

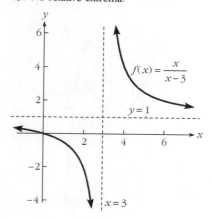

**21.** Relative maximum point is $(0, 0)$. Relative minimum point is $(4, 8)$.
**23.** Relative minimum point is $(0, 0)$.
**25.** No relative extrema.
**27. (a)** $C(x) = 2500 + 12x + \dfrac{x}{10000}$
**(b)** $\overline{C}(x) = \dfrac{2500}{x} + 12 + \dfrac{x}{10000}$
**(c)** Average daily cost per unit is minimized when daily production is 5000 units. The minimum average daily cost per unit is $13 per unit.
**29. (a)** Relative maximum point is $\left(10, \frac{3400}{3}\right)$. Relative minimum point is $\left(20, \frac{2900}{3}\right)$.
**(b)**

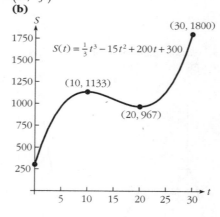

**31. (a)** Relative maximum point is approximately $(199.25, 29.88)$.
**(b)**

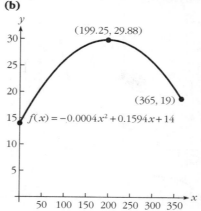

**33.**

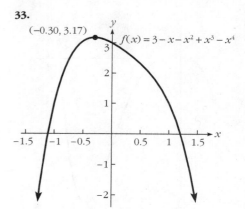

Relative maximum point is approximately $(-0.30, 3.17)$.

**35.**

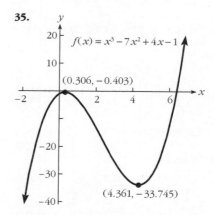

Relative maximum point is approximately $(0.306, -0.403)$. Relative minimum point is approximately $(4.361, -33.745)$.

**37.**

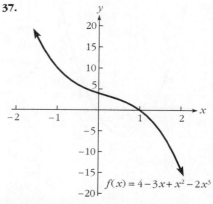

No relative extrema.

## 4.3 EXERCISES

**Short Answer**

1. up  3. (c)  5. (d)  7. (c)
9. true

**Exercises**

1. (a) Concave up on $(b, c)$ and on $(d, e)$. (b) Concave down on $(a, b)$, on $(c, d)$, and on $(e, f)$. (c) $b$, $d$, and $e$.
3. Concave down on $(-\infty, \infty)$.
5. Concave up on $(-\infty, 0)$ and concave down on $(0, \infty)$. Point of inflection is $(0, 3)$.
7. Concave down on $\left(-\infty, -\frac{5}{4}\right)$ and concave up on $\left(-\frac{5}{4}, \infty\right)$. Point of inflection is $\left(-\frac{5}{4}, \frac{265}{8}\right)$.
9. Concave down on $\left(-\infty, \frac{3}{2}\right)$ and concave up on $\left(\frac{3}{2}, \infty\right)$. Point of inflection is $\left(\frac{3}{2}, 0\right)$.
11. Concave down on $(-\infty, -1)$ and on $(1, \infty)$, concave up on $(-1, 1)$. Points of inflection are $(-1, 5)$ and $(1, 5)$.
13. Concave up on $(-\infty, 0)$ and on $\left(\frac{8}{3}, \infty\right)$, concave down on $\left(0, \frac{8}{3}\right)$. Points of inflection are $(0, 5)$ and $\left(\frac{8}{3}, -\frac{187}{81}\right)$.
15. Concave up on $(-\infty, 0)$ and concave down on $(0, \infty)$. Point of inflection is $(0, 0)$.
17. Concave down on $(-\infty, \infty)$.
19. Concave down on $\left(-\infty, -\frac{5}{2}\right)$ and on $\left(-\frac{5}{2}, \infty\right)$.
21. Concave down on $(-\infty, -3)$ and concave up on $(-3, \infty)$.
23. Concave down on $(-\infty, -1)$ and concave up on $(-1, \infty)$.
25. $\left(-\frac{5}{2}, \frac{65}{4}\right)$ is a relative maximum point.
27. $\left(-\frac{1}{6}, \frac{811}{108}\right)$ is a relative maximum point and $(3, -56)$ is a relative minimum point.
29. $\left(-\frac{1}{3}, -\frac{22}{27}\right)$ is a relative maximum point and $(1, -2)$ is a relative minimum point.
31. $\left(-\frac{5}{2}, 0\right)$ is a relative maximum point and $\left(-\frac{5}{6}, -\frac{250}{27}\right)$ is a relative minimum point.
33. $(0, 0)$ is a relative maximum point and $(\pm\sqrt{2}, -4)$ are relative minimum points.
35. $(-2, 6)$ and $(0, 6)$ are relative maximum points and $(-1, 5)$ is a relative minimum point.
37. $(-\sqrt{3}/9, -0.3849)$ is a relative minimum point and $(\sqrt{3}/9, 0.3849)$ is a relative maximum point. The $y$-coordinates of the given points are approximate.
39. No relative extrema.
41. No relative extrema.
43. $(0, 0)$ is a relative minimum point.
45. (a) $R(x) = -0.0001x^3 + 500x$; the graph of $R(x)$ is concave down on $(0, \infty)$ and so there are no points of inflection.
(b) $P(x) = -0.0001x^3 + 490x - 1000$; the graph of $P(x)$ is concave down on $(0, \infty)$ and so there are no points of inflection.
47. $C(x)$ is concave down on the interval $(0, 10)$ and concave up on the interval $(10, \infty)$. Thus the point of inflection is $(10, 1560)$ and the point of diminishing returns is 10 (thousand units).
49. (a) $S(x)$ is concave up on the interval $\left(4, \frac{50}{3}\right)$ and concave down on the interval $\left(\frac{50}{3}, 29\right)$, so the point $\left(\frac{50}{3}, \frac{710000}{27}\right)$ is a point of inflection. (b) $S(x)$ is increasing at an increasing rate on the interval $\left(4, \frac{50}{3}\right)$ and is increasing at a decreasing rate on the interval $\left(\frac{50}{3}, 29\right)$. (c) The point of diminishing returns is 50/3 thousand dollars, or about $16,667.
51. (a) $N(t)$ is concave up on the interval $(0, 25)$ and concave down on the interval $(25, 50)$, so the point of inflection is $(25, 80.125)$. (b) $N(t)$ is increasing at an increasing rate on the interval $(0, 25)$ and increasing at a decreasing rate on the interval $(25, 50)$. (c) 25 minutes
53. $f(x)$ is concave up on $(-\infty, 1)$ and concave down on $(1, \infty)$, so the $x$-coordinate of the point of inflection is 1.
55. Concave down on $(-\infty, -0.22)$ and $(0, 1.42)$, concave up on $(-0.22, 0)$ and $(1.42, \infty)$. Approximate $x$-coordinates of the points of inflection are $-0.22$ and $1.42$.

## 4.4 EXERCISES

**Short Answer**

1. true  3. 4  5. (a)  7. (c)
9. 0

**Exercises**

1. Absolute maximum is $f(4) = 20$. Absolute minimum is $f\left(-\frac{1}{2}\right) = -\frac{1}{4}$.
3. Absolute maximum is $f(2) = -1$. Absolute minimum is $f(0) = -5$.
5. Absolute maximum is $f(3) = 17$. Absolute minimum is $f(0) = -4$.
7. Absolute maximum is $f\left(-\frac{1}{4}\right) = \frac{97}{8}$. Absolute minimum is $f(2) = 2$.
9. Absolute maximum is $f(2) = 4$. Absolute minimum is $f(-2) = -12$.
11. Absolute maximum is $f(1) = 9$. Absolute minimum is $f(0) = f(3) = 5$.
13. Absolute maximum is $f(1) = 2$. Absolute minimum is $f(5) = -94$.
15. Absolute minimum is $f(-1) = -3$. No absolute maximum.
17. Absolute maximum is $f(2) = 4$. No absolute minimum.
19. Absolute minimum is $f(-1) = -6$. No absolute maximum.
21. Absolute maximum is $f(0) = 16$. Absolute minimum is $f(2) = 0$.
23. Absolute maximum is $f(3) = 251$. Absolute minimum is $f(-2) = -24$.
25. Absolute maximum is $f(1) = \frac{1}{2}$. Absolute minimum is $f(-1) = -\frac{1}{2}$.
27. No absolute maximum and no absolute minimum.
29. No absolute maximum and no absolute minimum.
31. Maximum profit of about $106,111 monthly is achieved with a sales level of about 33,333 units monthly.
33. (a) Absolute maximum revenue is $R(2000) = 40000$. (b) Absolute maximum profit is $P(750) = 4625$.
35. Absolute maximum occurs at $x = 0.66$ (approximately), absolute minimum occurs at $x = 4$.
37. No absolute extrema on $(0, \infty)$.

## 4.5 EXERCISES

**Short Answer**

1. $-6$  3. true  5. 2  7. no
9. (c)

**Exercises**

1. $x$-intercepts are $-1, 0, 1$, $y$-intercept is 0. Relative maximum point is $(-\sqrt{3}/3, 2\sqrt{3}/9)$, relative minimum point is $(\sqrt{3}/3, -2\sqrt{3}/9)$. Concave down on $(-\infty, 0)$, concave up on $(0, \infty)$, so point of inflection is $(0, 0)$.

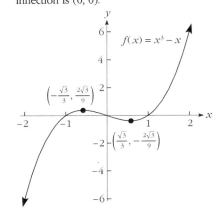

**3.** $y$-intercept is $-1$. Relative minimum point is $(0, -1)$, relative maximum point is $\left(\frac{4}{3}, \frac{5}{27}\right)$. Concave up on $\left(-\infty, \frac{2}{3}\right)$, concave down on $\left(\frac{2}{3}, \infty\right)$, so point of inflection is $\left(\frac{2}{3}, -\frac{11}{27}\right)$.

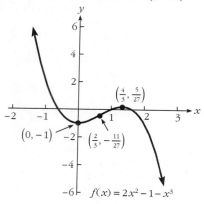

**5.** $y$-intercept is 5. Relative maximum point is $\left(\frac{-1-\sqrt{5}}{2}, f\left(\frac{-1-\sqrt{5}}{2}\right)\right)$ or approximately $(-1.62, 14.09)$. Relative minimum point is $\left(\frac{-1+\sqrt{5}}{2}, f\left(\frac{-1+\sqrt{5}}{2}\right)\right)$ or approximately $(0.62, 2.91)$. Concave down on $\left(-\infty, -\frac{1}{2}\right)$, concave up on $\left(-\frac{1}{2}, \infty\right)$, so point of inflection is $\left(-\frac{1}{2}, \frac{17}{2}\right)$.

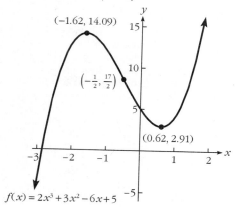

**7.** $y$-intercept is 3. Relative minimum point is $\left(\frac{1}{3}, \frac{77}{27}\right)$, relative maximum point is $(1, 3)$. Concave up on $\left(-\infty, \frac{2}{3}\right)$, concave down on $\left(\frac{2}{3}, \infty\right)$, so point of inflection is $\left(\frac{2}{3}, \frac{79}{27}\right)$.

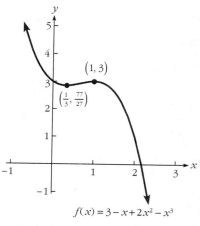

**9.** $x$-intercepts are 0 and $\sqrt[3]{4}$, $y$-intercept is 0. Relative minimum point is $(1, -3)$. Concave up on $(-\infty, 0)$ and $(0, \infty)$.

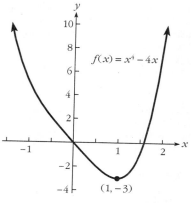

**11.** $y$-intercept is $-2$. Relative minimum points are $(\pm 2, -18)$, relative maximum point is $(0, -2)$. Concave up on $(-\infty, -2\sqrt{3}/3)$ and $(2\sqrt{3}/3, \infty)$, concave down on $(-2\sqrt{3}/3, 2\sqrt{3}/3)$, so points of inflection are $\left(\pm 2\sqrt{3}/3, \frac{-98}{9}\right)$ or approximately $(\pm 1.15, -10.89)$.

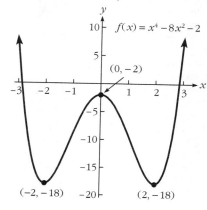

**13.** $y$-intercept is $-1$. Relative minimum point is $(3, -28)$. Concave up on $(-\infty, 0)$ and on $(2, \infty)$, concave down on $(0, 2)$. Points of inflection are $(0, -1)$ and $(2, -17)$.

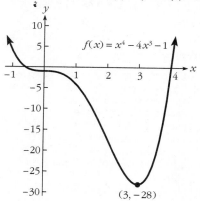

**15.** $y$-intercept is $-6$. Relative minimum points are $(0, -6)$ and $(4, -646)$. Relative maximum point is $\left(\frac{1}{3}, -\frac{139}{27}\right)$. Approximate intervals of concavity: concave up on $(-\infty, 0.16)$ and on $(2.73, \infty)$, concave down on $(0.16, 2.73)$. Approximate points of inflection: $(0.16, -5.58)$ and $(2.73, -384)$.

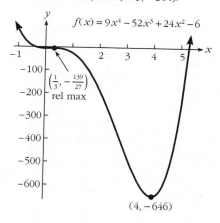

**17.** $y$-intercept is 0, $x$-intercepts are 0 and $\pm \sqrt[4]{40}$. Approximate relative maximum point is $(\sqrt[4]{8}, 53.82)$, approximate relative minimum point is $(-\sqrt[4]{8}, -53.82)$. Concave up on $(-\infty, 0)$, concave down on $(0, \infty)$, so point of inflection is $(0, 0)$.

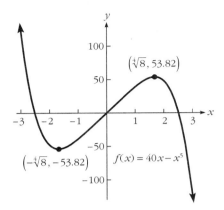

**19.** $y$-intercept is $-1$. Vertical asymptote is $x = 1$, horizontal asymptote is $y = 0$. Concave down on $(-\infty, 1)$, concave up on $(1, \infty)$.

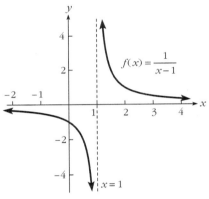

**21.** $y$-intercept is 0, $x$-intercept is 0. Vertical asymptote is $x = -3$, horizontal asymptote is $y = 2$. Concave up on $(-\infty, -3)$, concave down on $(-3, \infty)$.

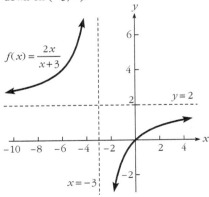

**23.** $x$-intercepts are $\pm 2$. Vertical asymptote is $x = 0$, horizontal asymptote is $y = 1$. Concave down on $(-\infty, 0)$ and on $(0, \infty)$.

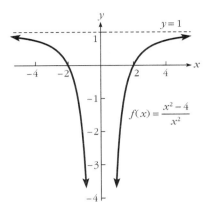

**25.** $x$-intercept is 0, $y$-intercept is 0. Vertical asymptote is $x = 3$, horizontal asymptote is $y = 0$. Relative minimum point is $\left(-3, -\frac{1}{12}\right)$. Concave down on $(-\infty, -6)$, concave up on $(-6, 3)$ and on $(3, \infty)$, so point of inflection is $\left(-6, -\frac{2}{27}\right)$.

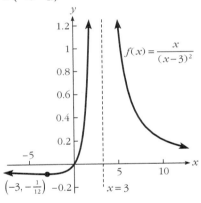

**27.** $x$-intercept is 2, $y$-intercept is $-\frac{2}{3}$. Vertical asymptote is $x = -3$, horizontal asymptote is $y = 1$. Concave up on $(-\infty, -3)$, concave down on $(-3, \infty)$.

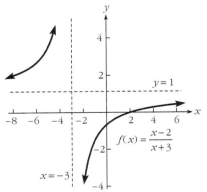

**29.** $x$-intercept is 0, $y$-intercept is 0. Horizontal asymptote is $y = 0$. Relative minimum point is $\left(-1, -\frac{1}{2}\right)$, relative maximum point is $\left(1, \frac{1}{2}\right)$. Concave down on $(-\infty, -\sqrt{3})$ and $(0, \sqrt{3})$, concave up on $(-\sqrt{3}, 0)$ and on $(\sqrt{3}, \infty)$, so one point of inflection is $(0, 0)$, and two other approximate points of inflection are $(\sqrt{3}, 0.433)$ and $(-\sqrt{3}, -0.433)$.

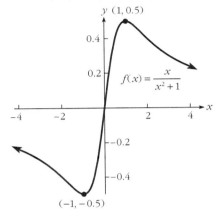

**31.** $x$-intercepts are 0 and 4, $y$-intercept is 0. Relative minimum point is $(1, -2)$. Concave up on $(0, \infty)$.

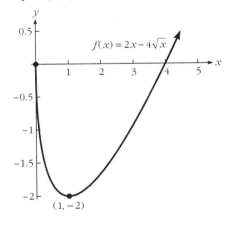

**33.** $x$-intercept is $-1$, $y$-intercept is 1. Concave up on $(-1, \infty)$.

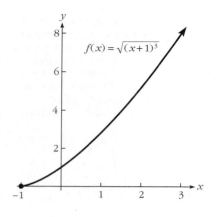

**35.** y-intercept is $-1$. Relative minimum point is $\left(\frac{64}{9}, -\frac{283}{27}\right)$. Concave up on $(0, \infty)$.

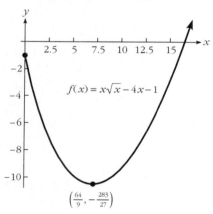

**37. (a)**

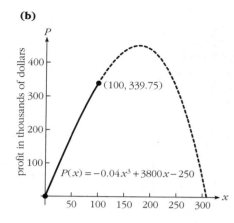

**(b)**

**39.**

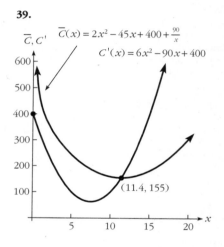

**41.** The x-intercepts are approximately $-1.4$, $1.1$, and $5.1$. Also, $x = 4.1$ is the approximate x-coordinate of a relative minimum and $x = -0.7$ that of a relative maximum. Points of inflection occur at points with x-coordinates approximately equal to $-0.3$, $0.2$, and $3.1$.

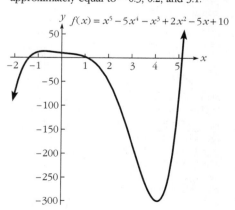

**43.** The x-intercepts are approximately $-3.2$, $-0.7$, $-0.3$, and $5.6$. Also, $x = -2.4$ and $x = 3.9$ are the approximate x-coordinates of relative minima, and $x = -0.5$ that of a relative maximum. Points of inflection occur at points with x-coordinates approximately equal to $-1.5$ and $2.2$.

$f(x) = x^4 - 1.4x^3 - 20.11x^2 - 18.424x - 3.7632$

**45.** The x-intercepts are approximately $-6.7$ and $-0.3$. There are no relative extrema.

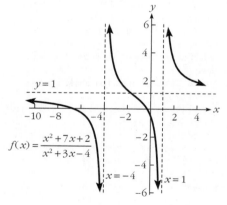

**47.** The x-intercepts are approximately $-4.6$ and $2.3$. There are no relative extrema. Points of inflection occur at points with x-coordinates approximately equal to $-4$ and $4.4$.

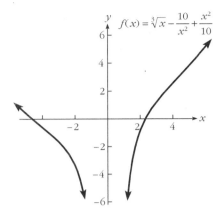

## 4.6 EXERCISES

### Exercises

1. Length and width are both 100 ft.
3. (a) 200 million dollars. (b) Maximum monthly profit is $102,500,000 when the production level is 15,000 units per month and the price is $12,500 per unit.
5. The base is 2 ft by 2 ft and the height is 1 ft.   7. 125,000 ft²
9. Each side should be exactly $20\sqrt{3}$ ft long.
11. Each wire should be cut in half so that the length of each of the resulting pieces is 12 in.   13. $36.50
15. To maximize yield per acre at 1800 lb of pecans, 30 trees should be planted.
17. (a) 14 in. by 14 in. by 28 in. (b) The radius of the cylinder is $28/\pi$ in. and the height is 28 in. (c) Each edge of the cube is $\frac{84}{5}$ in.
19. (a) 447,214 bolts (b) about $53,541
21. $\frac{1}{2}\sqrt{5}$ hr
23. At the optimal speed of $10\sqrt{10}$ mph, gas will cost $15.81 for the trip.
25. Spend no money on advertising; put the entire 50¢/pie into improved ingredients to produce weekly sales of 1150 pies.
27. The distance across each semicircular portion is $1/(\pi + 2)$ and the length of each of the other parallel straight sides is $1/(\pi + 2)$.
29. The maximum revenue is obtained when 600 members join.
31. $46 + 8\sqrt{15}$ in²
33. If $h$ is the diameter of the log, then $w = \sqrt{3}h/3$ and $d = \sqrt{6}h/3$.
35. 2 hr 40 min after the shift begins
37. The largest enclosed area has 100 ft of fencing parallel to the house and 50 ft of fencing on each of the two sides perpendicular to the house.   39. $(\frac{3}{2}, \frac{5}{2})$.

41. Approximately 2:45 P.M.
43. Run the pipe underwater to a point $8\sqrt{5}/5$ miles down the beach from $Q$ and from there over land to $R$.

## 4.7 EXERCISES

### Short Answer

1. 101st   3. 250   5. true   7. (b)
9. (c)

### Exercises

1. $C'(100) = 62, R'(100) = 20, P'(100) = -42; \Delta C = 62.25, \Delta R = 19.90, \Delta P = -42.35$
3. $C'(50) = 55, R'(50) = 9400, P'(50) = 9345; \Delta C = 55.40, \Delta R = 9249, \Delta P = 9193.60$
5. (a) $P(x) = 35x - 0.12x^2 - 200$; $P'(x) = 35 - 0.24x$ (b) $P'(100) = 11$, so the profit from manufacturing and selling the 101st unit is about $11. (c) approximately 145.83 units
7. (a) $149.50 (b) $150
9. (a) $R(x) = 150x - 0.1x^2$; $P(x) = 123x - 0.15x^2 - 500$ (b) $C'(x) = 0.1x + 27$; $R'(x) = 150 - 0.2x$ (c) $C'(400) = $67 is the approximate cost of producing the 401st unit; $R'(400) = $70 is the approximate revenue derived from the sale of the 401st unit; $C'(420) = $69 is the approximate cost of producing the 421st unit; $R'(420) = $66 is the approximate revenue derived from the sale of the 421st unit. (d) 410 units
11. (a) 120 (b) 120
13. The economic order quantity is 90 cases per order and the minimum yearly inventory cost is $630.
15. The economic lot size is 5 lots per year and the minimum yearly total of manufacturing and storage costs is $960,000.
17. The average elasticity of demand is $-0.15$. In this case a 66.6̄6̄% increase in price resulted in only a 10% decrease in demand, producing an elasticity of demand of $-0.10/0.666\overline{6} = -0.15$.
19. $E(p) = -p/(400 - p)$ and $E(100) = -0.3\overline{3}$, so demand is inelastic. A change in price of 1% produces a change in demand of approximately 0.3̄3̄% when price is $100. Further, $E(200) = -1$, so demand has unit elasticity. A change in price of 1% produces a change in demand of approximately 1% when price is $200. Finally, $E(300) = -3$, so demand is elastic. A change in price of 1% produces a change in demand of approximately 3% when price is $300.

21. $E(p) = -2p^2/(2500 - p^2)$ and $E(20) \approx -0.381$, so demand is inelastic. A change in price of 1% produces a change in demand of approximately 0.381% when price is $20. Next, $E(30) = -1.125$, so demand is elastic. A change in price of 1% produces a change in demand of approximately 1.125% when price is $30. Last, $E(40) = -3.5\overline{5}$, so demand is elastic. A change in price of 1% produces a change in demand of approximately 3.5̄5̄% when price is $40.
23. Demand is inelastic for $0 \le p < 625$ and it is elastic for $625 < p < 1250$. When $p = 625$ demand has unit elasticity.
25. Demand is inelastic for $0 \le p < 57.735$ and it is elastic for $57.735 < p < 100$. When $p = 57.735$ demand has unit elasticity.
27. Demand is inelastic for $0 \le p < 31.498$ and it is elastic for $31.498 < p < 50$. When $p = 31.498$ demand has unit elasticity.
29. Demand is inelastic for $p > 0$.
31. $R(p) = 4000p - 8p^2$ and maximum revenue is $R(250) = 500000$. Demand has unit elasticity when $p = 250$.
33. $R(p) = 150p - 0.2p^3$ and maximum revenue is $R(5\sqrt{10}) = 500\sqrt{10}$. Demand has unit elasticity when $p = 5\sqrt{10}$.
35. $R(p) = 50p - 10p^{3/2}$ and maximum revenue is $R(\frac{100}{9}) = \frac{5000}{27}$. Demand has unit elasticity when $p = \frac{100}{9}$.
37. $-9$   39. $-\frac{11}{18}$   41. $-\frac{387}{14}$

43.

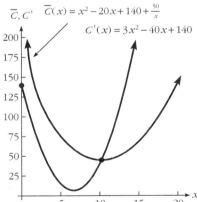

Average cost is a minimum when the production level is approximately 10,238 units.

45. (a) $0 \le p < 36.39$ (b) Demand is inelastic for $0 \le p < 22.85$ and is elastic for $22.85 < p < 36.39$. When $p = 22.85$ demand has unit elasticity.

## 4.8 EXERCISES

### Short Answer

**1.** decreasing  **3.** (d)  **5.** false
**7.** true  **9.** (b)  **11.** 2  **13.** (b)
**15.** −5  **17.** false  **19.** $x - 25$
**21.** 600  **23.** (d)

### Exercises

**1.** decreasing on $(-\infty, 2)$ and increasing on $(2, \infty)$
**3.** increasing on $(-\infty, -2)$ and on $(6, \infty)$, decreasing on $(-2, 6)$
**5.** decreasing on $(-\infty, 0)$ and on $(0, \infty)$
**7.** Relative maximum point is $(0, 0)$, relative minimum point is $\left(\frac{2}{3}, -\frac{4}{27}\right)$.
**9.** Relative maximum point is $(-2, 192)$, relative minimum point is $(7, -537)$.
**11.** Relative maximum point is $(0, 4)$.
**13.** Relative minimum points are $(-1, -3)$ and $(3, -131)$, relative maximum point is $(0, 4)$.  **15.** No relative extrema.
**17.** Concave up on $(-\infty, -\sqrt{2})$ and on $(\sqrt{2}, \infty)$ and concave down on $(-\sqrt{2}, \sqrt{2})$, so points of inflection are $(\pm\sqrt{2}, -20)$.
**19.** Concave up on $\left(-\infty, -\frac{1}{12}\right)$ and concave down on $\left(-\frac{1}{12}, \infty\right)$, so point of inflection is $\left(-\frac{1}{12}, \frac{2177}{216}\right)$.
**21.** Concave up on $(-\infty, 0)$ and concave down on $(0, \infty)$, so point of inflection is $(0, -2)$.
**23.** Absolute maximum is $f(1) = 4$, absolute minimum is $f(0) = -3$.
**25.** Absolute maximum is $g(2) = 16$, absolute minimum is $g(1) = -1$.
**27.** No absolute extrema.
**29.** $y$-intercept is $-6$. Relative maximum point is $(2, 22)$, relative minimum point is $(3, 21)$. Concave down on $\left(-\infty, \frac{5}{2}\right)$ and concave up on $\left(\frac{5}{2}, \infty\right)$, so point of inflection is $\left(\frac{5}{2}, \frac{43}{2}\right)$.

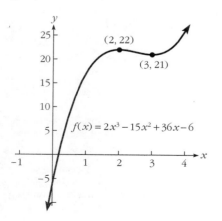

**31.** $x$- and $y$-intercepts are 0. Relative maximum point is $\left(-2, \frac{1}{2}\right)$, relative minimum point is $\left(2, -\frac{1}{2}\right)$. Concave up on $(-\infty, -2\sqrt{3})$ and on $(0, 2\sqrt{3})$, concave down on $(-2\sqrt{3}, 0)$ and on $(2\sqrt{3}, \infty)$, so points of inflection are $\left(-2\sqrt{3}, \frac{1}{4}\sqrt{3}\right)$, $(0, 0)$, and $\left(2\sqrt{3}, -\frac{1}{4}\sqrt{3}\right)$.

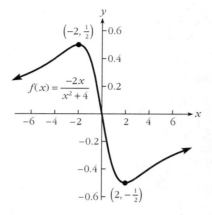

**33.** Profit is maximized at approximately $7045.61, when sales are approximately 676.89 units.
**35.** $C'(10) = 200$, $R'(10) = 0$, $P'(10) = -200$
**37.** Radius of the top and bottom is $2\sqrt[3]{36}/3$ in. and the height is $4\sqrt[3]{36}$ in.
**39.** The owner should charge $8 to obtain the maximum revenue of $2560.
**41.** The circle has a radius of length $25/[2(\pi + 4)]$ in. and the square has a side of length $25/(\pi + 4)$ in.
**43.** (a) approximately 5070  (b) in $\frac{40}{21}$ years
**45.** Minors of parents who earn $50,000 per year have the minimum arrest rate of 200 per year.
**47.** Demand is inelastic for $0 \leq p < 125$ and is elastic for $125 < p < 250$. When $p = 125$ demand has unit elasticity.

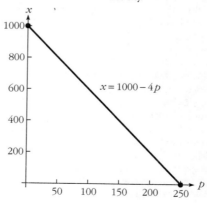

**49.** Demand is inelastic for $p > 0$.

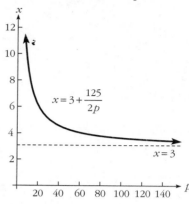

**51.** The revenue function is $R(p) = 10000p - 20p^2$ and the maximum revenue of $1,250,000 occurs when $p = $250$ and demand has unit elasticity.
**53.** The revenue function is $R(p) = 91125p - p^4$ and the maximum revenue of $1,937,424 occurs when price is approximately $28.35 and demand has unit elasticity.
**55.** $x = 1.1$ is the approximate $x$-coordinate of a relative minimum and $x = 2.3$ that of a relative maximum. Points of inflection occur at points with $x$-coordinates approximately equal to $-0.5$, $0.4$, and $1.8$.

## 5.1 EXERCISES

### Short Answer

**1.** (b)  **3.** false  **5.** $(0, 1)$  **7.** true
**9.** $20000e^{0.07t}$

### Exercises

**1.** 4  **3.** 1500  **5.** 7  **7.** 432
**9.** $3^{0.75}$  **11.** 2089 8135  **13.** $-2.3501$
**15.** 0.1869  **17.** 16.2425  **19.** 0.0226
**21.**

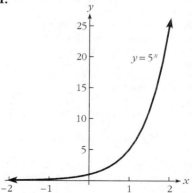

**23.**
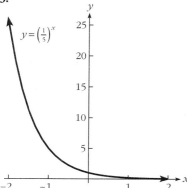

**29.**
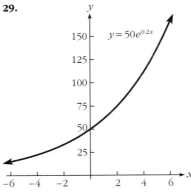

**55.**
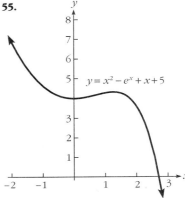

**25.**
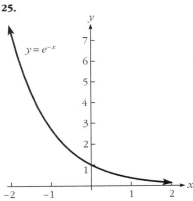

**31.** $32,772.33   **33. (a)** $151,106.87
**(b)** $152,010.55  **(c)** $152,190.02
**35.** $8,250
**37. (a)** $38,362.32  **(b)** $38,720.76
**39.** You will earn about $923.93 more interest with 9.5% interest compounded quarterly.
**41.** $6376.28   **43.** $24,980.05
**45. (a)** approximately 21.62
**(b)** $\lim_{t \to \infty} N(t) = 25$: this means that after a very large number of hours of training, an average worker will be able to assemble 25 components per hour.
**(c)**

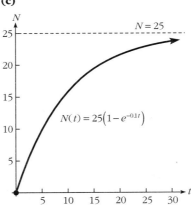

**47. (a)** 3,547,669  **(b)** 5,997,188
**49. (a)** Its value after 5 years is $23,618.33; after 10 years it is $11,156.51; after 40 years it is $123.94.  **(b)** $50,000  **(c)** $V(t) = 50000 - 1246.90t$  **(d)** The required values after 5, 10, and 40 years are $43,765.50, $37,531, and $124, respectively.
**51. (a)** The fraction operating after 5 years is about 0.6065; the fraction operating after 10 years is about 0.3679.  **(b)** 0.1813
**53. (a)** approximately 26.2%  **(b)** 6 weeks

**57.**

**59.** 2.71   **61.** 0.15 and 4.83

### 5.2 EXERCISES

**Short Answer**

**1.** 4   **3.** $e^0 = 1$   **5.** (d)   **7.** false
**9.** ln 10

**Exercises**

**1.** $2^4 = 16$   **3.** $2^{1/2} = \sqrt{2}$
**5.** $10^{-1} = \frac{1}{10}$   **7.** $\log_5 625 = 4$
**9.** $\log_{10} 0.00001 = -5$
**11.** $\log_{16} 64 = \frac{3}{2}$   **13.** 2   **15.** 5
**17.** $-3$   **19.** $\frac{1}{2}$   **21.** 2   **23.** $-2$
**25.** 1.2788   **27.** 0.2386   **29.** $-0.3567$
**31.** 1.1744

**27.**
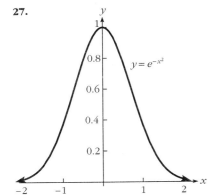

**33.**
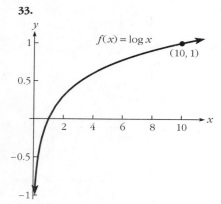

**35.**
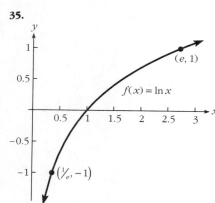

**37.**
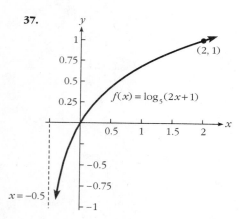

**39.** $\log 7 - \log 5$
**41.** $\ln x + \ln y - \ln 2$
**43.** $\log 3 + \log x - 2\log(x+1)$
**45.** $\frac{1}{3}\ln(2x+1)$   **47.** $\ln(2x^2)$
**49.** $\ln\left[\dfrac{(x-1)\sqrt{x+1}}{x}\right]$
**51.** $\ln\left[\dfrac{(x-1)^4(x^2+1)^2}{(x+1)^3}\right]$

**53.** $\ln\sqrt[3]{\dfrac{2x^2-5x}{2x^2+7}}$   **55.** $\dfrac{\ln 5}{2}$
**57.** $\dfrac{\ln 36}{2\ln 3}$   **59.** $\dfrac{\ln(0.35)}{4}$
**61.** $\dfrac{\ln 10}{1-2\ln 10}$   **63.** $-2$ and $1$
**65.** no solution   **67.** about 7.296 years
**69.** $\frac{1}{10}(\ln 2) \approx 6.93\%$
**71. (a)** about 8.751 years **(b)** about 13.870 years
**73. (a)** $3546.29 **(b)**

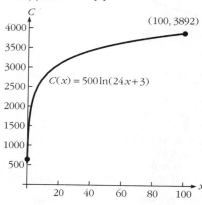

**(c)** $70.93

**75. (a)** $R(x) = \dfrac{25x}{\ln(x+1.2)}$ **(b)** $6577.65
**(c)** $8808.58
**77. (a)** $13,918.54 **(b)** $P(x) = \sqrt{x}\ln(x+5) - 0.0001x^2 - 0.1x - 5$
**(c)** $5136.51
**79. (a)** 54.89% **(b)** 21.82 months
**81. (a)** $10^{8.2} \approx 158{,}489{,}319$ **(b)** $10^{1.3} \approx 20$
**85.**
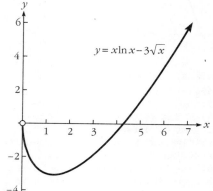

**87.**
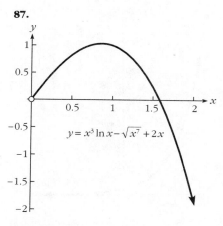

**89.** 0.96, 4.87   **91.** 0.48, 2.71

## 5.3 EXERCISES

### Short Answer
**1.** true   **3.** 1
**5.** $-1/x^2$, negative, down   **7.** (c)
**9.** true

### Exercises
**1.** $3e^x$   **3.** $4e^{4x}$   **5.** $-35e^{-5x}$
**7.** $-1.32e^{-0.22x}$   **9.** $(2-2x)e^{2x-x^2}$
**11.** $-e^{\sqrt{x}}/\sqrt{x}$   **13.** $3e^{-4x} - 12xe^{-4x}$
**15.** $\frac{1}{2}e^x - \frac{1}{2}e^{-x}$   **17.** $2(\ln 5)5^{2x}$
**19.** $3^x[1+(\ln 3)x]$
**21.** $2^x[(\ln 2)x - 1]/(5x^2)$
**23.** $2^x e^{x^2}(2x + \ln 2)$
**25.** $[1000\ln(1.02)](1.02)^x$
**27.** $5/x$   **29.** $3/(3x+4)$
**31.** $-7(3x^2+7)/(x^3+7x)$   **33.** $2/x$
**35.** $1/(2x\sqrt{\ln x})$   **37.** $1/(x\ln x)$
**39.** $x/(x^2+1)$   **41.** $(1-\ln x)/x^2$
**43.** $2x/(3x^2+48)$
**45.** $29/[(7x+3)(2x+5)]$
**47.** $2/(4-x^2)$
**49.** $\dfrac{x+2\sqrt{x^2+1}}{x^2+1+2x\sqrt{x^2+1}}$
**51.** $\dfrac{2}{2x+3} - \dfrac{1}{2x} - \dfrac{2}{6x+3}$
**53.** $\dfrac{x+1}{x^2+2x} - \dfrac{3}{x+1} - \dfrac{8}{2x-1}$
**55.** $\dfrac{12}{(\ln 3)x}$   **57.** $\dfrac{x}{\ln 10} + 2x\log x$
**59.** $\dfrac{1}{\ln 5} \cdot \dfrac{1+9x^2\sqrt[3]{x^2}}{3(x+x^3\sqrt[3]{x^2})}$
**61.** $\dfrac{1}{\ln 10}\left[\dfrac{-7}{(x+2)(x-5)}\right]$

**63.** $e^x(1 + \ln x + x \ln x)$

**65.** $\dfrac{e^x + e^{-x} - (x \ln x)(e^x - e^{-x})}{2x(e^x + e^{-x})^2}$

**67.** $36e^6$     **69.** $\dfrac{23(\ln 3)(\ln 46) - 12}{621(\ln 3)}$

**71.** The relative minimum point is (0, 1); concave up on the interval $(-\infty, \infty)$.

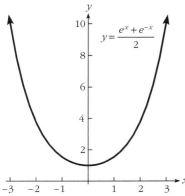

**73.** The relative minimum point is $(-1, -1/e)$. Concave down on the interval $(-\infty, -2)$ and concave up on the interval $(-2, \infty)$, so the point of inflection is $(-2, -2/e^2)$.

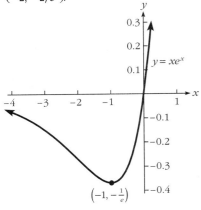

**75.** The relative maximum point is (0, 1). Concave up on the intervals $(-\infty, -\sqrt{2}/2)$ and $(\sqrt{2}/2, \infty)$, concave down on the interval $(-\sqrt{2}/2, \sqrt{2}/2)$, so the points $(\pm\sqrt{2}/2, 1/\sqrt{e})$ are points of inflection.

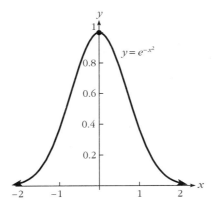

**77.** The relative minimum point is approximately (0.3679, −0.1598). Concave up on the interval $(0, \infty)$.

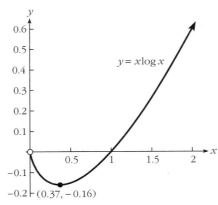

**79.** (a) $C'(x) = \dfrac{4x + 15}{2x^2 + 15x}$

(b) $\overline{C}(x) = \dfrac{\ln(2x^2 + 15x)}{x} + \dfrac{1200}{x}$

(c) $\overline{C}'(x) = [-2396x^2 - 17985x - (2x^2 + 15x)\ln(2x^2 + 15x)]/(2x^4 + 15x^3)$

**81.** The rate of change after seven days is $80e^{-1.4}$ or about 19.73 units per day. The rate of change after 14 days is $80e^{-2.8}$ or about 4.86 units per day.

**83.** The rate of change when $t = 6$ is $900e^{-1.2}$ or about 271.07 units per month. The rate of change when $t = 12$ is $900e^{-2.4}$ or about 81.65 units per month.

**85.** A production level of $e^6 \approx 403.43$ units yields the minimum average cost of $149.50 per unit.

**87.** It is decreasing at a rate of 14.49% per week.

**89.** The number of people who have had the flu is increasing at the rate of 11.92 people per day.

**91.** (a) about 51.99 units of sales per thousand dollars of advertising (b) about 29.90 units of sales per thousand dollars of advertising

**93.** (a) $150 - 100e^{-2}$ or about 136.47 units (b) $20e^{-4}$ or about 0.37 units per day

**95.** approximately 1214.9 units

**97.** approximately 25,276 units

## 5.4 EXERCISES

**Short Answer**

**1.** $Q_0 e^{0.05t}$    **3.** false    **5.** (a)    **7.** (d)
**9.** (d)

**Exercises**

**1.** $A(t) = 100e^{0.01t}$
**3.** $A(t) = 500(1 - 0.9e^{0.003t})$
**5.** $A(t) = 400/(1 + 3e^{-0.8t})$
**7.** $268,277.80
**9.** (a) 7.296 years (b) 12.556 years (c) 12.323 years
**11.** $14,067.62
**13.** (a) $x = 24414e^{-0.22314p}$ (b) approximately 6400 units
**15.** approximately 32.1 million
**17.** approximately 13.16 months
**19.** (a) approximately 17.21 g (b) approximately 60.37 days
**21.** approximately 189.84 g
**23.** approximately 3560 years
**25.** approximately 4.86%    **27.** $(\ln N)/k$
**29.** (a) $A(t) = 40(1 - e^{-0.18326t})$ (b) approximately 88.91%
**31.** (a) $A(t) = 350\left(1 - \tfrac{5}{7}e^{-0.22907t}\right)$ (b) approximately 7 weeks
**33.** about 69.9°
**35.** approximately $38,194.22
**37.** (b) $M$ (c) $MCke^{-kt} < 0$ (d) $-MCk^2e^{-kt} < 0$
**39.** (a) $A(t) = 15/(1 + 14e^{-0.38363t})$ (b) approximately 7.674 million
**41.** (a) 374 students (b) approximately 7.8 days
**43.** 6277    **47.** 6.88 hours
**49.** approximately 7.82 days
**51.** approximately 3.712 years

**53. (a)**

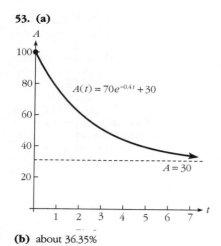

**(b)** about 36.35%

## 5.5 EXERCISES

### Short Answer

**1.** false  **3.** true  **5.** true  **7.** (a)
**9.** (c)  **11.** (d)  **13.** (d)  **15.** (c)

### Exercises

**1.** 4  **3.** 7  **5.**

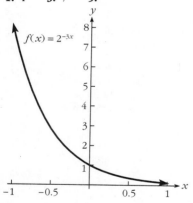

**7.**

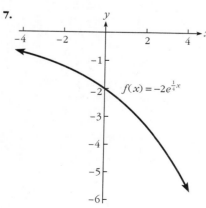

**9. (a)** $14,025.52 **(b)** $14,147.78 **(c)** $14,176.25 **(d)** $14,190.68
**11.** 2  **13.** $\frac{1}{3}$  **15.** 0.8047
**17.** 0.6671  **19.** 1.8295

**21.**

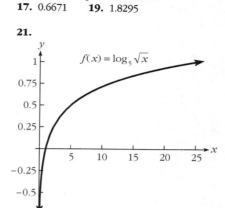

**23.**

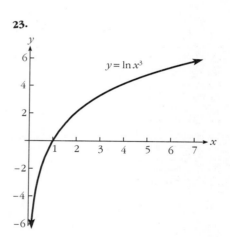

**25.** 4 ln 6
**27.** $(4 \ln 10 + 3)/(7 - \ln 10)$
**29.** $\pm e\sqrt[10]{3}$
**31. (a)** about 8.008 years **(b)** about 7.923 years **(c)** about 7.922 years
**33.** $-84e^{-7x+2}$  **35.** $30xe^{3x^2} - 21x^2$
**37.** $e^{\sqrt[3]{x}}\left(\frac{1}{3}\sqrt[3]{x} + 1\right)$  **39.** $-e^x/(e^x-1)^2$
**41.** $(\ln 2)2^x - 2e^{2x}$  **43.** 0

**45.** $\dfrac{2x + 5}{2x^2 + 10x + 14}$

**47.** $\dfrac{x}{x^2 + 15} - \dfrac{5x^4 + 7}{3x^5 + 21x + 9}$

**49.** $\dfrac{3x^3}{(\ln 5)(3x + 7)} + 3x^2 \log_5(3x + 7)$

**51.** The relative maximum point $\left(\frac{1}{5}, 1/(5e)\right)$. Concave down on the interval $\left(-\infty, \frac{2}{5}\right)$ and concave up on the interval $\left(\frac{2}{5}, \infty\right)$, so the point of inflection is $\left(\frac{2}{5}, 2/(5e^2)\right)$.

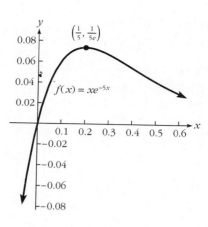

**53.** The relative minimum point is $(1/e^2, -1/e^2)$; concave up on $(0, \infty)$.

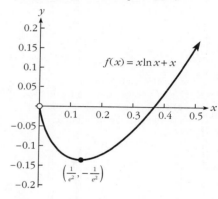

**55.** 0.16 g
**57.** approximately 19,030 years old
**59.** approximately 36.62 years
**61.** approximately 1642
**63.**

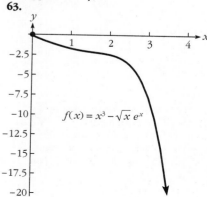

**65.**

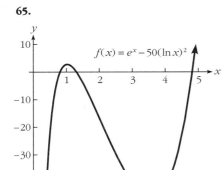

**67.**

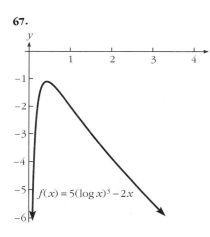

**69.** 0.35 and 3.11.

## 6.1 EXERCISES

**Short Answer**

**1.** false **3.** true **5.** $\frac{1}{2}xz^2$
**7.** false **9.** 1 **11.** 0

**Exercises**

**1.** $4x + C$ **3.** $\frac{1}{2}x^2 + C$
**5.** $\frac{5}{4}x^4 + C$ **7.** $\frac{2}{3}x^3 - \frac{7}{2}x^2 + C$
**9.** $\frac{4}{3}z^{3/2} + C$ **11.** $\frac{1}{4}y^4 - 4y^3 - 4y + C$
**13.** $\frac{1}{5}x^5 - \frac{5}{4}x^4 + x^3 + x^2 + x + C$
**15.** $\frac{1}{2}u^2 + \ln|u| - 2e^u + C$
**17.** $\frac{-3}{x} - \frac{1}{x^2} + x + C$
**19.** $\frac{2}{3}t^3 - \frac{4}{3t^3} + C$
**21.** $2e^x - \ln|x| + C$
**23.** $\frac{4}{3}x^{3/2} - \frac{15}{4}x^{4/3} + C$
**25.** $2x^{1/2} - 3x^{2/3} + C$
**27.** $\frac{1}{3}x^3 - \frac{5}{2}x^2 + C$
**29.** $\frac{5}{6}x^3 - \frac{7}{4}x^2 + \frac{1}{2}\ln|x| + C$
**31.** $\frac{5}{9}u^{9/5} - \frac{21}{5}u^{5/3} + C$
**33.** $\frac{4}{3}x^3 + 2x^2 + x + C$
**35.** $\frac{1}{2}x^4 - 3x + 2$
**37.** $2\ln|x| + \frac{2}{x^2} - \frac{3}{2} - 2\ln 2$
**39.** $\frac{1}{5}x^5 - \frac{5}{6}x^{6/5} + 3x - 3$
**41.** $C(x) = 0.03x^3 - 0.1x^2 + 2x + 2.5$
**43.** $V(t) = 11250 - 750t$
**45.** (a) $R(x) = 150x - \frac{1}{8}x^2$ (b) 600
**47.** 1100
**49.** (a) $s(t) = -16t^2 + 64t + 6$ (b) 70 ft
**51.** 38

## 6.2 EXERCISES

**Short Answer**

**1.** true **3.** $x^4$ **5.** false **7.** (c)
**9.** (b)

**Exercises**

**1.** $\frac{1}{3}(3x + 1)^3 + C$ **3.** $\frac{1}{6}(2x^3 - 7)^6 + C$
**5.** $\frac{1}{10}(x^2 - 5)^5 + C$ **7.** $e^{x^3} + C$
**9.** $-\frac{1}{2}e^{1-x^2} + C$ **11.** $\frac{1}{5}(7x - 5)^5 + C$
**13.** $\frac{1}{20}(2x - 3)^{10} + C$
**15.** $\frac{1}{48}(2x^3 + 3)^8 + C$
**17.** $\frac{2}{15}(x^5 + 16)^{3/2} + C$
**19.** $-\frac{1}{2}e^{4-x^2} + C$ **21.** $-3e^{1/x} + C$
**23.** $-e^{6-x^5} + C$ **25.** $\frac{1}{15}(7 + e^{3x})^5 + C$
**27.** $\frac{1}{3}(x^2 + 1)^{3/2} + C$
**29.** $\frac{4}{9}(3x - 5)^{1/2} + C$
**31.** $\frac{1}{3}(x^2 + 4x)^{3/2} + C$
**33.** $\frac{3}{4}(1 + 2x^2)^{4/3} + C$
**35.** $\frac{-1}{6(1 + y^2)^3} + C$ **37.** $\frac{-1}{x^2 + x + 4} + C$
**39.** $\frac{-4}{3(1 + \sqrt{x})^3} + C$ **41.** $\frac{1}{5}(\ln x)^5 + C$
**43.** $-\frac{1}{\ln x} + C$ **45.** $\frac{1}{2}[\ln(3x)]^2 + C$
**47.** $\ln|x + 1| + C$
**49.** $\frac{3}{2}\ln(x^2 + 5) + C$
**51.** $\ln|\ln x| + C$
**53.** $\frac{1}{2}[\ln(3 + e^{2x})]^2 + C$
**55.** $x + 1 - \ln|x + 1| + C$
**57.** $\frac{2}{25}(6 - 5x) - \frac{12}{25}\ln|6 - 5x| + C$
**59.** $\frac{1}{2}x^2 - x + 2\ln|x + 1| + C$
**61.** $\frac{2}{3}(\ln x + 1)^{3/2} + C$
**63.** It cannot be done by substitution.
**65.** It cannot be done by substitution.
**67.** $\ln|x| + C$
**69.** $2\sqrt{x^2 + 2500} + 1900$
**71.** about 523,195 gal
**73.** $P(x) = 5\sqrt{x^2 + 100} - 50\sqrt{10}$
**75.** approximately 22.629 million
**77.** approximately 33.52 g
**79.** (a) 50.64% (b) 47.73%

## 6.3 EXERCISES

**Short Answer**

**1.** $G(d) - G(c)$ **3.** $e - 1$
**5.** family of functions **7.** 8, 27
**9.** false

**Exercises**

**1.** 9 **3.** $\frac{25}{2}$ **5.** 8 **7.** $\frac{26}{3}$
**9.** $-\frac{104}{3}$ **11.** $-6$ **13.** $\frac{25}{6}$ **15.** $\frac{106}{35}$
**17.** $3e - 3$ **19.** $e^2 - e^4$ **21.** $6\ln 2$
**23.** $\frac{11}{8}$ **25.** $\frac{32}{3}$ **27.** $-\frac{1836}{5}$
**29.** $\frac{2}{5}[(26)^{3/2} - (11)^{3/2}]$
**31.** $\frac{1}{3}[(15)^{3/2} - (3)^{3/2}]$
**33.** $\frac{7448}{3}$ **35.** $\ln 3$ **37.** $\frac{15}{4}$
**39.** $1 - e^2$ **41.** $\frac{1}{6}\ln(\frac{31}{7})$
**43.** $\frac{3}{4}[(e + 5)^{4/3} - 6^{4/3}]$ **45.** $2708.33
**47.** approximately 1.919 billion gallons
**49.** 9620 people
**51.** approximately $7981.03 **53.** 14.4 gal
**55.** $176,000 **57.** 75 vehicles

## 6.4 EXERCISES

**Short Answer**

**1.** product **3.** $x^3 \, dx$
**5.** $(\ln x)^3, x^2 \, dx$
**7.** $x^2, (3x - 4)^3$ **9.** true

**Exercises**

**1.** $\frac{1}{2}x^2 \ln x - \frac{1}{4}x^2 + C$
**3.** $\frac{1}{4}x^4 \ln x - \frac{1}{16}x^4 + C$ **5.** $\frac{1}{9}(2e^3 + 1)$
**7.** $\frac{1}{4}xe^{-4x} - \frac{11}{16}e^{-4x} + C$
**9.** $-\frac{1}{4}x^2e^{-4x} - \frac{1}{8}xe^{-4x} - \frac{1}{32}e^{-4x} + C$
**11.** $\frac{7}{9}e^3(5e^3 - 2)$
**13.** $\frac{1}{36}(6x - 5)^5(6x + 1) + C$
**15.** $\frac{2}{15}(2x + 1)^{3/2}(3x - 1) + C$
**17.** $\frac{9}{700}(5x + 3)^{4/3}(20x - 9) + C$ **19.** $\frac{68}{35}$
**21.** $x(\ln x)^3 - 3x(\ln x)^2 + 6x \ln x - 6x + C$
**23.** $120 \ln 2 - \frac{56}{3}$ **25.** $(x^2 - 1)e^{x^2} + C$
**27.** 4 **29.** $\frac{1}{18}e^{-3x}(12x - 5)$
**31.** $\frac{1}{2}x \ln x - \frac{1}{2}x + C$ **33.** $\frac{1}{4}e^{x^4} + C$
**35.** $\frac{5}{16}(x - 3)^{16/5} + \frac{30}{11}(x - 3)^{11/5} + \frac{15}{2}(x - 3)^{6/5} + C$
**37.** $\frac{4}{9}$ **39.** approximately $9,775.02
**41.** approximately 22.31%
**43.** approximately 34,578 tons

## 6.5 EXERCISES

**Short Answer**

**1.** 12 **3.** 23 **5.** 18 **7.** 30
**9.** 24

## Exercises

1. $\frac{1}{3}\ln|2+3x|+C$
3. $\frac{3}{8}[1-2x-\ln|2-4x|]+C$
5. $\frac{1}{27}(6x+4)\sqrt{3x-1}+C$
7. $-\frac{1}{50}\ln|1-25x^2|+C$
9. $\frac{1}{9}\ln\left|\frac{5+x}{4-x}\right|+C$
11. $\frac{1}{2}\sqrt{16-x^2}-2\ln\left|\frac{4+\sqrt{16-x^2}}{x}\right|+C$
13. $\frac{1}{2}x-\frac{1}{3}\ln|2-3x|+C$
15. $\frac{1}{12}(2x+9)\sqrt{4x-9}+C$
17. $\frac{1}{100}x(50x^2-9)\sqrt{25x^2-9}-\frac{81}{500}\ln|5x+\sqrt{25x^2-9}|+C$
19. $\frac{7}{60}\ln\left|\frac{6x-5}{6x+5}\right|+C$
21. $\frac{(10x^4-6x^2+3)}{140}(3+4x^2)^{3/2}+C$
23. $-\frac{1}{75}(5x^2+8)\sqrt{4-5x^2}+C$
25. $(-\sqrt{e^{2x}-16}/e^x)+\ln|e^x+\sqrt{e^{2x}-16}|+C$
27. $\frac{1}{2}x^4e^{x^2}-x^2e^{x^2}+e^{x^2}+C$
29. $-\frac{1}{2}\ln(\frac{5}{3})$  31. $\frac{2}{9}\ln(1280)-\frac{4}{3}$
33. $-\frac{1}{15}\sqrt{61}+\frac{2}{3}\ln(10+\sqrt{61})+\frac{5}{12}-\frac{2}{3}\ln 13$
35. $\ln(1+\sqrt{5})-\ln(2+2\sqrt{2})$
37. $\frac{1}{5}\ln\left(\frac{e-1}{3e+2}\right)-\frac{1}{5}\ln\left(\frac{e^2-1}{3e^2+2}\right)$
39. approximately 489.2415
41. $2619.72
43. There will be an increase of approximately $24,787.12.

## 6.6 EXERCISES

### Short Answer

1. $-5$   3. true   5. true   7. (b)
9. 1   11. $3x^2\,dx, \frac{1}{5}e^{5x}$   13. 29
15. 18

### Exercises

1. $\frac{1}{3}x^3-\frac{7}{2}x^2+C$   3. $\frac{2}{3}x^{3/2}-\frac{9}{4}x^{4/3}+C$
5. $\frac{1}{2}x^3-\frac{1}{2}x^2+\frac{3}{2}\ln|x|+C$
7. $\frac{5}{7}x^{7/5}-\frac{2}{5}x^{5/2}+C$
9. $\frac{1}{5}xe^{5x}-\frac{1}{25}e^{5x}+C$
11. $-\frac{2}{9}(5-3x^2)^{3/2}+C$
13. $\frac{1}{8}[x(2x^2+5)\sqrt{x^2+5}]-25\ln|x+\sqrt{x^2+5}|+C$
15. $\frac{1}{2}x^2(\ln x)^3-\frac{3}{4}x^2(\ln x)^2+\frac{3}{2}x\ln x-\frac{3}{2}x$
17. $\frac{14}{1875}(15x+2)(1-5x)^{3/2}+C$
19. $\frac{1}{\sqrt{5}}\ln\left|\frac{\sqrt{5}+x}{\sqrt{5}-x}\right|+C$
21. $\frac{1}{240}(\ln x)^5+C$
23. $\frac{7}{8}\ln(4x^2+5)+C$
25. $\frac{1}{3}(e^{2x}-5)^{3/2}+C$

27. $5\ln|\ln x+3|+C$
29. 2   31. $18-\frac{10}{3}\sqrt{5}$
33. $\frac{1}{2}e^2-(1/2e)$   35. $-\frac{1}{9}(2e^3+1)$
37. approximately 15.1256
39. $6\ln\left(\frac{\sqrt{29}+2}{\sqrt{29}-2}\right)$
41. $\frac{1}{3}(2e^2+1)^3-\frac{1}{3}(2e+1)^3-\frac{4}{9}e^6-e^4+\frac{4}{9}e^3+e-\frac{1}{6}\approx 1042.4491$
43. $\frac{2}{3}(\ln 3+2)^{3/2}-\frac{2}{3}(\ln 2+2)^{3/2}$
45. $\ln(\frac{4}{3})$   47. $(1/\ln 2)-\frac{1}{2}$
49. (a) $C(x)=\frac{1}{10}(3x-5)(2x+5)^{3/2}+1000+\frac{5}{2}\sqrt{5}$  (b) $87,592.61
51. approximately 53.03 weeks
53. $3258.70   55. $15,339.39
57. $s(t)=\sqrt{4t^2+9}-3\ln\left|\frac{3+\sqrt{4t^2+9}}{2t}\right|+3\ln 2-5$

## 7.1 EXERCISES

### Short Answer

1. true   3. (c)   5. true
7. $g(x)=x$   9. 0, 1

### Exercises

1. $\frac{1}{2}$   3. 12   5. 15   7. $\frac{37}{3}$
9. $2\ln 3$   11. $\frac{5}{2}$   13. $\frac{1}{3}$   15. 48
17. 1   19. 36   21. $\frac{4}{3}$   23. $\frac{1}{3}$
25. $\frac{4000}{3}$   27. $\frac{1}{2}$   29. $\frac{5137}{96}$
31. $\frac{1}{24}(4\sqrt{2}-1)$   33. $\frac{99}{4}$
35. $e^2+(1/e)-3$   37. $e-1-\frac{1}{2}\ln 2$
39. $\frac{343}{6}$   41. $\frac{1331}{54}$   43. $\frac{37}{12}$
45. $\frac{1}{4}e^2+5-2e-2\ln 2$
47. (a) $C(x)=0.01x^2+2000$
(b) $252,000  (c) $250,000
(d)

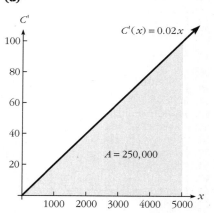

49. (a) $P(x)=4.25x+0.001x^2-1500$
(b) $2875  (c) $4375
(d)

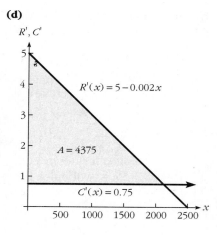

51. (a) 16,000 people
(b)

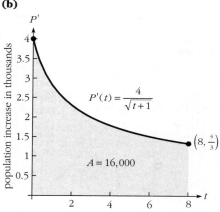

53. (a) $(6\ln 6-2\ln 2-4)$ ft $\approx 5.3643$ ft
(b)

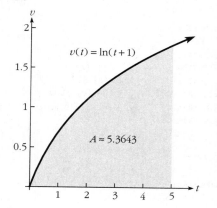

**55. (a)** Approximately 9.333 million dollars
**(b)**

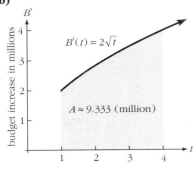

**57.** 1.7   **59.** 1.9   **61.** 2.0   **63.** 77.3

## 7.2 EXERCISES

**Short Answer**

**1.** rectangles   **3.** $\frac{5}{2}$   **5.** (a)   **7.** 25
**9.** true

**Exercises**

**1.** $\frac{7}{3}$   **3.** $\frac{4}{3}$   **5.** 8   **7.** $e - 1$
**9.** $\frac{3}{2}$ ft/sec   **11.** $85,583.33   **13.** 49.88 g
**15.** approximately 40.2 wpm
**17.** approximately 13.388 million gallons
**19.** approximately 2.00

## 7.3 EXERCISES

**Short Answer**

**1.** $20   **3.** true   **5.** $100   **7.** true
**9.** (c)

**Exercises**

**1. (a)** $2560 **(b)** $1920 **(c)** $640
**3. (a)** $733.33 **(b)** $200 **(c)** $533.33
**5. (a)** $2583.33 **(b)** $2250 **(c)** $333.33
**7. (a)** $3505.62 **(b)** $790 **(c)** $2715.62
**9. (a)** $400 **(b)** $240 **(c)** $160
**11. (a)** $4800 **(b)** $6600 **(c)** $1800
**13. (a)** $46.67 **(b)** $60 **(c)** $13.33
**15. (a)** $86,666.67 **(b)** $120,000
**(c)** $33,333.33
**17. (a)** $15,569.67 **(b)** $46,674
**(c)** $31,104.33
**19. (a)** $9.87 **(b)** $23.98 **(c)** $14.11
**21.** consumer's surplus = $320; producer's surplus = $320
**23.** consumer's surplus = $575.06; producer's surplus = $250
**25.** $1,203,692.82   **27.** $4458.58/yr
**29. (a)** $194,959.25 **(b)** $366,057.55
**31.** $94,281.94
**33.** consumer's surplus ≈ $13,542.40; producer's surplus ≈ $1319.22
**35.** consumer's surplus ≈ $3.42; producer's surplus ≈ $2.81

## 7.4 EXERCISES

**Short Answer**

**1.** $\frac{1}{2}, \frac{1}{4}$   **3.** false   **5.** 1.5   **7.** (d)
**9.** (d)

**Exercises**

**1. (a)** 8 **(b)** 12 **(c)** 10
**3. (a)** $\frac{451}{216}$ **(b)** $\frac{559}{216}$ **(c)** $\frac{1007}{432}$
**5. (a)** 6.32195 **(b)** 8.4677 **(c)** 7.3146
**7. (a)** 1.1889 **(b)** 1.0558 **(c)** 1.1165
**9. (a)** 4.8975 **(b)** 6.3518 **(c)** 5.6393
**11.** 10   **13.** 2.33796   **15.** 7.3948
**17.** 1.1224   **19.** 5.6246   **21.** 10
**23.** 2.3333   **25.** 7.3398   **27.** 1.1185
**29.** 5.6280
**31.** The left-hand sum is 0.725372, the right-hand sum is 0.662872, and the mid-point sum is 0.692661. The exact value is ln 2 ≈ 0.693147.
**33. (a)** 1.3877 **(b)** 0.025 **(c)** 14
**35. (a)** 1.5018 **(b)** $\frac{1}{3072}$ **(c)** 5104
**37. (a)** 0.5004 **(b)** $\frac{1}{384}$ **(c)** 12
**39.** 0.57   **41.** approximately 94 words
**43.** approximately 5,322
**45.** approximately $61.30

## 7.5 EXERCISES

**Short Answer**

**1.** true   **3.** 2   **5.** 9   **7.** diverges
**9.** $25,000

**Exercises**

**1.** $\frac{1}{8}$   **3.** diverges   **5.** diverges
**7.** $-1$   **9.** 1   **11.** diverges
**13.** diverges   **15.** 5   **17.** $-\frac{3}{2}$
**19.** diverges   **21.** $\frac{1}{4}$   **23.** 0   **25.** 0
**27.** $2/e$   **29.** $\frac{1}{4}$   **31.** 1
**33.** diverges   **35.** 1   **37.** 1

**39. (a)** $250,000 **(b)** $187,500
**41.** $125,000   **43.** $300,000
**45.** 250,000 gal

## 7.6 EXERCISES

**Short Answer**

**1.** false   **3.** $-\int_{1/2}^{1} \ln x \, dx$   **5.** false
**7.** $5000   **9.** (b)   **11.** true
**13.** true   **15.** (a)

**Exercises**

**1.** $\frac{32}{3}$   **3.** $\frac{1}{2}$   **5.** $\frac{32}{3}$   **7.** $\frac{237}{12}$   **9.** 10
**11.** $-\frac{5}{4}$   **13.** $50,000   **15.** $500
**17.** consumer's surplus = $50; producer's surplus = $150
**19.** consumer's surplus = $88; producer's surplus = $42.67
**21. (a)** 2.0932 **(b)** 2.9763 **(c)** 2.6962
**23. (a)** 1.47997 **(b)** 1.47997 **(c)** 1.5005
**25.** 2.5348   **27.** 1.47997   **29.** 2.6067
**31.** 1.4938
**33. (a)** 0.50455 **(b)** $\sqrt{3}/1728$ **(c)** 6
**35.** $1/4e^{16}$   **37.** $-1$   **39.** $1/(e + 1)$
**41.** $30,803.72   **43.** 1995
**45.** approximately 0.227 ft/sec/sec
**47.** $115,011.14   **49.** Approximately 220
**51.** 575   **53.** 12.8
**55.** consumer's surplus = $34.18; producer's surplus = $1.44

## 8.1 EXERCISES

**Short Answer**

**1.** $x, w, V$   **3.** true   **5.** true
**7.** (c)   **9.** $x^2 + z^2 = 16$

**Exercises**

**1. (a)** $-8$ **(b)** $-32$ **(c)** 25
**3. (a)** 29 **(b)** 24 **(c)** 8
**5. (a)** $\sqrt{2}$ **(b)** $2 + \ln 2$ **(c)** $2 + e\sqrt{2}$
**7. (a)** $\frac{1}{2}$ **(b)** $\frac{5}{13}$ **(c)** $a/2$ if $a \neq 0$
**9. (a)** 30 **(b)** 10 **(c)** 35
**11.** all ordered pairs of real numbers $(x, y)$
**13.** all ordered pairs of real numbers $(x, y)$ for which $x \geq 0$
**15.** all ordered pairs of real numbers $(x, y)$ for which $x \neq -2y$
**17.** all ordered triples of real numbers $(x, y, z)$ for which $x \neq z$

**A34** ANSWERS TO ODD-NUMBERED EXERCISES

**19.**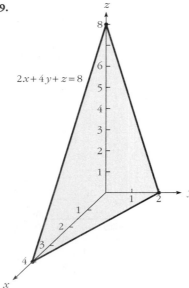

**25.**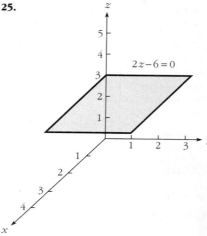

**27.** $3\sqrt{2}$  **29.** $\sqrt{51}$
**31.** $\sqrt{(a+1)^2 + (b+2)^2 + (c+3)^2}$
**33.** $x^2 + y^2 + z^2 = 16$
**35.** $(x+1)^2 + (y-2)^2 + (z-4)^2 = 9$

**37.**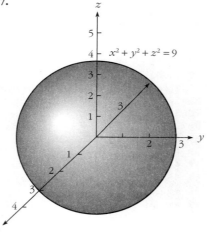

**41.** Trace in $x = 0$ is $z = 25 - y^2$; trace in $y = 0$ is $z = 25 - x^2$; trace in $z = 0$ is $x^2 + y^2 = 25$; trace in $z = 25$ is $x^2 + y^2 = 0$.

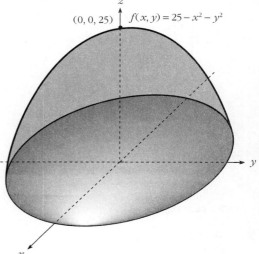

**21.**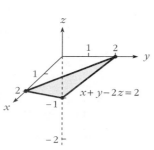

**23.**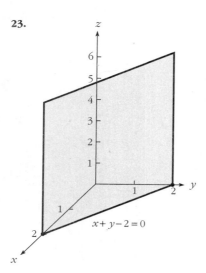

**39.**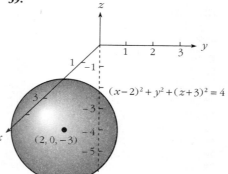

**43.** Trace in $x = 0$ is $z = y^2 + 1$; trace in $y = 0$ is $z = 2x^2 + 1$; trace in $z = 1$ is $2x^2 + y^2 = 0$; trace in $z = 4$ is $2x^2 + y^2 = 4$.

**45.** (a) $C(x, y) = 8x + 14y + 5000$
(b) $58,000
**47.** (a) 404  (b) It increases.
**49.** (a) 22400 units  (b) 44800 units

## 8.2 EXERCISES

**Short Answer**

**1.** true  **3.** false  **5.** (b)
**7.** product $A$  **9.** 0

## Exercises

1. $f_x = 2x; f_y = 2y$
3. $f_x = 4y + 6xy^2 - 1; f_y = 4x + 6x^2y$
5. $f_x = \frac{1}{y}; f_y = -\frac{x}{y^2} + 4y^3$
7. $f_x = \frac{x}{\sqrt{x^2+y^2}}; f_y = \frac{y}{\sqrt{x^2+y^2}}$
9. $f_x = 2\sqrt{y} - \frac{1}{2\sqrt{x}}; f_y = \frac{x}{\sqrt{y}}$
11. $f_x = \frac{7y}{(2x+3y)^2}; f_y = \frac{-7x}{(2x+3y)^2}$
13. $f_x = 2e^{2x-y}; f_y = -e^{2x-y}$
15. $f_x = ye^x + \frac{1}{x+2y}; f_y = e^x + \frac{2}{x+2y}$
17. $f_x(1,2) = 48, f_y(-3,4) = 60$
19. $\left.\frac{\partial z}{\partial x}\right|_{(1,3)} = \frac{87}{28}\sqrt{28}; \left.\frac{\partial z}{\partial y}\right|_{(-2,4)} = -\frac{100}{13}\sqrt{13}$
21. $-\frac{155}{48}$   23. $-1$
25. $f_{xx} = 0; f_{yy} = -6, f_{xy} = f_{yx} = 1$
27. $f_{xx} = 6xy; f_{yy} = -6xy; f_{xy} = f_{yx} = 3x^2 - 3y^2$
29. $f_{xx} = 0; f_{yy} = 2x/y^3; f_{xy} = f_{yx} = -1/y^2$
31. $f_{xx} = -1/(5y\sqrt{(2x+1)^3}); f_{yy} = 2\sqrt{2x+1}/(5y^3); f_{xy} = f_{yx} = -1/(5y^2\sqrt{2x+1})$
33. $f_{xx} = 1/x; f_{yy} = xe^y; f_{xy} = f_{yx} = e^y$
35. (a) 2 (b) 12 (c) 18
37. (a) 1 (b) $(8e^2 - 1)/(16e)$ (c) $-\frac{1}{108}$
39. (a) $C_x(200, 150) = 20$ is the approximate cost of producing an additional unit of product $A$ when the production level is 200 units of $A$ and 150 units of $B$. This is the same as the exact cost of producing the 201st unit of $A$. (b) $C_y(200, 150) = 25$ is the approximate cost of producing an additional unit of product $B$ when the production level is 200 units of $A$ and 150 units of $B$. This is the same as the exact cost of producing the 151st unit of $B$.
41. (a) $P(x, y) = 45x + 30y + 0.001xy - 6000$ (b) $C_x(300, 250) = 65$ is the approximate cost of producing another set of golf clubs when the production level is 300 sets of clubs and 250 bags. Thus this is the approximate cost of producing the 301st set of clubs. (c) $R_x(300, 250) = \$110.25$ is the approximate revenue generated from the sale of the 301st set of clubs. (d) $P_x(300, 250) = \$45.25$ is the approximate profit derived from the manufacture and sale of the 301st set of clubs.
43. (a) 30 (b) $\frac{80}{81}$
45. (a) $-0.5$ (b) 0.005 (c) $\frac{1}{48}$ (d) $\frac{3}{130}$: When the list price of the player is $150, advertising is $2500 per month, and the price of the competing CD players are $144 and $169, an increase of $1 in the list price of the player will cause an approximate decline in demand of $\frac{1}{2}$ unit, an increase of $1 in advertising will cause an approximate increase in demand of 0.005 unit, an increase of $1 in the price of the first competitor's player will cause an approximate increase in demand of $\frac{1}{48}$ unit, and an increase of $1 in the price of the second competitor's player will cause an approximate increase in demand of $\frac{3}{130}$ unit.
47. (a) $\partial x/\partial s = 20$, so for every $1 increase in the price of brand $B$, weekly demand for brand $A$ will increase by about 20 containers; $\partial y/\partial p = 15$, so for every $1 increase in the price of brand $A$, weekly demand for brand $B$ will increase by about 15 containers. (b) substitutes

## 8.3 EXERCISES

### Short Answer

1. false   3. $(-1, 3)$   5. saddle point
7. $(0, 0, 0)$   9. false

### Exercises

1. relative minimum point: $(0, 0, 5)$
3. relative maximum point: $\left(\frac{1}{2}, -1, \frac{47}{4}\right)$
5. relative minimum point: $(-1, 2, -5)$
7. saddle point: $(-2, 0, 16)$; relative minimum point: $(2, 0, -16)$
9. saddle point: $(0, 0, 7)$; relative minimum point: $\left(\frac{3}{4}, \frac{9}{8}, \frac{421}{64}\right)$
11. saddle points: $(0, -2, 3)$ and $(1, 2, 1)$
13. saddle point: $(0, 0, -2)$; relative maximum point: $\left(-\frac{1}{6}, -\frac{1}{12}, -\frac{863}{432}\right)$
15. saddle point: $(4, 3, -2)$
17. saddle point: $(0, 0, 4)$
19. 40,000 units of product $A$ and 20,000 units of product $B$
21. $37.45 each for Cutall and $33.48 each for Sawzall
23. $18,000 worth of labor and $18,000 worth of capital
25. Produce 163.3 units at factory $A$ and 116.9 units at factory $B$.
27. The minimum cost is approximately $2.71.   29. $\left(\frac{10}{57}, \frac{145}{19}\right)$

## 8.4 EXERCISES

### Short Answer

1. true   3. $x^2 + y^2 + 4 + \lambda(y - 3x - 5)$
5. 6   7. false
9. The maximum value is $f(0, 0) = 3$.

### Exercises

1. The constrained minimum is $f\left(\frac{4}{9}, \frac{16}{9}\right) = \frac{77}{9}$.
3. The constrained maximum is $f(4\sqrt{2}, 4\sqrt{2}) = f(-4\sqrt{2}, -4\sqrt{2}) = 64$, and the constrained minimum is $f(-4\sqrt{2}, 4\sqrt{2}) = f(4\sqrt{2}, -4\sqrt{2}) = -64$.
5. The constrained maximum is $f\left(\frac{1}{2}, \frac{3}{2}\sqrt{11}\right) = f\left(\frac{1}{2}, -\frac{3}{2}\sqrt{11}\right) = \frac{101}{4}$ and the constrained minimum is $f(-5, 0) = -5$.
7. The constrained maximum is $f(22, -2) = 404$.
9. The constrained minimum is $f(3, 3) = f(-3, -3) = 18$.
11. The constrained maximum is $f(2, 2) = f(2, -2) = f(-2, 2) = f(-2, -2) = 16$, and the constrained minimum is $f(0, \pm 2\sqrt{2}) = f(\pm 2\sqrt{2}, 0) = 0$.
13. The maximum product is 36 when both of the numbers are 6.
15. The constrained minimum is $f\left(\frac{6}{7}, -\frac{12}{7}, -\frac{18}{7}\right) = \frac{72}{7}$.
17. The constrained maximum is $f(3, 4, 5) = 2$.
19. The maximum product is 8000 when each of the three numbers is 20.
21. The company should produce 300 units at factory 1 and 400 units at factory 2.
23. labor: 320 units; capital: 240 units. The marginal productivity of money is approximately 27.27 units.
25. $3\frac{1}{3}$ lb of meat and $3\frac{1}{3}$ lb of vegetables.
27. The length is 40 in. and the two sides that yield the girth are each 20 in.
29.

## A36 ANSWERS TO ODD-NUMBERED EXERCISES

**31.**

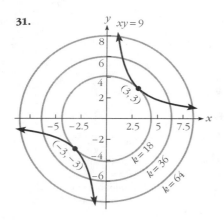

**3.** $\frac{1}{3}(y^2 + x)^{3/2} + C(x)$  **5.** $\frac{1}{2}e^{2xy} + C(y)$
**7.** $8x^2 + 4x$  **9.** $\ln\sqrt{\frac{y^2 + 9}{y^2 + 4}}$
**11.** $x + x^2 - 3x^3 - 3x^4$
**13.** $(\sqrt{y} + 2)(\sqrt{y} + 2) - y$
**15.** 3  **17.** 2  **19.** 2  **21.** $\frac{3}{2}$
**23.** $\frac{9}{2}$  **25.** $\frac{1}{6}\sqrt{2} - \frac{1}{3}$  **27.** $2 - e$
**29.** $\frac{1}{2}e^5 - \frac{1}{2}e^3 - e$  **31.** 78  **33.** $\frac{80}{3}$
**35.** 40  **37.** $-\frac{1}{28}$  **39.** $-\frac{1}{28}$  **41.** 0
**43.** $\frac{1}{4}(e - 1)$  **45.** 9  **47.** $\frac{115}{6}$
**49.** $\frac{261}{35}$  **51.** $\frac{250}{3}$  **53.** 208 ft$^3$
**55.** approximately 6499.08 units
**57.** \$8,375,000
**59.** $\frac{25}{9}$ deer per square mile  **61.** 5.2
**63.** 0.28

### 8.7 EXERCISES

**Short Answer**

**1.** $e + 1$  **3.** (d)  **5.** $-1/x^2$
**7.** true  **9.** true  **11.** false
**13.** true  **15.** (a)

**Exercises**

**1.** the set of all ordered pairs $(x, y)$ such that $x^2 + y^2 \le 25$
**3.** the set of all ordered pairs $(x, y)$ with $x \ne 0$ and $y \ne 1$
**5.**

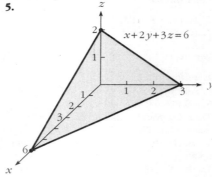

**7.**

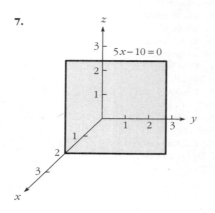

**9.** $(x + 1)^2 + (y - 2)^2 + (z + 4)^2 = 41$
**11.** Container $A$ has the larger surface area. The cost of materials is about \$2.51 for container $A$ and about \$2.36 for container $B$.
**13.** $\frac{\partial z}{\partial x} = \frac{x}{2\sqrt{x + y}} + \sqrt{x + y} - \frac{\sqrt{2}\,y}{2\sqrt{x}}$;
$\frac{\partial z}{\partial y} = \frac{x}{2\sqrt{x + y}} - \sqrt{2x}$
**15.** $\frac{\partial z}{\partial x} = \frac{2(y^2 - x^2 + xy)}{(x^2 + y^2)^2}$;
$\frac{\partial z}{\partial x} = \frac{y^2 - x^2 - 4xy}{(x^2 + y^2)^2}$
**17.** $f_x(0, 1) = 2$; $f_y(0, 4) = \frac{1}{4}$
**19.** $\frac{\partial z}{\partial x}\bigg|_{(-1,2)} = -1$; $\frac{\partial z}{\partial y}\bigg|_{(-1,2)} = -1$
**21.** $f_{xx} = -120x$; $f_{yy} = (2x/y^3) - 2$; $f_{xy} = f_{yx} = -1/y^2$
**23.** $f_{xx} = \frac{-4}{(2x + 3y)^2} - \frac{1}{\sqrt{(2x + 3y)^3}}$;
$f_{yy} = \frac{-9}{(2x + 3y)^2} - \frac{9}{4\sqrt{(2x + 3y)^3}}$;
$f_{xy} = f_{yx} = \frac{-6}{(2x + 3y)^2} - \frac{3}{\sqrt{(2x + 3y)^3}}$
**25.** (a) $-2$ (b) $\frac{5}{36}$ (c) $-4$
**27.** (a) $P_x(x, y) = 10 - 0.002y$: when production and sales of products $A$ and $B$ are $x$ and $y$ units per month, respectively, production and sales of an additional unit of product $A$ will contribute approximately $10 - 0.002y$ dollars to profit. (b) $P_y(x, y) = 15 - 0.002x$: an increase of one unit in the production and sales of product $B$ from an initial production and sales level of $x$ units of product $A$ and $y$ units of product $B$ will contribute approximately $15 - 0.002x$ dollars to profit.
**29.** saddle point: $(0, 0, 0)$
**31.** relative minimum point: $\left(-\frac{3}{11}, \frac{7}{11}, \frac{6}{11}\right)$
**33.** saddle point: $(0, 0, 1)$
**35.** The maximum profit of \$1,103,333.33 is derived from the sale of 6000 units of product $A$ and 10000 units of product $B$.
**37.** The length, width, and height should each be 9 in.
**39.** The constrained maximum is
$f\left(\frac{5}{2}\sqrt{2}, \frac{5}{2}\sqrt{2}\right) = f\left(-\frac{5}{2}\sqrt{2}, -\frac{5}{2}\sqrt{2}\right) = 50$
and the constrained minimum is
$f\left(\frac{5}{2}\sqrt{2}, -\frac{5}{2}\sqrt{2}\right) = f\left(-\frac{5}{2}\sqrt{2}, \frac{5}{2}\sqrt{2}\right) = -50$.
**41.** The constrained maximum is
$f\left(-\frac{7}{5}, \frac{34}{5}\right) = \frac{409}{5}$.
**43.** The constrained maximum is
$f\left(\frac{3}{7}\sqrt{14}, \frac{9}{14}\sqrt{14}, \frac{3}{14}\sqrt{14}\right) = 3\sqrt{14}$, and the constrained minimum is $f\left(-\frac{3}{7}\sqrt{14}, -\frac{9}{14}\sqrt{14}, -\frac{3}{14}\sqrt{14}\right) = -3\sqrt{14}$.
**45.** $y = -0.499x + 40.092$
**47.** $y = 77.667x - 493.467$
**49.** (a) $y = 2.547x + 26.826$ (b) \$65,031
**51.** $-\frac{1}{3}$  **53.** $\frac{104}{105}$  **55.** $-\frac{83}{198}$  **57.** $\frac{149}{2}$
**59.** $\frac{7}{4}$  **61.** 3

### 8.5 EXERCISES

**Short Answer**

**1.** false  **3.** $f_m = 330m + 42b - 690$
**5.** 4.7205  **7.** 16, 28, 90, 158
**9.** positive

**Exercises:** In these answers, $m$ and $b$ are rounded to three decimal places.

**1.** $y = 1.129x + 12.226$
**3.** $y = 0.417x + 10.583$
**5.** $y = 1.224x + 2.367$
**7.** $y = -1.963x + 10.935$
**9.** $y = 2.523x - 3.533$
**11.** $y = -0.738x + 114.900$
**13.** $y = 1.952x + 5.963$
**15.** (a) 9.082 (b) 14.128 (c) 46.927  Parts (a) and (b) should be accurate.
**17.** (a) 4.2 (b) 100.14 (c) 96.45  Parts (b) and (c) should be accurate.
**19.** (a) 6.3534 (b) 8.891 (c) 16.3086
Parts (a) and (b) should be accurate.
**21.** (a) $y = -32.8x + 206.8$ (b) 39,520
**23.** (a) $y = 13.6x + 759$ (b) \$827,000
**25.** (a) $y = 1.861x - 1.896$ (b) about 147 units
**27.** (a) $y = 0.0454x - 1.091$ (b) about 3.0 (c) approximately 68
**29.** (a) $y = 10x - 10.8$ (b) $y = 4.286x^2 - 16.214x + 20.7$ (c) the line

### 8.6 EXERCISES

**Short Answer**

**1.** true  **3.** false  **5.** $dx\,dy$  **7.** false
**9.** $\iint\limits_R dA = \int_c^d \int_{b_1(y)}^{b_2(y)} dx\,dy$

**Exercises**

**1.** $\frac{1}{2}x^2y + xy + C(y)$

**THE UNIVERSITY STORE**
**678-2011**

# PROTECT YOUR INVESTMENT
# SAVE YOUR RECEIPT

TEXT REFUND POLICY

A full refund will be given on new or used books provided . . .

1. You present your cash register receipt. NO RECEIPT—NO REFUND
2. New books must be in mint, clean condition. Do not write in new books.
3. The University Store price label is on the book.
4. Wrapped course materials must not be unwrapped or opened.
5. Fall and Spring semester last day for a full refund is two weeks from the start of the semester. Dates will be posted. No refunds on texts bought after last refund date.
6. Summer semester last day for a full refund is one week from start of the semester. Dates will be posted.
7. The manager reserves the right to make the decision on the condition or saleability of the merchandise.

PS 2/92-1126

Memphis State University is an Equal Opportunity/Affirmative Action University. It is committed to education of a non-racially identifiable student body.

**THE UNIVERSITY STORE**
**678-2011**

# PROTECT YOUR INVESTMENT
# SAVE YOUR RECEIPT

TEXT REFUND POLICY

A full refund will be given on new or used books provided . . . .

1. You present your cash register receipt. NO RECEIPT—NO REFUND
2. New books must be in mint, clean condition. Do not write in new books.
3. The University Store price label is on the book.
4. Wrapped course materials must not be unwrapped or opened.
5. Fall and Spring semester last day for a full refund is two weeks from the start of the semester. Dates will be posted. No refunds on texts bought after last refund date.
6. Summer semester last day for a full refund is one week from start of the semester. Dates will be posted.
7. The manager reserves the right to make the decision on the condition or saleability of the merchandise.

*Memphis State University is an Equal Opportunity/Affirmative Action University. It is committed to education of a non-racially identifiable student body.*

## 9.1 EXERCISES

**Short Answer**

1. 1   3. 2   5. false
7. initial-value problem   9. $k\sqrt[3]{x}$

**Exercises**

17. $y = 2e^{-3x}$   19. no solution
21. $y = \frac{3}{2}e^x - \frac{1}{2}e^{-x}$   23. $y = 2xe^x - e$
25. $y = \frac{1}{2}x^2 + C$
27. $y = -e^{-x} + x + C$
29. $y = \frac{2}{3}x^{3/2} - 2x + C$
31. $y = xe^x - e^x + C$
33. $y = \frac{1}{6}e^{3x} + \frac{1}{4}x^2 + C$
35. $y = \frac{4}{15}x^{5/2} + C_1 x + C_2$
37. $y = 3x^2 + 1$
39. $y = 4\ln|x| - 2x + 5$
41. $y = \frac{1}{4}x^{4/3} - \frac{5}{4}$   43. $y = \frac{1}{2}x^2 + \frac{1}{2}x - 3$
45. $y = -xe^{-x} - 3e^{-x} + 5$
47. (a) $dy/dt = 2000e^{4t}$; $y(0) = 5000$
(b) $y(t) = 500e^{4t} + 4500$
49. $N = (15000/r) - 500$   51. 2300

## 9.2 EXERCISES

**Short Answer**

1. separable   3. not separable
5. separable   7. $y = \sqrt[3]{Cx^2 + 1}$
9. $y = Ke^x$

**Exercises**

1. $y = \sqrt[3]{x^3 + C}$
3. $y = 1/(C - x^3)$, $y = 0$
5. $y = \sqrt{2x^4 + C}$ or $y = -\sqrt{2x^4 + C}$
7. $y = \sqrt[3]{(Cx - 9)/x}$
9. $y = \sqrt[3]{3e^x + 3x + C}$
11. $y = Ke^{x^2} - 2$
13. $y = \sqrt[3]{\frac{3}{2}x^2 + 8}$
15. $y = \sqrt[3]{3x^3 - 30}$
17. $y = \sqrt[4]{4e^{x^2} + 252}$
19. $y = e^{(3/2)x^2 - (2/3)x^{3/2} + (1/6)}$
21. $y = e^{(1/2)x^2 + x - (3/2)} - 1$
23. $3y^2 + 6y = 2x^3 + C$
25. $2y^3 + 6y = 12\ln|x| - 3x^2 + C$
27. $y^3 + 3y = 3xe^x - 3e^x + C$
29. $4ye^y - 4e^y = 2x^2 \ln x - x^2 + C$
31. $y^2 + 4y = 2x^2 - 2$
33. $y^3 + y = x + 3\ln|x| + 1$
35. $e^y + y = xe^x - e^x + 2$
37. $x^2 + y^2 = 25$   39. $x^2 - 9y^2 = 1$

## 9.3 EXERCISES

**Short Answer**

1. $e^{-x}$   3. true   5. true
7. $x^2 + C$   9. is not

**Exercises**

1. $y = xe^{-2x} + Ce^{-2x}$   3. $y = \frac{3}{4} + Ce^{-2x^2}$
5. $y = 2 + Ce^{-1/4 x^2}$
7. $y = (C/x) - (1/x^2)$   9. $y = Ce^{2x} - 2$
11. $y = 1 + Ce^{1/x}$   13. $y = \frac{1}{3}e^{5x} + Ce^{2x}$
15. $y = 3 + Ce^{-x^2 - x}$
17. $y = \frac{2}{3}(2x + 1) + C(2x + 1)^{-1/2}$
19. $y = \frac{1}{6}xe^{2x} + e^{2x}$
21. $y = \frac{1}{2}e^x - \frac{1}{2}e^{-x}$   23. $y = x^5$
25. $y = 2(x^2 + 4)^{-1/2}$
27. $y = Ce^{-1/x} - 2$   29. $y = 4 + Cx^{-1}$

## 9.4 EXERCISES

**Short Answer**

1. unlimited exponential   3. 250
5. true   7. $0.04A - 50$
9. $5A/(100 + 2t)$

**Exercises**

1. (a) approximately 10.13 g
(b) approximately 24.85 hr
3. approximately 6.931 yr
5. approximately 13.85 g
7. (a) approximately 117.89 g  (b) approximately 5.30 g  (c) approximately 12.30 hr
9. (a) approximately 207.08°
(b) approximately 32.55 min
11. (a) approximately 4,596,000
(b) approximately 58.4 weeks
13. after about 68.81 days
15. $p = k \ln|s| + C$
17. $N(t) = 1200/(1 + 239e^{-1.38421t})$
19. (a) $A(t) = 15789.47 - 789.47e^{0.095t}$
(b) approximately 31.534 yr  (c) $47,301
21. $1599.91   23. $357.75
25. (a) $A(t) = 4(100 + t) - 40000(100 + t)^{-1}$, $0 \le t \le 400$
(b) approximately 212.31 lb
27. (a) approximately 657.14 kg
(b) approximately 1.88 kg/l

## 9.5 EXERCISES

**Short Answer**

1. false   3. $k\sqrt{x}$   5. true
7. $e^{-(1/2)x^2}$   9. true   11. 100

**Exercises**

1. yes   3. no   5. no
7. $y = \ln|x| + C$   9. $y^2 = e^x + C$
11. $y = 2\ln|x + 2| + C$
13. $y = -\frac{1}{2}x - \frac{1}{4} + Ce^{2x}$
15. $y^2 = x^2 - (2/x) + C$
17. $y = 2\ln|x| + 2$

19. $y - \frac{2}{y} = \ln|x| + \frac{1}{x} - \frac{3}{2} - \ln 2$
21. $y - \frac{1}{2}\ln|y| = \ln|x| - x + 2$
23. $y = \frac{1}{6}x^4 + \frac{5}{6}x^{-2}$
25. $y = 2e^{-x} \ln|x + 1| - e^{-x}$
27. (a) $p = 200/(1 + 4e^{-0.28768t})$
(b) $177.51  (c) approximately 4.82 months
29. approximately 33
31. approximately 9.808 yr   33. $379.15
35. yes

## 10.1 EXERCISES

**Short Answer**

1. 4   3. false   5. 0.3   7. 100
9. false

**Exercises**

1. $S = \{HHH, HHT, HTH, THH, HTT, THT, TTH, TTT\}$   3. $\frac{1}{8}$

5.

| $x$ | 0 | 1 | 2 | 3 |
|---|---|---|---|---|
| $P(X = x)$ | $\frac{1}{8}$ | $\frac{3}{8}$ | $\frac{3}{8}$ | $\frac{1}{8}$ |

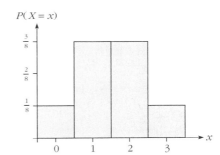

7. $V(x) = \frac{3}{4}$, $\sigma_x = \sqrt{3}/2$
9. (a) $\{(3, 4)\}$ (b) $\{(1, 2), (2, 2), (3, 2), (4, 2)\}$ (c) $\{(2, 2), (2, 4), (2, 6), (4, 2), (4, 4), (4, 6), (6, 2), (6, 4), (6, 6)\}$ (d) $\{(4, 1), (4, 2), (4, 3), (4, 4), (5, 1), (5, 2), (5, 3), (5, 4), (6, 1), (6, 2), (6, 3), (6, 4)\}$
(e) $\{(2, 1), (2, 3), (2, 5), (4, 1), (4, 3), (4, 5), (6, 1), (6, 3), (6, 5), (1, 2), (1, 4), (1, 6), (3, 2), (3, 4), (3, 6), (5, 2), (5, 4), (5, 6)\}$
11. (a) $\frac{3}{36}$ (b) $\frac{1}{9}$ (c) $\frac{1}{4}$ (d) $\frac{1}{3}$ (e) $\frac{1}{2}$
13. 7
15. $S = \{(5, 5), (5, 10), (5, 15), (5, 20), (10, 5), (10, 10), (10, 15), (10, 20), (15, 5), (15, 10), (15, 15), (15, 20), (20, 5), (20, 10), (20, 15), (20, 20)\}$   17. $\frac{1}{16}$

**19.**

| x | 10 | 15 | 20 | 25 | 30 | 35 | 40 |
|---|---|---|---|---|---|---|---|
| $P(X=x)$ | $\frac{1}{16}$ | $\frac{2}{16}$ | $\frac{3}{16}$ | $\frac{4}{16}$ | $\frac{3}{16}$ | $\frac{2}{16}$ | $\frac{1}{16}$ |

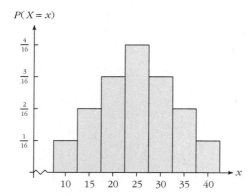

**21.** $\frac{125}{2}$
**23.** (a) $\{(10, 5)\}$ (b) $\{(10, 5), (10, 15), (15, 5), (20, 5), (20, 15)\}$ (c) $\{(5, 10), (10, 5)\}$ (d) $\{(10, 20), (20, 20)\}$ (e) $S$
**25.** (a) $\frac{1}{12}$ (b) $\frac{5}{12}$ (c) $\frac{1}{6}$ (d) $\frac{1}{6}$ (e) 1
**27.** 25
**29.** $S = \{G_1G_2, G_1G_3, G_1G_4, G_1G_5, G_1G_6, G_2G_3, G_2G_4, G_2G_5, G_2G_6, G_3G_4, G_3G_5, G_3G_6, G_4G_5, G_4G_6, G_5G_6, G_1D_1, G_2D_1, G_3D_1, G_4D_1, G_5D_1, G_6D_1, G_1D_2, G_2D_2, G_3D_2, G_4D_2, G_5D_2, G_6D_2, D_1D_2\}$ **31.** $\frac{1}{28}$

**33.**

| x | 0 | 1 | 2 |
|---|---|---|---|
| $P(X=x)$ | $\frac{15}{28}$ | $\frac{12}{28}$ | $\frac{1}{28}$ |

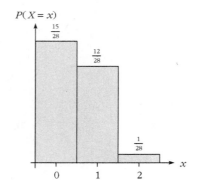

**35.** approximately 0.5669
**37.** (a) $\frac{39}{16}$ (b) not fair
**39.** (a) $-\frac{9199}{10000}$ (b) not fair
**41.** option $A$
**43.** Your expected annual loss is $170.

## 10.2 EXERCISES

### Short Answer

**1.** discrete  **3.** continuous  **5.** 0
**7.** true  **9.** 1

### Exercises

**1.**

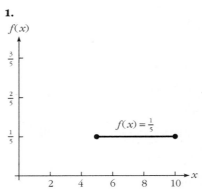

**3.**

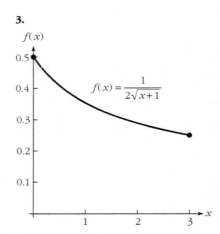

**5.**

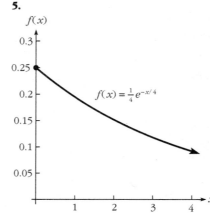

**7.** (a) $\frac{1}{5}$ (b) $\frac{4}{5}$
**9.** (a) $\sqrt{3} - 1$ (b) $2 - \sqrt{3}$
**11.** (a) $1 - (1/e)$ (b) approximately 0.5244
**13.** (a) $F(x) = \begin{cases} 0 & \text{if } x < 5 \\ \frac{1}{5}x - 1 & \text{if } 5 \leq x \leq 10 \\ 1 & \text{if } x > 10 \end{cases}$
(b) (i) 0.8 (ii) 0.4
**15.** (a) $F(x) = \begin{cases} 0 & \text{if } x < 0 \\ \sqrt{x+1} - 1 & \text{if } 0 \leq x \leq 3 \\ 1 & \text{if } x > 3 \end{cases}$
(b) (i) $2 - \frac{1}{2}\sqrt{10}$ (ii) $\sqrt{3} - \sqrt{2}$
**17.** (a) $F(x) = \begin{cases} 0 & \text{if } x < 0 \\ 1 - e^{-(1/4)x} & \text{if } x \geq 0 \end{cases}$
(b) (i) approximately 0.9179 (ii) approximately 0.2865  **19.** $\frac{1}{48}$  **21.** $\frac{1}{9}$  **23.** $\frac{3}{64}$
**25.** 1  **27.** $f(x) = \begin{cases} \frac{1}{2} & \text{if } 1 \leq x \leq 3 \\ 0 & \text{otherwise} \end{cases}$
**29.** $f(x) = \begin{cases} 132x^{10}(1-x) & \text{if } 0 \leq x \leq 1 \\ 0 & \text{otherwise} \end{cases}$
**31.** $\frac{2}{3}$  **33.** $\frac{1}{5}$  **35.** (a) 0.75 (b) 0.01
**37.** (a) $\frac{5}{4}$ (b) $\frac{1}{36}$  **39.** approximately 0.95
**41.** (a)

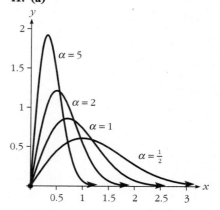

(b)

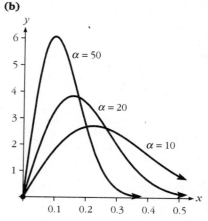

## 10.3 EXERCISES

### Short Answer

**1.** 15  **3.** at least 93.75%  **5.** false
**7.** ±5  **9.** (a)

### Exercises

**1.** $E(X) = \frac{5}{2}$, $V(X) = \frac{25}{12}$, $\sigma = \frac{5}{6}\sqrt{3}$
**3.** $E(X) = \frac{8}{3}$, $V(X) = \frac{8}{9}$, $\sigma = \frac{2}{3}\sqrt{2}$
**5.** $E(X) = \frac{4}{3}$, $V(X) = \frac{2}{75}$, $\sigma = \frac{1}{15}\sqrt{6}$
**7.** $E(X) = \frac{3}{2}$, $V(X) = \frac{3}{4}$, $\sigma = \frac{1}{2}\sqrt{3}$
**9.** $E(X) = \frac{1}{4}$, $V(X) = \frac{1}{16}$, $\sigma = \frac{1}{4}$
**11.** $E(X) = \frac{27}{8}$, $V(X) = \frac{67}{320}$, $\sigma = \frac{1}{40}\sqrt{335}$
**13.** $E(X) = \frac{4}{3}$, $V(X) = \frac{34}{45}$, $\sigma = \frac{1}{15}\sqrt{170}$
**15.** $3\sqrt{2}$   **17.** $\frac{1}{2}\sqrt[3]{4}$   **19.** $\frac{9}{4}$
**21.** $2\sqrt{2}$   **23.** $2 - \frac{1}{2}\sqrt{10}$   **25.** $\frac{1}{4}\ln 2$
**27.** $\frac{5}{4}\sqrt{10} - 1$   **29.** 3.5 min
**31. (a)** 10 days **(b)** 8.79 days **(c)** 50
**33.** $\frac{11}{31}$   **35. (a)** [11, 19] **(b)** [9.84, 20.16]
**37.** 2 yr

## 10.4 EXERCISES

### Short Answer

**1.** true   **3.** $P(X \geq 105)$   **5.** true

### Exercises

**1.** 0.6915   **3.** 0.3745   **5.** 0.8907
**7.** 0.8181   **9.** 0.0863   **11.** 0.5
**13.** 0.733   **15.** 0.6826   **17.** 0.0132
**19.** 0.1587   **21.** 0.9306   **23.** 0.9931
**25.** 0.4838
**27. (a)** almost zero **(b)** 0.4772 **(c)** 0.0228
**29. (a)** 0.0918 **(b)** 0.4972 **(c)** 0.0475
**31.** 8372   **33. (a)** 0.0038 **(b)** 0.0918
**35. (a)** 1587 **(b)** 7745 **(c)** 228
**39.**

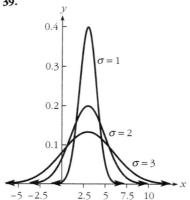

## 10.5 EXERCISES

### Short Answer

**1.** true   **3.** 0.4   **5.** (b)   **7.** $\frac{3}{8}$
**9.** 2   **11.** 8   **13.** true

### Exercises

**1.** $S = \{G_1G_2, G_1G_3, G_1G_4, G_1G_5, G_2G_3, G_2G_4,$
$G_2G_5, G_3G_4, G_3G_5, G_4G_5, G_1D, G_2D, G_3D,$
$G_4D, G_5D\}$   **3.** $\frac{1}{15}$
**5.** $P(X = 1) = \frac{1}{3}$, $P(X = 2) = \frac{2}{3}$

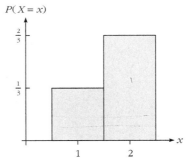

**7.** $\frac{2}{9}$   **9.** $A$   **15. (a)** $\frac{3}{4}$ **(b)** $\frac{2}{5}$
**17. (a)** $1 - \frac{3}{8}\sqrt{3}$ **(b)** $\frac{1}{4}\sqrt{2}$
**19. (a)** $F(x) = \begin{cases} 0 & \text{if } x < 1 \\ \dfrac{3x - 3}{2x} & \text{if } 1 \leq x \leq 3 \\ 1 & \text{if } x > 3 \end{cases}$
**(b)** (i) $\frac{1}{4}$ (ii) $\frac{3}{2}$ **(c)** $E(X) = \frac{3}{2}\ln 3$;
$V(X) = 3 - \frac{9}{4}(\ln 3)^2$
**21. (a)** $F(x) = \begin{cases} 0 & \text{if } x < 0 \\ \frac{1}{8}x^{3/2} & \text{if } 0 \leq x \leq 4 \\ 1 & \text{if } x > 4 \end{cases}$
**(b)** (i) $\frac{1}{8}(3\sqrt{3} - 1)$ (ii) $2\sqrt[3]{2}$ **(c)** $E(X) = \frac{12}{5}$; $V(X) = \frac{192}{175}$   **23.** $\frac{2}{9}$   **25.** $\frac{13}{16}$
**27.** 0.2877   **29.** 0.3355
**31.** 0.0475   **33.** 456

# INDEX

## A

Absolute extremum, 260–69
Absolute maximum, 260
Absolute minimum, 260
Acceleration, 181
Actual consumer expenditure, 456–70
Actual producer revenue, 461–64
Antiderivative, 386–94, 552
Antidifferentiation, 386, 552
Approximation:
   of a definite integral, 449–52, 470–79
   using differentials, 209–14
Area:
   between curves, 441–45
   under a curve, 438–41
Area theorem, 438
Asymptote, 85–87, 275–76
Average cost, 305–6
Average elasticity of demand, 310
Average rate of change, 134
Average value of a function, 452–53, 562–64
Average velocity, 133

## B

Base:
   of an exponential function, 330
   of a logarithmic function, 343
Break-even point, 30

## C

Capitalized value, 487–88
Carbon-14, 369
Carrying capacity, 372
Chain rule:
   for differentiation, 166–74
   for integration, 397–98
Change-of-base formula, 348
Chebyshev's theorem, 642–43
Cobb-Douglas production function, 506
Complementary products, 510–11
Composition of functions, 23–25, 166
Compound interest, 336
Concave down, 248–52
Concave up, 248–52
Constant multiple rule, 157
Constant rule, 156
Constrained optimization, 527–36
Constraint, 527
Consumer's surplus, 456–60
Consumer's willingness to pay, 456–60

Continuity:
   on an interval, 102–3
   at a point, 95–102
Continuous function, 97
Continuous random variable, 631–37
Continuous stream of income, 463–68
Continuously compounded interest, 338
Convergence of an improper integral, 484
Cost:
   average, 16
   fixed, 22
   function, 22
   marginal, 300
   ordering, 307–8
   variable, 22
Critical number:
   First-order, 226
   Second-order, 250
Critical point, 518
Cumulative probability distribution function, 636
Curve sketching, 271–83

## D

Data points, 539
Decay constant, 365
Decreasing function, 222–32
Definite integral:
   applications of, 456–68
   approximation of, 470–79
   as an area, 438–46
   as the limit of a Riemann sum, 449–54
   as a net change, 406–12
   properties of, 411
Delta notation, 33
Demand:
   function, 48
   schedule, 48
Dependent variable, 4
Depreciation, 43
Derivative:
   definition of, 118, 128
   geometric interpretation of, 119, 509
   notation for, 118, 122, 163, 177–78
   of order $n$, 177
   partial, 506–14
   second-order partial, 512–14
Deviation, 539
Difference quotient, 128
Differentiability, 125–28
Differential:
   of area, 554

of $x$, 208
of $y$, 209
Differential equations:
   applications of, 599–608
   explicit solutions of, 578, 587
   of the form $dy/dx = g(x)$, 581
   general solution of, 579
   implicit solution of, 587
   initial conditions for, 581
   integrating factor for, 594
   linear, 593–97, 604–8
   order of, 578
   parameters of, 579
   particular solution of, 579
   separable, 586–91
   singular solution of, 588
Differentiation:
   of composite functions, 166–70
   of exponential functions, 353–57
   implicit, 186–94
   of logarithmic functions, 357–60
   partial, 506–14
   of power functions, 170–74
   of products, 160
   of quotients, 160
   of sums, 158
Diffusion curve, 371
Diminishing returns, 246–47
Discontinuous function, 95–96
Discrete random variable, 621
Discriminant of $f(x, y)$, 520
Distance formula, 500
Divergent improper integral, 484
Domain, 3, 5, 498
Double integral, 551–64
Doubling time, 367

## E

$e$ (the number), 334
Ebbinghaus model, 378
Economic lot size, 309
Economic order quantity, 307
Elastic, 311
Elasticity of demand:
   average, 310
   point, 311
Error:
   for Simpson's rule, 477
   for the Trapezoidal rule, 475
Euler, Leonhard, 334
Event, 618

Expected value of a random variable, 623–24, 641–46
Exponential decay, 332, 364–73, 599–604
Exponential function, 328–38
Exponential growth, 331, 364–73, 599–604
Exponents (laws of), 329
Extrapolation, 48, 541
Extreme Value Theorem, 262

## F

Family of solutions, 579
First derivative test, 238–39
First octant, 499
Fixed cost, 22
Function(s):
   algebra of, 22–27
   average value of, 452–53, 562–64
   composition of, 23–25
   continuous, 95–103
   decreasing, 222–33
   definition of, 3
   derivative of, 118
   domain of, 3, 5, 498
   exponential, 328–38
   graph of, 9, 499
   implicit, 186
   increasing, 222–32
   inverse, 344
   limit of, 67
   linear, 30–43
   logarithmic, 342–49
   of one variable, 3
   quadratic, 47–57
   range of, 3, 12, 498
   of two variables, 498
   vertical line test for, 17
Functional notation, 4
Fundamental Theorem of Calculus, 451
Future value, 340, 465–66

## G

General solution of a differential equation, 579
Generalized power rule, 170–74
Gompertz growth law, 611
Graph:
   of a function of one variable, 9–18
   of a function of two variables, 499–500, 502–3
Growth constant, 365
Growth models, 364–73

## H

Half-life, 367–68
Higher-order derivatives, 177–83, 512–14
Horizontal asymptote, 85, 276
Horizontal line, 37

## I

Implicit differentiation, 186–94
Improper integrals, 483–88
Increasing function, 222–32
Indefinite integral, 386–94
Independent variable, 4
Index of summation, 542
Inelastic, 311
Inflection point, 252–54
Initial condition, 581
Initial value problem, 581
Instantaneous rate of change, 141
Instantaneous velocity, 139
Integral:
   convergent, 484
   definite, 406–12
   divergent, 484
   double, 551–64
   improper, 483–88
   indefinite, 385–86
   iterated, 553
   properties of, 388–89, 411
   sign, 387
   table, 424–26
Integrand, 387, 554
Integration:
   basic formulas, 388–89, 424
   constant of, 387
   chain rule for, 397–98
   lower limit of, 407
   numerical, 470–79
   partial, 551–52
   by parts, 415–22
   reduction formulas, 429
   by substitution, 398–403, 409–11
   upper limit of, 407
   using a table, 424–29
   variable of, 387
Integrating factor, 594
Intercepts, 32, 50–51, 499–500
Interest:
   compound, 336
   continuously compounded, 338
   simple, 334
Intermediate value theorem, 227
Inverse functions, 344
Inventory, 307–8
Iterated integral, 553

## L

Lagrange multiplier, 528
Learning curve, 370
Least squares line, 541
Left-hand limit, 69–71
Leibniz, Gottfried, 109
Leibniz notation, 109, 122, 164, 178
Level curve, 529
Limit(s):
   defintion of, 67
   infinite, 86–92
   at infinity, 80–85
   of integration, 407
   left-hand, 69–71
   one-sided, 69–71
   properties of, 72
   right hand, 69–71
Limited exponential decay, 370, 599, 601–2
Limited exponential growth, 370–71, 599, 601–2
Linear function, 30–43
Line(s):
   general form of the equation of, 39
   horizontal, 37
   parallel, 40–41
   perpendicular, 40–41
   point slope form of the equation of, 38
   of regression, 541
   of symmetry, 50–52
   slope of, 35
   slope-intercept form of the equation of, 37–38
   two-point form of the equation of, 38
   vertical, 37
Local extremum, 235
Logarithm:
   change of base, 348
   common, 343
   natural, 343
   properties of, 344
Logarithmic function, 342–49
Logistic growth, 371–72, 599, 602–4
Lower limit of integration, 407, 485–86

## M

Marginal cost, 141, 301–2
Marginal productivity of capital, 511
Marginal productivity of labor, 511
Marginal productivity of money, 535
Marginal profit, 32, 142, 301–2
Marginal revenue, 301–2
Market equilibrium, 462
Maximum:
   absolute, 260
   constrained, 527
   relative, 235–36
Mean, 623–24, 641–46
Median, 646–47
Method of least squares, 539–46
Midpoint Riemann sum, 471
Minimum:
   absolute, 260
   constrained, 527
   relative, 235–36
Mixed partial derivatives, 514

## N

Natural logarithm, 343
Net change, 407, 445
Newton's law of cooling, 376

Normal distribution:
 density function, 650
 graph of, 651
 standard, 652
Normal equations, 542
Numerical integration:
 Riemann sums, 470–73
 Simpson's rule, 476–78
 trapezoidal rule, 473–76

## O

Objective function, 527
Octant, 499
One-sided limit, 69–71
One-to-one function, 333
Optimization:
 constrained, 527–36
 of a function of one variable, 286–96
 of a function of two variables, 522–24
Order of integration, 556
Ordered pair, 9

## P

Parabola, 50–52
Parallel lines, 40–41
Partial antidifferentiation, 552
Partial derivative:
 first-order, 506–12
 second-order, 512–14
Partial integration, 551–52
Partial solution, 579
Perpendicular lines, 40–41
Plane, 499–500
Point elasticity of demand, 311
Point of diminishing returns, 246–47
Point of inflection, 252–54
Point-slope form, 38
Polynomial function, 271–75
Power rule, 156
Present value, 340, 467–68
Principal, 176, 334
Probability:
 density function, 632–33
 distribution, 621
 distribution function, 635–36
 histogram, 621
 of an event, 619–20
Producer's surplus, 461–64
Producer's willingness to accept, 461–64
Product rule, 160
Production function, 206, 362, 506, 516
Profit:
 function, 22
 marginal, 32, 142, 301–2

## Q

Quadratic equation, 52
Quadratic formula, 52

Quadratic function, 47–57
Quotient rule, 160

## R

Radioactive decay, 367–69
Random experiment, 618
Random variable:
 continuous, 631–32
 discrete, 621
Range of a function, 3, 12, 498
Rate of change:
 average, 134
 instantaneous, 141
Rational function, 82–83, 275–80
Rectangular coordinate plane, 9
Related rates, 196–204
Region of integration, 554
Regression line, 541
Relative extremum:
 of a function of one variable, 235–43
 of a function of two variables, 517–24
Relative maximum, 235–36, 518
Relative minimum, 235–36, 518
Revenue function, 22, 314–15
Reversing the order of integration, 559
Richter scale, 352
Riemann sum:
 general, 450–51, 471
 left-hand, 471
 midpoint, 471
 right-hand, 471
Right-hand limit, 69–71

## S

Saddle point, 519
Sample point, 618
Sample space, 618
Scatter diagram, 539
Secant line, 108
Second derivative, 177–83
Second derivative test:
 for absolute extrema, 267
 for relative extrema, 255
Second-order critical number, 250
Second-order partial derivative, 512–14
Second-partials test, 520
Separable differential equation, 586–91
Sigma notation, 542
Sign diagram, 228
Simple event, 618
Simple interest, 334
Simpson's rule, 476–78
Singular solution, 588
Slope, 35
Slope-intercept form, 37–38
Sphere, 501
Standard deviation, 626–27, 641–46
Standard normal curve, 652
Straight-line depreciation, 43

Substitute products, 510–11
Substitution, 397–403
Sum of functions, 22
Sum or difference rule, 158
Supply, 46
Surface, 499
Symmetry:
 line of, 50
 about the line $y = x$, 343–44

## T

Tangent line, 106–16
Tangent line approximation, 209–13
Test value, 228
Trace, 502
Trapezoidal rule, 473–76
Two-point form, 38
Type I region, 557
Type II region, 557

## U

Unlimited exponential decay, 365–66, 599–600
Unlimited exponential growth, 365, 367–70, 599–600
Uniform probability density function, 634
Unit elasticity of demand, 311
Upper limit of integration, 407, 484, 486

## V

Variable, 4
Variable cost, 22
Variance, 626–27, 641–46
Velocity:
 average, 133
 instantaneous, 139, 181
Vertex, 50–52
Vertical asymptote, 86–87, 276
Vertical line, 37
Vertical line test, 17
Volume, 555, 560–62

## X

$x$-intercept, 30

## Y

$y$-intercept, 32
Young's theorem, 514

## Z

$z$-intercept, 500

## Photo Credits

**1** Werner H. Muller, Peter Arnold, Inc.; **8** Paul Silverman, Fundamental Photographs; **65** Ellis Herwig, Stock, Boston; **153** Harry Haralambou, Peter Arnold, Inc.; **155** Richard Megna, Fundamental Photographs; **220** Heinz Plenge, Peter Arnold, Inc.; **221** Joel Gordon; **299** Peter Simon, Stock, Boston; **327** G.M. Medori, International Stock Photography; **377** Jean-Claude Lejeune, Stock, Boston; **385** Paul Silverman, Fundamental Photographs; **435** Owen Franken, Stock, Boston; **437** George W. Gardner, Stock, Boston; **496** Erika Stone, Peter Arnold, Inc.; **497** Ellis Herwig, Stock, Boston; **549** John Zoiner, International Stock Photography; **577** Joel Gordon; **614** Cary Wolinsky, Stock, Boston; **617** Malcolm S. Kirk, Peter Arnold, Inc.; **662** Spencer Grant, Stock, Boston.